A. Brahmananda Reddy · S. Nagini ·
Valentina E. Balas · K. Srujan Raju
Editors

Proceedings of Third International Conference on Advances in Computer Engineering and Communication Systems

ICACECS 2022

Springer

Editors
A. Brahmananda Reddy
Department of Computer Science
and Engineering
VNR VJIET
Hyderabad, India

S. Nagini
Department of Computer Science
and Engineering and Computer Science
and Business Systems
VNR VJIET
Hyderabad, India

Valentina E. Balas
Department of Automatics and Applied
Software
"Aurel Vlaicu" University of Arad
Arad, Romania

K. Srujan Raju
Department of Computer Science
and Engineering
CMR Technical Campus
Hyderabad, India

ISSN 2367-3370 ISSN 2367-3389 (electronic)
Lecture Notes in Networks and Systems
ISBN 978-981-19-9227-8 ISBN 978-981-19-9228-5 (eBook)
https://doi.org/10.1007/978-981-19-9228-5

This Springer imprint is published by the registered company Springer Nature Singapore Pte Ltd.
The registered company address is: 152 Beach Road, #21-01/04 Gateway East, Singapore 189721, Singapore

Committees—ICACECS-2022

Chief Patrons

Sri. D. Suresh Babu, President, Vignana Jyothi Society
Er. J. S. Rao, General Secretary, Vignana Jyothi Society

Patrons

Dr. C. D. Naidu, Principal, VNR VJIET
Dr. B. Chennakesava Rao, Director, Advancement, VNR VJIET
Dr. A. Subhananda Rao, Director, Research and Development, VNR VJIET

Program Chairs

Dr. A. Brahmananda Reddy, Department of CSE, VNR VJIET, Hyderabad, India
Dr. S. Nagini, Department of CSE, VNR VJIET, Hyderabad, India

Program Co-chairs

Dr. D. N. Vasundhara, Department of CSE, VNR VJIET, Hyderabad, India
Dr. G. S. Ramesh, Department of CSE, VNR VJIET, Hyderabad, India

General Chairs

Dr. C. Kiran Mai, Professor, Department of CSE, VNR VJIET, Hyderabad, India
Dr. B. V. Kiranmayee, Professor, Department of CSE, VNR VJIET, Hyderabad, India
Dr. P. V. Siva Kumar, Associate Professor, Department of CSE, VNR VJIET, Hyderabad, India

Honorary Chairs

Prof. Lakhmi C. Jain, University of Technology Sydney, Australia, and Founder of KES International
Dr. A. Govardhan, Professor in CSE and Rector JNTUH, Hyderabad, India

Editorial Board

Dr. A. Brahmananda Reddy, Associate Professor, VNR VJIET, Hyderabad, India
Dr. S. Nagini, VNR VJIET, Professor and Head, Hyderabad, India
Prof. Valentina E. Balas, Aurel Vlaicu University of Arad, Romania
Dr. K. Srujan Raju, Professor, CMR Technical Campus, Hyderabad, India

International Advisory Committee

Dr. Aynur Unal, Director, Member of the Executive Team, UK
Dr. Margarita N. Favorskaya, Siberian State Aerospace University of Science and Technology, Russian federation
Dr. Raghava Rao Mukkamala, CBDA, CBS, Denmark
Dr. Vishnu Pendyala, San Jose State University, USA
Dr. Radhakrishnan Palanikumar, King Khalid University, Abha, Kingdom of Saudi Arabia
Dr. Pawan Lingras, Saint Mary's University, Canada
Md. Qayyum, King Khalid University, Abha, Kingdom of Saudi Arabia
Dr. Rakhee, The University of the West Indies Mona, Jamaica
Dr. Sorayya Bibi Malek, University Malaya, Kuala Lumpur, Malaysia
Dr. Md. Rafiqul Islam, IIUM, Malaysia
Dr. Aisha Hasan Abdalla Hasim, IIUM, Malaysia

National Advisory Committee

Dr. Rajeev Srivastava IIT BHU, Varanasi, India
Dr. Suresh Chandra Satapathy, KIIT University, Bhubaneswar, India
Dr. Rakesh Matam, IIIT Guwahati, India
Dr. K. P. Supreethi, JNTUH, Hyderabad, India
Dr. P. Premchand, Professor, Osmania University, Hyderabad, India
Shri. Sudhakar Challaoalli, CSI-Hyderabad, India
Dr. U. Srinivasulu Reddy, NIT Tiruchirappalli, India
Dr. V. Kamakshi Prasad, JNTUH, Hyderabad, India
Dr. A. Sureshbabu, JNTUA, Anantapuramu, India
Dr. Y. Padmasai, VNR VJIET, Hyderabad, India
Dr. G. Ramesh Chandra, VNR VJIET, Hyderabad, India
Dr. B. V. Ram Naresh Yadav, JNTUH CES, Hyderabad, India
Dr. G. Suresh Reddy, VNR VJIET, Hyderabad, India
Dr. N. Sandhya, VNR VJIET, Hyderabad, India
Dr. K. Subrahmanyam, KL University, Vijayawada, India
Dr. T. V. Rajini Kanth, Dean, SNIST, Hyderabad, India
Dr. B. Vishnu Vardhan, JNTUHCEM, Manthani, India
Dr. S. Vasundra, JNTUA, Anantapuramu, India
Dr. C. Shoba Bindhu, JNTUA, Anantapuramu, India
Dr. V. A. Narayana, Principal, CMRCET, Hyderabad, India
Dr. Sidharth Dabhade, NIELIT, Aurangabad, India
Dr. P. Neelakantan, VNR VJIET, Hyderabad, India
Dr. M. Raja Sekar, VNR VJIET, Hyderabad, India
Dr. K. Vijaya Kumar, CMRCET, Hyderabad, India
Dr. P. Vijaya Pal Reddy, MEC, Hyderabad, India
Dr. S. Sampath, AIT, Chikkamagaluru, Karnataka, India
Dr. R. Manjula Sri, VNR VJIET, Hyderabad, India
Dr. D. Srinivasa Rao, VNR VJIET, Hyderabad, India
Dr. A. Mallika, VNR VJIET, Hyderabad, India
Dr. Ravichander Janapati, SR University, Warangal, India
Dr. G. Srinivasa Gupta, VNR VJIET, Hyderabad, India
Dr. N. Mangathayaru, VNR VJIET, Hyderabad, India
Dr. T. Srinivasa Rao, VNR VJIET, Hyderabad, India
Dr. T. Jayashree, VNR VJIET, Hyderabad, India
Anil Sukheja, Scientist "E" ISRO Ahmadabad, India
Dr. B. Sailaja, Principal Scientist, ICAR-IIRR, Hyderabad
Dr. K. Anuradha, VNR VJIET, Hyderabad, India
Dr. Poonam Upadhyay, VNR VJIET, Hyderabad, India

Technical Committee

Dr. Margarita N. Favorskaya, Siberian State Aerospace University of Science and Technology, Russian Federation
Dr. Raghava Rao Mukkamala, CBDA, CBS, Denmark
Dr. Vishnu Pendyala, San Jose State University, USA
Dr. Asad Ata, Malaysia Institute for Supply Chain Innovation, Malaysia
Dr. Nurul Fariza Binti Zulkurnain, IIUM, Malaysia
Dr. Pilli Emmanuel Subhakar, MNIT, Jaipur, India
Dr. A. P. Siva Kumar, JNTUCEA, Andhra Pradesh, India
Dr. Subodh Srivastava, NIT Patna, India
Dr. K. Suresh Babu, SIT, JNTUH, Hyderabad, India
Dr. C. Kiranmai, VNR VJIET, Hyderabad, India
Dr. B. V. Kiranmayee, VNR VJIET, Hyderabad, India
Dr. Padmaja Joshi, C-DAC Mumbai, India
Dr. Adiraju Prashanth Rao, Anurag College of Engineering, Hyderabad, India
Dr. G. Ramesh Chandra, VNR VJIET, Hyderabad, India
Dr. K. Purnachand, BVRIT Narasapur, Hyderabad, India
Dr. Korra Lakshman Naik, Scientist-D, NIELIT, Aurangabad, India
Dr. Konda Srinivas, CMRTC, Hyderabad, India
Dr. K. Seetha Ram Babu, Chairman CSI-Hyderabad Chapter, India
Dr. P. V. Siva Kumar, VNR VJIET, Hyderabad, India
Dr. T. Kishore Kumar, NIT Warangal, India
Dr. D. N. Vasundhara, VNR VJIET, Hyderabad, India
Dr. P. Dileep Kumar Reddy, SV College of Engineering, Tirupati
Dr. M. Ramakanth Reddy, Samskruti CET, Hyderabad
Dr. Talluri Sunil Kumar, VNR VJIET, Hyderabad, India
Dr. Deepak Sukheja, VNR VJIET, Hyderabad, India
Dr. P. Subhash, VNR VJIET, Hyderabad, India
Dr. K. Srinivas, VNR VJIET, Hyderabad, India
Dr. Sagar Yeruva, VNR VJIET, Hyderabad, India
Dr. A. Kousar Nikhath, VNR VJIET, Hyderabad, India
Dr. G. Madhu, VNR VJIET, Hyderabad, India
Dr. V. Radhakrishna, VNR VJIET, Hyderabad, India
Dr. G. Rajesh Kumar, VNR VJIET, Hyderabad, India
Dr. B. V. Seshu Kumari, VNR VJIET, Hyderabad, India
Dr. A. Srinivasa Rao, VNR VJIET, Hyderabad, India
Dr. Harshavardhan Awari, VNR VJIET, Hyderabad, India
Dr. M. Ravi Kanth, VNR VJIET, Hyderabad, India
Dr. Ch. Suresh, VNR VJIET, Hyderabad, India
Dr. K. Venkata Ramana, VNR VJIET, Hyderabad, India

Program Committee

Dr. S. Nagini, Department of CSE, VNR VJIET, Hyderabad, India
Dr. B. V. Kiranmayee, Department of CSE, VNR VJIET, Hyderabad, India
Dr. C. Kiran Mai, Department of CSE, VNR VJIET, Hyderabad, India
Dr. G. Ramesh Chandra, Department of CSE, VNR VJIET, Hyderabad, India
Dr. P. Neelakantan, Department of CSE, VNR VJIET, Hyderabad, India
Dr. V. Baby, Department of CSE, VNR VJIET, Hyderabad, India
Dr. P. V. Siva Kumar, Department of CSE, VNR VJIET, Hyderabad, India
Dr. Malige Gangappa, Department of CSE, VNR VJIET, Hyderabad, India
Dr. A. Brahmananda Reddy, Department of CSE, VNR VJIET, Hyderabad, India
Dr. Deepak Sukheja, Department of CSE, VNR VJIET, Hyderabad, India
Dr. D. N. Vasundhara, Department of CSE, VNR VJIET, Hyderabad, India
Dr. G. S. Ramesh, Department of CSE, VNR VJIET, Hyderabad, India
Dr. M. Ravi Kanth, Department of CSE, VNR VJIET, Hyderabad, India
Dr. K. Srinivas, Department of CSE, VNR VJIET, Hyderabad, India
Dr. Chalumuru Suresh, Department of CSE, VNR VJIET, Hyderabad, India
Dr. K. Venkata Ramana, Department of CSE, VNR VJIET, Hyderabad, India

Organizing Committee

Faculty and Staff from Department of CSE, VNR VJIET, Hyderabad, India

Conference Proceedings Committee

Dr. A. Brahmananda Reddy, Department of CSE, VNR VJIET, Hyderabad, India
Dr. D. N. Vasundhara, Department of CSE, VNR VJIET, Hyderabad, India
Mrs. A. Madhavi, Department of CSE, VNR VJIET, Hyderabad, India
Dr. N. V. Sailaja, Department of CSE, VNR VJIET, Hyderabad, India
Mrs. L. Indira, Department of CSE, VNR VJIET, Hyderabad, India
Mrs. K. Jhansi Laksmi Bai, Department of CSE, VNR VJIET, Hyderabad, India
Mrs. S. Nyemeesha, Department of CSE, VNR VJIET, Hyderabad, India
Mrs. G. Laxmi Deepthi, Department of CSE, VNR VJIET, Hyderabad, India
Mrs. A. Katyayani, Department of CSE, VNR VJIET, Hyderabad, India

Review Committee

Dr. Mohammed Shafiul Alam Khan, University of Dhaka, Dhaka, Bangladesh
Dr. Radhakrishnan Palanikumar, King Khalid University, Saudi Arabia
Dr. Ravi Vadapalli, HPCC, Texas Tech University, USA
Dr. Jagdish Shivhare, Distinguished Scientist and Member-IRBA, USA
Dr. Junyuan Zeng, The University of Texas at Dallas, USA
Dr. Neeraj Mittal, The University of Texas at Dallas, USA
Dr. Asad Ata, Malaysia Institute for Supply Chain Innovation, Malaysia
Dr. Nurul Fariza Binti Zulkurnain, IIUM, Malaysia
Dr. Shailendra Shukla, MNNIT, Allahabad, India
Dr. Sumit Mishra, IIIT Guwahati, Assam, India
Dr. Vivek Tiwari, IIIT Raipur, India
Dr. Muralidhar Kulkarni, NIT-Suratkal, Mangalore, Karnataka, India
Dr. S. Raghavan, NIT Tiruchirappalli, India
Dr. Kaustuv Nag, IIIT Guwahati, Assam, India
Dr. Angshuman Jana, IIIT Guwahati, Assam
Dr. Dr. Mohammed Misbahuddin, C-DAC, Bangalore, India
Dr. Subhodh Srivastava, NIT Patna, India
Dr. A. Srnivasa Rao, VNR VJIET, Hyderabad, India
Dr. K. Prasanna Lakshmi, GRIET, Hyderabad, India
Dr. P. Vijyapal Reddy, Matrusri Engineering College, Hyderabad, India
Dr. Srinivas Konda, CMRTC, Hyderabad, India
Dr. K. F. Bharati, JNTUA, Anantapuramu, Andhra Pradesh, India
Dr. N. Rama Subramanian, NIT, Tiruchirappalli, India
Dr. L. Anjaneyulu, NIT Warangal, India
Dr. P. Sreehari Rao, NIT Warangal, India
Dr. B. Ravindhar Reddy, AITS, India
Dr. K. Purna Chand, BVRIT Narsapur, Hyderabad, India
Dr. Kamakshaiah, GCET, Hyderabad
Dr. Kousar Nikhath, VNR VJIET, Hyderabad, India
Dr. Korra Lakshman, NIELIT, Aurangabad, India
Dr. V. Baby, VNR VJIET, Hyderabad, India
Dr. M. Gangappa, VNR VJIET, Hyderabad, India
Dr. Merugu Suresh, CMRCET, Hyderabad, India
Dr. G. Nagaraju, VNR VJIET, Hyderabad, India
Dr. V. Nagaveni, AIT, Bangalore, India
Dr. Deepak Sukheja, VNR VJIET, Hyderabad, India
Dr. Adiraju Prashanth Rao, Anurag University, Hyderabad, India
Dr. Mohd. Qayyum, King Khalid University, Abha, Saudi Arabia
Dr. V. Raman, Vardhaman College of Engineering, Hyderabad, India
Dr. K. Srinivas, VNR VJIET, Hyderabad, India
Dr. V. Venkateshwarlu, Vaagdevi College of Engineering, Warangal, India
Dr. CH. Suresh, VNR VJIET, Hyderabad, India

Dr. P. Lalitha Surya Kumari, KL University Hyderabad, Hyderabad, India
Dr. K. Bheemalingappa, VNR VJIET, Hyderabad, India
Dr. K. Prasanna, AITS, Rajampet, Andhra Pradesh
Dr. Shaheen, MVJ College of Engineering, Bangalore, Karnataka
Dr. G. Malini Devi, GNITS, Hyderabad
Dr. Megha Bushan, DIT University, Dehradun
Dr. D. Sunitha, KITS, Huzurabad, Karimnagar
Dr. R. Vijaya Saraswathi, VNR VJIET, Hyderabad, India
Dr. D. Sudheer, VNR VJIET, Hyderabad, India
Dr. Kranthi Kumar Singamaneni, GITAM, Visakhapatnam
Mrs. Radhika Patthi, VNR VJIET, Hyderabad, India
Mr. R. Kranthi Kumar, VNR VJIET, Hyderabad, India
Dr. J. Jeyabharathi, KL University, Vijayawada
Dr. M. Veeresha, Santhiram Engineering College, Nandyal, JNTUA
Dr. K. Govardhan Reddy, G. Pulla Reddy Engineering College, Kurnool
Dr. P. Venateswara Rao, VNR VJIET, Hyderabad, India
Dr. K. Shyam Sunder Reddy, Vasavi College of Engineering, Hyderabad
Dr. S. Ramakrishna, Baptla Engineering College, Baptla, Guntur, Andhra Pradesh
Dr. E. Sudarshan, SRITW, Warangal
Dr. R. Suneetha Rani, QIS College of Engineering and Technology, Ongole
Dr. C. Raghavendra, CVR College of Engineering, Hyderabad
Dr. R. Vijay Prakash, SR University, Warangal
Mrs. N. Sarika, VNR VJIET, Hyderabad, India
Dr. P. Pavan Kumar, ICFAITECH, Hyderabad
Dr. D. Ramesh, Kakatiya University, Warangal
Dr. S. V. Vasantha, MVSR Engineering College, Hyderabad
Dr. G. S. Ramesh, VNR VJIET, Hyderabad
Dr. Vasavi Ravuri, VNR VJIET, Hyderabad
Mr. K. G. Nandha Kumar, Presidency College, Bangalore
Dr. Sridhar Iyer, KLE Technological University MSSCET, Belagavi
Dr. S. Sudheer Mangalampalli, VIT- Andhra Pradesh University
Dr. Gaurav Kumar, Nigam Jaypee Institute of Information Technology, Noida
Mr. Gurpreet Singh Chhabra, GITAM School of Technology, Visakhapatnam
Mrs. Sonali Batra, MRIIRS, Faridabad
Dr. Vivek Upadhyaya, Poornima University, Jaipur
Dr. P. Mohamed Fathimal, SRMIST, Vadapalani Campus
Dr. Sagar Yeruva, VNR VJIET, Hyderabad
Dr. R. Senthil Kumar, Dr. N. G. P. Arts and Science College, Coimbatore
Dr. Kollati Vijaya Kumar, Vignan's Institute of Engineering for Women
Dr. T. Ananth Kumar, IFET College of Engineering, Tamil Nadu
Dr. Dattatray P. Gandhmal, NB Navale Sinhgad College of Engineering, Solapur
Dr. A. Harshavardhan, VNR VJIET, Hyderabad
Dr. S. Rama Krishna, Bapatla Engineering College (Autonomous)
Dr. Chaitanya Kulkarni, VPKBIET, Baramati
Dr. R. Delshi Howsalya Devi, KVCET

Dr. G. Rajavikram, VITS, Deshmukhi
Mr. P. Pramod Kumar, SR University, Warangal
Dr. P. Muthu, SRMIST, Kattankulathur
Dr. Anubhav Kumar, PUIT, Naini, Uttar Pradesh, India
Dr. Awanit Kumar, Sangum University, Rajasthan
Dr. Baswaraj Donagapure, Vasavi College of Engineering, Hyderabad
Dr. Umashankar Ghugar, GITAM, Vishakapatnam
Mrs. Monica Gahlawat, L J University
Dr. Tatireddy Subba Reddy, BVRIT, Narsapur
Dr. Sreekanth Puli, GITAM, Visakhapatnam
Dr. Yogesh Jadhav, Nmims University, Navi Mumbai
Mr. Ravi Aavula, Guru Nanak Institutions Technical Campus
Mr. A. Balaram, CMIT, Hyderabad
Dr. B. Sateesh Kumar, JNTUH UCEJ, Jagtial
Dr. Neelamadhab Padhy, GIET University
Dr. Debashreet Das, VIT, Vellore
Mr. P. Sakthivel, KIT, Salem
Dr. K. Murali Gopal, GIET University, Gunupur
Dr. Suman Kumar Swarnikar, Raipur, Chhattisgarh
Dr. Ranju S. Kartha, Rajagiri SET, Cochin
Dr. Roshan Gangurde Ajeenkya, DY Patil University, Pune
Dr. R. Thirumalai Selvi, Government Arts College, Chennai
Dr. Surekha Paneerselvam, Amrita School of Engineering, Bengaluru
Dr. T. Sudhakar, Lovely Professional University, Punjab
Mr. Soubraylu Sivakumar, SRFET
Dr. P. Subhash, VNR VJIET Hyderabad
Dr. Manu Gupta, SNIST, Hyderabad
Mr. Naina Narang, GITAM, Vishakhapatnam
Mr. K. Chandra Shekar, GNIT, Hyderabad
Mr. K. Ravikanth, RGUKT, Basar
Dr. Vempaty Prashanthi, VNR VJIET, Hyderabad
Mrs. Talakoti Mamatha, SNIST, Hyderabad
Dr. D. Sharada Mani, NRIIT, Guntur
Dr. T. V. Nagaraju, MLRIT, Dundigal
Dr. Vijaykumar B. Gadhavi, SSIT, Gandhinagar
Dr. G. N. Beena Bethel, GREIT, Hyderabad
Mrs. R. N. Ashlin Deepa, GRIET, Hyderabad
Mr. N. Balaji, NMAMIT, Nitte, Karkala, Udupi, Karnataka
Dr. M. Sridevi, CVRCE, Ibrahimpatnam, Hyderabad
Dr. G. Revathy, Sastra Deemed University, Thanjavur
Mr. Keshav Kaushik, University of PES, Dehradun
Dr. R. Suneetha Rani, BVRITW Hyderabad
Dr. M. Yeeresha, Dr. KVSRIT
Dr. Ghantasala Venu Gopal, PBR VITS, Kavali
Dr. L. Chandra Sekhar Reddy, CMRCET, Medchal, Hyderabad

Dr. G. Rajesh, NHCE, Bangalore
Mr. Ashutosh Kumar, IT Gopeshwar, Chamoli, Uttarakhand
Mrs. Divya Mohan, KLU, Vijayawada
Dr. S. Harihara Gopalan, SREC, Coimbatore
Mr. Yerragudipadu Subbarayudu, GRIET, Hyderabad
Mr. Mrutyunjaya S. Yalawar, CMREC, Hyderabad
Mr. K. Venkateshwara Rao, CMRCET, Medchal, Hyderabad
Dr. Gaddam Venu Gopal, VBIT, Ghatkesar, Hyderabad
Dr. Suresh Yadlapati, PVPSIT
Mr. J. Briskilal, SRMIST, KTR
Dr. G. Lakshmeeswari, Gitam University, Visakhapatnam
Dr. Padmaja, Gitam University, Visakhapatnam
Dr. P. Vijayapal Reddy, MEC, Hyderabad
Dr. Sailesh Iyer, Rai School of engineering, Rai University
Mr. Ramesh Dadi, SR University, Warangal
Dr. S. V. Yasantha, Vardhaman College of Engineering
Dr. Nagaratna P. Hegde, Vasavi College of Engineering
Mr. N. Sandeep Chaitanya, VNR VJIET, Hyderabad
Mr. J. Vamsinath, VNR VJIET, Hyderabad
Dr. K. Selvani Deepthi, ANITS, Visakhapatnam
Dr. A. Srinivasa Rao, VNR VJIET, Hyderabad
Dr. Shanti Verma, LJ University, Ahmedabad
Mr. M. Parameswar, CMRCET, Medchal, Hyderabad
Dr. D. B. Jagannadha Rao, Malla Reddy University, Hyderabad
Mr. Awaneesh Kumar Yadav, IIT Roorkee, Uttarakhand, India
Mr. Vivek Velleti, NIT Warangal, India
Dr. S. Rajkumar, VIT, Vellore
Mrs. T. Naga Lakshmi, KLU, Vijayawada, Andhra Pradesh
Mrs. K. Sai Madhuri, VNR VJIET, Hyderabad
Mrs. Anshul Khairwa, VIT, Vellore
Mr. B. Ravinder Reddy, SNITS
Mr. Rama Gaikwad, APCER
Dr. P. Guna Sundari, BMC for Women
Dr. L. Malliga, MREC for Women, Hyderabad

Preface

The field of computer engineering and communication systems is growing in leaps and bounds than any other time in the history. The interplay of computing and communication technologies ranging from artificial intelligence, machine learning, large-scale data mining, big data, data analytics, intelligent and communication systems, Internet of Things (IoT) applications and smart computing handles practical converged data into a wide variety of computer applications and has a significant role to play in the connotation of computer engineering with communication systems. The International Conference on Advances in Computer Engineering and Communication Systems (ICACECS-2022) is themed around Smart Innovations in Mezzanine Technologies: "Data Analytics, Intelligent Systems, Intelligent Computing, Networks and Communication Systems," thereby celebrating the emerging technology trends in computer engineering and communication systems and will create an arena for collaborative innovation. ICACECS-2022 is a 3rd international conference in a successive series of conferences consecutively in the month of August 2020 and August 2021 titled with ICACECS-2020 and ICACECS-2021, in collaboration with Springer Nature International as publishing partner.

In continuation of this legacy, the Department of Computer Science and Engineering, VNR Vignana Jyothi Institute of Engineering and Technology is organizing a virtual mode of the 3rd International Conference on "Advances in Computer Engineering and Communication Systems" from 11th to 12th August 2022 (ICACECS-2022) in collaboration with Springer Nature International as publication partner which brought into the best possibilities of the future is termed "Presencing", and it is a part of the core philosophy at Vignana Jyothi.

ICACECS-2022 is a unique forum bringing together the research scholars/scientists from the different countries to participate and transform the Research Landscape of the globe and carve a road map for implementation. It provides a valuable networking opportunity and brings a new era for the research scholars, students, professors, industries providing insights into the recent trends and developments in the field of computer science with a special focus on Mezzanine technologies.

ICACECS-2022 has received tremendous response with a great number of manuscripts of 439 from around the globe. Out of 441 manuscripts, 59 manuscripts are accepted for the presentations and will be published in the conference Springer proceedings of "Lecture Notes in Networks and Systems (LNNS)" (https://link.springer.com/bookseries/15179) a Scopus-indexed series.

The book volume focuses on thoroughly refereed post-conference enlargements and reviews on the advanced topics in artificial intelligence, machine learning, data mining and big data computing, cloud computing, Internet on Things, distributed computing and smart systems. We assure that we have put in every effort to ensure that the participatory experience in the e-Conference will not feel like a compromise but will add value in its own stride and enable more people to participate. We hope that the conference shall serve its purpose for the best exposition and progress of science, engineering, technology, and society.

Hyderabad, India
August 2022

Dr. A. Brahmananda Reddy
Corresponding Editor and Conference
Chair ICACECS-2022

Acknowledgemets

Team ICACECS-2022 acknowledges the support extended by Springer Nature International, for their continuous sustenance in publishing the proceedings of ICACECS-2022 in a Scopus-indexed series of "Lecture Notes in Networks and Systems (LNNS)".

We thank all the authors for their contributions and timely response. A special thanks to our reviewers who invested their valuable time and conscientious evaluation of the submissions for the best outcomes.

We express our profound gratitude for the inspiring and informative presentations of our keynote speakers on the frontline technologies, creating the curiosity to explore more on research trends in computer science and communication systems.

The cooperation extended by the sessions chair is immense, elevating the presentation skills amongst the participants and the valuable insights are enables them to diversify solutions and approaches.

We would like to thank our chief guests and guests of honour for accepting our invitation and making it to the e-conference, furthering our spirit to achieve more in providing technology solutions to the society—involve and evolve towards innovations and incubations leads to entrepreneurs.

Thanks to the founding members of Vignana Jyothi for their wisdom and social responsibility, who firmly believe that all technology must serve the urgent need of advancing the society, at micro- and/or macro levels.

Our sincere thanks are extended to all the chief patrons, patrons, chairpersons, members of the editorial board, eminent members of the programme committee and advisory committee for their guidance and support and to the enthusiastic people among the technical committee for their coordination and help in execution.

We would like to extend our unlimited appreciation for the amazing work done by our self-reliant and motivated team of CSE faculty and non-teaching staff. The amazing dedication and effort of the team enabled to reach our goal and put up this show with an ease.

Dr. A. Brahmananda Reddy
Dr. S. Nagini
Dr. D. N. Vasundhara
Dr. G. S. Ramesh

Contents

Editors and Contributors

About the Editors

Dr. A. Brahmananda Reddy is Associate Professor in the Department of Computer Science and Engineering at VNR Vignana Jyothi Institute of Engineering and Technology (VNR VJIET), Hyderabad. Dr. Reddy holds a Ph.D. from Jawaharlal Nehru Technological University Anantapuramu (JNTUA), Anantapuramu, in the area of Text Mining, M.Tech.—Computer Science in 2007 from JNTUCEA, Anantapuramu, and B.Tech. in 2004 in Computer Science and Engineering. Dr. Reddy has over 15 years of experience in the field of academic research and technological education, has more than 25 research papers published in various reputed national/international conferences and journals listed in Scopus, Web of Science, IEEE, Inderscience and Springer Proceedings etc., and also edited two books published by Springer Book Proceedings. He is Member of IEEE and Lifetime Member of ISTE and CSI. Dr. Reddy's research interests include data mining, text mining, natural language processing, semantic Web and social networks, machine learning, deep learning, and image processing. He is conducting many seminars/workshops/FDPs for the benefit of students and faculty in and out by the eminent personalities from various reputed institutions and delivered guest lectures on various topics in various academic institutions. Dr. Reddy guided many UG and PG projects in various mezzanine technologies and in charge for students industrial visits, establishing MoUs between institute and industries to reducing the gap between academia and industry. Dr. Reddy is Jury Member for the Smart India Hackathon Grand Finale (SIH 2018, SIH 2019, and SIH 2020) Software Edition and Evaluator for Toycathon-2021 organized by AICTE and MHRD in association with state governments and reputed industries and performed as Session Chairs for various international conferences, and also Reviewer for various reputed international conferences and Journals.

Dr. S. Nagini working as Professor and Head, Department of Computer Science and Engineering, VNR VJIET, has over 22 years of experience in the field of academic research and technological education. She has a multi-disciplinary approach due to varied roles taken up, including but not limited to teaching, research, and administration. She has organized several programs and workshops for the students to provide hands-on experience and bridge the gap between academics and industry—data structures, Linux fundamentals, unified modeling language, block chain technologies, and many more. She has successfully executed one UGC minor project on "Electricity Theft Analysis". She has published more than 29+ research papers which are indexed by Scopus and with good number of citations. Also, she has published four patents. She initiated a coding club "TURING HUT" whose mission is to create a diverse and inclusive tech community to empower the future generation of responsible students through numerous hackathons and coding competitions. She was Faculty Coordinator for annual technical fest convergence, at VNR VJIET during 2018. She is Member of several committees at VNR VJIET Academic Council, IAC, and at department level, she is Member of DAC, DDC, and BoS. Dr. Nagini is honored with "Top Performer" and "Award of Excellence" organized by IIT Bombay and sponsored by SAP India Pvt. Ltd. She is certified by NPTEL as "Top Performing Mentor".

Dr. Valentina E. Balas is currently Full Professor in the Department of Automatics and Applied Software at the Faculty of Engineering, "Aurel Vlaicu" University of Arad, Romania. She holds a Ph.D. Cum Laude, in Applied Electronics and Telecommunications from Polytechnic University of Timisoara. Dr. Balas is Author of more than 350 research papers in refereed journals and international conferences. Her research interests are in intelligent systems, fuzzy control, soft computing, smart sensors, information fusion, modeling, and simulation. She is Editor-in-Chief to *International Journal of Advanced Intelligence Paradigms* (IJAIP) and to *International Journal of Computational Systems Engineering* (IJCSysE), Member in Editorial Board of several national and international journals, and is Evaluator Expert for national, international projects, and Ph.D. thesis.

Dr. Balas is Director of Intelligent Systems Research Centre in Aurel Vlaicu University of Arad and Director of the Department of International Relations, Programs and Projects in the same university. She served as General Chair of the International Workshop Soft Computing and Applications (SOFA) in nine editions organized in the interval 2005–2020 and held in Romania and Hungary. Dr. Balas participated in many international conferences as Organizer, Honorary Chair, Session Chair, Member in Steering, Advisory or International Program Committees, and Keynote Speaker. She has published many books as an editor.

Now she is working in a national project with EU funding support: BioCell-NanoART = Novel Bio-inspired Cellular Nano-Architectures—For Digital Integrated Circuits, 3M Euro from National Authority for Scientific Research and Innovation. She is Member of European Society for Fuzzy Logic and Technology (EUSFLAT), Member of Society for Industrial and Applied Mathematics (SIAM), and Senior Member IEEE, Member in Technical Committee—Fuzzy Systems (IEEE Computational Intelligence Society), Chair of the Task Force 14 in Technical Committee—Emergent Technologies (IEEE CIS), and Member in Technical Committee—Soft Computing (IEEE SMCS).

Dr. Balas was past Vice-president (responsible with Awards) of IFSA—International Fuzzy Systems Association Council (2013–2015), is Joint Secretary of the Governing Council of Forum for Interdisciplinary Mathematics (FIM), A Multidisciplinary Academic Body, India, and Recipient of the "Tudor Tanasescu" Prize from the Romanian Academy for contributions in the field of soft computing methods (2019). She edited more than 140 reputed Book volumes.

Dr. K. Srujan Raju is currently working as Dean Student Welfare and Heading Department of Computer Science and Engineering at CMR Technical Campus. He obtained his Doctorate in Computer Science in the area of network security. He has more than 20 years of experience in academics and research. His research interest areas include computer networks, information security, data mining, cognitive radio networks and image processing, and other programming languages. Dr. Raju is presently working on two projects funded by Government of India under CSRI and NSTMIS, has also filed 7 patents and 1 copyright at Indian Patent Office, edited more than 14 books published by Springer Book Proceedings of AISC series, LAIS series, and other which are indexed by Scopus also authored books in *C Programming and Data Structure*, *Exploring to Internet*, *Hacking Secrets*, and contributed chapters in various books and published more than 30 papers in reputed peer-reviewed journals and international conferences. Dr. Raju was invited as Session Chair, Keynote Speaker, Technical Program Committee Member, Track Manager, and Reviewer for many national and international conferences also appointed as Subject Expert by CEPTAM DRDO—Delhi and CDAC. He has undergone specific training conducted by Wipro Mission 10X and NITTTR-Chennai which helped his involvement with students which is very conducive for solving their day-to-day problems. He has guided various student clubs for activities ranging from photography to hackathon. He mentored more than 100 students for incubating cutting-edge solutions. He has organized many conferences, FDPs, workshops, and symposiums. He has established the Centre of Excellence in IoT, Data Analytics. Dr. Raju is Member of various professional bodies and received Significant Contributor Award and Active Young Member Award from Computer Society of India also served as Management Committee Member, State Student Coordinator, and Secretary of CSI—Hyderabad Chapter.

Contributors

Esraa Abdalla Abdelhamid Faculty of Computer and Information Science, Ainshams University, Cairo, Egypt

D. Abdus Subhahan Department of CSE, BV Raju Institute of Technology, Hyderabad, India

Neha Adawadkar Department of Computer Engineering, Progressive Education Society's Modern College of Engineering, Pune, India

Sukriti Adhikary Department of Computer Science and Engineering, Meghnad Saha Institute of Technology, Kolkata, India

T. Adilakshmi Vasavi College of Engineering, Hyderabad, India

Nimmi Agarwal Department of Computer Science and Engineering, VNR Vignana Jyothi Institute of Engineering and Technology, Hyderabad, Telangana, India

Mahmood Zayid K. Al Ajmi Electrical and Electronics Section, Engineering Department, University of Technology and Applied Sciences-Al Musanna, Al Muladdah, Sultanate of Oman

Fatma Saleh H. AlBuradai Electrical and Electronics Section, Engineering Department, University of Technology and Applied Sciences-Al Musanna, Al Muladdah, Sultanate of Oman

B. Arathi Computer Science and Engineering, Kamala Institute of Technology and Science, Huzurabad, Karimnagar, Telangana, India

Mostafa Aref Faculty of Computer and Information Science, Ainshams University, Cairo, Egypt

M. S. Arpitha Department of Computer Science and Engineering, Vidyavardhaka College of Engineering, Mysuru, Karnataka, India

Md. Ashrafuzzaman Department of Biomedical Engineering, Military Institute of Science and Technology (MIST), Dhaka, Bangladesh

S. K. Ayesha Banu CSE Department, Vignan's Nirula Institute of Technology and Science for Women, Guntur, India

Sachin Babar Department of Computer Engineering, Sinhgad Institute of Technology, Pune, MS, India

G. L. Anand Babu University College of Engineering, Osmania University, Hyderabad, India

Srinivasu Badugu Department of CSE, Stanley College of Engineering and Technology for Women, Abids, Hyderabad, India

Dhanya Sri Bala Department of Computer Science and Engineering, VNR VJIET, Hyderabad, India

Vamsi Krishna Balusu Department of CSE, VNR Vignana Jyothi Institute of Engineering and Technology, Hyderabad, India

Samir Kumar Bandyopadhyay Lincoln University College, Petaling Jaya, Selangor Darul Ehsan, Malaysia

Ananta Banik Department of Biomedical Engineering, Military Institute of Science and Technology (MIST), Dhaka, Bangladesh

Srikanth Bethu Department of CSE, CVR College of Engineering, Hyderabad, Telangana, India;
Department of CSE, NIT Warangal, Warangal, Telangana, India

Deepali Arun Bhanage Symbiosis Institute of Technology, Symbiosis International (Deemed University), Pune, India;
Department of Computer Engineering, Pimpri Chinchwad Education Trust's, Pimpri Chinchwad College of Engineering, Pune, India

S. Bhargavi Latha Department of CSE, GRIET, Hyderabad, Telangana, India

Shripad Bhatlawande Department of E & TC Engineering, Vishwakarma Institute of Technology, Pune, India

Saurabh Bhattacharya Department of Computer Applications, NIT Raipur, Raipur, Chhattisgarh, India

Debnath Bhattacharyya Department of Computer Science and Engineering, Koneru Lakshmaiah Education Foundation, Vaddeswaram, Guntur, Andhra Pradesh, India

Shankar Nayak Bhukya Department of CSE (Data Science), CMR Technical Campus, Hyderabad, Telangana, India

Swapnil Biswas Department of Computer Science and Engineering, Military Institute of Science and Technology, Dhaka, Bangladesh

Payal Bose GLA University, Mathura, India

Debasmita Chakraborty Department of Computer Science and Engineering, Meghnad Saha Institute of Technology, Kolkata, India

Roneel Chand Fiji National University, Suva, Fiji

Soumajit Chatterjee Department of Computer Science and Engineering, Meghnad Saha Institute of Technology, Kolkata, India

Roshani Chavan Department of Computer Engineering, Progressive Education Society's Modern College of Engineering, Pune, India

Moshfiq-Us-Saleheen Chowdhury Department of Electrical, Electronic and Communication Engineering, Military Institute of Science and Technology (MIST), Dhaka, Bangladesh

A. K. Damodaram Department of Mechanical Engineering, Sree Vidyanikethan Engineering College (Autonomous), Tirupati, AP, India

Pratik Desai Department of ECE, RV College of Engineering, Bengaluru, India

S. P. Devika Department of Computer Science and Engineering, Vidyavardhaka College of Engineering, Mysuru, Karnataka, India

Gaurav Dhiman Department of Computer Science, Government Bikram College of Commerce, Patiala, Punjab, India;
Department of Computer Science and Engineering, University Centre for Research and Development, Chandigarh University, Gharuan, Mohali, India;
Department of Computer Science and Engineering, Graphic Era Deemed to be University, Dehradun, India

Junhua Ding Department of Information Science, University of North Texas, Denton, TX, USA

Uma N. Dulhare Computer Science and Artificial Intelligence, Muffakham Jah College of Engineering and Technology, Hyderabad, Telangana, India

Shawni Dutta Lincoln University College, Petaling Jaya, Selangor Darul Ehsan, Malaysia

Vijay Kumar Dwivedi United College of Engineering and Research, Naini, Prayagraj, U.P., India

Jannatul Ferdaus Department of Biomedical Engineering, Military Institute of Science and Technology (MIST), Dhaka, Bangladesh

Tejasri Gangireddy Department of Computer Science and Engineering, VNR VJIET, Hyderabad, India

Krishna Priya V. S. Garimella Vasavi College of Engineering, Hyderabad, India

Steven Fredrick Gilbert Department of Computer Science and Engineering, College of Engineering Guindy, Anna University, Chennai, Tamil Nadu, India

Ravi Chandana Golla Department of Computer Science and Engineering, VNR VJIET, Hyderabad, India

Shiramshetty Gouthami Osmania University, Hyderabad, Telangana, India

Naga Rama K. Guduri Electrical and Electronics Section, Engineering Department, University of Technology and Applied Sciences-Al Musanna, Al Muladdah, Sultanate of Oman

Praveen Kumar Gundaram C R Rao Advanced Institute of Mathematics, Statistics and Computer Science (AIMSCS), University of Hyderabad Campus, Hyderabad, India;
Acharya Nagarjuna University, Guntur, Andhra Pradesh, India

Indu Gurrala Department of Computer Science and Engineering, VNR Vignana Jyothi Institute of Engineering and Technology, Hyderabad, Telangana, India

Ganesan Hari Krishnan Government College of Technology, Coimbatore, India

Darshan Hegde Department of ECE, RV College of Engineering, Bengaluru, India

Nagarathna P. Hegde Vasavi College of Engineering, Hyderabad, India

Nagaratna P. Hegde Vasavi College of Engineering, Hyderabad, Telangana, India

Sally Ismail Faculty of Computer and Information Science, Ainshams University, Cairo, Egypt

Suhasini Itkar Department of Computer Engineering, Progressive Education Society's Modern College of Engineering, Pune, India

Amita Jain Department of Computer Science and Engineering, Netaji Subhas University of Technology, Delhi, India

Ch. Janakamma Osmania University, Hyderabad, India

Midhunchakkaravarthy Janarthanan Lincoln University College, Petaling Jaya, Selangor Darul Ehsan, Malaysia

S. D. Jani Nasima CSE Department, Vignan's Nirula Institute of Technology and Science for Women, Guntur, India

Bharti Joshi Department of Computer Engineering, RAIT, D. Y. Patil University, Nerul, Navi Mumbai, India

Ira Joshi University of Delhi, New Delhi, India

Manuja Joshi Department of E & TC Engineering, Vishwakarma Institute of Technology, Pune, India

Saadman Kabir Department of Biomedical Engineering, Military Institute of Science and Technology (MIST), Dhaka, Bangladesh

T. Kalaiselvi Department of Computer Science and Applications, The Gandhigram Rural Institute (Deemed to be University), Dindigul, Tamilnadu, India

Suresh Kallam Department of CSE, Sree Vidyanikethan Engineering College, Tirupati, AP, India

Komal Karishma The University of the South Pacific, Suva, Fiji

Amandeep Kaur Department of Computer Science and Engineering, University Centre for Research and Development, Chandigarh University, Gharuan, Mohali, India

Prabhjot Kaur Department of Computer Science and Engineering, National Institute of Technology, Uttarakhand, Srinagar, India

Nikhat Raza Khan Department of Computer Science and Engineering, IES College of Technology, Bhopal, M.P., India

S. Kirubakaran Department of CSE, CMR College of Engineering and Technology, Hyderabad, India

Chandrakant Kokane Department of Computer Engineering, Smt. Kashibai College of Engineering, Affiliated to Savitribai Phule Pune University, Pune, MS, India

Mandala Vamsi Krishna Department of Computer Science and Engineering, Vignan's Institute of Information Technology (A), Visakhapatnam, Andhra Pradesh, India

B. P. Deepak Kumar Department of CSE, CMR Technical Campus, Hyderabad, India

Budati Suresh Kumar Department of ECE, Chirala Engineering College, Chirala, India

Chitranjan Kumar Department of Computer Science and Information Technology, Mahatma Gandhi Central University, Motihari, Bihar, India

G. Pavan Kumar Department of CSE, CMR Technical Campus, Hyderabad, India

Nitin Kumar Department of Computer Science and Engineering, National Institute of Technology, Uttarakhand, Srinagar, India

Rakesh Kumar Department of Mathematics, Lovely Professional University, Phagwara, Punjab, India

Vipin Kumar Department of Computer Science and Information Technology, Mahatma Gandhi Central University, Motihari, Bihar, India

Suresh Kumar Kanaparthi Department of CSE-DS, Malla Reddy University, Hyderabad, Telangana, India

Taduvai Siva Likhitha Department of Computer Science and Engineering, Vignan's Nirula Institute of Technology and Science for Women, Guntur, India

Vadlapudi Likitha Department of Computer Science and Engineering, Vignan's Institute of Information Technology (A), Visakhapatnam, Andhra Pradesh, India

Sonal Linda University of Delhi, New Delhi, India

Mudassir Lone Department of Mathematical Sciences, Baba Ghulam Shah Badshah University, Rajouri, J&K, India

K. Reddy Madhavi CSE, Sree Vidyanikethan Engineering College, Tirupati, AP, India;
School of Computing, Mohan Babu University, Tirupati, India

G. S. Mahalakshmi Department of Computer Science and Engineering, College of Engineering Guindy, Anna University, Chennai, Tamil Nadu, India

Parikshit Mahalle Department of Artificial Intelligence and Data Science, Vishwakarma Institute of Information Technology, Pune, MS, India

Souvik Maitra National Institute of Technology Agartala, Agartala, India

Sharat Chandra Manchi Sarapu Vasavi College of Engineering, Hyderabad, India

Meghana Manda Department of CSE, VNR Vignana Jyothi Institute of Engineering and Technology, Hyderabad, India

Vaishnavi Mhamane Department of E&TC Engineering, Vishwakarma Institute of Technology, Pune, India

Hafeena Mohammad Department of CSE (AI&ML), CMR Technical Campus, Hyderabad, Telangana, India

Sallauddin Mohmmad JNTU, Hyderabad, India;
School of Computer Science and Artificial Intelligence, SR University, Warangal, TS, India

P. Mounika CMR Technical Campus, Hyderabad, India

Ramya Sri Myakala Department of Computer Science and Engineering, VNR Vignana Jyothi Institute of Engineering and Technology, Hyderabad, Telangana, India

Ausaaf Nabi University of Delhi, New Delhi, India

Grande. NagaJyothi ECE, Golden Valley Integrated Campus, Madanapalli, AP, India

Grande Nagajyothi ECE, Golden Valley Integrated Campus, Madanapalli, India

Sakshi Nagare Department of E & TC Engineering, Vishwakarma Institute of Technology, Pune, India

Analene Montesines Nagayo Electrical and Electronics Section, Engineering Department, University of Technology and Applied Sciences-Al Musanna, Al Muladdah, Sultanate of Oman

Anusha Nagula Department of Computer Science and Engineering, VNR VJIET, Hyderabad, India

Md. Najim Alam C R Rao Advanced Institute of Mathematics, Statistics and Computer Science (AIMSCS), University of Hyderabad Campus, Hyderabad, India;
School of Computer and Information Sciences, University of Hyderabad, Hyderabad, India

Shaik Najiya Department of Computer Science and Engineering, Vignan's Nirula Institute of Technology and Science for Women, Guntur, India

Thirupathi Rao Nakka Department of Computer Science and Engineering, Vignan's Institute of Information Technology (A), Visakhapatnam, Andhra Pradesh, India

Ananya Nandyala Department of CSE, VNR Vignana Jyothi Institute of Engineering and Technology, Hyderabad, India

Jonnadula Narasimharao Department of CSE, CMR Technical Campus, Hyderabad, India

Eali. Stephen Neal Joshua Department of Computer Science and Engineering, Vignan's Institute of Information Technology (A), Visakhapatnam, Andhra Pradesh, India;
Department of Computer Science and Engineering, Gitam (Deemed to be University), Visakhapatnam, India

Bokka Nishanth Department of Computer Science and Engineering, Vignan's Institute of Information Technology (A), Visakhapatnam, Andhra Pradesh, India

Syeda Ajbina Nusrat Department of Computer Science and Engineering, Military Institute of Science and Technology, Dhaka, Bangladesh

Anand Pande Department of E&TC Engineering, Vishwakarma Institute of Technology, Pune, India

Manju Pandey Department of Computer Applications, NIT Raipur, Raipur, Chhattisgarh, India

Surya Pandey Department of Computer Science and Engineering, New Horizon College of Engineering, Bangalore, India

Atharv Parbalkar Department of E&TC Engineering, Vishwakarma Institute of Technology, Pune, India

Abhishek Patil Department of E & TC Engineering, Vishwakarma Institute of Technology, Pune, India

Chaitra Patwardhan Department of Computer Engineering, Progressive Education Society's Modern College of Engineering, Pune, India

Ambika Vishal Pawar Symbiosis Institute of Technology, Symbiosis International (Deemed University), Pune, India

M. R. Pooja Department of Computer Science and Engineering, Vidyavardhaka College of Engineering, Mysuru, Karnataka, India

Pooja Department of Mathematics, Lovely Professional University, Phagwara, Punjab, India

Tejaswi Potluri Department of Computer Science and Engineering, VFSTR University, Guntur, India

N. Pranay CSE Department, UCOE, Osmania University, Hyderabad, Telangana, India

Y. Prasanthi CSE Department, Vignan's Nirula Institute of Technology and Science for Women, Guntur, India

C. H. Pravallika CSE Department, Vignan's Nirula Institute of Technology and Science for Women, Guntur, India

V. Prithviraj Vellore Institute of Technology, Vellore, India

R. J. Priyankasingh Department of ISE, JSS Science and Technology University, Mysore, India

V. Radha Centre for Cloud Computing, Institute for Development and Research in Banking Technology, Hyderabad, India

Krishna Raghuwaiya The University of the South Pacific, Suva, Fiji

Jai Raj The University of the South Pacific, Suva, Fiji

Thiyagarajan Rajasenbagam Government College of Technology, Coimbatore, India

Sujatha Rajkumar Vellore Institute of Technology, Vellore, India

G. Ramesh Chandra Department of Computer Science and Engineering, VNR VJIET, Hyderabad, India

B. Swaroopa Rani Department of CSE (AI&ML), CMR Technical Campus, Hyderabad, Telangana, India

Somiya Rani Department of Computer Science and Engineering, NSUT East Campus (Formerly AIACTR), Guru Gobind Singh Indraprastha University, Delhi, India

Shriya Rapolu Department of Computer Science and Engineering, VNR Vignana Jyothi Institute of Engineering and Technology, Hyderabad, Telangana, India

K. Ranjith Reddy Department of CSE, CMR Technical Campus, Hyderabad, India

Renuka Vasavi College of Engineering, Hyderabad, India

G. Roja CSE Department, UCOE, Osmania University, Hyderabad, Telangana, India

Dwijen Rudrapal National Institute of Technology Agartala, Agartala, India

M. Sahithi CSE Department, Vignan's Nirula Institute of Technology and Science for Women, Guntur, India

Talasila Dileep Eeswara Sai Department of Computer Science and Engineering, Vignan's Institute of Information Technology (A), Visakhapatnam, Andhra Pradesh, India

Archana G. Said Department of Computer Engineering, RAIT, D. Y. Patil University, Nerul, Navi Mumbai, India

Venkata Sailaja Department of CSE, VNR Vignana Jyothi Institute of Engineering and Technology, Hyderabad, India

Suresh Kumar Sanampudi Department of Information Technology, JNTUH College of Engineering, Jagtial, TS, India

V. Sandhya GITAM School of Technology Bengaluru, Bengaluru, India

Devdas Saraswat Department of Computer Science and Engineering, IES University, Bhopal, India

Niharika Sarkar Department of Computer Science and Engineering, Meghnad Saha Institute of Technology, Kolkata, India

G. Satyavathy Professor & Head, Department of Computer Science with Data Analytics, KPR College of Arts Science and Research, Coimbatore, India

S. Sendhilkumar Department of Information Science and Technology, College of Engineering Guindy, Anna University, Chennai, Tamil Nadu, India

Diganta Sengupta Department of Computer Science and Engineering, Meghnad Saha Institute of Technology, Kolkata, India;
Department of Computer Science and Business Systems, Meghnad Saha Institute of Technology, Kolkata, India

P. V. Shalini Department of Computer Science and Engineering, National Institute of Technology, Warangal, India;
Centre for Cloud Computing, Institute for Development and Research in Banking Technology, Hyderabad, India

Swati Shilaskar Department of E & TC Engineering, Vishwakarma Institute of Technology, Pune, India

K. Shilpa Department of CSE, CMR Technical Campus, Hyderabad, India

Nidhi Shukla United College of Engineering and Research, Naini, Prayagraj, U.P., India

K. Shyamala CSE Department, UCOE, Osmania University, Hyderabad, Telangana, India

Ashutosh Kumar Singh United College of Engineering and Research, Naini, Prayagraj, U.P., India

Maheep Singh Department of Computer Science and Engineering, National Institute of Technology, Uttarakhand, Srinagar, India

Venkatrama Phani Kumar Sistla Department of Computer Science and Engineering, VFSTR University, Guntur, India

Shivali Slathia Department of Mathematics, Lovely Professional University, Phagwara, Punjab, India

K. B. Sowmya Department of ECE, RV College of Engineering, Bengaluru, India

Malladi Sravya Department of Computer Science and Engineering, Vignan's Nirula Institute of Technology and Science for Women, Guntur, India

P. Sriramakrishnan Department of Computer Applications, Kalasalingam Academy of Research and Education, Krishnankoil, Tamilnadu, India

V. Sujatha Department of Computer Science and Engineering, Vignan's Nirula Institute of Technology and Science for Women, Guntur, India

Naheed Sultana Department of IT, Stanley College of Engineering & Technology, Hyderabad, India

J. Suman CSE Department, B V Raju Institute of Technology, Narsapur, Medak, Telangana, India

Gokul Sunilkumar Department of Computer Science and Engineering, College of Engineering Guindy, Anna University, Chennai, Tamil Nadu, India

Gurram Sunitha Department of CSE, School of Computing, Mohan Babu University, Tirupati, India

M. Sunitha Vasavi College of Engineering, Hyderabad, India

M. P. Swapna Research Scholar, Department of Computer Science, Sri Ramakrishna College of Arts and Science for Women, Coimbatore, India

S. Syedsafi Department of Computer Applications, Kalasalingam Academy of Research and Education, Krishnankoil, Tamilnadu, India

Naresh Tangudu IT Department, Aditya Institute of Technology and Management, Tekkali, Srikakulam, AP, India

P. Tarakanadha Reddy School of Computer and Information Sciences, University of Hyderabad, Hyderabad, India;
Centre for Cloud Computing, Institute for Development and Research in Banking Technology, Hyderabad, India

Noshin Tasnim Department of Computer Science and Engineering, Military Institute of Science and Technology, Dhaka, Bangladesh

Peravali Tejaswini Department of Computer Science and Engineering, Vignan's Nirula Institute of Technology and Science for Women, Guntur, India

Advait Thakur Department of Computer Engineering, Progressive Education Society's Modern College of Engineering, Pune, India

Aliya Thaseen Department of IT, Shadan Womens College of Engineering and Technology, Hyderabad, India

Raheem Unnisa Department of CSE, CMR Technical Campus, Hyderabad, India

G. Vani Department of CSE, Sreenidhi Institute of Science and Technology, Hyderabad, India

H. Y. Vani Department of ISE, JSS Science and Technology University, Mysore, India

Jito Vanualailai The University of the South Pacific, Suva, Fiji

A. Veerender Department of CSE—Data Science, CMR Technical Campus, Hyderabad, India

Sireesha Vikkurty Vasavi College of Engineering, Hyderabad, India

Ravi Vinayakumar Center for Artificial Intelligence, Prince Mohammad Bin Fahd University, Khobar, Saudi Arabia

Wattana Viriyasitavat Business Information Technology Division, Department of Statistics, Faculty of Commerce and Accountancy, Chulalongkorn University, Bangkok, Thailand

R. Hitesh Sai Vittal Hyundai Mobis, Hyderabad, India

Sandya Vooradi CMR Technical Campus, Hyderabad, India

Kusum Yadav College of Computer Science and Engineering, University of Ha'il, Ha'il, Kingdom of Saudi Arabia

Kairamkonda Poorna Yamini Department of CSE, VNR Vignana Jyothi Institute of Engineering and Technology, Hyderabad, India

Satyam Yedlla CMR Technical Campus, Hyderabad, India

Meghana Yelamarthi Department of CSE, VNR Vignana Jyothi Institute of Engineering and Technology, Hyderabad, India

Nagendar Yerukala C R Rao Advanced Institute of Mathematics, Statistics and Computer Science (AIMSCS), University of Hyderabad Campus, Hyderabad, India

Sagar Yeruva Department of Computer Science and Engineering, VNR Vignana Jyothi Institute of Engineering and Technology, Hyderabad, Telangana, India

Enneaontology: Toward an Enneagram Personality Detection

Esraa Abdalla Abdelhamid, Sally Ismail, and Mostafa Aref

Abstract Personality prediction catches research attention nowadays. In social media, attracting more users means getting more advertisements. Enneagram is a personality model which is used by psychiatrists. Enneagram is utilized to understand patients' personalities. This knowledge helps them to give the right support. The current method to realize Enneagram is questionnaire based. Humans feel boring to do long questionnaire. Enneagram personality detection system is required. There is not any knowledge representation for the Enneagram. Enneaontology provides an ontology for Enneagram. Enneaontology contains seven classes and nine objects. These classes are Enneagram, key motivation, fear, feature, desire, problem and best classes. These objects are reformer, helper, achiever, individualist, investigator, loyalist, enthusiast, challenger and peacemaker. Enneaontology is designed relative to METHONTOLOGY principles. Enneaontology is evaluated with Enneagram personality detection application. The promising results verify Enneaontology. Enneaontology is the first Enneagram ontology.

Keywords Ontology · Enneagram · Knowledge representation · Personality model · Artificial intelligence · Psychology

1 Introduction

Personality means behavioral, temperamental, emotional and mental characteristics that distinguish a person's individuality (Kaushal and Patwardhan 2018). Personality detection is a hot research topic. Understanding personality can be used in various fields. In social media, understanding the personality aids to attract users and getting more advertisement. Personality recognition is a difficult and complex activity for humans since it demands expert's knowledge. Questionnaire-based tests are the traditional way to know personality. People are unwilling to complete assessment test as it is long and time-wasting (Farnadi et al. 2016).

E. A. Abdelhamid (✉) · S. Ismail · M. Aref
Faculty of Computer and Information Science, Ainshams University, Cairo, Egypt
e-mail: esraa_cs@hotmail.com

A. B. Reddy et al. (eds.), *Proceedings of Third International Conference on Advances in Computer Engineering and Communication Systems*, Lecture Notes in Networks and Systems 612, https://doi.org/10.1007/978-981-19-9228-5_1

Enneagram is a personality model which describe human behavior. The types are reformer, helper, achiever, individualist, investigator, loyalist, enthusiast, challenger and peacemaker. Enneagram is a psychological development with origins in spiritual knowledge from a lot of different old traditions (Riso and Hudson 1999). Enneagram describes person's attitude in a detailed manner. Many universities in the United States are studying the Enneagram in medical, psychology, education, the arts and business (Demir et al. 2020). The Enneagram has the benefit of providing a comprehensive picture of human behavior. The Enneagram may teach a person about his or her strengths, weaknesses and areas for progress (Bland 2010). Enneagram assists in identifying person's directions like desires and fears. Enneagram can give a specific advices and recommendations according to person's personality type. The Enneagram is a useful tool for improving relationships with family, friends and workplace (Baron and Wagele 2009). Enneagram also has other usages. Enneagram assists the counselor in detecting client behavior which enhances healing and growth (Matise 2019). Since 1970, the Enneagram has been applied by psychiatrists (Alexander and Schnipke 2020). Psychiatrists utilize the Enneagram to determine the personality of their patients. This expertise enables them to provide psychological assistance to the patient.

Personality identification is a relatively new research area. The majority of studies do not employ ontology because machine learning is extensively applied. Ontology is used as a knowledge representation in some systems. A text classification ontology based on a three-factor personality model is presented. The model also employs linguistic feature analysis, personality questionnaire and machine learning methods (Sewwandi et al. 2017). A technique for measuring personality was suggested. Texts were classified based on the big five personality traits. An ontology model was used to describe the knowledge. Beside the ontology, other methods were also used (Alamsyah et al. 2019, 2020). A job application system used Twitter text to do a personality measurement based on an ontology model. Big five personality traits ontology identifying is used in order to evaluate user messages (Geovanni et al. 2021).

The absence of knowledge representation is one of the challenges in developing automated personality recognition. The present research aims at improving the use of more convenient methods, preprocessing approaches and the implementation of additional personality models (Ong et al. 2017). Ontology is a knowledge representation technique. Ontology connects concepts, relationships and rules in a certain domain. Ontologies are made up of classes, objects, functions, axioms and relations (Gómez et al. 2006). Ontologies have a lot of benefits including: sharing knowledge perception, separating knowledge from application, reusing knowledge and assisting with knowledge clearance and mapping (Noy et al. 2001). These advantages support developing ontologies. This results in machines that can infer knowledge in a specific area (Berners-Lee et al. 2001). This can be utilized in wide range of applications and industries. The advantage of a good ontology is that supports in the knowledge management systems and semantic development (Raad and Cruz 2015). A methodology is required to build ontology based on it. METHONTOLOGY is a technique for developing ontologies with standards (Fernandez et al. 1997). The METHON-

TOLOGY technique is used in the design. Specification, knowledge acquisition, conceptualization, integration, implementation, evaluation and documentation are the processes of constructing an ontology.

Enneaontology's design is explained in the next sections. The design of Enneaontology is divided into seven classes and nine objects. Enneaontology is developed by METHONTOLOGY and applied to application-based evaluation. The ontology's design is going to be explained. Enneaontology is partially presented as seven classes and three objects (Abdelhamid et al. 2022). Enneaontology's previous design and remaining objects are presented.

2 Building

Enneaontology is the first Enneagram ontology. Enneaontology applies METHONTOLOGY methodology. Enneaontology's objective is to create an automated personality identification system for the Enneagram. Scope has each personality type's basic properties. Questions that arise in the process of developing the ontology are as follows: What are the Enneagram's primary classes? How can these classes be grouped together? What are the connections between these classes? How can these relationships be connected using object properties? and What entities were chosen to personality detection? Information's reference is Enneagram institution's formal description (Enneagram official institute. https://www.enneagraminstitute.com/). Knowledge acquisition is based on text informal and formal analysis. Enneaontology's conceptualization composes of classes and object relations. Enneaontology employs semi-formal format in order to state concepts abstractly. Enneaontology is not integrated with other ontologies. Enneaontology is written in RDF/XML format. For implementation, Protege 5 is utilized. Protege provides many features and options (Horridge et al. 2004). Enneaontology is evaluated using the METHONTOLOGY framework. Using this paradigm, Enneaontology is evaluated according to METHONTOLOGY. Enneaontology is accurate, relatively complete and consistent. Ontology correctness accurately reflects the scope domain. Partially Complete ontology covers concepts within the scope of the ontology's planning usage. Consistency refers to the fact which all concepts are implemented in a similar way. All classes and object attributes in the Conceptual description adhere to the same rules. Documentation consists of this paper and the previous one (Abdelhamid et al. 2022).

There are many types of evaluations for the ontology like expert-based evaluation and application-based evaluation. Expert-based evaluations are expensive and require a lot of time. Domain experts have high costs so the method is not a practical solution to evaluate ontology when the required output is a single task (Sekandar 2018). Application-based evaluations measure the ontology according to application's results which the ontology is part of. Good ontology achieves useful outcomes in the application which intends to provide specific task (Brank et al. 2005). In this method, the application is the real assessment of the ontology. The difficulty of evaluating an ontology can be minimized to that of evaluating a particular task (Sekandar

2018). Application-based evaluation is applied on Enneaontology. The main target of the Enneaontology is to detect the personality. This goal is achieved through task-based application. This method is more suitable as expert-based evaluation is expensive. The task of personality detection is achieved through a system. Several case studies and results have conducted. The output of the system provides positive results.

3 Design

Enneaontology composed of Enneagram, desire, key motivation, fear, feature, best and problem classes. Enneagram personality type is expressed by the Enneagram class. The personality properties are demonstrated through the feature class. FName is an attribute of feature class that describes the names of features. Personality's fears are represented by fear class. FName is an attribute of fear class that represents the names of the fears. Personality's targets and aims are modeled by desire class. Desire class includes the attribute DName which contains the desired names. Personality's motivations drive person's actions which is modeled by key motivation class. The KMName property of the key motivation class represents the names of key motivations. Personality's challenges and difficulties are represented by the problem class. The PName property is part of the problem class. The outstanding personality properties are stated in the best class. When a person uses his or her qualifications well, his or her personality shines and become at his/her best. BName is mentioned as an attribute in best class. Each personality has features, desires, problems, key motivations, fears and best (Abdelhamid et al. 2022) as shown in Fig. 1.

Enneagram personalities are reformer, helper, achiever, individualist, investigator, loyalist, enthusiast, challenger and peacemaker. Every personality has special desires, fears, key motivations, features, problems and best. Each personality represents an object of Enneagram class. The first Enneagram type is reformer. Reformer is an

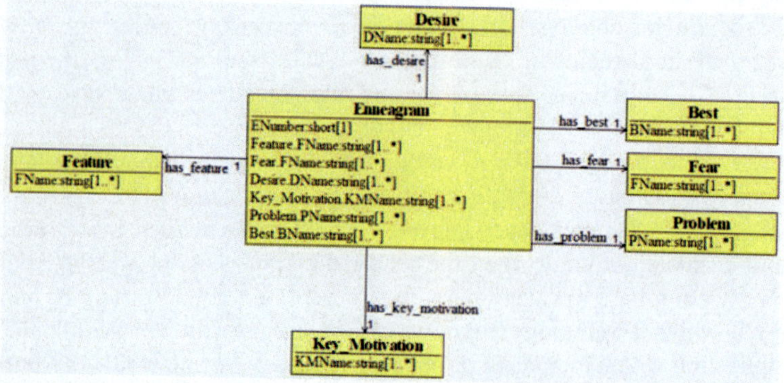

Fig. 1 Enneaontology class (Abdelhamid et al. 2022)

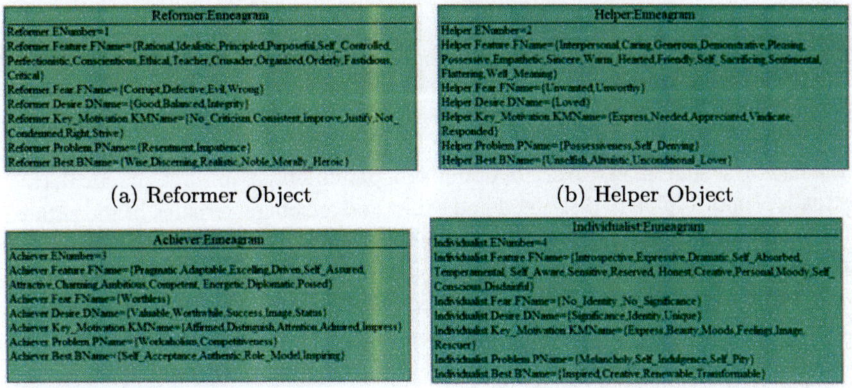

(a) Reformer Object

(b) Helper Object

(c) Achiever Object

(d) Individualist Object

Fig. 2 Reformer, helper, achiever and individualist objects

Enneagram object. The features of a reformer have many like rational, idealistic, perfectionistic, principled and conscientious (Enneagram types description. https://www.enneagraminstitute.com/type-descriptions) as seen in Fig. 2a.

Helper is the second Enneagram object. Caring, demonstrative, generous, possessive and empathetic are some characteristics of helpers (Enneagram types description. https://www.enneagraminstitute.com/type-descriptions) as shown in Fig. 2b.

The third Enneagram object is the achiever. Achiever has characteristics like ambitious, adaptable, driven, charming and image focused (Enneagram types description. https://www.enneagraminstitute.com/type-descriptions) as illustrated in Fig. 2c.

Individualist is the fourth object of the Enneagram. Individualist characteristics include introspective, expressive, dramatic, self-absorbed, temperamental, self-aware, sensitive, reserved, honest, creative, personal, moody, self-conscious and disdainful. Individualist anxieties include having no identity and no significance. Individualist aspirations include having significance, identity and unique. Individualist motives include to express, have beauty, moods, feelings, image and rescuer. Individualist issues include melancholy, self-indulgence and self-pity. Individualist best are to be inspired, creative, renewable and transformable as shown in Fig. 2d.

Investigator is the fifth object of the Enneagram. The characteristics of an investigator include cerebral, perceptive, innovative, secretive, isolated, alert, insightful, curious, independent, inventive, detached, strung, intense and preoccupied. Fears of the investigator are to be useless, helpless and incapable. The desires of the investigator are to be capable and competent. The essential motives of an investigator are to have knowledge, understand, figured out and defending. The issues of the investigator include eccentricity, nihilism and isolation. Investigator best are visionary, pioneers, ahead of time and discoverer as shown in Fig. 3a.

Loyalist is the sixth object of the Enneagram. Engaging, responsible, anxious, suspicious, committed, reliable, hardworking, trustworthy, troubleshooters, cautious, indecisive, defensive, evasive, reactive, defiant, rebellious, cooperative and complainer are characteristics of loyalists. Loyalist worries are to be unsupported and unguided. Security and support are top priorities for loyalists. The primary motives of loyalists are certitude and reassurance. Loyalist issues include self-doubt and suspicion. Loyalist's best are stable, self-reliant and championing as shown in Fig. 3b.

Enthusiast is the seventh object of the Enneagram. Busy, spontaneous, versatile, acquisitive, scattered, extroverted, optimistic, playful, high-spirited, practical, over-extended, undisciplined, distracted and exhausted are characteristics of the enthusiast. Fears of the enthusiast include to be deprived and in pain. Enthusiast demands are to be satisfied, be content, have variety and be fulfilled. Enthusiast's primary motivations are to have freedom, happiness, be excited, be occupied, have experience. Impatience and impulsiveness are common problems among enthusiasts. Enthusiast's best are focused, appreciative, joyous and satisfied as shown in Fig. 3c.

Challenger is the eighth object of the Enneagram. Powerful, dominating, self-confident, decisive, willful, confrontational, strong, assertive, protective, resourceful, straight, egocentric, domineering and intimidating are challenger characteristics. Fears of challengers are to be harmed and controlled. The desires of the challenger are to protect and control. The primary motives of the challenger are to be self-reliant, have strength, be important and dominate. Challenger problems are tempers and vulnerable. Challenger best are self-control, to improve, heroic, magnanimous and inspiring as shown in Fig. 3d.

Peacemaker is the ninth object of the Enneagram. Easygoing, self-effacing, receptive, reassuring, agreeable, complacent, accepting, trusting, stable, creative, optimistic, supportive, complacent and simplifying are all characteristics of a peacemaker. The worries of a peacemaker are loss and separation. The desires of a peacemaker are to have stability and peace. The fundamental motives of a peacemaker are to have harmony and preserve. Inertia and stubbornness are issues for peacemakers. Peacemaker best are indomitable, embracing and healing as shown in Fig. 3e.

4 Discussion

The Enneagram basic characteristic is now modeled using METHONTOLOGY. Enneaontology is evaluated using application-based evaluation. Enneaontology present features, key motivations, best, problems, fears and desires for each type. Enneaontology gives a benefit to develop Enneagram personality detection systems as Enneaontology is the first Enneagram ontology. Enneaontology maps knowledge representation as a part of a system.

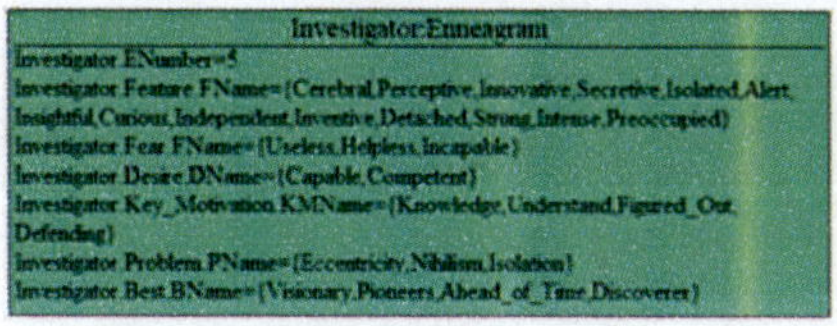

(a) Investigator Object

(b) Loyalist Object

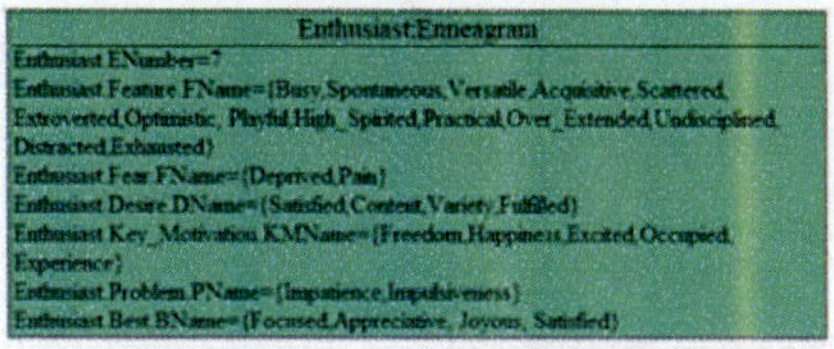

(c) Enthusiast Object

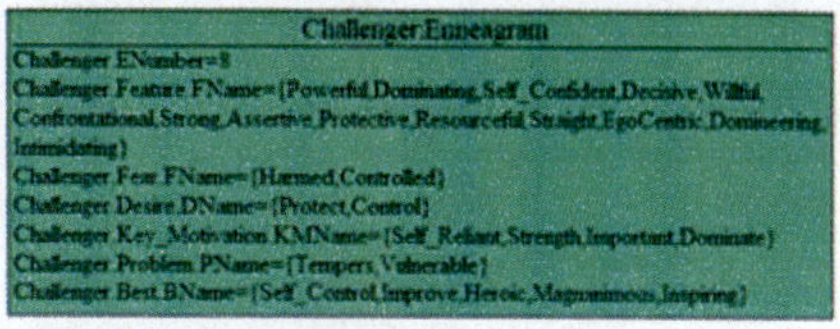

(d) Challenger Object

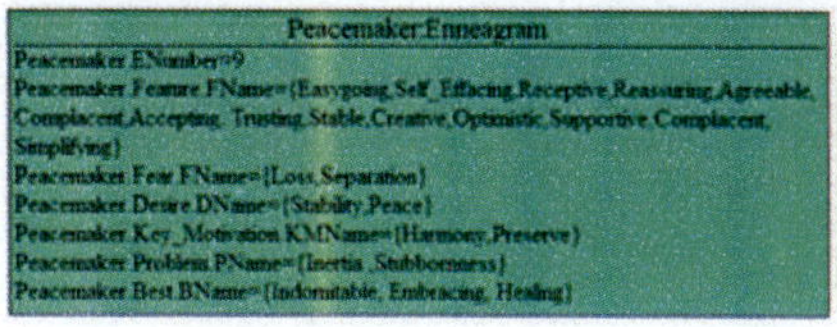

(e) Peacemaker Object

Fig. 3 Investigator, loyalist, enthusiast, challenger and peacemaker objects

5 Conclusion

Enneaontology is developed by METHONTOLOGY design principles. Enneaontology is evaluated by application-based evaluation. Enneaontology composes of seven classes: feature, desire, key motivation, fear, best, problem and Enneagram classes. Enneaontology's objects are nine: reformer, helper, achiever, individualist, investigator, loyalist, enthusiast, challenger and peacemaker objects. Several case studies and results are conducted. The promising results of application verify Enneaontology.

Enneaontology is the ontology first one to model the Enneagram. Enneaontology can be used by other researchers. This research encourages other researchers to create other Enneagram ontologies. In future work, a system will be applied to determine the personality state in the different development levels. Enneagram can predict suicide attempts and disorders in the lowest development levels.

References

E. Abdelhamid, S. Ismail,, M. Aref, Enneaontology: a proposed enneagram ontology, in *2nd International Conference on Ubiquitous Computing and Intelligent Information Systems. Smart Innovation, Systems and Technologies* (Springer, 2022), pp. 541–549

A. Alamsyah, M.F. Rachman, C.S. Hudaya, R.P. Putra, A.I. Rifkyano, F. Nurwianti, A progress on the personality measurement model using ontology based on social media text, in *2019 International Conference on Information Management and Technology (ICIMTech)*, vol. 1 (IEEE, 2019), pp. 581–586

A. Alamsyah, S. Widiyanesti, R.D. Putra, P.K. Sari, Personality measurement design for ontology based platform using social media text. Adv. Sci. Technol. Eng. Syst. **5**(3), 100–107 (2020)

M. Alexander, B. Schnipke, The enneagram: a primer for psychiatry residents. Am. J. Psychiatry Res. J. (2020)

R. Baron, E. Wagele, *The Enneagram Made Easy: Discover the 9 Types of People* (Harper Collins, 2009)

T. Berners-Lee, J. Hendler, O. Lassila, The semantic web. Sci. Am. **284**(5), 34–43 (2001)

A.M. Bland, The enneagram: a review of the empirical and transformational literature. J. Humanist. Couns. Educ. Dev. **49**(1), 16–31 (2010)

J. Brank, M. Grobelnik, D. Mladenic, A survey of ontology evaluation techniques, in *Proceedings of the Conference on Data Mining and Data Warehouses (SiKDD 2005)* (Citeseer, Ljubljana, Slovenia, 2005), pp. 166–170

A. Demir, O. Rakhmanov, K. Tastan, S. Dane, Z. Akturk, Development and validation of the nile personality assessment tool based on enneagram. J. Res. Med. Dent. Sci. **8**(4), 24–32 (2020)

Enneagram official institute. https://www.enneagraminstitute.com/

Enneagram types description. https://www.enneagraminstitute.com/type-descriptions

G. Farnadi, G. Sitaraman, S. Sushmita, F. Celli, M. Kosinski, D. Stillwell, S. Davalos, M.F. Moens, M. De Cock, Computational personality recognition in social media. User Model. User Adap. Inter. **26**(2), 109–142 (2016)

M. Fernandez, A. Gomez-Perez, N. Juristo, Methontology: from ontological art towards ontological engineering, in *Proceedings of the AAAI97 Spring Symposium Series on Ontological Engineering* (Stanford, USA, 1997), pp. 33–40

Y.M.F. Geovanni, A. Alamsyah, N. Dudija et al., Identifying personality of the new job applicants using the ontology model on twitter data, in *2021 2nd International Conference on ICT for Rural Development (IC-ICTRuDev)* (IEEE, 2021), pp. 1–5

A. Gómez-Pérez, M. Fernández-López, O. Corcho, *Ontological Engineering: With Examples from the Areas of Knowledge Management, e-Commerce and the Semantic Web* (Springer Science & Business Media, 2006)

M. Horridge, H. Knublauch, A. Rector, R. Stevens, C. Wroe, A practical guide to building owl ontologies using the protégé-owl plugin and co-ode tools edition 1.0. University of Manchester (2004)

V. Kaushal, M. Patwardhan, Emerging trends in personality identification using online social networks—a literature survey. ACM Trans. Knowl. Discov. Data (TKDD) **12**(2), 1–30 (2018)

M. Matise, The enneagram: an enhancement to family therapy. Contemp. Fam. Ther. **41**(1), 68–78 (2019)

N.F. Noy, D.L. McGuinness et al.: *Ontology Development 101: A Guide to Creating Your First Ontology* (2001)

V. Ong, A.D. Rahmanto, W. Williem, D. Suhartono, Exploring personality prediction from text on social media: a literature review. Internetworking Indones. **9**(1), 65–70 (2017)

J. Raad, C. Cruz, A survey on ontology evaluation methods, in *Proceedings of the International Conference on Knowledge Engineering and Ontology Development, Part of the 7th International Joint Conference on Knowledge Discovery, Knowledge Engineering and Knowledge Management* (2015)

D.R. Riso, R. Hudson, *The Wisdom of the Enneagram: The Complete Guide to Psychological and Spiritual Growth for the Nine Personality Types* (Bantam, 1999)

K. Sekandar, A quality measure for automatic ontology evaluation and improvement. Master's thesis (2018)

D. Sewwandi, K., Perera, S., Sandaruwan, O., Lakchani, A., Nugaliyadde, S., Thelijjagoda, Linguistic features based personality recognition using social media data, in *2017 6th National Conference on Technology and Management (NCTM)* (IEEE, 2017), pp. 63–68

Anomalies Detection on Contemporary Industrial Internet of Things Data for Securing Crucial Devices

Saurabh Bhattacharya and Manju Pandey

Abstract The industrial Internet of things (IIoT) is a framework that enables companies to improve their processes and increase their efficiency by using smart sensors and actuators. It is also used to collect data from various industrial devices and make use of this data to improve their operations. With the help of connected sensors and actuators, companies can save time. Due to the increasing number of industrial Internet of things (IIoT) devices, the data generated by these networks is constantly being analyzed. However, due to the high dimensionality of the data, it can cause fragmentation. Also, datasets collected by IIoT nodes are prone to displaying anomalous events. Unlike traditional networks, which are usually composed of several applications and protocols, IIoT networks are different. They require more stringent security measures to protect the integrity and confidentiality of their data. Due to the unique features of IIoT networks, they are prone to attack. This paper briefly describes the various requirements and challenges faced when it comes to protecting these networks. It also presents evaluation of various machine learning classifiers to detect anomalous behaviors in these networks. Due to the nature of the data that these networks contain anomalies, their analysis is critical to prevent attacks. In this paper, we analyze the usage of various machine learning algorithms to check the efficiency and accuracy of detecting anomalies in data generated by crucial IIoT devices.

Keywords Industrial Internet of things (IIoT) · Anomaly detection · Machine learning

---

S. Bhattacharya (✉) · M. Pandey
Department of Computer Applications, NIT Raipur, Raipur, Chhattisgarh, India
e-mail: babu.saurabh@gmail.com

M. Pandey
e-mail: mpandey.mca@nitrr.ac.in

A. B. Reddy et al. (eds.), *Proceedings of Third International Conference on Advances in Computer Engineering and Communication Systems*, Lecture Notes in Networks and Systems 612, https://doi.org/10.1007/978-981-19-9228-5_2

1 Introduction

The industrial Internet of things is a broad term that refers to the various technologies and services that are being used in the production and operation of industrial equipment and systems. These include machine-to-machine communication, big data, and artificial intelligence (Dhar Dwivedi et al. 2021). The IIoT goes beyond the traditional Internet of things concept by connecting various industrial equipment and systems. It is distinct from the concept of the IoT due to the way it combines the various aspects of information technology and operational technology. Combining operational technology (OT) and information technology (IT) enables industries to improve efficiency and reduce costs. For instance, they can save time and money by monitoring and controlling the various physical infrastructures in an industrial facility. The IIoT is expected to play a leading role in the fourth industrial revolution (Industry 4.0), which is based on the use of big data and artificial intelligence. The IIoT can provide real-time insights into industrial facilities through the data collected by sensors.

Through the IIoT, machines can now perform previously impossible tasks that previously could only be performed by humans. It can also help transform cities into smart cities and factories into smart factories. Consistent data transmission between various smart devices and machines provides enterprises with valuable insights that can help them improve their operations and finance. This data can also inform them about potential issues and provide them with actionable information. Through the IIoT, businesses can now collect and analyze vast amounts of data faster. This will allow to improve their performance and better understand their operations.

The remaining part of the paper is as follows: Sect. 2 describes literature review, Sect. 3 describes various security concerns and challenges in implementing the IIoT, Sect. 4 describes anomaly in IIoT data, Sect. 5 describes methodology, Sect. 6 describes the various results, and Sect. 7 concludes the paper.

2 Literature Review

We present an overview of literature related to anomaly detection in IIoT work. Ikram et al. (2022) aim to improve the search performance of the MNSWOA-IPM-RF by introducing a minimal features extraction model. Zhan et al. (2021) propose a method for developing a hierarchical representation of time series anomaly detection called HR-AD. Selim et al. (2021) focus on developing a machine learning system that can analyze and classify IIoT-based water infrastructures using various machine learning algorithms. Yan et al. (2020) aim to develop a new algorithm that can efficiently mine the data from the Internet of things (IoT) for untrustworthy information. Shah and Tiwari (2018) explore various techniques that can be utilized to detect anomalous behaviors in IIoT data collected from various engines. Wang et al. (2021) proposed a new federated learning method to improve the security and privacy of industrial

Internet of things (IIoT) applications, and a new type of anomaly detection strategy for IIoT using the federated learning method is proposed.

You and Tang (2021) described the performance of various PSOstrategies studied to determine their suitability for the IIoT environment. Zaminkar and Fotohi 2020) proposed a method that automatically detects the presence of sinkholes in the Internet of things using the RPL routing protocol. Radanliev et al. (2021) presented a study that aims to identify a framework that will enable the continuous improvement of AI by automating various tasks and systems. Genge et al. (2019) aim to develop a methodology that can detect abnormal behaviors in the IIoT in the context of its aging state.

Han et al. (2021) proposed the concept of the CCoV which was proposed to provide a measure of the value of the information that the IIoT nodes produce.

Kißkalt et al. (2018) The implementation of condition monitoring in production machinery can be challenging due to the various factors affecting communication between different systems. Zhou and Guo (2018) Due to the unique features of IIoT networks, they are prone to attack. It also presents various methods that are designed to detect anomalous behaviors in these networks. Zhou et al. (2021). The VLSTM model was introduced to introduce a generic framework that can be used to create a variety of networks and procedures related to the estimation and compression of data.

After an initial overview of related work, we examine specific prior work on anomaly detection in IIoT with machine learning and motivate us to perform the following work.

3 Various Security Concerns and Challenges in Implementing the IIoT

The industrial Internet of things can transform how businesses operate, but its implementation can expose them to risks. Many companies are not yet fully prepared to handle the complex issues associated with the IIoT. Due to the complexity of the IIoT's integration into the Internet, many organizations are not yet ready to fully implement it. They need to consider various factors such as availability, scalability, and security. Despite the potential benefits of the IIoT, many organizations still have outdated systems and processes. This can keep them from fully implementing the technology. Due to the increasing number of connected devices, many security vulnerabilities could be exploited. Due to the rise of the Internet of things, many security concerns have emerged. One of these is the failure to secure devices properly. This can be caused by a variety of factors, such as lack of proper authentication. The threat landscape has also changed due to the number of connected devices. Unsecure IIoT systems can lead to financial loss and operational issues for organizations. They can also expose them to various security risks.

While productivity is crucial for an IIoT system, security should also be a priority. Without proper protection, the system could be susceptible to exploitation. Having a security operations center can help organizations monitor and respond to threats. It can help prevent unauthorized users from accessing the system. Security operations centers (SOCs) can also help facilities monitor and respond to potential threats.

Having a comprehensive security stack built into an IIoT system can help organizations seamlessly operate their operations. It can help them manage the various devices and applications part of the system. The network area of an IIoT system is divided into various parts. One of these is the gateway, which is responsible for receiving and storing data from the devices. The other is the control center, which is the most critical part of the system. This can help prevent hackers from accessing the data stored on the system's servers.

4 Anomaly in IIoT Data

Anomaly refers to the unusual behaviors that occur in datasets. While it can be interpreted as an instance of random data that no longer fits the usual pattern, anomalies are groups or correlated with other datasets.

4.1 Types of Anomalies

Point anomaly: An anomaly is a point that can be considered different from the rest of the data. For instance, if a large cluster of points is outside a typical region, it might be considered an anomaly. Contextual anomaly: A contextual anomaly can be considered only in a specific context; an example is a data point's latitude and longitude, which can be used to determine its location. Collective anomaly: A collective anomaly occurs when a group of data points appears to be anomalous compared to the rest of the dataset. For instance, a script might be considered anomalous if its actions are performed in a sequence.

4.2 Methods to Resolve Anomalies

Unsupervised Anomaly Detection: Unsupervised anomaly detection is a method that does not require training data. Instead, it considers the data type and the statistical difference between the standard and anomalous samples. Then, it clusters the data using a similarity measure and considers the data points that are far away from the cluster as anomalies.

Supervised Anomaly Detection: Supervised anomaly detection is a process that involves labeling a dataset and constructing a predictive model to classify its data points. This process requires a dataset that contains both normal and anomalous samples. The most commonly used algorithms are supervised neural networks and K-nearest neighbors.

5 Methodology

5.1 Dataset

We use TON_IOT (Moustafa 2021) dataset for IIOT anomaly detection. The datasets were collected from various sources such as telemetry data, operating systems data, and network traffic datasets. They were created using a large-scale network designed and built by the ADFA, SEIT, and the Cyber Range.

The goal of the testbed was to establish an industry 4.0 network that includes the IIoT and the IoT. It was deployed using a combination of virtual machines and Kali and Linux operating systems hosts. Various attack techniques, such as denial of service (DoS), ransomware, and distributed denial of access (DDoA), are used against Web applications and computer systems connected to the Internet of things (IoT). Therefore, the datasets were collected in parallel to collect various cyberattack events. Figure 1 represents the dataset in tree viewer form.

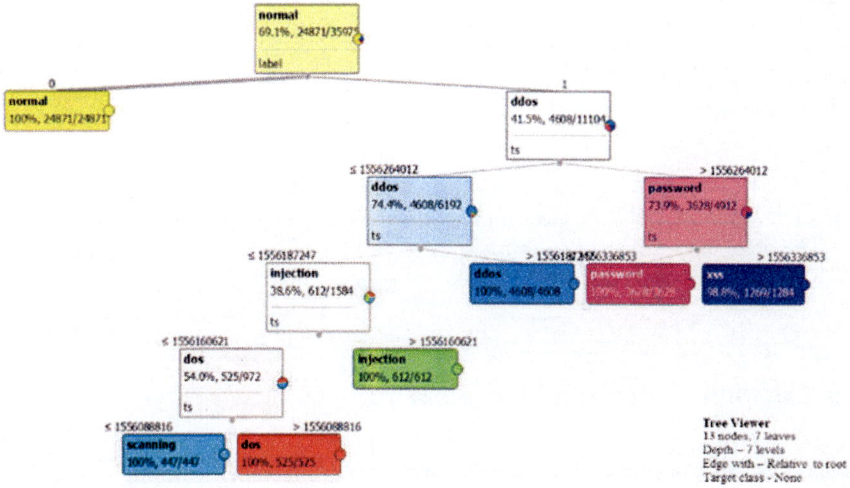

Fig. 1 Tree view of dataset

5.2 Classification Algorithm

We used five various classifiers to check the accuracy of detecting an anomaly in the available dataset.

5.3 Random Forest

The random forest is a type of machine learning that takes the average of the various decision trees in a dataset and predicts the outcome of the given task. It does so by taking the predictions from each tree and using most of them to generate the final output. The obtained results are as follows: AUC = 100%, classification accuracy = 99.99%, $F1 = 99.99\%$, precision = 99.99%, recall = 99.99%

5.4 Logistic Regression

A popular technique in machine learning for predicting a variable's output is logistic regression. Instead of giving a specific value, the method predicts the output of the variable by taking into account its independent variables. This method can perform various tasks, such as analyzing data and identifying the most effective ones. The obtained results are as follows: AUC = 83.14%, classification accuracy = 62.70%, $F1 = 58.69\%$, precision = 55.21%, recall = 62.70%

5.5 Stochastic Gradient Descent (SGD)

A stratified system or process is a random probability-based process that can generate a gradient. In most cases, the descent is performed on a dataset instead of the whole set. The obtained results are as follows: AUC = 99.94%, classification accuracy = 99.95%, $F1 = 99.93\%$, precision = 99.91%, recall = 99.95%

5.6 Support Vector Machine (SVM)

Support vector machine is a simple supervised machine learning algorithm commonly used for classification and regression. It finds a boundary between the data types that it encounters. In SVM, we plot the data items in a dataset in an N-dimension space. Then, find the optimal hyperplane for separating the data. Unfortunately, this method only works for binary classification. There are various techniques to perform multi-class problems. The obtained results are as follows: AUC = 99.99%, classification accuracy = 99.84%, $F1$ = 99.82%, precision = 99.80%, recall = 99.84%

5.7 *Naïve Bayes*

The Naive Bayes algorithm is based on the Bayes theorem. It is a simple and most effective method for building fast machine learning models. It uses a probabilistic approach to classify objects. The obtained results are as follows: AUC = 98.61%, classification accuracy = 75.82%, $F1$ = 82.80%, precision = 95.81%, recall = 75.82%

6 Results

6.1 *Receiver Operating Characteristic (ROC) Curve*

ROC analysis is a tool that enables users to identify optimal models and discard suboptimal ones. It is a direct and natural way to analyze the diagnostic decision-making process. The receiver operating characteristic curve is a plot that shows the diagnostic capabilities of a binary classification system. It plots the true-positive rate against the false-positive rate at various threshold conditions. Figure 2 represents the ROC curve.

6.2 *Evaluation Parameters*

Machine learning model evaluation is a process that involves analyzing the performance of a model. Here, five different parameters are used to evaluate the various classifiers, namely classification accuracy (CA), $F1$-score, precision, recall, and area under curve (AUC). Figure 3 shows random forest outperform as compared to other classifiers.

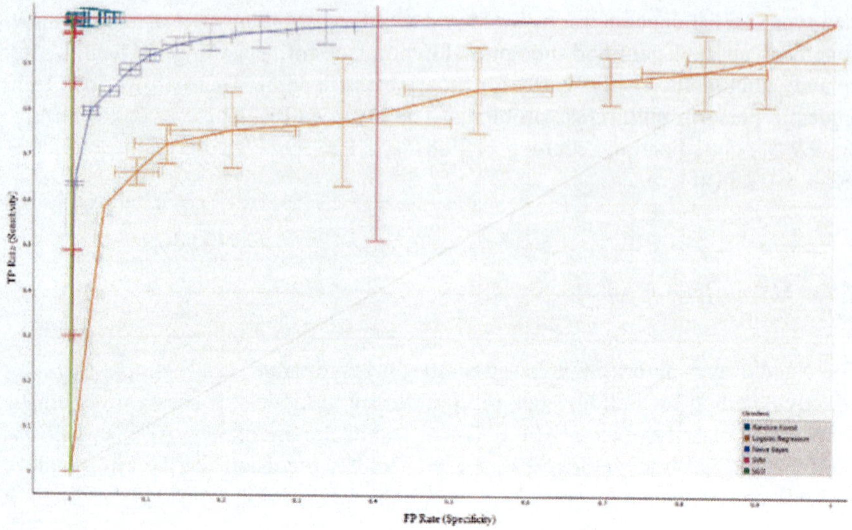

Fig. 2 ROC curve

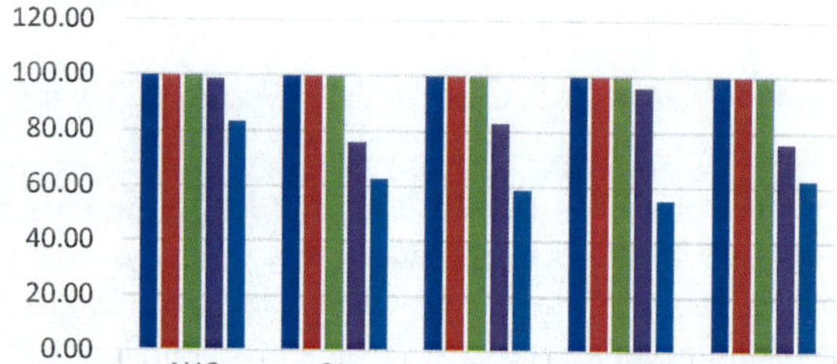

	AUC	CA	F1	Precision	Recall
SVM	99.99	99.84	99.82	99.80	99.84
SGD	99.94	99.95	99.93	99.91	99.95
Random Forest	100.00	99.99	99.99	99.99	99.99
Naive Bayes	98.61	75.82	82.80	95.81	75.82
Logistic Regression	83.14	62.70	58.69	55.21	62.70

■ SVM ■ SGD ■ Random Forest ■ Naive Bayes ■ Logistic Regression

Fig. 3 Comparison of various classifiers

7 Conclusion

The rapid emergence of the industrial Internet of things has raised the need for more effective security measures. This article aims to provide an in-depth analysis of this technology's various aspects of machine learning security. The outcome of the analysis led to the conclusion that out of the five classifiers, random forest provides high accuracy, almost 100% (Fig. 3), when detecting anomalous data. However, in terms of execution time, random forest is the most appropriate choice for detecting anomalous data. In addition, though the data in the dataset is small, the result may vary when tested with vast amounts of data.

References

A. Dhar Dwivedi, R. Singh, K. Kaushik, R. Rao Mukkamala, W.S. Alnumay, Blockchain and artificial intelligence for 5G-enabled internet of things: challenges, opportunities, and solutions. Trans. Emerg. Telecommun. Technol., 1–19 (2021). https://doi.org/10.1002/ett.4329

B. Genge, P. Haller, C. Enachescu, Anomaly detection in aging industrial internet of things. IEEE Access **7**, 74217–74230 (2019). https://doi.org/10.1109/ACCESS.2019.2920699

G. Han, J. Tu, L. Liu, M. Martinez-Garcia, Y. Peng, Anomaly detection based on multidimensional data processing for protecting vital devices in 6g-enabled massive IIoT. IEEE Internet Things J. **8**(7), 5219–5229 (2021). https://doi.org/10.1109/JIOT.2021.3051935

S.T. Ikram, V. Priya, B. Anbarasu, X. Cheng, M.R. Ghalib, A. Shankar, *Prediction of IIoT Traffic Using a Modified Whale Optimization Approach Integrated with Random Forest Classifier*, no. 0123456789 (Springer US, 2022)

D. Kißkalt, H. Fleischmann, S. Kreitlein, M. Knott, J. Franke, A novel approach for data-driven process and condition monitoring systems on the example of mill-turn centers. Prod. Eng. **12**(3–4), 525–533 (2018). https://doi.org/10.1007/s11740-018-0797-0

N. Moustafa, The TON_IoT datasets | UNSW research (2021) [Online]. Available: https://research.unsw.edu.au/projects/toniot-datasets

P. Radanliev, D. De Roure, M. Van Kleek, O. Santos, U. Ani, Artificial intelligence in cyber physical systems. AI Soc. **36**(3), 783–796 (2021). https://doi.org/10.1007/s00146-020-01049-0

G.E.I. Selim, E.E.D. Hemdan, A.M. Shehata, N.A. El-Fishawy, Anomaly events classification and detection system in critical industrial internet of things infrastructure using machine learning algorithms. Multimed. Tools Appl. **80**(8), 12619–12640 (2021). https://doi.org/10.1007/s11042-020-10354-1

G. Shah, A. Tiwari, Anomaly detection in IIoT: a case study using machine learning, in *International Conference Proceeding Series* (2018), pp. 295–300. https://doi.org/10.1145/3152494.3156816

X. Wang et al., Towards accurate anomaly detection in industrial internet-of-things using hierarchical federated learning. IEEE Internet Things J. 4662(c):1–10 (2021). https://doi.org/10.1109/JIOT.2021.3074382

X. Yan, Y. Xu, X. Xing, B. Cui, Z. Guo, T. Guo, Trustworthy network anomaly detection based on an adaptive learning rate and momentum in IIoT. IEEE Trans. Ind. Inform. **16**(9), 6182–6192 (2020). https://doi.org/10.1109/TII.2020.2975227

Q. You, B. Tang, Efficient task offloading using particle swarm optimization algorithm in edge computing for industrial internet of things. J. Cloud Comput. **10**(1) (2021). https://doi.org/10.1186/s13677-021-00256-4

M. Zaminkar, R. Fotohi, SoS-RPL: securing internet of things against sinkhole attack using RPL protocol-based node rating and ranking mechanism. Wirel. Pers. Commun. **114**(2), 1287–1312 (2020). https://doi.org/10.1007/s11277-020-07421-z

P. Zhan et al., Temporal anomaly detection on IIoT-enabled manufacturing. J. Intell. Manuf. **32**(6), 1669–1678 (2021). https://doi.org/10.1007/s10845-021-01768-1

X. Zhou, Y. Hu, W. Liang, J. Ma, Q. Jin, Variational LSTM enhanced anomaly detection for industrial big data. IEEE Trans. Ind. Inform. **17**(5), 3469–3477 (2021). https://doi.org/10.1109/TII.2020.3022432

L. Zhou, H. Guo, Anomaly detection methods for IIoT networks, in *Proceedings of the 2018 IEEE International Conference on Service Operations and Logistics, and Informatics, SOLI 2018* (2018), pp. 214–219. https://doi.org/10.1109/SOLI.2018.8476769

Automatic Sentiment Analysis Scalability Prediction for Information Extraction Using SentiStrength Algorithm

Shiramshetty Gouthami and Nagaratna P. Hegde

Abstract The social media platforms enable their users to provide feedback and voice complaints about the services and goods they have used. Sentiment analysis is a powerful tool that can help the software industry and company to better evaluate user needs and cater the software in a way to maximize the sales potential. One of the studying areas in natural language process (NLP) is sentiment analysis which is concerned for identifying the opinion or mood within a text. For extracting the information from the social big data, an automatic procedure is essential for decision-makers and marketers. For satisfying this requirement, an automatic sentiment analysis saleability prediction for information extraction using SentiStrength algorithm is presented in this paper. From consumers, data is collected through feedback forms on software product. Presented algorithm validity is proven through comparing the contrast rule-based sentiment analysis (CRbSA), general word counting, and extraction algorithms well-known sentiment information. Accuracy and processing time are two parameters used to analyze the performance of SentiStrength algorithm, and these values are 81.5% as accuracy and 15 ms as processing time. In a marketing system, this algorithm is employed for extracting the satisfaction of customers in particular, it serves as a warning mechanism for unfavorable remarks.

Keywords Natural language processing (NLP) · SentiStrength · Sentiment analysis · Scalability · CRbSA · Reviews

S. Gouthami (✉)
Osmania University, Hyderabad, Telangana, India
e-mail: gouthami.shiramshetty@gmail.com

N. P. Hegde (✉)
Vasavi College of Engineering, Hyderabad, Telangana, India

A. B. Reddy et al. (eds.), *Proceedings of Third International Conference on Advances in Computer Engineering and Communication Systems*, Lecture Notes in Networks and Systems 612, https://doi.org/10.1007/978-981-19-9228-5_3

1 Introduction

Currently, the majority of large enterprises are working hard to increase customer satisfaction because it helps for making better decisions as well as creating better strategies (Jie 2021). With the rapid increase of advent technologies like Internet of things (IoT), social network platforms, and cloud computing enable Internet users to provide feedback on services and goods (Çataltaş et al. 2020). In this case, the feedback of user refers the satisfaction of customer provided in the digital form. The challenges in these social marketing fields concern how this feedback information is used for improving their services and products. Every organization looks to maximize its profit through understanding customer needs and meeting them accordingly. These software companies then rely upon structured data in the form of feedback forms which require customer to rate their satisfaction on a numeric range to evaluate the areas which need improvement. Therefore, organizations should analyze the untapped potential of unstructured data (the quantity of which is exponentially greater as well captures the emotions of customers in words) in the form of customer reviews to enhance and predict the scalability of the product. Always "thinking nature of people" is a vital data part to everyone particularly in decision-making. The sentiment analysis is utilized for identifying the data polarity by NLP (Razzaq et al. 2019). In addition, it is defined as opinion mining as it analyzes the user attitude or opinion. The sentiment analysis technique is used to analyze how a user can think related to that specific topic. This can help for finding the document total contextual polarity and sentiments on various levels. It can categorize the entire document which determines the user words or expressions response in the whole document. Several companies have been utilized the sentiment analysis for observing their services and products (Zvarevashe and Olugbara 2018). Sentiments are depicted as emoticons, ideas, or opinions hued or incited by the emoticons. The sentiment analysis is an actual NLP task which can examine the sentiments, feelings, or attitudes of individuals toward the particular elements. It can be an inspection of utilizing a machine for selecting an appraisal polarity if it is negative, positive, or neutral (Solangi et al. 2018). This sentiment analysis is utilized for analyzing the polarity of data and accessing the input text positive and negative comments. The primary goal of sentiment analysis is to evaluate numerous items that are the subject of online discussion. The issue is extracting the information to a specific brand or product and classifying the opinions based on different product features. Based on the same process, this can involved feature extraction and classifier training. The product features have to be determined. These features are classified as negative and positive. The requirement of sentiment analysis is arising for the products because of unavailability of single, combined evaluation of several goods. The customer can come across negative and positive reviews and opinions, however, integrating these varying opinions for getting a deterministic result will be difficult. The customer is able to make an informed decision about a product prior to purchasing it because to this sentiment analysis of product reviews. This examination allows them for picking the right brand and product that can suit to their needs. Customers can choose wisely by using the characteristics and their evaluations (Srivats Athindran et al. 2018). This research aiming for the contribution of extracted information relies on SentiStrength algorithm. First based on NLP, the text mining is utilized for solving the unstructured information (Sintoris and Vergidis 2017). From such information, the object semantic meaning is determined. Then, the structures of sentence can be analyzed for categorizing the 3 groups: compound, complex, and simple. Eventually, every sentence is processed by the word counting for sentiment identification.

2 Literature Review

Alengadan and Khan (2017) suggested an aspect-based opinion mining method. Customer feedback is used to identify sentiment based on how satisfied customers are with product components or sub-customers (e.g., mobile phone display). It can provide more data to the marketers and support customer opinions on the products. However, it consumes more time for investigating the huge number of feature contrasts. Moharasar and Ho (2016) presented a new two-stage semi-supervised architecture for exploiting abundant annotated clinical text for automatically detecting the temporal data and gradually increases the annotated corpora size; this can lead to improvement in the accuracy of temporal data extraction. Firstly, possible features are generated from annotated corpora and select the significant features. At last, presented model is trained and evaluated with the selected features. This approach has attained 81.34% of F-measure and 75% of temporal data extraction. Lovelin et al. (2015) demonstrated an investigation on the reviews social media sentiment analysis. Massive volumes of data will be presented over the Web for Internet users with the rapidly increasing Web technologies. In addition, it became the place for ideas exchanging and online learning. The data collection has become more significant for finding the people thinking about any product, services, or organizations. This work mainly focused on different approaches and techniques to classify the opinion from the datasets of social media and its future aspects. Cernian et al. (2015) presented a semantic technique for the application of sentiment analysis and is based on the lexical resources of SentiWordNet. Effective connection among these aspects depends on opinion mining tools that automatically processed the reviews and opinions of consumers about the services or products. This validation has proved that it achieved 61% of application average success rate. Tan et al. (2015) described a rule-based sentiment analysis to classify the financial news articles polarity. This system has utilized a predefined polarity lexicon for the differentiation of negative and positive financial news stories. A dataset of manually annotated financial news articles collected from several online financial news sources is used to assess the sentiment analyzer's performance. The results demonstrated that this sentiment analyzer has achieved 75.6% of F-score for positive as well as negative classifications. Bafna and Toshniwal (2013) introduced a similar technique as well as maintains the same principal; frequently, the attributes of the product appear more than word phrases or other words in the customer reviews of a product. Hence, association rule mining can be utilized for extracting the nouns that appears more often. However, this rule mining might generate several features which are not represents the product features in real. Thus, they mentioned a characteristic or probabilistic power equation for removing the entire features of candidates that do not represent the actual features. Usha and Indra Devi (2013) presented a novel approach, namely combined sentiment topic (CST) approach, for identifying the topics and sentiments from the text. This approach works premised on the Gibbs sampling method. The objective of this analysis is identifying the emotional states in the online text. Unlike the supervised techniques for opinion mining that frequently fails to produce better performance while shifting to other fields, CST unsupervised nature made it as highly portable for other fields. This CST approach has better performance than the earlier supervised models. Ding et al. (2008) focus on product reviews from customers. They examined the issue of identifying the (positive, negative, or neutral) semantic orientations of opinions stated on product aspects in reviews. By utilizing outside facts and the linguistic standards of natural language expressions, they have suggested a comprehensive lexicon-based strategy to solving the problem. This method enables the system to tackle opinion words that depend on context, which provide significant challenges for the existing algorithms. Based on the suggested technique, a system called opinion observer has been put into place. The suggested technique is extremely effective, according to experimental outcomes employing a benchmark product review dataset and few supplementary reviews.

3 Sentiment Analysis Saleability Prediction for Information Extraction

The architecture of the automatic sentiment analysis saleability prediction for information extraction using the SentiStrength algorithm is represented in Fig. 1.

The data has been collected from consumers through feedback forms on software product. These forms were forwarded to consumers and consist of six questions. For the first five questions, the consumers give feedback in a text field. The last question asks the consumers about their likelihood to buy the product on a scale of 1–10 where 1 being ‘not at all likely’ to 10 being ‘definitely going to buy’. The dataset consists of 10,178 feedback forms gathered through software company’s consumers regarding several software products.

Preprocessing of collected data comprises of tokenization, cleaning the data, removing stop words, and lastly lemmatization. Cleaning the data removing noise is an essential part of the process. Text contains noise in the form of punctuation, abbreviations, etc., that play no role in the analysis of the sentiment of the text and hence, need to remove for making the process efficient and faster. Stop Words Removal: All languages contain stop words. These are words which are repeated and used regularly for instance ‘of’, ‘for’, ‘to’, ‘a’. The existence of stop words usually has no

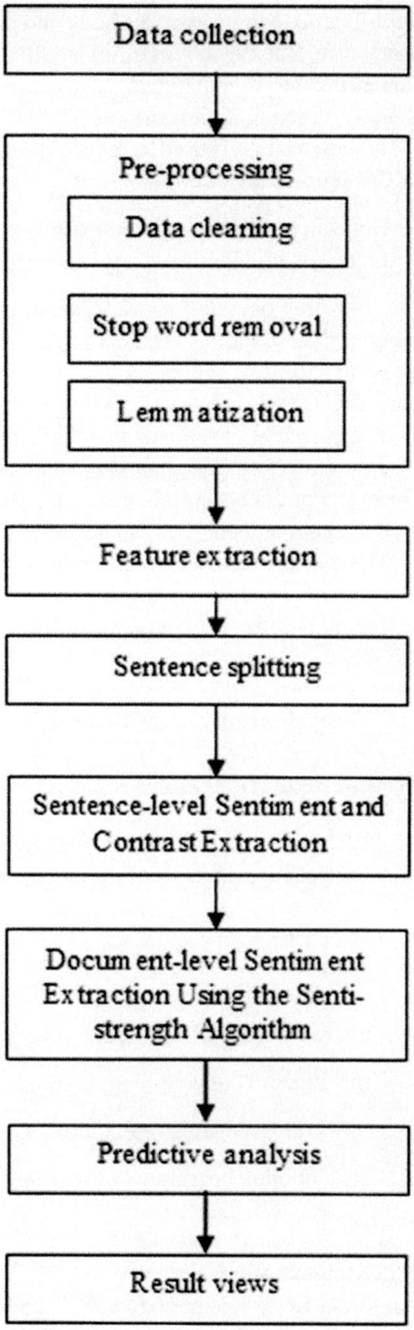

Fig. 1 System architecture overview

contribution to the message's body and particularly does not affect the implication of a sentence. But they do occupy a significant space in the size of the message. Besides increasing the computation overhead, they result in the accumulation of small errors, thereby effecting final result.

Lemmatization is used to reduce the token to its root word or lemma. The benefits of this process are twofold:

- Helps in reducing the size of text being processed
- Increased probability of a given word.

After preprocessing the data, features are extracted by using feature extraction module. The dividing sentences with regular expression model divide the string material into string sentences. This sentence-splitting component's goal is to enhance performance while extracting sentiments from the data.

The objective of sentence-level sentimental extraction is improving the accuracy of data collection for attaining better accuracy in terms of sentiment extraction. General word counting algorithm is utilized for discovering the contrast words as well as sentiments in every sentence.

The final process of extracting the sentiment data from the feedbacks of customer is document level sentiment extraction by the SentiStrength algorithm. A sentence sentiment is relies on the contrast. The SentiStrength algorithm will re-summarize the sentence-level sentiment extracted data and contrast for every sentence. Then, all sentences negative and positive percentages are calculated. At last, the document sentiment information is defined based on positive sentences average percentage.

SentiStrength Algorithm

1. Let R be the set of reviews, i.e., $R = \{r_1, r_2, r_3, r_n\}$, where each review *ri*.
2. Spilt each review *ri* into a sentence set S. S is composed of number of sentences, i.e., $S = \{S_1, S_2, S_{3,} S_m\}$. Each sentence S_j S.
3. For each sentence S_j S do,

   Extract opinion bearing words For each word w_q in the sentence do

   Traverse the lexicon to get

   the numeric polarity score If (score = exist) SO (w_q) = recorded

   // SO, i.e., semantic orientation of

   each opinion bearing word is recorded

   end if

   Sentence Score (S_j) =) end for
4. Total Review Score (*ri*) = $\sum_{j=1}^{m}$ Sentiment score S_j

   // Total review score (TRS) now contains the polarity numeric score of the review
5. Positive_count = Count ("Positive")

6. Negative_count = Count ("Negative")
7. Positive_average = 100 * (Positive_count/(Positive_count + Negative_count)) %
8. IF (Positive_average > 55%) THEN
9. Sentiment = "Positive"
10. ELSE IF (Positive_average < 45%) THEN
11. Sentiment = "Negative"
12. ELSE Sentiment = "Neutral"
13. RETURN review set

Here, the sentiment is defined as positive if the average percentage of 'positive' is more than 55% and 'negative' if the average is less than 45%. In between 45 and 55% is referred as 'neutral'.

4 Experimental Results

The implementation of the system is done using Python in Spyder. Several libraries are used including Scikit-learn, TextBlob, Matplotlib, Pandas, Numpy, etc. The models (SentiStrength, general word counting, and CRbSA algorithms) are tested over a public dataset that contains three different transactions: negative, positive, and neutral, and each one contain 100 user feedback transactions of services and products from Amazon. While the SentiStrength has exploited the contrast words and conjunction words in the algorithm. For verifying the results of presented algorithm, the number of correctly extracted data is compared based on these three algorithms, and it is represented in Table 1.

Table 1 Comparative performance of three algorithms

Algorit hm	Sentiment	Extract number in 100 transactions		Total correct predictions
		Correct	Incorrect	
SentiStrength	Positive	84	19	173
	Negative	79	21	
	Neutral	10	90	
CRbSA	Positive	67	33	147
	Negative	68	32	
	Neutral	12	88	
Genera l word counting	Positive	63	37	130
	Negative	28	72	
	Neutral	39	61	

Table 2 Processing time and accuracy parameters analysis

Algorithm	Processing time for 100 transactions (ms)	Accuracy (%)
SentiStrength	15	81.5
CRbSA	24	67.5
General word counting	28	45.5

Accuracy and processing time are used as the measures to evaluate the performance of information extraction model. In this case, the accuracy is defined as the total success of sentiment analysis model, i.e., ratio of total number of correct predictions to the total predicted sentences reviews:

$$\text{Accuracy} = \frac{\text{TP} + \text{TN}}{\text{TP} + \text{TN} + \text{FN} + \text{FP}} \tag{1}$$

where TP defines true positive, FP indicates false positives, TN referred as true negative, and FN is referred as false negative. The time to process and overall accuracy of the three algorithms is represented in Table 2. Presented algorithm has high accuracy and faster processing time compared to general word counting and very slightly slower compared to CRbSA. As the general counting word algorithm depends only on the count sum of negative and positive words values. Figure 2 shows the graphical representation of different algorithms accuracy (Fig. 3).

From above results, it is clear that described SentiStrength algorithm is efficiently extracts the information than other two algorithms as CRbSA algorithm and general word counting in terms of accuracy and processing time.

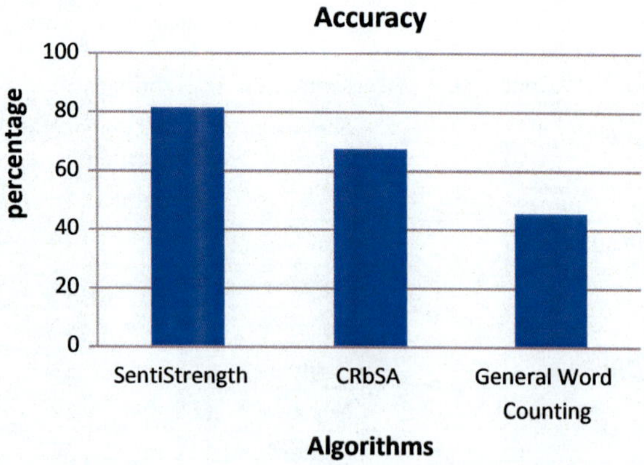

Fig. 2 Different algorithms accuracy value

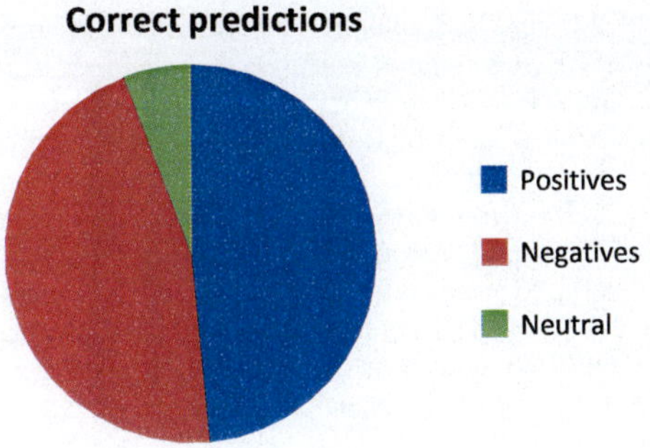

Fig. 3 Correct predictions using SentiStrength algorithm

5 Conclusion

The sentiment analysis has become essential for everyone who desires to make decisions like buying a product on Internet, etc. This sentiment analysis is helpful for various fields to calculate, identify, and express the sentiments. In this paper, automatic sentiment analysis saleability prediction for information extraction using the SentiStrength algorithm is described. SentiStrength algorithm extracts the sentiment data automatically from unstructured text information. Accuracy and processing time are two parameters used to analyze the performance for SentiStrength algorithm. CRbSA algorithm and general word counting are two algorithms used to compare the performance of SentiStrength algorithm. 81.5% of accuracy and 15 ms of processing time achieved with the SentiStrength algorithm which is better than remaining two algorithms.

References

B.B. Alengadan, S.S. Khan, A proposed system for modifying aspect based opinion mining for ranking of products, in *2017 Third International Conference on Sensing, Signal Processing and Security (ICSSS)* (2017), pp. 335–338

K. Bafna, D. Toshniwal, Feature based summarization of customers reviews of online products. Proc. Comput. Sci. **22**, 142–151 (2013)

M. Çataltaş, S. Doğramaci, S. Yumuşak, K. Öztoprak, Extraction of product defects and opinions from customer reviews by using text clustering and sentiment analysis, in *2020 IEEE International Conference on Big Data (Big Data)* (2020)

A. Cernian, V. Sgarciu, B. Martin, Sentiment analysis from product reviews using SentiWordNet as lexical resource, in *IEEE 7th International Conference on Electronics, Computers and Artificial Intelligence (ECAI)* (2015)

X. Ding, B. Liu, P.S. Yu, A holistic lexicon-based approach to opinion mining, in *Proceedings of the Conference on Web Search and Web Data Mining (WSDM'08)* (2008)

L. Jie, Customer satisfaction mining of tourism website based on social networking platform: a case study of Guizhou industrial heritage tourism route, in *2021 International Conference on Intelligent Transportation, Big Data and Smart City (ICITBS)* (2021)

M. Lovelin, P. Felciah, R. Anbuselvi, A study on sentiment analysis of social media reviews, in *IEEE International Conference on innovations in Information, Embedded and Communication Systems (ICIIECS)* (2015)
G. Moharasar, T.B. Ho, A semi-supervised approach for temporal information extraction from clinical text, in *2016 IEEE RIVF International Conference on Computing Communication Technologies, Research, Innovation, and Vision for the Future (RIVF)*, 2016, pp. 7–12
A. Razzaq, M. Asim, Z. Ali, S. Qadri, I. Mumtaz, D.M. Khan, Q. Niaz, Text sentiment analysis using frequency-based vigorous features. China Commun. **16**(12) (2019)
K. Sintoris, K. Vergidis, Extracting business process models using natural language processing (NLP) techniques, in *2017 IEEE 19th Conference on Business Informatics (CBI)* (2017)
Y.A. Solangi, Z.A. Solangi, S. Aarain, A. Abro, G.A. Mallah, A. Shah, Review on natural language processing (NLP) and its toolkits for opinion mining and sentiment analysis, in *2018 IEEE 5th International Conference on Engineering Technologies and Applied Sciences (ICETAS)* (2018)
N. Srivats Athindran, S. Manikandaraj, R. Kamaleshwar, Comparative analysis of customer sentiments on competing brands using hybrid model approach, in *2018 3rd International Conference on Inventive Computation Technologies (ICICT)* (2018)
L.I. Tan, W.S. Phang, K.O. Chin, P. Anthony, Rule-based sentiment analysis for financial news, in *2015 IEEE International Conference on Systems, Man, and Cybernetics* (2015), pp. 1601–1606
M.S. Usha, M. Indra Devi, Analysis of sentiments using unsupervised learning techniques, in *IEEE International Conference on Information Communication and Embedded Systems* (2013)
K. Zvarevashe, O.O. Olugbara, A framework for sentiment analysis with opinion mining of hotel reviews, in *2018 Conference on Information Communications Technology and Society (ICTAS)* (2018)

Vegetable Plant Leaf Image Classification Using Machine Learning Models

Chitranjan Kumar and Vipin Kumar

Abstract Vegetables are rich in minerals, vitamins, and calcium. It benefits our body in many ways. To know about the vegetable, firstly need to identify and their classification. The identification of the vegetables can be done with the help of their images. In this paper, 25 categories of vegetable leaf samples are collected, containing 7226 total RGB images. These images go through the machine learning process to train various models. Classifier such as Decision Tree (DT), Linear Regression (LR), Support Vector Machine (SVM), Multilayer Perceptron (MLP), *K*-Nearest Neighbors (*K*NN), and Naïve Bayes (NB) are used. After training the models, models are evaluated through accuracy (Acc), precision (Pr), recall (Re), and F1-score (F1-S). The result shows that MLP has performed better than another classifier with an accuracy of 90.68%.

Keywords Image pre-processing · Vegetable leaf · RGB image · Computer vision · Machine learning · Classification · Performance

1 Introduction

Life depends on agriculture and it plays a vital role in our life. Vegetables are exceptionally fundamental in our life as they give a wellspring of energy and conquer health issues (Toor et al. 2021). Thus, it is essential to recognize vegetable leaves precisely and conveniently. Vegetable leaves can be broadly assembled by the possibility of their fundamental causal administrator, either overpowering or non-irresistible. Machine learning algorithm methods are used to identify various vegetables (Tripathi and Maktedar 2021). It is difficult for a natural eye to recognize the names of many kinds of plants based on their leaves. Different machine learning models are trained to get the desired results. It starts with data collection, where a minimum of 250 samples are collected of each vegetable plant with their fixed size.

C. Kumar · V. Kumar (✉)
Department of Computer Science and Information Technology, Mahatma Gandhi Central University, Motihari, Bihar, India
e-mail: rt.vipink@gmail.com

A. B. Reddy et al. (eds.), *Proceedings of Third International Conference on Advances in Computer Engineering and Communication Systems*, Lecture Notes in Networks and Systems 612, https://doi.org/10.1007/978-981-19-9228-5_4

The recognition of images is done by machine learning (Kaur et al. 2022) due to its reach. Many features of the leaf are analyzed to predict the category of vegetables, including shape, size, colcolorsargins, veining, and sap. Vegetable leaves that are collected from different areas are fitted to the machine for applying the machine learning models. The models are trained, and then tested, and performance is measured with the help of Acc, Pr, Re, and F1-S (Agarwal et al. 2021). The supervised models such as DT, MLP, LR, *K*NN, SVM, and NB are being used to do the vegetable leave classification (Hameed et al. 2018).

An image is a representation of several pixels, represented across height and width. Pixels can be Grayscale, or RGB (Padmavathi and Thangadurai 2016). Grayscale contains pixel values between 0 and 255, where 0 indicates black and 255 indicates white. In RGB, a pixel contains three integers between 0 and 255, which shows the intensity of red, green, and blue. RGBA is the upgraded version of RGB, where *A* adds the alpha field. After knowing about the image, we can discuss image processing (Egmont-Petersen et al. 2002). Image processing converts the image into a digital form through which certain operations can be performed (Salvi et al. 2021). It shows the 2D signal representation when various methods are applied. Image processing contains various features such as visualization, recognition, sharpening, pattern recognition, and retrieval (Behura 2021).

Image processing involves various steps. Image is the first step toward image processing in which pre-processing of the image is done. Image enhancement works on brightness and contrast. Image restoration clears the noise and makes the appearance suitabled. Image compression is for reducing the size of the image so that data can be faded (Valikhani et al. 2021).

After having collected a dataset of 25 categories in which each category contains a minimum of 250 images. The images are captured from various fields, and image enhancement was done on the system. The dimension of each image is 250×250. The leaf images are collected from the market, agricultural land, and Krishi Vigyaan Kendra Piprakothi, Bihar. The RGB pictures were taken from a smartphone with camera quality of 16 MP. These data were stored as a dataset, and classifiers were applied. To crop the image we used a cropping technique (batch image cropper) to crop the images simultaneously after this all are resized to 250×250. Then various models are applied to get the desired results of the classification.

The novelty of the proposed research work

- *To the best of my knowledge, this is the first various vegetable plant leaf image dataset, consisting of 7226 RGB photos for 25 different vegetable plant types.*
- *The best performer was successfully found based on Acc, Pr, Re, and F1-S, and support performance parameters in a comparative assessment of the classification performance of the classifiers.*
- *The heavily misclassified leaf was studied using textual information, and visual representations of misclassified leaf categories were found to be similar.*

This paper is structured as follows: Sect. 2 includes a literature review in which some previous studies are mentioned. In Sect. 3, the proposed methodology is introduced as the study objectives. Section 4 shows the briefing about the dataset and the design of experiments. The results of experiment and analysis of the proposed work and their full incomplete are presented in Sect. 5, were reviewed the experimental outcomes. And at last, Sect. 6 contains the conclusion and future research directions closed with a discussion of current and prospective research projects.

2 Literature Review

In the paper (Tulshan and Raul 2019; Panchal et al. 2019) authors proposed the idea of identifying diseases in vegetable leaves using machine learning algorithms. They used RGB shading pictures of leaves and converted them into grayscale. The Image Segmentation method is applied and further divided into categories i.e., Segmentation by region and segmentation by the threshold, and the part having data of more than 50% is considered. The region-based k-mean segmentation method is applied and taken as input, and its output is in cluster form. Gray level co-occurrence (GLCM) is used to produce highlights dependent on grayscale images. Further co-event highlights are used in the SVM and KNN classifier. In k-mean clustering, the focus is to determine the Euclidean distances of similar properties. This divides the whole dataset into various types of clusters. After entering the cluster number, the disease name can be detected using the KNN classifier. In this paper (Jasim and Al-Tuwaijari 2020), authors Marwan Adnan Jasim and Jamal Mustafa Al-Tuwaijari used deep learning techniques to find the diseases in vegetable leaves using convolutional neural network (CNN) methods. A dataset of leaves containing images like tomatoes and potatoes and diseases containing 15 classes were classified. The images are of RGB color datasets and classified different types of diseases. The images are passed through the network using a CNN structure design. The CNN model architecture contains four layers, and each layer has different filters. The filters extract the corresponding disease classification and detect the plant species' diseases. ML algorithms are also utilized in different tasks of real-life like multi-view learning (Kumar and Minz 2015; Kumar 2015; Kumar and Minz 2017; Kumar et al. 2021), Air quality (Ahmad and Kumar 2022).

In the paper (Varshney et al. 2021; Rashmi and Shetty 2021) authors proposed the idea of Plant Disease Detection using machine learning algorithms. It also goes through the multiple data sources used to examine these strategies. The paper includes a summary of the various strategies and their benefits and drawbacks. In general, the data obtained by these approaches indicate that they perform well in the overall classification. The paper details the ML techniques involved in classifying and detecting plant diseases detection. It also discusses the many data sources used in researching these strategies. The research findings suggest that the strategy worked successfully in the majority of situations. The study describes the many methods used to categorize and detect plant diseases. The research results suggest that the strategies worked successfully in the majority of cases. They also demonstrated that data were significantly dependent on the successful deployment of the approach. This research aimed to look at the many components of ML algorithms in agriculture. There are other plant leaf classifications being performed using ML-like leaf classification (Kumar et al. 2022; Kumar and Kumar 2022; Aman and Kumar 2022; Ojha and Kumar 2022).

Disease detection in plants is one of the most essential concerns in agricultural image processing. In this paper (Madhavan et al. 2021), The authors are working on developing a system that can precisely diagnose and categorize illnesses in pomegranate plants. The framework employs image processing techniques such as picture capture, image scaling, image enhancement, image segmentation, ROI extraction (region of interest), and feature extraction.

In this paper (Babu et al. 2021), the authors focus on agriculture as the backbone of India as the majority population depends on agriculture or its plays a vital role in the Indian economy also. Still, it is currently experiencing several challenges such as disease, seed selection, water constraints, etc. ML models and image processing can be used to diagnose diseases. A training set can readily recognize sick plants. This research project aims to create a machine having to learn a framework in disease identification and classification that is appropriate for the Indian agricultural industry. This research (Verma and Mehta 2017) introduces an artificial intelligence-based autonomous plant leaf disease classification and detection system for quickly detecting illness, categorizing it, and conducting the necessary treatments to cure it. This strategy of ours aims to increase crop production in agriculture.

In (Sachar and Kumar 2021) the author presents a variety of computer vision algorithms for automatically identifying plants using images of leaves. Botanists use the gathered data to identify different plant species and utilize their medical or other characteristics. Although the human participation increased, less number of plants are available; yet their automated conservation may be the product of identification. The leaf may be photographed with any digital device which has camera in it. A tripod is used to support the camera. Dust, shadows, or other leaves can hide the leaves. In real-world applications based on automated plant identification, even leaves from plants that appear to be identical may be effectively detected in a variety of environmental circumstances. This research provides a contemporary overview of numerous leaf extraction methods that are classified based on leaf features used, as well as their advantages and disadvantages. In addition, we compare and discuss the various classifiers used in the identification method. The essay concludes with a discussion of potential areas to improve and future investigation.

3 Proposed Methodology

India is among the leading producer of vegetable. It is difficult to identify vegetables without technical expertise where the leaves are the critical features of plants. Therefore, it is required to had a device which can detect the vegetables-based leaf image by using technique of Artificial Intelligence. It will find way to detect the appropriate way to identify vegetable.

Objectives of this study's are:

- *To build the image dataset, different categories of vegetable plant leaves were collected.*
- *Cropping, scaling, and correcting are examples of dataset pre-processing.*
- *To use traditional machine learning methods on the newly obtained image dataset.*
- *To acquire the results and analyze them, including the highest performing classifier's confusion matrix.*

This paper aims to make the dataset of multiple vegetable leaf images, and pre-process the dataset with various techniques. This process included model training and selection a huge amount of training labeled data and ML models is used to train machine learning algorithms as each model has different features as we know LR computation is used to forecast a binary result either something happens or nothing happens. This can be expressed as Yes/No and so on, NB calculates the likelihood of a data coming into a category, *K*NN is a pattern recognition technique that uses training datasets to determine the *k* closest relatives in future samples, and a decision tree is a supervised learning technique that is ideal for classification issues since it can organize classes precisely and MLP Classifier does classification using an underlying neural network and to determine the effectiveness of each classifier, use the ML algorithm. Machine learning Models are used in image processing procedures to maximize quality for tasks involving computer vision such as adding creativity filters, changing a picture for the highest quality, or enhancing certain image attributes. A stepwise approach to the work is discussed in this section.

3.1 Dataset Preparation

An RGB image is collected and loaded into the system to read the dataset. Resizing of the image takes place in the initial stage. Images are converted into 250×250 sizes for processing. There are 25 categories of vegetable leaves with 7226 RGB images.

Let's say $V = \{V_i, y_i\}_{i=1}^{k}$ is sample of all the RGB image which has been captured. Where, V_i represents that the associated label for every ith sample in the dataset $y_i \in Y$ i.e., $Y = \{y_i\}_{i=1}^{k}$ for $i = \{1, 2, 3, \ldots, k\}$ and the instances in the dataset is given by k. A matrix $M \in R^{(h \times w \times b)}$ a be used to describe each sample $V_i \in V$. M_{iJ} represents the output in the matrix (M), row (i), and column (j). The height, width, and batch size are represented by h, w, or b. The required scaled dataset $\left\{V_i^t, y_i\right\}_{i=1}^{k}$ may be constructed by using the resized function $fx()$ in Eq. (1)

$$V^t = fx\left(\left\{V_i^r, y_i\right\}_{i=1}^{k}, p \times q\right) \tag{1}$$

3.2 Feature Extraction Using PCA

The principal component analysis (PCA) is used to extract data from the images (Wu et al. 2018). It removes of then the irrelevant data without losing the main content. The reduction of data leads to putting less effort into the machine and increases the speed of the learning process. Let us suppose that $(m-1)$ is the principal component where Eigenvectors are connected with the largest Eigenvalue of $(m-1)$. As S is symmetric, the spectrum theory says that we can make an orthonormal Eigen basis of $(m-1)$ dimensional space of R^D with the help of an eigenvector. Subtraction of $(m-1)$. Takes place for the (mth) principal component which is represented as $b_1, \ldots, b_{m-1}$. Other components minimize the left information. It results in a new matrix Eq. (2).

$$\hat{X} := X - \sum_{i=1}^{m-1} b_1 b_i^t X = X - B_{m-1} X \tag{2}$$

where $[x_1, \ldots, x_n] \in R^{D \times N}$ where column vectors are represented as data points and $B_{m-1} := X - \sum_{i=1}^{m-1} b_i b_i^T$. Is a project matrix that highlights the sunspace spanned by $b_1, \ldots, b_{m-1}$.

3.3 Splitting Dataset into the Test, Train, and Validate

The dataset must first be shuffled before training, validating, and testing the split (Joseph and Vakayil 2022). The algorithm divides each class picture into three stages: training, validation, and testing. The training set has 60% sample from each category, while the validation and test sets contain 20% sample from each category. Let $X = (X_1, \ldots, X_P)$ be the dataset and k be the number of samples. It is V^t split into the validation, train, and test data which are represented as $T_{\text{train}} \subset V^t$, $V_{\text{validation}} \subset V^t$ and $T_{\text{test}} \subset V^t$. Where $V^t = T_{\text{train}} \cup V_{\text{validation}} \cup T_{\text{test}}$ and $\{T_{\text{train}} \cap V_{\text{validation}}\} \cup \{T_{\text{train}} \cap T_{\text{test}}\} \cup \{V_{\text{validation}} \cap T_{\text{test}}\} = \varnothing$ width, Height, and channel of leaf image are assigned in tensor T. The ith picture of the dataset $\{V_i^t \in V^t\}$ expressed as $V_i^t = [H, W, C, i]$, where $i \in \{1, 2, 3, \ldots, b\}$. The batch size is denoted by b, As seen in the Eq. (3).

$$[X_T, Y_T] \tag{3}$$

where X_T represents training dataset whereas Y_T is testing dataset.

3.4 Using Machine Learning Classifiers to Analyze Datasets

The data analysis uses supervised learning to work with a set of label $\{(x_1, y_1), (x_2, y_2), \ldots, (x_N, y_N)\}$. Every element X_i among N is referred to as a feature vector. In turn, a one-dimensional array is an ordered and indexed series of values. The vector's dimensionality is defined as the length of that series of values, D. A feature vector is a matrix having values ranging from 1 to D in each dimension j. Each such value is referred to as a feature and is symbolized by the symbol $x^{(j)}$. X_k, for every k from 1 to N. The label y_i can be an element from one of a finite set of classes $\{1, 2, \ldots, N\}$.

In this classification (Simon et al. 2011), the labels are now D-dimensional vector y_i, whereas the predictions are D-dimensional vectors $\hat{y}_i$, with $\hat{y}_{i,j}$ values ranging from 0 to 1 in each dimension j. The loss for predicting one label y_i is defined as follows:

$$£_i = \sum_{j=1}^{D} \left[-y_{i,j} \times \log_2 \left(\hat{y_{i,j}} \right) - \left(1 - y_{i,j} \right) \times \log_2 \left(1 - \hat{y}_{i,j} \right) \right] \quad (4)$$

The total of single example losses is typically used to create the cost function for categorizing the whole training set.

$$£ = \sum_{i=1}^{N} £_i \quad (5)$$

In classification, the output layer consists of D logistic units with values in the range (0, 1), but their sum is (0, D).

4 Experiments

There are 7226 RGB images of leaves in total in the collection data, which are divided into 25 classes of vegetable plant leaf images. As shown in Fig. 1, numerous types of vegetables with plant-based leafs were used, including brinjal, onion, cauliflower, and beetroot, and others. There are a minimum of 250 images in each class. During the experiments, these images in each category are taken on a white background to eliminate anonymous or unusual data. The table provides a comprehensive dataset description, which includes a leaf picture of each class as well as its common name, scientific name, and the number of image dataset in each class, directly below each sample image. The picture dataset is separated into three sections: training, testing, and validation. The training dataset has more than 150 samples in each group, whereas the validation and test datasets each contain 50 samples.

Fig. 1 Collection of all 25 vegetable plant leaves classes with image data, the common name, botanical (scientific name), and the number of plant leaf images taken in each class

Fig. 2 Flow chart diagram of vegetable leaf image classification all the steps

4.1 Experimental Design

Figure 2 depicts the experimental flowchart of vegetable leaf image, which depicts the entire procedure from data collection to vegetable plant classification. The data set is collection is the first phase and gathered about a minimum of 250 images of leaves in each of the 25 categories of vegetable plants for this research. After collecting the dataset from various places, the data is pre-processed, which includes cropping and maintaining the size of the images. Images are converted into 250 × 250 size for evaluation. The images are in the form of RGB format. The images are captured through a phone camera whose pixel is 16 MP. It is done cluster-based. The cluster-based technique gives us a better result after the feature extraction. It extracts the image property and helps the classifier give an accurate effect. The classifier used here is DT, MLP, LR, *K*NN, SVM, and NB. The performance of the classifiers is measured through Acc, Pr, Re, and F1-S. The results are then visualized through a bar plot and box plot.

4.2 Experimental Setup

System specification: For research following software and hardware have been used: Our implemented high-level Python 3.9.7 64-bit programming language, IDE-Anaconda Navigator 2.1.1, and runs on the Google Collaboration(CoLab) Platform using a Jupyter Notebook 3.2 Windows 10 operating system is used in which 64 GB RAM is present, and it also includes GPU of Quadro RTX 4000. For training, testing, and validation, the dataset is divided into three parts: 60%, 20%, and 20%. In this experiment, the same parameter settings are used for each dataset, and the performance of each classifier is assessed. The whole process was carried out using an HP Z6 G4 workstation.

5 Result and Analysis

5.1 Description of Results

Bar plot of classifiers: Fig. 4 it shows the classification accuracy of different classifiers on the dataset. The six classifiers are DT, MLP, LR, *K*NN, SVM, and NB. The plot's x-axis shows the classifiers, whereas the plot's y-axis is accurate. Percentage of accuracies are put on the top of each bar of the classifier. As seen from the bar plot, the accuracy of the MLP classifier is performing highest accuracy at 90.68%, and the SVM classifier is closest to the MLP classifier. The lowest accuracy is the NB classifier. MLP classifier performed better than any other classifier.

Box plot of classifiers: Fig. 4b shows the box plot of classifiers with 9 iterations, The four evaluation criteria include min maximization, max maximization, median maximization, and the box size. MLP has the maximum min, and it also has the maximum max. The box size is the smallest, indicating that MLP has performed better than any other. Max and min values are also the highest, which says the classifier is best. MLP is again performing best among others. So, we will choose it for our further research.

Confusion matrix using heat-map: Fig. 3 is the confusion matrix using a heat map for the vegetable plant leaves dataset (Fernandez et al. 2017; Soler i Recasens 2022). A scale from 0 to 0–4.5 is present on the rightmost side, which shows the correlation among different classifiers. If the value is high or the scale value is 4.5, it shows that the two variables are highly correlated. Similarly, if the scale is low or 0, there is no correlation between the two variables. Now, in Fig. 3. The diagonal is highly correlated, meaning that the actual output and the predicted output are correctly predicted. The classifier has done a good work of classifying the data, but if the diagonal cells' color is darker (closer to 0), there have been a lot of misclassifications. But, except for this diagonal cell, the story will be exactly the reverse. That's a lot darker (closer to 0). The color indicates that there is an association between that class and the corresponding other vegetable plant leaf classes.

5.2 Analysis of Results

Using a bar plot to compare classifications: As shown by the bar plot in Fig. 4a, LR has 83.14%, *K*NN has 72.64%, SVM has 87.3%, DT has 48.68%, MLP has 90.68%, and NB has 21.98% classification accuracy. So, if MLP is used, which has the highest classification accuracy, this classifier will give the best result for the classification of vegetable plants leaf when compared to other classifiers. The lowest performance has been observed for NB, Because of its low accuracy, the NB will produce the poorest categorization report. This is expected because the probabilistic approach has not learned effectively for sparse data which is highly possible in image datasets.

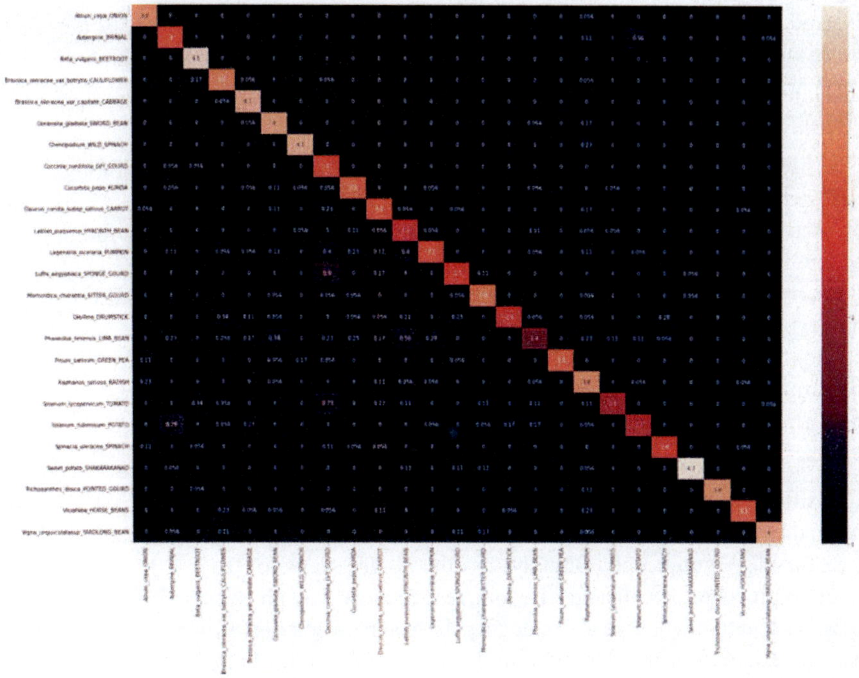

Fig. 3 Confusion matrix using heat map of classifiers

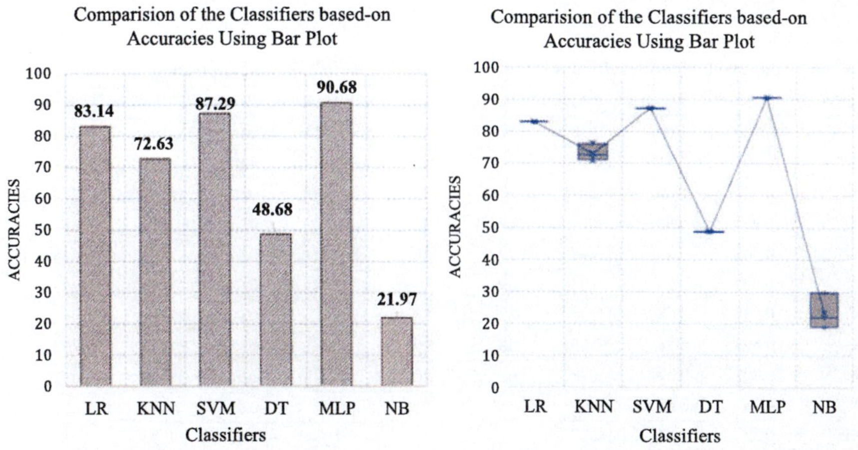

(a). Bar plot of ML algorithms classification performance based on accuracy measures

(b). Box plot of ML algorithms classification performance based on accuracy measures

Fig. 4 **a** Bar plot and **b** box plot given LR, *K*NN, SVM, DT, MLP and NB classifiers

Instead of this, MLP, LR, and SVM algorithms can handle the sparsity better than other algorithms. And, in these three algorithms, MLP has performed the best. The comparative analysis demonstrates that MLP has the best performance based on accuracy metrics.

Classification comparison using box plot: Fig. 4b shows the bar plot of the classifiers, which may affect the outlier performance with multiple iterations. A box plot of each is used to observe the distribution of accuracies. Max Maximization, Min Maximization, Size of a box plot, and the median of the box plots are the evaluation criteria for a box plot analysis. In comparison to other classifiers, MLP has the highest maximum and minimum values, the highest median value, and the smallest box size. So, based on all four box plot evaluation criteria, the MLP classifier performs best. Similarly, Super Vector Machine (SVM) and Linear Regression (LR) perform second and third over box plot measurements, indicating that MLP has the best performance among the other classifiers.

Classification comparison based on Pr, Re, and F1-s performance: Table 1 displays the performance of the MLP classifier. As shown Table 1, vegetable plant leaf Category-0 Allium Cepa (ONION) has the highest value of 0.97, 1.00, and 0.99 in terms of precision, recall, and F1-score, respectively. The Phaseolus Limensis (LIMA BEAN) vegetable plant leaf in category-15 has the worst performance, with the lowest Pr, Re, and F1-s based on support. Based on consent, the version of category-21 is highest, and category-7 has performed worst. So it is observed that the MLP classifier performs best in this category. The heat map in Fig. 3 validates these findings.

Table 1 Performance of the MLP classifier based on Pr, Re, F1-S, and support

Category	Precision	Recall	F1-score	Support	Category	Precision	Recall	F1-score	Support
0.	0.97	1.00	0.99	78	12.	0.93	0.90	0.92	63
1.	0.89	0.81	0.85	68	13.	0.94	0.95	0.94	78
2.	0.94	1.00	0.97	81	14.	0.98	0.95	0.97	64
3.	0.97	0.96	0.96	67	15.	0.74	0.75	0.74	64
4.	0.96	0.99	0.97	79	16.	0.89	0.92	0.91	64
5.	0.89	0.99	0.94	69	17.	0.94	0.94	0.94	78
6.	0.93	0.98	0.95	82	18.	0.91	0.91	0.91	82
7.	0.89	0.88	0.89	58	19.	0.83	0.78	0.80	73
8.	0.90	0.87	0.88	61	20.	0.97	0.98	0.98	63
9.	0.94	0.88	0.91	75	21.	0.94	0.97	0.97	85
10.	0.87	0.87	0.87	71	22.	0.93	0.96	0.96	65
11.	0.95	0.88	0.91	80	23.	0.97	0.96	0.96	61
					24.	0.96	0.94	0.94	71
Accuracy								0.92	1772
Macro Avg.						0.92	0.92	0.92	1772
Weighted Avg.						0.92	0.92	0.92	1772

Table 2 The maximum misclassification was observed in the two categories of leaves with similar views

Category 12 *SPONGE GOURD*	Category 07 *LVY GOURD*

Performance analysis of the best and the worst performers based on confusion matrix using heat-map: Fig. 3 this shows that category-0, which includes Allium Cepa (ONION), has the best categorization. The misclassification is highest in the cell corresponding to category-12 Luffa Aegyptiaca (SPONGE GOURD) on the x-axis and category-07 Coccinia Cordifolia (LVY GOURD) on the y-axis, with a value of 0.9. Their leaf images are very similar, as shown in Table 2.

Conclusion of the above analysis: From the results during this study, MLP classifiers have outperformed other models with highest accuracy shown in bar plot and box plot. SVM and LR classifiers come in second and third place for accuracy measurements, respectively. As a result, Pr, Re, and F1-S are the most important measures for analyzing classification performance, indicating that category-0 (ONION) performs best in the performance of the MLP classifiers.

6 Conclusion

Image processing is used to classify the various 25 categories of vegetable plant leaves images. For each of the 25 classes of vegetable plants, A minimum of 250 images were carefully gathered. As a result, this study's dataset is unique. To begin the categorization, the shape feature is extracted. Second, several approaches such as DT, MLP, LR, *K*NN, SVM, and NB were used to obtain classification accuracies. Based on these findings, Box plots, bar graphs, and heat maps were used in the analysis. Using the Multi-Layer Perceptron (MLP) classifier of the machine learning model, it obtained an accuracy of 90.68%. The farmers will get benefited directly after the implementation of this model in-app. This will boost the farming sector, leading to a boost in the Indian economy. Through this paper, it is seen that MLP classifiers performed better than any other classifier. I have used five classifiers and analyzed their performance through precision, recall, and f1-scores. For visualization, I used a heat map and box plot, which gives us a clear view of their classifier performance. I finalized this result by doing hyper parameter tuning and each classifier and checked their best effect.

In the future, the deep learning model will be used to the same dataset of 25 different vegetable plants to reach greater accuracy than the top ML algorithms, and the correlation of some vegetable plant classes, such as LIMA BEAN, TOMATO, POTATO, and others will benefit from proper data analysis. Other methods can be investigated in order to select the best classifier. Deep learning and other advanced methods can be utilized for significantly misclassified groups. More kinds of vegetable plant images can be gathered for categorization.

References

A. Agarwal, P. Sharma, M. Alshehri, A.A. Mohamed, O. Alfarraj, Classification model for accuracy and intrusion detection using machine learning approach. PeerJ Comput. Sci. **7**, e437 (2021)

N. Ahmad, V. Kumar, Deep learning for air quality prediction after Covid-19 pandemic based on pollutant and metrological data, in *The 4th International Conference on Communication and Information Processing (ICCIP-2022), SSRN Conference Series* (Open Access, Elsevier, 2022)

B.K. Aman, V. Kumar, Flower leaf classification using machine learning techniques, in *Third International Conference on Intelligent Computing, Instrumentation and Control Technologies (ICICICT-2022)* (IEEE Explore, 2022)

V.S. Babu, R.S. Kumar, R. Sunder, A comparative study on disease detection of plants using machine learning techniques, in *2021 7th International Conference on Advanced Computing and Communication Systems (ICACCS)*, vol. 1 (IEEE, 2021), pp. 1937–1941

A. Behura, The cluster analysis and feature selection: perspective of machine learning and image processing, in *Data Analytics in Bioinformatics: A Machine Learning Perspective* (2021), pp. 249–280

M. Egmont-Petersen, D. de Ridder, H. Handels, Pattern Recogn. Lett. **35**, 2279–2301 (2002)

N.F. Fernandez, G.W. Gundersen, A. Rahman, M.L. Grimes, K. Rikova, P. Hornbeck, A. Ma'ayan, Clustergrammer, a web-based heatmap visualization and analysis tool for high-dimensional biological data. Sci. data **4**(1), 1–12 (2017)

K. Hameed, D. Chai, A. Rassau, A comprehensive review of fruit and vegetable classification techniques. Image Vis. Comput. **80**, 24–44 (2018)

M.A. Jasim, J.M. Al-Tuwaijari, Plant leaf diseases detection and classification using image processing and deep learning techniques, in *2020 International Conference on Computer Science and Software Engineering (CSASE)* (IEEE, 2020), pp. 259–265

V.R. Joseph, A. Vakayil, Split: an optimal method for data splitting. Technometrics **64**(2), 166–176 (2022)

R. Kaur, A. Jain, S. Kumar, Optimization classification of sunflower recognition through machine learning. Mater. Today Proc. **51**, 207–211 (2022)

V. Kumar, Multi-view ensemble learning using optimal feature set partitioning: an extended experiments and analysis in low dimensional scenario. Proc. Comput. Sci. **58**, 499–506 (2015)

G. Kumar, V. Kumar, Herbal plants leaf image classification using deep learning models based on augmentation approach, in *The 4th International Conference on Communication and Information Processing (ICCIP-2022), SSRN Conference Series* (Open Access, Elsevier, 2022)

V. Kumar, S. Minz, Multi-view ensemble learning: a supervised feature set partitioning for high dimensional data classification, in *Proceedings of the Third International Symposium on Women in Computing and Informatics* (2015), pp. 31–37

V. Kumar, S. Minz, An optimal multi-view ensemble learning for high dimensional data classification using constrained particle swarm optimization, in *International Conference on Information, Communication and Computing Technology* (Springer, Singapore, 2017), pp. 363–378

V. Kumar, P.S.S. Aydav, S. Minz, Multi-view ensemble learning using multi-objective particle swarm optimization for high dimensional data classification. J. King Saud Univ. Comput. Inf. Sci. (2021)

G. Kumar, V. Kumar, A.K. Hritik, Herbal plants leaf image classification using machine learning approach, in *International Conference on Intelligent Systems and Smart Infrastructure (ICISSI-2022)* (CRC Press, Taylor & Francis Group, 2022)

M.V. Madhavan, D.N.H. Thanh, A. Khamparia, S. Pande, R. Malik, D. Gupta, Recognition and classification of pomegranate leaves diseases by image processing and machine learning techniques. Comput. Mater. Cont. **66**(3), 2939–2955 (2021)

A. Ojha, V. Kumar, Image classification of ornamental plants leaf using machine learning algorithms, in *4th International Conference on Inventive Research in Computing Application (ICIRCA-2022)* (IEEE Explore, 2022)

K. Padmavathi, K. Thangadurai, Implementation of RGB and grayscale images in plant leaves disease detection-comparative study. Indian J. Sci. Technol. **9**(6), 1–6 (2016)

P. Panchal, V.C. Raman, S. Mantri, Plant diseases detection and classification using machine learning models, in *2019 4th International Conference on Computational Systems and Information Technology for Sustainable Solution (CSITSS)*, vol. 4 (IEEE, 2019), pp. 1–6

N. Rashmi, C. Shetty, A machine learning technique for identification of plant diseases in leaves, in *2021 6th International Conference on Inventive Computation Technologies (ICICT)* (IEEE, 2021), pp. 481–484

S. Sachar, A. Kumar, Survey of feature extraction and classification techniques to identify plant through leaves. Exp. Syst. Appl. **167**, 114181 (2021)

M. Salvi, U.R. Acharya, F. Molinari, K.M. Meiburger, The impact of pre-and post-image processing techniques on deep learning frameworks: a comprehensive review for digital pathology image analysis. Comput. Biol. Med. **128**, 104129 (2021)

G.J. Simon, V. Kumar, P.W. Li, A simple statistical model and association rule filtering for classification, in *Proceedings of the 17th ACM SIGKDD International Conference on Knowledge Discovery and Data Mining* (2011), pp. 823–831

A. Soler i Recasens, Data capture and processing system to display a heat map. Bachelor's thesis. Universitat Politècnica de Catalunya (2022)

M.D. Toor, M. Adnan, F.U. Rehman, R. Tahir, M.S. Saeed, A.U. Khan, V. Pareek, Nutrients and their importance in agriculture crop production. A review. Ind. J. Pure Appl. Biosci. **9**(1), 1–6 (2021)

M.K. Tripathi, D.D. Maktedar, Detection of various categories of fruits and vegetables through various descriptors using machine learning techniques. Int. J. Comput. Intell. Stud. **10**(1), 36–73 (2021)

A.S. Tulshan, N. Raul, Plant leaf disease detection using machine learning, in *2019 10th International Conference on Computing, Communication and Networking Technologies (ICCCNT)* (IEEE, 2019), pp. 1–6

A. Valikhani, A. Jaberi Jahromi, S. Pouyanfar, I.M. Mantawy, A. Azizinamini, Machine learning and image processing approaches for estimating concrete surface roughness using basic cameras. Comput. Aided Civ. Infrastruct. Eng. **36**(2), 213–226 (2021)

D. Varshney, B. Babukhanwala, J. Khan, D. Saxena,, A. Kumar Singh, Machine learning techniques for plant disease detection, in *2021 5th International Conference on Trends in Electronics and Informatics (ICOEI)* (IEEE, 2021), pp. 1574–1581

A. Verma, S. Mehta, A comparative study of ensemble learning methods for classification in bioinformatics, in *2017 7th International Conference on Cloud Computing, Data Science and Engineering-Confluence* (IEEE, 2017), pp. 155–158

S.X. Wu, H.T. Wai, L. Li, A. Scaglione, A review of distributed algorithms for principal component analysis. Proc. IEEE **106**(8), 1321–1340 (2018)

A Novel SARS-COV-2 Variant Omicron Disseminating Evaluation

Shawni Dutta, Samir Kumar Bandyopadhyay, Midhunchakkaravarthy Janarthanan, and Payal Bose

Abstract Omicron is a relatively new form of COVID-19 that has created an unavoidable and life-threatening situation to the entire world since late 2021. Absence of appropriate vaccination, medication, the epidemiological cycle has become more complex. This study primarily concentrates on the analysis of genome sequence for COVID-19 variants. To conduct such analysis, two datasets are collected from Kaggle and GISAID. Using these datasets, the globally existing genome sequences are identified and insights regarding the countries that are carrying significantly higher genome sequence count are provided. This investigation analyzes the world-wide virus variants and further identifies that the United States and United Kingdom are the countries where proper inspection should be provided because of the genome sequence count. An adequate idea regarding the mutations of the Omicron virus is also considered in this study. To address this issue, recent genome sequence data ranging from February, 2022 to 10th March, 2022 is analyzed to understand how the latest arrival, Omicron, is perturbing the world. This study emphasizes on the constant surveillance of genome sequences among all the countries which in turn will benefit the health care professionals and frontline healthcare workers as well as the Governments can take necessary policies and precautions to combat such pandemic.

Keywords COVID-19 · Omicron variant · Genome sequence · World Health Organization · COVID-19 pandemic

1 Introduction

The Omicron variant of COVID has sparked widespread worry around the world. According to the Centers for Disease Control and Prevention, or CDC, the variant

S. Dutta (✉) · S. K. Bandyopadhyay · M. Janarthanan
Lincoln University College, Petaling Jaya, Selangor Darul Ehsan, Malaysia
e-mail: shawnidutta83@gmail.com

P. Bose
GLA University, Mathura, India

A. B. Reddy et al. (eds.), *Proceedings of Third International Conference on Advances in Computer Engineering and Communication Systems*, Lecture Notes in Networks and Systems 612, https://doi.org/10.1007/978-981-19-9228-5_5

Omicron was initially revealed in Botswana, South Africa on November 11, 2021. If virus genome sequencing is accomplished comprehensively, epidemiologists and public health administrators can apprehend the spreading characteristics of virus and evaluate the innovation capacity. The analysis of entire genome sequencing data has yielded numerous critical results regarding this virus. More sequence information from samples from around the world will be required to develop effective techniques to govern and prohibit COVID-19 outbreaks. Researchers across the world are working together to develop and disseminate this essential knowledge, which will be used to aid in disease diagnosis and safe handling. Throughout the pandemic, it is critical that pathogen sequence data be disseminated in public archives. Throughout the outbreak, the World Health Organization (WHO) actively encourages public access to sequencing data to guide population health and research decisions (Hemarajata 2020). According to the researchers, the variation Omicron represents a new arrival sequence. It was first discovered in South Africa and Botswana in early November 2021; however, the sequence testing was found in earlier evidence from people in England, later in the United States, South Africa, and Nigeria in November. By December, researchers had completed a detailed examination of the genetic diversity in millions of sequenced genomes, and also how aggressively the virus spread throughout populations. Based on that result, this sequence predicts that it appeared in the last week of September or very first week October of 2021. It is hypothesized that Omicron originated from Gauteng, a highly populated metropolitan zone located between Johannesburg and Pretoria. Then, it spread to other countries, including the neighboring nation of Botswana in southern Africa (Lundberg et al. 2022). Several varieties of SARS-CoV-2 evolved throughout the COVID-19 epidemic, with varying infectiousness and intensity. The Omicron (B.1.1.529) variation was discovered in a Botswana specimen and was officially reported by South Africa. WHO classified Omicron as a variation of concern at the end of November due its community spreading nature. This variant has been confirmed in 133 countries with available data and became the most common lineage internationally, accounting for 85% of mutant cases reported in late January. Between May and December 2021, the main variation in England was Delta (B.1.617.2). While in England, Omicron instances have swiftly multiplied at the middle of the 2nd week of January 2022. During this time, Omicron occurrences accounted for more than 99% of other variants (Nyberg et al. 2022).

There are two significant distinctions between Omicron and prior SARS-CoV-2 virus strains located in November 2019. The Delta variation became prevalent and superseded the alpha variant because it was 40–60% more transferrable. The Omicron form is now more transmissible than the Delta variant. Since human behavior and vaccination frequencies are continuously changing, it is hard to quantify how much more highly contagious one variety is than another. These characteristics, along with transmissibility, influence how quickly a virus spreads in a population (Hay et al. 2022). Running nose, headaches, mild to high fever, dizziness, fatigue, and muscle pain are the most familiar symptoms of this virus. According to preliminary research, Omicron is more powerful than earlier forms at propagating in the respiratory system, including the nose, throat, and mouth, making it much more identical to a respiratory syncytial virus. As the Omicron variant of SARS-CoV-2 maintains vulnerability in its global spread, vaccine manufacturers are investing in clinical trials with COVID-19 doses adapted to the highly transmissible variant. However, a slew of preliminary research indicates that Omicron-specific boosters offer no benefit over a third dosage of conventional immunizations (Hawman et al. 2022).

The study is motivated to ensure the following criterions as its objectives:

Analyze the genome sequences to monitor the worldwide effect of different genome sequences. A full-fledged genome sequence-wise analysis will provide an insight regarding the highly affected countries.

Analyzing genome sequence will enable us to track Omicron virus transmission dynamics. This will specifically address how the virus mutates and spreads among humans.

Omicron is one of the highly contagious variants of the coronavirus family. This variant started to evolve from 24th November, 2021 and spreaded all over the world (Chu 2021). Figure 1 depicts the global distribution of the Omicron variant in a chronological approach until January 2022.

2 Related Works

There are quite a few research articles that primarily encompass COVID-19-related transmission dynamics. Gambhir. E et al. investigated the trend along with the pattern of COVID-19 transmission in India using machine learning models. They investigated data from India's Ministry of Health and Family Welfare to anticipate the rise in COVID-19 cases over the following 60 days with an efficiency of 93% (Gambhir et al. 2020). For explaining and forecasting the actions of the COVID-19 contagion in Italy, a parameter-varying variant of the suspected-infected-recovered-deceased (SIRD) model has been developed. The recommended COVID-19 modeling technique can provide accurate short- and long-term forecasts, as well as helpful information on contagion mechanisms (Calafiore et al. 2020). Group of optimized and multisource selection (GROOMS) is an approach proposed by Fong et al. (Fong et al. 2020), which assembles an ensemble of five groups of outbreak forecasting methodologies. In this paper, a polynomial neural network with corrective feedback (PNN + cf) is integrated and applied. In terms of the root mean squared error (RMSE), experimental data shows that the collective strategy PNN + cf outperforms former methodologies significantly. A detailed review work on understanding of vaccines, immunization processes regarding the COVID-19 vaccination was provided in Pollard and Bijker (2021). While comparing Omicron to Delta variant, a study (Mallapaty 2022) has made following benchmarks:

According to Mallapaty (2022), when infected with Omicron, there is a 48% higher likelihood of passing the virus on than when impaired with Delta. Even with fewer opportunities for recurrent and sustained contact to the virus, the transmission capability of Omicron is more vulnerable outside the home, as the risk of contracting the virus is more than double that of Delta.

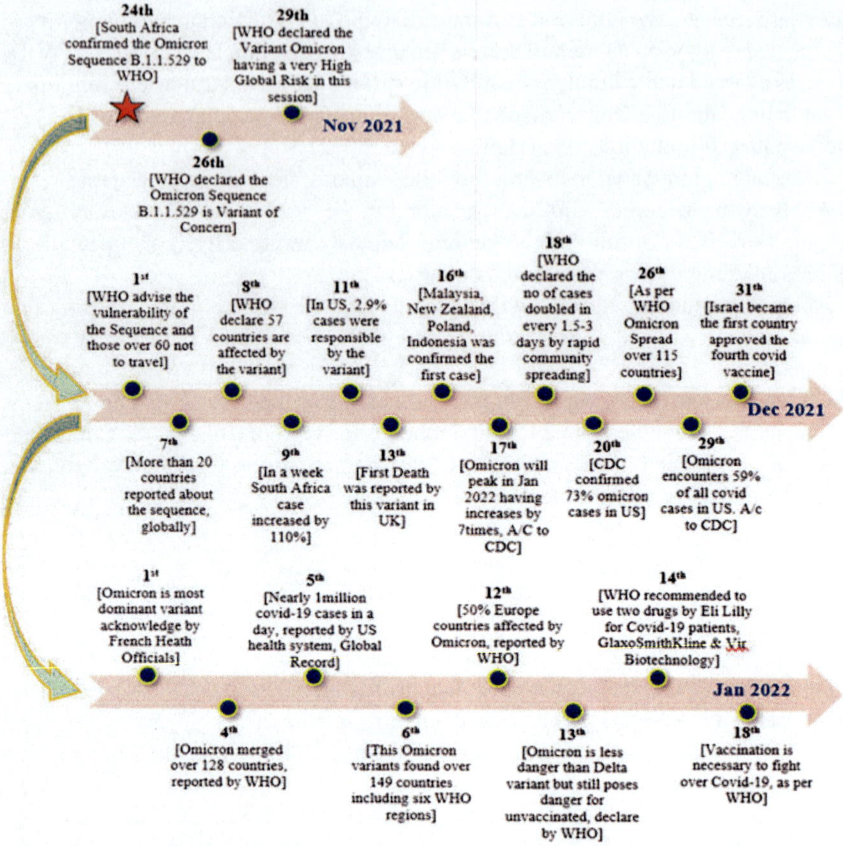

Fig. 1 Global distribution of the Omicron variant in a chronological approach

This research further exemplifies that the unvaccinated domestic members were 23% more in danger to infection with Omicron than with Delta. The unvaccinated people are more expected to spread Omicron variant as compared to the people who have received three doses of vaccination. However, this investigation further concluded vaccines are not a way of blocking the transmission.

3 Background

The process of determining the arrangement of DNA nucleotides, or compounds, in a genetic sequence order of As, Cs, Gs, and Ts that constitute an organism's DNA—is known as genomic information or genome sequence. The genetic sequence will be a helpful alternative for specialists, allowing them to locate genes much more efficiently and swiftly. Even while scientists are still discovering how to evaluate genome sequences, they still include some signals about where genes are located. This procedure, known as genomic sequencing, is used by scientists to analyze the genetic information which are existent in an organism or virus. Sequences from samples can be evaluated to assist experts in tracking the transmission of a virus, how it changes, and how those changes may affect population health (Houldcroft et al. 2017; Callaway 2020). Since novel variants emerge incessantly, genetic data has helped governments to take swift and well-versed populace healthcare judgments since the pandemic's inception. The significance of genetic sequencing for public health emergencies extends beyond simply learning when to take critical procedures such as lock down and immunization. As these are more highly infectious varieties, it can assist counties in preparing for future surges and taking essential steps such as expanding oxygen supply, providing more hospital wards, arranging the best effective vaccines, or ramping up monitoring (Kaggle 2022).

3.1 Dataset Details

The first dataset for the new COVID-19 variant Omicron has been obtained from the Kaggle archive (GISAID, https://www.gisaid.org/). Monitoring the evolution of the novel Omicron COVID-19 variation is the topic of the collection. The dataset has six properties, which are described in detail below:

'location'—Countries for which the various information are shared;
'date'—the date of record keeping;
'variant'—Virus variant associated with this data entry;
'num_sequences'—the quantity of analyzed sequences (for the nation, variation, and timestamp);
'perc_sequences'—a proportion of the whole amount of sequences (for the nation, variant, and date);
'numsequencestotal'—overall amount of sequences (countries, variants, and dates).

The second dataset was obtained from the GISAID Web data provider (WHO 2022). GISAID is a global research project and main contributor that was launched in 2008 with the purpose of providing free access to genomic information of influenza viruses and the COVID-19 coronavirus. On January 10, 2020, the full-fledged genome sequences of SARS-CoV-2 were first made publicly presented on GISAID. It supported the universal reactions to the outbreak, including the production of the first vaccines and clinical experiments to identify SARS-CoV-2. GISAID enables genetic forensics and instantaneous monitoring to track the spread of novel COVID-19 virus genotypes around the world.

4 Experimental Result

Any virus variant can be made by a series of mutations. RNA-based viruses, including SARS-CoV-2 and influenza, change considerably more quickly than DNA-based viruses. Every time SARS-CoV-2 replicates, the virus has the potential to evolve. During the year 2020, several COVID-19 variants were found with different genome sequences. However, during the time period of 2021 to January, 2022, more COVID-19 variants were reported. The different sequence statistics during the year 2020 and 2021 are shown in Figs. 2 and 3 separately. As revealed by Fig. 1, the Omicron variant started its impact during 2021. It is quite evident that the Omicron form of COVID has prompted a wave of concerns around the world. As discussed by the Centers for Disease Control and Prevention (CDC), the variant Omicron was first exposed in Botswana, South Africa on November 11, 2021 (Huamán-Saavedra 2022).

Genomic and transcriptomic data can be used to track viruses. Genomic analysis should be modeled based on gathering sufficient sequencing data from systems. They are implemented to detect novel variations and levels of popularity in circulating mutations. This process is popularly termed as genome surveillance. To support the process, the extreme mutated countries and low mutated countries during the time period from November, 2021 to January, 2022 are highlighted in Figs. 4 and 5,

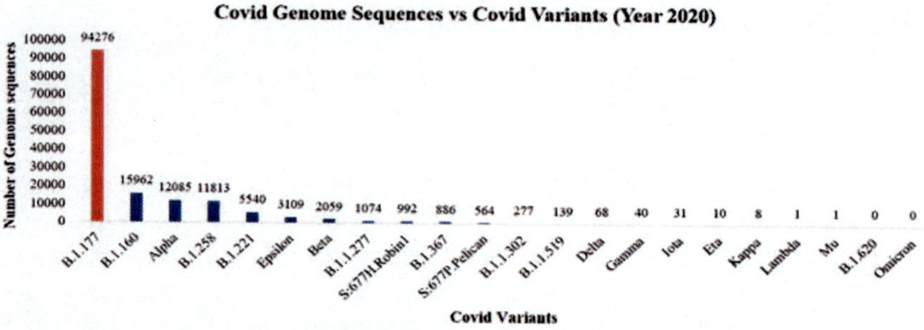

Fig. 2 Genome sequence and COVID-19 variants statistics (Year 2020)

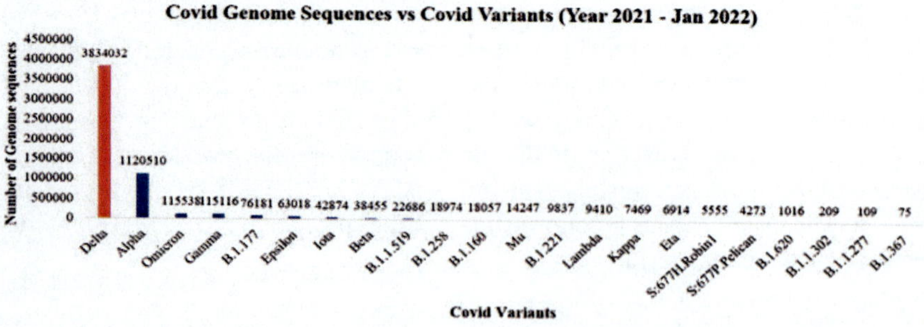

Fig. 3 Genome sequence and COVID-19 variants statistics (Year 2021)

respectively. As depicted in Fig. 4, the United States and the United Kingdom are the most mutated countries, whereas Canada, Brazil, Spain, Italy, France, India, Sweden, Japan, Turkey, and Denmark are other extreme mutated regions. Figure 5 displays the countries that are having low Omicron mutations. In spite of being low mutated countries, as marked in Fig. 5, Egypt, Jambia, Mozambique, Guatemala, Jordan, Bosnia, and Herzegovina are the countries where many genome sequences were found. Hence, proper genome surveillance is necessary for the aforementioned countries. In Fig. 5, countries marked in deep orange color also require close monitoring to the sequences. In the present day, numerous sequences of SARS-CoV2 virus are available, and Omicron is one of them. The figures, namely 4, are describing how the Omicron sequence affected several countries in the world. All the worldwide countries along with their Omicron sequences are depicted in Fig. 4 and 5. Figures 4 and 5 reveals that the countries such as the United States, United Kingdom, Denmark, and Germany are the highest Omicron genome sequence holders. The countries having relatively higher genome sequences are marked in red color as shown in Figs. 4 and 5. The blue and green marked countries are also having moderate genome sequences, and hence, proper monitoring can assist in tackling as well as identifying further genome sequences (Figs. 6 and 7).

WHO's Global Influenza Surveillance and Response System (GISRS) has released a data sharing platform GISAID. This platform shares different influenza viruses as well as coronavirus-related data. As available in this platform, only Omicron-related genome sequences are extracted within the time period of February, 2022 to 1st week of March, 2022. Figure 8 describes how developed countries such as the United Kingdom, United States, Denmark, and Germany are the top 4 countries that are carrying the higher genome sequence count. Time-based analysis between Nov, 2021–Jan, 2022 and Feb, 2022–Mar, 2022 for the most affected countries due to Omicron genome sequence is presented in Table 1.

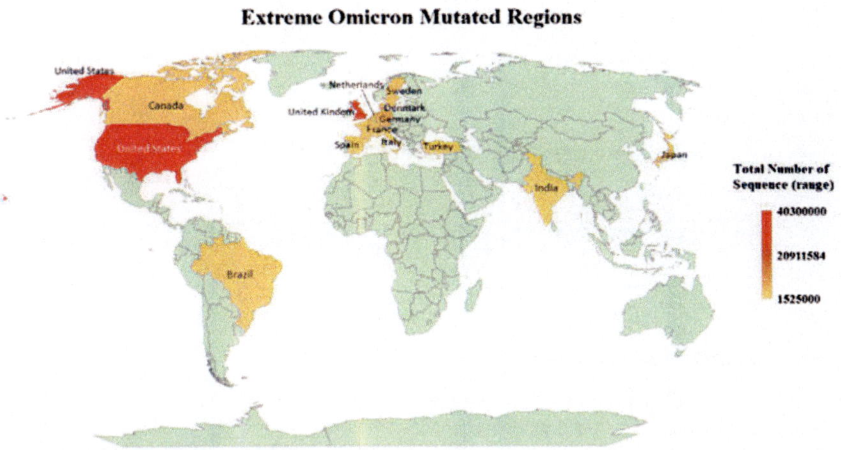

Fig. 4 Extreme Omicron mutated worldwide countries

Fig. 5 Low Omicron mutated worldwide countries

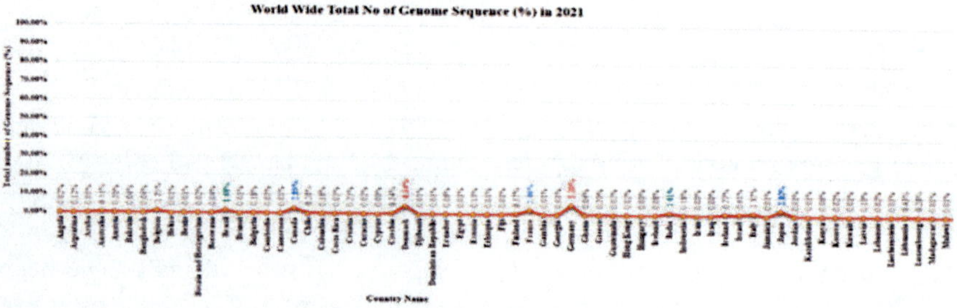

Fig. 6 Country-wise Omicron sequence distribution

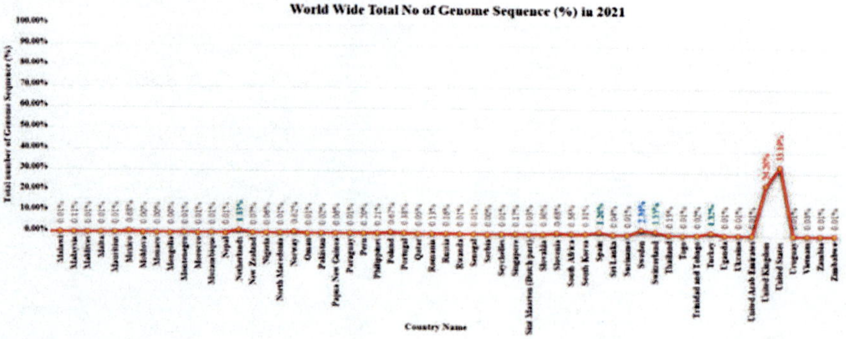

Fig. 7 Country-wise Omicron sequence distribution

Table 1 Most affected countries due to Omicron variant

Time period	Country name	Omicron genome sequence (%)
Nov, 2021–Jan, 2022	United States	33.19
	United Kingdom	24.2
	Germany	5.2
	Denmark	4.34
Feb, 2022–Mar, 2022	United Kingdom	99.6
	United States	94.7
	Denmark	93.9
	Germany	92.7
	France	90.4

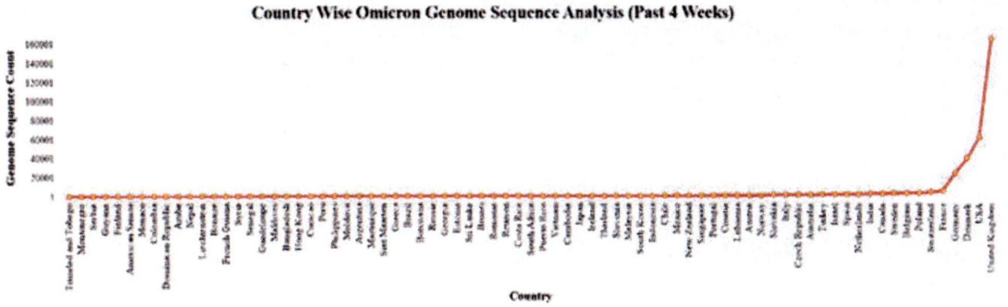

Fig. 8 Country-wise genome sequence analysis (February, 2022 to 10th March, 2022)

5 Discussions

This study conducts an exploratory data analysis based on genome sequence to track different virus variants. Different virus variants should be monitored and analyzed so that the nature of the corresponding virus can easily be assessed. This step will essentially assist the researchers to invent proper medicines, vaccines, and other necessary guidelines to combat the deadly disease. From this genome sequence analysis-based study, it is quite evident that the United States and United Kingdom are the two topmost countries having the most Omicron genome sequences. Hence, a proper genome surveillance-based system is necessary for these countries. However, other countries should also be closely monitored to identify further virus variants. A particular virus variant may have a heterogeneous impact on different countries. In fact, based on time, a variant may become more dangerous and can take numerous human lives. Thus, a worldwide monitoring system will help in assessing the spreading nature of the corresponding coronavirus variants. As a result, adequate health and related infrastructure can be provided to fight against the on-going pandemic.

6 Conclusion

Genome sequencing can aid scientists in determining the virus's spreading nature, statistics of virus's impact, and determining the prevalence of a variant in the population. The sequence approach can also be used to analyze the treatment's efficiency against viral variations. Because of the advantages indicated above, our study has mostly concentrated on genomic sequence analysis. Analyzing the number of genome sequences will highlight which countries are more vulnerable to virus propagation. As a result, in those high-risk countries, adequate genome sequence surveillance should be implemented. According to the findings of the current study, higher Omicron genome sequences were discovered in Denmark, the United Kingdom, Germany, and the United States. As a result, all of these countries will require a suitable monitoring system to identify more genome sequences as well as statistics on variation dissemination. The other countries, on the other hand, should conduct continual genome sequence analysis methods to ensure that no novel sequences go undetected.

References

G.C. Calafiore, C. Novara, C. Possieri, A time-varying SIRD model for the COVID-19 contagion in Italy. Annu. Rev. Control. **50**, 361–372 (2020). https://doi.org/10.1016/j.arcontrol.2020.10.005

E. Callaway, The coronavirus is mutating—does it matter? Nature **585**(7824), 174–177 (2020). https://doi.org/10.1038/d41586-020-02544-6

J. Chu, A statistical analysis of the novel coronavirus (COVID-19) in Italy and Spain. PLoS ONE **16**(3), e0249037 (2021). https://doi.org/10.1371/journal.pone.0249037

S.J. Fong, G. Li, N. Dey, R. Gonzalez-Crespo, E. Herrera-Viedma, Finding an accurate early forecasting model from small dataset: a case of 2019-nCoV novel coronavirus outbreak. Int. J. Interact. Multimedia Artif. Intell. **6**(1), 132 (2020). https://doi.org/10.9781/ijimai.2020.02.002

E. Gambhir, R. Jain, A. Gupta, U. Tomer, Regression analysis of COVID-19 using machine learning algorithms, in *2020 International Conference on Smart Electronics and Communication (ICOSEC)* (2020). https://doi.org/10.1109/icosec49089.2020.9215356

GISAID—initiative. GISAID. https://www.gisaid.org/

D.W. Hawman, K. Meade-White, C. Clancy, J. Archer, T. Hinkley, S.S. Leventhal, D. Rao, A. Stamper, M., R. Rosenke, K. Krieger, S. Randall, A.P. Khandhar, L. Hao, T.Y. Hsiang, A.L. Greninger, M. Gale, P. Berglund, D.H. Fuller, J.H. Erasmus, et al.: Replicating RNA platform enables rapid response to the SARS-CoV-2 Omicron variant and elicits enhanced protection in naïve hamsters compared to ancestral vaccine. Preprint at bioRxiv (2022). https://doi.org/10.1101/2022.01.31.478520

J.A. Hay, S.M. Kissler, J.R. Fauver, C. Mack, C.G. Tai, R.M. Samant, S. Connelly, D.J. Anderson, G. Khullar, M. MacKay, M. Patel, S. Kelly, A. Manhertz, I. Eiter, D. Salgado, T. Baker, B. Howard, J.T. Dudley, C.E. Mason, Y.H. Grad et al., Viral dynamics and duration of PCR positivity of the SARS-CoV-2 Omicron variant. medRxiv Preprint (2022). https://doi.org/10.1101/2022.01.13.22269257

P. Hemarajata, *SARS-CoV-2 Sequencing Data: The Devil Is in the Genomic Detail* (ASM.Org., 2020). https://asm.org/Articles/2020/October/SARS-CoV-2-Sequencing-Data-The-Devil-Is-in-the-Gen. Accessed 28 Oct 2020

C.J. Houldcroft, M.A. Beale, J. Breuer, Clinical and biological insights from viral genome sequencing. Nat. Rev. Microbiol. **15**(3), 183–192 (2017). https://doi.org/10.1038/nrmicro.2016.182

J.J. Huamán-Saavedra, The Omicron variant of SARS-CoV-2. Rev. Méd. Trujillo **17**(1), 3–4 (2022). https://doi.org/10.17268/rmt.2022.v17i1.4256

A.L. Lundberg, R. Lorenzo-Redondo, E.A. Ozer, C.A. Hawkins, J.F. Hultquist, S.B. Welch, P.V. Prasad, J.F. Oehmke, C.J. Achenbach, R.L. Murphy, J.I. White, R.J. Havey, L.A. Post, Has Omicron changed the evolution of the pandemic? JMIR Publ. Health Surveill. **8**(1), e35763 (2022). https://doi.org/10.2196/35763

S. Mallapaty, COVID-19: how Omicron overtook delta in three charts. Nature (2022). https://doi.org/10.1038/d41586-022-00632-3

T. Nyberg, M. Ferguson, S.G. Nash, H.H. Webster, S. Flaxman, N. Andrews, W. Hinsley, J.L. Bernal, M. Kall, S. Bhatt, P. Blomquist, A. Zaidi, E. Volz, N.A. Aziz, K. Harman, S. Funk, S. Abbott, R. Hope, A. Charlett, S. Thelwall et al., Comparative analysis of the risks of hospitalisation and death associated with SARS-CoV-2 Omicron (B.1.1.529) and delta (B.1.617.2) variants in England: a cohort study. Lancet (2022). https://doi.org/10.1016/s0140-6736(22)00462-7

Omicron daily cases by country (COVID-19 variant). Kaggle. https://www.kaggle.com/datasets/yamqwe/omicron-covid19-variant-daily-cases. Accessed 16 Feb 2022

A.J. Pollard, E.M. Bijker, A guide to vaccinology: from basic principles to new developments. Nat. Rev. Immunol. **21**, 83–100 (2021). https://doi.org/10.1038/s41577-020-00479-7

Why genomic sequencing is crucial in COVID-19 response: WHO | Regional Office for Africa. https://www.afro.who.int/news/why-genomic-sequencing-crucial-covid-19-response. Accessed 18 Mar 2022

Cryptanalysis of Tiny Encryption Algorithm Based on SMT Solvers Using HPC

Md. Najim Alam, Praveen Kumar Gundaram, and Nagendar Yerukala

Abstract Tiny Encryption Algorithm (TEA) is a block cipher algorithm that uses symmetric key of 128-bits. It performs 32 rounds for encryption/decryption. TEA uses addition modulo 2^{32}, XoR, and shift operations in its algorithm. We convert TEA algorithm into nonlinear system of equations and solve the system using Satisfiability Modulo Theory solvers (SMT) on a desktop computer and also on high performance computing (HPC) facility. As solving system of equations is NP-complete problem, we tried solving the system for various number of rounds out of 32 rounds. The solver $Z_3(py)$, a Satisfiability Modulo Theories (SMT) solver, has been chosen to perform algebraic cryptanalysis. We could solve the system up to 5th round and found the actual secret key successfully among few solutions of high probable keys which we got from solver with in 15,576.34 min using HPC. Nonlinearity will increase as number of round increases, so solving high nonlinear system is very difficult. Our aim is to solve a specific kind of nonlinear system of equations. We tried to recover the partial keys for rounds greater than 5. Results of the key recovery are present in this paper.

Keywords Block cipher · Tiny encryption algorithm · Satisfiability Modulo Theories (SMT) solver · Algebraic cryptanalysis · High performance computing

Md. Najim Alam · P. K. Gundaram · N. Yerukala (✉)
C R Rao Advanced Institute of Mathematics, Statistics and Computer Science (AIMSCS), University of Hyderabad Campus, Hyderabad 500046, India
e-mail: nagendar@crraoaimscs.res.in

P. K. Gundaram
e-mail: praveenkumar@crraoaimscs.res.in

Md. Najim Alam
School of Computer and Information Sciences,University of Hyderabad, Hyderabad 500046, India

P. K. Gundaram
Acharya Nagarjuna University, Guntur, Andhra Pradesh, India

A. B. Reddy et al. (eds.), *Proceedings of Third International Conference on Advances in Computer Engineering and Communication Systems*, Lecture Notes in Networks and Systems 612, https://doi.org/10.1007/978-981-19-9228-5_6

1 Introduction

Security is becoming more critical as computer systems become more prevalent and complicated. The key component of any mechanism that protect exchange of data over wired/wireless channel is a cryptographic techniques and protocols. The principles, procedures, and methods for keeping messages secure are referred to as cryptography. There are number of cryptographic algorithm and protocols available in the literature. Each having its own pros and cons.

The study of how to attack cryptographic mechanisms is known as cryptanalysis. There are various attack mechanism such as algebraic, differential, correlation, and related key which exists in the literature for attack/cryptanalyze the cryptographic algorithm. Aim of these attacks is to find the secret key/plaintext (message). Asymmetric or public key cryptography and symmetric or secret key cryptography are two types of key-based encryption algorithms. David J. Wheeler and Roger M. Needham developed a symmetric key encryption algorithm called Tiny Encryption Algorithm (TEA) which is a Feistel-type block cipher, at Cambridge University computer laboratory in 1994 (Wheeler and Needham 1994) and first presented in 1994 at Fast Software Encryption (FSE).

Aim of this paper is to apply algebraic attack on TEA by generating system of equations and solve those using SMT solvers and try to recover the key. High performance computing (HPC) has been used to get the result faster and we could get the key up to 5th round results which are shown in this paper.

2 Tiny Encryption Algorithm

TEA performs 32 simple rounds to provide security rather than few complex rounds (Wheeler and Needham 1994). TEA is a Feistel cipher. It takes 128-bits key, 64-bits plaintext data as a block and some constant delta (0x9e3779b9) which has very good meaning in mathematical world and perform 32 rounds and produce 64-bits ciphertext. TEA does not use any predefined table or any kind of S-boxes to do substitution and permutation but with the help of addition, XOR and shifting TEA provide substitution, whereas permutation achieves by swapping the result at every round. One important property of block cipher is Shannon's confusion and diffusion which also achieve by same operation ADD, XOR, and SHIFT. TEA does not use any key scheduling algorithm, instead it uses constant delta to make the sub-keys complex.

TEA block cipher (Wheeler and Needham 1994) encrypts (decrypts) 64-bit plaintext/ message (ciphertext) at a time and outputs corresponding ciphertext (plaintext) of 64-bit length after performing 32 rounds using a 128-bit secret key, and a constant Delta = 0x9e3779b9. Full description of TEA algorithm can be found from original paper on TEA (Wheeler and Needham 1994). Round function of encryption of TEA algorithm can be understood by Fig. 1.

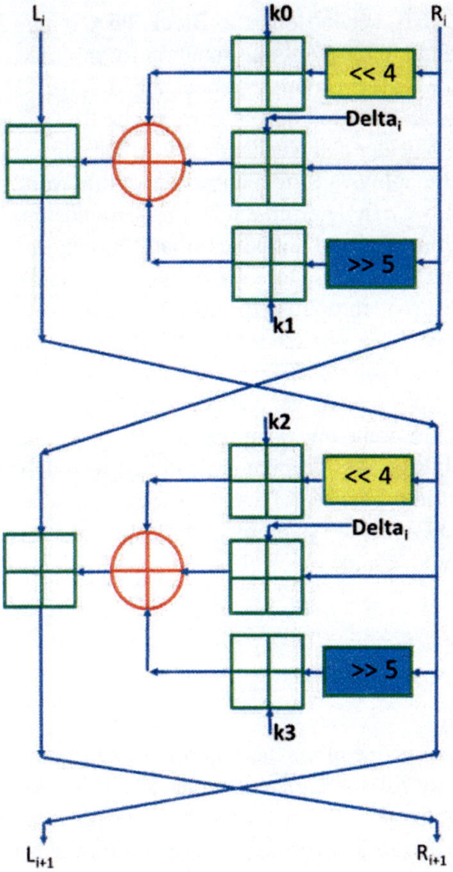

Fig. 1 TEA encryption round

3 Related Study on TEA

Because of its simple and easy design implementation, the TEA was extensively researched and subjected to a various cryptanalysis/attacks. Kelsey et al. (1997) found out that the practical key size of TEA to be 126-bits. Xbox has made use of TEA based on its simple and easy design implementation, because of 126-bit key size, it leads to an attack on Xbox. With two 2^{23} plaintexts and two 2^{32} time (Shoeb and Gupta 2012), Wagner, Kelsey, and Schneier come up with related key cryptanalysis on TEA. Following these findings, David J. Wheeler and Roger M. Needham modified TEA to produce Extended TEA (XTEA), Block TEA in 1996 (Wheeler et al. 1996). While XTEA has the same number of rounds, block size, and key size as the TEA, second one the Block TEA is more flexible because it uses the XTEA (Sekar et al. 2011) round function for multiple iterations to perform encryption/decryption. Later, David J. Wheeler and Roger M. Needham devised corrected one the Block TEA, to address flaws in the Block TEA, and published it in a research article (Wheeler and Needham 1994). This cipher operates on variable length messages and employs an imbalanced Feistel network. Based on the block size of the algorithm, we can determine how many rounds required to make cipher not easily breakable, but it should not be less than six rounds. In Sekar et al. (2011), it has been shown that the entire Block TEA has been attacked, as well as various weaknesses in XTEA also identified. In the last decade, several cryptanalysis findings on the TEA group have been published. It was stated in Sekar et al. (2011) that a low power consumption design of XTEA implementation may be more appropriate for low resource situations than Advance Encryption Standard (AES). Because of its simple design, it is been come under numerous equivalent and related key attacks. David J Wheeler and Roger M. Needham in 1997 corrected the original TEA and it leads to two new algorithms called XTEA and Block TEA where XTEA (Hong et al. 2003; Moon et al. 2002) has the same block size with some changes in operations but Block TEA provide variable block size for each iteration.

4 Algebraic Cryptanalysis

Algebraic cryptanalysis is one of the methods of cryptanalysis to break the cipher to get the key/plaintext by solving algebraic equations. It is a known plaintext attack. Attacker chose some plaintext (P) and corresponding ciphertext (C). Keys kept secret (unknown). Based on these known (P, C) pairs, generate system of equations and solve for the key. To perform algebraic cryptanalysis:

- Represent encryption operation as system of equations.
- Substitute some known values form (P, C) pairs.
- Solve for the key/plaintext.

Solving nonlinear system of equations is NP-complete. But our aim is not to solve generalized nonlinear system. Aim is to solve our specific system using various methods to recover the key .There are various methods to solve system of multivariate equations: (1) Linearization method, (2) Extended linearization, (3) Gaussian Elimination technique, (4) Groebner Basis Method, (5) SAT/SMT solver, etc. In this paper, SMT solver has been chosen to solve nonlinear equations for our approach to get faster results.

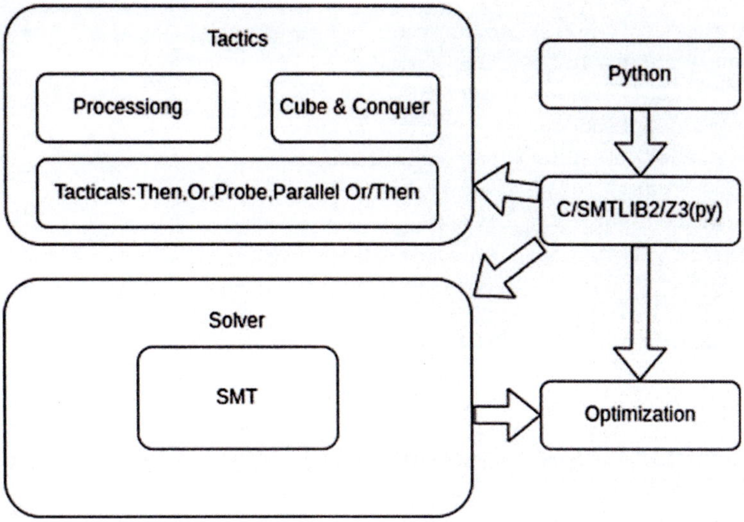

Fig. 2 Execution flow of Z3 solver

4.1 SMT Solver

Set of decision for logical first-order equations in terms of various background theories such as uninterpreted functions, arrays, arithmetic, and bit-vectors comes under, the Satisfiability Modulo Theories (SMT) problem. There are various SMT solvers (Brummayer and Robert 2010) such as Boolector, Yices, Optimathsat, and Z3 present in the literature. Microsoft has made available for free Z3, a modern and improved SMT solver. It is used in a wide range of verification applications and software analysis and we can also use it to solve our system of equations without actually taking care of the Boolean level difficulty. To perform algebraic attack on TEA, the SMT solver Z3 has been chosen for our approach. Z3 is a fast SMT solver with specialized background theory solution techniques. Z3 can resolve not only Boolean formulas (e.g., $x \vee y =$ True) but also non-Boolean formulations (e.g., $x^2 - 2y + 5 < 0$) in comparison with the SAT solver.

Figure 2 shows overall working components of Z3 version 4.8. Complete details of Z3 component can be referred from Bjørner et al. (2018). Example program on Z3 where we declare two Boolean variable and applying logical Or operation and try to see whether it is satisfiable or not if satisfiable than print any possible value for variable $p1$ and $p2$. In result, one can see that it is sat means satisfiable, and one possible value for variables is $p1 =$ True and $p2 =$ False.

```
Example:
from z3 import *
p1,p2 = Bool("p1 p2") #declare Boolean variable
```

```
phi = Or (p1, p2)    #construct a formula
print(phi) #printing the formula
s = Solver() #create SMT object
s.add(phi) #add formula to the solver
r = s.check() #Check satisfiability
if r == sat:
    print("sat, s.model())
else:
    print("unsat")
```

Output: sat with p1 = True, p2 = False.

5 Algebraic Cryptanalysis on TEA

TEA has simple and easy cipher operations that can be easily translated into system of equations. We choose Z3 SMT solver to solve our TEA equations. MathSAT, CryptoSMT, Yices, Boolector, Z3, OpenSMT, STP, Spear, and CVC3 are currently available SMT solvers. The following approach has been used for generating and solving using Z3 by algebraic cryptanalysis (Gundaram et al. 2021; Sahu et al. 2020).

Algorithm 1: Algebraic Cryptanalysis in Z3(py) format

1. **Inputs:** Plaintext-Ciphertext (P, C) and Constant delta = (0X9E379B9)
2. **Output:** Key (K)
3. Declare the variables (X0, X1,. . . ,Y0, Y1,. . . ,K0, K1, K2,, K3, etc) in Z3(py)
4. Initialize Z3 SMT solver
5. Add known value to variables declared in step 3
6. Write cipher description as an algebraic equations in terms of key variables
   $X_{i+1} = X_i + ((Y_i << 4) + k0) \oplus (Y_i + Delta_i) \oplus ((Y_i >> 5) + k1)$
   $Y_{i+1} = Y_i + ((X_{i+1} << 4) + k2) \oplus (X_{i+1} + Delta_i) \oplus ((X_{i+1} >> 5) + k3)$
   where, $Delta_i$ is constant, X_i and Y_i are 32-bit palintext, X_{i+1} and Y_{i+1} are 32-bit ciphertext and (k_0, k_1, k_2, k_3) are key variables
7. Provide equations generated in step 6 to Z3 SMT solver.
8. Check for Satisfiability
9. **if UNSAT** then
10. return **Erorr :** Equation not satisfiable or invalid input constraints
11. **else:** continue
12. **end if**
13. Generate the model for satisfiable constraints
14. Find all possible assignments of keys using generated model in step 14
15. Find the correct key from all possible candidate keys by encrypting other pairs of plaintext and compare with the respective ciphertext.

5.1 Model Generation in Z3(Py) Format

The following 32-bit variables are used for TEA Z3 model to perform operations under modulo 2^{32}.

Inputs are plaintext $X_i(x_0, x_1, x_2, \ldots, x_{31})$, Y_i $(y_0, y_1, \ldots, y_{31})$, ciphertext X_{i+1} $(x_0, x_1, \ldots, x_{31})$, Y_{i+1} $(y_0, y_1, \ldots, y_{31})$. Key variables are K_0 $(k_0, k_1, \ldots, k_{31})$, K_1 $(k_{32}, k_{33}, \ldots, k_{63})$, K_2 $(k_{64}, k_{65}, \ldots, k_{95})$ and K_3 $(k_{96}, k_{97}, k_{98}, \ldots, k_{127},)$. Constant $Delta$ is provided at every round in multiple of itself $Delta\ _i$ $(d_0, d_1, \ldots, d_{31})$.

The following is the approach for solving the equations. The plaintext data and its associated ciphertext are made fixed first. Write equations in Z3 format by fixing plaintext and ciphertext. This is checked for all possible keys, out of that one key will definitely be the original key from which plaintext has been encrypted.

In this section, we show how to perform algebraic cryptanalysis on TEA using Algorithm 1 based on the Z3 SMT solver.

1. Variable declaration Z3(py):
   Plaintext declaration: x0, y0 = BitVecs('x0 y0', 32)
   Ciphertext declaration: x31, y31 = BitVecs('x31 y31', 32)
2. Initialize solver: s = Solver()
3. Assign plaintext sub-block to variables x0 and y0
   s.add(x0, y0 == 0x12345678, 0x33333333)
4. Declaration of Z3Py variables for master sub-keys:
   k0, k1, k2, k3 = BitVecs('k0 k1 k2 k3', 32)
5. Create a ciphertext expression in terms of unknown key bits by following the encryption method E and the plaintext as known Z3(py) variables, produce the Boolean expression corresponding to output Ciphertext bits as given in Sect. 5.3. The following are the equations derived after one round:

$$X_1 = X_0 + ((Y_0 \ll 4) + k0) \oplus (Y_0 + Delta_0) \oplus ((Y_0 \gg 5) + k1)$$
$$Y_1 = Y_0 + ((X_1 \ll 4) + k2) \oplus (X_1 + Delta_0) \oplus ((X_1 \gg 5) + k3)$$

   The first round's output is utilized as input to build equations for the following rounds. Because the size of the equations grows rapidly in successive rounds, it is impractical to write all the equations in this paper.
6. Equate the system of equations for the ciphertext calculated with key as unknown with original ciphertext to find the unknown: The example given cipher text C = 0x423c29fb, 0xd0a3664 divided into two equal sub-blocks and assign these ciphertext value to variable X_{31}, Y_{31} as below.

$$\text{s.add(x31, y31 == 0x423c29fb, 0xd0a3664)}$$

7. Examine the algebraic equations generated in earlier Step for Satisfiability after equating with original ciphertext.
8. Using the model() function, generate all potential satisfying assignments K_0, K_1, $K2$, K_3. All of these assignments are possible keys for the cipher.

9. Encrypt other pairs of plaintext to find the correct key among all possible keys and compare with the respective ciphertext to get the Master key K.

6 Cryptanalysis Results of TEA

Table 1 gives comparison of results using desktop computer and HPC up to 5th round of TEA. Till five rounds of TEA, our attack requires one plaintext ciphertext pair to recover the key. For a given one pair of plaintext and ciphertext corresponding to 5th round, it took 15,576.345 min to recover the Reduced Round (5th round) full key using HPC. It is giving 70 possible keys. For given one pair of plaintext and ciphertext, and partial known key (from 64-bits to 96-bits), remaining partial key is recovered using desktop. Results containing number of possible keys and execution time till 6th round are given in Table 2. We were able to recover 32/64-bit partial keys of 6th round provided the input as given plaintext ciphertext pairs and along with the partial 32/64-bit key out of 128-bit as input.

Table 1 TEA results Normal-PC versus HPC

Round No.	Time taken by desktop PC	Time taken by HPC	No. of possible keys obtained
2	52.64 s	1.754 s	1024
3	278.24 s	9.274 s	1024
4	–	1037.5 min	1024
5	–	15,576.345 min	69

Table 2 TEA 6th round table

Round No.	Plaintext and ciphertext	Known partial keys	No. of possible keys	Time taken (s)
2	X_0, Y_0, X_2, Y_2	K_1, K_2, K_3	1	0.12
2	X_0, Y_0, X_2, Y_2	K_0, K_1	1024	53.34
3	X_0, Y_0, X_3, Y_3	K_1, K_2, K_3	1	0.37
3	X_0, Y_0, X_3, Y_3	K_0, K_1	8	90.27
4	X_0, Y_0, X_4, Y_4	K_1, K_2, K_3	1	0.06
4	X_0, Y_0, X_4, Y_4	K_0, K_1	2	4017
5	X_0, Y_0, X_5, Y_5	K_1, K_2, K_3	1	20.4
6	X_0, Y_0, X_6, Y_6	K_1, K_2, K_3	1	3175.54

Desktop Computer Configuration: Ubuntu 20.04 LTS 64-bit, AMD A10 processor at 1.8 GHz ×8, 8GB RAM and 2TB Hard Disk.

HPC Configuration: CentOS Linux 7 (Core) 64 bit, Intel (R) Xeon (R) E5-2680 v4 at 2.40 GHz, Core per socket: 32.

7 Conclusion

Algebraic cryptanalysis on block cipher TEA using SMT solver Z3 has been performed in this paper. Firstly, we modeled the TEA in Z3(py) format, and to solve the cipher expression, we have used Z3 which is one of the best SMT solvers for cryptanalysis to use under known plaintext attack. By analyzing other SMT solver (Gundaram et al. 2021)-based algebraic cryptanalysis on different ciphers, we found out that Z3 is best suitable for our approach. We could cryptanalyze fifth round TEA using Z3 SMT solver that only requires two pairs of plaintext and ciphertext within 259.605 h. If we want to go for more number of rounds, till 6th round We were able to recover 32-bits out of a full 128-bits key. There is potential to create a customized efficient SMT solver that supports cryptanalysis theories and accurately reflects cryptographic primitive operations. Parallel SMT solvers can be used to speed up the SMT attack. To cryptanalyze a larger number of TEA rounds, algebraic attack can be combined with other attacks like MITM, differential to get the better results compare to stand alone SMT solver.

References

N. Bjørner, L. de Moura, L. Nachmanson, C.M. Wintersteiger, Programming Z3, in *International Summer School on Engineering Trustworthy Software Systems* (Springer, 2018), pp. 148–201

R. Brummayer, *Efficient SMT Solving for Bit-Vectors and the Extensional Theory of Arrays* (Trauner, 2010)

P.K. Gundaram, A.N. Tentu, N.B. Muppalaneni, Performance of various SMT solvers in cryptanalysis, in *2021 International Conference on Computing, Communication, and Intelligent Systems (ICCCIS)*, pp. 298–303 (IEEE, 2021)

S. Hong, D. Hong, Y. Ko, D. Chang, W. Lee, S. Lee, Differential cryptanalysis of TEA and XTEA, in *International Conference on Information Security and Cryptology* (Springer, 2003), pp. 402–417

J. Kelsey, B. Schneier, D. Wagner, Related-key cryptanalysis of 3-WAY, Biham-DES,CAST, DES-X, NewDES, RC2, and TEA, in *Information and Communications Security. ICICS 1997*. Lecture Notes in Computer Science, vol. 1334, ed. by Y. Han, T. Okamoto, S. Qing (Springer, Berlin, Heidelberg). https://doi.org/10.1007/BFb0028479

D. Moon, K. Hwang, W. Lee, S. Lee, J. Lim, Impossible differential cryptanalysis of reduced round XTEA and TEA, in *International Workshop on Fast Software Encryption* (Springer, 2002), pp. 49–60

Rajashekarappa, K.M. Sunjiv Soyjaudah, K.A. Sumithra Devi, Study on cryptanalysis of the tiny encryption algorithm. Int. J. Innov. Technol. Expl. Eng. (IJITEE) **2**, 88–91 (2013)

H.K. Sahu, N.R. Pillai, I. Gupta et al., SMT solver-based cryptanalysis of block ciphers. SN Comput. Sci. **1**, 169 (2020). https://doi.org/10.1007/s42979-020-00181-4

G. Sekar, N. Mouha, V. Velichkov, B. Preneel, Meet-in-the-middle attacks on reduced-round XTEA, in *Cryptographers' Track at the RSA Conference* (Springer, 2011), pp. 250–267

M. Shoeb, V.K. Gupta, A crypt analysis of the tiny encryption algorithm in key generation. Int. J. Commun. **10** (2012)

D.J. Wheeler, R.M. Needham, TEA, a tiny encryption algorithm, in *International Workshop on Fast Software Encryption* (Springer, 1994), pp. 363–366

Diagnosis of Pulmonary Diseases from Chest X-ray Using Deep Learning Approaches

Chaitra Patwardhan, Advait Thakur, Neha Adawadkar, Roshani Chavan, and Suhasini Itkar

Abstract In recent years, a severe pandemic has struck worldwide with the utmost shutter, enforcing a lot of stress in the medical industry. Moreover, the increasing population has brought to light that the work bestowed upon the healthcare specialists needs to be reduced. Medical images like chest X-rays are of utmost importance for the diagnosis of diseases such as pneumonia, COVID-19, thorax, and many more. Various manual image analysis techniques are time-consuming and not always efficient. Deep learning models for neural networks are capable of finding hidden patterns, assisting the experts in specified fields. Therefore, collaborating these medical images with deep learning techniques has paved the path for enormous applications leading to the reduction of pressure embarked upon the health industry. This paper demonstrates an approach for automatic lung diagnosing of COVID-19 (coronavirus) and thorax diseases from given CXR images, using deep learning techniques. The previously proposed model uses the concept of ResNet-18, ResNet-50, and Xception algorithms. This model gives the highest accuracy of 98% without segmentation and 95% with segmentation. Whereas, the proposed model uses CNN and CLAHE algorithms which achieves an accuracy of 99.22% without segmentation and 98.39% with segmentation. Therefore, this model will be able to provide assistance to health workforces and minimize manual errors precisely.

Keywords Convolutional neural network (CNN) · Deep learning · Chest X-ray (CXR) · Contrast limited adaptive histogram equalization (CLAHE)

1 Introduction

In late 2019, the outburst of COVID-19 resulted in numerous deaths and emphasized various aspects of individual's health. The rise in the COVID-19 infected cases spread globally. With a sequence of instances, it was declared that coronavirus became a

C. Patwardhan (✉) · A. Thakur · N. Adawadkar · R. Chavan · S. Itkar
Department of Computer Engineering, Progressive Education Society's Modern College of Engineering, Pune, India
e-mail: chaitrapatwardhan@gmail.com

A. B. Reddy et al. (eds.), *Proceedings of Third International Conference on Advances in Computer Engineering and Communication Systems*, Lecture Notes in Networks and Systems 612, https://doi.org/10.1007/978-981-19-9228-5_7

pandemic with a tremendously high threat that would affect millions of lives around the world, India being one of the countries to experience the same. Following the rise of COVID-19 cases, from June 2020, India became the third-worst affected country. Therefore, these circumstances brought to attention that doctors and professionals require some assistance.

Chest radiograph, also known as chest X-rays (CXRs), is a very common method for diagnosing any pulmonary disease. Professionals for many years have been relying on manual analysis of chest X-rays, which is quite time-consuming and may result in some subjective errors. In current years, deep learning techniques have become a choice for the tasks of image analysis and have made a high impact in the healthcare field. Various deep learning approaches are described to make improvements in segmentation quality and accuracy. However, the prevailing methods shortfall in robustness to the deviation of the target. For helping professionals to make an accurate disease diagnosis, segmentation of crucial objects and extracting meaningful features from specific areas is necessary. Therefore, automatic diagnosis of diseases from CXR images is necessary, which can be achieved through image segmentation. Among various deep learning techniques, CNN-based models have attained huge success in an extended range of tasks. Moreover, CNN algorithms have much more applications in artificial intelligence, such as image segmentation, object detection, and recognition.

In this paper, a system for automatic diagnosis of COVID-19 and thorax diseases is demonstrated. The image is first pre-processed by applying cropping techniques followed by CLAHE for better localization and equalization. By using CLAHE, it prevents the over-amplification of images. Furthermore, the pre-processed images are provided to the multi-class classification model, which works quite well with instances obtained from various sources. Finally, these segmented images give a precise and accurate diagnosis of a disease in a split second.

2 Related Works

Researchers are using deep learning techniques for identifying diseases from medical images like CTs and CXRs. Let us discuss some of the work done by various researchers for identifying patterns and discovering pulmonary diseases from the CXRs.

Yang et al. (2020) gives a brief introduction about the novel coronavirus and calls to attention the significance of imaging in its diagnosis. In (https://doi.org/10.1007/s00330-020-06827-4Olaf), Xception architecture which is similar to Inception V3 is developed which efficiently uses the model parameters. Chollet (2017) proposes a binary classifier using support vector machine (SVM) which classifies a CXR as either normal or abnormal giving an accuracy of 82.1% and area under the ROC curve as 88.5%. In Jaeger et al. (2014), a U-Net network is calculative trained with the use of data augmentation by using elastic deformations for effective use of annotated samples. The segmentation results for ISBI cell tracking challenge with average intersection over union (IOU) for 'PhC-U373' dataset is 92% and for 'DIC-HeLa' dataset is 77.5%. In Ronneberger et al. (2015a), a framework for dynamic sequences of 2D cardiac magnetic resonance (MR) images is reconstructed using the under-sampled data with the use of deep cascade of CNN. The model reconstructs most of the anatomical structures more precisely by the previously learned data. The system in Schlemper et al. (2017) re-purposes the networks which are trained on image classification to semantic segmentation by using 'atrous convolution' with unsampled filters for the purpose of feature extraction. User can explicitly manage the resolution where upon feature responses are determined and maximize the field view of filters to subsume large context without increasing the number of parameters or the amount of computation. Chen et al. (2018) adapts contemporary classification networks into fully convolutional networks and transfers their learned representations by fine-tuning to the segmentation task. In Long et al. (2015), an architecture to diagnose COVID-19, thorax, and other diseases that affect a patient's lung from CXR using ResNet-18, ResNet-50, and Xception model is used. This ResNet-18 model gives an accuracy of 94%, while ResNet-50 gives accuracy of 93%, and the Xception model is having an accuracy of 95%. In FUIQA (Kittiworapanya and Pasupa 2020), lookup-based CNN (LCNN) and deep CNN are used together for analysis, wherein LCNN gives overall accuracy of 92%, and the DCNN gives overall accuracy of around 97%.

Wu et al. 2017) uses CNN to reconstruct and assesses the quality of the images. Feature learning and regression are integrated into an optimization process, leading to a constructive system for assessing image quality. In Kang et al. (2014), U-Net is used for the segmentation of biomedical images. In Ronneberger et al. (2015b), an automated analysis method for cardiovascular magnetic resonance (CMR) images built upon a fully convolutional network (FCN). UK Biobank dataset is used for training and evaluation purpose. This model has a dice metric of 0.93. In Bai et al. (2018), convolutional recurrent neural network (CRNN) model is used for image quality assessment that reconstructs high-quality cardiac MR images from highly under-sampled *k*-space data by jointly exploiting the dependencies of the temporal sequences as well as the iterative nature of the traditional optimization algorithms. This model gives a minimum time of 3 s. In Qin et al. (2018), the authors have proposed the use of VGG16 for breast cancer detection giving an overall accuracy of 94.77%. In Yongbin et al. (2020), implementation of FCN, ResNet, U-Net, encoder/decoder, and CNN is done giving an overall accuracy of 75% to 80%. Quality assessment of Echocardiograms, Ali et al. (2020) developed a regression model to quantify the condition of echo images. Stochastic gradient descent algorithm is used to train the loss function of the model. Convolution and fully connected stages are used to design the model. In Abdi et al. (2017), basic operations are done to eliminate known distortions from images. Distorted and reference signals are scaled and aligned, then metrics of digital values need to be converted into pixel intensity using nonlinear pointed transformations. Discrete cosine transform or separable wavelet transform methods are used for quality assessment methods.

3 System Architecture

The data used is the Deep Learning and AI Summer/Winter School (DLAI) Hackathon Phase 3—Multiclass COVID-19 Chest X-ray Challenge dataset (Wang et al. 2004). This data is collected from multiple sources. There are three classes present in the dataset thorax diseases, COVID-19, and clear (Fig. 1).

3.1 Pre-processing

All the RGB X-ray images were converted to monochrome images, to match the requirements of CNN networks. Since OpenCV needs the images to be of same dimensions, resizing is done so that the calculation of pixel values of original images can be done easily. The images were resized with dimensions 64 * 64 px to regularize the dimensions of all the images in the dataset. INTER_NEAREST, INTER_LINEAR, INTER_AREA, etc., are the five such interpolation methods provided by OpenCV. In this paper, INTER_AREA has been used for reducing the size of the image (shrinking) with the help of pixel area relation for resampling. The dataset is then cleaved into 90% and 10% for training and testing dataset, respectively. Range rotation was done in which a range was fixed to randomly rotate the images between the degrees of 0–180. Along with this, width_shift and height_shift were taken with which we can translate the images vertically and horizontally randomly. Rescaling is the method in which a value will multiply the data before any other processing.

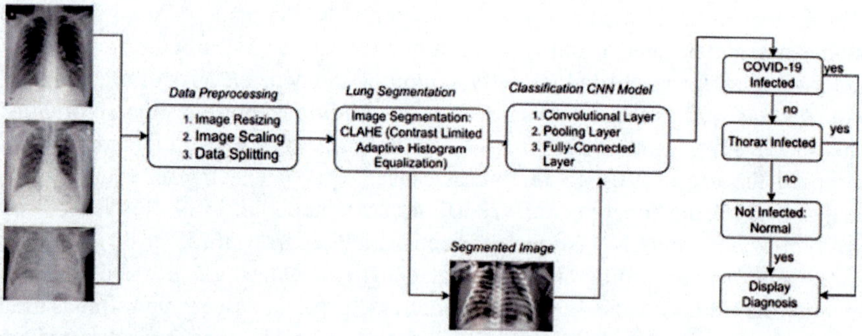

Fig. 1 Proposed chest X-ray diagnosis framework

3.2 Segmentation

The procedure of dividing an image into multiple sets of pixels in the medical images is known as segmentation. Segmentation is used to make things simpler and transform the representation of medical images into a meaning full subject.

For the dataset used, the challenging problem for segmentation is the unavailability of the masks for the lung and its region of interest. The main requirement for segmenting the CXR images is the masks needed, and since this dataset does not contain these masks, in the experiment, we have used a variant of adaptive histogram equalization, contrast limited adaptive histogram equalization (CLAHE) in which the contrast amplification is limited (https://www.kaggle.com/jonathanchan/dlai3-hackathon-phase3-covid19-cxr-challenge/version/1). In CLAHE, the slope of transformation function states the contrast amplification of a given pixel value. The entire image is divided into small regions called tiles on which the CLAHE operates. The neighboring tiles are then combined using interpolation to remove the artificial boundaries (https://www.kaggle.com/jonathanchan/dlai3-hackathon-phase3-covid19-cxr-challenge/version/1). CLAHE can be applied to color images and grayscale images, where in color images, when CLAHE is applied on the RGB model, new artifact colors are introduced different than the original images, and on the HSV images, the contrast of the images not only increases but the outcomes are much better.

There are two main important parameters for CLAHE which are clipLimit that sets the threshold for contrast limiting whose default value is 40 and tileGridSize that places the number of tiles in the row and column which is set to 8 * 8 by default.

CLAHE Algorithm

Step 1: Unclear or blurred image is acquired.

Step 2: Parameters like clipLimit and distribution parameter along with number of regions in the row and column are obtained in the enhancement process.

Step 3: Processes are applied over the tile which are gained by dividing the original image into different regions.

Step 4: Clipped histogram along with gray level mapping is generated, and they interpolate gray level mapping which are beneficial to create enhanced image (Fig. 2).

3.3 Classification

In this step, two commonly used state-of-the art CNN-based models are used. A 54 layered CNN ResNet model is coded and along with that an Inception pretrained model is used. This Inception model was pretrained using ImageNet dataset. When the depth of CNN layer increases, a residual learning model uses skip connections

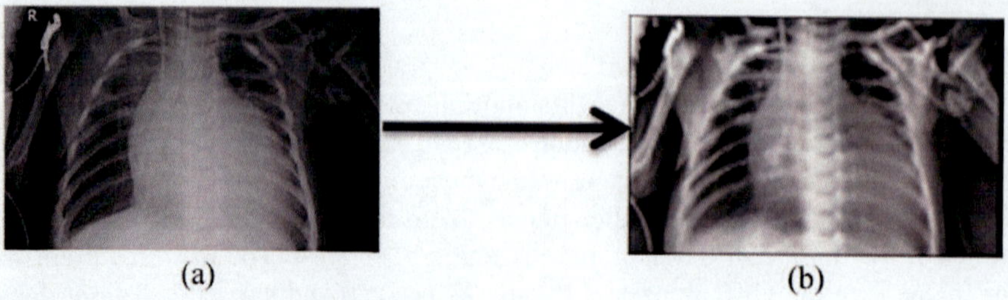

Fig. 2 **a** Original image and **b** CLAHE applied image

to solve the gradient vanishing problem. The CNN models categorize CXR images into three classes of thorax, COVID-19, and normal (not infected).

The another type of deep neural network is an Inception network consisting Inception modules which repeats the components. Pretrained model is the trained models to find solution of similar typed problems. The pretrained models save the efforts to build the model from the very beginning. Pretrained models help to work faster as it reduces efforts for training and testing purpose to build architecture of a model. The dataset division was performed using split folders library such that 9 folders were made. In the pretrained model, image size of 150 * 150 is used. For training purpose, the 30 number of epochs were used along with 75 steps per epoch.

System consists of the main proposed model, i.e., CNN ResNet model with 54 convolutional layers, where ReLU activation and batch normalization operations are performed after each convolutional layer. Yadav et al. (2014) Output of the previous layer is normalized in batch normalization. The activation function scales the input layer by using normalization. The independent learning is encouraged after every layer. Numbers of epochs were up to its maximum limit 100.

Rectified linear unit (ReLU) is one of common function used by neurons. The major advantage of ReLU is that is it very easy to calculate. The activation function of 'ReLU' and 'softmax' was used. Both the models were compiled with 0.001 learning rate and 'categorical_crossentropy' loss function.

4 Experimental Results

The dataset (Wang et al. 2004) is divided into 3736 thorax diseases, 1408 clear, and 363 COVID-19 giving us a total of 5507 images. The chest X-rays with absence of thorax disease and may have occurrences of other unknown disease patterns are also included in clear label. The CXR images differ in contrasts, sizes, views, and ratios. Some CXR images focus exactly on the lung region while some stretch from neck up to the stomach. To address these problems, pre-processing needs to be done before assigning to the model. After pre-processing, 551 images were used for testing purpose and 4956 images used for training purpose.

In the experiment, $F1$-score, accuracy, precision, and recall are used for evaluation purpose. In classification problems, the number of correct predictions made by the model over all predictions is called as accuracy.

$$\text{Accuracy} = \frac{\text{TP} + \text{TN}}{\text{TP} + \text{FN} + \text{FP} + \text{TN}}$$

Precision is the ratio of the quality of the predictions that are positive, i.e., TP and all the positives. Recall is the measure of the correctly identifying TP. The harmonic mean of precision and recall is known as $F1$-score.

$$\text{Precision} = \frac{\text{TP}}{\text{TP} + \text{FP}}$$

$$\text{Recall} = \frac{\text{TP}}{\text{TP} + \text{FN}}$$

$$F1\ \text{Score} = \frac{2 * \text{precision} * \text{recall}}{\text{precision} + \text{recall}}$$

From Table 1, it is evident that the proposed model from the paper gives the best results as compared to that of Xception (Long et al. 2015) when segmentation methods are applied. The above results are acquired from the performance of the training dataset (Fig. 3).

In graph (a) and (b), the number of epochs is same 0–200, but the accuracy of the models is different. The training accuracy ranges from 98 to 99 and testing accuracy goes from 97 to 95 and then again increases to 97 for the proposed model. The training accuracy varies from 88 to 90 while testing accuracy ranges between 88 and 90 for Inception. While from the graph (b), it can be recognized that the loss of proposed model is comparatively less than that of Inception.

Table 1 Classification experiment results

Method	Models	Accuracy (%)	Precision (%)	Recall (%)	$F1$-score (%)
Without segmentation	Proposed model	99.22	96.89	96.85	96.86
	Inception	90.35	93.21	92.87	92.94
	Xception	98	98	96	97
	ResNet-18	98	98	95	96
	ResNet-50	97	97	95	96
With segmentation	Proposed model	98.39	97.16	96.97	96.94
	Inception	87.74	91.11	90.63	90.74
	Xception	95	94	90	92
	ResNet-18	94	90	89	89
	ResNet-50	93	87	90	88

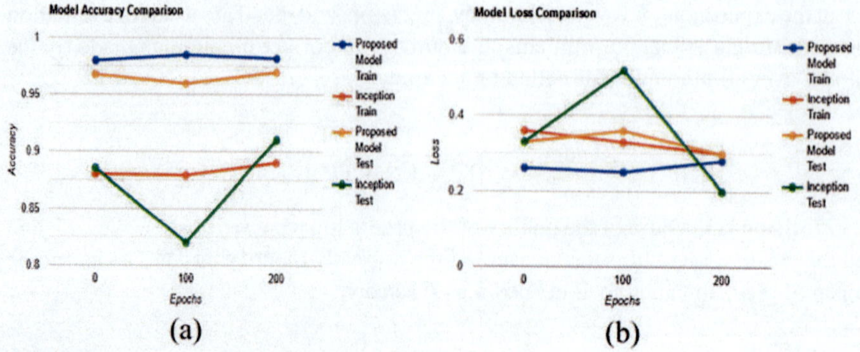

Fig. 3 Graphs showing comparisons for **a** model accuracy and **b** model loss for proposed model and inception with applied CLAHE

5 Conclusion

During the critical phase of pandemic, rapid and accurate diagnosis of a disease is an extreme necessity. As thorax and COVID-19 are severely affecting our lungs, it becomes challenging to predict correct disease with very thin line of differences. The accuracy of implemented Inception model is 89.06% while the model implemented using CNN outsmarts the Inception model with an accuracy of 99%. The implemented model outsmarts the previous implemented models, starting with implementation of CLAHE. The increase in convolutional layers and change of activation function and loss function increases overall performance of the model.

The future scope can include diagnosis of various diseases related to different body organs for disease diagnosis which are crucial to dissect and require a lot of human effort and time. Our paper solely focuses on the diagnosis of pulmonary disease from chest X-rays, but in successive trials for successive methods, more enhanced methods can be used for diagnosis in images other than chest X-rays.

References

A.H. Abdi, C. Luong, T. Tsang, G. Allan, S. Nouranian, J. Jue, D. Hawley, S. Fleming, K. Gin, J. Swift et al., Automatic quality assessment of apical four-chamber echocardiograms using deep convolutional neural networks. Proc. SPIE **10133**, 101 330S-1 (2017)

M. Ali, S.O. Gilani, A. Waris, K. Zafar, M. Jamil, Brain tumour image segmentation using deep networks. IEEE Open Access J **8** (2020)

W. Bai, M. Sinclair, G. Tarroni, O. Oktay, M. Rajchl, G. Vaillant, A. M. Lee, N. Aung, E. Lukaschuk, M.M. Sanghvi, Automated cardiovascular magnetic resonance image analysis with fully convolutional networks. J. Cardiovasc. Magn. Reson. **20**(1), 65 (2018)

L.-C. Chen, G. Papandreou, I. Kokkinos, K. Murphy, A.L. Yuille, Deeplab: semantic image segmentation with deep convolutional nets, atrous convolution, and fully connected CRFs. IEEE Trans. Pattern Anal. Mach. Intell. **40**(4), 834–848 (2018)

F. Chollet, Xception: deep learning with depthwise separable convolutions. In: *Proceedings of the IEEE Conference on Computer Vision and Pattern Recognition (CVPR 2017)* (IEEE, Honolulu, HI, 2017), pp. 1800–1807. https://doi.org/10.1109/CVPR.2017.195
https://doi.org/10.1007/s00330-020-06827-4Olaf
https://www.kaggle.com/jonathanchan/dlai3-hackathon-phase3-covid19-cxr-challenge/version/1
S. Jaeger, A. Karargyris, S. Candemir, L. Folio, J. Siegelman, F. Callaghan, Z. Xue, K. Palaniappan, R.K. Singh, S. Antani, G. Thoma, Y. Wang, P. Lu, C.J. McDonald, Automatic tuberculosis screening using chest radiographs. IEEE Trans. Med. Imaging **33**(2), 233–245 (2014)
L. Kang, P. Ye, Y. Li, D. Doermann, Convolutional neural networks for no-reference image quality assessment, in *Proceedings of the IEEE Conference on Computer Vision and Pattern Recognition* (2014), pp. 1733–1740
P. Kittiworapanya, K. Pasupa, *An Image Segment Based Classification for Chest XRay Image* (ACM, 2020). https://doi.org/10.1145/3429210.3429227
J. Long, E. Shelhamer, T. Darrell, Fully convolutional networks for semantic segmentation. Proc. CVPR, 3431–3440 (2015)
C. Qin, J. Schlemper, J. Caballero, A.N. Price, J.V. Hajnal, D. Rueckert, Convolutional recurrent neural networks for dynamic MR image reconstruction. IEEE Trans. Med. Imaging **38**(1), 280–290 (2018)
O. Ronneberger, P. Fischer, T. Brox, U-net: convolutional networks for biomedical image segmentation, in *Proceedings of the International Conference on Medical Image Computing and Computer-Assisted Intervention (MICCAI 2015a)* (Springer, Munich, Germany, 2015a), pp. 234–241. https://doi.org/10.1007/978-3-319-24574-4_28
O. Ronneberger, P. Fischer, T. Brox, *U-Net: Convolutional Networks for Biomedical Image Segmentation|| in MICCAI* (Springer, 2015b), pp. 234–241
J. Schlemper, J. Caballero, J.V. Hajnal, A.N. Price, D. Rueckert, A deep cascade of convolutional neural networks for dynamic MR image reconstruction. IEEE Trans. Med. Imaging **37**(2), 491–503 (2017)
Z. Wang, A.C. Bovik, H.R. Sheikh, E.P. Simoncelli, Image quality assessment: From error visibility to structural similarity. IEEE Trans. Image Process. **13**(4), 600–612 (2004)
L. Wu, J.-Z. Cheng, S. Li, B. Lei, T. Wang, D. Ni, FUIQA: fetal ultrasound image quality assessment with deep convolutional networks. IEEE Trans. Cybernet. **47**(5), 1336–1349 (2017)
G. Yadav, S. Maheshwari, A. Agarwal, Contrast limited adaptive histogram equalization-based enhancement for real time video system, in *International Conference on Advances in Computing, Communications and Informatics (ICACCI, 2014)*. https://doi.org/10.1109/icacci.2014.6968381
W. Yang, A. Sirajuddin, X. Zhang, G. Liu, Z. Teng, S. Zhao, M. Lu, The role of imaging in 2019 novel coronavirus pneumonia (COVID-19). Eur. Radiol. **30**(9), 4874–4882 (2020)
Y. Yu, E. Favour, P. Mazumder, Convolutional neural network design for breast cancer medical image classification, in *2020 IEEE 20th International Conference on Communication Technology*

Performance Analysis of CNN for Patch-Based Sclera–Periocular Biometrics

V. Sandhya and Nagarathna P. Hegde

Abstract Image classification is a process of categorizing images based on features. Features in an image could be identified as a change in pixel intensity or an edge. A color image has the pixel values represented using R, G, and B. Multiple such images are labeled and used for image classification. The challenging part is to identify the features in which such images are a complex task. CNN is a widely used image processing algorithm particularly for image classification. The three layers of CNN—convolution layer, pooling layer, and fully connected layers—can be applied to an image for image processing problems such as image recognition, object detection, and segmentation. In the proposed system, CNN is applied on patch-based sclera–periocular images. The model has shown an accuracy of 99.3% for patch-based images. The model was trained on image patches of size 100 × 100, 50 × 50, and 25 × 25.

Keywords CNN · Patch · Sclera–periocular

1 Introduction

In a conventional image processing system, features are extracted using edge detection, line detection, and corner detection methods. For a gray scale image to classify images into two classes, Otsu's threshold-based classification can be applied. Canny edge detection algorithm detects the edges in an image. Hough transform is used to identify lines in an image. For text to extract from an image, corner detection is applied. Gabor filters are used for feature extraction. The input image is passed through a Gabor filter that can identify edge and texture changes. Scale invariant feature transformation (SIFT) extracts local features by identifying key points known

V. Sandhya (✉)
GITAM School of Technology Bengaluru, Bengaluru, India
e-mail: svinayak@gitam.edu

N. P. Hegde
Vasavi College of Engineering, Hyderabad, India
e-mail: nagaratnaph@staff.vce.ac.in

A. B. Reddy et al. (eds.), *Proceedings of Third International Conference on Advances in Computer Engineering and Communication Systems*, Lecture Notes in Networks and Systems 612, https://doi.org/10.1007/978-981-19-9228-5_8

Fig. 1 Deep learning classifier

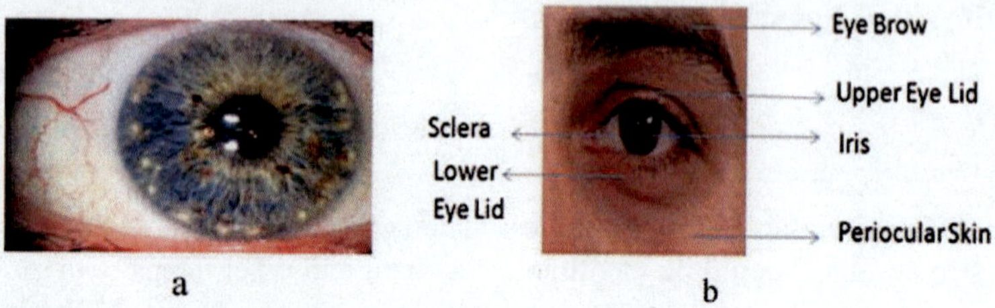

Fig. 2 **a** Sclera. **b** Periocular

as descriptors. These key points are used for model training in classification. Identifying rotation or scale invariant descriptors are important in the SIFT algorithm. For a small training data set, traditional image processing techniques can be applied as it occupies less memory. If the data set is large and to extract complex features of an image, CNN is used.

CNN is a deep neural network algorithm most widely used for image processing techniques such as image classification, image segmentation, and object detection. This algorithm has three basic operations for image processing—convolution, pooling, and passing through the dense layers. Figure 1 shows a deep learning-based classifier. The input image is passed through N layers. Each layer extracts specific features that can be used for classification of images. The extracted features are passed through a classifier to identify the input images based on labels.

Sclera is the white part of the eye that surrounds the iris and has blood vessel patterns. Periocular is the surrounding part of the eye that has eyebrows, eyelids, and skin texture. Figure 2a shows the sclera, and Fig. 2b shows the periocular images of eye.

2 Related Work

In Kao et al. (2020), the author introduced a method, patch-based neural network for brain tumor segmentation. The author applied data pre-processing technique such as clipping of non-zero voxels to remove outliers. A modified DeepMedic architecture was applied. In medical image, analysis segmentation plays an important role. Minaee et al. (2020) carried out various segmentation techniques such as fully convolutional pixel labeling networks, recurrent networks, encoder–decoder, pyramid-based approaches, and generative models. In Gayathri et al. (2020), the author proposed CNN model to extract features of retinal fundus images to classify using SVM, AdaBoost, Naive Bayes, and random forest. The performance of the CNN is 99.59% for multi-class classification. Deep learning models are used to train large number of samples. The given image is trained in forward pass (Sekoua et al. 2019) with a small number of samples and image patches using a light weighted network to a densely connected networks. In Kumari and Seeja (2019), the author explained the trade-off between face and iris biometric which requires user cooperation. Application of periocular biometrics and a detailed study of future research was carreied out. Hossain and Sajib (2019), Hesamian et al. (2019) explains about various deep learning methods for medical image segmentation such as CNN, 2D CNN, 3D CNN, and FCN. The author reviewed a study of periocular biometrics (Huang and Basanta 2019) based on existing databases, algorithms, segmentation methods, features for recognition, the discriminative regions of the periocular area, impact of plastic surgery, etc. Due to the traditional hand crafted segmentation limits in sclera extraction, Tiong et al. (2019) proposed CNN for image segmentation with noise removal using conditional random fields. A fully automatic minutiae extractor known as MinutiaNet (Hui and Lu 2019) using deep neural networks to estimate minutiae score map and minutiae orientation-based on CNN generates minutiae sets that have better precision and recall values. The discriminative features of the face are extracted using local binary pattern (LBP) in Roy et al. (2018), and city block distance is used as a classifier. The many challenges in sclera segmentation are iris gaze directions (Proenc and Neves 2018), image capturing distance, and differences in lighting conditions and proposed blood vessel enhancement and feature extraction methods to increase the adaptability to noisy sclera vessel deformations. To increase the accuracy of a biometric system, multimodal biometrics is used. In Rot et al. (2018), the author proposes a face and iris multimodal biometric system using matching score level and feature level for feature extraction. The multimodal system is implemented using LBP (Dong et al. 2018) feature extraction method. SVM is used to improve the recognition performance of the proposed system. A patch-based evaluation of image segmentation (PEIS) to Sharma et al. (2017) assess the quality of segmentation based on patch identification and patch displacement. In Hou et al. (2016), Chen et al. (2016), Zhang et al. (2006), Mechrez et al. (2016), and Ledig et al. (2014), the author proposed a novel deep learning framework for SIP-SegNet semantic segmentation of sclera, iris, and pupil traits in an unconstrained scenario by suppressing the periocular region.

3 Patch-Based Sclera–Periocular Biometrics

In patch-based segmentation, the given image is divided into segments of specified patch size. Patch can be of different shapes such as square, rectangle, or ellipse. Let image I be represented by height h and width was $h \times w$. Let the size of the patch be p as patch height and q as patch width. The number of patches that will be generated for given image is

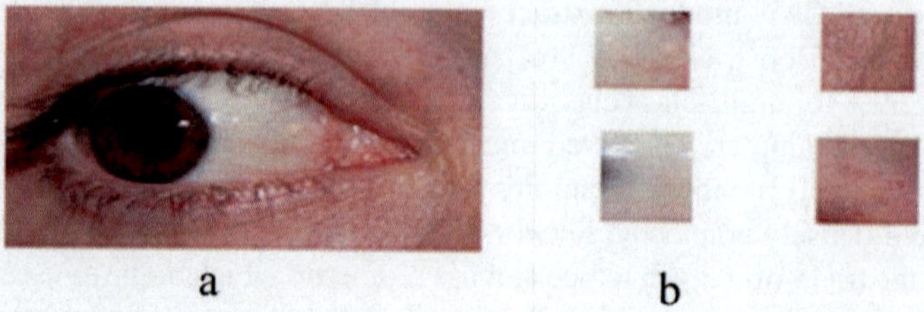

Fig. 3 a Eye. b Patches

$$N = \frac{h}{p} \times \frac{w}{q} \tag{1}$$

Figure 3a shows the given eye images, and Fig. 3b shows the sclera patch and periocular patch for the given eye image.

To generate the patches for the given image I, Algorithm 1 is applied. Mean pixel intensities are calculated for each patch. If the mean patch intensity is zero which is an indication of black patch.

Algorithm 1 For a given image I = {p1, p2, …, pm}, patches are generated as output.

1. Specify the threshold values t1 and t2.
2. For each patch of the image I.
3. Calculate mean pixel intensity, if the value is less than t1 or greater than t2, then remove the patch Pi from the image I folder.
4. Steps 2 and 3 are repeated for all patches of the image.

4 CNN Model

The proposed CNN model for sclera–periocular data set is as shown in Fig. 4. The model has a convolution layer followed by a max pooling layer and at the last layer a fully connected layer to classify the input images according to the labels. The first layer has three convolution layers with filters or kernels of size 3 × 3 with a depth of 16. Second convolution layer has three Conv2D layers of filter size 3 × 3 with a depth of 32. The third layer has again three Conv2D layers of filter size 3 × 3 with a depth of 64. The last layer has three Conv2D layers of filter size 3 × 3 with a depth of 128.

5 Sclera–Periocular Data Set

In the proposed model, the images are taken from three data sets such as VISOB 2.0, UBRIS, and SBVPI. Each subject has a left image and a right image. The ten

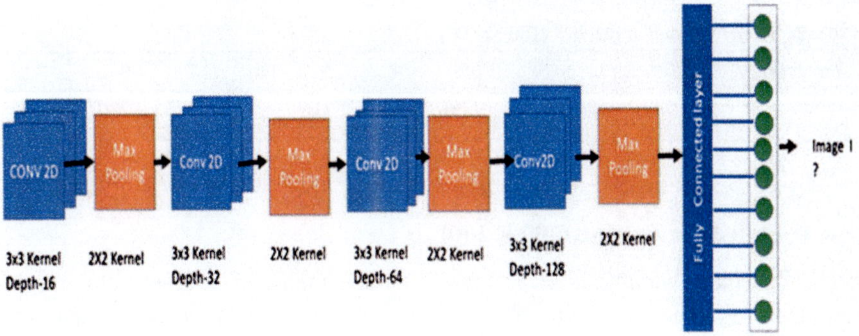

Fig. 4 Proposed CNN model

subjects are chosen on different features like images with makeup, with spectacle, with occluded sclera part, occluded periocular part, and difference in brightness. The number of images for this study has a total of 20 images with 2 images for each subject (Fig. 5).

Each subject's eye image was divided into patches with different sizes. Initially, the model was trained on patch size of 100 × 100, then with patch size of 50 × 50, and finally with a patch size of 25 × 25.

Sclera–Periocular Patches (50 x 50)

The images are collected from three different data sets. Images are resized to a shape of 3000 × 1700. Patch size is chosen as 50 × 50. Each subject has a left eye and a right eye image. Table 1 shows the class wise distribution of Sclera–periocular patches for a patch size of 50 × 50.

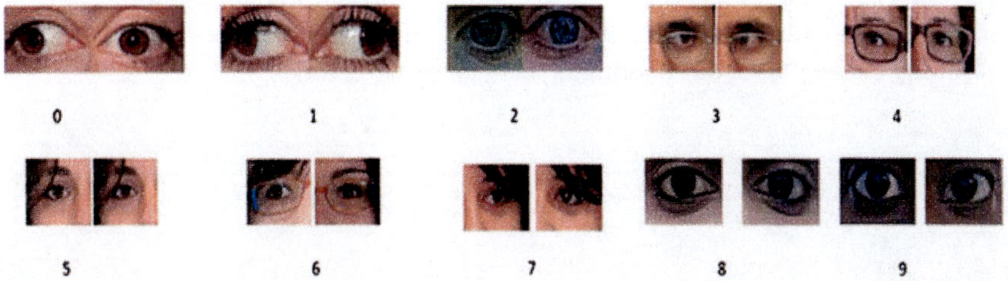

Fig. 5 Challenging data set

Table 1 Sclera–periocular patches (50 × 50)

Class label	O	1	2	3	4	5	6	7	8	9
Training patches	2156	2256	993	2770	3332	2376	1625	2030	1419	2150
Validation patches	538	563	248	692	832	594	406	507	354	537
Test patches	586	413	1544	713	516	437	313	425	468	301

Table 2 Sclera–periocular patches (100 × 100)

Class label	0	1	2	3	4	5	6	7	8	9
Training patches	537	596	514	790	954	716	392	525	280	623
Test patches	141	166	132	187	214	136	182	172	119	126

Sclera–Periocular Patches (100 x 100)

Each subject has two images such as first is left eye image, and the second is a right eye image eye. The total count of the patches is 1020. In the proposed system, the image patches are trained corresponding to sclera and periocular components only. After discarding the other patches of the other eye image features such as Iris and eye brow, we get the resultant data set for each subject. Table 2 lists the number of training and test set patches for the ten classes of patch size 100 × 100.

6 Results and Analysis

Training CNN Model on Sclera–Periocular Patches (100 x 100)

The sclera–periocular data set is trained using the proposed CNN model. The training is performed on 4702 patches. The model is trained using NVIDIA-SMI 495.44 CUDA version 11. 2 Persistence-M GPU which is used. The training is iterated for 100 epochs with a batch size of 16, around 261 batches. The execution time for each batch is 8 s at the rate of 68 ms/step.

Figure 6a depicts the progress of the CNN model for the sclera–periocular patches. The training curve is a smooth one, whereas the validation accuracy has few fluctuations. This is due to the data set having images from three different image data sets. The less gap between the training and validation accuracy curve indicates less overfitting as the model has pooling layers and dense layers.

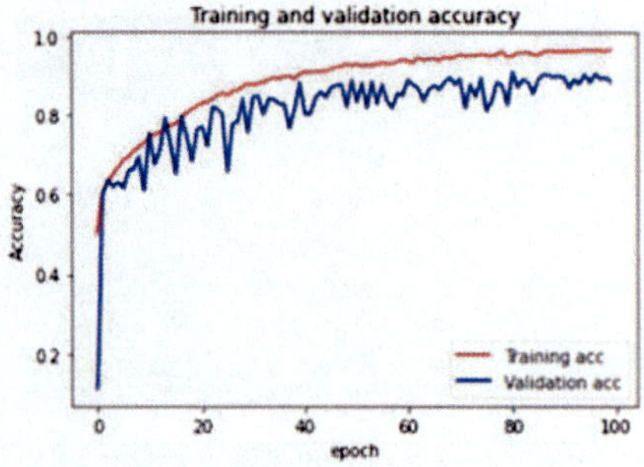

a

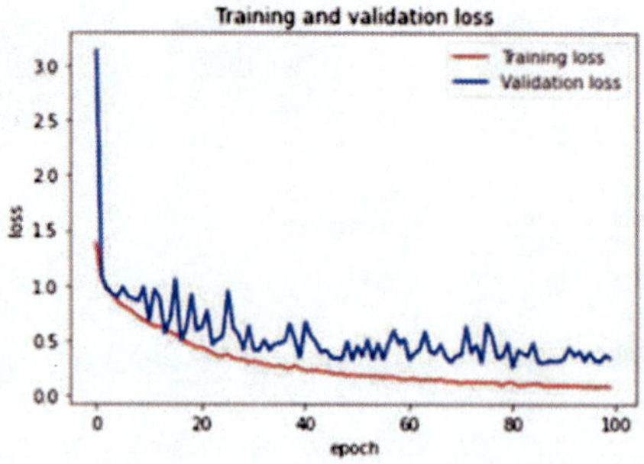

b

Fig. 6 **a** Accuracy. **b** Loss

Figure 6b shows the proposed CNN model training and validation loss for the sclera–periocular patch data set. From the curve, the validation loss curve is decreasing continuously but with few variations. Up on weight updates in each iteration for 100 epochs using categorical loss function, the validation loss tends to decrease. The training loss curve is a smooth one and tends to decrease for each epoch.

Confusion Matrix

This matrix represents how the sclera–periocular patches are classified. From the given 10 classes, Fig. 7 shows how many patches of the class are correctly classified with respect to their predicted Image patches. The score for class 10 is 122, that is, image patches of class 10 are classified as predicted class 10 for 122 patches. Then, 12 patches of image class 10 are classified as class 2. Patches of class label 1 are correctly classified as class 1 for 142 patches and incorrectly classified as class label 9 for 23 patches. Patches of class label 4 classified most correctly with 209 patches and incorrectly classified as class label 8 only for 14 patches and class label 7 for 16 patches.

Classification Report

The classification report is a metric to determine the performance of the proposed CNN model that lists the details of true predictions using precision, recall scores for the test data set. From Table 3, metrics for each class label in terms of true and predicted labels are generated for each class label. The precision value indicates the true predictions for all positive predictions. In Table 3, class label 10 has precision value 91%. Recall score identifies the true positives for the model. Here for class label 10, the recall value is 87%, that is, the model is able to classify correctly 87%. The F1-score represents the good recall and precision values. The support values

Fig. 7 Confusion matrix

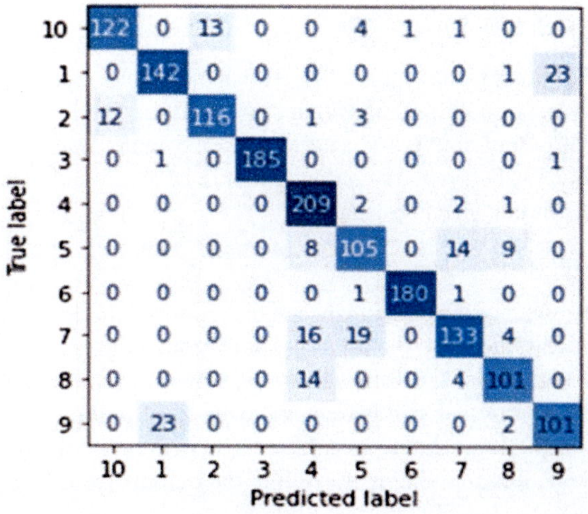

Table 3 CNN classification report

Image labels	Precision	Recall	F1 score	Support
10	0.91	0.87	0.89	141
1	0.86	0.86	0.86	166
2	0.9	0.88	0.89	132
3	1	0.99	0.99	187
4	0.84	0.98	0.9	214
3	0.78	0.77	0.78	136
6	0.99	0.99	0.99	182
7	0.86	0.77	0.81	172
8	0.86	0.85	0.85	119
9	0.81	0.8	0.8	126

represent the number of samples considered for classification. The class 10 label has 141 image patches.

7 Comparison of CNN for Different Patch Sizes

The data set size must be very large in an image classification problem to apply a deep learning technique. In the proposed CNN model, experimentation has been performed on different patch sizes. All the images in the proposed data set are of size 3000 × 1700. In the fusion of sclera–periocular systems, initially, a patch size of 100 is chosen. The actual image size is not considered since it may occupy more RAM space during training, and also patches of sclera and periocular are needed.

Patch Size 100

The input patch size is 100 × 100 in the first layer, after applying CONV2D, followed by max pooling, the resultant layer output size is 50 × 50. In the next three layers, the output size is reduced to 25 × 25 using max pooling layer. In the next three CONV2D layers, the output patch is 12 × 12, and then after applying max pooling, the output size is 6 × 6.

Patch Size 50

To enhance the performance of the proposed model and to accomplish better accuracy in classification, the input patch is reduced to 50 × 50. Applying CONV2D in the first three layers, followed by max pooling, the resultant layer output size is 25 × 25. The output size is reduced to 12 × 12 in the next three layers followed by max pooling layer. In the next three CONV2D layers the of the output patch is 6 × 6 and then after applying max pooling the output size is 3 × 3.

Patch Size 25

To check the optimal patch size of the proposed model, the patch size is again reduced to 25 × 25 as input to the first CONV2D layer. The output size after the first three convolution layers is 12 × 12. In the second layer, the output size is reduced to 6 × 6 in the next three layers followed by max pooling layer. In the last three CONV2D layers, the output patch is 3 × 3, and then after applying max pooling, the output size is 1 × 1.

Patch Size 10

To find out whether the model accuracy increases by still reducing the patch size to 10. Following the layer wise output, if the patch size 10 × 10 as input to the proposed model, then the output sizes are 5 × 5 in next layer, 2 × 2 in the following layer, and 1 × 1 in the last convolution layer. But max pooling layer succeeds CONV2D layer with a filter size of 2 × 2. The input patch size is 1 × 1. The model does not accept and generates an error.

Comparison of Patch Size

The patch wise accuracy values are 97.88%, 98.22%, and 99.3%. Patches of size 100 × 100, 50 × 50, and 25 × 25 are trained using the proposed CNN model as depicted in Fig. 8. Accuracy values of the three patch size models are compared. The accuracy increases as the patch size decreases, simultaneously the number of layers in the proposed CNN can accept up to only patches of size 20.

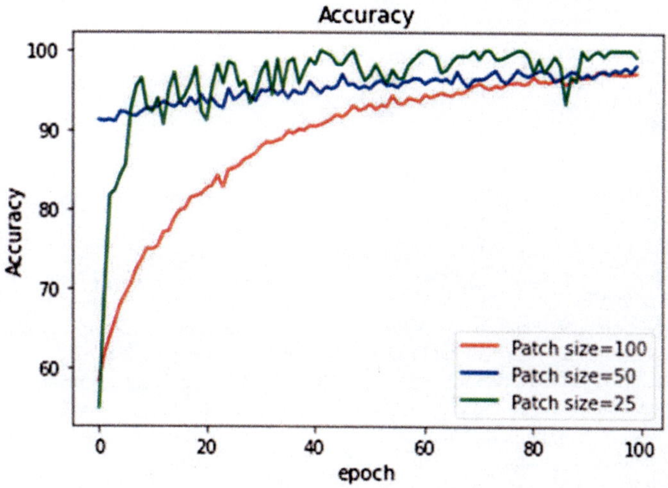

Fig. 8 Patch wise accuracy

Optimal Patch Size

Experiments on three different patch sizes are performed on the proposed CNN model. The accuracy of the model improved by reducing the patch size, but at the same time, the model is able to fit only up to patch size 20×20. If the patch size is still reduced, the last max pooling layer will not be able to accept the input patch. Patch size started with 100×100 or at least half of the eye image size as we require more patches to train the CNN. Patch size output cannot be less than the last layer max pooling size that is less than 2.

8 Conclusion

In this proposed sclera–periocular patch-based biometric system, the CNN model performance is improved with the decrease in patch size. As the patch size is decreased, the number of patches increases in the data set. For a large data sets, CNN would be the best model for image classification. The accuracy has been improved with reduction in patch size. The optimal patch size depends on the layers in the proposed model. A CNN model with layer invariant patch-based classification could be a future scope for research.

Acknowledgements We like to express gratitude to the University of Missouri-Kansas City for providing access to the **VISOB** data set and the Department of Computer Science, the University of Beira Interior, for providing access to the **UBIPr** and SBVPI data set.

References

G.H. Chen, D. Shah, P. Golland, A latent source model for patch-based image segmentation. Med. Image Comput. Comput. Assist. Interv. (2016)

W. Dong, H. Zhou, D. Xu, A new sclera segmentation and vessel extraction method for sclera recognition, in *International Conference on Communication Software and Networks (ICCSN)* (2018), pp. 552–556

S. Gayathri, V.P. Gopi, P. Palanisamy, A lightweight CNN for diabetic retinopathy classification from fundus images. Biomed. Signal Process. Control **62**, 102–115 (2020)

M.H. Hesamian, W. Jia, X. He, P. Kennedy, Deep learning techniques for medical image segmentation: achievements and challenges. J. Digit. Imaging (2019)

M.A. Hossain, M.S.A. Sajib, Classification of image using convolutional neural network (CNN). Glob. J. Comput. Sci. Technol.: D Neural Artif. Intell. **2**, 1–7 (2019)

L. Hou, D. Samaras, T.M. Kurc, Y. Gao, J.E. Davis, J.H. Saltz, Patch-based convolutional neural network for whole slide tissue image classification, in *2016 IEEE Conference on Computer Vision and Pattern Recognition (CVPR)*

Y.-P. Huang, H. Basanta, Bird image retrieval and recognition using a deep learning platform. EEE Access **7**, 66980–66989 (2019)

M.Q. Hui, Y. Lu, Multimodal biometrics based on convolution neural networks by two-layer fusion, in *2019 12th International Congress on Image and Signal Processing, Biomedical Engineering and Informatics(CISP-BMEI)* (2019), pp. 1–6

P.-Y. Kao, S. Shailja, J. Jiang, A. Zhang, A. Khan, J.W. Chen, B.S. Manjunath, Improving patch-based convolutional neural networks for MRI brain tumor segmentation by leveraging location information. Front. Neuro Sci. (2020)

P. Kumari, K.R. Seeja, Periocular biometrics: a survey. J. King Saud Univ. Comput. Inf. Sci. 1–12 (2019)

C. Ledig, W. Shi, W. Bai, D. Rueckert, *Patch-Based Evaluation of Image Segmentation* (2014)

R. Mechrez, J. Goldberger, H. Greenspan, Patch-based segmentation with spatial consistency: application to MS lesions in brain MRI. Int. J. Biomed. Imaging (2016)

S. Minaee, Y. Boykov, F. Porikli, A. Plaza, N. Kehtarnavaz, D. Terzopoulos, Image segmentation using deep learning: a survey. arXiv:2001.05566 [cs.CV] (2020)

H. Proenc, J.C. Neves, Deep-PRWIS: periocular recognition without the iris and sclera using deep learning frameworks. IEEE Trans. Inf. Forens. Secur. **13**(4), 888–896 (2018)

P. Rot, Z. Emersic, V. Struc, P. Pee, Deep multi-class eye segmentation for ocular biometrics, in *IEEE International Work Conference on Bioinspired Intelligence* (2018), pp. 1–8

K. Roy, D. Banik, D. Bhattacharjee, M. Nasipuri, Patch-based system for classification of breast histology images using deep learning. Comput. Med. Imaging Graph. (2018)

T.B. Sekoua, M. Hidanc, J. Olivicr, H. Cardot, From patch to image segmentation using fully convolutional networks—application to retinal images. Comput. Med. Image Graph. (CMIG) (2019)

A. Sharma, X. Liu, X. Yang, D. Shi, A patch-based convolutional neural network for remote sensing image classification. Neural Netw. **95** (2017)

L.C.O. Tiong, Y. Lee, A.B.J. Teoh, Periocular recognition in the wild: implementation of RGB-OCLBCP dual-stream CNN. Appl. Sci. **9**(13), 2709 (2019)

L. Zhang, X. Wang, N. Penwarden, Q. Ji, An image segmentation framework based on patch segmentation fusion, in *International Conference on Pattern Recognition* (IEEE, 2006)

Reliable Transmission of Multimedia Data Over Wireless Sensor Networks

Ch. Janakamma and Nagaratna P. Hegde

Abstract Wireless multimedia sensor networks (WMSNs) are developed as a new class of wireless sensor networks to satisfy the strict quality of service (QoS) necessities of emerging applications. Nowadays, there occurs a large requirement to broadcast the video over the network. Therefore, it is necessary to design a routing protocol along with bandwidth estimation and compression approaches. In this paper, the congestion aware tunicate swarm algorithm (CATSA) is proposed for an effective routing over the WMSN. The passive available bandwidth estimation (PABE) and hybrid compression approach (HCA) are used to improve the QoS. The PABE is used to estimate the bandwidth of the routing path followed by the HCA which is used to compress the data for minimizing the resources. In that, HCA includes a lossless lifting wavelet transform (LWT) and lossy deep learning-based compression (DLC), whereas these approaches are applied according to the estimated bandwidth. The performance of the CATSA-PABE-HCA method is analyzed by means of packet drop, end-to-end delay (EED), and peak signal noise ratio (PSNR). The existing researches HRDSS and HSSFF are used to analyze the CATSA-PABE-HCA. The PSNR of the CATSA-PABE-HCA for 150 Kbps of bit rate is 2400 which is high than the HRDSS and HSSFF.

Keywords Congestion aware tunicate swarm algorithm · Hybrid compression approach · Passive available bandwidth estimation · Quality of service · Wireless multimedia sensor networks

Ch. Janakamma (✉)
Osmania University, Hyderabad, India
e-mail: janaki.chawalam@gmail.com

N. P. Hegde
Vasavi College of Engineering, Hyderabad, India
e-mail: nagaratnaph@staff.vce.ac.in

A. B. Reddy et al. (eds.), *Proceedings of Third International Conference on Advances in Computer Engineering and Communication Systems*, Lecture Notes in Networks and Systems 612, https://doi.org/10.1007/978-981-19-9228-5_9

1 Introduction

In general, the wireless sensor networks (WSNs) have low-cost sensors which can wirelessly forward the physical phenomena for essential calculation (Gorgich and Tabatabaei 2021; Ezhilarasi and Krishnaveni 2019). The modern generation of multimedia sensor nodes has the extensive capacity to capture the multimedia and scalar information from the sensing fields (Genta et al. 2019). Generally, the WMSN is developed of a huge amount of inexpensive, less power, and multi-functional nodes (Hasan et al. 2017; Li et al. 2020). These WMSNs are mostly used in unattended or hostile environments (Yu and Lu 2019). Moreover, the WMSN requires higher bandwidth than the typical WSN (Habib and Moh 2019). Network bandwidth computation is important for various Internet applications and protocols, specifically those participating in the huge file transmission, and those participating in real-time quality of service (QoS) restrictions, e.g., streaming media (Salcedo et al. 2018).

An estimation of bandwidth is an important approach to admission control for QoS in both wired and wireless networks (Chaudhari and Biradar 2015). The multimedia data requires huge bandwidth and energy which affects the network performances (Tian and Liao 2020). The solution to overcome the aforementioned disadvantage is compressing the data for reducing the size (Heng et al. 2017; Banerjee and Das Bit 2019; Banerjee et al. 2019). However, the congestion occurred in WMSN because more resources are required to perform the continuous flows and a huge amount of data transfer. The occurred congestion creates queuing delays and packet drops (Sumathi and Pandiaraja 2020). The existing routing protocols for WSN are not appropriate for the transmission of multimedia data such as audio and video streams (Han et al. 2016).

The main contributions of this research paper are given as follows: The congestion aware routing is accomplished by proposing the CATSA that used to improve data delivery. According to the estimated bandwidth using PABE, the LWT and DNN-based compression are applied over the input multimedia data. Therefore, the transmission of compressed multimedia data over an optimal path is used to improve the QoS.

The remaining paper is organized as follows: Sect. 2 provides the related work about the WMSN. An explanation of the CATSA-PABE-HCA is given in Sect. 3. Section 4 provides the outcomes of the CATSA-PABE-HCA. Further, the conclusion of this research is made in Sect. 5.

2 Related Work

Alqahtani (2021) presented the ant colony optimization (ACO)-based multipath routing protocol for enhancing the QoS in WMSN. The congestion over the network was not considered by this ACO routing. Suseela et al. (2020) developed the priority-based multipath routing algorithm to enhance the network performances. But, the delay was increased according to the increment in the number of nodes. Aswale and Ghorpade (2021) presented link quality and an interference aware cross-layer multipath routing for handling the multimedia data over the WMSN. The signal-to-noise ratio can create a noisy link quality evaluation because of the radio miscalibration.

Ambareesh and Madheswari (2021) developed a routing approach based on hybrid red deer salp swarm (HRDSS). The multiple objectives considered in this HRDSS were ETX, memory, delay, and packet drop. Srinivasa Gowda and Annamalai (Srinivasa Gowda and Annamalai 2021) presented the hybrid salp swarm firefly (HSSFF) to perform routing over the WMSN. The HSSFF generated the route with less EED and ETX. The packet drop of the WMSN was affected because of the congestion over the WMSN.

3 CATSA-PABE-HCA Method

In this CATSA-PABE-HCA method, the QoS of the data transmission is improved while transmitting the multimedia data over the sensor networks. This CATSA-PABE-HCA method has three important phases such as path identification using CATSA, bandwidth estimation using PABE, and compression using HCA. The proposed CATSA-based routing is used to avoid collisions while transmitting the multimedia data. The block diagram for the CATSA-PABE-HCA method is shown in Fig. 1.

3.1 *Routing Using CATSA*

Generally, the TSA (Kaur et al. 2020) mimics the weird behaviors of tunicates done in oceans, especially the jet-drive and swarm intelligence of their searching process. The population of CATSA is initialized as the possible paths between the source

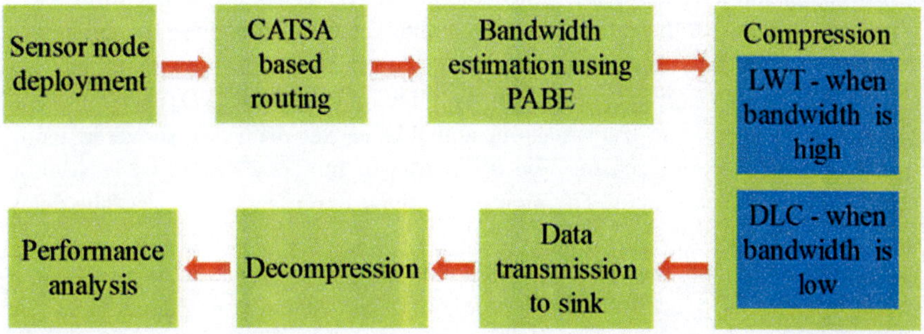

Fig. 1 Block diagram of the CATSA-PABE-HCA method

and the sink. Next, these possible paths are given as input to the CATSA to find the appropriate path.

Iterative process of CATSA. An optimal path WMSN is identified using CATSA based on the queue utilization (QU), packet drop, memory, delay, and ETX for improving the QoS. The tunicate is moved in the direction of the current best solution, once the conflict is avoided among the neighbors. This helps to reach the optimal solution. The optimal location of a best agent is identified using Eq. (1).

$$\overrightarrow{\text{PD}} = \left| X_{\text{best}} - r_{\text{rand}}.\overrightarrow{P_P}(x) \right| \tag{1}$$

where, the $\overrightarrow{\text{PD}}$ indicates the distance among the optimal solution (X_{best}) and location of tunicate $\overrightarrow{P_P}(x)$; the current iteration is represented as x, and the random number generated in the range of [0, 1] is r_{rand}. The tunicate's position is maintained near to the optimal solution by calculating the tunicate's position as shown in Eq. (2).

$$\overrightarrow{P_P}(x) = \begin{cases} X_{\text{best}} + \vec{A}.\overrightarrow{\text{PD}}, & \text{if } r_{\text{rand}} \geq 0.5 \\ X_{\text{best}} - \vec{A}.\overrightarrow{\text{PD}}, & \text{if } r_{\text{rand}} < 0.5 \end{cases} \tag{2}$$

where vector of new agent location is represented as $\vec{A}$.

The updated positions of optimal route based on the X_{best} are resulted from Eq. (2). The location of the current tunicate is updated based on the locations of two tunicates for modeling the swarming behavior that is given in Eq. (3).

$$\overrightarrow{P_P}(x+1) = \frac{\overrightarrow{P_P}(x) + \overrightarrow{P_P}(x+1)}{2 + c_1} \tag{3}$$

In this phase, an optimal solution X_{best} is discovered according to the cost function which is derived in the following section.

Cost function derivation. The cost function used in this CATSA includes queue utilization, packet drop, memory, delay, and ETX. The derived cost function is used to improve the QoS of the network. QU is defined as the ratio among the amount of occupied packets in the queue of ith node to the total amount of available packets in ith a node which is expressed in Eq. (4). The packet drop (PD) is required to be less for effective data transmission, and this packet drop is discovered using Eq. (5). For an effective routing, the utilization of memory should be less while transmitting the video frames. The memory usage is expressed in Eq. (6). The delay achieved through the derived path should be less for effective data transmission. In general, the delay includes four different delays such as propagation delay (De_{prop}), processing delay (De_{proc}), transmission delay (De_{tran}), and queuing delay (De_{queu}). Equation (7) expresses the delay obtained using the route. ETX shown in Eq. (8) is the relation between the amount of transmissions and the retransmissions required for the effective data transmission.

$$\mathrm{QU} = \frac{\mathrm{QU}_{\mathrm{occupied}}(i)}{\mathrm{QU}_{\mathrm{total}}} \tag{4}$$

$$\mathrm{PD} = 1 - \prod_{N \epsilon P} \mathrm{Min}\left[\frac{\varepsilon(N)}{\sum_{S \epsilon NN} \varepsilon(S)}\right] \tag{5}$$

where the number of nodes is N; the transmission path is denoted as P; the rate of packet drop for N nodes is denoted as $\varepsilon(N)$, and the group of neighbor nodes is denoted as NN.

$$MU = \prod_{N \epsilon P} \mathrm{Min}\left[\frac{\mu(N)}{\sum_{S \epsilon NN} \mu(S)}\right] \tag{6}$$

where the memory value for the N nodes is denoted as $\mu(N)$

$$\mathrm{Delay} = \mathrm{Min} \sum_{EE=1}^{\mathrm{HP}} \mathrm{De}_{\mathrm{prop}} + \mathrm{De}_{\mathrm{proc}} + \mathrm{De}_{\mathrm{tran}} + \mathrm{De}_{\mathrm{queu}} \tag{7}$$

$$ETX = \frac{1}{\mathrm{FD} \times \mathrm{RD}} \tag{8}$$

where the forward data delivery is denoted as FD, and reverse data delivery or acknowledgment from the destination is denoted as RD.

The aforementioned multiple cost functions are converted into single cost metric values as shown in Eq. (9) which used to discover the optimal path.

$$\mathrm{Cost} = \delta_1 \times QU + \delta_2 \times \mathrm{PD} + \delta_3 \times MU + \delta_4 \times \mathrm{Delay} + \delta_5 \times ETX \tag{9}$$

Next, the bandwidth of the detected path is estimated by using PABE.

3.2 Passive Available Bandwidth Estimation

In this section, the bandwidth for the possible links is estimated using the PABE. This is used to estimate the available bandwidth of the discovered link as shown in Eq. (10), and accordingly, compression techniques are enabled to achieve reliable communication.

$$AB = \frac{T_{\mathrm{observe}} - T_{tx} - T_{rx} - T_{\mathrm{sensing}}}{T_{\mathrm{observe}}} \times \frac{L}{T} \tag{10}$$

where the first part of equation denotes the IdleRatio, while the second part denotes the $\mathrm{Capacity}_{\mathrm{effective}}$; the time period for idle, transmission, receiving, and sending states is denoted as T_{idle}, T_{tx}, T_{rx}, and T_{sensing}, respectively, and the time period for evaluating the time ratio for the idle channel is T_{observe}; amount of bits transferred through the link is denoted as L, and time taken to transmit the bits is denoted as T. The calculation of available bandwidth is enhanced by concentrating on the estimation accuracy of the channel idle time ratio and distance between the sender and receiver.

3.3 Hybrid Compression for an Effective Data Transmission

In this proposed method, a hybrid compression is developed where it contains lossless and lossy compression approaches. These compression approaches are developed for an effective video data transmission over WMSN. In that, the images frames are taken from the video, and it is compressed and transmitted through the detected path. If the bandwidth of the detected link from CATSA is high, a lossless compression, i.e., LWT (Zhang et al. 2020) applied over the input video frames. Otherwise, the lossy compression, i.e., DLC (Dua et al. 2021) is applied, when the bandwidth of the detected link CATSA is low.

4 Results and Discussion

The design and implementation of this CATSA-PABE-HCA are done in MATLAB R2018a software. The main objective of this research is to enhance the QoS during multimedia data broadcasting over the WMSN. The simulation parameters of the CATSA-PABE-HCA method are represented in Table 1.

The CATSA-PABE-HCA is analyzed by means of QoS parameters such as packet loss, EED, and PSNR. Here, the CATSA-PABE-HCA method is evaluated with existing researches of HRDSS (Ambareesh and Madheswari 2021) and HSSFF (Srinivasa Gowda and Annamalai 2021). Figures 2, 3 and 4 show the performance analysis of the packet loss, EED, and PSNR for CATSA-PABE-HCA with HRDSS (Ambareesh and Madheswari 2021) and HSSFF (Srinivasa Gowda and Annamalai 2021). From the figures, it is known that the CATSA-PABE-HCA achieves less packet loss, less EED, and high PSNR than the HRDSS (Ambareesh and Madheswari 2021)

Table 1 Simulation parameters

Parameter	Value
Area	1000 m × 1000 m
Amount of nodes	100
Amount of sink and source	1–1
Rate of source data	128 kbps
Size of the packet	2325 bytes

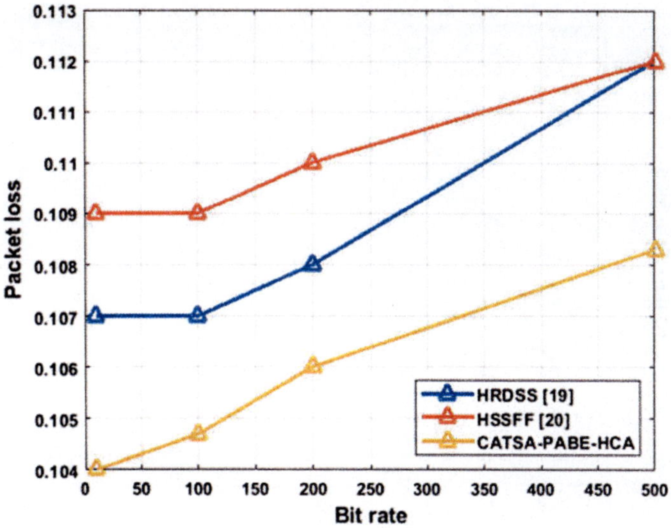

Fig. 2 Comparison of packet loss versus bit rate

and HSSFF (Srinivasa Gowda and Annamalai 2021). The designed congestion aware routing of CATSA-PABE-HCA is used to minimize the packet loss that occurred because of collision and also compression according to the bandwidth used to minimize the losses. Next, the consideration of queue utilization and delay in CATSA are used to minimize the delay over the network. Further, HCA-based compression according to the estimated bandwidth of the selected path is used to improve the PSNR of CATSA-PABE-HCA.

5 Conclusion

In this paper, the collisions during the video frame transmission are avoided by using the CATSA. The PABE is used for identifying the available bandwidth for the route selected from the CATSA. Here, the distance is considered in the PABE for improving the bandwidth calculation. According to the estimated bandwidth, the HCA uses LWT and CNN. If the bandwidth is high, an LWT-based lossless compression is applied over the input video frames. Otherwise, the CNN is applied to compress the input video frames. Therefore, this reliable transmission over the WMSN is used to improve the QoS. The PSNR of the CATSA-PABE-HCA for 150 Kbps of bit rate is 2400 which is high than the HRDSS and HSSFF. In future, the novel optimization algorithm-based routing can be used for improving the performances of the WMSNs.

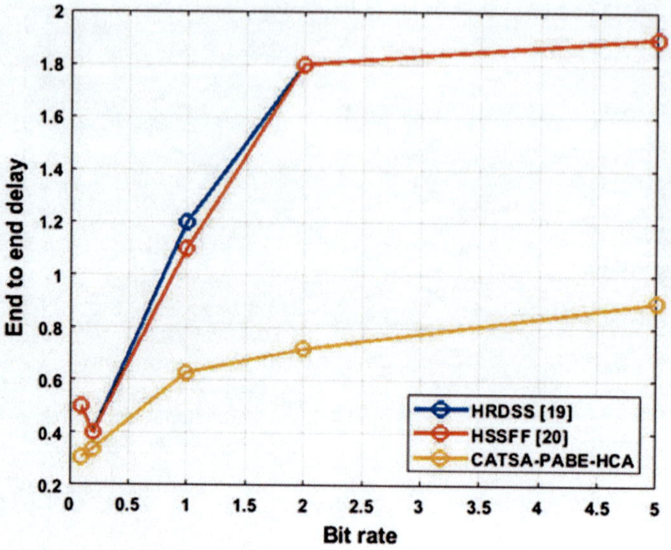

Fig. 3 Comparison of EED versus bit rate

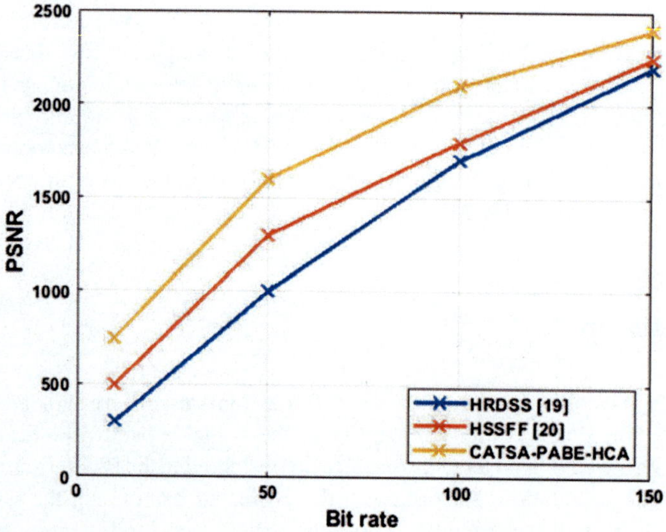

Fig. 4 Comparison of PSNR versus bit rate

References

A.S. Alqahtani, Improve the QoS using multi-path routing protocol for wireless multimedia sensor network. Environ. Technol. Innov. **24**, 101850 (2021)

S. Ambareesh, A.N. Madheswari: HRDSS-WMSN: a multi-objective function for optimal routing protocol in wireless multimedia sensor networks using hybrid red deer salp swarm algorithm. Wireless Pers. Commun. 119(1):117–146

S. Aswale, V.R. Ghorpade, Geographic multipath routing based on triangle link quality metric with minimum inter-path interference for wireless multimedia sensor networks. J. King Saud Univ. Comput. Inf. Sci. **33**(1), 33–44 (2021)

R. Banerjee, S. Das Bit, An energy efficient image compression scheme for wireless multimedia sensor network using curve fitting technique. Wireless Netw. **25**(1), 167–183 (2019)

R. Banerjee, S. Chatterjee, S. Das Bit (2019) Performance of a partial discrete wavelet transform based path merging compression technique for wireless multimedia sensor networks. Wireless Pers. Commun. **104**(1), 57–71

S.S. Chaudhari, R.C. Biradar, Survey of bandwidth estimation techniques in communication networks. Wireless Pers. Commun. **83**(2), 1425–1476 (2015)

Y. Dua, R.S. Singh, K. Parwani, S. Lunagariya, V. Kumar, Convolution neural network based lossy compression of hyperspectral images. Sig. Process. Image Commun. **95**, 116255 (2021)

M. Ezhilarasi, V. Krishnaveni, An evolutionary multipath energy-efficient routing protocol (EMEER) for network lifetime enhancement in wireless sensor networks. Soft. Comput. **23**(18), 8367–8377 (2019)

A. Genta, D.K. Lobiyal, J.H. Abawajy, Energy efficient multipath routing algorithm for wireless multimedia sensor network. Sensors **19**(17), 3642 (2019)

S. Gorgich, S. Tabatabaei, Proposing an energy-aware routing protocol by using a fish swarm optimization algorithm in WSN (wireless sensor networks). Wireless Pers. Commun. **119**(3), 1935–1955 (2021)

M.A. Habib, S. Moh, Robust evolutionary-game-based routing for wireless multimedia sensor networks. Sensors **19**(16), 3544 (2019)

G. Han, J. Jiang, M. Guizani, J.J.C. Rodrigues, Green routing protocols for wireless multimedia sensor networks. IEEE Wireless Commun. **23**(6), 140–146 (2016)

M.Z. Hasan, F. Al-Turjman, H. Al-Rizzo, Optimized multi-constrained quality-of-service multipath routing approach for multimedia sensor networks. IEEE Sens. J. **17**(7), 2298–2309 (2017)

S. Heng, C. So-In, T.G. Nguyen, Distributed image compression architecture over wireless multimedia sensor networks. Wireless Commun. Mob. Comput. (2017)

S. Kaur, L.K. Awasthi, A.L. Sangal, G. Dhiman, Tunicate swarm algorithm: a new bio-inspired based metaheuristic paradigm for global optimization. Eng. Appl. Artif. Intell. **90**, 103541 (2020)

H. Li, Q. Qi, J. Liu, P. Zhao, Y. Yang, Mobile wireless multimedia sensor networks image compression task collaboration based on dynamic alliance. IEEE Access **8**, 86024–86037 (2020)

D. Salcedo, C. Guerrero, R. Martinez, Available bandwidth estimation tools metrics, approaches and performance (2018)

A. Srinivasa Gowda, N.M. Annamalai, Hybrid salp swarm–firefly algorithm-based routing protocol in wireless multimedia sensor networks. Int. J. Commun. Syst. **34**(3), e4633 (2021)

K. Sumathi, P. Pandiaraja, Dynamic alternate buffer switching and congestion control in wireless multimedia sensor networks. Peer Peer Netw. Appl. **13**(6), 2001–2010 (2020)

S. Suseela, R. Eswari, S. Nickolas, M. Saravanan, QoS optimization through PBMR algorithm in multipath wireless multimedia sensor networks. Peer Peer Netw. Appl. **13**(4), 1248–1259 (2020)

Y. Tian, R. Liao, Multinode collaborative image compression algorithm for wireless multimedia sensor networks based on LBT. IEEE Sens. J. **20**(20), 12065–12073 (2020)

Z. Yu, B. Lu, A multipath routing protocol using congestion control in wireless multimedia sensor networks. Peer Peer Netw. Appl. **12**(6), 1585–1593 (2019)

H. Zhang, X.Q. Wang, Y.J. Sun, X.Y. Wang, A novel method for lossless image compression and encryption based on LWT, SPIHT and cellular automata. Sig. Process. Image Commun. **84**, 115829 (2020)

Leader-Follower Based Low-Degree Formation Control of Fixed-Wing Unmanned Aerial Vehicles in 3D

Roneel Chand, Krishna Raghuwaiya, Jito Vanualailai, and Jai Raj

Abstract The motion planner of several fixed-wing unmanned aerial vehicles (UAVs) using a leader-follower strategy is presented in this paper, using the architecture of the Lyapunov-based control scheme (LbCS). Here, we use the leader-follower formation type, where the flock must move in accordance with its leader through a congested workplace to fulfill its predetermined goals. When there are obstacles, the formation gets slightly distorted and rejoins at a safer location while preserving its low degree of formation. In our scenario, the obstacles will be spherical stationary obstacles, and each moving aircraft in the swarm will serve as an obstacle for every other member (moving obstacles). The Lyapunov-based control scheme will be used to create a series of nonlinear acceleration-based controllers to successfully achieve this task. The controllers will facilitate and ensure the flock's capacity to coordinate its travel in a well-planned manner, converge to its intended destination, and do so while keeping a low degree of formation and avoiding obstacles. The effectiveness of the suggested nonlinear acceleration control principles will be proved by computer simulations of a virtual scenario. This will reveal the simplicity and effectiveness of the control technique used. The paper finally ends with a conclusion and future recommendations.

Keywords Lyapunov · UAVs · Controllers · Low-degree formation

R. Chand (✉)
Fiji National University, Suva, Fiji
e-mail: roneel.chand@fnu.ac.fj

K. Raghuwaiya · J. Vanualailai · J. Raj
The University of the South Pacific, Suva, Fiji
e-mail: krishna.raghuwaiya@usp.ac.fj

J. Vanualailai
e-mail: jito.vanualailai@usp.ac.fj

J. Raj
e-mail: jai.raj@usp.ac.fj

A. B. Reddy et al. (eds.), *Proceedings of Third International Conference on Advances in Computer Engineering and Communication Systems*, Lecture Notes in Networks and Systems 612, https://doi.org/10.1007/978-981-19-9228-5_10

1 Introduction

Transportation-related industries all across the world rely heavily on linked and autonomous vehicles. These are typically utilized for tasks that are too dangerous or impossible for men to perform. Due to the fact that numerous robots can complete cooperative activities more successfully, such as security and cooperative tasks, the employment of multiple robots has been increasingly popular in recent years (Ben Hamida et al. 2015; Festag 2014). In addition, using numerous robots is less expensive than using a single specialized robot (Festag 2014).

For the control of cooperative intelligent transport systems (C-ITSs), a number of applications have already been proposed, including collaborative mapping, mapping and planning, flocking and schooling, search and rescue operations, target finding, and so on (Arrichiello 2006; Balch and Hybinette 2000; De La Cruz and Carelli 2008; Raj et al. 2021). Also in today's real-world, application of work hugely depends on C-ITSs to work autonomously in a stable configuration while avoiding obstacles and collisions.

As given in De La Cruz and Carelli (2008), Arrichiello (2006), Balch and Hybinette (2000), and Raj et al. (2021), numerous control mechanisms for robotic control have recently been proposed. In addition, Sharma developed an artificial potential field (APF) in (Sharma, 2008), which uses the architecture of the Lyapunov-based control scheme (LbCS) approach. This approach creates the attractive and repulsive potential functions which controls the robotic formations. We adopt this strategy to control the C-ITS for the case of this paper.

Raghuwaiya, in Raghuwaiya et al. (2018), further extended this control technique using LbCS and developed a new leader-follower scheme that maintains locally rigid formation through out the journey. The novelty of this method is that it is able to create continuous nonlinear controller control laws which has the ability to convert locally rigid forms of nonholonomic system with dynamic constraints. Compared to merely cooperative agents, the new leader-follower scheme's virtual connectedness makes a variety of jobs easier while still avoiding obstacles (Raj et al. 2020a, b; Sharma 2008; Sharma et al. 2017, 2011; Vanualailai et al. 2013).

For the case of this paper, the flock will be controlled in formation using the leader-follower approach described in Raghuwaiya et al. (2018). A motion planner for the flock traveling in the workspace (Crombie 1997; Mamino and Viglietta 2016; Shojaei 2017) will be created using LbCS. To successfully converge on its pre-defined target, the flock must carefully plan its travel while avoiding potential obstacles and maintaining a locally rigid formation (Mamino and Viglietta 2016; Shojaei 2017; Vanualailai et al. 2013). In our scenario, the obstacles will be spherical stationary obstacles, and each moving aircraft in the swarm will serve as an obstacle for every other member. Then, we will apply the direct method of Lyapunov to develop continuous acceleration-based controllers, which will provide stability to our system, using artificial potential fields. The research given in this paper is an extension of that found in Raghuwaiya et al. (2018), where we move the research from two dimensions to three.

The rest of the work done in this paper are as follows: We define the robot model together with leader-follower-based formation scheme in Sect. 2; Sect. 3 looks at the artificial potential field functions taking kinodynamic constraints into consideration; we define the acceleration-based control laws in Sect. 4; Sect. 5 explains the stability analysis of the given robotic system; computer simulations of a virtual scenario are given in Sect. 6 which also shows the effectivenesss of the designed controllers; and the paper is finally concluded in Sect. 7 outlining possible future work to be done in this area.

2 Vehicle Model

In this part, we construct a simple kinematic model for the formation control of several fixed-wing aircraft using the leader-follower method. We shall take into account fixed-wing aircraft in the Euclidean plane, n, $n \in \mathbb{N}$. $\mathcal{A}_1$ will serve as the leader, while $\mathcal{A}i$ for $i = 2, \dots, n$, will serve as the robot followers. As given in Fig. 1, the reference point in 2-dimensional space is given by (x_i, y_i), which represents the Cartesian coordinates of our fixed-wing aircraft. L_i and l_i give the overall length and complete wingspan of the aircraft, respectively, as given in Fig. 1. Furthermore, θ_i represents the orientation of the aircraft which is calculated with respective to z_1-axis. Next, the nomenclature of Sharma (2008) is adopted, and the spherical zones shielding the aircraft is built. This is done to ensure aircraft safely pass around an obstacle. Furthermore provided, we have *clearance parameters* $\epsilon 1 > 0$ and $\epsilon 2 > 0$, and the aircraft is enclosed in a protective sphere which has center of (x_i, y_i, z_i) and radius of:

$$r_i = \frac{\sqrt{(L_i + 2\epsilon_1)^2 + (l_i + 2\epsilon_2)^2}}{2}. \tag{1}$$

Next, we assume no slippage condition (i.e., $\dot{x}_i \sin\theta_i - \dot{y}_i \cos\theta_i = 0$) together with pure rolling (i.e., $\dot{x}_i \cos\theta_i + \dot{y}_i \sin\theta_i = v_i$) of the aircraft in this scenario. This will generate *nonholonomic constraints* on the given system. Next, we adopt the kinematics of the system from (Sharma 2008), which takes care of the nonholonomic constraints of the system. This is used for our case for $i \in \{1, \dots, n\}$, presented below as the kinematics model of our system.

$$\begin{aligned}
\dot{x}_i &= v_i \cos\theta_i - \frac{L_i}{2}\omega_i \sin\theta_i, \\
\dot{y}_i &= v_i \sin\theta_i + \frac{L_i}{2}\omega_i \cos\theta_i, \\
\dot{z}_i &= 0, \quad \dot{\theta}_i = \frac{v_i}{L_i}\tan\phi_i = \omega_i, \\
\dot{v}_i &= \sigma_{i1}, \quad \dot{\omega}_i = \sigma_{i2}, \quad \dot{\mu}_i = \sigma_{i3}.
\end{aligned} \tag{2}$$

Additionally, we assume $\phi_i = \theta_i$ without losing generality. Now, with reference to system (2), v_i, ω_i, and μ_i are the instantaneous translational and rotational velocities, whereas σ_{i1}, σ_{i2}, and σ_{i3} represent the instantaneous translational and rotational accelerations of the ith robot. Next, $\dot{z}_i$ is assumed to be 0, which means that the aircraft cannot change its altitude during the course of its flight, or it will travel at a fixed altitude. We also assume that the swarm is already at a height, before their time of flight. The kinematic model of the system (2) now is a description of the instantaneous velocities and accelerations of $\mathcal{A}_i$. Next, we use vector $\mathbf{x}_i = (x_i, y_i, z_i, \theta_i, v_i, \omega_i, \mu_i) \in \mathbb{R}^7$ to redefine the position (x_i, y_i), orientation, θ_i, together with the velocities (v_i, ω_i, μ_i) of $\mathcal{A}_i$ at time $t \geq 0$.

2.1 *Leader-Follower-Based Formation Scheme*

As discussed in introduction, Raghuwaiya in Raghuwaiya et al. (2018) designed a new model for leader-follower-based formation of multiple vehicles, which takes locally rigid formation into consideration. This model will be used for our case. He defined two reference frames in this model. The first reference frame, which is body frame, will be fixed to the leader's rotating body, $\mathcal{A}1$, and secondly, a space frame, similar to one in Kang et al. (2004), is defined, which we let to be an inertial frame. Moreover, cartesian coordinate system $(X - Y)$ is introduced, which is fixed on the leader's body, initially developed in the (Raghuwaiya et al. 2016) scheme. This idea relies on an instantaneous co-rotating frame of reference to guarantee that $(X - Y)$ rotates when the leader, $\mathcal{A}_1$, does.

Fig. 1 Kinematic model of the fixed-wing UAV

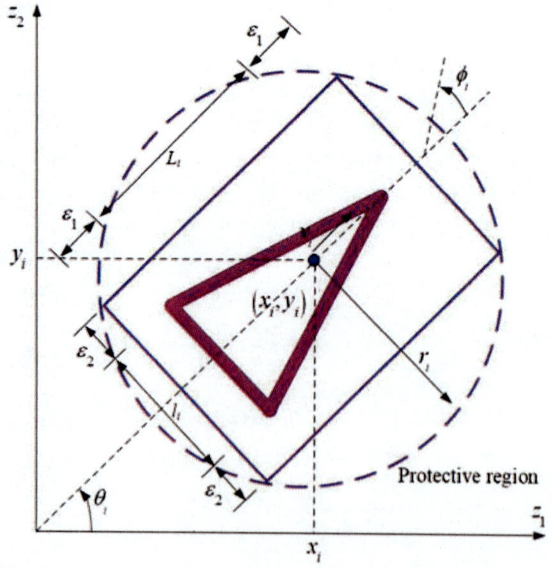

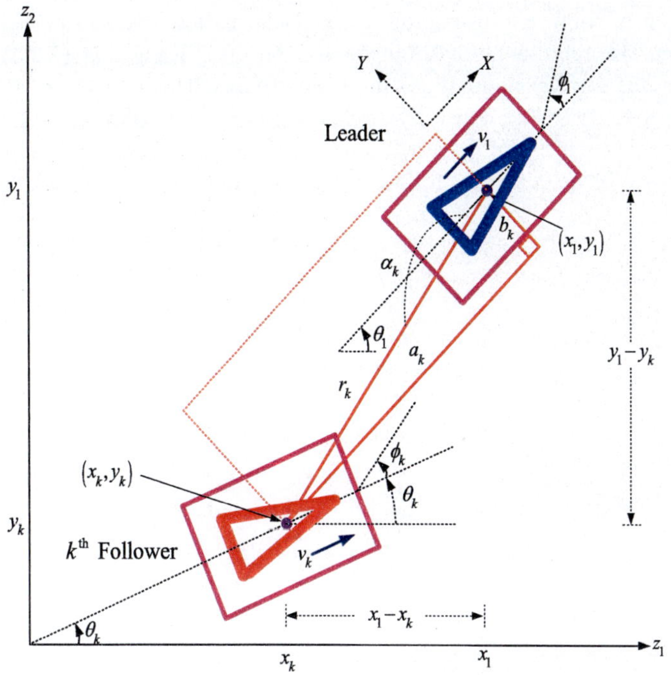

Fig. 2 Proposed scheme utilizing a rotation of axes with the body frame fixed at the leader robot

Additionally, we fix the leader robot's origin at (x_1, y_1) and designate it as the co-rotating frame of reference. Next, a rotational axis is set up perpendicular to the plane of the motion leader. This guarantees that the rate of rotation of the leader about (x_1, y_1) and the rate of rotation of the selected rotating frame of reference at any given point t are equal. Next, as given in Fig. 2, r_{1k} gives the straight line distance between the leader and the kth follower with respect to the reference point of both leader and followers. The measured angle between X axis and the above described straight line is further represented by α_{1k}. Next, taking into consideration the above definitions and leader's position, the followers robots kth position will always be unique if we fix (r_{1k}, α_{1k}), as given in Fig. 2. Next, we adopt and define the shape of the formation of mobile robots as in Raghuwaiya et al. (2016) as $\zeta = [\zeta_{12}, \zeta_{13}, \ldots, \zeta_{1n}]^T$, where $\zeta_{1k} = [r_{1k}, \alpha_{1k}]^T$ for $k \in \{2, \ldots, n\}$.

We then introduce the definition as it is stated in Raghuwaiya et al. (2016).

Definition 1 Let $2r_k < r_{1k}^d < 2r_k + \varsigma$ where for $\varsigma > 0$, and $k \in \{2, \ldots, n\}$, the group of mobile robots make a $\zeta^d = [\zeta_{12}^d, \zeta_{13}^d, \ldots, \zeta_{1n}^d]^T$ formation, if $\exists T > 0, \forall t > T$:

$$r_{1k} = r_{1k}^d \quad \text{and} \quad \alpha_{1k} = \alpha_{1k}^d.$$

The position of followers in relation to the leader is now represented using polar coordinates. This representation's drawback is that, as in Kang et al. (2004), it occasionally causes specific singularities in the controllers. This needs to be prevented, thus in order to do so, we consider the position of the kth follower by taking into account its relative distance from the leader, $\mathcal{A}_1$, along the specified X and Y directions. As a result, we define the following as stated in Raghuwaiya et al. (2016):

$$\begin{aligned} A_k &= -(x_1 - x_k)\cos\theta_1 - (y_1 - y_k)\sin\theta_1, \\ B_k &= (x_1 - x_k)\sin\theta_1 - (y_1 - y_k)\cos\theta_1. \end{aligned} \tag{3}$$

for $k \in \{2, \ldots, n\}$. A_k and B_k defines kth followers relative position with reference to $(X - Y)$ coordinate system. Furthermore, to make sure that the follower's position is unique, we again fix A_k and B_k, and it should be known. As a result, one should know the design parameters a_k to get a desired formation. In this way, our control objective would be to achieve $A_k \longrightarrow a_k$ and $B_k \longrightarrow b_k$, i.e., $r_{1k} \longrightarrow r_{1k}^d$ and $\alpha_{1k} \longrightarrow, \alpha_{1k}^d$, where $r_{1k}^d = \sqrt{a_k^2 + b_k^2}$ and $\alpha_{1k}^d = \tan\left(\frac{a_k}{b_k}\right)$, which will result in maintaining local rigidity of any formations (Raghuwaiya et al. 2016).

3 Artificial Potential Field Function

Here, under kinodynamic constraints in a predetermined workspace, we derive collision-free trajectories for the fixed-wing aircraft system. These functions, which will be created in the following section, will be added together to create the system's artificial potential field.

3.1 *Attractive Potential Field Functions*

Attraction to Target We provide leader a target and allow leader to direct the overall movement of the vehicle system. The followers move in accordance with the leader as it goes toward the target destination, keeping the ideal formation. In order to properly converge at the target's center, leader $\mathcal{A}_1$ must first move from an initial location to its pre-defined target while avoiding any obstacles that may be in its way. We begin with the definition given in Raghuwaiya et al. (2016):

Definition 2 The leader robot, $\mathcal{A}_1$, is assigned a stationary target which is assumed to be a sphere with center (p_{11}, p_{12}, p_{13}) with radius rt.

According to the definition given above, the leader, $\mathcal{A}_1$, must travel in the direction of its predetermined destination, and the follower vehicles must rigorously follow the leader while retaining a specific position, leading to a locally rigid formation.

According to Sharma (2008), the attractive function for the leader $\mathcal{A}_1$ to its target is constructed as follows and measures its convergence to the target with the inclusion of velocity components.

$$V_1(\mathbf{x}) = \frac{1}{2}\left[(x_1 - p_{11})^2 + (y_1 - p_{12})^2 + (z_1 - p_{13})^2 + v_1^2 + \omega_1^2 + \mu_1^2\right]. \quad (4)$$

Next, we introduce the following function to guarantee that the follower vehicles, $\mathcal{A}_i$ for $i = 2, \ldots, n$, keep their desired position in relation to the leader, $\mathcal{A}_1$.

$$V_i(\mathbf{x}) = \frac{1}{2}\left[(A_i - a_i)^2 + (B_i - b_i)^2 + (z_i - p_{13})^2 + (\theta_i - \theta_1)^2 + v_i^2 + \omega_i^2 + \mu_i^2\right]. \quad (5)$$

Both a_i and b_i values will be distinct based on how the intended configuration should be formed.

Auxiliary Function The following auxiliary function, which ensures the leader successfully converges to its target, takes the form of

$$G_1(\mathbf{x}) = \frac{1}{2}\left[(x_1 - p_{11})^2 + (y_1 - p_{12})^2 + (z_1 - p_{13})^2 + (\theta_1 - p_{14})^2\right]. \quad (6)$$

where p_{14} is the prescribed final orientation of the leader robot, $\mathcal{A}_1$ and

$$G_i(\mathbf{x}) = \frac{1}{2}\left[(A_i - a_i)^2 + (B_i - b_i)^2 + (z_i - p_{13})^2\right]. \quad (7)$$

represents the auxiliary function for follower vehicles for $i = 2, \ldots, n$. The given auxiliary functions ensure that the controllers get zero at the target of the leader. The auxiliary function and the given potential field functions are later multiplied together.

3.2 *Repulsive Potential Field Functions*

Generally, we have presence of unknown agents, such as obstacles in our environment. This should be successfully avoided by the aircraft in order to converge to their goals. For this to take place, we design respective obstacle avoidance function, which calculates the Euclidean distances between the robots and obstacles present in our workspace. According to Sharma (2008), these repulsive potential fields function is an inverse function that transforms an avoidance function into the denominator and a control parameter into the numerator.

3.3 Fixed Obstacles

We fix solid obstacles inside the workspace's confines using $q \in \mathbb{N}$. We let the lth obstacle be a sphere wit center of (o_{l1}, o_{l2}, o_{l3}) and radius ro_l. For the aircraft to avoid these obstacles, we formulate the avoidance function as given below:

$$FO_{il}(\mathbf{x}) = \frac{1}{2}\left[(x_i - o_{l1})^2 + (y_i - o_{l2})^2 + (z_i - o_{l3})^2 - (ro_l + r_i)^2\right]. \quad (8)$$

where $i = 1, \ldots, n$, and $l = 1, \ldots, q$.

3.4 Moving Obstacles

Any MRS that has multiple aircraft in it creates moving obstacles for all other aircraft in the system because of their presence. All aircraft must move away from other aircraft in the workspace, which must be viewed as moving obstacles, in order to progress toward their destination. The following avoidance function is designed for the aircraft, $\mathcal{A}_i$ to avoid the aircraft $\mathcal{A}_j$.

$$MO_{ij}(\mathbf{x}) = \frac{1}{2}\left[(x_i - x_j)^2 + (y_i - y_j)^2 + (z_i - z_j)^2 - (2r_i)^2\right]. \quad (9)$$

for $i, j = 1, \ldots, n$, $j \neq i$.

3.5 Workspace Limitations

We design a specific 3D framework using dimensions η_1 by η_2 by η_3 to be the workspace of our aircraft. In our LbCS, these boundaries are considered as *fixed obstacles*. For the aircraft to clear these boundaries, we formulate the following potential function:

$$\begin{aligned} W_{i1}(\mathbf{x}) &= (x_i - r_i), \\ W_{i2}(\mathbf{x}) &= (\eta_2 - (y_i + r_i)), \\ W_{i3}(\mathbf{x}) &= (\eta_1 - (x_i + r_i)), \\ W_{i4}(\mathbf{x}) &= (y_i - r_i), \\ W_{i5}(\mathbf{x}) &= (z_i - r_i), \\ W_{i6}(\mathbf{x}) &= (\eta_3 - (z_i + r_i)). \end{aligned} \quad (10)$$

for $i = 1, \ldots, n$.

These avoidance functions will then be added to the control rules so that the movement of the aircraft remains within the given boundaries of the workspace.

Combining these functions with the proper tuning parameters, later, it will result in repulsive potential field functions.

3.6 Dynamic Constraints

Practically, limits are imposed on the yaw angle of the ith autonomous aircraft due to mechanical singularities and aircrafts translational speed which is limited due to safety concerns (Sharma 2008). Due to this, we take the following into consideration;

(i) $|v_i| < v_{\max}$, letting $v_{\max}$ to be the ith aircraft *maximal achievable speed*;
(ii) $|\omega_i| < \frac{v_{\max}}{|\rho_{\min}|}$, where $\rho_{\min} := \frac{L_i}{\tan(\phi_{\max})}$. We have this condition due to the aircrafts yaw angle boundness. That is, $|\phi_i| \leq \phi_{\max}$, where $\phi_{\max}$ is the *maximal yaw angle of the aircraft.*

Next, we let the values of $v_{\max}$, and $\phi_{\max}$, to be same for simplicity reasons. Furthermore, we design corresponding artificial obstacle for each dynamics constraint using the LbCS. Hence, we consider the following functions into consideration:

$$\begin{aligned}
U_{i1}(\mathbf{x}) &= \frac{1}{2}\left((v_{\max})^2 - (v_i)^2\right), \\
U_{i2}(\mathbf{x}) &= \frac{1}{2}\left(\left(\frac{v_{\max}}{|\rho_{\min}|}\right)^2 - (\omega_i)^2\right), \\
U_{i3}(\mathbf{x}) &= \frac{1}{2}\left((\mu_{\max})^2 - (\mu_i)^2\right).
\end{aligned} \tag{11}$$

for $i = 1, \ldots, n$.

4 Design of the Acceleration-Based Controllers

4.1 Lyapunov Function

The total potentials, also known as a Lyapunov function, must be constructed next. To do this, we must first define the *control parameters* for $i = 1, \ldots, n$, that will be used in the repulsive potential functions. This will be as follows:

(i) $\alpha_{il} > 0$, $l = 1, \ldots, q$, for the collision avoidance of q spherical-shaped obstacles.
(ii) $\beta_{is} > 0$, $s = 1, \ldots, 3$, for the avoidance of the artificial obstacles from dynamic constraints.
(iii) $\eta_{ij} > 0$, $j = 1, \ldots, n$, $i \neq j$, for the collision avoidance between any two vehicles.

(iv) $\kappa_{ip} > 0$, $p = 1, \ldots, 6$, for the avoidance of the lane boundaries.

Next, using the above *control parameters*, we design the Lyapunov function with two components, namely the attractive and repulsive field functions:

$$L(\mathbf{x}) = \sum_{i=1}^{n} \left[V_i(\mathbf{x}) + G_i(\mathbf{x}) \left(\sum_{l=1}^{q} \frac{\alpha_{il}}{FO_{il}(\mathbf{x})} + \sum_{s=1}^{3} \frac{\beta_{is}}{U_{is}(\mathbf{x})} + \sum_{p=1}^{6} \frac{\kappa_{ip}}{W_{ip}(\mathbf{x})} \right) \right] + \sum_{i=1}^{n} G_i(\mathbf{x}) \left(\sum_{\substack{j=1\\ j\neq i}}^{n} \frac{\eta_{ij}}{MO_{ij}(\mathbf{x})} \right). \tag{12}$$

4.2 Nonlinear Acceleration Controllers

Next, we design the feedback controllers noting that the functions f_{ik} to g_{ij} for $i = 1, j = 1, \ldots, 3$ and $k = 1, \ldots, 4$ are defined (on suppressing $\mathbf{x}$):

$$f_{11} = \left[1 + \sum_{l=1}^{q} \frac{\alpha_{1l}}{FO_{1l}} + \sum_{s=1}^{3} \frac{\beta_{1s}}{U_{1s}} + \sum_{p=1}^{6} \frac{\kappa_{1p}}{W_{1p}} + \sum_{\substack{j=1\\ j\neq i}}^{n} \frac{\eta_{1j}}{MO_{1j}} \right] (x_1 - p_{11})$$
$$- G_1 \left[\sum_{l=1}^{q} \frac{\alpha_{1l}}{FO_{1l}^2} (x_1 - o_{l1}) - 2 \sum_{\substack{j=1\\ j\neq i}}^{n} \frac{\eta_{1j}}{MO_{1j}^2} (x_1 - x_j) \right] - \frac{G_1 \beta_{11}}{(W_{11})^2} + \frac{G_1 \beta_{13}}{(W_{13})^2},$$

$$f_{12} = \left[1 + \sum_{l=1}^{q} \frac{\alpha_{1l}}{FO_{1l}} + \sum_{s=1}^{3} \frac{\beta_{1s}}{U_{1s}} + \sum_{p=1}^{6} \frac{\kappa_{1p}}{W_{1p}} + \sum_{\substack{j=1\\ j\neq i}}^{n} \frac{\eta_{1j}}{MO_{1j}} \right] (y_1 - p_{12})$$
$$- G_1 \left[\sum_{l=1}^{q} \frac{\alpha_{1l}}{FO_{1l}^2} (y_1 - o_{l2}) - 2 \sum_{\substack{j=1\\ j\neq i}}^{n} \frac{\eta_{1j}}{MO_{1j}^2} (y_1 - y_j) \right] - \frac{G_1 \beta_{12}}{(W_{12})^2} + \frac{G_1 \beta_{14}}{(W_{14})^2},$$

$$f_{13} = \left[1 + \sum_{l=1}^{q} \frac{\alpha_{1l}}{FO_{1l}} + \sum_{s=1}^{3} \frac{\beta_{1s}}{U_{1s}} + \sum_{p=1}^{6} \frac{\kappa_{1p}}{W_{1p}} + \sum_{\substack{j=1 \\ j \neq i}}^{n} \frac{\eta_{1j}}{MO_{1j}}\right](z_1 - p_{13})$$

$$- G_1\left[\sum_{l=1}^{q} \frac{\alpha_{1l}}{FO_{1l}^2}(z_1 - o_{l3}) - 2\sum_{\substack{j=1 \\ j \neq i}}^{n} \frac{\eta_{1j}}{MO_{1j}^2}(z_1 - z_j)\right] - \frac{G_1\beta_{15}}{(W_{15})^2} + \frac{G_1\beta_{16}}{(W_{16})^2},$$

$$f_{14} = \left[\sum_{l=1}^{q} \frac{\alpha_{1l}}{FO_{1l}} + \sum_{s=1}^{3} \frac{\beta_{1s}}{U_{1s}} + \sum_{p=1}^{6} \frac{\kappa_{1p}}{W_{1p}} + \sum_{\substack{j=1 \\ j \neq i}}^{n} \frac{\eta_{1j}}{MO_{1j}}\right](\theta_i - p_{14}),$$

$$g_{11} = 1 + G_1\frac{\beta_{11}}{U_{11}^2}, \quad g_{12} = 1 + G_1\frac{\beta_{12}}{U_{12}^2}, \quad g_{13} = 1 + G_1\frac{\beta_{13}}{U_{13}^2}.$$

and for $i = 2, \ldots, n$.

$$f_{i1} = \left[1 + \sum_{l=1}^{q} \frac{\alpha_{il}}{FO_{il}} + \sum_{s=1}^{3} \frac{\beta_{is}}{U_{is}} + \sum_{p=1}^{6} \frac{\kappa_{ip}}{W_{ip}} + \sum_{\substack{j=1 \\ j \neq i}}^{n} \frac{\eta_{ij}}{MO_{ij}}\right](-(A_i - a_i)\cos\theta_1$$

$$+ (B_i - b_i)\sin\theta_1)$$

$$- G_i\left[\sum_{l=1}^{q} \frac{\alpha_{il}}{FO_{il}^2}(x_i - o_{l1}) - 2\sum_{\substack{j=1 \\ j \neq i}}^{n} \frac{\eta_{ij}}{MO_{ij}^2}(x_i - x_j)\right] - \frac{G_i\beta_{i1}}{(W_{i1})^2} + \frac{G_i\beta_{i1}}{(W_{i3})^2},$$

$$f_{i2} = \left[1 + \sum_{l=1}^{q} \frac{\alpha_{il}}{FO_{il}} + \sum_{s=1}^{3} \frac{\beta_{is}}{U_{is}} + \sum_{p=1}^{6} \frac{\kappa_{ip}}{W_{ip}} + \sum_{\substack{j=1 \\ j \neq i}}^{n} \frac{\eta_{ij}}{MO_{ij}}\right](-(A_i - a_i)\sin\theta_1$$

$$+ (B_i - b_i)\cos\theta_1)$$

$$- G_i\left[\sum_{l=1}^{q} \frac{\alpha_{il}}{FO_{il}^2}(y_i - o_{l2}) - 2\sum_{\substack{j=1 \\ j \neq i}}^{n} \frac{\eta_{ij}}{MO_{ij}^2}(y_i - y_j)\right] - \frac{G_i\beta_{i2}}{(W_{i2})^2} + \frac{G_i\beta_{i4}}{(W_{i4})^2},$$

$$f_{i3} = \left[1 + \sum_{l=1}^{q} \frac{\alpha_{il}}{FO_{il}} + \sum_{s=1}^{3} \frac{\beta_{is}}{U_{is}} + \sum_{p=1}^{6} \frac{\kappa_{ip}}{W_{ip}} + \sum_{\substack{j=1 \\ j \neq i}}^{n} \frac{\eta_{ij}}{MO_{ij}}\right](z_i - p_{13})$$

$$- G_i\left[\sum_{l=1}^{q} \frac{\alpha_{il}}{FO_{il}^2}(z_i - o_{l3}) - 2\sum_{\substack{j=1 \\ j \neq i}}^{n} \frac{\eta_{ij}}{MO_{ij}^2}(z_i - z_j)\right] - \frac{G_i\beta_{i5}}{(W_{i5})^2} + \frac{G_i\beta_{i6}}{(W_{i6})^2},$$

$$f_{i4} = \left[\sum_{l=1}^{q} \frac{\alpha_{il}}{FO_{il}} + \sum_{s=1}^{3} \frac{\beta_{is}}{U_{is}} + \sum_{p=1}^{6} \frac{\kappa_{ip}}{W_{ip}} + \sum_{\substack{j=1 \\ j \neq i}}^{n} \frac{\eta_{ij}}{MO_{ij}}\right](\theta_i - \theta_1),$$

$$g_{i1} = 1 + G_i\frac{\beta_{i1}}{U_{i1}^2}, \quad g_{i2} = 1 + G_i\frac{\beta_{i2}}{U_{i2}^2}, \quad g_{i3} = 1 + G_i\frac{\beta_{i3}}{U_{i3}^2}.$$

The behavior of the Lyapunov function $L(\mathbf{x})$ given in (12) with its terms specified in (4)–(11) depends on the following 2 properties. Firstly, the function gains it minimum value as the robots approach their desired target, and secondly, the value increases to infinity when the robots encounter obstacles. Next, we design the theorem similar to one in Raghuwaiya et al. (2016).

Theorem 1 *Let the ODEs given in system (2) control the motion of group of n mobile robots in the C-ITS. The main objective here is to create and maintain a pre-defined formation, generate aircraft movement in a pre-defined workspace, and safely reach at the pre-defined target while maintaining pre-defined formation. Achieving the pre-determined targets, limiting the workspace, and taking the kinodynamic constraints are among the subtask that needs to be considered. According to the LbCS of the system (2), the following continuous time-invariant acceleration control laws can be produced using the attracting and repulsive potential field functions which ensures stability of system (2) in Lyapunov sense.*

$$\begin{aligned}
\sigma_{i1} &= -\frac{\delta_{i1} v_i + f_{i1}\cos\theta_i + f_{i2}\sin\theta_i}{g_{i1}}, \\
\sigma_{i2} &= -\frac{\delta_{i2}\omega_i + \frac{L_i}{2}(f_{i2}\cos\theta_i - f_{i1}\sin\theta_i) + f_{i4}}{g_{i2}}, \\
\sigma_{i3} &= -\frac{\delta_{i3}\mu_i + f_{i3}}{g_{i3}}.
\end{aligned} \tag{13}$$

where $i = 1, \ldots, n$, where $\delta_{i1} > 0$, $\delta_{i2} > 0$, and $\delta_{i3} > 0$ are defined as the convergence parameters.

5 Stability Analysis

Here, using the Lyapunov's direct method and the theorem of Raghuwaiya et al. (2016), we provide the proof of the mathematical stability of the system (2).

Theorem 2 *Let (p_{11}, p_{12}, p_{13}) be the position of the target of the leader, and p_{i3} for $i = 1, \ldots, n$, be the prescribed final orientations of the follower robots. Let p_{i1}, p_{i2}, and p_{i3} satisfy*

$$\begin{aligned}
a_i &= -(p_{11} - p_{i1})\cos\theta_1 - (p_{12} - p_{i2})\sin\theta_1, \\
b_i &= (p_{11} - p_{i1})\sin\theta_1 - (p_{12} - p_{i2})\cos\theta_1.
\end{aligned}$$

for any given a_i,b_i, and c_i for $i = 2, \ldots, n$. If $\mathbf{x}_{ei} \in \mathbb{R}^{7n}$ is an equilibrium point for (2), then $\mathbf{x}_e = (x_{e1}, x_{e2}, \ldots, x_{en}) \in D(L_{(2)}(\mathbf{x}))$ is a stable equilibrium point of system (2).

Proof The following is simple to verify, for $i \in \{1, \dots, n\}$:

1. $L(\mathbf{x})$ is defined, continuous, and positive over the domain $D(L(\mathbf{x})) = \{\mathbf{x} \in \mathbb{R}^{4n} : FO_{il}(\mathbf{x}) > 0, l = 1, \dots, q;\ MO_{ij}(\mathbf{x}) > 0,\ j = 1, \dots, n,\ j \neq i;\ W_{ip}(\mathbf{x}) > 0, p = 1, \dots, 6;\ U_{is}(\mathbf{x}) > 0, s = 1, \dots, 3\}$,
2. $L(\mathbf{x}_e) = 0, \dot{L}_{(2)}(\mathbf{x}_e) = 0$,
3. $L(\mathbf{x}) > 0 \forall \mathbf{x} \in D(L(\mathbf{x}))/\mathbf{x}_e$,
4. $\dot{L}_{(2)}(\mathbf{x}) = -\sum_{i=1}^{n} \left(\delta_{i1} v_i^2 + \delta_{i2} \omega_i^2 + +\delta_{i3} \mu_i^2\right) \leq 0, \forall \mathbf{x} \in D(L(\mathbf{x}))$,
5. $L(\mathbf{x}) \in C^1\left(D(L(\mathbf{x}))\right)$.

Hence, $L(\mathbf{x})$ is classified as a Lyapunov function for system (2), and $\mathbf{x}_e$ is a stable equilibrium point in the sense of Lyapunov. □

6 Simulation Results and Discussion

A virtual scenario for two robots forming a *Column* is shown in this section. We look at the movement of two aircraft in a *Column Formation* when there are spherical obstacles in their path. Figure 3 illustrates how the follower should stand in relation to the leader, $\mathcal{A}_1$. In order to prevent inter-collision, follower aircraft must maintain its desired separation from the leader and its orientation with respect to the leader as the leader moves toward its target. With this, the structure will stay locally rigid the entire way. Figure 4 shows the development of the Lyapunov function (red) and its derivative (blue). Assuming all the units have been taken into account, Table 1 provides the correct configurations of the two aircrafts along with additional parameters required

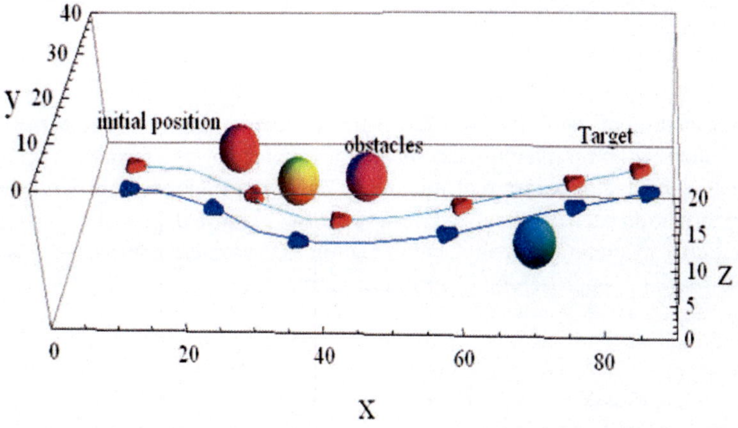

Fig. 3 Positions of $\mathcal{A}_1$ and $\mathcal{A}_2$ in column formation. Red designates the leader, and blue designates the follower. The system's trajectory is shown by the cyan and blue lines. In the event of obstacles, the system distorts slightly before reestablishing itself

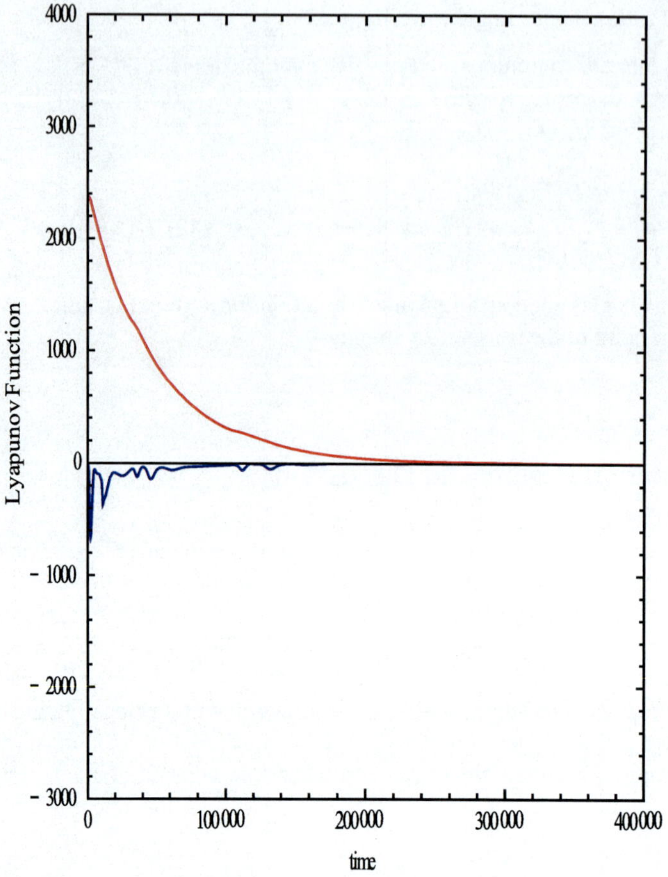

Fig. 4 Evolution of $L(\mathbf{x})$ in red and $\dot{L}_{(1)}(\mathbf{x})$ in blue

for the simulation of Fig. 2. Figure 5 shows the relative separation between the leader and the follower. The graph unmistakably demonstrates how the formation is slightly deformed by obstacles before returning to the desired formation. This demonstrates how the formation keeps its localized rigid structure. Figure 6 depicts the translational and rotational accelerations, which begin to drop as the system approaches obstacles and then rise later, approaching to zero near target.

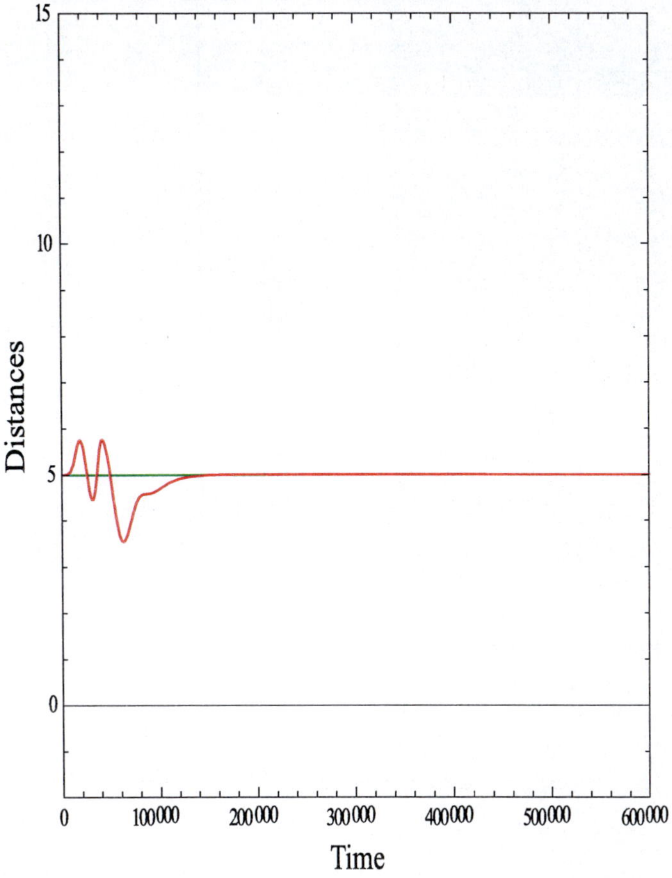

Fig. 5 Relative distance error for a_1 in red

7 Conclusion

The research effectively provided a set of nonlinear control principles utilizing LbCS. The provided controllers are made to move the system in a leader-follower manner while maintaining a locally rigid formation. The leader's body has been fixed, and the position of the follower has been assigned specifically in relation to the leader using Cartesian coordinate systems. All of this is predicated on the idea of a frame of reference that rotates instantly, totally avoiding the singularities that arise from polar representation. The acceleration-based controllers are additionally created to guarantee convergence of the entire system to its equilibrium state while avoiding obstacles, giving practicable trajectories. The follower vehicle followed and kept the pre-defined leader in place, clearly maintaining the locally rigid formation. Additionally, the assumption of no slippage and pure rolling has been examined in order to

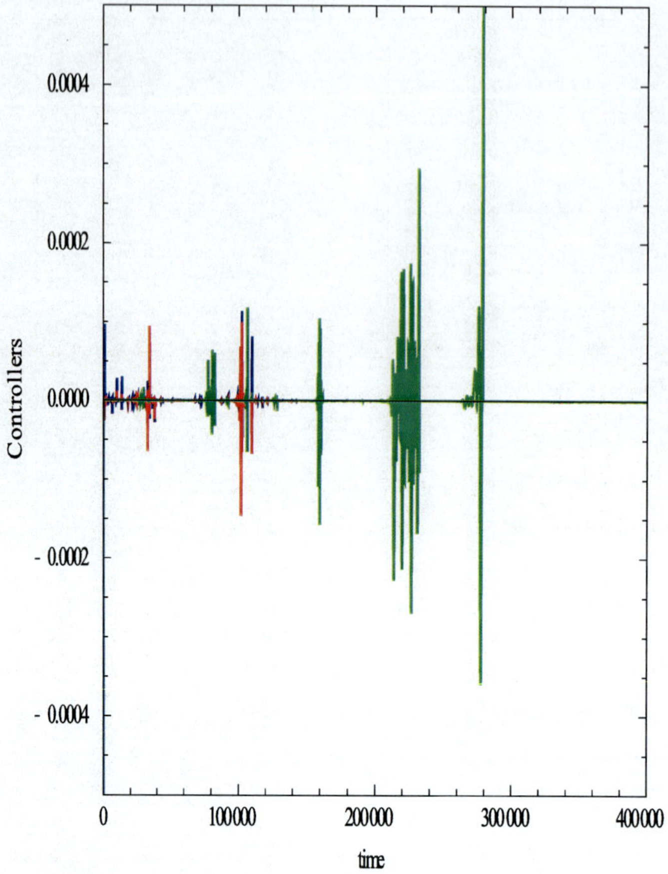

Fig. 6 Evolution of acceleration σ_{11} in blue, σ_{12} in green, and σ_{13} in red

establish nonholonomic constraints. Finally, to demonstrate the effectiveness of the method, computer simulations of a virtual scenario have been run. Future work will introduce cylinder- and rectangle-shaped obstacles and analyze stability dynamically using a leader-follower strategy.

Table 1 Numerical values of initial and final states, with constraints and parameters for Fig. 3

Rectangular position of leader	$(x_1, y_1, z_1) = (10, 20, 10)$
Velocities	$v_i = 0.5$, $\omega_i = 0$, $\nu_i = 0.05$ for $i = 1, 2$
Angular positions	$\theta_i = 0$, for $i = 1, \ldots, n$
Relative distance	$(a_2, b_2, c_2) = (0, 5, 0)$
Dimensions of robots	$L_i = 1.6$, $l_i = 1.2$ for $i = 1, \ldots, n$
Leader's target	$(p_{11}, p_{12}, p_{13}) = (85, 20, 10)$, $rt = 0.6$
Final orientations	$p_{14} = 0$
Max. translational velocity	$v_{\max} = 5$, $\nu_{\max} = 1$
Clearance parameters	$\epsilon_1 = 0.1$, $\epsilon_2 = 0.05$
Fixed obstacles	$(o_{11}, o_{12}, o_{13}) =$ $(25, 24, 10)$, $(o_{21}, o_{22}, o_{23}) = (35, 17, 10)$
	$(o_{31}, o_{32}, o_{33}) =$ $(45, 18, 10)$, $(o_{41}, o_{42}, o_{43}) = (70, 5, 10)$
	$ro_l = 3$ for $l = 1, \ldots, 4$
Collision avoidance	$\varphi_{ij} = 0.00001$, for $i, j = 1, 2$, $j \neq i$,
	$\alpha_{il} = 0.001$, for $i = 2, \ldots, n$, $l = 1, \ldots, q$
	$\alpha_{il} = 1$, for $i = 1, ..., 1$, $l = 1, \ldots, q$
Dynamics constraints	$\beta_{is} = 0.001$, for $i, s = 1, 3$,
Convergence	$\delta_{11} = 100{,}000$, $\delta_{12} = 50{,}000$, $\delta_{13} = 10{,}000$, $\delta_{i1} = 10{,}000$,
	$\delta_{i2} = 10{,}000$, $\delta_{i3} = 10{,}000$, $i = 2, \ldots, n$

References

F. Arrichiello, *Coordination control of multiple mobile robots* (Ingegneria Dell'informazione e Matematica Industriale, Dipartimento di Automazione, Elettromagnetismo, 2006)

T. Balch, M. Hybinette, Social potentials for scalable multi-robot formations, in *Proceedings 2000 ICRA. Millennium Conference. IEEE International Conference on Robotics and Automation. Symposia Proceedings (Cat. No. 00CH37065)*, vol. 1 (IEEE, 2000), pp. 73–80

E. Ben Hamida, H. Noura, W. Znaidi, Security of cooperative intelligent transport systems: standards, threats analysis and cryptographic countermeasures. Electronics **4**(3), 380–423 (2015)

D. Crombie, The examination and exploration of algorithms and complex behavior to realistically control multiple mobile robots. Master's thesis, Australian National University, 1997

C. De La Cruz, R. Carelli, Dynamic model based formation control and obstacle avoidance of multi-robot systems. Robotica **26**(3), 345–356 (2008)

A. Festag, Cooperative intelligent transport systems standards in Europe. IEEE Commun. Mag. **52**(12), 166–172 (2014)

W. Kang, N. Xi, J. Tan, Y. Wang, Formation control of multiple autonomous robots: theory and experimentation. Intell. Autom. Soft Comput. **10**(2), 1–17 (2004)

M. Mamino, G. Viglietta, Square formation by asynchronous oblivious robots. arXiv preprint arXiv:1605.06093 (2016)

K. Raghuwaiya, B. Sharma, J. Vanualailai, Cooperative control of multi-robot systems with a low-degree formation, in *Advanced Computer and Communication Engineering Technology* (Springer, 2016), pp. 233–249

K. Raghuwaiya, B. Sharma, J. Vanualailai, Leader-follower based locally rigid formation control. J. Adv. Transp. **2018** (2018)

J. Raj, K. Raghuwaiya, B. Sharma, J. Vanualailai, Motion control of a flock of 1-trailer robots with swarm avoidance. Robotica 1–26 (2021)

J. Raj, K.S. Raghuwaiya, J. Vanualailai, Novel Lyapunov-based autonomous controllers for quadrotors. IEEE Access **8**, 47393–47406 (2020a)

J. Raj, K. Raghuwaiya, J. Vanualailai, Collision avoidance of 3D rectangular planes by multiple cooperating autonomous agents. J. Adv. Transp. **1–13**(10), 2020 (2020b)

B. Sharma, *New Directions in the Applications of the Lyapunov-Based Control Scheme to the Findpath Problem*. PhD thesis, University of the South Pacific, Suva, Fiji Islands, July 2008

B. Sharma, J. Vanualailai, A. Prasad, A dø-strategy: facilitating dual-formation control of a virtually connected team. J. Adv. Transp. (2017)

B. Sharma, J. Vanualailai, A. Prasad, Formation control of a swarm of mobile manipulators. Rocky Mt. J. Math. 909–940 (2011)

K. Shojaei, Neural adaptive output feedback formation control of type (m, s) wheeled mobile robots. IET Control Theory Appl. **11**(4), 504–515 (2017)

J. Vanualailai, A. Sharan, B. Sharma, A swarm model for planar formations of multiple autonomous unmanned aerial vehicles, in *2013 IEEE International Symposium on Intelligent Control (ISIC)* (IEEE, 2013), pp. 206–211

Effective Load Balancing and Load Sharing in Multi-access Edge Computing for Extreme Network Congestion

Ausaaf Nabi, Ira Joshi, and Sonal Linda

Abstract A multi-access edge computing (MEC) plays a vital role in dealing with the exponential growth of data traffic to avoid network congestion. This is also essential for fulfilling the user requirements of computation resources. MEC network brings cloud functionalities to the edges that are in close proximity to the devices supporting multiple access points. Mobile devices that have resource constraints can save energy and enrich users' experience using the computation load balancing technique. This paper provides a comprehensive approach to load balancing and load sharing in a MEC environment. We proposed two algorithms for the MEC framework. First is the proactive load rationalization technique that accounts for the load as a function of chaos resolution through user load prediction. The second algorithm is used for the distribution of data across servers in a given geographical region which aims to achieve an equilibrium that ensures no server is overworked while other servers are left idle. Finally, a fine-grained computation-based MEC framework is developed using our proposed algorithms that is capable of load balancing and sharing in extreme network congestion. The ultimate goal is to provide optimal resource services to each user while balancing the load among the MEC servers. The simulation results demonstrate the efficiency of our methods in balancing load in the real-time scenario. Moreover, it has the ability to produce minimal load overhead during load failure.

Keywords MEC · Load balancing · Load sharing · Failure resolution · Base stations

1 Introduction

Multi-access edge computing (MEC) is defined as a cloud-based IT services environment at the network's edge. It ticks all the boxes by delivering ultra-low latency and high bandwidth while allowing apps to make use of real-time radio network data.

A. Nabi (✉) · I. Joshi · S. Linda
University of Delhi, 110021 New Delhi, India

A. B. Reddy et al. (eds.), *Proceedings of Third International Conference on Advances in Computer Engineering and Communication Systems*, Lecture Notes in Networks and Systems 612, https://doi.org/10.1007/978-981-19-9228-5_11

MEC infrastructures allow the software to be deployed simply for mobile functions or software-as-a-service (SaaS) applications that operate on a standardized virtualization platform at or near the network edge. With the exponential growth of big data and the high demand for real-time applications, a fast response time, i.e., less than 1 ms has left an indelible mark on the ICT sector (Chen et al. 2021a). As a consequence, resource management and latency difficulties have appeared in fulfilling such demands. Furthermore, the distance between the user equipment and the base station (BS) in the edge computing architecture creates a hurdle to meeting the needs of real-time applications (Mondal et al. 2020). The traffic traveling to virtual machines and container-based servers deployed in edge data centers must be handled. Continuous service is required with as little delay as feasible in the IoT and 5G era we are going. Load balancers can aid one in this endeavor. In each edge data center, several application instances should be running, and traffic to them should be distributed via load balancers to avoid servers from becoming overloaded.

In comparison with mobile cloud computing (MCC), MEC takes use of its close proximity to end-users to dramatically reduce job offloading transmission latency (Mirrokni et al. 2018). Local computing, edge computing, and cloud computing are used to construct a three-tier MEC architecture that utilizes edge and cloud servers. Due to the connectivity between the mobile devices and edge servers across wireless channels, offloading can access processing resources at the price of additional transmission latency and related energy consumption. Furthermore, load balancing enhancement techniques were discussed in Chen et al. (2021a) and Mondal et al. (2020) where the former discussed the impact of load estimation on load balancing in ultra-dense network sensors. While the latter talks about how game-theoretic approaches can be used in load balancing for non-cooperative cloudlets. In this paper, the semi-distributed offloading strategy is being used (Mondal et al. 2020), and the load sharing goal is to optimize job distribution in a MEC network in order to decrease the risk of failure. However, it has only been viewed as an issue with a single goal of minimizing execution time.

The major contributions of this paper are summarized as follows:

- Firstly, a proactive load rationalization scheme is proposed for load balancing while maintaining node failure tolerance in multi-access edge computing (MEC).
- Secondly, the proposed algorithm can address the problem of unexpected load overhead in the sparse network which occurs due to the highly dynamic environment of MEC.
- Lastly, a novel way of utilizing the software define network (SDN) node for MEC is introduced to deal with load balancing failure. It can perform central tasks offloading within a distributed system that will lead to the reduction of the cost of setting up a specialized SDN server node for task management.

The remainder of this paper is organized as follows. Section 2 presents the related work. In Sect. 3, we present the system model, problem formulation, and proposed algorithm. In Sect. 4, we describe the experimentation and simulation results. Finally, we conclude this paper in Sect. 5.

2 Related Work

MEC has been drawing attention from academia and industry and provides facilities in terms of computation and storage at the network edge. It allows computing and caching capabilities to be brought closer to users which is in line with the requirements of several new and emerging applications, such as vehicular applications. Recently, MEC researchers focus on decreasing energy consumption and latency via offloading computation-intensive and latency-sensitive operations.

Zhang et al. have introduced the idea of minimizing system energy usage while guaranteeing the latency criteria are satisfied (Zhang et al. 2016). Binary offloading (Zhang et al. 2013) and partial offloading (Wang et al. 2016) for decreasing energy usage by manipulating CPU-cycle frequencies developed using dynamic voltage and frequency scaling approaches. The offloading problem was investigated in You et al. (2016) with the goal of minimizing the weighted total of mobile energy consumption for a multi-user MEC system while keeping compute delay in mind. In the same direction, computation offloading was examined in Yu et al. (2016) to reduce mobile device energy consumption. Li et al. proposed an optimization problem to minimize the weighted sum of the total delay and energy consumption of all SMDs in a multi-MECS–multi-SMD network via multi-dimensional optimization on offloading strategy (Li et al. 2020). The problem includes load balancing, computation resource allocation, and transmit power control. Yu et al. (2017) propose a new design of a scalable and dynamic load balancer, called SDLB, that satisfies the requirements of MEC. SDLB is claimed to be 4× to 10× faster than the widely used cloud-load balancer at that time. Inspired by the above work, we propose two algorithms for load balancing and load sharing in the MEC environment.

Among the listed papers, Tang et al. (2020) stands out as it discusses reliability in MEC offloading in V2X systems. The paper further goes on to devise a decision-based strategy for handling faults in MEC servers. Similar to MEC in nature, the clustering in WSN was discussed in Mazinani et al. (2021). This goes on to provide a semi-distributed clustering technique by using virtual grids.

3 Fault Tolerant Load Balancing and Sharing Scheme for Reduced Network Congestion

Due to the limited capacity of MEC, the servers in the MEC network are prone to task failures as the network congestion increases. The congestion impacts the load of the systems reducing efficiency and reliability. The algorithm deals with several types of fault models for MEC load balancing systems: node crash, link crash, network anomalies (delay, congestion, partitioning). We identified three major design goals: predictable latency, continued performance under scaling, and guaranteed delivery even in the presence of faults. In the following subsections, we discuss the system model, problem statement and propose fault tolerant load balancing and sharing scheme for reduced network congestion.

3.1 System Model

For the problem of fault tolerant load balancing, we considered a real-world network configuration. The information on base stations (BSs) is collected from the largest open-source telecom dataset, OpenCellID.[1] Then, we used that information to build a network topology and simulate it further. Initially, we assume that all BSs do not have a server. Therefore, BSs are clustered to get the MEC server location to provide the feasibility of the cost of network setup. The MEC servers M, i.e., $\mathcal{M} = \{1, 2, 3, \ldots, M\}$ has limited computation capacity, storage capacity, and bandwidth.

We assume that there exists a topology for the MEC servers for communication with one another. However, the topology has no effect on load balancing as long as there exists a spanning tree in topology. In the proposed system model, a discrete time is adopted which is indexed by time t, i.e., $T = \{0, 1, \ldots, t\}$. The operating period is split into equal lengths τ (small discrete time unit). The conditions for the MEC server and task arrival rate vary for each slot $t \in T$.

3.2 Problem Formulation

When task i arrives at the jth cluster, the cost $W_{i,j}$ is computed as the load incurred considering computation, storage, and bandwidth resources at edge nodes. Our goal is to maximize cost $W_{i,j}$, while performing load resolution, subsequently dealing with fault tolerance (Zhang et al. 2016). For brevity, we simplify the expression as follows:

$$W_{i,j} = \max\left|\frac{\text{Load}_{t,j}}{1 + \text{Load}_{a,j} + \text{Load}_{m,j}}\right| \tag{1}$$

subject to,

$$C1 : \lambda_{cc} + \lambda_{sc} + \lambda_{bc} = 1; \quad \lambda_{cc}, \lambda_{sc}, \lambda_{bc} \in (0, 1) \tag{2}$$

$$C2 : L_{a,j} < L_j \quad \forall j \tag{3}$$

$$C3 : i * F_{i,j} = 1 \quad \forall j \in \{1, 2, \ldots, N\} \tag{4}$$

[1] https://opencellid.org.

Hence, the server current average network resource can easily be calculated as

$$\text{Load}_{\text{canr}} = \text{Load}_{\text{max}} - \text{Residual}_i \tag{5}$$

The time for task computation is used as follows,

$$t_i^{\text{act},j} = \min\left\{t_i^{c,j} + \max\left\{t_i^{\text{off},j}, t_i^{\text{res},j}\right\}, t_{\max}^{T}\right\} \tag{6}$$

The problem formulated by Eq. (1) is subject to constraints Eqs. (2), (3), and (4). We are proposing a semi-distributed algorithm with the help of leader node election (Zhang et al. 2013). This permits us to solve our problem of balancing the resource allocation decisions (Carvalho et al. 2021), by edge nodes in almost real-time by calculating residual load from Eq. (5). Finally, we are using Eq. (6) as a sub-metric for experiment.

3.3 Proposed Framework

In this section, we describe our proposed event driven MEC resource sharing (ED-MRS) scheme for efficient load balancing and sharing. If any of the servers in a cluster suffers from an uneven load, the algorithm acts as a reactive load resolution mechanism that evenly distributes the load among the servers.

Our proposed model operates in three steps as follows: initialization phase, election phase, and action phase (see Fig. 1).

Initialization Phase: In this phase, we define clusters based on geolocation (Bouet and Conan 2017) and build a consistent hash ring (Mirrokni et al. 2018). Next, we extract a priori data and compute constants.

Election Phase: In this phase, we elect a virtual leader in every cluster that would be responsible for doing load balancing and computing the resource reallocation on

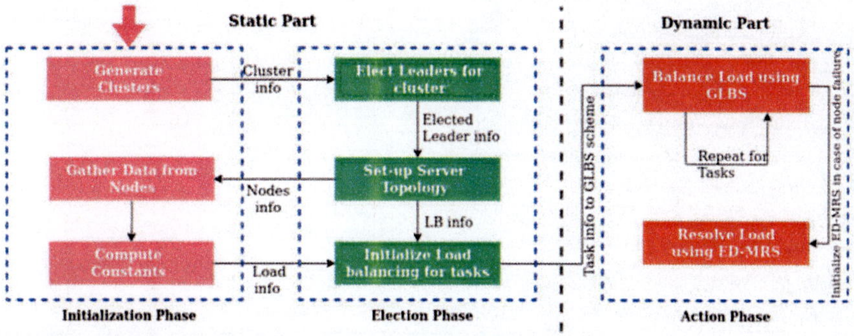

Fig. 1 Flow diagram of fault tolerant load balancing framework

failure. This will introduce a single point of command that will make the interprocess resource sharing easier and streamlined.

Action Phase: In this phase, the virtual leader nodes (VLNs) redirect the load using the GLBS algorithm. Further, it waits for the signal from any worker nodes in case a fault occurs. If the event is triggered, then the ED-MRS algorithm starts the task-offloading (Ma et al. 2020; Wu 2018; Chen et al. 2021b) based on the load to fellow MEC nodes by processing the current network. The process will repeat until the system settings change or the leader node collapses. In case the leader node collapses, then repeat step (3) in Algorithm 1 to re-elect the leader node.

Algorithm 1: Initialization

1. Initialize the MEC server M at t = 0
2. Using optimal clustering strategy create cluster
$\mathbf{C_k} \leftarrow \kappa$
3. ***for each*** C_i in C_k, ***do***:
4.   N = ElectLeaderNode(C_i)
5. $ND_i \Leftarrow getNodesData(C_i)$
6.   compute constants from eq (2 LoadFac.), (5. Cap)
7. ***endfor***
8. ***global*** $Residual[M] \leftarrow max(Cap)$
9. ***for*** t: = 0 to T ***do***:
10.   balance **GLBS(task(n,t))**, where
        n = Total task at time t
11.   ***if*** node M_i^k fails, ***do***:
12.   MutexLock(C_i) where $M_i^k \in C_i$
13.   Run **ED-MRS** algorithm
14.   MutexUnlock(C_i)
15.   ***endif***
16. ***endfor***

Algorithm 2: GLBS **T = O(N)**

global $Residual[M]$
for incoming requests i at $\mathbf{t = t_i}$, ***do***:
    compute load for task(n,t)
    Hash task based on location
    servers ← Load / $max(W_{i,j})$
    Find $\mathbf{Bins_j}$ from consistent hash
    j ← OffloadTask($\mathbf{task_i}$, $\mathbf{Bins_j}$, servers)
    $Residual[j] \leftarrow Residual[j] - Load$
    if $Residual[M] < L_{critical}$, ***do***:
        return M_i^k
endif
endfor

Algorithm 3: ED-MRS **T = O(MlogM + N)**

currentState $\leftarrow Residual[M]$
weights[M/2] $\leftarrow 0$
sort currentState
MEC-RHL $\leftarrow currentState[0, M/2]$
MEC-RLL $\leftarrow currentState[M/2+1, M]$
$L_{median} \leftarrow currentLoadatservercurrentState[M/2+1]]$
for i in **MEC-RHL, *do*:**
    $j \leftarrow$ **MEC-RLL**[N-i]
    weights[i] $\leftarrow max(W_{i,j})$
    dequeue tasks t until **weights**$[i] * Load_i < L_{median}$
        $Residual[i] \leftarrow Residual[i] +$ **transferLoad[i]*Load**$_i$
endfor
for all **tasks t, *do*:**
    select task slice from tasks, and **Bins**$_i$ from MEC-RLL
    offload (task slice, **Bins**$_i$)
endfor

In order to achieve faster load balancing, we start off by reducing our space to make the load balancing work faster (Carvalho et al. 2021). For that, we approximated the search space by clustering the servers into small unique clusters based on the physical coordinates. We used both hard and soft (fuzzy) clustering approaches but found out that soft clustering causes incompatibility in consistent hash in the algorithm. Hence, the model works well with hard clustering like the k-means clustering algorithm.

After the clustering is done, the load for each time instance T_n was transferred to the nearest cluster using the leader node. The leader node is elected at instance $t = 0$, which acts as a temporary load balancer. The temporary load balancer helps in reducing the cost of network setup and load balancing via the ED-MRS scheme even on failures. It does so by recreating a new leader node if the load balancer fails.

For every request that is applied to the load balancer, the load balancer is responsible for transferring the data to the nodes with the most residual load. In case, the load of any node in the cluster gets above the threshold capacity, i.e., residual becomes less than zero. All this operation is done with significant constant time complexity. Consequently, the total load balancing time complexity would be only dependent upon the number of tasks coming at T^{th} interval. Moreover, the server starts the reallocation procedure which reallocates the load into all nearby clusters defined by the same search space, i.e., the same cluster. Successfully reallocating load from MEC-RHL (relatively high load) to MEC-RLL (relatively low load) (Chen et al. 2021a) in the failed clusters. By the time, the clusters are reallocating the tasks, and the next tasks are passed onto the next nearest cluster. Consequently, giving the ability to simultaneously heal the server health reduces the failures.

4 Result and Analysis

This section describes the effectiveness of our proposed schemes for load balancing and load sharing through various experiments. This includes load estimation via user load prediction and observing average request time for tasks with respect to faults. The primary parameters for the experimental setup are shown in Table 1. The whole experimental environment and process were simulated in Python, running on a configuration with Intel Core i7-6650U 2.20 GHz processor, 8 GB RAM.

In this experiment, a MEC topology with 60 server nodes is taken into consideration as in Fig. 2. The number of tasks set by a randomized task scheduler range from 1 to 50 tasks per cycle. The acceptable time for the request completion is set at $t + 5$ cycle. The performance measures selected are the mean response time of a task and error rate resolution time. In each simulation run, 1000 to 50,000 tasks were generated depending upon the randomization of the scheduler. Extensive simulations have been performed to determine the impact of the parameters such as the system load, load overhead on failures, size, and topology of the network after failed nodes. Finally, the estimated load of the MEC server is exactly equal to its actual load for each iteration. Figures 3 and 4 representing the change in bandwidth and total resource consumption over time for $t = 30$ for 2 distant cooperative servers.

Table 1 Parameters and constants

Parameter constants	Symbol	Value
No. of BSs/MEC in the topology considered	M	60
Clusters in topology	C_k	5
Critical residual load	L_{critical}	0.05
Maximum task completion time	$t^T_{\max}$	6
Tasks for every iteration	T_{mac}	1–50

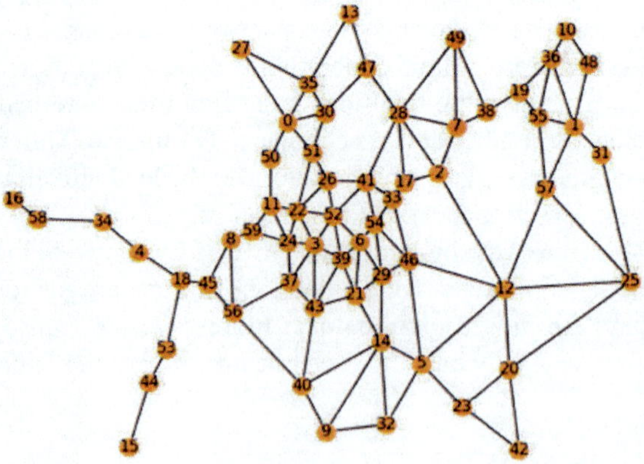

Fig. 2 Sampled graph from the data source

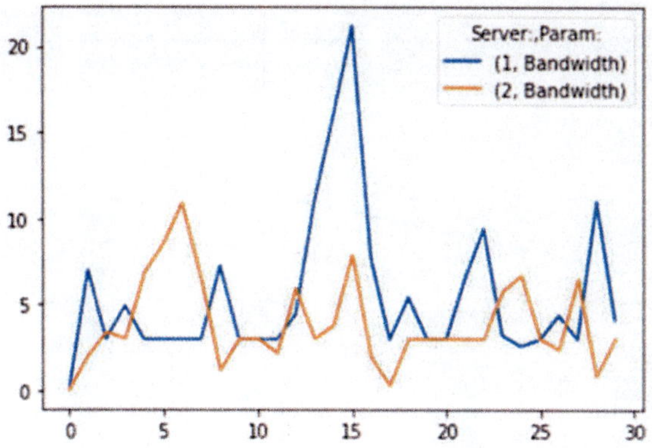

Fig. 3 Comparison of bandwidth of 2 instances at time $t \leq 30$

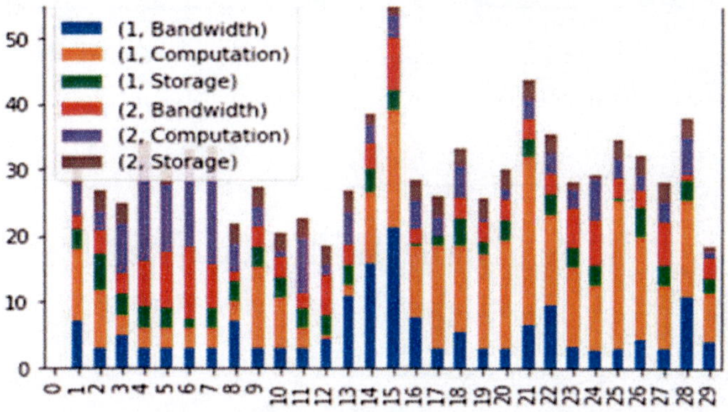

Fig. 4 Stacked representation for all the resources for the first 30 iterations

For the experiment, we run the simulation for 1000 distinct iterations, and the server load capacity changes for every iteration. In any instance, if the total load exceeds the threshold, then the fault occurs (see Fig. 6), i.e., the Residual$[M] < L_{\text{critical}}$. Then, the leader node will start the reallocation procedure within the cluster to rectify the fault, while all new tasks redirect to the next nearest cluster in the hash according to the GLBS scheme.

Hence, the algorithm increases load balancing overhead for fault resolution while trying to rectify errors. The green arrows (see Fig. 6) represent faults that lead to step increment in balancing overhead due to task reallocation. As a result, the maximum and average request time will change with respect to the number of faults as depicted in Fig. 5. The proposed scheme works well for optimizing the task's average request time even in extreme load balancing failures.

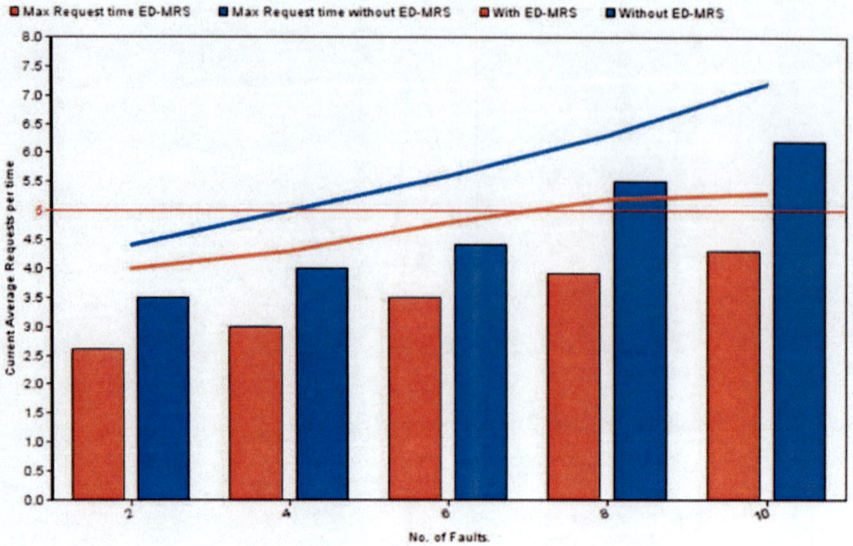

Fig. 5 Maximum and average request time with respect to the number of faults

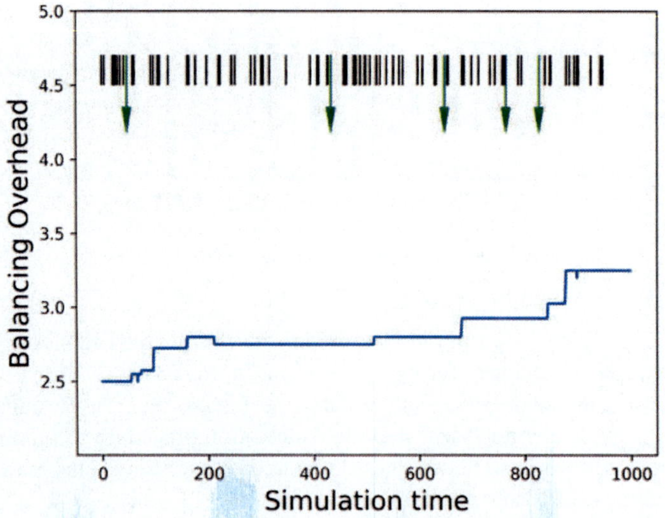

Fig. 6 Load balancing and load sharing with failure resolution using our proposed ED-MRS scheme

5 Conclusion and Future Directions

In this paper, we addressed the problem of fault tolerant load balancing in multi-access edge networks, where each user is served by several BSs cooperatively. Our proposed schemes, geolocation-based load balancing scheme (GLBS) and event driven MEC resolution strategy (ED-MRS), are used to optimize both delay and fault resolution. Initially, we describe the system model of a multi-access edge computing (MEC) network. Then, a cluster-based scheme is utilized for designing GLBS and ED-MRS algorithms to solve the problem of fault tolerant load balancing. Finally, the simulation results show the effectiveness of the proposed schemes. The future research directions would include more experiments on the action phase, considering larger MEC networks, and updating task offloading strategy using several distinct methods. Further, we will try to use different methods for virtual leader election.

References

M. Bouet, V. Conan, Geo-partitioning of MEC resources, in *Proceedings of the Workshop on Mobile Edge Communications*, pp. 43–48 (2017)

G. Carvalho, B. Cabral, V. Pereira, J. Bernardino, Edge computing: current trends, research challenges and future directions. Computing **103**(5), 993–1023 (2021)

W. Chen, Y. Zhu, J. Liu, Y. Chen, Enhancing mobile edge computing with efficient load balancing using load estimation in ultra-dense network. Sensors **21**(9), 3135 (2021a)

R. Chen, H. Lu, P. Ma, User-centric cooperative MEC service offloading, in 2021 IEEE wireless communications and networking conference (WCNC) (IEEE, 2021b), pp. 1–6

S.L. Li, J.B. Du, D.S. Zhai, X.L. Chu, F.R. Yu, Task offloading, load balancing, and resource allocation in MEC networks. IET Commun. **14**(9), 1451–1458 (2020)

S. Ma, S. Song, J. Zhao, L. Zhai, F. Yang, Joint network selection and service placement based on particle swarm optimization for multi-access edge computing. IEEE Access **8**, 160871–160881 (2020)

A. Mazinani, S.M. Mazinani, S. Hasanabadi, FSCVG: a fuzzy semi-distributed clustering using virtual grids in WSN. Wireless Pers. Commun. **118**(2), 1017–1038 (2021)

V. Mirrokni, M. Thorup, M. Zadimoghaddam, Consistent hashing with bounded loads, in *Proceedings of the Twenty-Ninth Annual ACM-SIAM Symposium on Discrete Algorithms* (Society for Industrial and Applied Mathematics, 2018), pp. 587–604

S. Mondal, G. Das, E. Wong, A game-theoretic approach for non-cooperative load balancing among competing cloudlets. IEEE Open J. Commun. Soc. **1**, 226–241 (2020)

L. Tang, B. Tang, L. Tang, F. Guo, J. Zhang, Reliable mobile edge service offloading based on P2P distributed networks. Symmetry **12**(5), 821 (2020)

Y. Wang, M., Sheng, X. Wang, L. Wang, J. Li, Mobile-edge computing: partial computation offloading using dynamic voltage scaling. IEEE Trans. Commun. **64**(10), 4268–4282 (2016)

H. Wu, Multi-objective decision-making for mobile cloud offloading: a survey. IEEE Access **6**, 3962–3976 (2018)

C. You, K. Huang, H. Chae, B.H. Kim, Energy-efficient resource allocation for mobile-edge computation offloading. IEEE Trans. Wireless Commun. **16**(3), 1397–1411 (2016)

Y. Yu, J. Zhang, K.B. Letaief, Joint subcarrier and CPU time allocation for mobile edge computing, in *2016 IEEE Global Communications Conference (GLOBECOM)* (IEEE, 2016), pp. 1–6

Y. Yu, X. Li, C. Qian, SDLB: a scalable and dynamic software load balancer for fog and mobile edge computing, in *Proceedings of the Workshop on Mobile Edge Communications* (2017), pp. 55–60

W. Zhang, Y. Wen, K. Guan, D. Kilper, H. Luo, D.O. Wu, Energy-optimal mobile cloud computing under stochastic wireless channel. IEEE Trans. Wireless Commun. **12**(9), 4569–4581 (2013)

K. Zhang, Y. Mao, S. Leng, Q. Zhao, L. Li, X. Peng, L. Pan, S. Maharjan, Y. Zhang, Energy-efficient offloading for mobile edge computing in 5G heterogeneous networks. IEEE Access **4**, 5896–5907 (2016)

Prediction of Depression in Techies at Workplaces

Venkata Sailaja, Meghana Yelamarthi, Ananya Nandyala, Meghana Manda, Kairamkonda Poorna Yamini, and Vamsi Krishna Balusu

Abstract Mental health is nowadays a topic that is most often discussed when it comes to research but least frequently discussed when it comes to personal life. The expanding utilization of innovation will prompt a way of life of less actual work. Additionally, the constant pressure on a worker in the IT industry will make it more defenseless against mental issues. Employees in the tech industry are most vulnerable to such illness as this quick moving industry has huge stakes, which expect individuals to meet considerably better standards. So, it is of extreme importance to develop a prescient predictive model for automated diagnosis of mental illness. We aim to propose a model using Stacking Classifier with the help of feature selection for prediction which outperforms the existing models implemented.

Keywords Mental health · Depression · Workplace · Stacking classifier · Survey · Accuracy · Recall · Precision · $F1$ score measures

1 Introduction

Mental health connected issues such as depression and anxiety are predominant and normal these days in a lot of people and they suffer irrespective of any gender, age, physical well-being, etc. The problems affect a person's thinking, acting, retrospecting, and assessing capabilities. It is a parameter of estimating how one can handle stress and make rightful decisions with every step-in life. It is usually not widely discussed with others and out in public, and no proper awareness is there in society. Depression has many characteristics such as mood swings along with disturbed sleeping patterns and suicidal thoughts. Stressful work environments can also lead to a person's mental well-being affected. The working conditions are stressful in the

V. Sailaja (✉) · M. Yelamarthi · A. Nandyala · M. Manda · K. P. Yamini · V. K. Balusu
Department of CSE, VNR Vignana Jyothi Institute of Engineering and Technology, Hyderabad, India
e-mail: sailaja_nv@vnrvjiet.in

V. K. Balusu
e-mail: balusuv@uci.edu

A. B. Reddy et al. (eds.), *Proceedings of Third International Conference on Advances in Computer Engineering and Communication Systems*, Lecture Notes in Networks and Systems 612, https://doi.org/10.1007/978-981-19-9228-5_12

Tech industries where there's constant stress and pressure of meeting deadlines and delivering the work. A study shows that, in tech, over one-third of professionals admit they have issues with depression. Depression is a disagreeable condition of inward unrest, regularly joined by anxious conduct and so on. It is frequently joined by fretfulness, weakness, issues in fixation, and solid pressure. These conditions contrarily sway numerous areas of a person's working. Two significant difficulties in satisfactorily tending to Depression are distinguishing influenced people and guaranteeing proper and ideal treatment. Since Depression's side effects are inside experienced, it regularly goes undetected.

Early detection and analysis are principal to understanding and tending to dysfunctional behavior on a public level.

1.1 Causes of Depression

There are a few potential reasons for gloom which go from normal to fortuitous. The causes include

- A foundation set apart by drug or alcohol misuse can impact.
- Mind structure is a danger for Depression if the front-facing flap of the brain is less active.
- Certain conditions may put you at greater threat, like constant sickness, sleep deprivation, or attention deficit hyperactivity disorder.

1.2 Symptoms of Depression

Depression can be more than a steady attitude of pity or feeling "blue". Critical sadness can cause a grouping of signs. The signs of Depression can be differentiated among men, women, and youths in a sudden way.

- Temperament, like annoyance, forcefulness, peevishness, tension, anxiety.
- Conduct, like loss of interest, done discovering delight in most loved exercises, feeling tired effectively, and so on.
- Psychological capacities, like failure to focus, trouble finishing errands, deferred reactions during discussions.
- Lifestyle patterns, like sleep deprivation, extreme tiredness, not staying unconscious for the whole evening.

1.3 Treatments for Depression

Living with depression can be challenging, but treatments can be helpful in improving one's quality of life. One can talk to their doctor about possible ways to get out of it.

You may adequately treat depression with one treatment, or you may find that a mix of medicines works best. It's entirely expected to join clinical medicines and way of life treatments, like drugs, psychotherapy, light treatment, work out, way of life changes, taking care of yourself, and so on.

1.4 *Statistical Information*

According to the CDC, Mental health statistics show that Depression interferes with a person's ability to wrap up work assignments about 20% of the time and diminishes mental execution around 35% of the time. Only 57% of laborers who report mid-level hopelessness and 40% of the people who report outrageous serious depression.

According to a survey by WHO, Internationally, an expected 264 million individuals deal with depression, prime sources of inability, with many of these people also suffering from signs of anxiety. The assessed cost to the worldwide economy is US$ 1 trillion every year in lost profitability (Sailaja et al. 2021).

A study shows that, in tech, more than 33% of experts concede they have issues with depression. Specifically, 38.8% of tech aces reacting to a Blind overview led by top tech organizations, say they're discouraged.

2 Literature Review

Relationships between depression and the work environment, depressive problems are exceptionally predominant in the work environment. Using different algorithms and achieved the highest accuracy of around 80% (Bender and Farvolden 2008).

Depression has high pervasiveness and causes suffering, it is essential to foresee the beginning of depression among adults, and for that we can build an Artificial Intelligence-based prescient model for future expectation with tenfold CV giving accuracy of 87% (Na et al. 2020).

Depression is a psychological sickness that causes huge aggravations in day-by-day life. Depression is ordinarily connected with low temperament, serious medical issues, and significant socioeconomic burden. The Random Forest gave an accuracy of 83% (Cho et al. 2020).

Depression, the most recent plague of the advanced time, has consistently drawn the consideration of specialists to discover and assess the level, causes and anticipation using the predictive model Apriori Algorithm got the frequent item sets with 100% confidence (Jena and Kamila 2014).

In this paper AI alongside electroencephalogram (EEG) signal investigation is utilized. In this test pressure is added to the analysis dependent on the MIST. This utilizes the accompanying calculations as *K*NN, Decision Tree, Random Forest, etc. and has 81.25% as the most noteworthy precision with the Random Forest (Sandhya and Kantesaria 2019).

Depression gives irate upheavals, rest unsettling influences, low energy and helpless fixation and best-case scenario, can prompt self-destruction. On applying algorithms such as SVM, Bayes net, Cat boost. Using the Cat boost got accuracy of 85% (Shanthalakshmi Revathy et al. 2020).

Anxiety and depression are two of the mail emotional well-being disorders among seafarers. In this paper on using 5 ML classifiers such as Random Forest, Logistic Regression, CatBoost, Naïve Bayes, and SVM, utilizing CatBoost got the most agreeable measure, with an accuracy of 82.6% (Sau and Bhakta 2019).

In this paper, the author meant to use a thorough scope of patient-level analytic, conduct, and segment information, just as past visit history information from a statewide well-being data trade. Using a Decision model for high-risk patient groups yielded an accuracy of 86% (Sau and Ishita 2019).

In this paper, the author used the OSMI 2018 dataset and compared 9 models while using the same questionnaire approach with only 15 features, using Random Forest turned out to be the best, with an accuracy of around 81% (Kasthurirathne, et al. 2019).

In this paper, the author used data from the OSMI 2017 responses of techies. Different ML techniques were applied. Boosting had the highest accuracy among the models which is of 75% (Katti, et al. 2020).

In this paper, the author explored how Social Network Sites users' posts can help characterize them as per emotional wellness levels. Using SVM and Naïve Bayes classifiers gave 63% accuracy (Reddy et al. 2018).

The enthusiastic, mental, and social welfare assistance of an individual is uncovered by their psychological health. They used online datasets. Classification algorithms like Random Forest, Naïve Bayes, and Decision Tree got more accuracy of 82% (Aldarwish et al. 2017).

In this, collection of the dataset was through polls, asked the individual, posts via online media, text utilized all through verbal correspondence and demeanors on the face. In this research work is done with different classifiers, on using *K*NN with Twint tool got 89% accuracy (Laijawala et al. 2020).

3 Proposed Methodology

The essential aim of this paper is planning and building up a procedure for prediction of depression in tech workers using Stacking Classifiers. The proposed procedure implies 4 important steps, such as preprocessing, feature importance, selection, and classification. Firstly, preprocessing of the information is done. Next, feature importance is performed with the Xgboost Algorithm. Next, feature selection is done using the Random Forest. The proposed classifier, called Stacking Classifier, provides the prediction of depression. The method we proposed appears as the architectural view of the modules in the proposed framework (shown in Fig. 1). Python is used to carry out this execution. Then, it is evaluated in terms of accuracy, precision, *f*1 score,

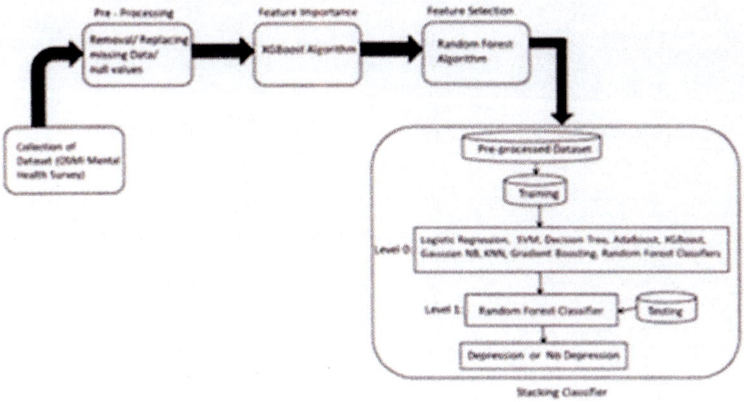

Fig. 1 Proposed technique–block diagram

recall to show the performance. This will be contrasted with the works that already exist.

3.1 Dataset

Dataset is gathered from a survey of tech workers. Open Sourcing Mental Illness is a non-benefit enterprise devoted to bringing issues to light, teaching, and giving assets to help mental health in the tech and open-source networks. OSMI Mental Health Survey Data (2018–2020) is taken along with the survey conducted by us by taking up similar questions (Narayanrao and Kumari (2020). Total dataset is of size 125 Kb. In the dataset missing data is taken care of by overlooking the highlights with over 80% of missing information.

3.2 Features

There are 25 features in total which include the following questionnaire as (shown in Fig. 2).

3.3 Preprocessing

Both the surveys have been combined to form the actual dataset. In this missing data, null data is handled.

country: Which country are you from?
state: Which state are you from?
age: What is your age?
gender: What is your gender?
family_history : Do you have a Family member who has a history of Mental illness?
treatment : Have you ever been diagnosed with a Mental health disorder?
work_interfere : Do you feel it interferes with your work if you have a mental health condition?
self_employed : Are you self employed?
no_of_employees : No. of employees in your organisation
tech_company :Is your employer primarily a tech company / organisation?
benefits : Does your employer provide Mental health benefits as part of healthcare coverage?
care_options : Do you know the alternatives for Mental medical care accessible under your manager given wellbeing inclusion?
wellness_program : Has your employer at any point talked about emotional well-being as a component of a health program for your workers?
seek_help : Does your employer give assets to become familiar with issues of emotional well-being and how to look for help?
anonymity : If you decide to exploit their wellsprings of emotional wellness or substance misuse, is your anonymity secured?
leave : How simple is taking clinical leave for a state of psychological well-being for you?
mental_health_consequence : Have you at any point talked about your psychological well-being with your employer?
physical_health_consequence : Overall, what amount of significance does your manager put on actual wellbeing?
coworkers : Would you feel great talking about an emotional wellness issue with your associates?
supervisor : Would you feel great talking about an emotional wellness issue with your direct supervisor(s)?
mental_health_interview : Would you bring up your Mental health with a potential employer in an interview
phys_health_interview :Would you be willing to bring up a physical health issue with a potential employer in an interview?
mental_health : Overall, how much importance does your employer place on Mental health?
obs_consequence : If you have revealed a Mental health disorder to a coworker or employee, has this ever impacted you or the relationship?

Outcome feature:
depressionLabel : Do you currently have a Mental health disorder?

Fig. 2 Features in the dataset

3.4 *Feature Importance*

The features significant for forecasting the depression are chosen with assistance of the Xgboost (shown in Fig. 3), advantage is that after the boosted trees are constructed, it gets the significant scores for every individual attribute. For the most part, it gives a f-score that demonstrates how valuable every attribute in the model. The most important attribute will have a high f-score (Narayanrao and Kumari 2020).

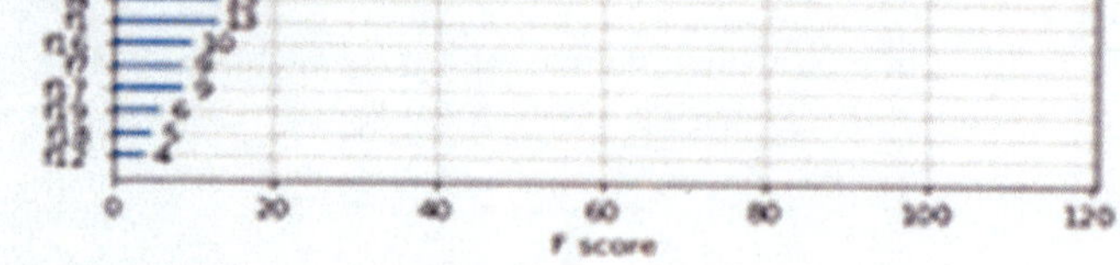

Fig. 3 F-score values of attributes using XGBoost

3.5 Feature Selection

Feature importance scores which have been collected from above step are considered and with the help of Random Forest Classifier we select the features which are significant for the prediction of depression, the benefit is that it is the tree-based methodology utilized by random forests normally positions by how well they improve the purity of the node. Thus, by pruning trees, we can get a set of most significant features.

3.6 Model Selection

The choice of the model is important, we use ML algorithms to obtain the best results. The managed method is utilized as a groundwork for an assortment of I/O pairs and to test their association pattern (Narayanrao and Kumari 2020). Dataset has 22 Dependent variables, and a nonmainstream variable is the aim or performance variable, that decides if patients have depression. Multiple classification algorithms are utilized in this project. Classifiers are a direction-based learning that helps in gaining knowledge. This is afterward used to recognize current discoveries. The dataset has been examined utilizing the accompanying classifiers:

(1) Stacking Classifier (2) Ada Boost (3) Gradient Boosting (4) Gaussian NB (5) SVM (6) KNN (7) Random Forest (8) Decision Tree (9) XGBoost (10) Logistic Regression.

Stacking Classifier

Stacking is a technique of ensemble learning which combines various models of classification via a meta-classifier. Every single model is trained on the complete training dataset; then, the meta-classifier fits accordingly on the O/P features of every single model in the ensemble. The purpose in stacking is to investigate a space of various models for a similar problem. The thought is that one can tackle a learning problem with various sorts of models which are fit for learning some part of the problem, however not the entire space of it. Thus, we can assemble numerous various learners and use them to fabricate an intermediary forecast, one prediction for each learned model. Then at that point we add a new model which learns from the intermediate predictions the same target. This final model is said to be stacked on the top of the others, hence the name. The advantages of using a Stacking Classifier is that it helps improve model performance and creates a robust model by reducing variance by combining the predictions of multiple models. The drawbacks are that it requires more memory and can take significantly longer to train than simpler models. Generating predictions using this might be slower and more computationally expensive.

Algorithm

1. Take the Survey dataset and perform preprocessing of the dataset
2. Utilizing Xgboost Algorithm get the importance levels of features
3. Utilizing Random Forest Algorithms, get the component significance and take out the less significant highlights

    3.1 Use RandomForestClassifier() to select the features
    3.2 fit(X, y) into the model

4. Take the Preprocessed dataset and X as input and y as output

    4.1 Create the pipeline
        4.1.1 Create the feature Union with the transformations with

        MinMaxScalr(), StandardScaler(), RobustScaler(), QuantileTransformer(n_quantiles = 100, output_distribution = 'normal'), KBinsDiscretizer(n_bins = 10, encode = 'ordinal', strategy = 'uniform'), PCA(n_components = 7), TruncatedSVD(n_components = 7)
        4.1.2 Create feature Selection using RandomForest as estimator using RFE()
    4.2 Train the StackingClassifier model with Training dataset(X)
        4.2.1 StackingClassifier Function, i.e., StackingClassifier() with following
            4.2.1.1 In this function fixing following values for the parameters estimators = level0, final_estimator = level1, cv = 5

            Where,

            level0 = [LogisticRegression(max_iter = 500), KNeighborsClassifier(), DecisionTreeClassifier(), SVC(), GaussianNB(), XGBClassifier(), AdaBoostClassifier(), GradientBoostingClassifier(), ExtraTreesClassifier(), LinearDiscriminantAnalysis(), QuadraticDiscriminantAnalysis(), HistGradientBoostingClassifier(max_iter = 100), RandomForestClassifier(n_estimators = 100, random_state = 0)]

            level1 = RandomForestClassifier(n_estimators = 100, random_state = 0).
            4.2.1.2 Putting away the arrival esteem into a variable
    4.3 Validate the prepared model with tenfold cross-validation
    4.4 Print the accuracy

The model is built using a Stacking Classifier, in which the preprocessed data is given to the model and the pipeline is created with the transformations and feature selection and this is given to that Stacking Classifier as accuracy 90.052%, recall 0.947, precision 0.915, $f1$score 0.904.

Ada-boost or Adaptive Boosting: It associates various classifiers so as to increment the accuracy of classifiers and is an iteration-based ensemble method. It frames a powerful classifier by associating multiple below-par performing classifiers to get a high veracity model. The results obtained as accuracy 75.1%, recall 0.726, precision 0.820, $f1$ score 0.735.

Gradient Boosting: It develops a prediction model in the form of a set of weak models. The model is built stage-wise same as other boosting methods, and it hypothesizes by granting enhancement of random differentiable loss functions. The results obtained as accuracy 80%, recall 0.765, precision 0.901, $f1$ score 0.790.

Gaussian Naive Bayes: Models that assign class labels to issue instances, communicated as vectors of factor esteems, where these class names are taken from some limited set. There isn't a single algorithm for instructing such classifiers, but a tribe of algorithms based on a typical principle: all Naive Bayes classifiers infer that a particular feature's value is sovereign of other features value, given the class variable. The results obtained as accuracy 51.6%, recall 0.997, precision 0.508, $f1$ score 0.692.

***K*-Nearest Neighbors (*K*NN)**: *K*NN is a non-parametric and sluggish learning algorithm. Accuracy is proportional to quality of the data. Using huge data, the slower the prediction. Requires more memory and becomes weak on having irrelevant features. The results obtained as accuracy 73.3%, recall 0.647, precision 0.756, $f1$ score 0.708.

Support Vector Machine (SVM): In this, we plot every information item as a point in n-dimensional space (where $n =$ no. of features) with the worth of each element as a specific coordinate. We classify by drawing the hyper-plane that separates 2 classes (Sailaja et al. 2021). The results obtained as accuracy 50.8%, recall 0.998, precision 0.502, $f1$score 0.668.

Random Forest Classifier: It is a supervised, flexible, simple to utilize algorithm. More trees, then it is more robust. It makes decision trees on arbitrarily chosen data samples, gets predictions from each tree and with help of voting it selects the best solution. The results obtained as accuracy 85.9%, recall 0.791, precision 0.760, $f1$score 0.843.

XGBoost Classifier: XGBoost is a gradient boosting to improve speed and performance. It is bound to overfit than bagging does (i.e., random forest) but with a powerful enough dataset and moderate hyperparameters The results obtained as accuracy 78.8%, recall 0.741, precision 0.805, $f1$score 0.771.

Logistic Regression: To discover the likelihood of success and failure events. Utilized when the dependent variable is paired. If the no. of observations are less than the no. of features it may provoke overfitting. The results obtained as accuracy 75.1%, recall 0.593, precision 0.718, $f1$score 0.670.

Decision Tree: Decision tree resolves the problem of ML by transforming the data into a tree structure representation. A slight change in the data can cause an immense change in the development of the decision tree causing shakiness. The results obtained as accuracy 81.7%, recall 0.735, precision 0.881, $f1$score 0.798.

4 Experimental Results and Discussion

Simulation results of the proposed model for the prediction of depression is discussed in this section. The classifier's performance is assessed utilizing different evaluation metrics.

4.1 Simulation Setup

The classifier is run in PYTHON 3.6 on a computer with an Intel I7 processor and 16 GB of RAM running Windows 10 operating system.

4.2 Evaluation Metrics

In this paper, models are assessed on accuracy, recall, precision, and $f1$ score as metrics. Evaluation metrics are as follows:

Accuracy = Number of Correct Predictions / Total Number of Predictions.

Precision = TP/(TP + FP) where TP = True Positive, FP = False Positive.

Recall = TP/(TP + FN) where TP = True Positive, FN = False Negative.

***F*1** = (2 * precision * recall) / (precision + recall).

4.3 Comparative Analysis

Represents the relative analysis of the different models for the prediction of depression at workplaces in tech workers.

The above outcomes show various models like Stacking Classifier, AdaBoost-Classifier, Gaussian NB, GradientBoosting, *K*NN, SVM, Logistic Regression, XGBoost, Decision Tree and RandomForestClassifier gives the accuracies as 90%, 75.1%,51.6%, 80%, 73.3%, 50.8%, 75.1%, 78.8%, 81.7%, and 85.6%, respectively (shown in Fig. 4). Comparison of recall versus precision also given as shown in Fig. 5. At last, the outcomes showed that the Stacking Classifier gives the best accuracy in prediction of depression in techies at workplaces.

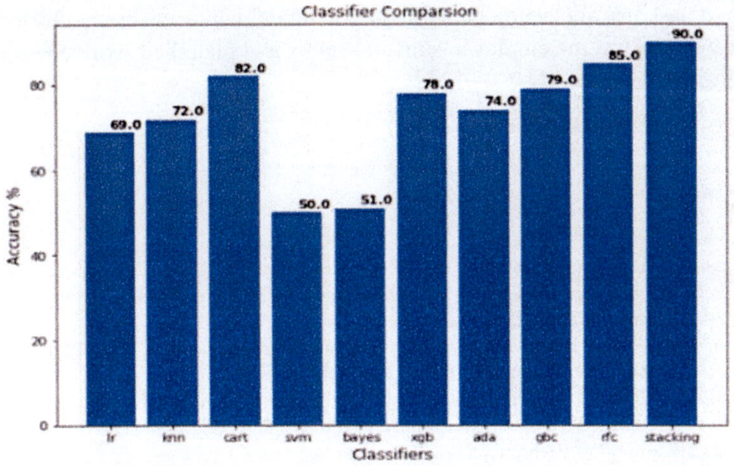

Fig. 4 Accuracy in bar graph

Fig. 5 Recall versus precision comparison

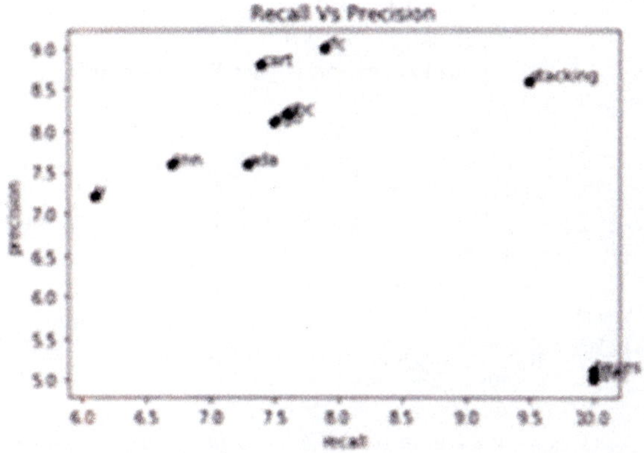

5 Conclusion

Depression is described as a mood disorder. It also can very well be understood as sensations of sadness, misery, anger or outrage that messes with a person's regular life. Science and innovation is changing at a quick speed that it is getting hard for the experts to stay informed concerning the forthcoming innovation alongside the everyday tasks of the work environment. IT experts experience a ton of stress, nervousness, wretchedness, and dejection relating to their workplace and uncover sensations of deficiency, bringing down confidence and disappointment, which brings about friendly, conjugal and medical conditions. This paper has presented an overview about depression, its causes and symptoms. This paper also presents the prediction of depression among IT professionals at early and beginning stages with

Stacking Classifier to develop great prediction model which enables software organizations to analyze the employee's mental status and plan their work environment accordingly.

References

M.M. Aldarwish, H.F. Ahmad, Predicting depression levels using social media posts, in *2017 IEEE 13th International Symposium on Autonomous Decentralized System (ISADS)* (Bangkok, 2017), pp. 277–280. https://doi.org/10.1109/ISADS.2017.41

A. Bender, P. Farvolden, Depression and the workplace: a progress report. Curr. Psychiatry Rep. **10**, 73–79 (2008). https://doi.org/10.1007/s11920-008-0013-6

S.-E. Cho, Z.W. Geem, K.-S. Na, Prediction of depression among medical check-ups of 433,190 patients: a nationwide population-based study. Psychiatry Res. **293**, 113474 (2020). ISSN 0165-1781. https://doi.org/10.1016/j.psychres.2020.113474

L. Jena, N.K. Kamila, A model for prediction of human depression using apriori algorithm, in *2014 International Conference on Information Technology* (Bhubaneswar, 2014), pp. 240–244. https://doi.org/10.1109/ICIT.2014.65

S.N. Kasthurirathne et al., Identification of patients in need of advanced care for depression using data extracted from a statewide health information exchange: a machine learning approach. J. Med. Int. Res. **21**(7), e13809 (2019)

P.U.S. Katti et al., Screening depression in IT industry using machine learning. Int. J. Progress. Res. Sci. Eng. **1**(5), 85–88 (2020)

V. Laijawala et al.: Classification algorithms based mental health prediction using data mining, in *2020 5th International Conference on Communication and Electronics Systems (ICCES)* (2020), pp 1174–1178

K.-S. Na, S.-E. Cho, Z.W. Geem, Y.-K. Kim, Predicting future onset of depression among community dwelling adults in the Republic of Korea using a machine learning algorithm, Neurosci. Lett. **721**, 134804 (2020). ISSN 0304-3940. https://doi.org/10.1016/j.neulet.2020.134804

P.V. Narayanrao, P.L.S. Kumari, Analysis of machine learning algorithms for predicting depression, in *2020 International Conference on Computer Science, Engineering and Applications (ICCSEA)* (Gunupur, India, 2020), pp. 1–4. https://doi.org/10.1109/ICCSEA49143.2020.9132963

U.S. Reddy, A.V. Thota, A. Dharun, Machine learning techniques for stress prediction in working employees, in *2018 IEEE International Conference on Computational Intelligence and Computing Research (ICCIC)* (Madurai, India, 2018), pp. 1–4. https://doi.org/10.1109/ICCIC.2018.8782395

N.V. Sailaja, M. Yelamarthi, Y.H. Chandana, P. Karadi, S. Yedla, Early detection of sepsis on clinical data using multi-layer perceptron, in C.K. Mai, A.B. Reddy, K.S. Raju (eds.) *Machine Learning Technologies and Applications. Algorithms for Intelligent Systems* (Springer, Singapore, 2021). https://doi.org/10.1007/978-981-33-4046-6_22

P. Sandhya, M. Kantesaria, Prediction of mental disorder for employees in IT industry. Int. J. Innov. Technol. Explor. Eng. (IJITEE) **8**(6S) (2019). ISSN: 2278-3075

A. Sau, I. Bhakta, Screening of anxiety and depression among seafarers using machine learning technology. Inf. Med. Unlock. **16**, 100228 (2019). ISSN 2352-9148. https://doi.org/10.1016/j.imu.2019.100228

J. Shanthalakshmi Revathy, N. Uma Maheswari, S. Sasikala, A model for predicting human depression using machine learning algorithm. Int. J. Innov. Technol. Explor. Eng. (IJITEE) **9**(3S) (2020). ISSN: 2278-3075. https://doi.org/10.35940/ijitee.C1086.0193S20

Experimental Validation of Mesa Sine Wave in Stock Price Prediction

Soumajit Chatterjee, Sukriti Adhikary, Debasmita Chakraborty, Niharika Sarkar, and Diganta Sengupta

Abstract Rising dissemination of learning algorithms in almost all spheres of life has been witnessed in the last 5 years. In this regard, stock market has provided a huge landscape for data science to introduce computational intelligence in otherwise traditional method of handling the global economy. This study evaluates whether Lead Mesa Sine Wave (LMSW) can be a good marker in stock price prediction. Our results reflect that LMSW cannot be used for stock price prediction. We validate our result using learning algorithms. Moreover we have also observed that the future price prediction using historical closing price data can be used as a dependable marker. Due to the time scale nature of the data, we have used recurrent neural network [specifically Long Short-Term Memory (LSTM)] for our prediction model design. The results from the prediction model exhibit better performance with respect to literary counterparts. We have used three publicly available datasets from Reliance, Infosys, and Grasim for the study. We claim two results in this study—LMSW cannot be used as stock price predictor, and LSTM can be used as a good predictor using historical closing price data.

S. Chatterjee · S. Adhikary · D. Chakraborty · N. Sarkar · D. Sengupta (✉)
Department of Computer Science and Engineering, Meghnad Saha Institute of Technology, Kolkata 700150, India
e-mail: sg.diganta@ieee.org

S. Chatterjee
e-mail: soumajit_c.cse2019@msit.edu.in

S. Adhikary
e-mail: sukriti_a.cse2019@msit.edu.in

D. Chakraborty
e-mail: debasmita_c.cse2019@msit.edu.in

N. Sarkar
e-mail: niharika_s.cse2019@msit.edu.in

D. Sengupta
Department of Computer Science and Business Systems, Meghnad Saha Institute of Technology, Kolkata 700150, India

A. B. Reddy et al. (eds.), *Proceedings of Third International Conference on Advances in Computer Engineering and Communication Systems*, Lecture Notes in Networks and Systems 612, https://doi.org/10.1007/978-981-19-9228-5_13

Keywords Mesa sine wave · LSTM · Stock price prediction · Learning algorithms · Closing price

1 Introduction

Stock market prediction has been a topic of computational interest for almost two decades. Technological advancements have recently added fuel to the global interest with loads of scope being witnessed in the use of machine learning (ML) and deep learning (DL) frameworks for stock analysis (Sharma et al. 2017; Kleina et al. 2020). One of the most popular research questions (RQ) being which learning algorithm suits best in determining the future trend of a stock. Certain proposals have been observed in literature which furnish new frameworks or improvised algorithms to deal with the RQ. In this study, we try to address the RQ using Lead Mesa Sine Wave (LMSW). Although LMSW being a technical indicator basically depicts the nature of the market whether the market exhibits cycle mode (CM) or trend mode (TM) (RTMath, Division of Deltix Inc. 2022), our quest for better alternatives for RQ initiated the present study. CM or TM is decided using two sine plots, judging the sinusoidal mode or wander mode of the plots. Readers are directed to RTMath, Division of Deltix Inc. (2022), Ameritrade (2022), and ForexRealm (2022) for complete understanding of the utilization of LMSW. The equations governing the LMSW as a CM or TM analyzer are provided in Eqs. (1)–(5).

$$\mathrm{Real}_k = \sum_{i=0}^{n-1}\left(\cos\left(\frac{360 * i}{n}\right) * \mathrm{close}_k\right) \tag{1}$$

$$\mathrm{Imag}_k = \sum_{i=0}^{n-1}\left(\sin\left(\frac{360 * i}{n}\right) * \mathrm{close}_k\right) \tag{2}$$

$$\mathrm{Phase}_k = \mathrm{atan}\left(\frac{\mathrm{Real}_k}{\mathrm{Imag}_k}\right) + 90 \tag{3}$$

$$\begin{aligned}
&\text{If } \mathrm{Real}_k > 0.001 \ \ \mathrm{Phase}_k = \ \arctan\left(\mathrm{Imag}_k/\mathrm{Real}_k\right)\\
&\text{else if } \mathrm{Imag}_k < 0 \text{ then } \mathrm{Phase}_k = \left(\frac{360^\circ}{2}\right) * -1\\
&\text{else } \mathrm{Phase}_k = \left(\frac{360^\circ}{2}\right) * 1\\
&\text{If } \mathrm{Real}_k < 0: \ \mathrm{Phase}_k = \mathrm{Phase}_k + 180^\circ\\
&\text{If}(\mathrm{Phase}_k < 0): \ \mathrm{Phase}_k = \mathrm{Phase}_k + 360^\circ\\
&\text{If}(\mathrm{Phase}_k > 360^\circ): \ \mathrm{Phase}_k = \mathrm{Phase}_k - 360^\circ
\end{aligned}$$

$$\mathrm{Sine}_k = \sin(\mathrm{Phase}_k) \tag{4}$$

$$\text{LeadSine}_k = \sin(\text{Phase}_k + 45) \tag{5}$$

where n = time during which the indicator is calculated. We would like to mention that to the best of our knowledge, study is the first study in experimental validation of LMSW for stock price prediction. The second part of the study involves Stock Market Prediction (SMP) using DL framework. Since the data happens to be a time series data, hence we used Recurrent Neural Network (RNN) for SMP. Minutely, we used the LSTM model for SMP.

The rest of the paper is organized as follows. Section 2 presents the existing notable literature in this domain. Our experimental analysis of LMSW viability for stock price prediction followed by proposed prediction model using LSTM, along with the experimental results is presented in Sect. 3. Section 4 concludes the paper.

2 Literature Survey

Use of artificial intelligence for SMP can be observed in the last four to five years. This has further led to the debate as to which algorithm suits better. In this regard, Wang et al. (2018) have presented a brief study. Certain amount of notable literature exists in use of ML and DL frameworks for proper stock price prediction. A comprehensive survey of such approaches can be observed in Sharma et al. (2017) and Kleina et al. (2020). Also the work by Nabipour et al. (2020) provide a comparative analysis of such frameworks with independent attention to use case application of the frameworks, or algorithms. Use of DL for SMP can be observed in Nabipour et al. (2020), Gao et al. (2020), Ingle and Deshmukh (2021), and Koratamaddi et al. (2021). Since the historical data in case of stock market is often time series data, considerable amount of RNN have been used till date, notably the studies in Idrees et al. (2019) and Li et al. (2019). One of the frequently used stock momentum indicators is the moving average convergence divergence (MACD) (Wan and Kim 2018). Analysis between two important factors in SMP lies in the momentum guided trend and the moving averages (Zakamulin and Giner 2020). In this study the authors have claimed that the moving average rule fares better in comparison to the momentum rule in SMP. Usually there two approaches in SMP, short term and long term. Shen and Shafiq (2020) have proposed a DL framework for short term SMP in which they have used an LSTM model in the penultimate step. In another study, Li et al. (2021) have done SMP taking into account unexpected incidents that can affect the trend.

3 Proposed Work

This study focuses on two parts. The first one is whether LMSW can predict future prices of stocks, and how well LSTM (DL RNN framework) can work in SMP. The next two sub-sections detail each of the two parts.

3.1 Experimental Validation of Lead Mesa Sine Wave for Stock Market Prediction

For the experimental validation, three stock datasets have been used from Reliance, Infosys, and Grasim. The data instances in each of the dataset are as follows: Grasim = 4903, Infosys = 6598, and Reliance = 6598. The view of the data fetched is presented in Table 1. Using Eqs. (4) and (5), the sine and the LMSW values are calculated and presented in Table 2 as per the algorithm presented in AL01. In this regard, we have taken a period of 2, and a phase angle of 45°. It can be observed that the values for LMSW lies in the range [−1, 1].

Table 1 View of the raw data

Date	Close	Date	Close
1-1-2020	100	7-1-2020	130
2-1-2020	105	8-1-2020	135
3-1-2020	110	9-1-2020	140
4-1-2020	115	10-1-2020	145
5-1-2020	120	…	…
6-1-2020	125		

Table 2 View of the data for sine and LMSW

Date	Close	Sine	LMSW
1-1-2020	100	–	–
2-1-2020	105	–	–
3-1-2020	110	0.78	– 0.88
4-1-2020	115	0.19	0.67
5-1-2020	120	– 0.90	0.96
6-1-2020	125	– 0.49	– 0.56
7-1-2020	130	0.84	– 0.86
8-1-2020	135	– 0.99	0.23
9-1-2020	140	0.71	– 0.52
10-1-2020	145	0.90	0.14
…	…	…	…

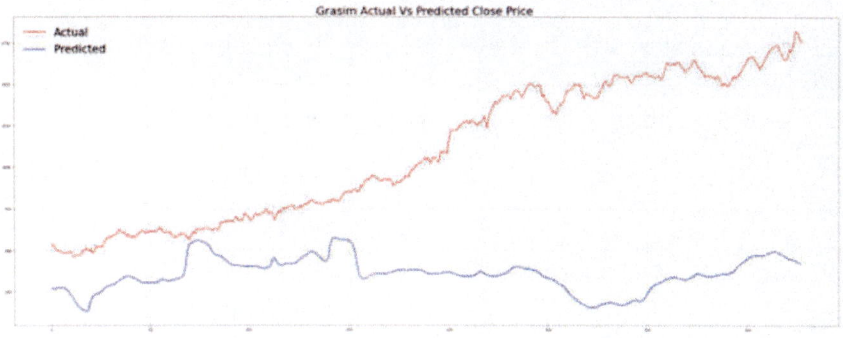

Fig. 1 Actual versus LMSW predicted price for Grasim using LMSW

AL01 Mesa Sine and Lead Sine

```
Define libraries
Begin Load_closing_prices_and_period_value
        Begin Iteration_over_closing_prices
              Calculate RealPart_and_ImagPart_for_closing_price_after_period
                Calculate Phase_from_RealPart_and_ImaginaryPart
                Check Phase_valid_or_not:   If not valid: Update Phase
                End Check
               Calculate Sine_from_Phase && Calculate Leadsine_from_Phase
          End Begin
End Begin
```

The prediction model used the data presented in Table 2 and generated the results as shown in Figs. 1, 2, and 3. The Red line and the Blue line in Figs. 1, 2, and 3 represent the actual and the predicted values, respectively. It can be observed from the three figures that the prediction of future stock prices by far don't follow the actual prices. Hence, we conclude that the LMSW does not serve as a trend indicator in predicting future stock prices.

3.2 *Stock Price Prediction using Historical Closing Values*

In this section, we propose a SMP framework based on LSTM. The block diagram for LSTM is presented in Fig. 4 with Fig. 5 presenting an LSTM memory cell. The complete workflow is presented in Fig. 6.

The choice of LSTM framework lies in the fact that the data is a time series data, and RNN is the best approach to deal with such data as has been observed technically and historically. As we have already observed that the LMSW values range [−1.1], hence the closing values were scaled in the range [0, 1]. The curated dataset thus obtained was split into three parts for training, validation, and testing purpose. The

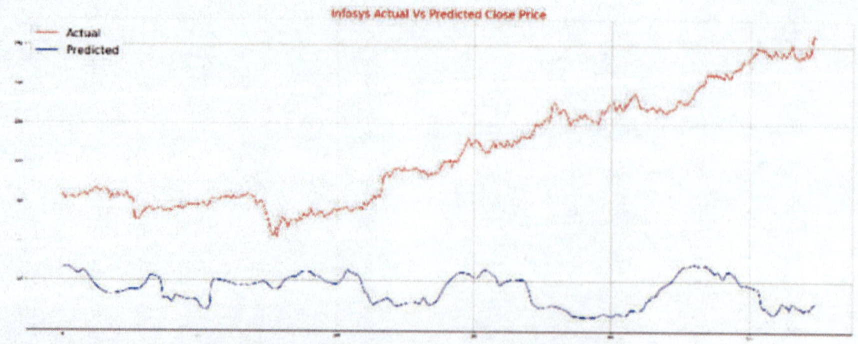

Fig. 2 Actual versus LMSW predicted price for Infosys using LMSW

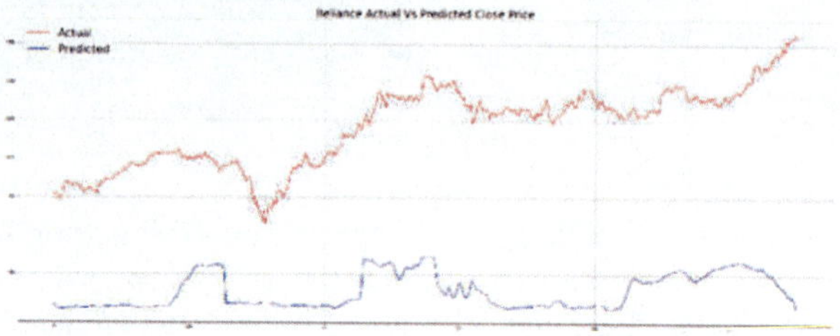

Fig. 3 Actual vs. LMSW predicted price for Reliance using LMSW

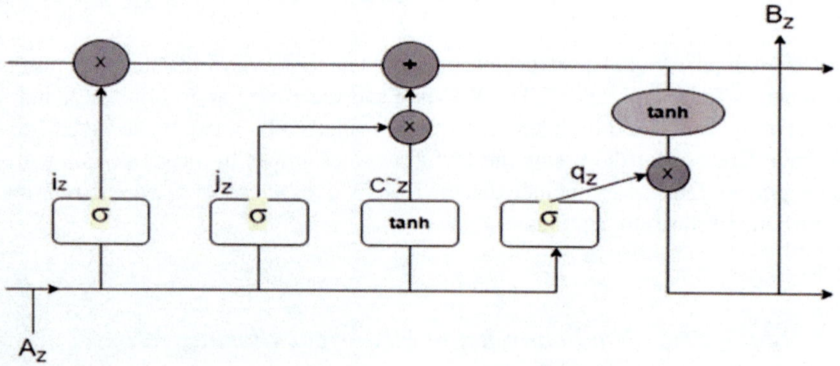

Fig. 4 Block diagram of LSTM

ratio of train:validate:test was 0.7:0.2:0.1. Thereafter a timestamp was created based on which the dataset was designed for training the LSTM model, i.e., if the timestamp = 4, then the first 4 closing prices as input, and the 5th closing price as the output.

$$\text{if(timestamp} == 4) : \text{input} = \text{CP}_i \forall i \in [1, 4] \ \&\& \ \text{output} = \text{CP}_5$$

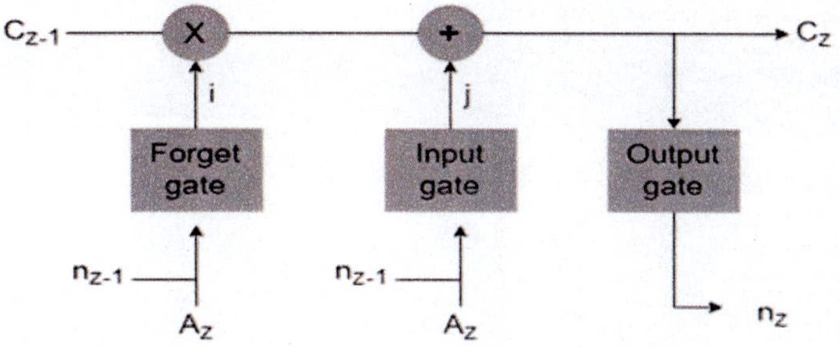

Fig. 5 LSTM memory cell

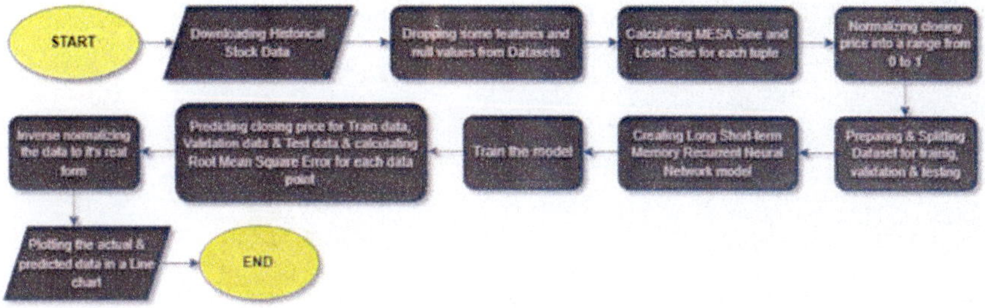

Fig. 6 Workflow diagram

$$\text{if(timestamp} == 5) : \text{input} = \text{CP}_i \forall i \in [1, 5] \;\&\&\; \text{output} = \text{CP}_6 \text{ and so on} \ldots$$

where $\text{CP}_i = i$th Closing Price. Therefore the training dataset thus generated is presented in Table 3. Figures 6, 7, and 8 present the actual and the predicted values with respect to our proposal. The figure itself depicts the accurateness of the model.

From Figs. 7, 8, and 9, it can be observed that the predicted values closely follow the actual value thereby validating the effectiveness of the model for all the three datasets. Further tuning of the model can generate even better predictions.

Table 3 Training dataset

100	105	110	115	120
105	110	115	120	125
110	115	120	125	130
115	120	125	130	135
120	125	130	135	140
125	130	135	140	145
…	…	…	…	

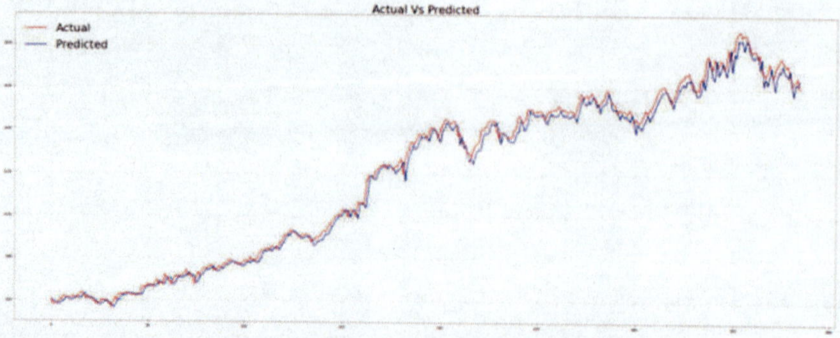

Fig. 7 Actual versus LMSW predicted price for Grasim using closing price

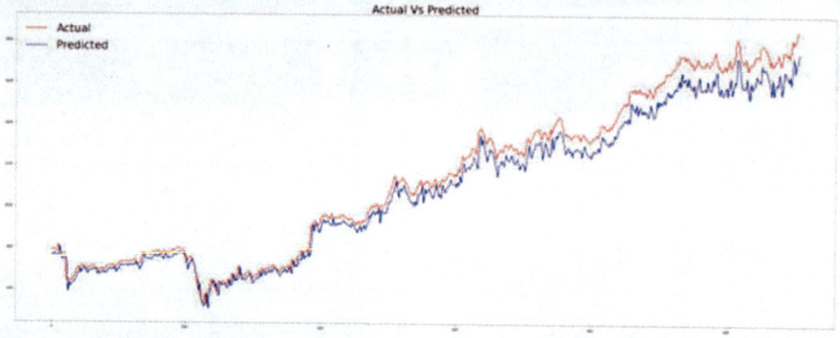

Fig. 8 Actual versus LMSW predicted price for Infosys using closing price

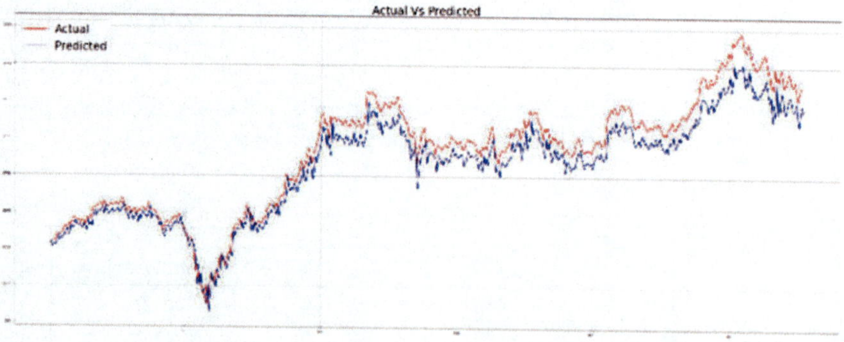

Fig. 9 Actual versus LMSW predicted price for Reliance using closing price

4 Conclusion

Although Lead Mesa Sine Wave is a pair of sine waves used for predicting whether the stock is in cyclic mode or trend mode, this study experimentally establishes the fact that Lead Mesa Sine Wave cannot be used as an indicator for future stock price prediction. Moreover prediction can be made much more efficiently using closing prices using Recurrent Neural Networks, typically LSTM. It has been observed that the predicted results using closing prices closely follow the actual price using the basic LSTM model.

The work in this study can be extended further to design more efficient networks/models using deep learning for future stock price prediction. Moreover, tuning the proposed model can be done to experimentally check if LSTM serves as a better model for prediction post tuning.

References

Ameritrade, MESASineWave (2022) [Online]. https://tlc.thinkorswim.com/center/reference/Tech-Indicators/studies-library/M-N/MESASineWave

ForexRealm, Mesa Sine Wave (2022) [Online]. https://www.forexrealm.com/technical-analysis/technical-indicators/mesa-sine-wave.html

P. Gao, R. Zhang, X. Yang, The application of stock index price prediction with neural network. Math. Comput. Appl. 1–16 (2020)

S.M. Idrees, M. Afshar Alam, P. Agarwal, A prediction approach for stock market volatility based on time series data. IEEE Access 17287–17298 (2019)

V. Ingle, S. Deshmukh, Ensemble deep learning framework for stock market data prediction (EDLF-DP). Global Transit. Proc. 47–66 (2021)

M. Kleina, F. Oscar Drozda, M. Augusto Mendes Marques, S. Teixeira Zavadzki, Computational intelligence techniques used for stock market prediction: a systematic review. IEEE Latin Am. Trans. 744–755 (2020)

P. Koratamaddi, K. Wadhwani, M. Gupta, S.G. Sanjeevi, Market sentiment-aware deep reinforcement learning approach for stock portfolio allocation. Eng. Sci. Technol. Int. J. 848–859 (2021)

P. Li, L. Zhang, M. Wen, Y. Chen, Stock market trend prediction using high-order information of time series. IEEE Access 28299–28308 (2019)

Z. Li, H. Zhang, S. Lyu, T. Jiang, One step ahead: a framework for detecting unexpected incidents and predicting the stock markets. IEEE Access 30292–30305 (2021)

M. Nabipour et al., Deep learning for stock market prediction. Entropy 1–23 (2020)

M. Nabipour, P. Nayyeri, H. Jabani, S. Shahab, A. Mosavi, Predicting stock market trends using machine learning and deep learning algorithms via continuous and binary data; a comparative analysis. IEEE Access 150199–150212 (2020)

RTMath, Division of Deltix Inc., MESASine Wave (2022) [Online]. https://rtmath.net/assets/docs/finanalysis/html/e1749c57-2542-4c8e-821f-b48ed8e0213e.htm

A. Sharma, D. Bhuriya, U. Singh, Survey of stock market prediction using machine learning approach, in *International Conference on Electronics, Communication and Aerospace Technology*, Indore (2017), pp. 506–509

J. Shen, M.O. Shafiq, Short-term stock market price trend prediction using a comprehensive deep learning system. J. Big Data 2–33 (2020)

J. Wan, J. Kim, Predicting stock price trend using MACD optimized by historical volatility. Math. Probl. Eng. **2018**, 1–12 (2018)

M. Wang et al., Which artificial intelligence algorithm better predicts the Chinese stock market. IEEE Access 48625–48633 (2018)

V. Zakamulin, J. Giner, Trend following with momentum versus moving averages: a tale of differences. Quant. Finance **20**(6), 985–1007 (2020)

Testbed Implementation of MAX LEACH Routing Protocol and Sinkhole Attack in WSN

J. Suman, K. Shyamala, G. Roja, and N. Pranay

Abstract The WSN is a deployment of small sensors which are powered by a small battery. These sensors sense their surroundings according to the application and transfer the data to the BS due to which the energy of these sensors is dissipated. The residual energy of the nodes gradually decreases and this effects the Lifetime of the network. To increase the performance and Lifetime of the WSN, a MAX LEACH Routing protocol was proposed. There are chances of security attacks externally like wear and tear of the nodes and theft of the nodes which are deployed in remote places. But more than that due to the presence of the malicious nodes in the network causes Internal attacks and more damage than the external security attacks. The WSNs are vulnerable to various types of security attacks in every layer. Out of them the most dangerous attack is the sinkhole attack. In this paper we implement the TESTBED of the MAX LEACH Routing protocol and also the Sinkhole attack on the MAX LEACH Routing protocol in the WSN.

Keywords WSN · Routing protocol · Residual energy · Security attacks · Compromised nodes · Sinkhole · MAX LEACH

1 Introduction

The WSN have sensor nodes that are powered by a small battery and they are not rechargeable. Depending upon the application these nodes are deployed in remote areas. The nodes in the network sense the surroundings and transfer the information to the BS with the help of the nearby nodes in the network. Energy is depleted when the nodes sense the surroundings and communicate between them and the BS. Another problem in these sensor networks is due to the malicious nodes in the network. They

J. Suman (✉)
CSE Department, B V Raju Institute of Technology, Narsapur, Medak, Telangana, India
e-mail: sumojon@gmail.com

K. Shyamala · G. Roja · N. Pranay
CSE Department, UCOE, Osmania University, Hyderabad, Telangana, India

A. B. Reddy et al. (eds.), *Proceedings of Third International Conference on Advances in Computer Engineering and Communication Systems*, Lecture Notes in Networks and Systems 612, https://doi.org/10.1007/978-981-19-9228-5_14

try to attack the network and snip the valuable information and also they disrupt the network completely. WSN is vulnerable to various attacks (Akyildiz et al. 2002). These attacks try to adopt new strategies depending upon the uniqueness and various characteristics of WSN. This paper is concentrated on Testbed Implementation of MAX LEACH Routing Protocol and the sinkhole attack in the MAX LEACH Routing protocol in the Network Layer. The paper is organized as follows; Sect. 2 gives the related work of MAX LEACH protocol and sinkhole attack in WSN. Section 3 gives the proposed work of Testbed Implementation of MAX LEACH routing protocol. Section 4 gives the Results and Analysis. Section 5 is the Sinkhole attack in the MAX LEACH protocol in WSN. Section 6 is the conclusion and future work of the proposed work.

2 Related Work

The authors in Suman et al. (2021) have proposed a MAX_LEACH Routing protocol to enhance the network lifetime in WSN. Initially there are 2 phases in this protocol. They are setup phase and steady state phase. In the setup phase, once the nodes are deployed randomly in the network according to the application. The first round is implemented based on the LEACH protocol. In the steady state phase the sensors sense their surroundings and transfer the data to the CH and from there to the BS. After the first round, the residual energy is calculated for all the nodes. Then the clusters are formed. The node which has high residual energy will be chosen as a CH for that particular round. This continues until the nodes are dead in the network.

2.1 Sinkhole Attack

Sinkhole attack in WSN is an active type of attack. The compromised node in the network acts as a sinkhole and tries to attract all the nodes towards it. This is done because the compromised node projects itself as a nearest node to the sink and sends the messages to all the nodes in the network. Once the nodes have sent their data to this compromised node, then it starts the attack by manipulating the data or dropping of the data which is to be sent to the base station (Fig. 1).

In Krontiris et al. (2008), authors have studied about the sinkhole attack in WSN and various solutions provided to detect the attack. A rule-based approach is used here to detect the sinkhole attack. Two rules were formed and implanted in Intrusion detection system (IDS). When any node violates the rules, the Intrusion detection system triggers an alarm but it does not provide node ID of compromised node. The first rule "for each overhead route update packet the ID of the sender must be different your node ID". The second rule "for each overhead route update packet the ID of the sender must be one of the nodes ID in your neighbours".

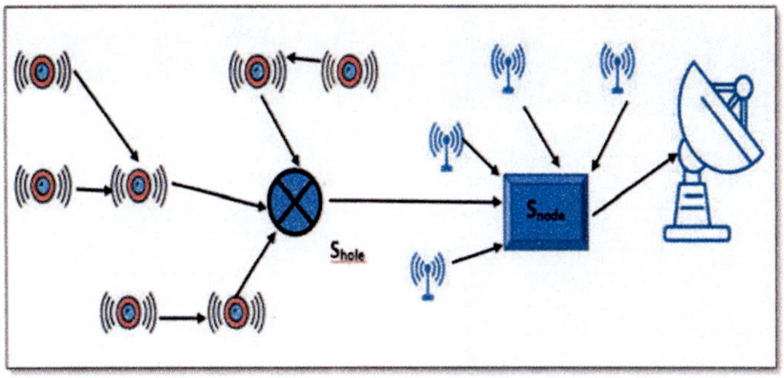

Fig. 1 Sinkhole attack in WSN

In Teng and Zhang (2010), authors have studied the sinkhole attack in TinyAODV protocol for WSN. They have considered the Hop count to the BS as a metric. The compromised node will send the RREP packets to their neighbouring node with less no of HOP count to attract the nodes. They have evaluated the delay per hop for every node which is in the path and found the malicious node in the network.

In Iqbal et al. (2014), authors have proposed an Intrusion detection system to find the intruder in the WSN using the LEACH protocol for routing purpose. Based on the packets received and transmitted they have calculated the Intrusion ratio. This value shows either the normal or malicious activity in the network. So when there is any malicious activity, the agent will alert the system so that the data transmission will be stopped. They have used TETCOS NETSIM for the simulation purpose.

In Jahandoust and Ghassemi (2017), authors have proposed an algorithm called "Hop Count-Based Detection Scheme for Sinkhole Attack (HCODESSA)". They have compared the same with another algorithm called Ibrahim. The performance is evaluated based on various parameters like the Detection probability, false positive rate and false negative rate. The simulation shows that the proposed algorithm has outperformed the Ibrahim algorithm. Sinkhole attack effects were also studied using ratios of packet delivery and packet loss which show that they were adverse effect on the performance of the network.

3 Testbed Implementation of MAX LEACH Routing Protocol

In MAX LEACH routing protocol, the nodes with the highest Residual Energy are chosen as CH in the network. In this manner the Network Stability and Network lifetime is improved to 50% (Suman et al. 2021).

$$R = \left(\frac{P}{1 - P \cdot \mathrm{mod}\left(r, \mathrm{round}\left(\frac{1}{p}\right)\right)} \right) \cdot (S(i) \cdot (\mathrm{Max\_Energy})) \tag{1}$$

where p stands for random probability, r stand for number of rounds, $S(i)$ is the remaining energy; Max_Energy is the total/highest energy of node. Energy dissipations are calculated by below equation:

$$\begin{aligned} S1(i) \cdot E = S1(i) \cdot E - (\mathrm{ETX} * (4000) + \mathrm{Emp} * 4000 \\ * (\mathrm{min\_dis} * \mathrm{min\_dis} * \mathrm{min\_dis} * \mathrm{min\_dis})) \end{aligned} \tag{2}$$

where ETX stand for transference energy and Min_dis is the minimum distance from the member node to CH.

In Fig. 2, the MAX LEACH Routing protocol is shown. There are 2 clusters with nodes N-1, N-2, N-3, N-4 and N-5 in cluster-1 and N-6, N-7, N-8, N-9 and N-10 in cluster-2. The Node N-5 in cluster-1 and the node N-8 in cluster-2 whose residual energy is high are chosen as CH for the second round. For Testbed implementation we have considered the Radio frequency model of IEEE 802.15.4 from the SENSEnuts. We have considered 10 coordinator nodes and one PAN coordinator node which act as the Base station/Sink. The pan coordinator node is only the radio module and doesn't have any Temperature and Light Sensors (TL). This pan coordinator node which acts as a Base station (BS/Sink) is connected directly to the computer and is powered continuously as shown in Fig. 4.

Figure 3 shows the coordinator node that is fixed with the Temperature and Light Sensors (TL) and is powered by two AA Batteries. Once the setup of nodes is done. The pan coordinator node and the coordinator nodes are flashed with the respective codes using the SENSEnuts Gateway module. For this purpose the code is written in the "C" language and to generate the binary file which is to be flashed on to the nodes is done using the make command. The binary file is flashed using the Device

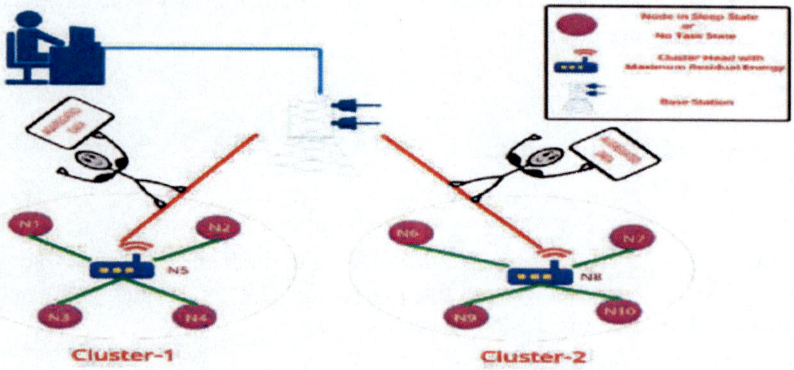

Fig. 2 MAX LEACH routing protocol in WSN (Guntuku et al. 2021)

Fig. 3 Coordinator node with the TL sensor from SENSEnuts

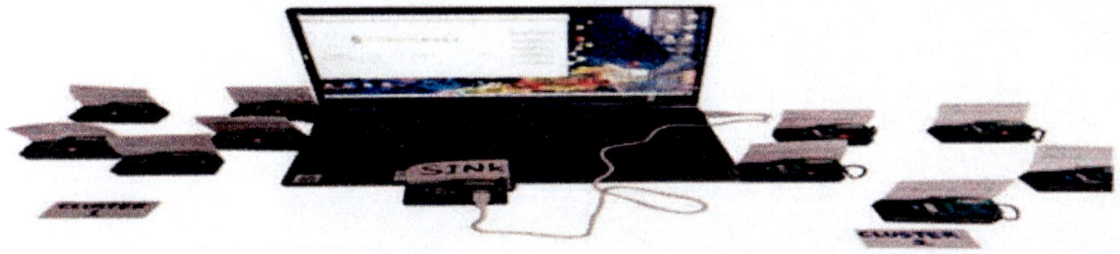

Fig. 4 Testbed setup of SENSEnuts pan coordinator and coordinator

Programmer in the GUI of SENSEnuts and are switched on as shown in Fig. 4. The pan coordinator node that is the sink node is directly connected to the computer for uninterrupted power supply.

4 Results and Analysis

In Fig. 5, five nodes are connected in a cluster-1 in round-1. It is clear that, the node-5 with the MAC ID: ODB0 is chosen as CH randomly in the round-1. The other nodes—1, 2, 3 and 4 will send their collected data to node-5 according to the TDMA schedule. Then the node-5 consolidates the data and sends it to the BS. This setup and steady state phase takes place in one round. The non-glowing node-5 is the CH node and the glowing nodes are the member nodes in that particular round.

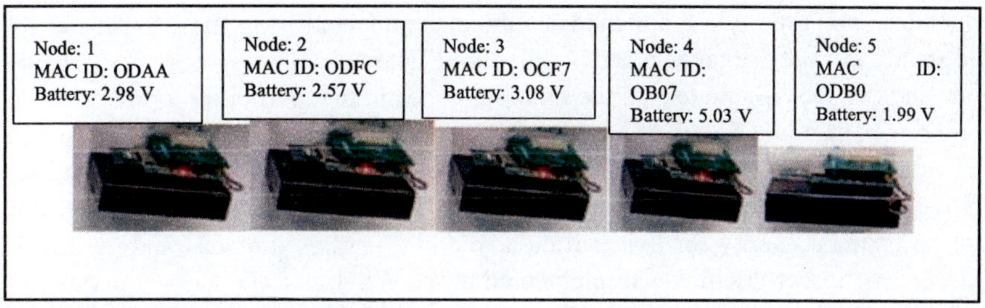

Fig. 5 Formation of cluster-1 in round-1

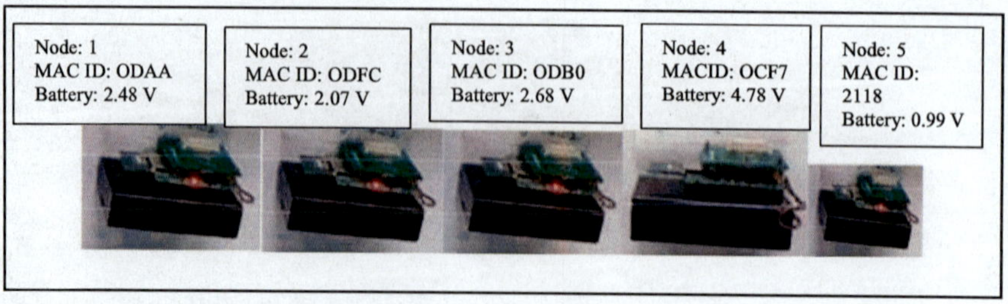

Fig. 6 Formation of cluster-1 in round-2

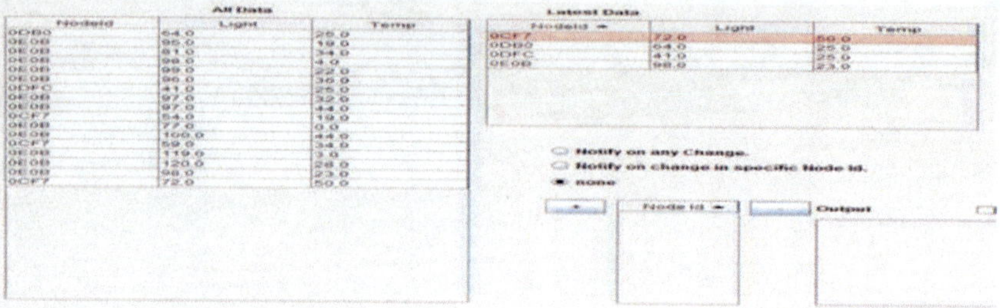

Fig. 7 Sense live data for cluster-1

In Fig. 6, the node 4 with MAC ID—OCF7 is chosen as CH for the second round. The other nodes 1, 2, 3 and 5 with MAC ID's–ODAA, ODFC, ODB0 and 2118 are the member nodes for that particular round-2. This shows that the node-4 with highest residual energy for that particular round-2 is chosen as CH and the MAX LEACH routing protocol was implemented in the WSN.

In Fig. 7, the Temperature and the Light data from the sensor nodes is received in the Sense Live module in the SENSEnuts Graphical User Interface for cluster-1 initially.

In Fig. 8, five nodes—6, 7, 8, 9 and 10 with MAC ID's–ODE3, OD98, OE08, ODB0 and OCED are connected in a cluster-2 in round-1. It is clear that, the node-8 with the MAC ID: OE08 is chosen as CH randomly in the round-1. The other nodes–6, 7, 9 and 10 will send their collected data to node-8 according to the TDMA schedule. Then the node-8 consolidates the data and sends it to the BS. This setup and steady state phase takes place in one round. The non-glowing node-8 is the CH node and the glowing nodes are the member nodes in that particular round.

In Fig. 9, the node-6 with MAC ID—ODE3 is chosen as CH for the second round. The other nodes 7, 8, 9 and 10 with MAC ID's—OD98, OE08, ODB0 and OCED are the member nodes for that particular round-2. This shows that the node-6 with highest residual energy for that particular round-2 is chosen as CH and the MAX LEACH routing protocol was implemented in the WSN.

In Fig. 10, the Temperature and the Light data from the sensor nodes is received in the Sense Live module in the SENSEnuts Graphical User Interface for cluster-2.

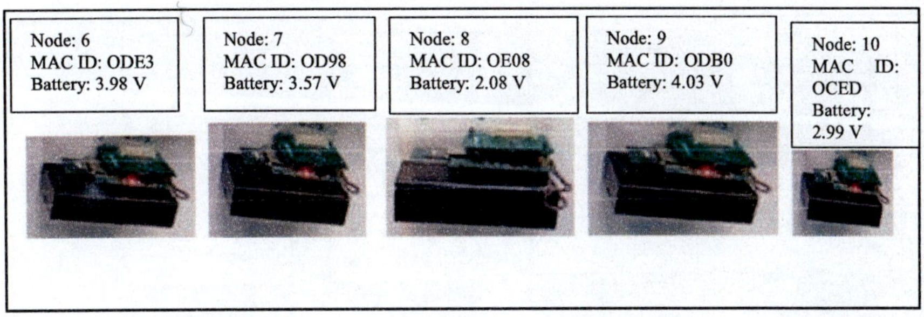

Fig. 8 Formation of cluster-2 in round-1

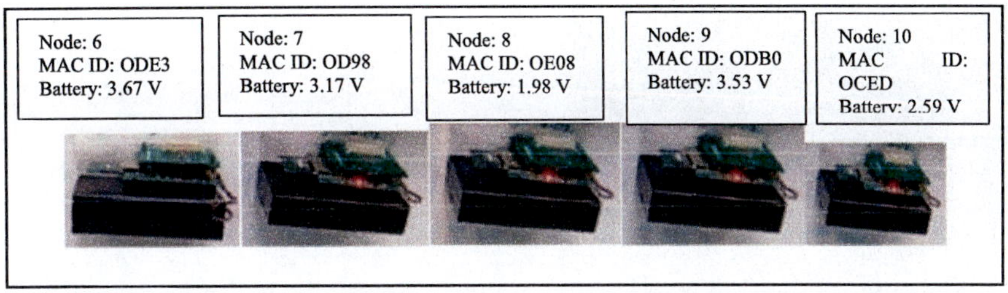

Fig. 9 Formation of cluster-2 in round-2

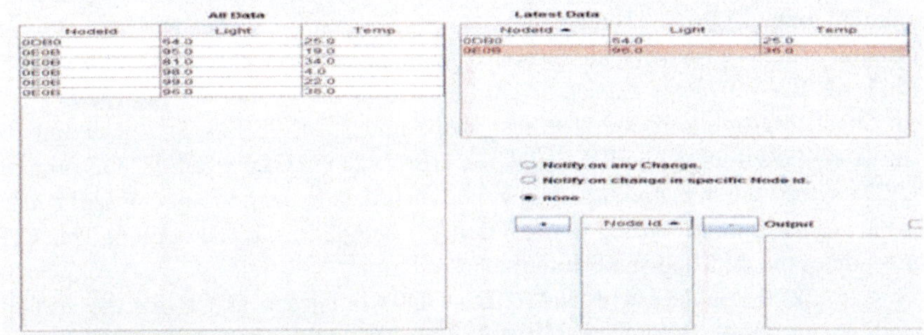

Fig. 10 Sense live data for cluster-2

5 Testbed Implementation of Sinkhole Attack in MAX LEACH

In the proposed work, the sinkhole attack in MAX LEACH routing protocol in WSN is implemented. All the nodes from 1 to 10 are flashed with the Coordinator code using SENSEnuts Graphical User Interface in the Devise Programmer module. In the same manner the PAN Coordinator node is also flashed with the code. Another node-11 with the MAC ID: 31D7 is flashed with the sinkhole attack code. Once this

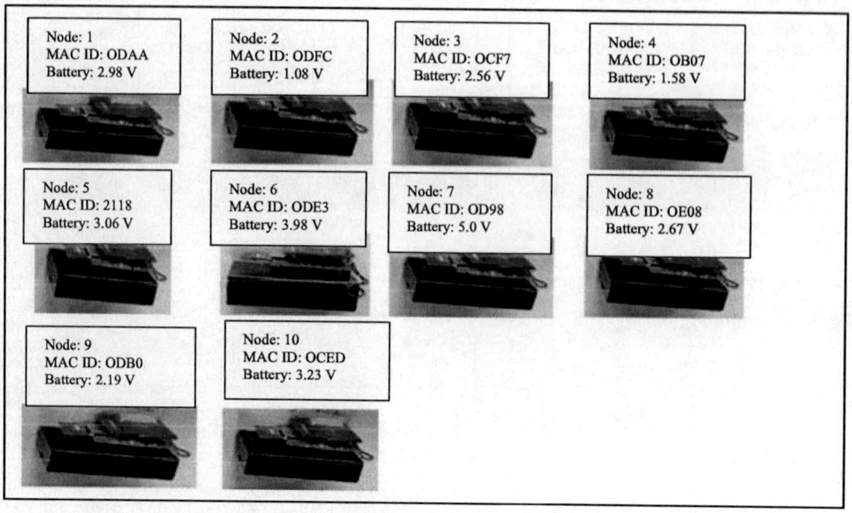

Fig. 11 Formation of cluster in round-1 before sink hole attack

flashing of the codes is done. The setup is done. Now all the nodes are switched on except the node-11 that is the sinkhole attack node. We can observe that the node-6 with MAC ID—ODE3 is chosen as CH for the round-1. The rest of the nodes are the member nodes for that round and they send their data to the CH. This happens according to the TDMA schedule.

In Fig. 11, we can see that all the nodes—1 to 10 are formed as cluster and node-6 with MAC ID—ODE3 is chosen as cluster head as its residual energy (3.98 V) is high for that particular round. The rest of the nodes—1, 2, 3, 4, 5, 7, 8, 9 and 10 with MAC ID's—ODAA, ODFC, OCF7, OB07, 2118, OD98, OE08, ODB0 and OCED are the members nodes and they send their data to the node-6. Once all nodes have sent their data according to the TDMA schedule to the CH node-6. This CH consolidates the data and sends it to the base station.

In Fig. 12, the node-7 with MAC ID—OD98 is chosen as CH for the second round. As this node's residual energy (4.95 V) is high when compared to all the other nodes in the network.

In Fig. 13, the sinkhole node with MAC ID—31D7 is switched on in the network, though its residual energy (2.08 V) is not high when compared to other nodes residual energy, we can observe that the sinkhole node is chosen as CH for the round-3. This is because the sinkhole node project itself as the node with high residual energy and attracts all the other nodes towards it.

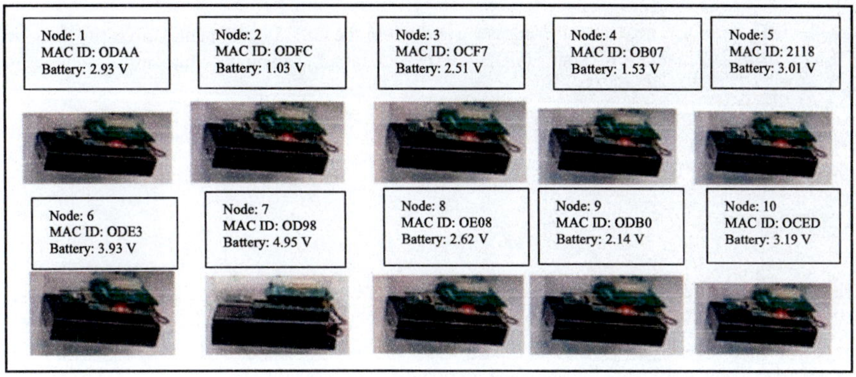

Fig. 12 Formation of cluster in round-2 before sink hole attack

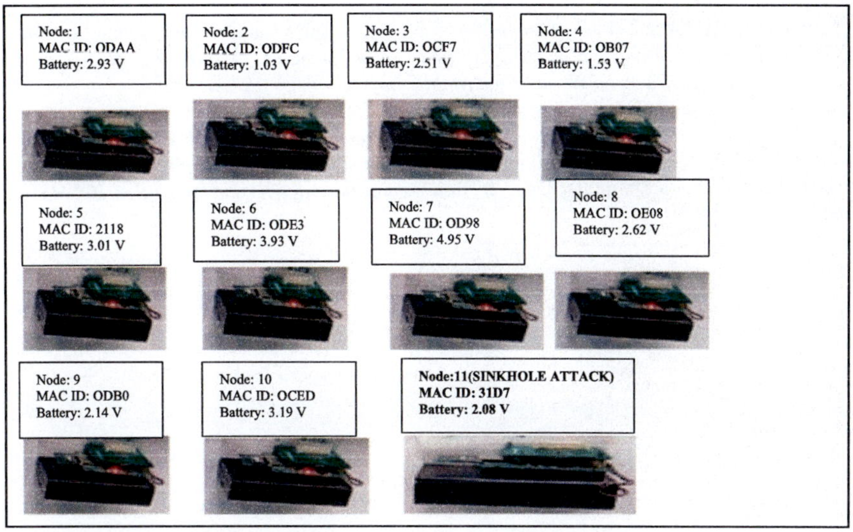

Fig. 13 Sink hole attack in the cluster

6 Conclusion and Future Work

In this paper, the Testbed Implementation of MAX LEACH Routing protocol and sinkhole attack in MAX LEACH protocol in WSN is done using 10 Radio Model of IEEE. 802.15.4 from the SENSEnuts. This work is in continuation of the work done in Suman et al. (2021) as we have proposed a MAX_LEACH Routing protocol to improve the Lifetime of the WSN's which is based on the maximum residual energy. The future work can be extended with more no of sensors on the testbed and observe the results.

Acknowledgements The research work is carried out in the CSE Department, University College of Engineering, Osmania University (UCE, OU), Hyderabad, Telangana State and is funded by DST-SERB (EEQ/2018/000623).

References

I.F. Akyildiz, W. Su, Y. Sankarasubramaniam, E. Cayirci, Wireless sensor networks: a survey. Comput. Netw. **38**(4), 393–422 (2002)

R. Guntuku, K. Shyamala, P. Nigam, Design and testbed implementation of MAX LEACH protocol using SENSEnuts, in *IEEE Bombay Section Signature Conference* (2021)

S. Iqbal, S.P. Aravind Srinivas, G. Sudharsan, S.S. Kashyap, Comparison of different attacks on LEACH protocol in WSN. Int. J. Electr. Electron. Data Commun. **8**(8), 16–19 (2014)

G. Jahandoust, F. Ghassemi, An adaptive sinkhole aware algorithm in wireless sensor networks. Ad Hoc Netw. (2017)

I. Krontiris, T. Dimitriou, T. Giannetsos, M. Mpasoukos, Intrusion detection sinkhole attacks in wireless sensor network, in *Networking and Communications, in 2008. WIMOB'08. IEEE International Conference on Wireless and Mobile Computing*, pp. 526–531 (2008)

J. Suman, K. Shyamala, G. Roja, Improving network lifetime in WSN based on maximum residual energy, in *2021 2nd International Conference for Emerging Technology (INCET)* (2021), pp. 1–5

L. Teng, Y. Zhang, Secure routing algorithm against sinkhole attack for mobile wireless sensor network, in *Second International Conference on Computer Modelling and Simulation, ICCMS'10*, vol. 4 (IEEE, 2010), pp. 79–82

An Adaptive Algorithm for Polysemous Words in Natural Language Processing

Chandrakant Kokane, Sachin Babar, and Parikshit Mahalle

Abstract In this research article the supervised adaptive approach for word sense disambiguation is discussed. The most critical and identified problem of natural language understanding is the lexical ambiguity. Lexical ambiguity is introduced by the polysemy words. This paper describes different state-of-the-art techniques to detect ambiguous words from the ambiguous sentence. The word embedding is an important phase of word sense disambiguation and which is succeeded by ambiguous word detection and processed by classification. The different embedding techniques are discussed here and the uniqueness of adaptive word embedding has been proved. There are two standard available datasets OMSTI and WordNet are used for the data processing. This article elaborates the newly generated dataset Adaptive-Lex for disambiguation. There are two important challenges in word sense disambiguation: identifying polysemy words without context information and constructing word embedding for polysemy words with highest sense values. These challenges are addressed with a complex network approach and adaptive word embedding technique. The performance evaluation classifiers are supervised neural network approaches such CNN and DNN.

Keywords Lexical ambiguity · Adaptive word embedding · Supervised learning · Word sense disambiguation

C. Kokane (✉)
Department of Computer Engineering, Smt. Kashibai College of Engineering, Affiliated to Savitribai Phule Pune University, Pune, MS, India
e-mail: cdkokane1992@gmail.com

S. Babar
Department of Computer Engineering, Sinhgad Institute of Technology, Pune, MS, India

P. Mahalle
Department of Artificial Intelligence and Data Science, Vishwakarma Institute of Information Technology, Pune, MS, India

A. B. Reddy et al. (eds.), *Proceedings of Third International Conference on Advances in Computer Engineering and Communication Systems*, Lecture Notes in Networks and Systems 612, https://doi.org/10.1007/978-981-19-9228-5_15

1 Introduction

Word sense disambiguation is a sub-domain of natural language processing which mainly concerns ambiguous words. The ambiguous words are also known as polysemy words. The word having more than one meaning is generally called as an ambiguous word.

Example: Statement-1 may I come in?
Statement-2 Let's meet in May.

In the above example both the statements containing "may" as an ambiguous word. The first statement "may" along with "?" indicates the person is asking for permission. In the second statement the "May" is the noun and its month. The motivation behind this research is the ambiguous words are the challenge in the field of information extraction and sentiment analysis. In information extraction if ambiguous query is processed by machine then there are hundred percent chances that the machine will generate ambiguous results (Kokane et al. 2020).

In Fig. 1, the architecture of natural language processing is categorized into two sections such as Natural Language Generation (NLG) and Natural Language Understanding (NLU). In NLU the machine understands the meaning of the given user query and accordingly the machine generates the results at NLG. The lexical ambiguity is the most crucial and identified problems at NLU (Kokane et al. 2020). There are three suggested machine learning approaches for word sense disambiguating as supervised machine learning, semi-supervised machine learning and unsupervised machine learning (Kokane and Babar 2019). In the proposed adaptive approach for word sense disambiguation is tested and validated with supervised machine learning model. The proposed adaptive approach is classified into adaptive training and intelligent testing. The goal of this research is to minimize the ambiguity in information extraction and machine translation.

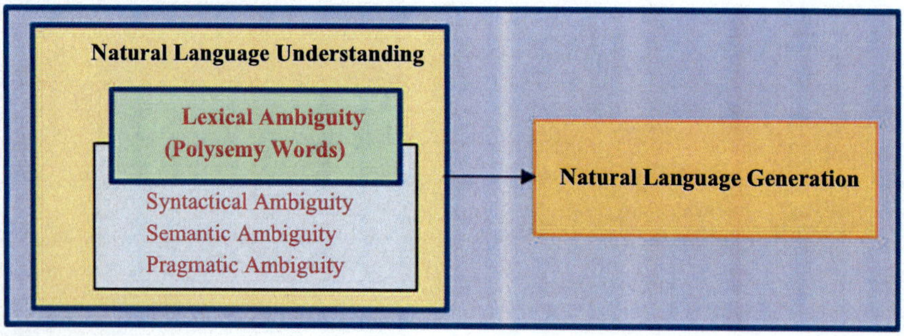

Fig. 1 Natural language processing

2 Related Work

The problem of word sense is effectively solved by the machine learning techniques. The review of available literature is carried out by considering the training and testing paradigms of supervised, unsupervised and semi-supervised machine learning techniques as.

2.1 Supervised Learning

The supervised algorithm for word sense disambiguation is discussed by Kokane et al. (2020). Authors have proposed deep neural network classifiers with normal word embedding. The word vector is further provided to deep neural network classifiers with seven hidden layers. The two standard datasets OMSTI and SemCor are used for the disambiguation. The accuracy of the discussed model is compared with deep neural network model. This approach generates superior performance for single polysemy words in the sentences (Kokane et al. 2020). The supervised recurrent neural network approach to address the problem of word sense disambiguation is proposed by Kokane and Babar (2019). Authors have introduced the unique technique for ambiguous word detection. The ambiguous sentence is tokenized and the sense key of every ambiguous word is retrieved from the lexical resource WordNet. The local generated data structure is processed with binary search technique for sense key matching. The close similar words of the ambiguous words are determined by using LSTM approach (Kokane and Babar 2019). The efficient model for single polysemy word detection is proposed by Rahman et al. (2019). Authors have used the freely available lexical resource WordNet for ambiguous word detection. The methodology is morally based on the available context information of the identified polysemy word (Rahman et al. 2019). The context information of every polysemy word is stored into a local generated information context table (ICT). This method is simple and in terms of training have less complexity. The methodology generates superior performance as compared with the Lex algorithm (Calvo et al. 2019; Iacobacci et al. 2016).

2.2 Unsupervised Learning

In this learning approach Gutiérrez et al. (2017) have spoken the issue of extracting the sense value of polysemy word. Authors have proposed page rank model to initialize sense values to the ambiguous words. The sense values of the ambiguous words are mapped with the help of freely available lexical resource WordNet that is machine readable dictionaries (Gutiérrez et al. 2017). The word sense disambiguation for regional Indian language proposed by the authors, Rao and Kallimani (2016). This discussed methodology is designed for regional Indian language. This research aims on lexical ambiguity mainly caused by the words having the same spelling and pronunciation but different meaning. In this approach like knowledge-based approach the ambiguous word can be matched with available standard dataset. This algorithm is applicable only for the sentences with multiple ambiguous words. The performance of this system is not up to the mark because linear searching is used for matching purposes (Rao and Kallimani 2016). The graph-based approach for polysemy words is proposed by Navigli and Lapata (2009) in this paper, authors have discussed unsupervised approach for word sense disambiguation. The discussed methodology there are two measures that are taken into the consideration as local measures and global measures. The graph is constructed first by considering the local and global features of ambiguous words. Authors have compared the performance of local verses global measures and found that system genres fine results for local measures. The best local measures that are affecting on the performance are Degree and PageRank of ambiguous words because these local measures are closely related (Sruthi Sankar et al. 2016; Navigli and Lapata 2009).

2.3 *Semi-supervised Learning*

The adaptive word sense disambiguation is proposed by Li et al. (2019). The discussed methodology is applicable for biomedical word sense disambiguation. The context embedding generated here based on Bi-LSTM gives superior results as compared to traditional approaches. The high quality context representation plays a vital role for getting more precise accuracy. Considering the performance of supervised and semi-supervised learning techniques the unlabeled tanning data gives superior performance for biomedical word sense disambiguation (Li et al. 2019).

2.4 *Challenges*

To identify ambiguous words from the context information is easy and lots of research is done on the same. But to disambiguate the ambitious words without any context information is the challenge identified.

Second challenge is the size of the word vector. The polysemy words with minimum sense values can be easily embedded into word vectors. The challenging task is generating word vectors for polysemy words having highest sense values like break having sense value "74".

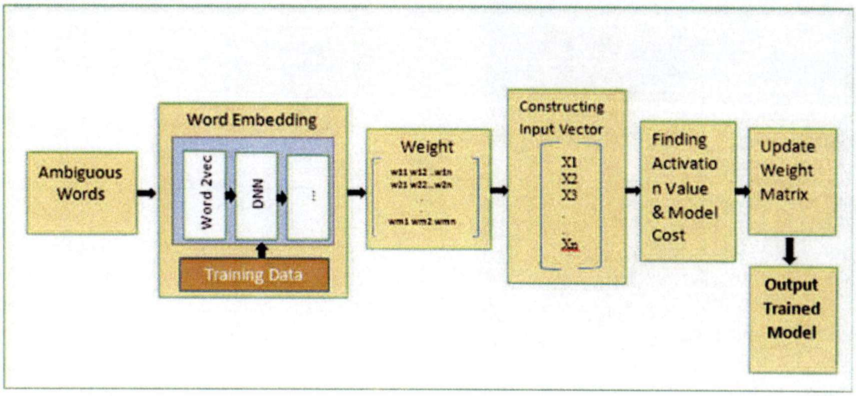

Fig. 2 Method of training

3 Methodology

The extreme computing environment is required for the processing; the OMSTI dataset because it requires more than 1.5 GB of storage on the local system and it is very difficult on the local system, so we processed the data with the help of Google Collaborators. For the same the OMSTI dataset is uploaded to Google cloud.

3.1 Training

The supervised learning approach is used for the training as shown in Fig. 2. The training is done for "528" ambitious words with respect to the sense values (Kokane and Babar 2019). Extracted-data files are formed for each polysemy word. The adaptive word embedding approach is used for constructing word vectors. At the time of tanning the semantic information is not considered because every ambiguous word is individually trained.

3.2 Testing

In the testing approach there should be successful classification of the ambiguous sentence as presented in Fig. 3. The input ambiguous user query or input text is taken from the user with the help of user_ui.py file. The user_ui.py is a python file generated graphical user interface for the testing module. After accepting input from the user the preprocessing.py file will perform all preprocessing tasks that are splitting sentence, punctuation removal, tokenization, stemming and lemmatization. After this step we will get the words the input sentence into the purest form, the

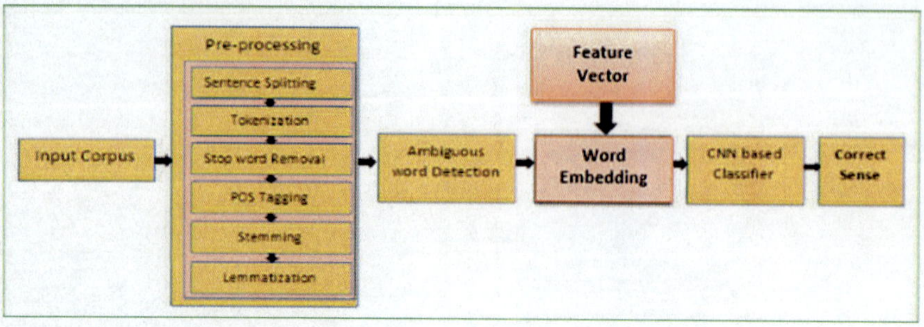

Fig. 3 Method of testing

purest form of words without any tense or suffix. After preprocessing the ambiguous words are detected with the help of WordNet (Kokane et al. 2020).

3.3 System Architecture

Figure 4 describes the system architecture of word sense disambiguation with convolutional neural network as a classifier. The ambiguous user query is accepted as an input here. The user query will be further pre-processed. The preprocessing includes sequence of tokenization, removing of stop words, stemming and lemmatization (Kokane et al. 2020). After preprocessing, the output will be the purest form of the input query from which we are going to identify the ambiguous words. The ambiguous word is detected by standard available lexical dataset WordNet. The word vector is constructed with respect to ambiguous word and the same is provided as input to CNN. CNN generates the closest sense of ambiguous word which is further mapped with WordNet to get exact meaning of polysemous word.

3.4 Construction of Word Embedding

The adaptive word embedding method is used while constructing the input vector with size 10 generates efficient input with cosine similarities addresses the second challenge. The input word vector is created for the ambiguous word with max_similarity and stored_context parameters. Table 1 shows the max_similarity of the ambiguous word break. Top ten max_similarity values are taken to the consideration here. The breaking and shatter are very close to the ambiguous word break that's why they are having higher values for max_similarity. The complex network graph of the max_similarity values with the closed vertex describing the highest max_similarity value that addressing the first challenge.

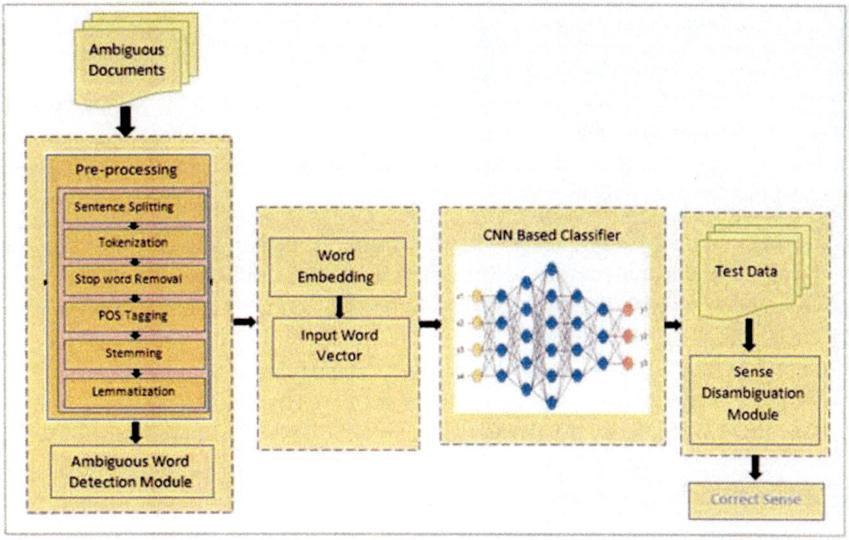

Fig. 4 System architecture

Table 1 Construction of word embedding for "break"

Breaking	0.678
Shatter	0.614
Broken	0.599
Smashed	0.557
Smashe	0.502
Smash	0.501
Crack	0.501
Tear	0.491
Push	0.486
Fall	0.476

4 Experimentation Details

In this section the experimentation details are elaborated along with the available state-of-the-art datasets. The experimentation is done with two available standard datasets such as: OMSTI, SemCor and Adaptive-Lex as.

4.1 Datasets

The performance of the adaptive word sense disambiguation model is compared with the existing deep neural network model. By considering the evaluation parameters, the performance of convolutional neural network model is far better for the OMSTI and adapted Lex datasets, the details of datasets as,

a. **OMSTI**: One Million Sense Tagged Instances is the huge dataset for the word sense disambiguation because it contains one million sense instances of ambiguous words. The size is very large and requires a high configuration environment for processing.
b. **SemCor**: The SemCor dataset is manually tagged and it is more accurate than OMSTI. All the resources of SemCor are annotated by using the lexical resource WordNet. WordNet is the freely available large lexical resource for the Word Sense Disambiguation.
c. **Adaptive-**Lex: The dataset is created by the authors for word sense disambiguation. The feature of this dataset is the ambitious words are represented with possible semantic similar values.

5 Result and Discussion

The novelty of this work is applying adaptive convolutional neural network models for classification with adaptive word embedding. The neural network model is constructed and tested for accuracy and the same is evaluated using the Keras neural network API. Before presenting a contrast of the experimental results with other state-of-the-art models, Table 2 provides the general idea about the results. The input statement is processed and provided some selective information about the polysemy words. Table 2 shows how different parameters of ambiguous words are taken into consideration.

Table 2 shows the stages of the adaptive CNN model. The input sentence is pre-processed and an ambiguous word is identified first. As shown in Table 2 the may

Table 2 Result of convolutional neural network model

Sentence	After preprocessing	Ambiguous word	Identified sense value
May I come in?	May I come in	"May"	10
Let's meet in may	Meet may	"May"	10
Bank need donor's	Bank need donor	"Bank"	18
I am going to bank	I go bank	"Bank"	18

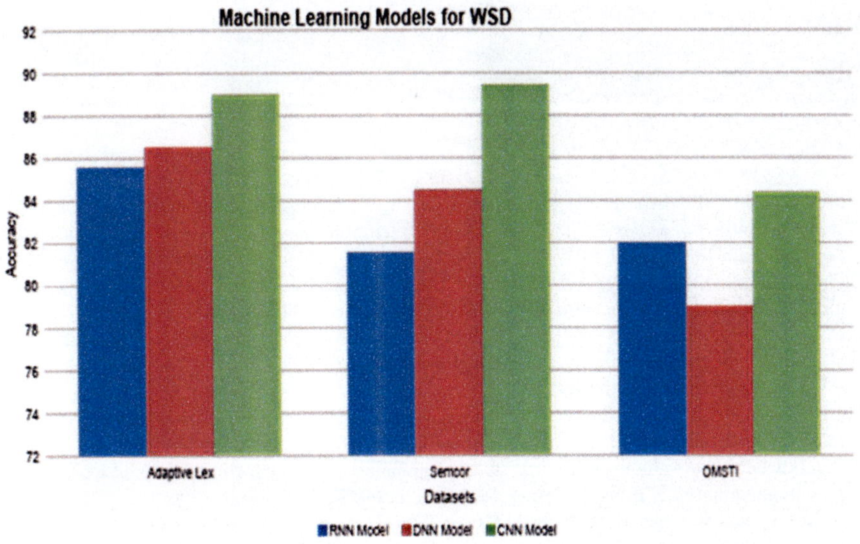

Fig. 5 Performance analysis of WDS

is an ambiguous word and as per the SemCor the sense value of the bank is "ten" (Fig. 5) (Calvo et al. 2019; Gutiérrez et al. 2017).

6 Conclusion and Future Work

This work has been presented with supervised convolutional neural networks as a classifier and tested with OMSTI, SemCor and Adaptive-Lex datasets. The discussed adaptive word embedding is inspired by the closest values of the words. The article also explores the utility of Adaptive-Lex for developing word embeddings for context-free polysemy words along with knowledge-based lexical resource (WordNet). The fact that the addition of the adaptive word embedding with complex network approach improves the results significantly. Another improvement that can contribute greatly is disambiguation of large documents instead of single sentences. The discussed complex network for representing polysemy words with respect to semantic similarities can be used for word sense disambiguation in large documents by representing every sentence as an isolated node. As shown in graph-1 the achieved average training accuracy of adaptive convolutional neural network models is 87.7%.

References

H. Calvo et al., Toward universal word sense disambiguation using deep neural networks. IEEE Access **7**, 60264–60275 (2019)

Y. Gutiérrez, S. Vázquez, A. Montoyo, Spreading semantic information by word sense disambiguation. Knowl.-Based Syst. **132**, 47–61 (2017). ISSN: 0950-7051

I. Iacobacci, M.T. Pilehvar, R. Navigli, Embedding for word sense disambiguation: an evaluation study, in *2016 Proceedings of the 54th Annual Meeting of the Association for Computational Linguistics* (2016), pp. 897–907

C.D. Kokane, S.D. Babar, Supervised word sense disambiguation with recurrent neural network model. Int. J. Eng. Adv. Technol. (IJEAT) **9**(2) (2019). ISSN: 2249-8958

C.D. Kokane, S.D. Babar, P.N. Mahalle, An adaptive algorithm for lexical ambiguity in word sense disambiguation, in *Proceeding of First Doctoral Symposium on Natural Computing Research: DSNCR 2020*, vol. 169 (Springer, 2021)

Z. Li, F. Yang, Y. Luo, Context embedding based on bi-LSTM in semi-supervised biomedical word sense disambiguation. IEEE Access **7**, 72928–72935 (2019)

R. Navigli, M. Lapata, An experimental study of graph connectivity for unsupervised word sense disambiguation. IEEE Trans. Pattern Anal. Mach. Intell. **32**(4), 678–692 (2009)

M.M. Rahman, S.A. Khan, K.M. Azharul Hasan, Word sense disambiguation by context detection, in *2019 4th International Conference on Electrical Information and Communication Technology (EICT)* (IEEE, 2019)

R. Rao, J.S. Kallimani, Analysis of polysemy words in Kannada sentences based on parts of speech, in *2016 International Conference on Advances in Computing, Communications and Informatics (ICACCI)* (IEEE, 2016)

K.P. Sruthi Sankar, P.C. Reghu Raj, V. Jayan, Unsupervised approach to word sense disambiguation in Malayalam. Procedia Technol. **24**, 1507–1513 (2016). ISSN: 2212-0173

CS-FA Nature Inspired Algorithm-Based Robust Video Watermarking

Srikanth Bethu, S. Bhargavi Latha, Suresh Kumar Kanaparthi, D. Abdus Subhahan, and G. Vani

Abstract This paper illustrates a hybrid algorithm that is amalgamation of two bio-inspired algorithms such as Cuckoo search and Firefly optimization algorithms are incorporated to find optimize scaling factor intended for the watermark insertion. This combination is considered as one of the utmost usages in the recent past. The foremost benefit is obtained by combining several features of both the algorithms. SSIM and BER are used to evaluate fitness function in this optimization technique. Contemplating above opportunities, a novel video watermarking method is proposed by using CS-FA in DWT, SVD domain. Besides, the watermark security is also improvised using secret sharing method. Experimental tests reveal that the video watermarking approach suggested has a reasonable imperceptibility and an improved robustness against attacks.

Keywords Optimization · Cuckoo search algorithm · Video watermarking · Firefly algorithm · Secret sharing

---

S. Bethu (✉)
Department of CSE, CVR College of Engineering, Hyderabad, Telangana 501510, India
e-mail: srikanthbethu@cvr.ac.in

Department of CSE, NIT Warangal, Warangal, Telangana 506004, India

S. Bhargavi Latha
Department of CSE, GRIET, Hyderabad, Telangana 500090, India

S. Kumar Kanaparthi
Department of CSE-DS, Malla Reddy University, Hyderabad, Telangana 500100, India
e-mail: drsureshk@mallareddyuniversity.ac.in

D. Abdus Subhahan
Department of CSE, BV Raju Institute of Technology, Hyderabad 501301, India

G. Vani
Department of CSE, Sreenidhi Institute of Science and Technology, Hyderabad 501301, India
e-mail: vanig@sreenidhi.edu.in

A. B. Reddy et al. (eds.), *Proceedings of Third International Conference on Advances in Computer Engineering and Communication Systems*, Lecture Notes in Networks and Systems 612, https://doi.org/10.1007/978-981-19-9228-5_16

1 Introduction

Robustness and Invisibility are the two core traits that must be considered for development of a watermark technique. Video watermarking is segregated in to two distinct types. The first being Spatial domain technique and the other one is Transform domain method. The embedding of the watermark is attained by tweaking the coefficients in the Transform Domain Technology. Over last few years, it's been observed that the following techniques such as DCT, DFT, DWT and matrix decompositions are often used in the video watermarking. Again, in the side, the common matrix decomposition will be encompassed of QR decomposition, (SVD) Singular decomposition and Schur decomposition. In the area of watermarking, there are both benefits and drawbacks of any specific decomposition and specific transform techniques. In case, if one transform is combined with another, then this combined transform will equilibrize the barriers of an individual transform and overall, this will boost the efficiency of the watermarking process. DWT watermarking strategies would have certain benefits including multi-resolution representation, improvised energy compression. On other side, the vulnerability to image processing attacks is very strong and whereas geometric attacks exhibit feeble robustness.

Matrix decomposition methods can also be used for watermarking in which the geometric properties are dig out from the given video-frame and these features used for the insertion of watermark. So, it provides more robustness against geometric attacks. In the similar way, in watermarking whenever the amalgamation the DWT and matrix decomposition are applied then it will take the benefits of the applied methods together and hence determines robustness on image processing and geometric attacks (Makbol et al. 2016; Bao and Ma 2005; Mehta et al. 2016; Jane and Elbasi 2014). Although such conventional methods deliver decent computing speed but lacks in balancing the robustness and invisibility automatically.

The steadiness among the two important parameters, robustness and imperceptibility is realized via optimization techniques. BPSO optimization algorithm is used in the implementation of video watermarking by Prathik et al. (2013). The optimal frames for incorporating the watermark are chosen using BPSO but the method needs to remember the embedding frame order sequence for the extraction purpose.

Genetic algorithm was used by Takore et al. (2016) to enhance robustness. Sake and Tirumala (2017) successfully embed the watermark using the bee colony technique as an optimizer. To calculate the scaling factor for embedding Aditya et al. (2017) leverages Cuckoo search algorithm and scene change detection is used the frames selection for framing.

Deb and Yang have developed Firefly and Cuckoo search (CS) metaheuristic algorithms (Yang and Deb 2009). The firefly algorithm is inspired to grab attention of other fireflies from their flashing behavior. The Cuckoo search algorithm is stirred from the postulation of restricted brood parasitism conduct of Cuckoo species that utilizes nest of other birds to lay eggs. It is inspired on the postulation of constrained brood parasitism Both FA and CS are very effective in providing the solutions to global optimization problems. Firefly algorithm takes less time to generate near optimum value when compared to Cuckoo search. In recent studies, it has been hypothesized that CS is more effective than FA. By considering the benefits of both algorithms, a novel hybrid algorithm CS-FA is proposed in Elkhechafi et al. (2018).

2 Materials and Methods

2.1 Discrete Wavelet Transforms

Owing the benefits of the DWT like De-noising, compression and many more, the usage of DWT was introduced into watermarking. DWT when compared with the other techniques such as DFT and DCT epitomizes any signal with a series of essential functions in a well manner and flexibly with the use of filter banks. DWT signifies a specified image with the sequence of wavelets in a multi-resolution pattern for effective analysis and also the signal may also be displayed or analyzed. A given image is split in to four sub-bands if it is applied to DWT.

2.2 Singular Value Decomposition (SVD)

It is a recognized, accepted and standard technique which involves diversified functions from resolving linear least-squares analysis (LSA) to calculation of pseudo matrix inverse. By converting a frame or image matrix into $U * S * V$, where the diagonal matrix (S) carries the image's singular values in decreasing order, SVD reduces the complexity of image or video processing.

2.3 Hybrid CS-FA Algorithm

The two algorithms, Cuckoo search and Firefly algorithm have demonstrated remarkable results in working out solutions for optimization complications (Kanagaraj et al. 2013). It's been noted from the recent past research that the CS illustrated better results that of FA in multi-modal optimization. Whereas FA surpass CS algorithm when time taken for the generation of optimum solution is taken into the consideration (Arora and Singh 2013). A hybrid algorithm CS-FA is developed which incorporates the reproduction strategy of CS and the attraction concept in FA as mentioned in Algorithm 1.

Algorithm 1 CS-FA Nature Inspired Algorithm

Algorithm: Pseudo-code CS-FA
Begin: Separate all the values into two groups: G1, G2

Initialize the populations of two groups G1 and G2
Calculate fitness value of each value
Repeat
    Do in Parallel
    Carry out CS operation in G1
    Carry out FA operation in G2
    End do in parallel
    Update the global best in both populations
    Combine the two groups together, then reorganize them into new groups at random: G1 and G2
    Calculate each particle's fitness value
Until a terminate-condition is satisfied
End: Post-Process results and visualization

2.4 *Secret Sharing*

By using secret sharing scheme (Hsieh et al. 2007), we can provide efficient security to the watermark where one pixel is converted in to two 2 * 2 shared blocks, one is talk about as Principal share (P-share) and Complementary share (C-share). Features of the original frame would be a base for the obtained share image. The frame selected is separating up into size of 32 * 32 blocks. Every 32 * 32 bock is applied to 3-level DWT results to four sub-bands of 4 * 4 size and SVD is used on all these sub-bands to accomplish singular values. Average of all sub-band's first singular values is determined. The feature type (*n*) would be dependent on the mean value. A P-share image is generated by using the watermark and feature share image. The 2 * 2 share block generation is represented by using the mapping Table 1. During decoding processing XOR operations are employed on the principle share image and this can be illustrated in Fig. 1. The redundant noise which is generated by the pixel expansion during the secret share generation phase can be corrected using watermark reduction process. The reduction process is explained in Fig. 1.

Table 1 Accuracy for the proposed method

Video	SSIM	BER	PSNR
Sports	0.9906	0.31	53.65
Nature	0.9978	0	55.92
Cartoon	0.9968	0	54.56
News	0.9956	0	55.63

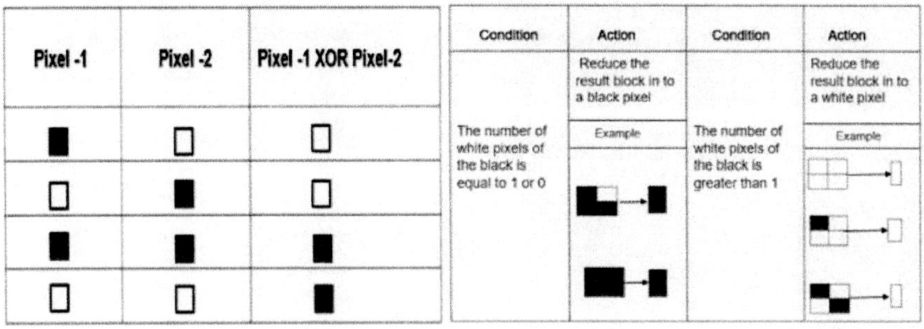

Fig. 1 XOR rule and watermark reduction scheme

3 Proposed Algorithm

The embedding phase consists of 3 stages and they are generation stage, discovering a proper scaling factor, P-share embedding stage. All of the watermark bits are placed into each and every frame of the video to demonstrate robustness against frame drop attacks (Fig. 2).

Due to its accessibility characteristics R Channel is chosen for the embedding phase. The entire embedding procedure is illustrated in Fig. 3. The replacement of

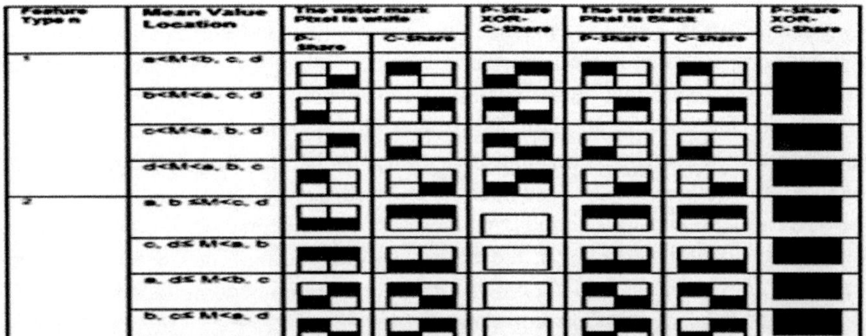

Fig. 2 Mapping table

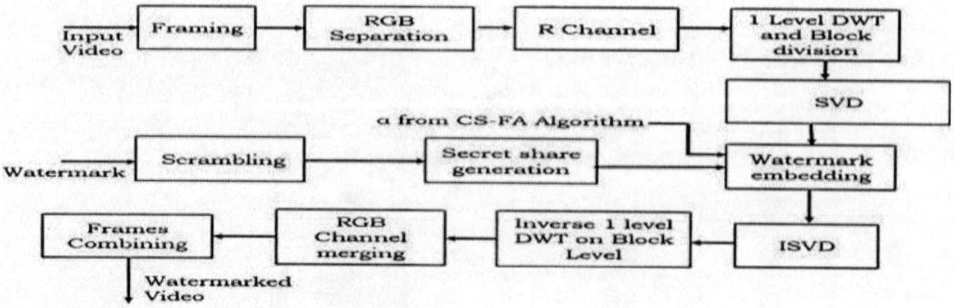

Fig. 3 Block diagram for watermarking embedding

singular values with watermark bits is used to replace singular components according to the formula below.

$$\mathrm{Sv}(1,1) = \mathrm{Sv}(1,1) + \mathrm{alpha} * \mathrm{wmbit}(k) \tag{1}$$

3.1 *Achieving Optimal Scaling Factor Using CS-FA Algorithm*

The main factors that have fascinated consciousness from most researchers in digital watermarking techniques are robustness and invisibility. On the other side of the story, it is being said that their Invisibility and robustness are very unreliable and interrelated. The need of the hour is to determine a solid solution that can bring equilibrium between robustness and invisibility. In the recent past, the researchers have shown interest on the NIA's (Nature Inspired Algorithms) to achieve balance in between them. Generally, the NIA's are used to decide embedding position of a watermark and strength factor of the watermark embedding. The CS-FA technique is utilized in this approach to get the proper watermark embedding factor.

PSNR, SSIM, NC and Corr-correlation coefficient are some of the reliability or performance metrics used to assess invisibility. In the same way robustness is decided in terms of SSIM, BER and NC, however PSNR is used to decide the similarity amid watermarked and original video frames, whilst BER, NC and SSIM metrics are used to know the similarity between two given frames.

3.2 *Mathematical Model*

$$\mathrm{MSE} = \frac{\sum_{m,n}\left[I_{\mathrm{img1}}(p,q) - I_{\mathrm{img2}}(p,q)\right]^2}{p * q} \tag{2}$$

$$\mathrm{PSNR}(I_{\mathrm{img1}}, I_{\mathrm{img2}}) = 10\log_{10}\left(\frac{R*R}{\mathrm{PSE}}\right) \tag{3}$$

$$\mathrm{NC}(I_{\mathrm{img1}}, I_{\mathrm{img2}}) = \frac{\sum_{m,n} I_{\mathrm{img1}}(p,q) * I_{\mathrm{img2}}(p,q)}{\sqrt{\sum_{p,q} I_{\mathrm{img2}}(p,q)^2} * \sqrt{\sum_{p,q} I_{\mathrm{img2}}(p,q)^2}} \tag{4}$$

where I_{img1} and I_{img2} are two frames with same size (p * q). $R = 255$ if frame is 8 bit image.

$$\mathrm{SSIM}(I_{\mathrm{img1}}, I_{\mathrm{img2}}) = \frac{(2\mu_1\mu_2 + c_1)(2\sigma_1^2 + c_2)}{(\mu_1^2 + \mu_2^2 + C_1)((\sigma_1^2 + \sigma_2^2 + C_2)} \tag{5}$$

μ_1 parameter and σ_1 parameter are average and variance of frame I_{img1}, μ_2 parameter and σ_2 parameter are average and variance of frame I_{img2}. The covariance is σ_1^2 of frames I_{img1} and I_{img2}. The variables C_1 and C_2 have been used to keep division stable.

$$\mathrm{BER} = \frac{\sum_{p,q}\left[I_{\mathrm{img1}}(p,q) \oplus I_{\mathrm{img2}}(p,q)\right]}{p*q} \tag{6}$$

The ex-or operation is signified by the symbol $\oplus$. In effort to accomplish watermark equilibrium and robustness, several researchers have suggested numerous concepts of optimized watermarking.

$$F = \mathrm{PSNR}(X_{\mathrm{OF}}, X_{\mathrm{WF}}i') + \Phi * [\mathrm{NC}(w_{\mathrm{OF}}, w'_{\mathrm{WF}}i) + \sum_{i=1}^{N} \mathrm{NC}(w_{\mathrm{OF}}, w'_{\mathrm{WF}}i) \tag{7}$$

$$F = \mathrm{PSNR}(X_{\mathrm{OF}}, X_{\mathrm{WF}}i') + 30 \cdot \sum_{i=1}^{N} \mathrm{BER}(w_{\mathrm{OF}}, w'_{\mathrm{WF}}i) \tag{8}$$

$$F = \mathrm{PSNR}(X_{\mathrm{OF}}, X_{\mathrm{WF}}i')/100 + \sum_{i=1}^{N} \mathrm{NC}(w_{\mathrm{OF}}, w'_{\mathrm{WF}}i) \tag{9}$$

$$F = \frac{N}{\sum_{i=1}^{N} \mathrm{Corr}(w, w'i)} - \mathrm{Corr}(w_{\mathrm{OF}}, w'_{\mathrm{WF}}i) \tag{10}$$

$$F = 10 \cdot \left[1 - \mathrm{SSIM}(X_{\mathrm{OF}}, X_{\mathrm{WF}}i')\right] + \frac{1}{N}\sum_{i=1}^{N} \mathrm{BER}(w_{\mathrm{OF}}, w'_{\mathrm{WF}}i) \tag{11}$$

This proposed algorithm utilizes the amalgamation of SSIM and BER as an objective function.

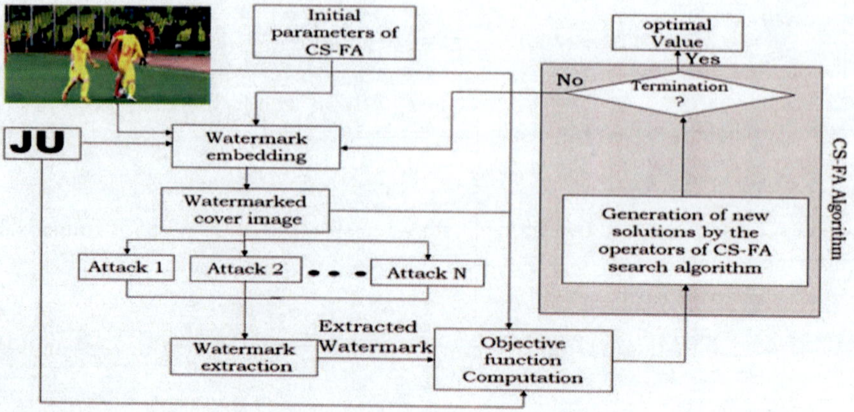

Fig. 4 Cuckoo search (CS) algorithm work flow process and watermark extraction process

$$F = \left[1 - \text{SSIM}\left(X_{\text{of}}, X_{\text{wf}}i'\right)\right] + \Phi \cdot \frac{1}{N} \sum_{i=1}^{N} \text{BER}\left(w_{\text{of}}, w'_{\text{wf}}i\right) \quad (12)$$

where weight factor is denoted by Φ, 'X_{of}' states to original frame and '$X_{\text{WF}}i$' watermarked frame. 'w_{of} and $w'_{\text{wf}}i$' refers to the original watermark and extracted watermark under ith attack. The number of attacks is denoted by the letter N. The main motives for deciding the objective function (Eq. 12) with the use of SSIM and BER are given by Papakostas et al. (2014) demonstrated that the image quality can be measured efficiently by SSIM than PSNR. Robustness can be measured efficiently by using BER than SSIM and NC. Although, Eq. (11) seems to be more likely as Eq. (12) that's recommended in this paper but differs in terms of the weights. The proposed method (Eq. 12) mainly focusses on robustness whereas Eq. (11) talks about imperceptibility. In this method, per the suggestion made by Guo and Li (2017) owing the below facts the weight factor Φ is considered as 30. The key focus in the proposed method is on robustness, to achieve this weight factor of robustness must be greater than watermark invisibility. To achieve the imperceptibility where SSIM (Xo, Xw) is greater than 0.9, the weight factor of robustness is considered a low compared to the referred value. Video watermarking approach using CS-FA, DWT and SVD is exemplified in the following using the flow chart in Fig. 4.

4 Results and Discussion

A range attacks are used out for assessing the performance and robustness of the proposed method. Figure 5 shows video frames and Fig. 6 shows a sample input watermark, a scrambled watermark and the respective secret shares. The initial parameters for the CS and FA are $n = 10$, pa $= 0.25$, lb $= 8$ and ub $= 15$. $\alpha = 1.0$, $\beta_0 = 1$,

$\gamma = 0.01$, the maximum iterations are set as 20, and $\phi = 30$. The scaling factor is adjusted from 8 to 15 to run both the algorithms. Figure 7 shows the video frames and their watermark.

In this simulation experiments, PSNR, SSIM and BER are used to measure imperceptibility. Watermarked frame and its corresponding extracted watermarks are shown in Fig. 7. Numerical values of imperceptibility for a range of types of videos are represented in Table 1. Figure 8 shows graphs comparisons of CS-FA algorithms with CS and FA.

In these experiments, to review the robustness of the watermark numerous attacks were brought out on different watermarked videos which normally happens at the time of transferring the video content through Internet. BER is used as metric to assess robustness. The kind of attack and average bit error rate for range of attacks are presented in Table 2.

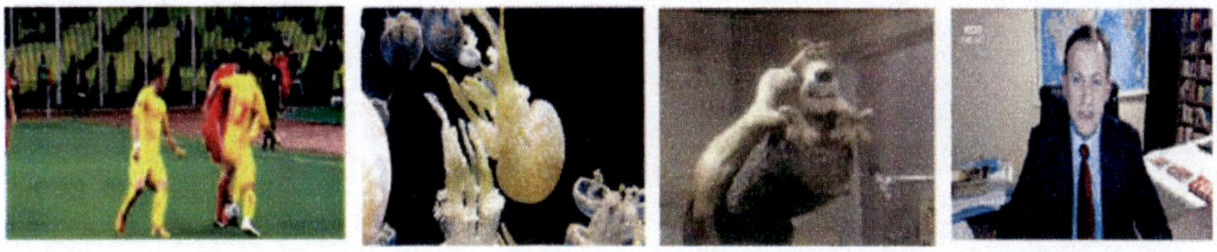

Fig. 5 Sample video frames

Fig. 6 Watermark and shares

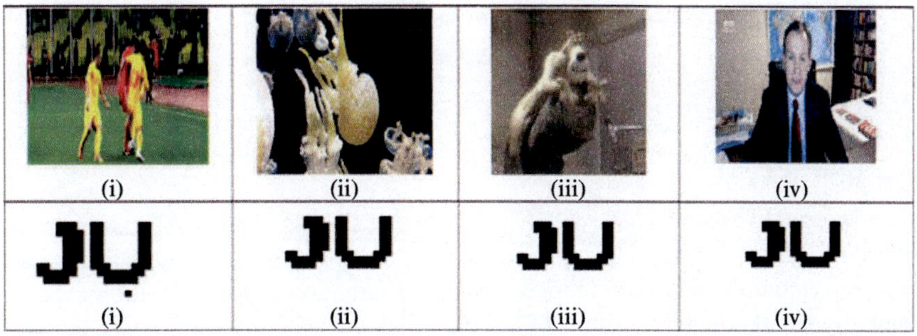

Fig. 7 Watermark frame and result ant extracted watermark

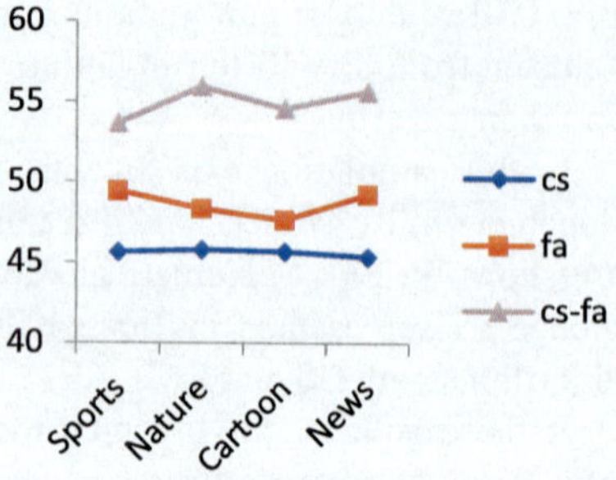

Fig. 8 Comparison of graphs

Table 2 Avg bit errors

Index of attack	Sports video	Nature video	Cartoon video	News video
Attack name	17	12.42	0.556	9.52
Resize	28.4	37.86	31	40.53
Salt and pepper noise	0	0	0	0
Gaussian LPF	8.7	9.32	9.87	9.7
Cropping	0.23	0	0	0
JPEG compression	7.53	8.74	3.6	7.34
Sharpening filter	5.7	0.58	0	1.0
Motion filter	7.64	7	6.85	7.65

Table 3 Proposed method performance

Metric	BDWTSVD (Sake and Tirumala 2016)	Level-3 DWT (Shukla and Sharma 2018)	DWT (Kadu et al. 2016)	Proposed method
PSNR	44.68	45.8	45.44	54.36
SSIM	0.94	0.938	0.928	0.998

The method is assessed by comparing PSNR and SSIM to the recent past methods and the same is mentioned in Table 3 [Sports Video is chosen].

5 Conclusion

In the proposed method, i.e., the CS-FA, secret sharing, DWT-SVD domain is propounded for watermarking a given video. The features like robustness and invisibility are examined by excogitating videos of different types for simulation experiments. By having the high values for PSNR, SSIM the watermarked frames demonstrates improved visual quality. To illustrate the robustness of the suggested proposal, the embedded watermarked frames are extracted by using different kinds of attacks on the frame. The method's perception is also carried out by evaluating PSNR, SSIM and BER against contemporary methods. Our technique outperforms theirs, according to the results.

References

K. Aditya, A. Choudhary, A. Raj, M. Sing, Video watermarking based on Cuckoo search and scene change detection, in *2017 International Conference on Intelligent Sustainable Systems (ICISS)*, Palladam (2017), pp. 420–424

S. Arora, S. Singh, A conceptual comparison of firefly algorithm, bat algorithm and cuckoo search, in *2013 International Conference on Control Computing Communication & Materials (Allahabad 2013)* (IEEE Press, Piscataway, 2013), pp. 1–4

P. Bao, X. Ma, Image adaptive watermarking using wavelet domain singular value decomposition. IEEE Trans. Circuits Syst. Video Technol. **15**(1), 96–102 (2005)

M. Elkhechafi, H. Hachimi, Y. Elkettani, A new hybrid cuckoo search and firefly optimization. Monte Carlo Methods Appl. (2018)

Y. Guo, B. Li, Optimised blind image watermarking method based on firefly algorithm in DWT-QR transform domain. IET Image Process. **11**(6), 406–415 (2017)

S.L. Hsieh, J.J. Jian, I.J. Tsai, B.Y. Huang, A color image watermarking scheme based on secret sharing and wavelet transform, in *Proceedings of IEEE International Conference on Systems, Man and Cybernetics*, Montreal, Quebec, Canada (2007), pp. 2143–2148

O. Jane, E. Elbasi, A new approach of nonblind watermarking methods based on DWT and SVD via LU decomposition. Turk. J. Electr. Eng. Comput. Sci. **22**(5), 1354–1366 (2014)

S. Kadu, C. Naveen, V.R. Satpute, A.G. Keskar, Discrete wavelet transform based video watermarking technique, in *2016 International Conference on Microelectronics, Computing and Communications (Micro Com)*, Durgapur (2016), pp. 1–6

G. Kanagaraj, S.G. Ponnambalam, N. Jawahar, A hybrid cuckoo search and genetic algorithm for reliability-redundancy allocation problems. Comput. Ind. Eng. **66**(4), 1115–1124 (2013). https://doi.org/10.1016/j.cie.2013.08.003

N.M. Makbol, B.E. Khoo, T.H. Rassem, Lock-based discrete wavelet transform-singular value decomposition image watermarking scheme using human visual system characteristics. IET Image Process. **10**(1), 34–52 (2016)

R. Mehta, N. Rajpal, V.P. Vishwakarma, LWT-QR decomposition based robust and efficient image watermarking scheme using Lagrangian SVR. Multimed. Tools Appl. **75**(7), 4129–4150 (2016)

G.A. Papakostas, E.D. Tsougenis, D.E. Koulouriotis, Moment-based local image watermarking via genetic optimization. Appl. Math. Comput. **227**, 222–236 (2014)

P. Prathik, R. Krishna, R.A. Nafde, K. Shreedarshan, An adaptive blind video watermarking technique based on SD-BPSO and DWT-SVD, in *Proceedings of International Conference on Computer Communication and Informatics*, Coimbatore, India (2013), pp. 1–15

A. Sake, R. Tirumala, A robust fusion of DWT and PCA for video watermarking using optimization technique. Int. J. Eng. Res. Electron. Commun. Eng. **4** (2017)

A. Sake, R. Tirumala, Bi-orthogonal wavelet transform based video watermarking using optimization techniques, in *2nd International Conference on Solar Energy Photovoltaic*, Ongole, Mater. Today: Proc. 5(1), pp. 1470–1477 (2018)

D. Shukla, M. Sharma, Robust scene-based digital video watermarking scheme using level-3 DWT: approach, evaluation, and experimentation. Radio Electron. Commun. Syst. **61**(1), 1–12 (2018)

T.T. Takore, P.R. Kumar, G.L. Devi, A modified blind image watermarking scheme based on DWT, DCT and SVD domain using GA to optimize robustness, in *2016 ICEEOT*, Chennai (2016), pp. 2725–2729

X.-S. Yang, S. Deb, Cuckoo search via Lévy flights, in *World Congress on Nature and Biologically Inspired Computing*, Coimbatore (IEEE Press, Piscataway, 2009), pp. 210–214

Investigation of MANET Routing Protocols to Enhance QoS for Communication

Devdas Saraswat and Nikhat Raza Khan

Abstract On-demand routing is a primary function of mobile ad hoc network, and routing broadcasting are utilized network capacity nearly half of network capacity because node frequently change their location and number of time call the route broadcasting methodology that increase the overhead as well as utilization of channel capacity of the network. Due to the large overhead various researchers are work in the field of overhead minimization some of work un-assign slot utilization so fully network bandwidth are utilized. But in our proposed work combine the routing, queuing and location mechanism and improve service quality of network. Proposed approach uses multipath routing with QoS measuring methodology that utilized un-assign slot as well as distributed load sharing mechanism which increase network reliability for ad hoc communication. Our proposed QoS measuring technique with enhanced routing protocol is to achieve better reliability. Un-assign slot utilization methodology in multipath environment get more reliable network.

Keywords MPDSR · QoS · MANET · Queue utilization · Channel access

1 Introduction

The MANET is a mobile ad hoc network that is decentralized. Because it does not rely on pre-existing equipment like wired network routers or managed wireless network access points, the network is ad hoc (infrastructure). Instead, each node contributes to routing by forwarding data to other nodes; hence, assessing a node's forward data is dynamically dependent on network connection and the routing algorithm. Wireless ad hoc mobile networks are dynamic, self-configuring networks in which nodes wander freely. Multicast routing, rather than only unicast or broadcast, is required by

D. Saraswat (✉)
Department of Computer Science and Engineering, IES University, Bhopal, India
e-mail: devdas_saraswat@rediffmail.com

N. R. Khan
Department of Computer Science and Engineering, IES College of Technology, Bhopal, M.P., India

A. B. Reddy et al. (eds.), *Proceedings of Third International Conference on Advances in Computer Engineering and Communication Systems*, Lecture Notes in Networks and Systems 612, https://doi.org/10.1007/978-981-19-9228-5_17

definition in a true MANET. Any device in a MANET has the freedom to go in any direction and, as a result, will regularly modify its links to other devices. Each one must be a router since it must forward traffic that is unrelated to its intended purpose. The most challenging part of building a MANET is supplying each device with the data it needs to properly route traffic on a continuous basis. MANET can function as a stand-alone system or as a component of a larger network. They create a highly dynamic, autonomous architecture when one or more distinct transceivers are present between nodes. As multimedia apps grow increasingly common, quality of service (QoS) is becoming one of the most enticing characteristics of mobile ad hoc networks (MANETs). However, QoS support in MANETs is difficult to provide due to node mobility, limited bandwidth, and the highly competitive nature of network topologies. The major purpose of this study is to improve MANET QoS by introducing three techniques: self-organization, corruption rate, and jitter level.

2 Literature Survey

Liang et al. (2021) "A dynamic source routing protocol based on path reliability and link monitoring repair". In this title we discuss the two most important aspects for mobile self-organizing networks that apply to drone dependability and stability are discussed in this book. We propose a dynamic source routing protocol based on path reliability and a monitoring repair mechanism to improve the communication quality of UAVs moving at high speeds (DSR-PM). Data transmission is carried out by the model by filtering the most reliable path. During transmission, link-state information is monitored, and broken links are repaired as soon as possible to ensure link communication stability and dependability, as well as increase data transmission efficiency.

Chen et al. (2020) "An Adaptive on-Demand Multipath Routing Protocol With QoS Support for High-Speed MANET". In this article, we'll look at a routing protocol called Topological Change Adaptive Ad hoc On-demand Multipath Distance Vector (TA-AOMDV) that can adapt to high-speed node movement while maintaining QoS. This protocol proposes a stable path selection technique that considers both node resources (residual energy, available bandwidth, and queue length) and the connection stability probability between nodes as path selection factors.

Priyambodo et al. (2021) "Performance Optimization of MANET Networks through Routing Protocol Analysis". This study looked at the effects of route request parameters like RREQ RETRIES and MAX RREQ TIMOUT on the Ad Hoc On-demand Distance Vector (AODV) protocol, which was then compared to the standard AODV performance Optimized Link State Routing (OLSR) protocols.

Jayabarathan et al. (2016) "QoS enhancement in MANETs using priority aware mechanism in DSR protocol". This phrase may be found in the book's title. A priority-aware (PA) technique is developed in this book to improve the QoS for MANET by allocating priority depending on data rates in the dynamic source routing (DSR) protocol.

Jara et al. (2021) "QSMVM: QoS-Aware and Social-Aware Multi metric Routing Protocol for Video-Streaming Services over MANETS". In this paper, we extend our earlier approach, the multipath multimedia dynamic source routing (MMDSR) protocol, by incorporating a social measure TS into the forwarding algorithm's judgments. The quality of service (QoS) and the level of confidence amongst users who constitute the forwarding channel in the MANET are found to be trade-offs. Our goal is to raise the trust metric while maintaining a high level of QoS.

Bujari (2021) "Location Dynamic Tabu Routing Protocol for MANETs". We examine how mobile ad hoc networks are presently enjoying a second youth in terms of research interest and societal advantages in this title. This has been facilitated by the vast number of applications made available by the ubiquitous and rising availability of smartphones, drones, sensors, and other tiny devices with communication and sensing capabilities.

Sirmollo (2021) "Mobility-Aware Routing Algorithm for Mobile Ad Hoc Networks". The main purpose of this title is to construct and enhance the mobility-aware routing algorithm (MARA) in order to improve the performance of the routing protocol in MANETs. The suggested technique allows mobile nodes to rebroadcast or remove received broadcasted messages. The decision is based on a combination of node speed, node distance, and node residual energy.

Mandhare et al. (2016) "Improving QoS of Mobile Ad hoc Network using Cache Update Scheme in Dynamic Source Routing Protocol", we suggest a new approach for cache updating based on the distributed route-cache update algorithm. In this we used distributed cache replacement strategy.

Johnson (2021) "QoS Improvement in MANET using Self Organized Balanced Optimization" by adding trust between social and QoS in this term. The major solution of the proposed method is to reduce the number of nodes that display various packet forwarding abuses and to figure out how to provide secure communication using the trust mechanism. QoS variables such as channel quality, residual energy, connection quality, and so on are used to compare the results.

Alfawaer (2017) "An enhanced Multipath Strategy in Mobile Ad hoc Routing Protocols". The main purpose of this research was to look at the characteristics of MANET multipath routing technologies. Two multipath routing protocols were investigated, and a simulation study using NS2 was undertaken to compare DSR and AODV in order to provide a better means of reaching the target while maintaining QoS.

Liang et al. (2021). They explore a dynamic source routing protocol based on path reliability and link monitoring repair mechanism in "A dynamic source routing protocol based on path reliability and link monitoring repair" (DSR-PM). The model handles data transfer by filtering the most reliable path. Link-state information is monitored throughout transmission, and damaged links are repaired as quickly as possible to assure link communication stability and reliability, as well as to improve data transmission efficiency.

3 Proposed Work

A mobile ad hoc network is made up of moveable nodes that create a network for communication between them. MANET employs a dynamic routing protocol to improve route construction; however it is unable to manage node mobility, which increases network overhead and instability. As a result, devise a method to reduce overhead while improving network quality of service.

Mobile ad hoc networks will fundamentally alter communication structures in the future, allowing end-users to communicate at their leisure. Its implementation is difficult because to several restrictions, such as restricted processing capabilities, limited node energy, route determination process, dynamic node motion, and so on. Most of the difficulties in earlier study were solved using various approaches, yet research is unending. The goal of the proposed study is to find the most energy-saving approach for establishing the route and improving network QoS. The proposed method route selection criteria are based on a number of factors, including starting energy, data transmission per packet power requirements, connection dependability, link expiration time, reception signal intensity, and distance between communicator nodes. During route selection, each node examines the intensity of the receiving signal as well as its distance from all of its neighbors, which aids in determining the best path from source to destination.

3.1 QoS Implementation Architecture

See Fig. 1.

3.2 MPDSR with QoS Decision System

Routing is a crucial aspect of any connection since real data is transmitted based on the chosen path following route discovery. This section explains how multipath dynamic source routing is used to estimate the QoS parameter. Multipath dynamic source routing works effectively when intermediary nodes are kept to a minimal, implying that DSR routing chooses fewer participant hops between source and destination. DSR routing performs well in terms of security and QoS, but because the topology of a mobile ad hoc network is continually changing and open, we cannot limit the number of source nodes that wish to connect at the same time; this would damage network performance. The performance degradation limit is dependent on channel use, queue usage, and energy consumption, all of which contribute to the network's quality of service. The suggested architecture depicts the whole network deployment for detecting quality of service deterioration and resolving the issue in MANETs using multipath dynamic source routing (MPDSR).

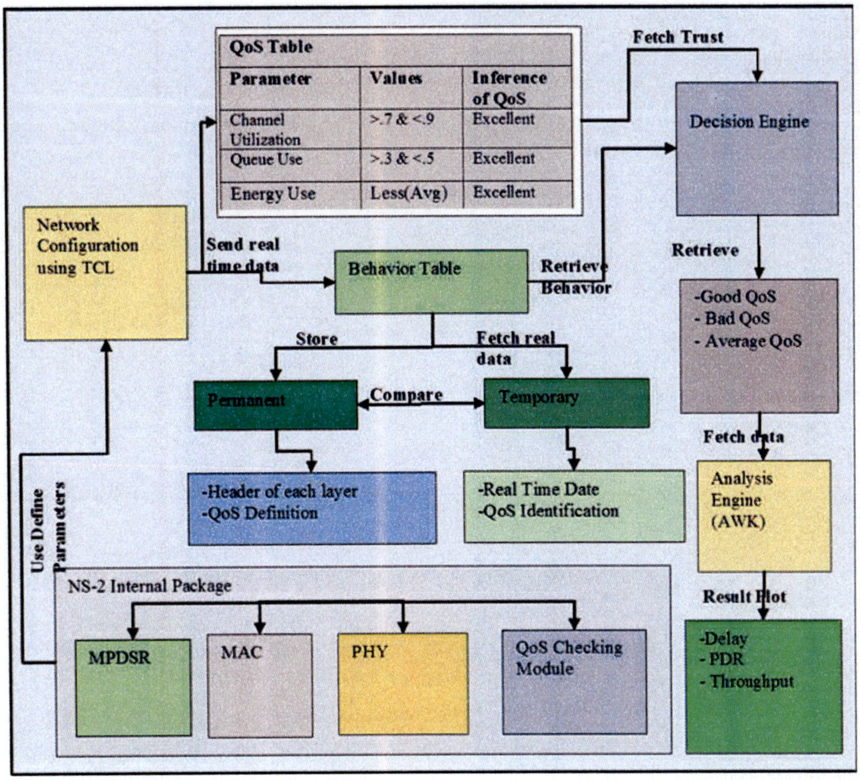

Fig. 1 QoS implementation architecture

4 Simulation Tool Overview

Network Simulator (Version 2), wide called NS2, is solely an occasion driven simulation tool that has tried helpful in finding out the dynamic nature of communication networks. Simulation of wired similarly as wireless network functions and protocols. Due to its flexibility and modular nature, NS2 has gained constant popularity in the networking research community since its birth in 1989. Since 1995 the Defense Advanced analysis comes Agency (DARPA) (Henderson et al. 2008) supported development of NS through the Virtual inhume Network work (VINT) project.

Table 1 Network simulation parameters

Parameters	Value
Area of simulation	1000 m × 1000 m
Antenna type	Omni-directional
Mobile nodes	100
Radio range (m)	550
Transferring mode	Multi-hop
Maximum speed (ms)	Random
Routing protocol	AOMDV, AODV, DSR
Transport layer	TCP, UDP
Traffic	CBR
Application layer	FTP
Simulation time (s)	500
Packet size	512 bytes
Area of simulation	1000 m × 1000 m
Antenna type	Omni-directional

4.1 Simulation Parameters

In Table 1 shows the simulation parameter which is used for analyzing the network activity, those parameters are number of nodes, routing protocol, transport protocol, traffic type, etc.

4.2 Performance Metrics

The performance metrics in case of proposed profile-based cluster communication is considered for performance measurement is mentioned below:

Packet Delivery Ratio. In the resultant graph represent the comparative analysis of percentage of data receives and gets AOMDV better result in terms of percentage of data receives (Fig. 2).

Throughput Analysis. Throughput is an analysis of per unit data receives in receiver. In the architecture calculate the throughput in terms of Kbps and conclude that AOMDV routing perform well with respect to throughput analysis (Fig. 3).

Normal Routing Load. Normal routing is a ratio of number of routing packet from number of data packet. While the NRL is higher it means network overhead is higher. The resultant graph represent AOMDV gives low normal routing load its means multipath routing perform with low overhead (Fig. 4).

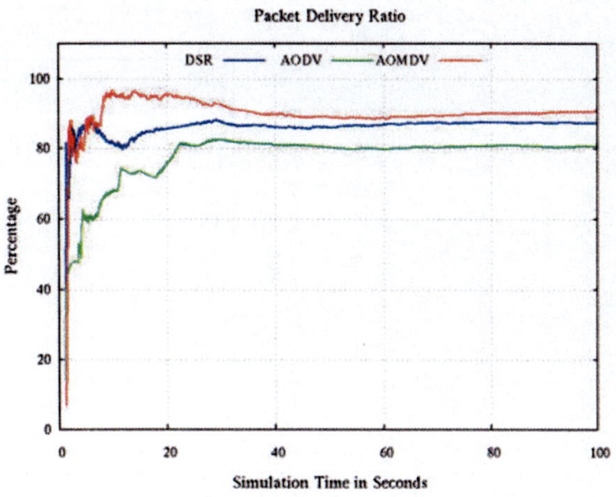

Fig. 2 Percentage of data receives

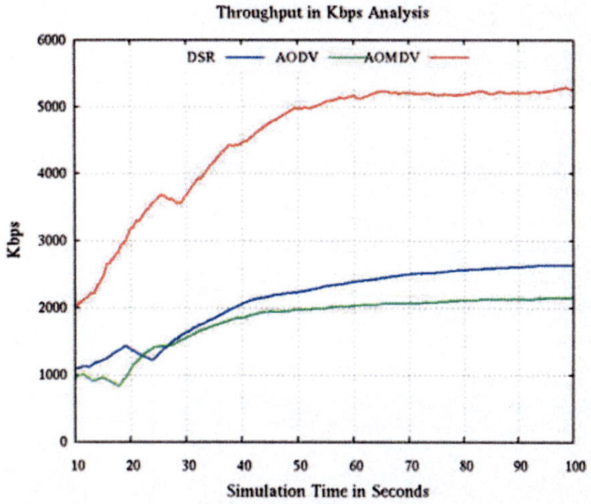

Fig. 3 Throughput analysis (Kbps)

End to End Delay (ms). Delay is important parameter for data transmission, while delay is lower it means communication is fast. In the resultant graph our result conclude that AOMDV routing gives low delay with respect to other routing protocol it means it provide fast communication (Fig. 5).

5 Conclusion

In this paper, the proposed approach for identifying the quality of service mechanism is feasible for MANET with low overhead. We also compare our result with the base MPDSR and the existing un-assigned slot mechanism. Our primary goal in

Fig. 4 Normal routing load

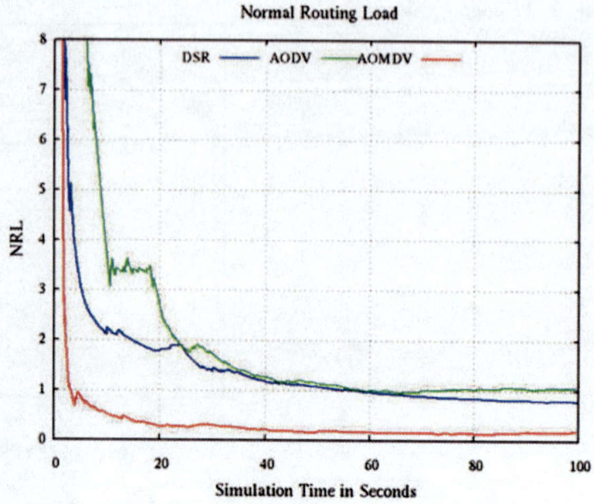

Fig. 5 End to end delay (ms)

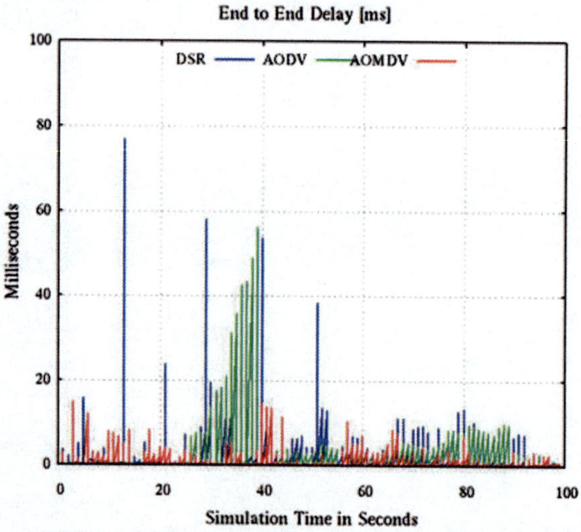

designing this dissertation is to provide reliable service to communicator nodes in a dynamic network with congestion and collisions. In our proposed multipath routing approach, senders are aware of the capacity of the link as well as free slots so that work minimizes the data drop and increases link bandwidth between intermediate nodes of the network. Another technique, namely location detection, is useful for local route repair while breaking the link, and that work decreases the routing overhead and further improves the quality of service in terms of delay, throughput, packet delivery ratio, etc.

References

Z.M. Alfawaer, An enhanced multipath strategy in mobile ad hoc routing protocols, in *2017 9th IEEE-GCC Conference and Exhibition (GCCCE)* (2017)

A. Bujari, Location dynamic tabu routing protocol for MANETs. Mob. Netw. Appl. (2021)

Z. Chen, W. Zhou et al., An adaptive on-demand multipath routing protocol with QoS support for high-speed MANET. IEEE Access (2020). Date of current version March 13, 2020

T.R. Henderson, M. Lacage, G.F. Riley, C. Dowell, J. Kopena, Network simulations with the ns-3 simulator. SIGCOMM demonstration (2008)

E.P. Jara et al., QSMVM: QoS-aware and social-aware multi metric routing protocol for video-streaming services over MANETS. Sensors (2021)

J.K. Jayabarathan et al., QoS enhancement in MANETs using priority aware mechanism in DSR protocol. EURASIP J. Wirel. Commun. Netw. (2016)

A.J. Johnson, QoS improvement in MANET using self organized balanced optimization. Turk. J. Comput. Math. Educ. **12**(11) (2021)

Q. Liang et al., A dynamic source routing protocol based on path reliability and link monitoring repair. PLoS One (2021)

V.V. Mandhare et al., Improving QoS of mobile ad-hoc network using cache update scheme in dynamic source routing protocol, in *7th International Conference on Communication, Computing and Virtualization* (2016)

T.K. Priyambodo et al., Performance optimization of MANET networks through routing protocol analysis. Computers (2021)

C.Z. Sirmollo, Mobility-aware routing algorithm for mobile ad hoc networks. Wirel. Commun. Mob. Comput. **2021** (2021)

Leader-Follower Based Control of Fixed-Wing Multi-Robot System (MRS) via Split-Rejoin Maneuvers in 3D

Roneel Chand, Krishna Raghuwaiya, Jito Vanualailai, and Jai Raj

Abstract In this research, we present a motion planner for numerous fixed-wing unmanned aerial vehicles (UAVs) using the Lyapunov-based control scheme (LbCS) architecture. To go to their predetermined target, various formations of multi-robot systems must navigate a workspace in the presence of obstacles with respect to their leaders. Every formation has a leader, and the followers follow their leaders' behavior. In the event of obstacles, the forms split and rejoin farther away. The flock moves toward its goal while navigating the surroundings in its predetermined pattern. The obstacles in our scenario will be spherical fixed obstacles, and the swarm's moving aircraft will act as an obstacle for all other members (moving obstacles). The particular formation will stay clear of other formations as well. To successfully complete this work, a group of nonlinear acceleration-based controllers using the Lyapunov-based control scheme are developed. The proposed controllers will make sure that the UAVs coordinate their motion in a well-planned manner and ensure that the aircraft converges to their intended destination while avoiding obstacles in their route. It has been examined how several predefined flock formations have been developed in computer-generated simulations of a virtual scenario. The results of the simulation demonstrate how well the nonlinear acceleration control principles are working. Finally, a conclusion and suggestions for further research in this area complete the paper.

Keywords Lyapunov · Fixed-wing UAVs · Multi-robot system · Flock · Split-rejoin

R. Chand (✉)
Fiji National University, Suva, Fiji
e-mail: roneel.chand@fnu.ac.fj

K. Raghuwaiya · J. Vanualailai · J. Raj
The University of the South Pacific, Suva, Fiji
e-mail: krishna.raghuwaiya@usp.ac.fj

J. Vanualailai
e-mail: jito.vanualailai@usp.ac.fj

J. Raj
e-mail: jai.raj@usp.ac.fj

A. B. Reddy et al. (eds.), *Proceedings of Third International Conference on Advances in Computer Engineering and Communication Systems*, Lecture Notes in Networks and Systems 612, https://doi.org/10.1007/978-981-19-9228-5_18

1 Introduction

The multi-robot system (MRS), which is receiving a lot of attention due to developments in technology and research, has emerged as one of the most effective tools for performing tasks that are unsafe for people to perform or are otherwise impractical. According to research, MRS is also more efficient than single or split robots, as well as cheaper and faster due to its ability to cover a large area quickly (Sharma 2008; Sharma et al. 2009). Due to the fact that robots must coordinate their motion in advance, most MRS navigation is done in a certain way (Raj et al. 2021; Sharma et al. 2009).

Multi-robot formation control has received a lot of attention, and intriguing control principles have been developed as a result of extensive research. Interestingly, natural phenomena such as schools of fish, herds of land animals, a flock of birds, and many others have been employed to successfully control the formation of the multi-robot system (MRS) (Gazi 2005; Raghuwaiya et al. 2011; Raj et al. 2020).

Recently, many control strategies for coordinating MRS behaviors have been put forth. According to Olfati Saber et al. (2003), it can be roughly divided into centralized and decentralized (hybrid) control methods. Robots use their own built-in sensors to collect data from the surroundings and the locations of other robots during decentralized control. This is then utilized when making control decisions. The leader-follower strategy, in which the leader acquires environmental data and directs the movement of the follower robots, is the foundation of centralized control. This strategy is more reliable and accurate, but if the leader fails, the entire control would collapse (Khaksar et al. 2019).

Olfati Saber et al. (2003) introduced and talked about a graphical theoretical framework for the algebraic definition of a formation on the notation of the graph. Later, an adaptive approach-based graph theory was employed by Guo. In his approach, Guo used LbCS to contain a moving target and achieve the necessary inter-robot formation (Guo et al. 2010). Additionally, hybrid systems made up of distributed smooth-time changing feedback control laws and Lagrangian system control were also suggested in the works of Chung and Slotine (2009), Toibero et al. (2007) and Khaksar et al. (2019).

For the case of this paper, we present a leader-follower multiple formation via split-rejoin technique. To control a collection of multi-robot systems, cooperative formation controllers will be created. Individual formation has its own leader who is responsible of defining their goals and assigning tasks. The leader also controls the overall journey from the starting configuration to the final target. The main task here is to guarantee that the whole set of mobile robots safely travel from the starting point to the final predefined target avoiding any obstacles present in the workspace. Upon meeting obstacles, the formation splits and rejoins at a safer location. The work to be done in this paper is an extension of work published in Raghuwaiya and Chand (2018) and Raghuwaiya et al. (2018). The rest of this work is organized as follows: Sect. 2 defines the model of the fixed-wing aircraft, whereas in Sect. 3, artificial potential field functions are defined; the acceleration-based control laws are derived in Sect. 4, followed by the stability analysis of the robotic system in Sect. 5; the effectiveness of the suggested controllers is shown through computer simulations in Sect. 6, and finally, Sect. 7 concludes the paper and outlines the future work in the area.

2 Vehicle Model

In this part, we derive a simple kinematic model for the formation control of several fixed-wing aircraft using the leader-follower method. We shall take into account fixed-wing aircraft in the Euclidean plane, $n, n \in \mathbb{N}$. Each formation's leader is represented by $\mathcal{A}_{h1}, h = 1, \dots, p$, while the follower robots are represented by $\mathcal{A}_{hi}$ for $i = 2, \dots, n$ and $h = 1, \dots, p$, respectively. Next, Cartesian coordinates (x_{hi}, y_{hi}), is used, show the two-dimensional reference point of the fixed-wing aircraft, as seen in Fig. 1. Furthermore, θ_{hi} with respect to z_1-axis gives aircraft's orientation, and L_{hi} and l_{hi} gives the aircraft's overall length and its entire wingspan, respectively. The pitch angle and rolling angle of the aircraft is ignored in this scenario. Furthermore, adapting the nomenclature of Raghuwaiya and Singh (2013), we construct a spherical region around the aircraft to protect it. This enables the fixed-wing aircraft to safely move past an obstacle present in the workspace. Furthermore, provided that the *clearance parameters* $\epsilon_1 > 0$ and $\epsilon_2 > 0$, we enclose the aircraft using a protective spherical region centered at (x_{hi}, y_{hi}, z_{hi}), with radius $r_{hi} = \frac{\sqrt{(L_{hi}+2\epsilon_1)^2+(l_{hi}+2\epsilon_2)^2}}{2}$. Next, we define the model of the system, adopted from Hwangbo et al. (2007) as:

Fig. 1 Kinematic model of the fixed-wing UAV

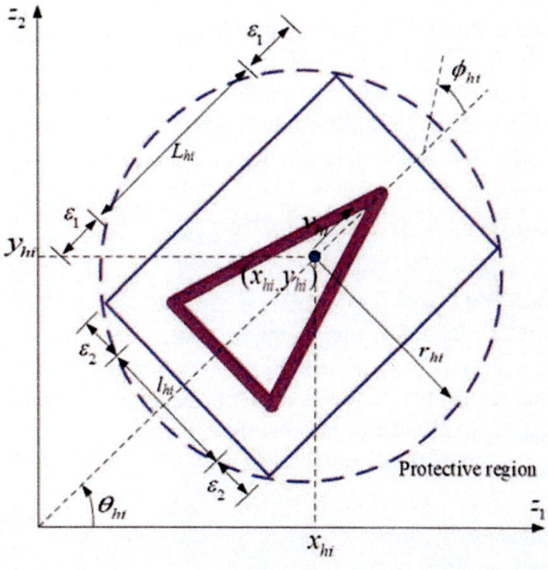

$$\begin{aligned} \dot{x_{hi}} &= v_{hi} \cos\theta_{hi} - \frac{L_{hi}}{2}\omega_{hi} \sin\theta_{hi}, \\ \dot{y_{hi}} &= v_{hi} \sin\theta_{hi} + \frac{L_{hi}}{2}\omega_{hi} \cos\theta_{hi}, \\ \dot{z_{hi}} = \nu_{hi}, \quad \dot{\theta_{hi}} &= \omega_{hi}, \quad \dot{v_{hi}} = \sigma_{hi1}, \quad \dot{\omega_{hi}} = \sigma_{hi2}, \quad \dot{\nu_{hi}} = \sigma_{hi3}. \end{aligned} \tag{1}$$

With reference to the system (1), v_{hi} is the aircraft's instantaneous translational velocity; ω_{hi} is the yaw rate, ν_{hi} is the rate of change in altitude, and $\sigma_{hi1}, \sigma_{hi2}, \sigma_{hi3}$ for $i = 1, \ldots, n$ are the instantaneous accelerations for $h = 1, \ldots, p$. The instantaneous velocities and accelerations of the fixed-wing aircraft are described in system (1) at this point. The state of the fixed-wing UAV can then be described by $\mathbf{x} = (x_{hi}, y_{hi}, z_{hi}, \theta_{hi}, v_{hi}, \omega_{hi}, \nu_{hi}) \in \mathbb{R}^7$. With the help of this acceleration-based control, the vehicles will move in the direction of the leader while avoiding obstacles and collisions based in its state in the prescribed formation.

3 Artificial Potential Field Function

Here, we wish to build the acceleration controllers, σ_{hi} for $i = 1, \ldots, 3$, and $h = 1, \ldots, p$, so that the fixed-wing aircraft safely navigates in the 3D workspace, avoiding obstacles and collisions simultaneously.

3.1 Attractive Potential Field Functions

Attraction to Target We provide the leader a stationary target in the form of a sphere that has the center (p_{11}, p_{12}, p_{13}) and radius r_t. The duty of the follower aircraft is to move with respective to their leader, as the leader moves toward it target maintaining a desirable relative position. We take into consideration an alluring prospective function for the leader's appeal to its target as:

$$V_{h1}(\mathbf{x}) = \frac{1}{2}\left[(x_{h1} - p_{11})^2 + (y_{h1} - p_{12})^2 + (z_{h1} - p_{13})^2 + v_{h1}^2 + \omega_{h1}^2 + \nu_{h1}^2\right], \tag{2}$$

for $h = 1 \ldots, p$.

Next, the following aircrafts, $i = 2, \ldots n$ and $h = 1, \ldots p$, with center $(p_{i1}, p_{i2}, p_{i3}) = (x_{11} - a_{hi}, y_{11} - b_{hi}, z_{11} - c_{hi})$ where a_{hi}, b_{hi} and c_{hi} are relative x, y, z distances of the followers from the leader of each formations, need to maintain their respective relative position with respective to the leader. To achieve this, we design the following function:

$$V_{hi}(\mathbf{x}) = \frac{1}{2}\left[(x_{hi} - (p_{i1}))^2 + (y_{hi} - (p_{i2}))^2 + (z_{hi} - (p_{i3}))^2 + v_{hi}^2 + \omega_{hi}^2 + \nu_{hi}^2\right]. \tag{3}$$

Auxiliary Function The vehicles must successfully converge to their destination, and in order to do this, we must build an auxiliary function that looks like:

$$G_{h1}(\mathbf{x}) = \frac{1}{2}\left[(x_{h1} - p_{11})^2 + (y_{h1} - p_{12})^2 + (z_{h1} - p_{13})^2\right], \tag{4}$$

for $h = 1, \ldots, p$, and

$$G_{hi}(\mathbf{x}) = \frac{1}{2}\left[(x_{hi} - p_{i1})^2 + (y_{hi} - p_{i2})^2 + (z_{hi} - p_{i3})^2\right], \tag{5}$$

for $i = 2, \ldots, n$ and $h = 1, \ldots, p$.

The given auxiliary functions guarantees the controllers zero out at the leader's target. These functions are then multiplied with the repulsive potential field function which will be developed in the following sections.

3.2 *Repulsive Potential Field Functions*

Generally, we have presence of unknown agents, such as obstacles in our natural environment. These needs to be avoided by the aircraft to successfully reach to their intended target. For this to take place, we construct a suitable obstacle avoidance function, which calculates the Euclidean distances between the obstacles and robots in the workspace.

Fixed Obstacles in the Workspace We fix solid obstacles inside the workspace's confines using $q \in \mathbb{N}$. In this case, we let the lth obstacle to be a sphere with a radius of ro_l and a center of (o_{l1}, o_{l2}, o_{l3}). The aircraft needs to avoid this to converge to its target, so for the aircraft to avoid the lth obstacle, we consider

$$FO_{hil}(\mathbf{x}) = \frac{1}{2}\left[(x_{hi} - o_{l1})^2 + (y_{hi} - o_{l2})^2 + (z_{hi} - o_{l3})^2 - (ro_l + r_{hi})^2\right], \tag{6}$$

where $l = 1, \ldots q$, $i = 1, \ldots, n$ and $h = 1, \ldots, p$ as an avoidance function.

Moving Obstacles in the Workspace

Minimum Inter-robot Obstacles: We need to take moving obstacles into account in our 3D workspace to create workable trajectories for the aircraft since we have presence of moving obstacles in the environment. It is necessary to avoid moving aircraft since they become moving obstacles for all other aircraft. We create the repulsive potential field function with the following formula for $i, j = 1, \ldots, n, j \neq i$, in order for vehicle B_i to avoid vehicle B_j:

$$MO_{hij}(\mathbf{x}) = \frac{1}{2}\left[(x_{hi} - x_{hj})^2 + (y_{hi} - y_{hj})^2 + (z_{hi} - z_{hj})^2 - (2r_{hi})^2\right]. \quad (7)$$

Inter-formation Avoidance: Since there are multiple formations in a cooperative system, it is essential that they avoid one another. To prevent collisions with any other formation structures at work, each formation structure is the system that is desired. For ith body of ψ_h to avoid the uth body of ψ_m, for $i, u = 1, \ldots, n, j \neq i$ and $h, m = 1, \ldots, p$ with $m \neq h$, we design and adopt:

$$DO_{himu}(\mathbf{x}) = \frac{1}{2}\left[(x_{hi} - x_{mu})^2 + (y_{hi} - y_{mu})^2 + (z_{hi} - z_{mu})^2 - (r_{hi} - r_{mu})^2\right]. (8)$$

Workspace Limitations We design a specific 3D framework using dimensions $\eta 1$ by $\eta 2$ by $\eta 3$ to be the workspace of our aircraft. In our LbCS, these boundaries are considered as *fixed obstacles*. For the aircraft to clear these boundaries, we formulate the following potential function for $i = 1, \ldots, n$ and $h = 1, \ldots, h$:

$$\begin{aligned}
W_{hi1}(\mathbf{x}) &= (x_{hi} - r_{hi}),\\
W_{hi2}(\mathbf{x}) &= (\eta_2 - (y_{hi} + r_{hi})),\\
W_{hi3}(\mathbf{x}) &= (\eta_1 - (x_{hi} + r_{hi})),\\
W_{hi4}(\mathbf{x}) &= (y_{hi} - r_{hi}),\\
W_{hi5}(\mathbf{x}) &= (z_{hi} - r_{hi}),\\
W_{hi6}(\mathbf{x}) &= (\eta_3 - (z_{hi} + r_{hi})).
\end{aligned} \quad (9)$$

These avoidance functions will then be added to the control rules so that the movement of the aircraft remains within the given boundaries of the workspace.

Dynamic Constraints Practically, limits are imposed on the yaw angle of the ith autonomous aircraft due to mechanical singularities, and aircrafts translational speed is limited due to safety concerns (Sharma 2008). Due to this, we take the following into consideration;

(i) $v_{\min} < v_{hi} < v_{\max}$ the aircraft while $v_{\max}$ is the *maximal achievable speed* of the aircraft,
(ii) $|\omega_{hi}| < \omega_{\max}$, where $\omega_{\max}$ is the maximum yaw rate,
(iii) $|\nu_{hi}| < \nu_{\max}$, where $\nu_{\max}$ is the maximum altitude change.

Hence, for $i = 1, \ldots, n$ and $h = 1, \ldots, p$, we consider the following functions:

$$
\begin{aligned}
U_{hi1}(\mathbf{x}) &= \frac{1}{2}\left[(v_{\max} - v_{hi})(v_{\max} + v_{hi})\right],\\
U_{hi2}(\mathbf{x}) &= \frac{1}{2}\left[(v_{hi} - v_{\min})(v_{hi} + v_{\min})\right],\\
U_{hi3}(\mathbf{x}) &= \frac{1}{2}\left[(\omega_{\max} - \omega_{hi})(\omega_{\max} + \omega_{h})\right],\\
U_{hi4}(\mathbf{x}) &= \frac{1}{2}\left[(\nu_{\max} - \nu_{hi})(\nu_{\max} + \nu_{hi})\right].
\end{aligned} \tag{10}
$$

These positive functions would make sure that the limitations put on the aircraft's yaw angle and speed were followed appropriately.

4 Design of the Acceleration Controllers

The nonlinear acceleration control laws for system (1) will be designed using LbCS as proposed in Sharma (2008).

4.1 Lyapunov Function

The total potentials, also known as a Lyapunov function, must be constructed next. To do this, we must first define the *control parameters* for $i = 1, \ldots, n$ and $h = 1 \ldots p$, that will be used in the repulsive potential functions. This will be as follows:

(i) $\alpha_{hil} > 0, l = 1, \ldots, q$, for avoidance of q sphere-liked obstacles.
(ii) $\beta_{hik} > 0, k = 1, \ldots, 6$, for workspace of aircraft treated as fixed obstacles.
(iii) $\gamma_{his} > 0, s = 1, \ldots, 4$, for avoidance of obstacles from dynamic constraints.
(iv) $\eta_{hij} > 0, j = 1, \ldots n, j \neq i$, for the collision avoidance between any two aircraft.
(v) $\kappa_{himu} > 0, m = 1, \ldots, p$, and $u = 1, \ldots, n$ with $m \neq h$ for inter formation avoidance.

$$
\begin{aligned}
L(\mathbf{x}) = \sum_{i=1}^{n}\sum_{h=1}^{p}\left[V_{hi}(\mathbf{x}) + G_{hi}(\mathbf{x})\left(\sum_{l=1}^{q}\frac{\alpha_{hil}}{FO_{hil}(\mathbf{x})} + \sum_{k=1}^{6}\frac{\beta_{hik}}{W_{hik}(\mathbf{x})} + \sum_{s=1}^{4}\frac{\gamma_{his}}{U_{his}(\mathbf{x})}\right)\right]\\
+\sum_{i=1}^{n}\sum_{h=1}^{p}\left[G_{hi}(\mathbf{x})\left(\sum_{\substack{j=1\\ j\neq i}}^{n}\frac{\eta_{hij}}{MO_{hij}(\mathbf{x})} + \sum_{\substack{m=1\\ h\neq m}}^{p}\frac{\kappa_{himu}}{D_{himu}(\mathbf{x})}\right)\right].
\end{aligned} \tag{11}
$$

4.2 Nonlinear Acceleration Controllers

The design of the feedback controllers begins by noting that the functions f_{hi} for $i = 1, \ldots, 6$ and $h = 1, \ldots, p$ are defined as (on suppressing $\mathbf{x}$):

$$f_{h11} = \left[1 + \sum_{l=1}^{q} \frac{\alpha_{h1l}}{FO_{h1l}} + \sum_{k=1}^{6} \frac{\beta_{h1k}}{W_{h1k}} + \sum_{s=1}^{4} \frac{\gamma_{h1s}}{U_{h1s}} + \sum_{j=2}^{n} \frac{\eta_{h1j}}{M_{h1j}} + \sum_{m=2}^{p} \frac{\kappa_{h1mu}}{D_{h1mu}}\right]$$

$$(x_{h1} - p_{11}) - \frac{G_{h1}\beta_{h11}}{(W_{h11})^2} + \frac{G_{h1}\beta_{h13}}{(W_{h13})^2}$$

$$-\sum_{i=2}^{n}\left[1 + \sum_{l=1}^{q} \frac{\alpha_{hil}}{(FO_{hil})} + \sum_{\substack{j=1,\\ j\neq i}}^{n} \frac{\eta_{hij}}{M_{hij}} + \sum_{s=1}^{4} \frac{\gamma_{his}}{U_{his}} + \sum_{\substack{m=1,\\ m\neq h}}^{p} \frac{\kappa_{himu}}{D_{himu}}\right]$$

$$(x_{hi} - (x_{h1} - a_{hi})) - G_{h1}\left[\sum_{l=1}^{q} \frac{\alpha_{h1l}}{(FO_{h1l})^2}(x_{h1} - o_{hl1})\right.$$

$$\left. -2\sum_{j=2}^{n} \frac{\eta_{h1j}}{(M_{h1j})^2}(x_{h1} - x_{hj}) - 2\sum_{m=2}^{p} \frac{\kappa_{h1mu}}{(D_{h1mu})^2}(x_{h1} - x_{mu})\right],$$

$$f_{h12} = \left[1 + \sum_{l=1}^{q} \frac{\alpha_{h1l}}{FO_{h1l}} + \sum_{k=1}^{6} \frac{\beta_{h1k}}{W_{h1k}} + \sum_{s=1}^{4} \frac{\gamma_{h1s}}{U_{h1s}} + \sum_{j=2}^{n} \frac{\eta_{h1j}}{M_{h1j}} + \sum_{m=2}^{p} \frac{\kappa_{h1mu}}{D_{h1mu}}\right]$$

$$(y_{h1} - p_{12}) - \frac{G_{h1}\beta_{h12}}{(W_{h12})^2} + \frac{G_{h1}\beta_{h14}}{(W_{h14})^2}$$

$$-\sum_{i=2}^{n}\left[1 + \sum_{l=1}^{q} \frac{\alpha_{hil}}{(FO_{hil})} + \sum_{\substack{j=1,\\ j\neq i}}^{n} \frac{\eta_{hij}}{M_{hij}} + \sum_{s=1}^{4} \frac{\gamma_{his}}{U_{his}} + \sum_{\substack{m=1,\\ m\neq h}}^{p} \frac{\kappa_{himu}}{D_{himu}}\right]$$

$$(y_{hi} - (y_{h1} - b_{hi})) - G_{h1}\left[\sum_{l=1}^{q} \frac{\alpha_{h1l}}{(FO_{h1l})^2}(y_{h1} - o_{l2})\right.$$

$$\left. -2\sum_{j=2}^{n} \frac{\eta_{h1j}}{(M_{h1j})^2}(y_{h1} - y_{hj}) - 2\sum_{m=2}^{p} \frac{\kappa_{h1mu}}{(D_{h1mu})^2}(y_{h1} - y_{mu})\right],$$

$$f_{h13} = \left[1 + \sum_{l=1}^{q} \frac{\alpha_{h1l}}{FO_{h1l}} + \sum_{k=1}^{6} \frac{\beta_{h1k}}{W_{h1k}} + \sum_{s=1}^{4} \frac{\gamma_{h1s}}{U_{h1s}} + \sum_{j=2}^{n} \frac{\eta_{h1j}}{M_{h1j}} + \sum_{m=2}^{p} \frac{\kappa_{h1mu}}{D_{h1mu}}\right]$$
$$(z_{h1} - p_{13}) - \frac{G_{h1}\beta_{h15}}{(W_{h15})^2} + \frac{G_{h1}\beta_{h16}}{(W_{h16})^2}$$
$$-\sum_{i=2}^{n}\left[1 + \sum_{l=1}^{q} \frac{\alpha_{hil}}{(FO_{hil})} + \sum_{\substack{j=1,\\ j\neq i}}^{n} \frac{\eta_{hij}}{M_{hij}} + \sum_{s=1}^{4} \frac{\gamma_{his}}{U_{his}} + \sum_{\substack{m=1,\\ m\neq h}}^{p} \frac{\kappa_{h1mu}}{D_{h1mu}}\right]$$
$$(z_{hi} - (z_{h1} - c_{hi})) - G_{h1}\left[\sum_{l=1}^{q} \frac{\alpha_{h1l}}{(FO_{h1l})^2}(z_{h1} - o_{l3})\right.$$
$$\left. -2\sum_{j=2}^{n} \frac{\eta_{h1j}}{(M_{h1j})^2}(z_{h1} - z_{hj}) - 2\sum_{m=2}^{p} \frac{\kappa_{h1mu}}{(D_{h1mu})^2}(z_{h1} - z_{mu})\right],$$
$$f_{h14} = 1 + G_{h1}\left[\frac{\gamma_{h11}}{(U_{h11})^2} - \frac{\gamma_{h12}}{(U_{h12})^2}\right],$$
$$f_{h15} = 1 + \frac{G_{h1}\gamma_{h13}}{(U_{h13})^2},$$
$$f_{h16} = 1 + \frac{G_{h1}\gamma_{h14}}{(U_{h14})^2},$$

for $i = 2, \ldots, n$ and $h = 1, \ldots, p$

$$f_{hi1} = \left[1 + \sum_{l=1}^{q} \frac{\alpha_{hil}}{FO_{hil}} + \sum_{k=1}^{6} \frac{\beta_{hik}}{W_{hik}} + \sum_{s=1}^{4} \frac{\gamma_{his}}{U_{his}} + \sum_{j=2}^{n} \frac{\eta_{hij}}{M_{hij}} + \sum_{m=2}^{p} \frac{\kappa_{himu}}{D_{himu}}\right]$$
$$(x_{hi} - (x_{h1} - a_{hi})) - \frac{G_{hi}\beta_{hi1}}{(W_{hi1})^2} + \frac{G_{hi}\beta_{hi3}}{(W_{hi3})^2}$$
$$-G_{hi}\left[\sum_{l=1}^{q} \frac{\alpha_{hil}}{(FO_{hil})^2}(x_{hi} - o_{l1}) - 2\sum_{j=1}^{n} \frac{\eta_{hij}}{(M_{hij})^2}(x_{hi} - x_{hj})\right.$$
$$\left. -2\sum_{m=2}^{p} \frac{\kappa_{himu}}{(D_{himu})^2}(x_{hi} - x_{mu})\right],$$

$$f_{hi2} = \left[1 + \sum_{l=1}^{q} \frac{\alpha_{hil}}{FO_{hil}} + \sum_{k=1}^{6} \frac{\beta_{hik}}{W_{hik}} + \sum_{s=1}^{4} \frac{\gamma_{his}}{U_{his}} + \sum_{j=1}^{n} \frac{\eta_{hij}}{M_{hij}} + \sum_{\substack{m=1,\\ m\neq h}}^{p} \frac{\kappa_{himu}}{D_{himu}}\right]$$
$$(y_{hi} - (y_{h1} - b_{hi})) - \frac{G_{hi}\beta_{hi2}}{(W_{hi2})^2} + \frac{G_{hi}\beta_{hi4}}{(W_{hi4})^2}$$
$$-G_{hi}\left[\sum_{l=1}^{q} \frac{\alpha_{hil}}{(FO_{hil})^2}(y_{hi} - o_{l2}) - 2\sum_{j=1}^{n} \frac{\eta_{hij}}{(M_{hij})^2}(y_{hi} - y_{hj})\right.$$
$$\left. -2\sum_{m=2}^{p} \frac{\kappa_{h1mu}}{(D_{h1mu})^2}(y_{hi} - y_{mu})\right],$$

$$f_{hi3} = \left[1 + \sum_{l=1}^{q} \frac{\alpha_{hil}}{FO_{hil}} + \sum_{k=1}^{6} \frac{\beta_{hik}}{W_{hik}} + \sum_{s=1}^{4} \frac{\gamma_{his}}{U_{his}} + \sum_{j=1}^{n} \frac{\eta_{hij}}{M_{hij}} + \sum_{\substack{m=1,\\ m\neq h}}^{p} \frac{\kappa_{himu}}{D_{himu}}\right]$$

$$(z_{hi} - (z_{h1} - c_{hi})) - \frac{G_{hi}\beta_{hi5}}{(W_{hi5})^2} + \frac{G_{hi}\beta_{hi6}}{(W_{hi6})^2}$$

$$-G_{hi}\left[\sum_{l=1}^{q} \frac{\alpha_{hil}}{(FO_{hil})^2}(z_{hi} - o_{l3}) - 2\sum_{j=1}^{n} \frac{\eta_{hij}}{(M_{hij})^2}(z_{hi} - z_{hj})\right.$$

$$\left.-2\sum_{m=2}^{p} \frac{\kappa_{himu}}{(D_{himu})^2}(z_{hi} - z_{mu})\right],$$

$$f_{hi4} = 1 + G_{hi}\left[\frac{\gamma_{hi1}}{(U_{hi1})^2} - \frac{\gamma_{hi2}}{(U_{hi2})^2}\right],$$

$$f_{hi5} = 1 + \frac{\gamma_{hi3}G_{hi}}{(U_{hi3})^2},$$

$$f_{hi6} = 1 + \frac{\gamma_{hi4}G_{hi}}{(U_{hi4})^2}.$$

So, we design the following theorem:

Theorem 1 *Let the ODEs given in system* (1) *control the motion of group of n mobile robots in the C-ITS. The main objective here is to create and maintain a predefined formation, generate aircraft movement in a predefined workspace, and safely reach at the predefined target while maintaining predefined formation. Achieving the predetermined targets, limiting the workspace and taking the kinodynamic constraints are among the subtask that needs to be considered. According to the LbCS of the system* (1)*, the following continuous time-invariant acceleration control laws can be produced using the attracting and repulsive potential field functions which ensures stability of system* (1) *in Lyapunov sense.*

$$\sigma_{hi1} = \frac{-(\delta_{hi1}v_i + f_{hi1}\cos\theta_{hi} + f_{hi2}\sin\theta_{hi})}{f_{hi4}},$$

$$\sigma_{hi2} = \frac{-(\delta_{hi2}\omega_{hi} + \frac{L}{2}f_{hi2}\cos\theta_{hi} - \frac{L}{2}f_{hi1}\sin\theta_{hi})}{f_{hi5}}, \quad (12)$$

$$\sigma_{hi3} = \frac{-(\delta_{hi3}v_{hi} + f_{hi3})}{f_{hi6}}.$$

where $\delta_{hi1} > 0$, $\delta_{hi2} > 0$ *and* $\delta_{hi3} > 0$ *for* $i = 1, \ldots, 3$ *and* $h = 1, \ldots, p$ *are constants commonly known as convergence parameters.*

5 Stability Analysis

Theorem 2 *Given that* $\mathbf{x}_{ehi} := (p_{hi1}, p_{hi2}, p_{hi3}, p_{hi4}, 0, 0, 0) \in \mathbb{R}^7$, *is an equilibrium point for* (1), *then* $\mathbf{x}_e = (x_{e11}, x_{e12}, \ldots, x_{ehn}) \in D(L(\mathbf{x}))$ *is a stable equilibrium point of system* (1).

Proof One can easily verify the following, for $i = 1, \ldots n$, and $h = 1, \ldots, p$:

1. $L(\mathbf{x})$ is defined, continuous and positive over the domain $D(L(\mathbf{x})) = \{\mathbf{x} \in \mathbb{R}^{7n} : FO_{hil}(\mathbf{x}) > 0, l = 1, \ldots, q, MO_{hij}(\mathbf{x}) > 0, i, j = 1, \ldots, n, i \neq j, DO_{himu}(\mathbf{x}) > 0, i, u = 1, \ldots, n$ and $h, m = 1, \ldots, p, h \neq m, U_{his}(\mathbf{x}) > 0, s = 1, \ldots, 4, W_{hik}(\mathbf{x}) > 0, k = 1, \ldots, 6;$
2. $L(\mathbf{x}_e) = 0;$
3. $L(\mathbf{x}) > 0 \;\; \forall \mathbf{x} \in D(L(\mathbf{x}))/\mathbf{x}_e.$

Next, take into account the time derivative of the potential Lyapunov function along a specific system (1) trajectory:

$$\dot{L}_{(1)}(\mathbf{x}) = \sum_{i=1}^{n}\sum_{i=h}^{p}[f_{hi1}\dot{x_{hi}} + f_{hi2}\dot{y_{hi}} + f_{hi3}\dot{z_{hi}} + f_{hi4}v_{hi}\dot{v_{hi}} + f_{hi5}\omega_{hi}\dot{\omega_{hi}} + f_{hi6}\nu_{hi}\dot{\nu_{hi}}].$$

Substituting the controllers given in (12) and the governing ODEs for system (1), we obtain the following semi-negative definite function

$$\dot{L}_{(1)}(\mathbf{x}) = -\left(\delta_{hi1}v_{hi}^2 + \delta_{hi2}\omega_{hi}^2 + \delta_{hi3}\nu_{hi}^2\right) \leq 0.$$

Thus, $\dot{L}_{(1)}(\mathbf{x}) \leq 0 \; \forall \mathbf{x} \in D(L(\mathbf{x}))$ and $\dot{L}_{(1)}(\mathbf{x}_e) = 0$. Finally, it can be easily verified that $L(\mathbf{x}) \in C^1(D(L(\mathbf{x})))$, which makes up the fifth and final criterion of a Lyapunov function. Hence, $L(\mathbf{x})$ is classified as a Lyapunov function for system (1), and $\mathbf{x}_e$ is a stable equilibrium point in the sense of Lyapunov. □

6 Simulation Results and Discussion

In this section, we simulate a virtual C-ITS scenario on our workspace to show how the suggested continuous time-invariant acceleration-based controllers perform in the context of the Lyapunov-based control scheme. We consider how two vehicle formations—one in a horizontal configuration and another in a vertical formation—move toward their target while encountering obstacles. In the presence of obstacles, the formation splits and reassembles at a safer position. Also, as the leaders of each formation move in the direction of their respective destinations, the followers keep the proper distance and orientation from the leaders. Additionally, as seen in Fig. 2,

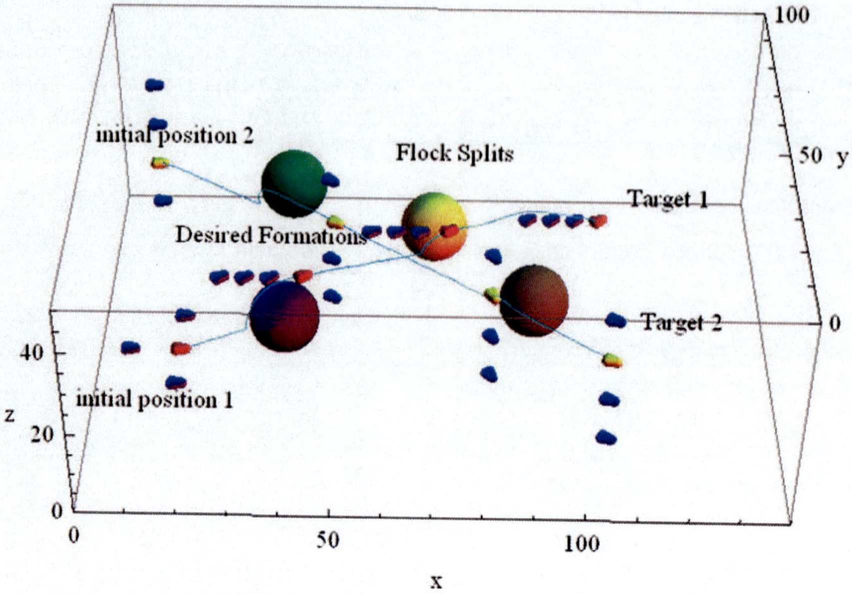

Fig. 2 In the presence of obstacles, two formations of fixed-wing UAVs are traveling in a horizontal and vertical direction, each with a leader (in red and yellow) and followers (in blue and purple). When faced with obstacles, the flock divides into the desired shape, rejoining afterward to avoid intercollision. The system's trajectory is shown by the cyan lines

the individual formation stays clear of intercollision, which is regarded as a moving obstacle in this situation. Table 1 lists the corresponding initial and final setups as well as the additional parameters needed for the simulation. Figure 3 depicts the Lyapunov function's (red) and its derivative's (blue) evolution along the system's trajectory. The translational and rotational velocities are also shown in Fig. 4, and they fall in the presence of obstacles before increasing and convergent to zero close to the goal.

Table 1 Numerical values of initial final states, constraints, and parameters for Fig. 2

	Initial configuration
Workspace	$\eta_1 = 200$, $\eta_2 = 100$, $\eta_3 = 50$
Rectangular positions	$(x_{11}, y_{11}, z_{11}) = (20, 75, 20)$, $(x_{12}, y_{12}, z_{12}) = (20, 85, 20)$
	$(x_{13}, y_{13}, z_{13}) = (20, 65, 20)$, $(x_{14}, y_{14}, z_{14}) = (10, 75, 20)$
	$(x_{21}, y_{21}, z_{21}) = (20, 25, 20)$, $(x_{22}, y_{22}, z_{22}) = (20, 15, 20)$
	$(x_{23}, y_{23}, z_{23}) = (20, 35, 20)$, $(x_{24}, y_{24}, z_{24}) = (10, 25, 20)$
Velocities	$v_{hi} = 1.5$, $\omega_{hi} = 0.5$, $\nu_{hi} = 0.5$ for $i = 1, \ldots, n$, $h = 1, \ldots, p$
Angular positions	$\theta_{ih} = 0$, for $i = 1, \ldots, n$, $h = 1, \ldots, p$
	Constraints and parameters
Dimensions of UAV	$L = 2$, $w = 1$
Leaders target	$(p_{11}, p_{12}, p_{13}) = (195, 30, 23)$, $rt_1 = 2$
	$(p_{21}, p_{22}, p_{23}) = (185, 80, 18)$, $rt_1 = 2$
Position of ghost targets	$(a_{12}, b_{12}, c_{12}) = (5, 0, 0)$, $(a_{13}, b_{13}, c_{13}) = (10, 0, 0)$
	$(a_{14}, b_{14}, c_{14}) = (15, 0, 0)$, $(a_{22}, b_{22}, c_{22}) = (5, 0, 0)$
	$(a_{23}, b_{23}, c_{23}) = (10, 0, 0)$, $(a_{24}, b_{24}, c_{24}) = (15, 0, 0)$
	$(o_{11}, o_{12}, o_{13}) = (80, 40, 20)$, $(o_{21}, o_{22}, o_{23}) = (100, 55, 25)$
Fixed obstacles	$(o_{31}, o_{32}, o_{33}) = (150, 34, 30)$, $(o_{41}, o_{42}, o_{43}) = (140, 75, 30)$
Max. velocities	$v_{\max} = 5$, $\omega_{\max} = 1$, $\nu_{\max}$=1
Min. velocity	$v_{\min} = 1$
Clearance parameters	$\epsilon_1 = 0.1$, $\epsilon_2 = 0.05$
	Control and convergence parameters
Collision avoidance	$\beta_{hij} = 0.001$, for $i, j = 1, \ldots, n$, $j \neq i$, $h = 1, \ldots, p$
	$\alpha_{hil} = 0.01$, for $i = 1, \ldots, n$, $l = 1, \ldots, q$, $h = 1, \ldots, p$
	$\kappa_{himu} = 0.01$, for $h = 1, \ldots p$, $i = 1, \ldots n$, $m = 1, \ldots p$, $u = 1, \ldots n$
Dynamics constraints	$\gamma_{his} = 0.001$, for $i = 1, \ldots, n$, $s = 1, \ldots, 3$, $h = 1, \ldots, p$
Convergence	$\delta_{1ik} = 4000$, $\delta_{1ik} = 100$, $\delta_{2ik} = 5000$, $\delta_{2ik} = 150$
	for $i = 1, \ldots, n$, $k = 1, \ldots, 3$

7 Conclusion

In order to solve motion planners for several formations of fixed-wing aircraft in an environment with obstacles, this paper discusses the split-rejoin method using the leader-follower strategy. Each formation in this scenario has a leader in charge of managing the motion. The remaining robots in the formation place themselves in relation to the leader as followers. The formation moved toward its predetermined destination while navigating the environment in the established formation. The flock divides and reunites at a safer location or a greater distance in the presence of unknown agents, such as obstacles. Additionally, the configuration avoided intercollision, which in this instance was seen as a moving obstacle. Finally, the control rules have been demonstrated, and their efficacy has been demonstrated using

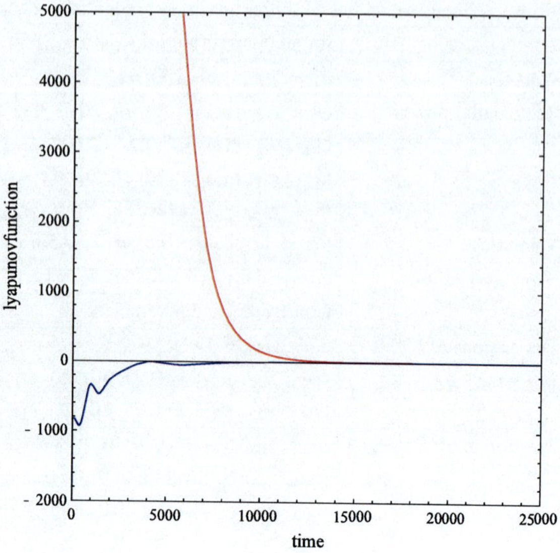

Fig. 3 Evolution of $L(\mathbf{x})$ in red and $\dot{L}_{(1)}(\mathbf{x})$ in blue

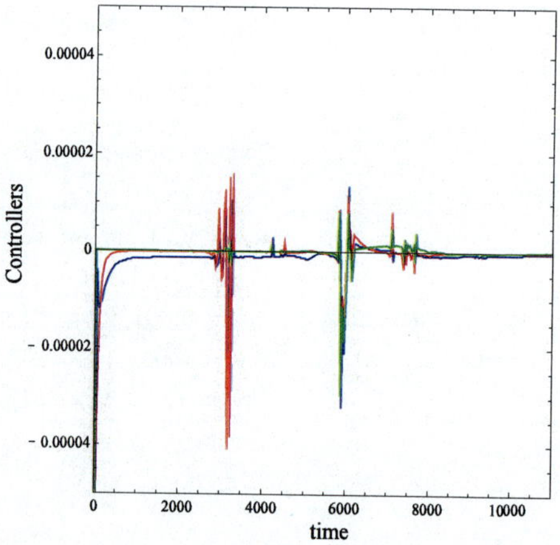

Fig. 4 Evolution of acceleration σ_{11} in blue, σ_{12} in green and σ_{13} in red

computer simulation of a virtual scenario. Future work will involve adding obstacles like cylindrical obstacles and rectangular obstacles. Other mechanical robots, such as quadrotors, can be used to apply the leader-follower strategy to the dynamic examination of stability.

References

S.-J. Chung, J.-J.E. Slotine, Cooperative robot control and concurrent synchronization of Lagrangian systems. IEEE Trans. Robot. **25**(3), 686–700 (2009)

V. Gazi, Swarm aggregations using artificial potentials and sliding-mode control. IEEE Trans. Robot. **21**(6), 1208–1214 (2005)

J. Guo, G. Yan, Z. Lin, Cooperative control synthesis for moving-target-enclosing with changing topologies, in *2010 IEEE International Conference on Robotics and Automation* (IEEE, 2010), pp. 1468–1473

M. Hwangbo, J. Kuffner, T. Kanade, Efficient two-phase 3D motion planning for small fixed-wing UAVs, in *Proceedings of the IEEE International Conference on Robotics and Automation*, Rome, Italy, Apr 2007, pp. 1035–1041

W. Khaksar, M.Z. Uddin, J. Torresen, Fuzzy motion planning for nonholonomic mobile robot navigation in unknown indoor environments. Int. J. Mech. Eng. Robot. Res. **8**(1), 6–11 (2019)

R. Olfati Saber, W.B. Dunbar, R.M. Murray, Cooperative control of multi-vehicle systems using cost graphs and optimization, in *Proceedings of the 2003 American Control Conference, 2003*, vol. 3 (IEEE, 2003), pp. 2217–2222

K. Raghuwaiya, R. Chand, 3D motion planning of a fixed-wing unmanned aerial vehicle, in *2018 5th Asia-Pacific World Congress on Computer Science and Engineering (APWC on CSE)* (IEEE, 2018), pp. 241–245

K. Raghuwaiya, B. Sharma, J. Vanualailai, Leader-follower based locally rigid formation control. J. Adv. Transp. **2018** (2018)

K. Raghuwaiya, S. Singh, B.N. Sharma, G.I. Lingam, Formation types of a flock of 1-trailer mobile robots, in *Proceedings of the 7th IMT-GT International Conference on Mathematics, Statistics and Its Applications* (2011), pp. 368–382

K. Raghuwaiya, S. Singh, Formation types of multiple steerable 1-trailer mobile robots via split/rejoin maneuvers. New Zeal. J. Math. **43**, 7–21 (2013)

J. Raj, K. Raghuwaiya, J. Vanualailai, Collision avoidance of 3D rectangular planes by multiple cooperating autonomous agents. J. Adv. Transp. **1–13**(10), 2020 (2020)

J. Raj, K. Raghuwaiya, B. Sharma, J. Vanualailai, Motion control of a flock of 1-trailer robots with swarm avoidance. Robotica **39**(11), 1926–1951 (2021)

B. Sharma, New directions in the applications of the Lyapunov-based control scheme to the findpath problem. PhD thesis, University of the South Pacific, Suva, Fiji Islands, July 2008

B.N. Sharma, J. Vanualailai, U. Chand, Flocking of multi-agents in constrained environments. Eur. J. Pure Appl. Math. **2**(3), 401–425 (2009)

J.M. Toibero, F. Roberti, P. Fiorini, R. Carelli, Hybrid formation control for non-holonomic wheeled mobile robots, in *Recent Progress in Robotics: Viable Robotic Service to Human* (Springer, 2007), pp. 21–34

MR Image Block-Based Brain Tumour Detection Using GLCM Texture Features and SVM

S. Syedsafi, P. Sriramakrishnan, and T. Kalaiselvi

Abstract A brain tumour is a deadly disease, and it is an unwanted cells development in the human brain. In medical technology, brain tumour detection and diagnosis increase the patient's life days. In this manuscript, an effective block-based brain tumour detection method is proposed for MR images. The proposed method has four steps: convert T2W images into 8 × 8 blocks, Feature extraction, Feature selection and Classification. This method uses the Brain Tumour Segmentation (BRATS)-2013 dataset with High-grade glioma (HGG) and Low-grade glioma (LGG) MRI multi-modal on T2-Weighted images. Feature extraction is achieved by GLCM texture features. The Chi-square test method is implemented to rank features in the feature selection process. Finally, the classification process is achieved by SVM, and it has two main phases training and testing. This method uses 90,000 blocks for training and 36,000 blocks for testing in the combination of HGG and LGG images. The blocks are classified into normal or tumour based on their features during the testing phase. The proposed block-based brain tumour detection method achieves 100% sensitivity, 100% specificity and 100% accuracy in the testing phase.

Keywords Brain tumour · GLCM features · Chi-square test · Classification · SVM

1 Introduction

In the human body brain is a vital organ, and it has only soft tissues (Amin et al. 2020). The human brain is affected by many types of diseases. Amongst them, brain tumour is one of the significant dangerous diseases. The most common signs of brain tumour are low concentration, severe headache and memory loss. The brain tumour

S. Syedsafi (✉) · P. Sriramakrishnan
Department of Computer Applications, Kalasalingam Academy of Research and Education, Krishnankoil, Tamilnadu 626126, India
e-mail: syedsafi.safi@gmail.com

T. Kalaiselvi
Department of Computer Science and Applications, The Gandhigram Rural Institute (Deemed to be University), Dindigul, Tamilnadu 624302, India

A. B. Reddy et al. (eds.), *Proceedings of Third International Conference on Advances in Computer Engineering and Communication Systems*, Lecture Notes in Networks and Systems 612, https://doi.org/10.1007/978-981-19-9228-5_19

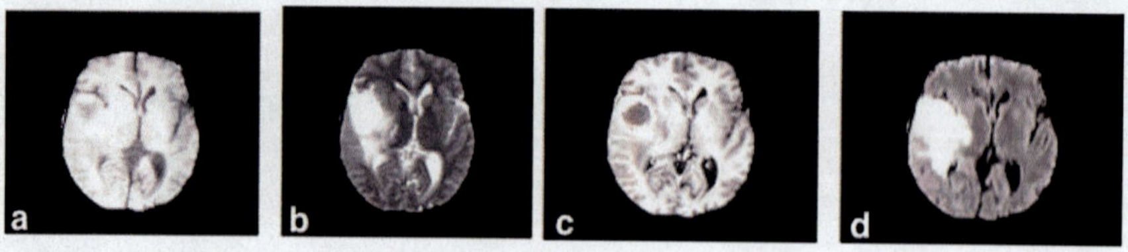

Fig. 1 MRI multi-sequence scans, **a** T1W, **b** T2W, **c** T1C, **d** FLAIR

has two types of classifications: primary and secondary. The primary type of tumour it starts and stays only in the brain, but the secondary type of brain tumour starts anywhere in the human body and moves towards the brain (Kalaiselvi et al. 2019).

Glioma is one of the brain tumour types coming under primary type brain tumour. The medical field gives more important to glioma type tumours because it mainly affects adults (Dupont et al. 2016). The WHO arrange the brain tumours in four grades from Grade I to Grade IV (Johnson et al. 2016). The I and II are coming under low grade type; III and IV are coming under high grade type. The high grade tumours are very danger and also in future low grade is may automatically converted to high grade tumour. So, glioma type tumour classification in the initial stage is very essential. Many scanning procedures are available in the medical field, like CT, PET, and MRI. Compared with other scanning procedures, MRI gives more detail about tumours using the radio frequency and magnetic field. It provides multi-sequence scans T1-weighted (T1W), T2-weighted (T2W), T1-weighted with gadolinium contrast enhancement (T1C), Fluid attenuated inversion recovery (FLAIR) these are shown in Fig. 1 (Kalaiselvi et al. 2016a).

Early detection of glioma type tumour is used to increase the survival days of the patients. The clinical experts do the tumour detection process, but manual detection of tumours in MRI scans takes a long time (Andjouh and Zidane 2018). Therefore, to solve this issue, automated detection is needed. Hence, this manuscript proposes the best automatic detection method for whether the MR image block has a tumour.

The proposed method has four steps: Convert T2W images into 8×8 blocks, Feature extraction, Feature selection and Classification. The classification work is completed using SVM, it classifies the MRI image blocks whether it has a tumour or not. The SVM has two phases: training and testing. These two phases use the 8×8 MR image blocks selected features as inputs. The SVM method classify the image blocks by 0 (normal) and 1 (tumour).

The remaining manuscript is planned as follows: Sect. 2 discuss a work-related to the proposed method, Sect. 3 discuss dataset and metrics, Sect. 4 proposes an efficient method for classifying the MRI image blocks, Sect. 5 discuss the outcomes of the proposed method. Eventually, Sect. 6 gives the conclusions of the manuscript.

2 Related Work

Kharrat et al. proposed an automatic brain tumour classification method using the SVM. In this method, the feature extraction process is performed using Discrete Wavelet Transmission (DWT) and spatial grey level depended matrix (SGLDM). Feature reduction is achieved by using simulated annealing (SA) (Kharrat et al. 2015). Vani et al. have proposed an automatic method for brain tumour classification using SVM. This approach used the cancer imaging archive dataset images. It used the threshold method and DWT to represent the image effectively (Vani et al. 2017).

Bangare et al. have proposed an automatic classification method for tumour using SVM. Initially, the MRI is transformed to a grayscale image with the help of threshold technique, next grayscale image is transformed to a binary (black and white) image. The feature selection process is done by using a Genetic algorithm (GA). The median, mod and mean operations are used to classify the tumour tissues (Bangare et al. 2017). Sukumar and Augustine developed a method for brain tumour classification. This method uses preprocessing work to remove the noises. This method used principal component analysis (PCA) and linear discriminant analysis (LDA) to perform the classification work (Athul Sukumar and Augustine 2017).

Deepika et al. have proposed an SVM-based automatic classification method for a brain tumour. It classifies the MR image into normal or abnormal. The feature extraction process is achieved by local binary pattern (LBP) it automatically removes the unwanted features (Deepika et al. 2019). Polepaka et al. have proposed an Interactive Diagnosis Support System for brain tumour classification using SVM. This method includes three major steps: preprocessing the MRI, extracting features, and classifying the images. It used a two-stage interactive SVM classification technique (Polepaka et al. 2020).

3 Dataset and Metrics Used

3.1 Dataset

In this manuscript, MRI-based BRATS2013 multi-modal dataset is used for SVM training and testing (Menze et al. 2015). It has 20 HGGs and 10 LGGs volumes. Each volume consists of T1, T2, T1C and FLAIR. The dataset also has a ground truth images. Every image has 240 × 240 pixels size. The proposed method used 100 images (T2W) in training phase and 40 images (T2W) in testing phase. The normal type MRI images did not have tumour tissues. The proposed method used the normal and tumour images for the training and testing process in the combination of HGGs and LGGs.

3.2 Metrics

This paper used three metrics to classify the MRI images. The classification output of SVM is calculated by using the following metrics sensitivity, specificity and accuracy. True positive (TP) indicates that the number of tumour blocks is perfectly classified. True negative (TN) indicates that the number of ordinary blocks is perfectly classified. False positive (FP) indicates the number of tumour block is mistakenly classified as normal. False negative (FN) indicates the number of normal blocks is wrongly classified as tumour. Sensitivity denotes the diagnostic test that perfectly identifies all the TP. Specificity denotes the diagnostic test that perfectly identifies all the TN. Accuracy denotes only the true results of either TP or TN in a group. The metric equation is given below (David and Kaufmann 2012; Kumar et al. 2016; Kalaiselvi et al. 2016b).

$$\text{Sensitivity } \% = \frac{\text{TP}}{\text{TP} + \text{FN}} \tag{1}$$

$$\text{Specificity } \% = \frac{\text{TN}}{\text{TN} + \text{FP}} \tag{2}$$

$$\text{Accuracy } \% = \frac{\text{TP} + \text{TN}}{\text{TP} + \text{TN} + \text{FP} + \text{FN}} \tag{3}$$

4 Proposed Method

The manuscript proposes an efficient technique to detect brain tumour. The T2W MRI images are converted into 8 × 8 blocks and extract the features from every block. The features selection is achieved by the Chi-square test method. It is used to find the best features (predictors) by their ranks, and selected predictors are given SVM input for the classification process. SVM has two main phases—training phase and the testing phase. The T2W 8 × 8 blocks selected features are input with class labels in the training phase. The class label 0 indicates the block is normal, 1 indicates the tumour. In the testing phase, T2W 8 × 8 block features labelled by SVM and compared with training blocks to determine accuracy. The proposed method workflow diagram is shown in Fig. 2.

4.1 Convert T2W Images into 8 x 8 Blocks

The proposed method uses only the T2W MRI images. In the BRATS2013 dataset, all T2W images are 240 × 240 pixels. T2W images are converted to 8 × 8 blocks.

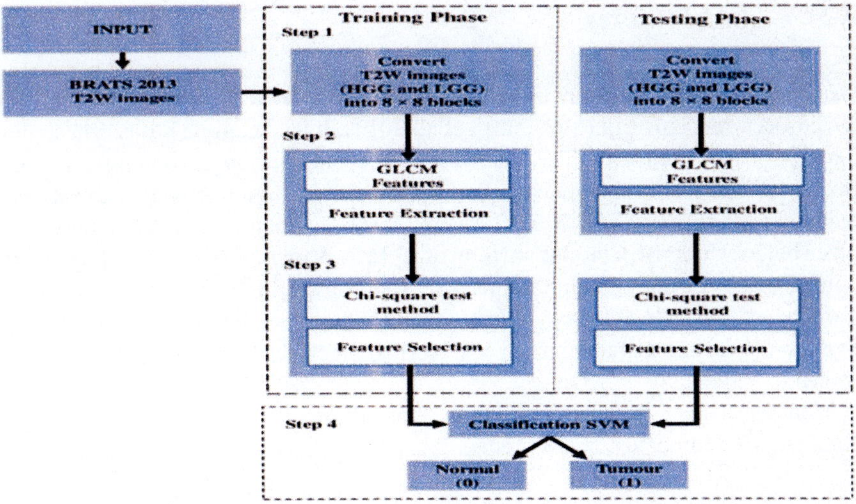

Fig. 2 Workflow diagram of the proposed method

Each block has a totally 64 pixels. So, finally, one single T2W image is converted to 900 blocks. Compared with other MRI image block sizes from 4×4 to 16×16 the 8×8 results are most relevant (Kalaiselvi et al. 2020). A single T2W image with 900 blocks wise conversion is shown in Fig. 3.

Feature Extraction. In the proposed method, grey level co-occurrence matrix (GLCM) is used to find the features from the 8×8 blocks of T2W MRI Images. It denotes the characteristics of the particular image block it is used to distinguish the tumour block from the normal block. In this article, the GLCM matrix finds fourteen features from each block.

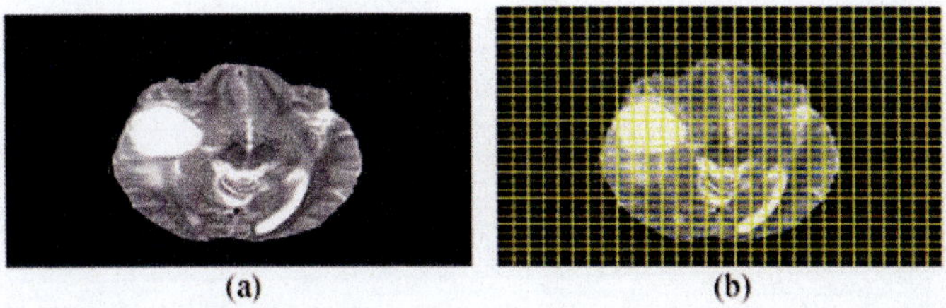

Fig. 3 **a** T2W **b** T2W with 8×8 blocks

4.2 GLCM Features

GLCM features define texture property and spatial features. These are taken out from the co-occurrence matrix (Mohanaiah et al. 2013). It has multiple projection angles like 0°, 45°, 90° and 135°. The GLCM must use only one angle to find the pattern from MRI images. In this manuscript, the 45° projection angle is used to find the features from the 8 × 8 T2W MRI image blocks. Extracting fourteen features are given below: Contrast, Correlation, Sum of squares: Variance, Energy, Homogeneity, Sum average, Entropy, Sum variance, Sum entropy, Difference variance, Cluster prominence, Difference entropy, Cluster shade and Inverse different moment (IDM). The GLCM feature extraction for 8 × 8 Normal block and Tumour block values are shown in Table 1.

Contrast defines the total local grey level difference of MRI image. The high contrast value can define noise or edge. It is predicted by the following equation.

$$\text{Contrast} = \sum_{i=1}^{M}\sum_{j=1}^{N}(X * Y) * \frac{G(i, j)}{R} \tag{4}$$

Here X and Y denote the tuples (rows) and attributes (columns) of the GLCM matrix. $G(i, j)$ denotes the values in created GLCM matrix.

Correlation is used in the particular MRI image pixel group set to find the joint probability event. The equation is given below.

Table 1 GLCM feature extraction for 8 × 8 normal and tumour blocks

S. No.	GLCM features for 8 × 8 T2W MR image block	Normal block	Tumour block
1	Contrast	0.4081	0
2	Correlation	0.7671	NaN
3	Energy	0.4385	1
4	Entropy	1.1390	− 2.22E−16
5	Homogeneity	0.8448	1
6	Sum of squares: variance	55.0755	63.7502
7	Sum average	15.1020	16
8	Sum variance	196.8529	256
9	Sum entropy	1.1390	− 2.22E−16
10	Cluster prominence	10.4210	0
11	Cluster shade	− 3.1707	0
12	Difference variance	0.4081	0
13	Difference entropy	0.7154	− 2.22E−16
14	Inverse different moment (IDM)	0.9938	1

$$\text{Correlation} = \frac{\sum_{I=1}^{M}\sum_{j=1}^{N}\frac{i*j*G(i,j)}{R} - \mu x * \mu y}{\text{o'}x * \text{o'}y} \tag{5}$$

Energy is used to differentiate the tumour parts from the normal parts in the MRI image. The equation is given below.

$$\text{Energy} = \sum_{i=1}^{M}\sum_{j=1}^{N} G(i, j)^2 \tag{6}$$

Entropy is used to calculate the degree of unordered in pixels in the MRI image. It is oppositely correlated with uniformity. The equation is given below.

$$\text{Entropy} = \sum_{i=1}^{Ng}\sum_{j=1}^{Ng} p(i, j)\log[p(i, j)] \tag{7}$$

Homogeneity defines the direct bond between every pixel in the MRI image. The equation is given below.

$$\text{Homogeneity} = \sum_{i=1}^{M}\sum_{j=1}^{N} \frac{G(i, j)}{1 + |i - j|} \tag{8}$$

Sum of squares variance is used to calculate the scatter of the grey level pixels in the MRI image. The equation is given below.

$$\text{Sum of squares variance} = \sum_{i=1}^{Ng}\sum_{j=1}^{Ng} (i - \mu)2p(i, j) \tag{9}$$

Sum average is used to calculate the mean of the grey pixels in the MRI images, the equation is given below.

$$\text{Sum average} = \sum_{i=2}^{2Ng} iPx + y(i) \tag{10}$$

Sum variance is used in concern to the mean, to calculate the scattering of grey pixels of MRI images. The equation is given below.

$$\text{Sum variance} = \sum_{i=2}^{2Ng}\left(i - \left[\sum_{i=2}^{2Ng} iPx + y\right]^{(i)}\right)^2 \tag{11}$$

Sum entropy is used in grey level sum distribution to calculate the unordered pixels in the MRI image. The equation is given below.

$$\text{Sum entropy} = \sum_{i=2}^{2Ng} Px + y^{(i)} \log\{Px + y^{(i)}\} \tag{12}$$

Cluster prominence is used to calculate the imbalance (Unser 1986). When the cluster prominence value is very high, the image has a low imbalance. The equation is given below.

$$\text{Cluster prominence} = \sum_{i=0}^{Ng-1} \sum_{j=0}^{Ng-1} (i + j - ux - uy)^4 p(i, j) \tag{13}$$

Cluster shade is used to calculate the bend of the matrix and also used to calculate the similar concepts of the perception (Unser 1986). The equation is given below.

$$\text{Cluster shade} = \sum_{i=0}^{Ng-1} \sum_{j=0}^{Ng-1} (i + j - ux - uy)^3 p(i, j) \tag{14}$$

Difference variance is used to calculate the scatter with mean of the various spreading of the grey pixels in MRI. The equation is given below.

$$\text{Difference variance} = \sum_{i=2}^{2Ng} \left(i - \left[\sum_{i=2}^{2Ng} i Px - y^{(i)} \right] \right)^2 \tag{15}$$

Difference entropy is used to calculate the unordered distinguish distribution of grey pixels in the MRI image. The equation is given below.

$$\text{Difference entropy} = \sum_{i=0}^{Ng-1} Px - y(i) \log\{Px - y(i)\} \tag{16}$$

Inverse different moment (IDM) is used to calculate the local homogeneity of the MRI image. The equation is given below.

$$\text{IDM} = \sum_{i=0}^{Ng-1} \sum_{j=0}^{Ng-1} \frac{1}{1 + (i - j)^2} p(i, j) \tag{17}$$

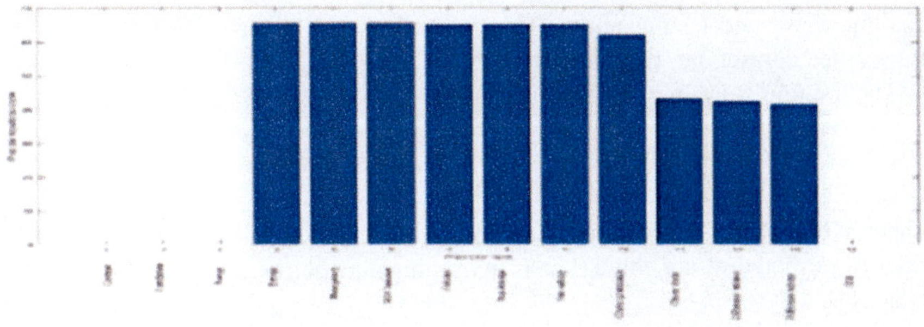

Fig. 4 Predictor rank for 14 GLCM Features using the chi-square test method

4.3 *Feature Selection*

This section uses the Chi-square test method for Feature selection (https://in.mathworks.com/help/stats/fscchi2.html#mw_3a4e15f8-e55d-4b64-b8d0-1253e2734904_head. Accessed 12 Jan 2021). The input has features (Predictors) variables, response variables and response variable names. The *idx* function is used to find the best predictors for the classification problems. First, the *idx* function creates a bar chart for the above fourteen features (Contrast to IDM) discussed in the feature extraction section. It gives the predictor ordered by their predictor importance and gives predictor rank. The empty predictor rank denotes that the feature has infinite values. The proposed method uses the first four rank predictors: Sum entropy, sum average, sum variance and sum of squares: variance. The above four rank predictor values give normal and tumour blocks more differentiated. The bar chart gives the details about the predictor rank shown in Fig. 4.

4.4 *Classification*

Classification is an important task to distinguish the tumour parts from the normal parts in the brain MRI image (Bezdek 1981). The classification performance is calculated by classification accuracy. It defines the ratio of how many images are perfectly classified from the total images. In this manuscript, the classification performance is calculated by SVM classifier.

4.5 *Support Vector Machine (SVM)*

Another name binary classification also calls SVM. The binary SVM has two types linear SVM and non-linear SVM. This SVM method consumes two important phases:

training phase and testing phase. This method uses a hyper-plane to classify the unspecified dataset into two classes.

The training set values are represented as,

$$T = \{x_1 y_1, x_2 y_2, x_3 y_3, \ldots, x_n y_n\} \tag{18}$$

where x represents the feature extraction values and n represents the number of features extracted. y_i are either 0 or 1. In training data points x_i come with their own categories y_i.

The hyper-plane is represented as,

$$f(x) = x!\beta + b = 0 \tag{19}$$

4.6 Training Phase

In the training phase normal and tumour T2-Weighted 50 HGGs and 50 LGGs MR images converted into 8 × 8 blocks. From Fig. 3 one T2W image is converted to 900 blocks, so a total of 50 HGG images are converted to 45,000 blocks, 50 LGG images are converted to 45,000 blocks and extract the fourteen features from every block. The Chi-square test method is used to rank all the fourteen features, the first four rank features are selected and their class labels normal (0), tumour (1) are taken as an input. The input data is given to the SVM classifier for the training phase. The normal blocks and tumour blocks are trained and create their patterns. When the new tumour block comes, the pattern will be updated automatically.

4.7 Testing Phase

In the testing phase, normal and tumour T2-Weighted 40 High-grade glioma images converted to 18,000 blocks and 40 Low-grade glioma images converted to 18,000 blocks, and fourteen features are extracted from every block. The first four rank features are taken without class labels and input for the testing phase. The testing blocks selected features are compared with training phase features to assign the class labels. The SVM classifier assigns labels by using the training phase knowledge.

5 Results and Discussion

In this method brain tumour classification is tested on the BRATS2013 dataset. MATLAB R2021a student version is used for classification with Intel® Core™ i3-1115G4@3.00GHz and 8 GB RAM as hardware specifications. The proposed method is used only the T2W images from the dataset. The proposed method uses the SVM machine learning method to do the classification process. This method uses the two-class labels normal (0) and tumour (1).

The proposed method in the HGG training phase correctly classifies 44,656 blocks as normal (0) and 328 blocks as tumours (1). In the LGG training phase, 44,632 blocks are normal (0) and 368 blocks are tumours (1) out of 90,000 T2W image blocks. This method in the HGG testing phase correctly classifies 17,916 blocks are normal (0) and 84 blocks are tumour (1), and in the LGG testing phase, 17,948 blocks are normal (0), and 52 blocks are tumour (1), out of over all 36,000 T2W MRI image blocks. This method used the support vectors to differentiate the normal blocks from the tumour blocks. The circle denotes the support vectors is shown in Fig. 5. Table 2 shows the detail about training and testing phases results.

Table 2 gives the details of sensitivity, specificity and accuracy for the training phase and testing phase results of HGG and LGG type T2W image blocks. In the HGG training phase, the sensitivity result is 95.34% because 16 tumours (1) blocks are wrongly classified as normal (0) blocks, so the accuracy is 99.96% but, in the LGG training phase, SVM perfectly classifies the normal blocks and tumour blocks so, it gives the 100% accuracy. In both HGG and LGG training and testing phases also perfectly classify the normal and tumour blocks so, both give 100% accuracy.

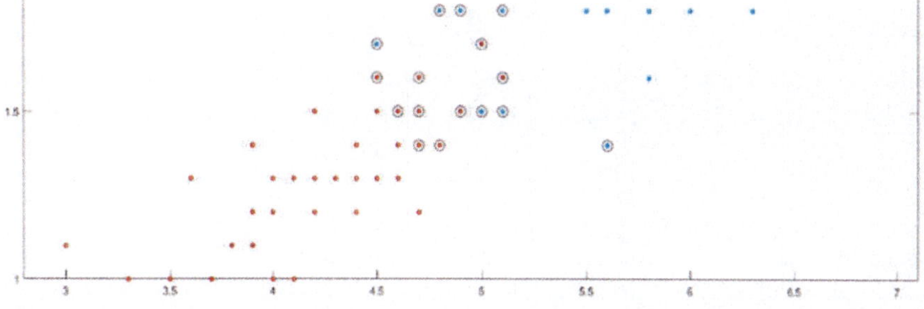

Fig. 5 SVM support vectors

Table 2 HGG and LGG 8 × 8 blocks of training and testing results

Dataset	Image used	Type	Training phase (8 × 8 block size)			Testing phase (8 × 8 block size)		
			Sensitivity%	Specificity%	Accuracy%	Sensitivity%	Specificity%	Accuracy%
BRATS2013	T2W	HGG (45,000 blocks in training) (18,000 blocks in testing)	100	95.34	99.96	100	100	100
		LGG (45,000 blocks in training) (18,000 blocks in testing)	100	100	100	100	100	100
Overall results (90,000 blocks in training) (36,000 blocks in testing)			100	97.67	99.98	**100**	**100**	**100**

Bold indicates the overall training and testing dataset got the final result of 100

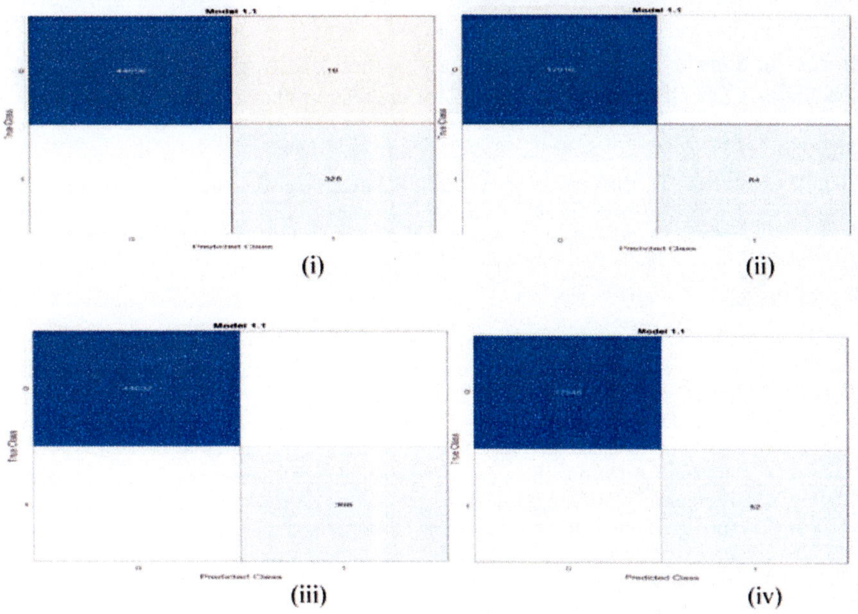

Fig. 6 Confusion matrix for training and testing blocks of HGG and LGG. **a** HGG training, **b** HGG testing, **c** LGG training, **d** LGG testing

The confusion matrix for HGG and LGG blocks in the training and testing phase is shown in Fig. 6.

In the above figure represents the confusion matrix for training and testing phase of the SVM classifier. The confusion matrix has two main classes: True class and Predicted class. True class is called by another name actual class, and Predicted class is called by another name tested class. Both classes have positive and negative values. Figure 6a and b denotes the training phase of SVM, and Fig. 6c and d denotes the testing phase of SVM. Finally, the proposed method gives the overall result for 90,000 blocks (HGG and LGG) in the training phase: sensitivity 100%, specificity 97.67%, accuracy 99.98%, and 36,000 blocks (HGG and LGG) testing phase is sensitivity 100%, specificity 100%, accuracy 100%.

6 Conclusion

This manuscript proposes a block-wise classification method for MR images using the BRATS2013 dataset. This method uses only T2W MR images and is converted to 8 × 8 blocks. The GLCM texture features are used to extract features from the blocks. The Chi-square test method is used for feature selection; it gives rank to all the features. The first four rank features are taken with the class labels and given as training inputs, and without class, labels are given as a testing input. The classification process is done by using the SVM. This method finally gives sensitivity 100%, specificity 100%, the accuracy of 100% for the testing phase.

Conflict of Interest The authors have no conflicts of interest regarding this investigation.

References

J. Amin, M. Sharif et al., A distinctive approach in brain tumor detection and classification using MRI. Pattern Recogn. Lett. **139**, 118–127 (2020). ISSN: 0167-8655

K.K. Andjouh, F. Zidane, Fully automated brain tumour segmentation system in 3D-MRI using symmetry analysis of brain and level sets. IET Image Process. **12**(11), 1964–1971 (2018)

A.M. Athul Sukumar, P. Augustine, Efficient brain tumor classification using PCA and SVM. Int. J. Res. Eng. IT Soc. Sci. **7**, 1–7 (2017)

S.L. Bangare et al., Brain tumor classification using mixed method approach, in *2017 International Conference on Information Communication and Embedded Systems (ICICES)*, pp. 1–4

J.C. Bezdek, Objective function clustering, in *Pattern Recognition with Fuzzy Objective Function Algorithms*. Advanced Applications in Pattern Recognition (Springer, Boston, MA, 1981). https://doi.org/10.1007/978-1-4757-0450-13

B.K. David, W.H. Kaufmann, *Praise of Programming Massively Parallel Processors: A Hands-on Approach*, 2nd edn. (Elsevier, 2012), pp. 1–514

K. Deepika et al., An efficient automatic brain tumor classification using LBP features and SVM-based classifier, in *Proceedings of International Conference on Computational Intelligence and Data Engineering*, Singapore (2019), pp. 163–170

C. Dupont, N. Betrouni et al., On image segmentation methods applied to glioblastoma: state of art and new trends. Innov. Res. BioMed. Eng. **37**(3), 131–143 (2016)

https://in.mathworks.com/help/stats/fscchi2.html#mw_3a4e15f8-e55d-4b64-b8d0-1253e2734904_head. Accessed 12 Jan 2021

D.R. Johnson, J.B. Guerin et al., 2016 updates to the WHO brain tumour classification system: what the radiologist needs to know. Radiographics **37**, 2164–2180 (2016)

T. Kalaiselvi, P. Kumarashankar, P. Sriramakrishnan, S. Karthigaiselvi, Brain tumor detection from multimodal MRI brain images using Pseudo coloring processes. Procedia Comput Sci **165**, 173–181 (2019)

T. Kalaiselvi et al., Three-phase automatic brain tumor diagnosis system using patches based updated run length region growing technique. J. Digit. Imaging **33**, 465–479 (2020)

T. Kalaiselvi, P. Sriramakrishnan, K. Somasundaram, Brain abnormality detection from MRI of human head scans using the bilateral symmetry property and histogram similarity measures, in *International Computer Science and Engineering Conference* (2016a), pp. 1–6

T. Kalaiselvi, P. Sriramakrishnan, P. Nagaraja, Brain tumor boundary detection by edge indication map using bi-modal fuzzy histogram thresholding technique from MRI T2-weighted, I.J. Image Graphics Signal Process. 51–59 (2016b)

A. Kharrat et al., MRI brain tumor classification using support vector machines and meta-heuristic method, in *Proceedings of the 2015 IEEE 15th International Conference on Intelligent Systems Design and Applications (ISDA)*, Marrakesh, Morocco, pp. 446–451 (2015)

R. Kumar et al., GPU-based level set method for MRI brain tumour segmentation using modified probabilistic clustering, in *Medical Imaging: Concepts, Methodologies, Tools, and Applications* (IGI Global, 2017), pp. 1053–1078

B.H. Menze et al., The Multimodal Brain Tumor Image Segmentation Benchmark (BRATS). IEEE Trans. Med. Imaging **34**, 1993–2024 (2015)

P. Mohanaiah et al., Image texture feature extraction using GLCM approach. Int. J. Sci. Res. Publ. **3**(5), 1–5 (2013)

S. Polepaka et al., IDSS-based Two stage classification of brain tumor using SVM. Health Technol. **10**, 249–258 (2020). https://doi.org/10.1007/s12553-018-00290-4

M. Unser, Sum and difference histograms for texture classification. IEEE Trans. Pattern Anal. Mach. Intell. **8**(1), 118–125 (1986). https://doi.org/10.1109/TPAMI.1986.4767760

N. Vani et al., Brain tumor classification using support vector machine. Int. Res. J. Eng. Technol. (IRJET) **4** (2017)

DFM: Deep Fusion Model for COVID-19 Vaccine Sentiment Analysis

Somiya Rani and Amita Jain

Abstract The coronavirus disease (COVID-19) pandemic has created a lot of healthcare concerns. Over the past two years, healthcare professionals worked hard to develop numerous vaccines to combat this virus which is truly remarkable. However, a large proportion of the global population is skeptical about the vaccines and the sudden emergence of the new strain of the virus is stirring up mixed emotions causing the use of opinion terms having varying polarities in different contexts which poses a challenge to predict the accurate sentiments from the user-generated data. In this work, a novel architecture namely a deep fusion model (DFM) with a meta-learning ensemble method is proposed for sentiment analysis of public opinions on COVID-19 vaccines and omicron variant on Twitter. The proposed model employed using natural language processing with deep learning models such as LSTM, GRU, CNN, and their various combinations. The purpose of this study is to understand the public opinion around COVID-19 vaccines and omicron variant through the proposed model. In addition, the experiment demonstrated effectiveness with an accuracy of up to 88% in comparison with state-of-the-art models.

Keywords COVID-19 · Coronavirus · Deep learning · Natural language processing · Sentiment analysis · Vaccine

---

S. Rani (✉)
Department of Computer Science and Engineering, NSUT East Campus (Formerly AIACTR), Guru Gobind Singh Indraprastha University, Delhi, India
e-mail: somiya1093@gmail.com

A. Jain
Department of Computer Science and Engineering, Netaji Subhas University of Technology, Delhi, India
e-mail: amita.jain@nsut.ac.in

A. B. Reddy et al. (eds.), *Proceedings of Third International Conference on Advances in Computer Engineering and Communication Systems*, Lecture Notes in Networks and Systems 612, https://doi.org/10.1007/978-981-19-9228-5_20

1 Introduction

COVID-19 outbreak was declared a pandemic by WHO (Gandhi et al. 2020). On December 31, the first case of COVID-19 was identified and confirmed in Wuhan, China. As of January 18, 2022, there have been 328,532,929 confirmed cases of COVID-19, including 5,542,359 deaths reported by WHO (https://covid19.who.int/). It is well established that vaccination has helped in slowing the coronavirus spread, as it increases immunity, provides memory to the immune system to combat the virus when infection occurs, and remains in the human body to fight against exposure to the virus in the future. Over the past year, many vaccines have been developed such as Oxford/AstraZeneca, Pfizer/BioNTech, Covaxin, Moderna, Sputnik V, Sinovac, Sinopharm, etc. Each vaccine has gone through multiple phases in a clinical trial, safety checks, and protocols, however, people are skeptical and fearful concerning vaccine safety. To fight COVID-19, we need to understand the necessity of vaccination. As of January 16, 2022, a total of 9,395,059,118 vaccine doses have been administered by WHO (https://covid19.who.int/).

Social media is one such platform where people interact with each other and talk about different topics. With natural language processing (NLP) using deep learning (DL) techniques on the user-generated textual data, we can understand the public sentiment regarding vaccines and their effectiveness against COVID-19 and the latest omicron variant which can help to identify the possible misconceptions and fears. Therefore, to achieve this objective, we have proposed a novel 2-tier deep fusion model (DFM) for sentiment analysis (SA) by employing long short-term memory (LSTM), gated recurrent unit (GRU), convolutional neural network (CNN), bidirectional LSTM (BiLSTM), stacked LSTM, stacked GRU, LSTM + CNN, and GRU + CNN models at tier-1, and meta-classifier at tier-2 which chooses the best DL classifier for predicting the final sentiment label. We have used two datasets to test the generalizability of the proposed model. The first dataset consists of COVID-19 vaccine tweets (Preda 2021). The second dataset consists of tweets related to the omicron variant which is collected through Twitter API. We refer to the vaccine tweets dataset as 'VD' and the omicron variant tweets dataset as 'OD'. Tweets are classified into three classes: positive, negative, or neutral.

The main contributions of this paper are as follows:

- Propose a novel deep fusion model that combines DL models with a meta-classifier to overcome the challenge of contextual polarity ambiguity within the user-generated data to identify the accurate sentiment polarity and enhance the classification of COVID-19 vaccines and omicron variant tweets.
- Several deep learning models, LSTM, GRU, CNN, BiLSTM, stacked LSTM, stacked GRU, LSTM + CNN, and GRU + CNN are employed and are combined with meta-classifier.
- Conduct experiments by comparing the performance of DFM with various baseline models to determine the effectiveness of the model.
- Exploratory data analysis of the datasets is presented to identify the meaningful patterns related to COVID-19 vaccines and omicron variant.

The remaining paper is organized as follows: in Sect. 2, a literature review is conducted. In Sect. 3, we thoroughly explained the proposed methodology. In Sect. 4, we discussed the experiment conducted, and the results obtained. Lastly, the paper is concluded in Sect. 5.

2 Literature Review

COVID-19 news has surpassed any other news on social media. Twitter is one of the most popular social media platforms with 500 COVID-19 related tweets sent daily. COVID-19 variant and the vaccine are among the hot topics since coronavirus emergence. Authors have utilized this data and performed SA and various other NLP tasks. For example, Li et al. (2020) analyze the tweets on COVID-19 to assess the mental health of the users. Eight emotions were used to classify the tweets with DL models. Similarly, Chandra and Krishna (2021) employ LSTM with global vector embedding and the BERT language model on tweets from India to investigate the emotion expressed in the tweets. Additionally, Wang et al. (2020) propose a method to classify the sentiments expressed in Weibo posts related to COVID-19 vaccines by using BERT and performing the topic modeling by using TF-IDF. Furthermore, Samuel et al. (2020) employ different ML classifiers on short and long tweets. The SA is also performed on news headlines. For example, Aslam et al. (2020) perform sentiment classification of global English news headlines. Fear, trust, anticipation, sadness, and anger were the main emotions evoked by news headlines and showed 52% negative, 30% positive, and 18% neutral sentiment. Another study was done by Imran et al. (2020), where LSTM is used to analyze and estimate the sentiment polarity of the tweets from citizens of different culture. Jelodar et al. (2020) used NLP-based LSTM RNN approach for deep sentiment classification and topic discovery of COVID-19 posts on Reddit. An accuracy of 81.15% was achieved for the sentiment classification task. Zhang et al. (2020) conducted SA on Chinese micro-blogs related to COVID-19. Sanders et al. (2021) perform SA on tweet clusters, and they developed a divisiveness score to assess the sentiment polarity of tweets.

3 Proposed Methodology

In this section, we discussed the data utilized in this study, pre-processing, and proposed methodology. The proposed deep fusion model is depicted in Fig. 1. We have given an in-depth description of each model and discussed the performance achieved in the experiment evaluation, result, and discussion section.

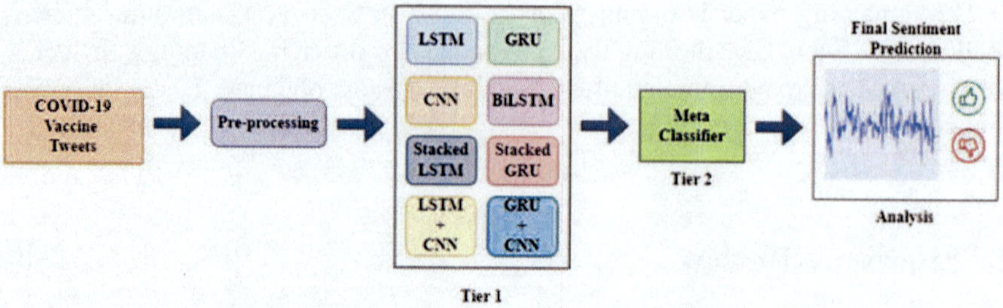

Fig. 1 Proposed deep fusion model (DFM)

3.1 Data and Pre-processing

The datasets, namely VD and OD, used in this study feature the tweets associated with the COVID-19 vaccine from December 20, 2020, to November 15, 2021 (Preda 2021), and tweets on the omicron variant from January 29, 2022, to February 1, 2022. The VD consists of 228,207 tweets and OD consists of 84,993 tweets. The OD tweets are collected from Twitter API using the terms "omicron", "omicron variant", "#omicron", and "third wave". DFM is trained using a tweet sentiment extraction dataset (Kaggle 2020). The collected data is first preprocessed to remove noise data. The tweets are tokenized to present each word of the sentence as a token, the stop words and numeric values are removed, each token is converted to its stem word and lemma, and any URL, punctuation marks, or duplicate tweets are removed from the tweets.

3.2 Methodology

The proposed DFM in the paper is based on NLP using DL models. The main idea behind the DFM is to fuse DL models and utilize the meta-classifier to opt for the best classifier for making the final sentiment prediction.

Deep fusion Model. Most studies related fusion of models are restricted to ML models. Therefore, we aimed for utilizing only DL models to learn more contextual features. Inspired by Sesmero et al. (2015) DL classifiers viz., LSTM, GRU, CNN, BiLSTM, stacked LSTM, stacked GRU, LSTM + CNN, and GRU + CNN are combined with a fusion strategy. The fusion of these classifiers is done to make an ensemble decision which ultimately leads to more accurate results in contrast to the individual model. The ensemble decision is considered more accurate as each of the classifiers learns the features differently and when individual predictions are combined the resultant model is considered more accurate (Sesmero et al. 2015). The meta-classifier at the second tier (tier-2), combines the outputs from the individual base classifier at the first tier (tier-1) and generates the result. Several efficient meta-learning techniques such as bagging, boosting, and stacking exist (Wolpert 1992; Zhang et al. 2014). In this study, a stacking technique is implemented where the gradient boost is used as the meta-classifier is trained on the output of the base classifier to fuse the predictions and make the final sentiment prediction.

Long Short-Term Memory. The first model we have applied is a unidirectional LSTM which can process variable length sequential data. LSTM takes word representations from the embedding layer and produces the input representations which are then fed into hidden layers to extract the features. A dense layer interprets the LSTM output and improves the model performance by overfitting. Then, a dense + softmax layer predicts the sentiment of the tweet. An LSTM is typically computed using Eqs. (1)–(5) as shown below

$$f_t = \sigma\left(w_f \cdot \left[h_{t-1}, x_t\right] + b_f\right) \tag{1}$$

$$i_t = \sigma\left(w_i \cdot \left[h_{t-1}, x_t\right] + b_i\right) \tag{2}$$

$$o_t = \sigma\left(w_o \cdot \left[h_{t-1}, x_t\right] + b_o\right) \tag{3}$$

$$c_t = f_t \odot c_{t-1} + i_t \odot \tanh\left(w_c \cdot \left[h_{t-1}, x_t\right] + b_c\right) \tag{4}$$

$$h_t = o_t \odot \tanh(c_t) \tag{5}$$

where f_t is the forget gate, i_t is the input gate, o_t is the output gate, h_t is the hidden state, c_t is the cell state, and x_t is input vector. $w_f, w_i, w_o \in \mathbb{R}^{d\times 2d}$ is the weighted matrices, and $b_f, b_i, b_o \in \mathbb{R}^{d}$ is the biases of LSTM to be learned during training, parameterizing the transformations of the forget, input, and output gates, respectively.

Gated Recurrent Unit. GRU is similar to LSTM with an additional memory unit. The embedding representations from the embedding layers is fed into the unidirectional GRU. Then, the dense layer works similarly as it does in LSTM. Finally, the dense + softmax layer predicts the sentiment polarities of the tweets. A GRU cell is computed using Eqs. (6)–(9) as shown below, where z_t is the update gate, r_t is the reset gate, and $\overline{h}_t$ is the candidate state.

$$z_t = \sigma(w_z x_t + U_z h_{t-1} + b_z) \tag{6}$$

$$r_t = \sigma(w_r x_t + U_r h_{t-1} + b_r) \tag{7}$$

$$\overline{h}_t = \tanh(w_h x_t + r_{t-1} \odot (U_r h_{t-2}) + b_h) \tag{8}$$

$$h_t = (1 - z_t) \odot h_{t-1} + z_t \odot \overline{h}_t \tag{9}$$

Convolutional Neural Network. CNN is another popular DL model widely used for SA. Conv1D is applied to learn the textual features. Firstly, the embedding layers generate the embeddings E_w for each word using Eq. (10), where S^w is the embedding matrix and v^w is a vector of size $|V_w|$ with value 1 at index w and zero in all positions. These embeddings are then fed into the Conv1D layer which produces feature map $M = [m_0, m_1, m_2, \ldots, m_s - h]$ by applying filter F of size $M \times h$ to the sub-matrices of input matrix using Eq. (11), where $i \in \{0, 1, \ldots, s - h\}$ and $S_{i:j}$ is a sub-matrix of S from i to j. Max-pooling is used to select the most significant feature (msf) as shown in Eq. (12).

$$E_w = S^w \cdot v^w \tag{10}$$

$$m_i = F \cdot S_{i:i+h-1} \tag{11}$$

$$\text{msf} = \max_{0 \le i \le s-h} \left(m_j\right) \tag{12}$$

Bidirectional LSTM (BiLSTM). Bidirectional is similar to unidirectional LSTM but with an additional capability of encoding the input sequence in both forward and backward directions. BiLSTM has more trainable parameters in comparison with LSTM and takes longer to train. However, it results in more meaningful output. Therefore, we used this as the fourth model.

Stacked Model. The DL models perform well with more layers as it provides an additional level of abstraction for feature learning and improves the performance of the model in comparison with the conventional DL models. For this reason, we opted for the stacked model which has two DL models and the output from both models is concatenated into a single vector. The vector is then processed with a dense layer and an output layer. We have created four variations of the DL models, i.e., stacked LSTM which consists of two LSTM layers, stacked GRU which consists of two GRU layers, LSTM + CNN consists of LSTM and CNN layer, and the last combination GRU + CNN consisting of GRU and CNN layer which allowed the models to learn the integration in the best possible way.

4 Experiment Evaluation, Results, and Discussion

The visualization of the VD and OD is provided in Figs. 2 and 3 through a color-coded word cloud of the tweets. Some common words frequently observed in the VD are "Vaccine", "Moderna", and "Covaxin". The most positive tweets were from the people who either received the first dose or were fully vaccinated. It was found that most of the negative tweets were re-tweeted from diplomats. The neutral tweets were informational and showcased day-to-day news headlines and guidelines. Word cloud of the frequent terms in OD shows words such as "omicron variant", "antiviral effect", "omicron surge", and "omicron wave". These words are frequently observed in the tweets showing peoples' concerns toward the omicron variant. Most of the positive tweets were observed from healthcare workers, medical practitioners, and pharmaceutical researchers. Positive tweets showed a high level of confidence in the vaccine's protection against the omicron variant. However, most tweets showed negative polarity as the people are not sure whether the vaccine protects from the omicron variant.

The methodology used in this paper is based on DL models namely LSTM, GRU, CNN, BiLSTM, stacked LSTM, stacked GRU, LSTM + CNN, and GRU + CNN for our novel deep fusion model for SA of the COVID-19 vaccine and omicron variant tweets. DFM has achieved great results in comparison with the baseline models and has shown effectiveness on VD and OD. The results achieved from the models are summarized in Table 1. The LSTM achieved 78.13% for VD and 79.21% accuracy for OD. The LSTM and GRU are performing almost similarly. The training time for GRU was comparatively less than LSTM. The BiLSTM performed better than LSTM and GRU and achieved an accuracy of 81.53% for VD and 80.13% for OD. We did not observe any performance improvement for CNN but the accuracy for stacked models achieved quite a high performance. This is due to the reason that the separate layer of the neural network is adding a level of abstraction and helps in extracting more complex features than the single-layer model. The combination models, LSTM + CNN and GRU + CNN showed drastic improvement in performance compared

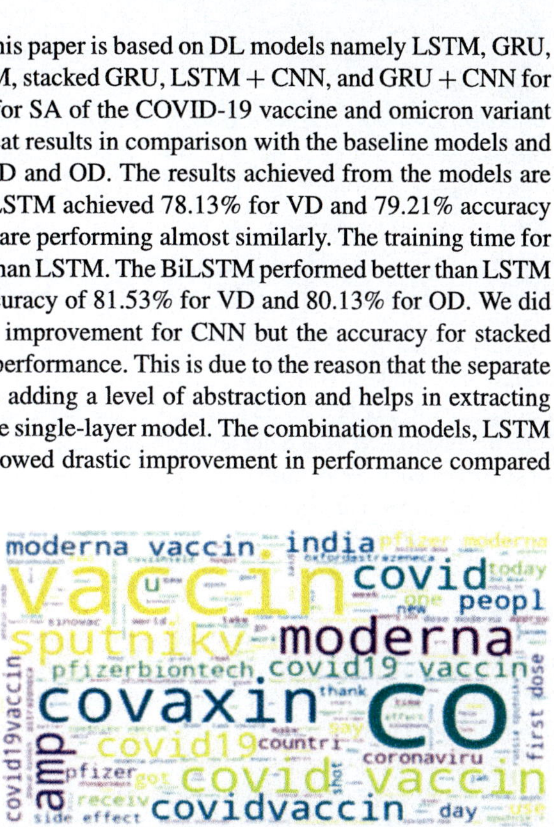

Fig. 2 Word cloud of the tweets in the VD

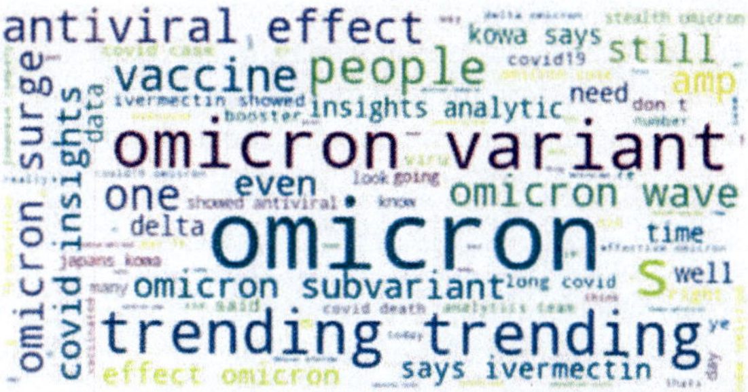

Fig. 3 Word cloud of the tweets in the OD

Table 1 Comparison of models in terms of accuracy, precision, and recall on VD and OD (%)

Dataset	VD			OD		
Model	Accuracy	Precision	Recall	Accuracy	Precision	Recall
LSTM	78.13	77.32	77.07	79.21	78.44	78.1
GRU	79.47	77.46	76.07	78.91	77.12	76.88
CNN	78.2	76.74	76.2	78.44	76.03	75.89
BiLSTM	81.53	79.9	78.13	80.13	79.1	77.99
Stacked LSTM	80.27	78.1	78.27	81.66	80.45	80.1
Stacked GRU	80.07	79.09	77.27	82.64	81.87	81.28
LSTM + CNN	85	84.89	83.07	84.87	83.98	82.45
GRU + CNN	83	81.65	82.33	84.89	83.13	83.49
BERT (Wang et al. 2020)	81.34	80.19	78.62	79.53	78.94	78.77
Naïve Bayes (Villavicencio et al. 2021)	78.47	77.21	76.56	81.75	80.31	80.67
DFM (proposed)	86.7	85.11	84.89	87.82	86.94	85.47

to baselines models. BERT proposed by Wang et al. (2020) achieved an accuracy of 81.34% on VD and 79.53% on OD. The naïve Bayes model achieved similar performance to LSTM and achieved an accuracy of 78.47% on VD and 81.75% on OD (Villavicencio et al. 2021). DFM has achieved outstanding performance among all the models with an accuracy of 86.70% for VD, and an accuracy of 87.82%.

5 Conclusion

This paper illustrates the application of the DFM for the sentiment analysis of public opinion toward COVID-19 vaccines and the omicron variant expressed on Twitter. DFM is implemented using four DL classifiers and their four different variations to fuse them with meta-classifier to enhance the model. The analysis of 228,207 COVID-19 vaccines and 84,944 omicron variant tweets provides insights into people's concerns about vaccines and omicron variants. Additionally, the findings from the experiments show the effectiveness of the models and achieved better performance in comparison with different baseline models. The results suggest that the model is performing well for sentiment analysis and can further be applied to other domains in the future. We observe that the BiLSTM, stacked models, and combination of models were better. LSTM, GRU, and CNN models were slightly weak due to the existence of bias in the training dataset which caused the misclassification of some tweets. In the future, we try removing these limitations by utilizing a larger dataset which will improve the accuracy. In addition, the model could be extended to topic modeling and fake tweet detection related to the COVID-19 vaccines and emerging infectious diseases in the future.

References

F. Aslam, T.M. Awan et al., Sentiments and emotions evoked by news headlines of coronavirus disease (COVID-19) outbreak. Hum. Soc. Sci. Commun. **7**(1), 1–9 (2020). https://doi.org/10.1057/s41599-020-0523-3

R. Chandra, A. Krishna, COVID-19 sentiment analysis via deep learning during the rise of novel cases. PLoS ONE **16**(8), e0255615 (2021). https://doi.org/10.1371/journal.pone.0255615

R.T. Gandhi, J.B. Lynch, C. Del Rio, Mild or moderate Covid-19. N. Engl. J. Med. **383**, 1757–1766 (2020). https://doi.org/10.1056/nejmcp2009249

A.S. Imran, S.M. Daudpota et al., Cross-cultural polarity and emotion detection using sentiment analysis and deep learning on COVID-19 related tweets. IEEE Access **8**, 181074–181090 (2020). https://doi.org/10.1109/ACCESS.2020.3027350

H. Jelodar, Y. Wang et al., Deep sentiment classification and topic discovery on novel coronavirus or covid-19 online discussions: NLP using LSTM recurrent neural network approach. IEEE J. Biomed. Health Inform. **24**(10), 2733–2742 (2020). https://doi.org/10.1109/JBHI.2020.3001216

Kaggle, Tweet Sentiment Extraction (2020). https://www.kaggle.com/c/tweetsentiment-extraction/overview. Accessed 13 Jan 2022

I. Li, Y. Li, T. Li et al., What are we depressed about when we talk about covid-19: mental health analysis on tweets using natural language processing, in *International Conference on Innovative Techniques and Applications of Artificial Intelligence* (2020), pp. 358–370. https://doi.org/10.1007/978-3-030-63799-6_27

G. Preda, COVID-19 All Vaccines Tweets (2021). https://doi.org/10.34740/KAGGLE/DSV/2845240. Accessed 13 Jan 2022

J. Samuel, G.G. Ali et al., Covid-19 public sentiment insights and machine learning for tweets classification. Information **11**(6), 314 (2020). https://doi.org/10.3390/info11060314

A.C. Sanders, R.C. White et al., Unmasking the conversation on masks: natural language processing for topical sentiment analysis of COVID-19 Twitter discourse. medRxiv 2020-08 (2021). https://doi.org/10.1101/2020.08.28.20183863

M.P. Sesmero, A.I. Ledezma, A. Sanchis, Generating ensembles of heterogeneous classifiers using stacked generalization. Wiley Interdiscip. Rev.: Data Min. Knowl. Discov. **5**(1), 21–34 (2015). https://doi.org/10.1002/widm.1143

C. Villavicencio, J.J. Macrohon et al., Twitter sentiment analysis towards covid-19 vaccines in the Philippines using naïve Bayes. Information **12**(5), 204 (2021). https://doi.org/10.3390/info12050204

T. Wang, K. Lu et al., COVID-19 sensing: negative sentiment analysis on social media in China via BERT model. IEEE Access **8**, 138162–138169 (2020). https://doi.org/10.1109/ACCESS.2020.3012595

WHO COVID-19 Dashboard. World Health Organization. https://covid19.who.int/. Accessed 18 Jan 2022

D.H. Wolpert, Stacked generalization. Neural Netw. **5**(2), 241–259 (1992). https://doi.org/10.1016/S0893-6080(05)80023-1

Y. Zhang, H. Zhang et al., A weighted voting classifier based on differential evolution. Abstr. Appl. Anal. (2014). https://doi.org/10.1155/2014/376950

H. Zhang, S. Sun et al., Sentiment classification for Chinese text based on interactive multitask learning. IEEE Access **8**, 129626–129635 (2020). https://doi.org/10.1109/ACCESS.2020.3007889

Driver Assistance System for Recognition of Brake and Parking Signal

Shripad Bhatlawande, Vaishnavi Mhamane, Anand Pande, Atharv Parbalkar, and Swati Shilaskar

Abstract This work presents a driver assistance system to detect brake and parking signals. Vehicle present ahead of the host vehicle performs sudden action such as applying brakes. The host vehicle's driver must respond in real time to avoid accidents or collisions. Detection of action performed by a leading vehicle using its taillights is implemented with computer vision and machine learning techniques. Features are extracted using scale invariant feature transform (SIFT) and accelerated-KAZE (AKAZE). The dimensions are reduced using K-means clustering and then by principal component analysis (PCA). Five classification models have been trained to evaluate the performance. Random forest classifier provided the highest accuracy of 82% among all classifiers. Voting classifier provides final prediction based on five classification model's output. It provided 81% accuracy.

Keywords Driver assistance system · Computer vision · Machine learning · Brake signal detection · Parking signal detection

S. Bhatlawande (✉) · V. Mhamane · A. Pande · A. Parbalkar · S. Shilaskar
Department of E&TC Engineering, Vishwakarma Institute of Technology, Pune 411037, India
e-mail: shripad.bhatlawande@vit.edu

V. Mhamane
e-mail: vaishnavi.mhamane20@vit.edu

A. Pande
e-mail: anand.pande20@vit.edu

A. Parbalkar
e-mail: atharv.parbalkar20@vit.edu

S. Shilaskar
e-mail: swati.shilaskar@vit.edu

A. B. Reddy et al. (eds.), *Proceedings of Third International Conference on Advances in Computer Engineering and Communication Systems*, Lecture Notes in Networks and Systems 612, https://doi.org/10.1007/978-981-19-9228-5_21

1 Introduction

The lead cause of many accidents is a human fault (Reliance General Insurance 2022). Distracted, drunk, speeding/reckless driving, and tailgating are the most common reasons behind the crashes. According to the American Automobile Association (AAA), 9.5% of crashes/accidents have happened due to drowsy driving. The National Safety Council (NSC) report indicates that about 100,000 crashes, 71,000 injuries, and 1550 fatalities are due to drowsy driving (Borelli and Kempken 2022). Advanced driver assistance systems have proven critical in preventing such incidents and accidents (Synopsys 2022). The system capable of detecting the front car's behavior is very helpful for overcoming the crash/accident rate.

2 Literature Survey

Preprocessing techniques such as boundary detection and noise removal are used in Gupta et al. (2016) carried further with image processing and pattern recognition using linear SVM and KNN classifiers. The presented system uses a driving simulator (Saito et al. 2016) to investigate the system's effectiveness by simultaneously checking driver status and performing safety control of the vehicle. The hierarchical framework is proposed in Cui et al. (2015) for detecting alert signals such as brake and turning signals in the daytime. Daytime detection of break signals is done in Chen et al. (2015), and it gives a detection rate of 79%. It works in 2 parts first detecting the taillight and then detecting its status. Detection of taillamps and headlamps at night is accomplished using blob detection in Rezaei et al. (2015), Humnabadkar et al. (2019) and Dave et al. (2016). Interconnected pixels having high luminance are taken as blobs. Detection of vehicles is done using optical cameras in Pradeep and Ramanathan (2018) which can detect vehicles having both the taillights and only one taillight functional. The proposed system (Zhang et al. 2015) detects the vehicles by taking red plane images obtained by simultaneous noise reduction, and edge enhancement processing is achieved using the ADV morphology function.

The exploited system has used a CMOS-based sensor and RGB color filter-based camera (O'Malley et al. 2011). This captures the rear side lamps of target vehicles by filtering out bright objects. In O'Malley et al. (2008), white regions near to red colored portions are searched to get symmetrical pairs. Bounding boxes are obtained using the constraints of aspect ratio. Another algorithm is proposed in Lee et al. (2014) which detects and tracks vehicles by calculating the intensity of diffused taillights. This reduces the effects of signals and streetlights. In Weis et al. (2017), sub-modalities are extracted using cognitive theory, which is similar to a graphical model. Nakagami-m distribution is used in Pillai et al. (2016) and Chen et al. (2012, 2014, 2017) for detection. In order to get the region of the interested segment, the scattering of taillights is used. Chen et al. (2017) provided 95% vehicle detection and 90% turn (left and right) signal detection accuracy. Blob properties analysis and convolutional neural networks (CNN) are used in Bogacki and Dlugosz (2019) to improve the accuracy.

The system exploited in Qing and Jo (2013) uses Gabor feature set for training BPNN and particle filter to track multiple vehicles and overcome vehicle occlusion. Brake light patterns (BLPs) are used for the detection of vehicles in Wang et al. (2016) by using a multilayer perception neural network. YOLOv3 (Li et al. 2020) algorithm is used for training and validating the datasets with Pascal VOC and Labeling tool. The system gives an accuracy of between 40 and 80%. YOLOv3-tiny in Alsfasser et al. (2019) is used as the backbone model and uses a feature fusion strategy to increase prediction layers. This system fails when the backlights are obscured by a specific threshold.

3 Methodology

The proposed system detects brake and parking signals for alerting the driver. The detection and classification system include a camera and a processor-based system. It detects and classifies brake signal and parking signal. Feedback is provided in the form of voice and text messages. Block diagram of the system is shown in Fig. 1.

3.1 Details of Dataset

The dataset consists of a total of 6207 images. It includes 3 classes, namely (i) brake signal, (ii) parking signal, and (iii) negative. It includes 60% images captured using a smartphone camera, and remaining 40% images are obtained from the Internet. Sample images from the dataset are shown in Fig. 2.

The dataset is divided into 4503 training images and 1704 testing images. The distribution of images among classes is listed in Table 1.

Fig. 1 Detection and classification of brake and parking signal

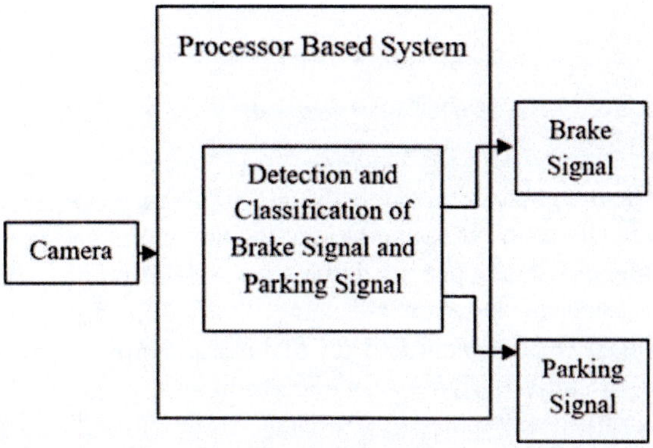

Fig. 2 Sample images from the dataset

Table 1 Distribution of dataset among the classes

Sr. No.	Classes	No. of training images	No. of testing images
1.	Brake signal	1503	600
2.	Parking signal	1500	600
3.	Negative images	1500	504
	Total no. of images	4503	1704
		6207	

3.2 Design and Implementation of System

The images are resized to 250 × 250 pixels and then converted to grayscale. Prewitt edge detection is applied to get the horizontal and vertical edges in the images. The images are then provided to a feature descriptor to extract the features.

Features are extracted using two feature descriptors, namely (i) SIFT and (ii) AKAZE. SIFT identified key features and provided a feature vector of size 2,023,926 × 128 after applying it on 4503 training images.

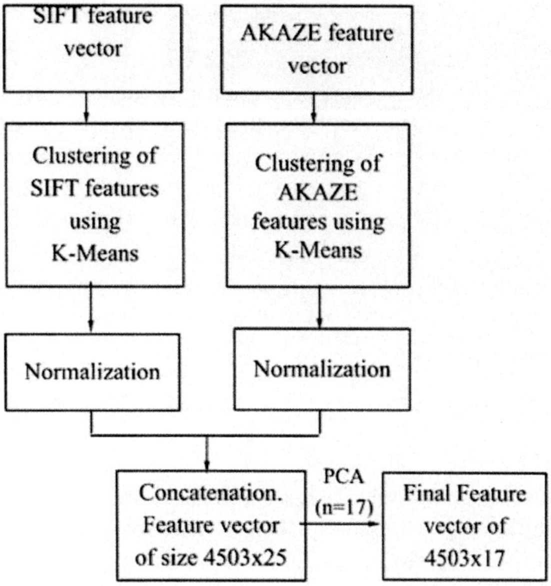

Fig. 3 Compilation of optimized feature vector

The AKAZE detector is based on the Hessian matrix determinant. Detector response maxima in spatial area is taken as a feature point. It provided a feature vector of size 576,419 × 61 after applying it on 4503 images.

The processing of large sized feature vector requires high computational cost. The dimension of the feature vector is reduced using K-means clustering and PCA. K-means clustering is applied on the SIFT feature vector of size 2,023,926 × 128 and the AKAZE feature vector of size 576,419 × 61. Euclidean distance measure is used to identify the clusters. Value of '*k*' (number of clusters) for SIFT and AKAZE feature vectors is 13 and 11, respectively. It is decided using the elbow curve. The feature vector for each image is obtained by applying the K-means model. This vector is normalized and concatenated to form a feature vector of size 1 × 24 for each image. Feature vectors of all images are appended to form a resultant vector of size 4503 × 24.

The highly correlated features are merged using PCA. The first 17 columns of the data represent 99% of the variation. PCA reduces 24 columns to 17 columns for training feature vectors. Optimized feature vector of size 4503 × 17 is obtained using Algorithm 1. The process of compilation of optimized feature vectors is shown in Fig. 3.

Algorithm 1 Computation of optimized feature vector

Input: Image dataset (4503 images)

Output: Feature Vector with reduced dimension (4503 × 17)

1. **For** each image in dataset

1. extract features—SIFT
2. extract features—AKAZE
3. Hist1 = Apply Pre-trained K-Means to SIFT features [$K = 13$]
4. Hist2 = Apply Pre-trained K-Means to AKAZE features [$K = 11$]
5. Hist1 = Normalize(Hist1) //normalize SIFT features
6. Hist2 = Normalize(Hist2) //normalize AKAZE features
7. Hist_new = concatenate(Hist1,Hist2) //Concatenate the features
8. FD.append(Hist_new)
9. End **For**
10. Final = PCA(FD, $n = 17$) //Optimized feature vector of size 4503×17

The training and testing data at a ratio of 80:20 are distributed. Classification is performed using 5 classifiers, namely (i) decision tree classifier, (ii) random forest classifier, (iii) K-nearest neighbor classifier, (iv) support vector machine (SVM), and (v) Gaussian NB classifier. The random forest classifier provided highest accuracy among all classifiers. It has several decision trees in various subparts and results from different decision trees which are used to decide the object's end class. It uses mean square error (MSE) in Eq. (1) to form a branch from each node.

$$\text{MSE} = \frac{1}{N}\sum_{i=1}^{N}(ki - li)x^2 \tag{1}$$

where N is the data points count,
ki is the value returned by model,
li is data point i actual value.

Voting-based classifier takes multiple models as parameters. The classifier predicts the result on the basis of maximum votes obtained for the same class by the base models. The soft voting classification is implemented to predict the probabilities of each class.

4 Results and Discussion

The model is trained and tested with a dataset of 6207 images. Performance of classifiers is evaluated based on parameters such as accuracy, precision, recall, and $F1$-score. These parameters for each classifier are listed in Table 2.

The accuracies obtained from decision tree, SVM, and KNN are 71%, 77%, and 77%, respectively. Random forest classifier provided highest accuracy of 82%. Soft voting classifier provided 81% accuracy. Region of convergence (ROC) curve shows area under the ROC curve (AUC) for each class. The ROC plot for the random forest classifier is shown in Fig. 4.

Table 2 Results obtained with classifiers

Classifiers	Performance parameters			
	Precision	Recall	$F1$-score	Accuracy
Decision tree	0.71	0.71	0.71	0.71
Random forest	0.83	0.83	0.82	0.82
KNN	0.79	0.77	0.77	0.77
Gaussian NB	0.69	0.70	0.68	0.69
SVM	0.77	0.77	0.77	0.77
Soft voting	0.82	0.82	0.81	0.81

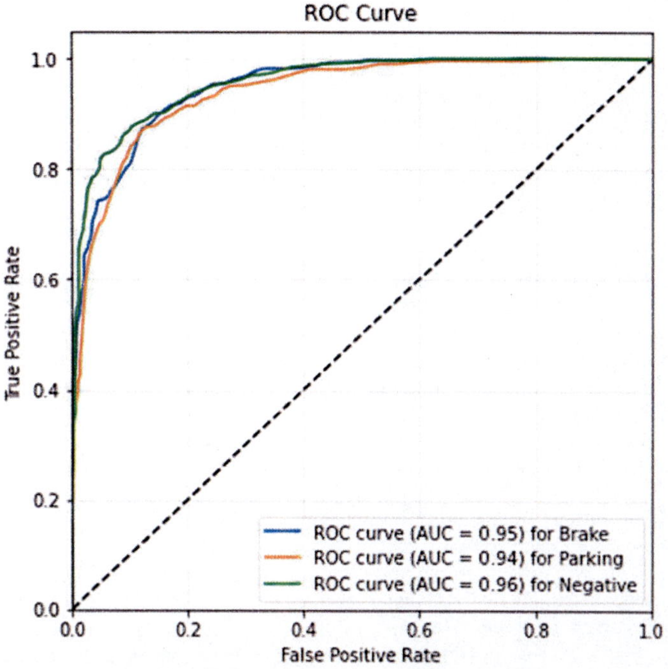

Fig. 4 ROC curve of random forest classifier

AUC of brake, parking, and negative classes for random forest classifier is 95%, 94%, and 96%, respectively. Gaussian NB is less reliable as its accuracy is 69%. The best accuracy is achieved with random forest classifier.

5 Conclusion

This paper presented a driver assistance system to detect brake and parking signals using machine learning and computer vision techniques. As per author's knowledge, this is the first system which detects brake signal and parking signal for driver assistance. This system is evaluated with 6 classifiers with highest accuracy of 82% obtained by random forest classifier. Improvisation to the system is in progress by training the model with a larger size dataset.

References

M. Alsfasser, M. Meuter, A. Kummert, Combinatorial use of optical tracker, Gaussian mixture PHD and group tracking for vehicle light tracking, in *IEEE Intelligent Vehicles Symposium* (2019), pp. 410–416

P. Bogacki, R. Dlugosz, Selected methods for increasing the accuracy of vehicle lights detection, in *24th International Conference on Methods and Models in Automation and Robotics* (2019), pp. 227–231

L. Borelli, M. Kempken, Drowsy driving statistics and facts (2022). [Online]. https://www.bankrate.com/insurance/car/drowsy-driving-statistics/. Accessed 5 May 2022

D.-Y. Chen, Y.-H. Lin, Y.-J. Peng, Nighttime brake-light detection by Nakagami imaging. IEEE Trans. Intell. Transp. Syst. **13**(4), 1627–1637 (2012)

D. Chen, Y. Peng, L. Chen, J. Hsieh, Nighttime turn signal detection by scatter modeling and reflectance-based direction recognition. IEEE Sens. J. **14**(7), 2317–2326 (2014)

H.-T. Chen, Y.-C. Wu, C.-C. Hsu, Daytime preceding vehicle brake light detection using monocular vision. IEEE Sens. J. **16**(1), 120–131 (2015)

L. Chen, X. Hu, T. Xu, H. Kuang, Q. Li, Turn signal detection during nighttime by CNN detector and perceptual hashing tracking. IEEE Trans. Intell. Transp. Syst. **18**(12), 3303–3314 (2017)

Z. Cui, S.-W. Yang, H.-M. Tsai, A vision-based hierarchical framework for autonomous front-vehicle taillights detection and signal recognition, in *IEEE 18th International Conference on Intelligent Transportation Systems* (IEEE, 2015), pp. 931–937

P. Dave, N.M. Gella, N. Saboo, A. Das, A novel algorithm for night time vehicle detection even with one non-functional taillight by CIOF (color inherited optical flow), in *International Conference on Pattern Recognition Systems* (2016), pp. 1–6

S. Gupta, Y.J. Singh, M. Kumar, Object detection using multiple shape-based features, in *Fourth International Conference on Parallel, Distributed and Grid Computing* (2016), pp. 433–437

A. Humnabadkar, O. Kulkarni, P.K. Rajani, Automation of vehicle identification at night using light source recognition, in *5th International Conference on Computing, Communication, Control and Automation* (2019), pp. 1–4

J.-D. Lee, J.-T. Wu, C.-H. Hsieh, J.-C. Chien, Close range vehicle detection and tracking by vehicle lights, in *11th IEEE International Conference on Advanced Video and Signal Based Surveillance* (2014), pp. 381–386

Q. Li, S. Garg, J. Nie, X. Li, R.W. Liu, Z. Cao, M.S. Hossain, A highly efficient vehicle taillight detection approach based on deep learning. IEEE Trans. Intell. Transp. Syst. **22**(7), 4716–4726 (2020)

R. O'Malley, M. Glavin, E. Jones, Vehicle detection at night based on tail-light detection, in *1st International Symposium on Vehicular Computing Systems*, Trinity College Dublin (2008)

R. O'Malley, E. Jones, M. Glavin, Rear-lamp vehicle detection and tracking in low-exposure color video for night conditions. IEEE Trans. Intell. Transp. Syst. **2**, 453–462 (2011)

S.S. Pillai, B. Radhakrishnan, L.P. Suresh, Detecting tail lights for analyzing traffic during night using image processing techniques, in *International Conference on Emerging Technological Trends* (2016), pp. 1–7

C.S. Pradeep, R. Ramanathan, An improved technique for night-time vehicle detection, in *International Conference on Advances in Computing, Communications, and Informatics* (2018), pp. 508–513

M. Qing, K.-H. Jo, A novel particle filter implementation for a multiple-vehicle detection and tracking system using tail light segmentation. Int. J. Control Autom. Syst. **11**(3), 577–585 (2013)

Reliance General Insurance, Common causes of road accidents in India (2022). [Online]. https://www.reliancegeneral.co.in/Insurance/Knowledge-Center/Blogs/Common-Causes-of-Road-Accidents-in-India.aspx. Accessed 5 May 2022

M. Rezaei, M. Terauchi, R. Klette, Robust vehicle detection and distance estimation under challenging lighting conditions. IEEE Trans. Intell. Transp. Syst. **16**(5), 2723–2743 (2015)

Y. Saito, M. Itoh, T. Inagaki, Driver assistance system with a dual control scheme: effectiveness of identifying driver drowsiness and preventing lane departure accidents. IEEE Trans. Hum.-Mach. Syst. **46**(5), 660–671 (2016)

Synopsys, What is ADAS (Advanced Driver Assistance Systems) (2022). [Online]. https://www.synopsys.com/automotive/what-is-adas.html. Accessed 5 May 2022

J.-G. Wang, L. Zhou, Y. Pan, S. Lee, Z. Song, B.S. Han, V.B. Saputra, Appearance-based brake-lights recognition using deep learning and vehicle detection, in *IEEE Intelligent Vehicles Symposium* (2016), pp. 815–820

T. Weis, M. Mundt, P. Harding, V. Ramesh, Anomaly detection for automotive visual signal transition estimation, in *IEEE 20th International Conference on Intelligent Transportation Systems* (2017), pp. 1–8

J. Zhang, J. Oh, J. Kim, Night time vehicle detection by using color information based on tail-light, in *15th International Conference on Control, Automation and Systems* (2015), pp. 1724–1727

DEEC-Based Meta-heuristic Unequal Clustering Scheme for Energy Sustenance of Heterogeneous Nodes in WSN

M. P. Swapna and G. Satyavathy

Abstract Wireless sensor network (WSN) is becoming denser and more heterogeneous each and every day, with the overwhelming impact of technologies like IoT. Managing the energy and maintaining the lifetime longevity of the heterogeneous WSN is a challenging task. Unequal clustering is one of the imperative solutions for maintaining energy efficiently, ideal for minimizing hotspot problem in WSN. Heterogeneity of nodes in WSN needs apposite attention to avoid energy depletion of low energy nodes. This paper proposes the novel algorithm, clustering of heterogeneous nodes with bat algorithm (CHNBA) that integrates the uniform energy dissemination protocol, distributed energy efficiency clustering (DEEC) with unequal clustering. The location of cluster head (CH) is optimized using the bio-inspired improved bat algorithm. DEEC has high ability to assure heterogeneity performance and clustering stability, while optimized unequal clustering ensures energy efficiency resulting in prolonged network lifetime. The effectiveness of the algorithm was experimented using MATLAB simulations. The results proved that the protocol, CHNBA, is effective for the energy sustenance of a network. It outperforms the existing clustering protocols in energy efficiency, network lifetime, throughput, packet delivery ratio and stability.

Keywords Heterogeneous WSN · Unequal clustering · Competition radius · Residual energy · Cluster head · Stability

---

M. P. Swapna (✉)
Research Scholar, Department of Computer Science, Sri Ramakrishna College of Arts and Science for Women, Coimbatore, India
e-mail: mpswapna77@gmail.com

G. Satyavathy
Professor & Head, Department of Computer Science with Data Analytics, KPR College of Arts Science and Research, Coimbatore, India

A. B. Reddy et al. (eds.), *Proceedings of Third International Conference on Advances in Computer Engineering and Communication Systems*, Lecture Notes in Networks and Systems 612, https://doi.org/10.1007/978-981-19-9228-5_22

1 Introduction

Wireless sensor networks (WSNs) are distributed adhoc networks having small, energy-constrained, inexpensive devices called sensor nodes. WSNs are deployed and are efficiently used in a broad range of areas like health care, traffic and environment monitoring, military, boundary and battlefield surveillance, etc. (Akyildiz et al. 2002; Mehmood et al. 2016; Mo et al. 2014). The energy of the sensor nodes in the network has a strong impact on the WSN's lifetime. A sensor node has one or more sensors, with a low power battery, a microcontroller assisted with external memory. WSNs are mainly deployed in hostile environments, where human intervention is arduous. The short duration of the battery-operated sensors and the vulnerability of nodes in the deserted and unmanned environment makes it necessary to find methods for energy retention and lifetime enhancement of the network. Heterogeneous wireless sensor network (HWSN) has sensors with different sizes and abilities, varied sensing capabilities, computational power, etc. The heterogeneity and mobility of WSNs is the most challenging research trend that requires attention and solution (Arjunan and Pothula 2017). The exponential growth of devices and sensors attached to IoT represents the real scenario. The heterogeneity among the power of nodes requires proper scrutiny, else would result in victimization of low energy nodes affecting the stability of HWSN. With the increased heterogeneity and complexity in WSN, the retention of energy and network lifetime requires extensive research. Though a wide variety of algorithms are proposed for clustering of sensor nodes, the solution to HWSN needs more probing. This work is a novel attempt to integrate unequal clustering optimized by bat algorithm with DEEC, a proved scheme for multi-level HWSN. The paper proposes, clustering of heterogeneous nodes with bat algorithm (CHNBA), an energy-efficient algorithm for large-scale HWSN.

The current section discussed briefly on the nature of wireless sensor networks. Section 2 is an eye opener on the related works of unequal clustering, heterogeneous WSN and optimization algorithms. Section 3 details the proposed protocol along with the base concepts and preliminaries, Sect. 4 elaborates the comparative analysis of the simulated results of proposed CHNBA algorithm, and finally, Sect. 5 concludes the research work with a light on future extension.

2 Related Works

"Energy balanced unequal clustering protocol" uses particle swarm optimization method as a solution to the problems faced by sink. This approach proposed maximum number of unequal clusters to form small group dimensions so that energy can be saved for the higher inter-cluster relay. EBUC's methodologies for inter-cluster-based multi-hop systems are inept. The scheme failed to have fault tolerance solutions to overcome the fast failure of overloaded CH (Kumar et al. 2009). "Approximation algorithm" (Yarinezhad and Hashemi 2019) is proposed for clustering data with virtual grid infrastructure. Apt paths are detected as routes among sink and cluster. LEACH-C utilizes residual energy. The sink selects the best CHs and maximizes the network lifetime through the virtual strengthening method. The scheme uses inter-cluster routing and leads to unnecessary power distribution, especially when several CHs are long away from the sink. Thus, CH consumes more energy when interacting with the sink and ultimately drains to death (Shi et al. 2012).

An optimization algorithm is proposed in Sahoo et al. (2020) that integrates PSO with efficient clustering and sink mobility, reduces energy utilization. Opposition-based bat algorithm (Saha et al. 2013), a novel BA method is effective in accuracy and conjunction velocity compared to differential evolution (DE), particle swarm optimization (PSO) and genetic algorithm (GA). Based on residual energy and energy consumption rate of each node, the algorithm BAUC (Sahoo and Amgoth 2021) optimizes CH selection to improve network performance. The algorithm is energy efficient and outperforms the existing evolutionary methods like improving particle swarm optimization (IPSO) includes a quicker convergence. "Reverse glowworm swarm optimization" (Chowdhury and De 2020) proposes green sensor network to improve the performance and energy efficiency. Fuzzy inference system is utilized to reduce the energy consumption and distance traversing. The simulation results show that the live sensor outperforms other protocols in execution. The role of optimization is prominent in networking (Ramkumar and Vadivel 2020a, b, 2021).

The energy efficiency of the EBAN and DEEC protocols in WSNs (Prakash et al. 2020; Qing et al. 2006) for computational link and energy heterogeneity was explored and experimented. For two-level heterogeneous wireless sensor networks, the SEP technique (Abbasi and Younis 2007) was proposed. Although SEP extends the stability time, it is ineffective for multi-level HWSNs. The DEEC scheme is mostly utilized for three-degree heterogeneity, and it outperforms Stable Election Protocols (SEPs) that use stochastic scheme detection for network lifespan extension.

3 Clustering of Heterogeneous Nodes with Bat Algorithm (CHNBA)

3.1 Preliminaries and Base Concepts

Unequal clustering is the most widely recommended energy-efficient technique for WSN. Unequal clustering (Smaragdakis et al. 2004) is a promising method to overcome the hotspot problem in wireless networks. The idea behind unequal clustering is having varied size clusters. This helps to balance the load and ensure that the CHs are all using the same amount of energy. Unequal clustering considers the inter- and intra-cluster traffic of the network. Uneven sized clusters are formed using competition radius (R_c). In heterogeneous WSN, the competition radius is computed by the formula (Yu et al. 2011b)

$$R_c = \left[1 - \propto \frac{d_{max} - d(s_i - DS)}{d_{max} - d_{min}} - \beta\left(1 - \frac{E_r}{E_{max}}\right)\right] R_{max} \tag{1}$$

where d_{max} and d_{min} specify the minimum and maximum distance between the sensor node and that of the base station. The initial residual energy and maximum residual energy are E_r and E_{max}, while R_{max} is the maximum radius.

Heterogeneity of Nodes in HWSN in a HWSN, nodes differ from one another, the energy of the nodes at the beginning varies from one another, which is contradictory to homogeneous WSN. A heterogeneous network with two levels of heterogeneity's total beginning energy is given as

$$E_{tot} = N(1 - m)E_i + NmE_i(1 + g) \tag{2}$$

where E_i is the initial energy of normal nodes, while m represents the advanced nodes in the network having 'g' fold increase in energy over other nodes. In a network with multi-level heterogeneity, initial energy of sensor nodes is spread at random and bounded to the values $[E_i, E_i\ (1 + g_{max})]$, where E_i is the lower bound and g_{max} represents the maximum energy. For a node si, the initial energy is $E_i\ (1 + g_i)$, g_i times more than the lower bound E_i. The energy of a multi-level HWSN at the initial stage is

$$E_{tot} = \sum_{i=1}^{N} E_i(1 + g_i) = E_i\left(N + \sum_{i=1}^{N} g_i\right) \tag{3}$$

The variation in initial energy of nodes must be a primary criteria, when clustering the nodes of HWSN.

DEEC for Heterogeneous WSN Distributed energy efficiency clustering (DEEC) is heterogeneous aware technique, a proved scheme for multi-level heterogeneous WSN. The algorithm draws a probability threshold of the initial energy and residual energy of a node. As the initial energy of nodes varies in a HWSN, the time span of a node being a cluster leader is examined. DEEC guarantees that energy is distributed evenly throughout the network, ensuring network stability and extending network lifespan.

Working procedure of DEEC protocol

1. Cluster heads (CHs) are elected considering the initial energy and residual energy of a node.
2. For a heterogeneous network, the weighted probability p_i of a node is given by

$$p_i = \frac{p_{op}N(1 + x)EN_i(rs)}{\left(N + \sum_{i=1}^{N} x_i\right)EN_r} \tag{4}$$

where p_{op} is the optimum probability, $EN_i(rs)$ is the residual energy of the node, EN_r the average energy at round r in the network.
3. The repetition of node si becoming a CH, called as CH_Rotation, is different for each node within a heterogeneous network due to typical initial energy of nodes. This fluctuates based on the residual energy $EN_i(rs)$ too. As a result, nodes with greater residual energy have an increased probability to become CHs, resulting in a more uniform energy distribution in the network thus maintaining stability.
4. Average energy is the mean of energy that each node must possess in the present round to retain the network alive. This guarantees that the network's and nodes' energy is spread evenly, and that all nodes perish at the same time. The average energy of ith round is given by

$$EN_i = \frac{1}{N} EN_{total} \left(1 - \frac{i}{R}\right) \quad (5)$$

where R is the sum of rounds in the network's lifetime.

Bio-inspired Bat Algorithm for Optimization of Search Space. The location of selected CH in a WSN can be optimized using metaheuristic approaches. Bat algorithm is based on a novel metaheuristic hunting procedure that mimics bat's. To arrive at optimal solution in the problem-solving space, the behavior of bat is considered as an evolutionary technique. Based on bat's echolocation tactic of identifying prey, Yang et al. (2010) specified the following rules

- Bats can clearly distinguish prey and other barriers using their echolocation ability to sense distance.
- When hunting for prey, bats fly with a velocity (v), minimum frequency (f_m), variable wavelength (λ), loudness (W_0), at a place (x). They immediately adjust the wavelength (or maybe frequency) of transmitted pulses. Depending on the closeness of the prey, they regulate the frequency or wavelength of emitted pulses.
- The variation in loudness ranges from A_0 to A_{min}, the large and the minimum constant values, respectively.

Considering a velocity v^t and position x^t in a solution space at the iteration t, the above three rules can be represented as equations

$$f = f_{mn} + (f_{mx} - f_{mn}) \times r_v \quad (6)$$

$$v_t = v_{t-1} + (x_{t-1} - x_b) f \quad (7)$$

$$x_t = x_{t-1} + v_t \quad (8)$$

where $r_v \in (0, 1)$ is the random vector from an even distribution, x_b is present best position, every bat is assigned a random frequency between $[f_{mn}, f_{mx}]$ with values $f_{mn} = 0$ and $f_{mx} = O(1)$ depending on the size of problem domain.

3.2 The Protocol—CHNBA

The proposed protocol clustering of heterogeneous nodes with bat algorithm (CHNBA) is an energy-efficient clustering scheme to prolong the network lifetime of HWSN. It is based on unequal clustering where CH selection is done using DEEC and the location of CH is optimized using bat algorithm.

Working Procedure of CHNBA. Considering a HWSN having 3-level heterogeneity with normal nodes (maximum count), advanced nodes and superior nodes (minimum count), the total energy of the HWSN is given by Eq. 9.

$$E_{\text{tot}} = \emptyset \times N \times E_1 + \emptyset^2 \times N \times E_2 + \left(1 - \emptyset - \emptyset^2\right) \times N \times E_3 \tag{9}$$

where E_1, E_2 and E_3 are the energy of the above specified heterogeneous categories of nodes, respectively. N indicates the number of nodes, while $\emptyset \times N$ is the minimum energy, $0 \leq \emptyset \leq 1$.

1. Compute the energy at the initial stage and residual energy of each node in HWSN.
2. The mean probability (P_i) of node i to be a CH in the round r is

$$P_i = P_{\text{opt}} \frac{E_{\text{r}}}{E_{\text{a}}} \tag{10}$$

   where P_{opt} is the optimal probability at the initial stage for each node to be the CH, E_{r} is the residual energy of node, E_{a} is the average energy at round r in the network.
3. Nodes with a lot of leftover energy (residual) have a better chance of becoming the CH in additional rounds.
4. All nodes of the HWSN will be CH exactly once in $\frac{1}{P_{\text{opt}}}\left(\emptyset + \emptyset^2 \times \frac{E_2}{E_1} + \left(1 - \emptyset - \emptyset^2\right) \times \frac{E_3}{E_1}\right)$ rounds and $\left(\emptyset + \emptyset^2 \times \frac{E_2}{E_1} + \left(1 - \emptyset - \emptyset^2\right) \times \frac{E_3}{E_1}\right) \times N \times P_n$ average CHs per rnd.
5. Assigning values to the optimal probability for each type of normal, advanced and superior nodes, we get weighted probability as given below

$$P_{\text{nr}} = \frac{P_{\text{opt}}(E_{\text{r}})}{\left(\emptyset + \emptyset^2 \times \frac{E_2}{E_1} + \left(1 - \emptyset - \emptyset^2\right) \times \frac{E_3}{E_1}\right)(E_{\text{a}})} \tag{11}$$

$$P_{\text{ad}} = \frac{P_{\text{opt}}(1 + \propto)(E_{\text{r}})}{\left(\emptyset + \emptyset^2 \times \frac{E_2}{E_1} + \left(1 - \emptyset - \emptyset^2\right) \times \frac{E_3}{E_1}\right)(E_{\text{a}})} \tag{12}$$

$$P_{\text{sp}} = \frac{P_{\text{opt}}(1 + \beta)(E_{\text{r}})}{\left(\emptyset + \emptyset^2 \times \frac{E_2}{E_1} + \left(1 - \emptyset - \emptyset^2\right) \times \frac{E_3}{E_1}\right)(E_{\text{a}})} \tag{13}$$

6. The probability threshold value (T_{si}) helps a node to determine itself whether to become a CH or not

$$T_{si} = \left\{ \begin{array}{cc} \frac{P_d}{1-P_d\left(r \bmod \frac{1}{P_d}\right)} & \text{if si} \in G \\ 0 & \text{otherwise} \end{array} \right\} \tag{14}$$

   where P_d is the percentage of total nodes that were initially CHs, G is the collection of nodes that are not CHs in the previous $1/P_d$ rounds. P_d can be replaced by P_{nr}, P_{ad} and P_{sp}, respectively, to arrive the probability threshold of normal, advanced and superior nodes, respectively.
7. Once the CHs are elected in each round, the optimal location of CH is determined using the bio-inspired BA, with frequency, loudness and pulse emission rate of the node as primary parameters.
8. The convergence of nodes toward the CH is faster, with appropriate sized unequal clusters based on competition radius.

In a HWSN, the probability of a node repeatedly becoming CH varies due to the difference in initial energy and balance energy of the node. The proposed protocol, CHNBA, with the attributes weighted probability and threshold value, assures uniform allocation of energy in the heterogeneous network, prolongs the network lifetime. The evaluation is segregated based on the type of nodes, CHNBA ensures that nodes with high remaining energy (residual) have a higher chance of becoming CH, and nodes with low residual energy do not die easily. Unequal clustering is an ideal solution for the hotspot problem faced by WSN, the integration of DEEC and unequal clustering with BA, enhances stability of the network and improves energy efficiency. The process flow of CHNBA is represented in Fig. 1.

4 Results and Discussions

The proposed algorithm, CHNBA was experimented in MATLAB-2019 simulator to determine the effectiveness of performance metrics (a) energy efficiency, (b) network lifetime, (c) throughput, (d) packet delivery ratio and (e) stability. CHNBA was analyzed with the existing protocols BAUC, WOTC and IPSO. The nodes in the network are heterogeneous with 52 normal nodes having an energy level of 0.5 J each, 27 advanced nodes with 1.48 J each and 21 super nodes with 1.68 J each. The simulations are carried out with 100 nodes that are deployed randomly in an area of 200×200 m^2 with a threshold distance of 85 m.

Figure 2 illustrates the energy exhaustion rate of CHNBA with X-axis having the count of rounds and the energy exhaustion percentage on the Y-axis. The ideal dissemination of heterogeneous nodes in the network, distributes the energy evenly, nodes with high residual energy after each round takes the chance of being CH, resulting in efficient energy retention in the network. Unequal clusters formed is an added effect for energy retention. When evaluated against the existing algorithms BAUC, WOTC

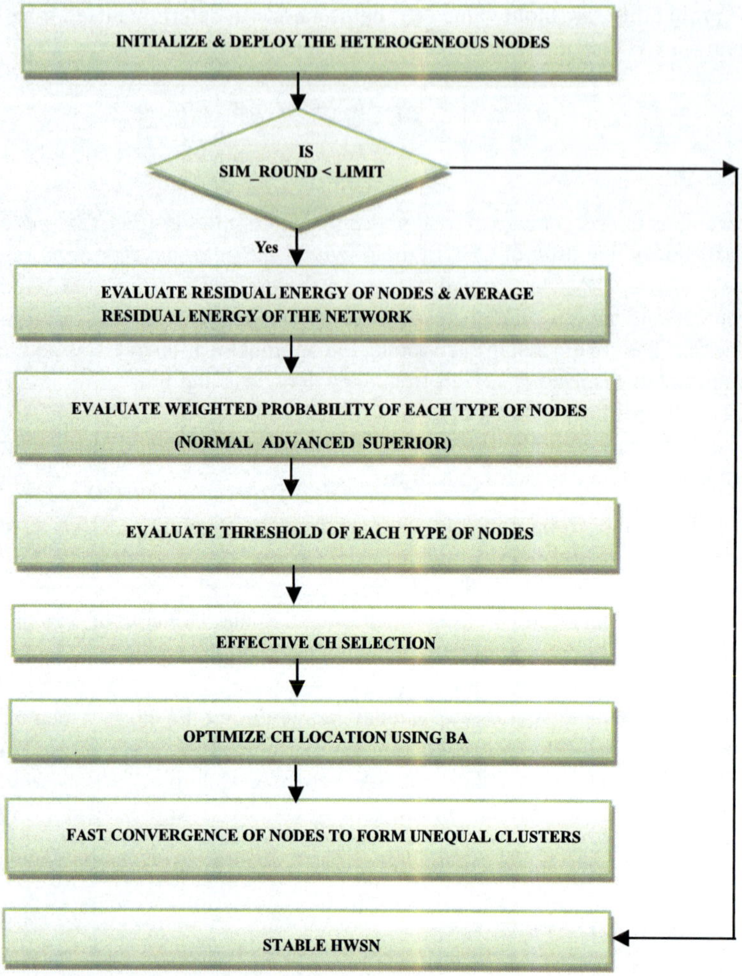

Fig. 1 Flowchart representing the process flow of CHNBA

and IPSO, the energy depletion rate of CHNBA is comparatively less. The network lifetime of CHNBA compared with BAUC, WOTC and IPSO is represented in Fig. 3. X-axis represents the time in milliseconds, and Y-axis has the count of nodes. The experimental results clearly demonstrate that CHNBA enhances the network lifetime to the maximum. The death of low energy nodes is eliminated by the DEEC scheme, uniform energy dissemination is assured by proposed protocol, CHNBA, this enhances the network lifetime. Figure 4 indicates an optimum throughput by the proposed algorithm, CHNBA. The comparative analysis with the throughput in X-axis and count of nodes in the Y-axis depicts the superior nature of CHNBA with that of the existing algorithms. The random allotment of heterogeneous nodes in the network and the uniform energy dissemination by CHNBA tends to enhance the overall throughput in the HWSN. To mitigate hotspot problem in HWSN, CHNBA uses unequal clustering, which is an added thrust to optimize the throughput of the network. The downs in the figure indicate the drop in throughput due to the collision that occurs when the node density is increased beyond 60. Figure 5 is an illustration of packet delivery ratio (PDR) of CHNBA, the *X*-axis indicates the density of nodes and the *Y*-axis has the PDR measured in percentage. Efficient energy retention capability and sustenance of network lifetime by the proposed algorithm CHNBA is the key for having good PDR percentage. Death of low energy nodes between the transmission is regulated in this protocol, this is an added input for improved PDR percentage of CHNBA, than the existing clustering protocols. The comparative analysis of stability of the network is illustrated in Fig. 6. The stability is measured by assessing the parameters, half node death and lifespan of the network. The collaborative effect of DEEC on heterogeneity performance and uneven sized clusters on hotspot problem enhances the stability of the HWSN. The stability level and the network lifetime of the proposed CHNBA is outstanding compared to the existing protocols BAUC, WOTC and IPSO.

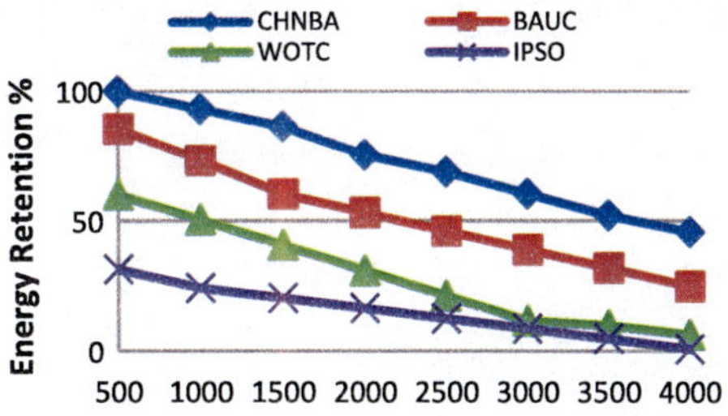

Fig. 2 Energy exhaustion analysis

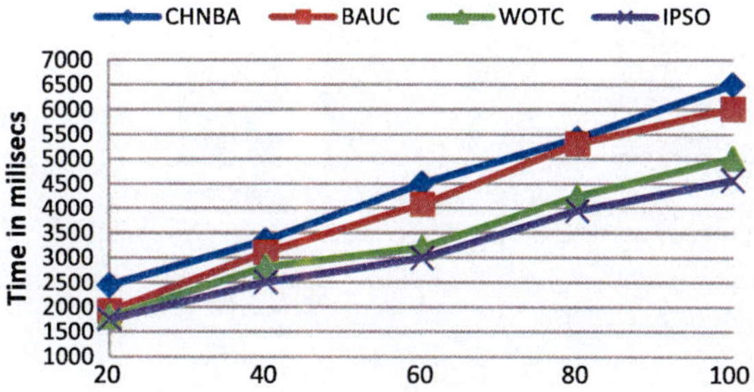

Fig. 3 Network lifetime in ms against the number of nodes

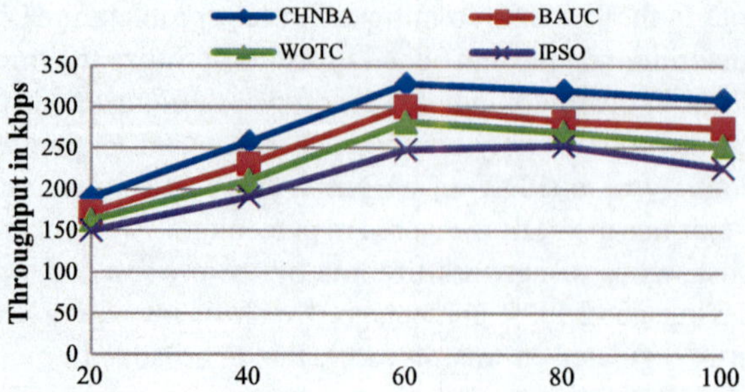

Fig. 4 Throughput in kbps against number of nodes

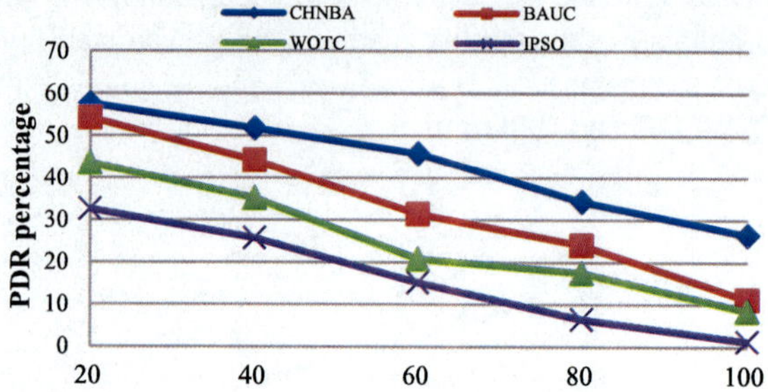

Fig. 5 PDR against node density

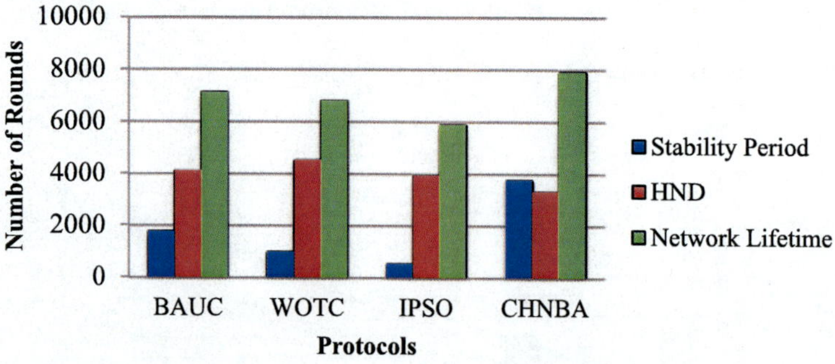

Fig. 6 Stability analysis of CHNBA

5 Conclusion

The heterogeneous model of WSN can have heterogeneity in computational power, energy and link capability of nodes in the network. This work proposed an efficient clustering solution, clustering of heterogeneous nodes using bat algorithm (CHNBA) based on the heterogeneous aware DEEC and unequal clustering. Residual energy of nodes is evaluated after each round to elect the best CHs, further it is optimized using improved bat algorithm. Unequal clusters are formed to reduce hotspot problem in the network and uniform energy distribution is assured. The simulation results of CHNBA experimented using MATLAB reveal that it is more efficient and superior compared to existing clustering protocols BAUC, WOTC and IPSO. This paper proposes on having a stable and energy-efficient HWSN using the novel clustering technique. The work could be extended to recommend an ideal routing scheme for selecting the reliable forwarder node, for the acquired data to reach the sink.

References

A.A. Abbasi, M. Younis, A survey on clustering algorithms for wireless sensor networks. Comput. Commun. **30**(14), 2826–2841 (2007)

I.F. Akyildiz, W. Su, Y. Sankarasubramaniam, E. Cayirci, Wireless sensor networks: a survey. Comput. Netw. **38**(4), 393–422 (2002)

S. Arjunan, S. Pothula, A survey on unequal clustering protocols in wireless sensor networks. J. King Saud Univ. Comput. Inf. Sci. (2017)

A. Chowdhury, D. De, FIS-RGSO: dynamic fuzzy inference system based reverse glowworm swarm optimization of energy and coverage in green mobile wireless sensor networks. Comput. Commun. **163**, 12–34 (2020)

D. Kumar, T.C. Aseri, R. Patel, EEHC: energy efficient heterogeneous clustered scheme for wireless sensor networks. Comput. Commun. **32**(4), 662–667 (2009)

A. Mehmood, J. Lloret, S. Sendra, A secure and low-energy zone-based wireless sensor networks routing protocol for pollution monitoring. Wireless Commun. Mob. Comput. **16**(17), 2869–2883 (2016)

B. Mo, W. Dong, C. Chen, J. Bu, Q. Wang, An efficient differencing algorithm for reprogramming wireless sensor networks. Ad Hoc Sens. Wireless Netw. **21**(3–4), 201–218 (2014)

A. Prakash, R. Mishra, A.S. Kushwaha, Exterior bound allotted advanced node distribution for energy efficient clustering (EBAN DEEC) protocol in WSN. Int. J. Res. Dev. Appl. Sci. Eng. (IJRDASE) **20**(1) (2020)

L. Qing, Q. Zhu, M. Wang, Design of a distributed energy-efficient clustering algorithm for heterogeneous wireless sensor networks. Comput. Commun. **29**, 2230–2237 (2006)

J. Ramkumar, R. Vadivel, Intelligent fish swarm inspired protocol (IFSIP) for dynamic ideal routing in cognitive radio ad-hoc networks. Int. J. Comput. Digit. Syst. **10**(1), 1063–1074 (2020a). https://doi.org/10.12785/ijcds/100196

J. Ramkumar, R. Vadivel, Improved wolf prey inspired protocol for routing in cognitive radio ad hoc networks. Int. J. Comput. Netw. Appl. **7**(5), 126–136 (2020b). https://doi.org/10.22247/ijcna/2020/202977

J. Ramkumar, R. Vadivel, Multi-adaptive routing protocol for internet of things based ad-hoc networks. Wireless Pers. Commun. 1–23 (2021). https://doi.org/10.1007/s11277-021-08495-z

S.K. Saha, R. Kar, D. Mandal, S.P. Ghoshal, V. Mukherjee, A new design method using opposition-based BAT algorithm for IIR system identification problem. Int. J. Bio-Inspir. Comput. **5**(2), 99–132 (2013)

B.M. Sahoo, T. Amgoth, An improved bat algorithm for unequal clustering in heterogeneous wireless sensor networks. SN Comput. Sci. **2**, 290 (2021). https://doi.org/10.1007/s42979-021-00665-x

B.M. Sahoo, T. Amgoth, H.M. Pandey, Particle swarm optimization based energy efficient clustering and sink mobility in heterogeneous wireless sensor network. Ad Hoc Netw. **106**, 102237 (2020)

S. Shi, X. Liu, X. Gu, An energy-efficiency optimized LEACH-C for wireless sensor networks, in *7th International Conference on Communications and Networking in China*, Aug 2012 (IEEE, 2012), pp. 487–492

G. Smaragdakis, I. Matta, A. Bestavros, SEP: a stable election protocol for clustered heterogeneous wireless sensor networks, in *Second International Workshop on Sensor and Actor Network Protocols and Applications (SANPA 2004)* (2004)

R. Yarinezhad, S.N. Hashemi, Solving the load balanced clustering and routing problems in WSNs with an fpt-approximation algorithm and a grid structure. Pervasive Mob. Comput. **58**, 101033 (2019)

Grid-Based Pathfinding Using Ant Colony Optimization Algorithm

Swapnil Biswas, Syeda Ajbina Nusrat, and Noshin Tasnim

Abstract The unique behavior of ants has inspired a number of methods, and ant colony optimization (ACO) method has been the most successful general-purpose optimization technique. The ant colony metaheuristic has been shown to be effective in solving complex problems such as combinatorial problems and NP-Hard problems, frequently producing the best solution in the shortest time. However, ACO has received insufficient attention as a way of solving problems with optimal solutions that can be found using other approaches. The pathfinding problem is without a doubt one of the most important areas of navigation and telecommunications. Ants release chemical substances known as a pheromone in the ground while searching for food to mark a path that is preferred by them, and other ants will get a hint about which path to follow from the strong smell of pheromone. The ACO algorithm uses a similar mechanism. So, in ACO, solutions to an optimization problem are being created by a group of artificial ants by exchanging information on their quality using a communication mechanism that is similar to actual ants. In this paper, an approach to implement the ACO algorithm for pathfinding in a 2D grid as the search space has been proposed.

Keywords Ant colony optimization · Pathfinding · 2D grid

1 Introduction

The concept of ant colony optimization (ACO) was proposed by an Italian researcher, Dorigo. The ACO algorithm has been widely used in combinatorial optimization problems (Dorigo et al. 1999). ACO is a searching algorithm that simulates biological group evolution and has been utilized in a variety of applications such as image processing, function optimization, and vehicle scheduling due to its positive feedback

S. Biswas (✉) · S. A. Nusrat · N. Tasnim
Department of Computer Science and Engineering, Military Institute of Science and Technology, Mirpur Cantonment, Dhaka 1216, Bangladesh
e-mail: swapnil.cse16@gmail.com

A. B. Reddy et al. (eds.), *Proceedings of Third International Conference on Advances in Computer Engineering and Communication Systems*, Lecture Notes in Networks and Systems 612, https://doi.org/10.1007/978-981-19-9228-5_23

mechanism, distributed computing, greedy search features and robustness, parallel processing (Dréo and Siarry 2004). ACO involves a group of artificial ants creating solutions to an optimization problem and exchanging information on their quality using a communication technique similar to that used by actual ants (Stützle and Hoos 2000). Artificial ants (for example, simulation agents) move across a two-dimensional space (grid). While exploring their surroundings, real ants leave pheromones that direct each other to food sources. Choice of next node is probabilistic in nature and depends on the pheromone level of a path. Du and Swamy (2016) simulated 'ants' keep track of their positions and the quality of their solutions in the same way, so that more ants find better solutions in subsequent simulation iterations.

The process of finding a path between two points in a given environment is described by pathfinding theory (Sharon et al. 2015). In most circumstances, the goal is to discover the shortest path that is optimal, cheapest, or simplest. In many applications: military simulations, network traffic, computer games, robot planning, rely on pathfinding. A grid (Zarembo and Kodors 2013) is often placed over a region, and a searching algorithm is used to determine the optimal (lowest cost) solution.

The purpose of this research is to dig the implementation and performance of ACO by taking all the parameters associated with algorithm via self made simulation in 2D grid.

A $m \times n$ 2D grid is similar to a matrix with m number of rows and n number of columns. A 2D grid can be represented as an undirected unweighted graph where each cell can be represented as the vertex of the graph. Two adjacent cells share a common undirected edge. There can be obstacles on some of the cells. The problem is to find the minimum number of empty cells that denotes the shortest path to move from one empty cell to another. Breadth-first search algorithm is widely used to solve this type of problem. However, Dijsktra's algorithm, Bellman Ford's algorithm are also used in the case of weighted graphs.

To introduce a metaheuristic approach in this field, ant colony optimization algorithm has been applied to find the shortest path from one cell to another in a 2D grid that defines the novelty of this work. At first, the 2D grid has been converted into an undirected unweighted graph, and then, the shortest path from one cell to another has been calculated for various numbers of ants and iterations. After that, a comparison has been illustrated between this approach and the widely used algorithms to compare the quality of the path generated by the ant colony optimization algorithm. In summary, the contribution of this work is as follows:

- Representing a 2D grid with some obstacles into an undirected unweighted graph.
- Applying ant colony optimization to find the shortest path from one empty cell to another in the 2D grid.
- Comparing the path quality generated by this approaches with the commonly used approaches.

2 Literature Survey

An article (Maniezzo and Carbonaro 2002) by Maniezzo et al. presents the current research within the ACO framework and focuses on a particular research line, the ANTS metaheuristic, discusses the details on the ACO algorithm where they present results recently obtained on the quadratic and on the frequency assignment problems. A work (Wang et al. 2014) by Wang et al. presents the process planning optimization problem for prismatic pieces using an upgraded ACO technique. A weighted graph is used for process planning problem, and a pheromone updating strategy is proposed which includes global update rule and local update rule.

Through the study of classic genetic algorithms and ant colony optimization, a research (Zhang and Lu 2012) by Zhang et al. presents a genetic ant algorithm. This algorithm starts with the results of the genetic algorithm to establish the pheromone distribution, then uses its rapid global convergence to find the best solution using the colony algorithm, which features parallelism, positive feedback, and excellent solution efficiency. A research (Mulani and Desai 2018) by Mulani et al. tries to establish the best possible solution in terms of its variable parameters. As a probabilistic algorithm, ACO is strongly reliant on its two constants and. This paper shows that if stagnation phase occurs, ACO algorithm puts the solution into local optima. And suggests to recover it by changing the values of. Another work (Ajeil et al. 2020) by Ajeil et al. shows a modification based on the age of the ant named aging-based ant colony optimization (ABACO) which is implemented to solve the path planning problem in grid-based static and dynamic environment. An article (Runka 2009) by Runka uses the evolutionary process present in genetic programming to improve the ant system algorithm's decision formula.

3 Experimental Setup

3.1 Pathfinding in a 2D Grid

A grid in two-dimensional space is similar to a $m \times n$ matrix where m is the number of rows and n is the number of columns. The maximum value of m has been considered as 30, and the maximum value of n has been considered as 100 in this paper. Rows are marked from 0 to $m - 1$, and columns are marked from 0 to $n - 1$. A cell is represented as Ci, j where i indicates the row number and j indicates the column number of that cell. A cell can be either empty or an obstacle. An empty cell allows to put an object on it where a cell with an obstacle does not allow. There are 2 special empty cells named source and destination. The type of a cell can be determined by its value. The following representation has been used to classify a cell in this work.

$$C_{i,j} = \begin{cases} \text{Obstacle} & C_{i,j} = -1 \\ \text{Empty} & C_{i,j} = 0 \\ \text{Source} & C_{i,j} = 1 \\ \text{Destination} & C_{i,j} = 2 \end{cases}$$

Each cell Ci, j can have at most four adjacent cells which are $Ci - 1, j, Ci + 1, j, Ci, j - 1, Ci, j + 1$. The boundary condition of Ci, j is defined with a Boolean function $b(i, j)$. False value of $b(i, j)$ indicates that the position of Ci, j is out of the grid.

$$b(i, j) = \begin{cases} \text{false} & i < 0 \text{ or } i > m - 1 \text{ or } j < 0 \text{ or } j > n - 1 \\ \text{true} & \text{otherwise} \end{cases}$$

1 is defined as the movement cost from one cell to any of its adjacent cells if both of the cells satisfy the boundary condition. Also, it is not possible to move an object from one cell to any of its adjacent if any of the cells contain an obstacle. Now, the goal of the pathfinding problem is to move an object from source to destination via a path in the 2D grid that is optimal or near optimal.

3.2 *Graph Representation of 2D Grid*

The 2D grid is represented by an undirected graph $G(V, E)$ where V represents the set of vertices and E represents the set of edges. Each cell on the grid is a vertex of G. A vertex can have at most 4 adjacents. The total number of vertices in the graph, $|V| = mn$. Each vertex is represented with a single integer x $(0 \leq x < mn)$ where a cell in a grid is identified by 2 integers i and j. A function f does the work here to take i and j as parameters and return the value of corresponding $x \cdot f(i, j) = ni + j$ where n is the number of columns of the grid.

If two cells are adjacent in the grid only in that case, an edge is possible between the corresponding pair of vertices. In the grid, the movement cost from one cell to any of its adjacent cells has been considered as 1, and so, in the graph representation, the weight of every edge is 1. S denotes the source vertex, and D denotes the destination vertex of the graph.

4 Proposed Method

In the 2D grid, all ants initially start their tour from source. Initially, a constant and equal amount pheromone is put in all the edges. A two-dimensional array of real numbers is maintained to store the amount of the pheromone at each edge. As the total number of vertices in the graph is mn, so the dimension of the pheromone matrix is $mn \times mn$. A list can be used to track the pheromone level, but for simplicity, two-dimensional array has been used for implementation. Initially, the pheromone level at all the edges is the same so that each ant can choose among the adjacent vertices with equal probability. In the grid, the movement of an ant has been classified in 2 categories

- **Forward movement**: A movement is classified as forward movement if an ant has started its journey from the source toward destination. A forward movement is completed when an ant reaches the destination. As initially all the ants start their journey from the source, so the initial movement is always a forward movement for all the ants.
- **Backward movement**: A movement is named as backward when the ant starts moving from destination to the source. A backward movement is completed when the ant reaches the source.

All the ants have a memory to track its visited vertices in the graph. This is done by an array of size *mn*. At the same time, the ant also keeps track of its last visited vertex. Also, the path followed by an ant for either forward or backward movement is tracked by an array of size.

The algorithm executes for a total *N* (uppercase) number of times. At each iteration, each ant is moved from its current vertex to any of its adjacent vertex which is not an obstacle or visited earlier. The choice of the next vertex is made based on the pheromone intensity of that edge and the heuristic value of that vertex. The heuristic value for a node is calculated by the cityblock distance from the destination. The cityblock distance of a cell Ci, j is calculated in the following way:

cityBlock$(i, j) = |p - i| + |q - j|$ where p denotes the row and q denotes the column of the destination cell. An ant can move in both forward and backward directions for multiple times in between *N* number of iterations. To control this type of movements, the array that tracks the visited vertices and the path followed by an ant is cleared after completing a forward or backward movement. After completing a forward of backward movement, a constant amount of pheromone is added on each edge of a path which has been followed by an ant. After completing a total *N* number of iterations, the path from source to destination is traced with the highest possible pheromone level to find the shortest path.

The remarkable functions and the pseudocode of ACO are described in the following sections.

4.1 Initialization

In this paper, the number of ants is denoted by *k*. First, an array of size *k* is declared where each element of the array represents an ant uniquely. The index of an array element decides the unique id of the corresponding ant. Each ant contains a visited array of size *mn* of Boolean type. Initially, all the elements of the visited array for each ant are set to false which means that none of the vertices is visited by that corresponding ant. Source is set as the current node for each ant in the initialization phase. A parent array of size mn is maintained for each ant to trace the path for forward or backward movement. Parent[i] = j represents that if vertex-j has been visited at tth iteration, then at $(t+1)$th iteration vertex-i has been visited. Initially, all the elements of the parent array for each ant are set to -1. After initializing the ants, the pheromone matrix is initialized. A constant amount of pheromone is put in all the edges of the graph. For this work, the amount of constant pheromone has been considered as 1.0. The following init function is called for k number of times to initialize all the ants.

Pseudocode for init(i, m, n, S)

```
ants[i].curNode := S
for j ← 0 to m ∗ n − 1 do
    ants[i].visited[j] := false
    ants[i].parent[j] := −1
```

Pseudocode for initPheromone(m, n, ph = 1.0)

```
for i ← 0 to m ∗ n − 1 do
    for j ← 0 to m ∗ n − 1 do
        pheromone[i][j] := ph
        pheromone[j][i] := ph
```

4.2 *Deciding Next Node*

At each iteration, all the ants move from their current vertex to the best probable adjacent vertex. The decision of the next vertex is made from the pheromone intensity of the edge and also the heuristic of the next node. Probability for all the adjacent vertices of the current node is calculated. The vertex with highest probability is selected for the next node. If an ant is currently located at u and the set of adjacents of u is denoted by A which are not visited by that ant and not an obstacle, then the probability of choosing $v \in A$ as the next node has been determined by the following method:

$$P(u, v) = \frac{\text{pheromone}[u][v] * \text{cityBlock}(v)}{\sum_{f \in A} \text{pheromone}[u][f] * \text{cityBlock}(f)}$$

If v is a vertex with an obstacle or visited[v] is true in that case $P(v)$ returns 0. When an ant completes a forward or backward journey, then the visited array is cleared to allow that ant to move through the same vertices.

4.3 Updating Pheromone

The pheromone for each edge is evaporated at each iteration like the real-life scenario. The pheromone of an edge is evaporated by the following way where i and j are the two endpoints of the edge and e is the evaporation factor: pheromone[i][j] = $(1 - e)$pheromone[i][j]. In this paper, e has been taken as 0.5. However, pheromone intensity is also updated when an ant completes either a forward or backward movement. Pheromone level of the complete path is increased by $1/L$ amount which is followed by the ant and L denotes the length of the path. A recursive function updateTrail does the work here.

Pseudocode for updateTrail(u, L = 1)

```
v := parent[u]
if v equals −1 then
    pheromone[u][v] := pheromone[u][v] + 1L
    pheromone[v][u] := pheromone[v][u] + 1L
updateTrail(v, L + 1)
```

If the ant completes a forward journey, then destination (D) is passed as u otherwise source (S) is passed as u.

5 Result

As ACO chooses the path based on probabilistic choice so as it is not guaranteed that the optimal path will be chosen by the algorithm. However, increasing the number of iterations and ant can lead the algorithm to choose a better or near optimal path. The function for updating the pheromone trail is inversely proportional to the length of the path. This leads the ants to choose that shorter path but not the optimal one.

The grid illustrated in Fig. 1 was taken for performing the ACO. In the figure, a dot (.) represents an empty cell; (–) represents a cell with an obstacle; (S) denotes the source, and (D) denotes the destination.

In Fig. 2, the path with highest pheromone level has been chosen illustrated by (*) sign for 1000 iterations and 500 ants. The length of the chosen path is 17 which is not the shortest.

In Fig. 3, the optimal path has been illustrated using breadth-first search (BFS) algorithm. The length of the shortest path is 15. Decreasing the number of iterations to 100 produced a path of length 19. The best path was found for 10,000 iterations and 750 ants where the path cost was 15 which is the shortest. But, it is not guaranteed that iterating for 10,000 times with 750 ants will always produce the best path.

Another experiment was performed on a completely empty cell of a 5×15 grid where the source node was located in (0, 0) location and the destination was located

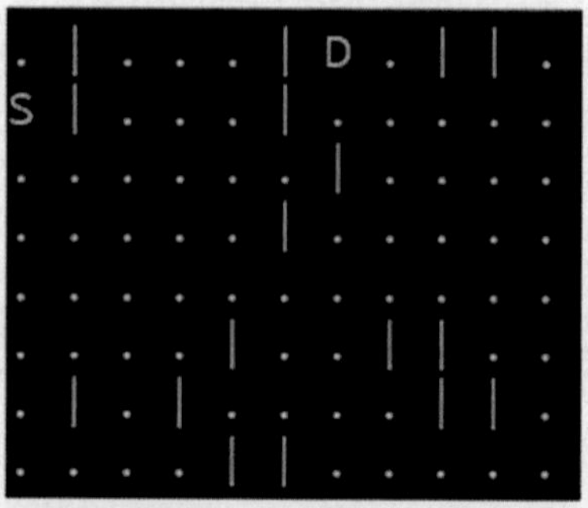

Fig. 1 Representation of the grid

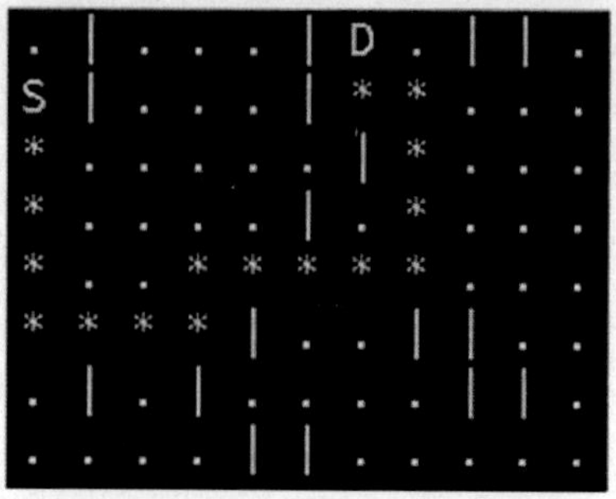

Fig. 2 ACO pathfinding

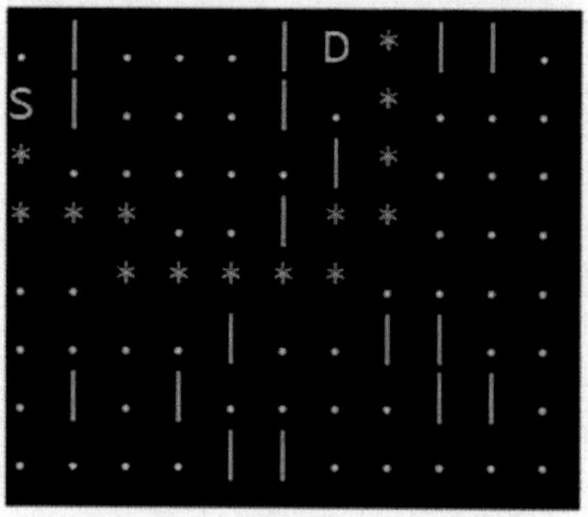

Fig. 3 ACO shortest pathfinding

in (0, 14). So, the length of the shortest path is 14. In this scenario, if the number of iteration is above 50 with an ant colony more than 10 ants always produces the optimal result. Now, considering the same scenario, but two obstacles in (0, 7) and (1, 1) position bring a drastic change in the previous observation. Figures 4, 5, 6, and 7, respectively, show result generated by the algorithm for 50 iterations, 10 ants, 100 iterations, 10 ants, 50 iterations, 100 ants, 100 iteration, 100 ants. The best path is found for 100 iterations with 100 ants which can claim that increasing both the number of ants and iterations can significantly improve the path quality as well decrease the performance of the algorithm considering the required amount of time.

In this work, the concept of static number of iterations has been used to terminate the algorithm. This concept has mainly been used to resolve some corner cases of

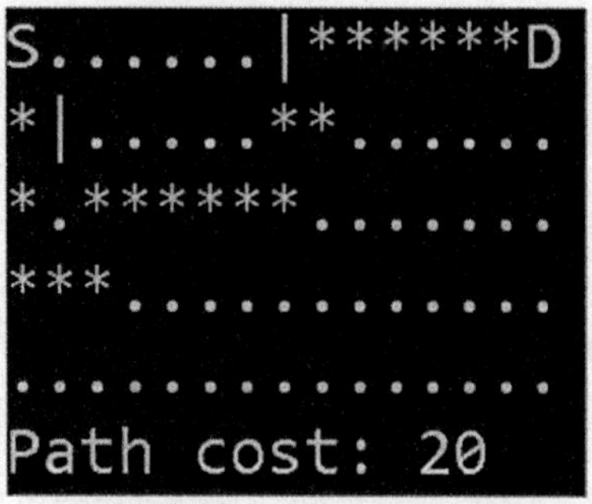

Fig. 4 Pathfinding (50 iterations, 10 ants)

Fig. 5 Pathfinding (100 iterations, 10 ants)

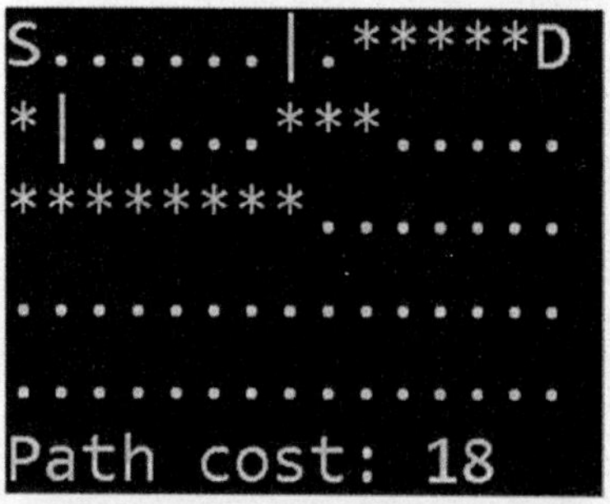

Fig. 6 Pathfinding (50 iterations, 100 ants)

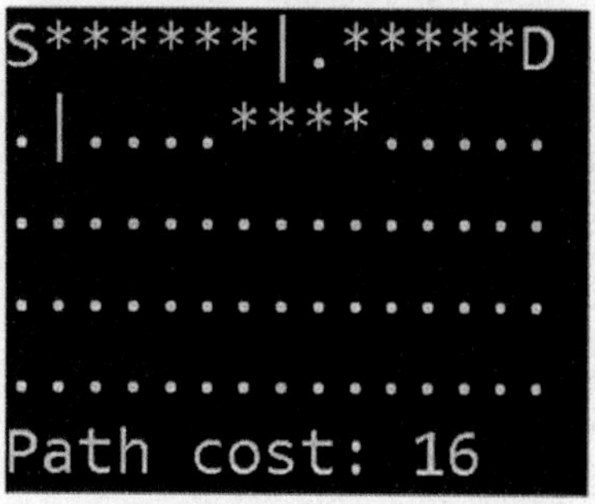

Fig. 7 Pathfinding (100 iterations, 100 ants)

the traditional ant colony algorithm. In the traditional ant colony algorithm, all the ants start their journey from the source, reach the destination, and then return to the source. When all the ants return to the source, then the algorithm terminates. But, if any ant stocks in any of the internal nodes except the source or destination, then the ant can not reach to either source or destination. In that case, the algorithm can lead to an infinite condition. But, the concept of fixed number of iterations solves this problem though the number of iterations influences the path quality. In this case, the algorithm is not dependent on the movements of the ants. No matter whether any ant stucks at any point or not the algorithm terminates after a fixed number of iterations. As a result, this approach is highly usable to find a path from source to destination in a 2D grid using ant colony optimization.

5.1 Time Complexity Analysis

The number of ants is denoted by k. So, for clearing each of the nodes in the search space, the complexity becomes $O(kmn)$ where m is the number of rows and n is the number of columns in the grid. So, mn represents the total number of nodes of the search space. For updating the pheromone values on each edge of the search space, the required time is $O(m^2n^2)$ as there are edges between every two nodes, and the total number of nodes is mn. Then, for the next part, the outer loop will continue as per user input (let t), and the inner loop will execute for k times. For each of the iterations, from each node, the next node is found, and if the goal reached, the pheromone level is updated for that path. The time complexity for this part is $O(tkm^2n^2)$. Total time complexity becomes $O(kmn) + O(m^2n^2) + O(tkm^2n^2)$. Ignoring the first two literals, it becomes $O(tkm^2n^2)$. The more the iteration number, the better the solution is. So, if iteration number (t) and the number of ants (k) increase, then the time complexity will increase accordingly.

6 Conclusion

Finding the shortest path in a road map network is one of the most common problems, and many of the static algorithms have been already used to solve this problem, but dynamic methods like ACO can also be a great asset to get a solution of pathfinding problems. Some future directions can be taken into consideration. This work does not specify the number of ants or iterations for generating the optimal path. As well the proposed method can not be applied on a weighted graph. Hence, it is required to develop techniques to apply the ant colony optimization algorithm on a weighted graph with a suggestion of optimal number of ants and iterations to generate the best quality path in a 2D grid from one cell to another. The next direction is related to the technique fixed number of iterations that has been used for avoiding the infinite loop condition. Hence, a technique can be introduced to avoid deadlock with a dynamic loop-breaking condition.

References

F.H. Ajeil, I.K. Ibraheem, A.T. Azar, A.J. Humaidi, Grid-based mobile robot path planning using aging-based ant colony optimization algorithm in static and dynamic environments. Sensors **20**(7), 1880 (2020)

M. Dorigo, G. Di Caro, L.M. Gambardella, Ant algorithms for discrete optimization. Artif. Life **5**(2), 137–172 (1999)

J. Dréo, P. Siarry, Continuous interacting ant colony algorithm based on dense hierarchy. Future Gener. Comput. Syst. **20**(5), 841–856 (2004)

K.-L. Du, M.N.S. Swamy, Ant colony optimization, in *Search and Optimization by Metaheuristics* (Springer, 2016), pp. 191–199

M.D.L.M. Gambardella, M.B.A. Martinoli, R.P.T. Stützle, Ant colony optimization and swarm intelligence, in *5th International Workshop* (Springer, 2006)

U. Jaiswal, S. Aggarwal, Ant colony optimization. Int. J. Sci. Eng. Res. **2**(7), 1–7 (2011)

V. Maniezzo, A. Carbonaro, Ant colony optimization: an overview, in *Essays and Surveys in Metaheuristics* (2002), pp. 469–492

M. Mulani, V.L. Desai, Design and implementation issues in ant colony optimization. Int. J. Appl. Eng. Res. **13**(16), 12877–12882 (2018)

A. Runka, Evolving an edge selection formula for ant colony optimization, in *Proceedings of the 11th Annual Conference on Genetic and Evolutionary Computation* (2009), pp. 1075–1082

G. Sharon, R. Stern, A. Felner, N.R. Sturtevant, Conflict-based search for optimal multi-agent pathfinding. Artif. Intell. **219**, 40–66 (2015)

K. Socha, M. Dorigo, Ant colony optimization for continuous domains. Eur. J. Oper. Res. **185**(3), 1155–1173 (2008)

T. Stützle, H.H. Hoos, Max–min ant system. Future Gener. Comput. Syst. **16**(8), 889–914 (2000)

J.F. Wang, X.L. Fan, H. Ding, An improved ant colony optimization approach for optimization of process planning. Sci. World J. **2014** (2014)

I. Zarembo, S. Kodors, Pathfinding algorithm efficiency analysis in 2D grid, in *ENVIRONMENT. TECHNOLOGIES. RESOURCES. Proceedings of the International Scientific and Practical Conference*, vol. 2 (2013), pp. 46–50

W. Zhang, T. Lu, The research of genetic ant colony algorithm and its application. Procedia Eng. **37**, 101–106 (2012)

Improving Classification-Based Log Analysis Using Vectorization Techniques

Deepali Arun Bhanage and Ambika Vishal Pawar

Abstract IT infrastructure components are exposed to miscellaneous anomalies or failures as they flourish swiftly in scale and usage. The identification of failures is possibly managed by system logs produced on the execution of logging statements. A recent highly advantageous technique is to observe the system's behavior and identify the anomalous log entries to take corrective actions. However, current methods focus on classifying logs but overlook the nature of data. This paper proposes the log analysis system contingent on natural language processing (NLP) techniques considering logs as natural language text. This model is trained through TF-IDF, polarity score, and Word2Vec as vectorization techniques and conventional machine learning classifiers, suitable to group records as per the assigned level. The efficacy of the proposed model was validated on various IT infrastructure logs. Experimental results demonstrate that sentiment analysis is possibly the encouraging technique for analyzing complex, huge, and irregular system logs.

Keywords Log analysis · NLP · ML · Classification

1 Introduction

IT infrastructure monitoring demands the accurate detection of failures using the log analysis approach to reduce the downtime in the system. Due to failure in any IT infrastructure component, there can be unavailability of services for some time which interrupts the execution flow of operation and results in financial as well as

D. A. Bhanage (✉) · A. V. Pawar
Symbiosis Institute of Technology, Symbiosis International (Deemed University), Pune 412115, India
e-mail: deepali.bhanage.phd2019@sitpune.edu.in

A. V. Pawar
e-mail: ambikap.imc@gmail.com

D. A. Bhanage
Department of Computer Engineering, Pimpri Chinchwad Education Trust's, Pimpri Chinchwad College of Engineering, Pune, India

A. B. Reddy et al. (eds.), *Proceedings of Third International Conference on Advances in Computer Engineering and Communication Systems*, Lecture Notes in Networks and Systems 612, https://doi.org/10.1007/978-981-19-9228-5_24

productivity losses (Das et al. 2018). If service interruption is observed due to not working components, we need to fix this issue by applying a reactive approach. In a reactive approach, remedial actions will be taken by analyzing the root cause of the failure. In contrast, proactive approaches are taken to fix future failures before they occur (Tan and Gu 2010). In continuously working IT infrastructures, failures can occur at any time and location; consequently, they must be monitored regularly. IT infrastructure monitoring involves tracking events in each component, alerts on abnormal conditions, preserving details about each activity, data mining to find the root cause, etc.

The system logs register the activities and events in the IT infrastructure's assets and components. Therefore, they supply worthwhile particulars which possibly employed to understand and maintain the system's state. This large amount of log data can be studied to observe the behavior pattern and detect strange situations in the system. Log examination be partly responsible for in failure identification, root cause analysis, and failure prediction to secure the reliability of the IT infrastructure (Bhanage et al. 2021).

System logs are unstructured textual data, and each system produces a massive amount of logs each minute. Traditionally, system administrators used to check logs with standard expression matching or keyword probing (like "error", "kill", and "fail") to identify failure (Wang et al. 2020). However, this process was highly counterproductive. Accordingly, it encourages the need for automatic log analysis.

In the present literature, a lot of work has been done to propose automatic log analysis. For further investigation, log messages are unstructured and must be converted to structured forms. Log parsers mainly aim to retrieve log templates from log messages. For example, a log entry for BGL system is "ciod: generated 128 core files for program /g/g24/germann2/SPaSM_mini/MEAM/r13" and extracted event template is "ciod: generated <*> core files for program <*>".

Even though the log templates are structured, they still hold data in natural language. Log templates must be presented in digital format to provide them as input to train ML or DL models. In the recent past, several researchers have started applying NLP approaches for log template vectorization like term frequency-inverse document frequency (TF-IDF) (Alharthi et al. 2021), word2vec (Wang et al. 2022; Yang et al. 2021), and bag-of-words (Meng et al. 2019a). These word embedding techniques convert log templated into vectors by understanding the input's sentiments. The generated vectors are further offered to ML (Alharthi et al. 2021) and DL (Platini et al. 2021; Meng et al. 2019b) models for identification or prediction of faults.

Accordingly, to withdraw the vectors from the log templates, understand the newly generated log entries and provide sentiment analysis of logs, we explore the vectorization by TF-IDF, polarity score, and word2vec, followed by training for ML classifiers. The principal endowment of this paper is outlined as follows:

- Word embedding is performed based on sentient analysis of log templates using TF-IDF, polarity score, and word2vec.

- Training using machine learning supervised classifiers to categorize log entries considering their level.
- Execution of trained models on various infrastructure logs such as Android, Hadoop, BGL, and OpenStack
- Performance examination of all the NLP and ML models using precision, recall, $F1$-score, and accuracy as evaluation metrics.

2 Related Work

In the recent years, researchers have suggested that the semantic analysis of textual logs can be more beneficial than the traditional analysis methods. Ample amount of time, NLP techniques are utilized for document processing to understand the sentiments of the writers. Sentiment analysis was applied to detect cyber-attacks by exploiting tweets. But NLP techniques were not utilized for log analysis. By virtue of the study conducted on the non-traditional use of NLP algorithms, Bertero et al. (2017) attempted to make use of Google's word2vec algorithm for log mining. Authors transformed logs into multi-dimensional vectors of features. Then, they provided classifiers like the random forest, MLP, and Gaussian NB to detect the stressed behavior in vending machine with approximately 90% accuracy. Wang et al. (2018) applied TF-IDF and Word2vec feature extraction techniques to transform words into vectors and claimed that Word2vec could enhance the effectiveness of semantic information of log than TF-IDF for anomaly detection. Ren et al. (2019) generated numerical semantic feature vectors by calculating and combining the semantic similarity values for filtered log events and then applying a deep convolutional neural network (CNN) algorithm for cluster systems log classification. Meng et al. (2019b) proposed template2Vec inspired by word2vec, the first model that considered semantic (synonyms and antonyms) and syntax information of log templates to determine anomalous log sequence. LogRobust (Zhang et al. 2019) is not dependent on simple occurrence details of the log template but converts each log template into a semantic vector of the fixed dimension that is proficient at apprehending semantic information embedded in the log template. LogTransfer (Chen et al. 2020) framework proposed to detect the anomaly in cross-system based on the concept of similarity in the pattern. In the preprocessing phase, fixed dimension vectors were developed using GloVe, a word representation technique, then forwarded to LSTM for detection. Huang et al. (2020) augmented the parameter vector encoder and log sequence encoder to build the anomaly detection model using LSTM. Also, it declared the performance and robustness of HitAnomaly on precarious logs. Wang et al. (2020) proposed LogEvent2vec offline feature extraction model, directly extracting the connection between log and vectorized log events. LogEvent2vec with TF-IDF and Naïve Bayes shows a significantly low computational time (30 min).

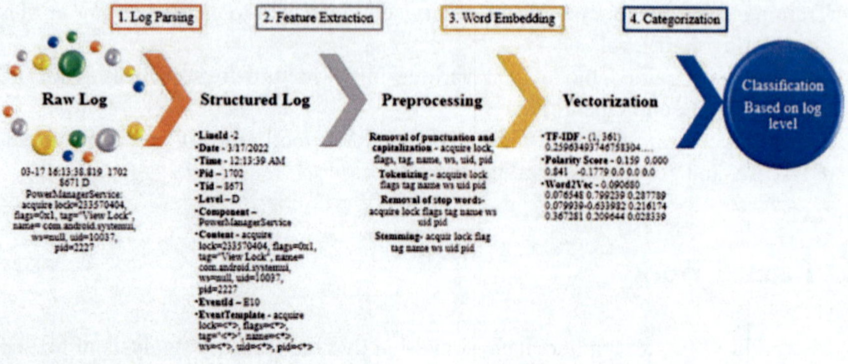

Fig. 1 Basic workflow of proposed system

3 Design of Log Analysis System

3.1 System Overview

Figure 1 exhibits the general workflow of the proposed system. The workflow mainly comprises of four steps, as (i) log parsing, (ii) feature extraction, (iii) word embedding, and (iv) classification of log records. The system generates logs to preserve the information about the events occurring during the communication between various components. The logs are collected from different infrastructures, and they are in raw format. The first and essential step is log parsing which transforms unstructured logs into a structured format. On parsing, logs are stored in a .csv file with different columns like LineId, Data, Time, Level, Content Event Template, etc. The "log template" is selected for further evaluation. Thus, log template can be intended as a feature of the whole log entry. The selected feature undergoes word embedding in the third step and represents text data in numerical forms. Various word embedding techniques such as TF-IDF, polarity score, and Word2Vec are applied to produce vectors for the log template. Finally, extracted vectors are supplied to the ML or DL models to classify logs as per the allocated log level. Logs get classified into different categories, which include "fail", "fatal", "error", etc. This automatic classification of logs is helpful to the system administrator in identifying faulty logs and taking corrective actions.

3.2 Data Acquisition

System logs carry all the particulars concerning the execution of each component and resources involved in the event's operation. Elements present in the system logs are confidential; thus, real-world log datasets are infrequent in public. This

Table 1 Statistics of datasets used for experimentation

Log data	Infrastructure type	Dataset size	Number of log records	Log templates
Android framework log	Mobile system logs	3.38 GB	30,348,042	76,923
Blue Gene/L supercomputer log	Supercomputer	708.76 MB	4,747,963	619
Hadoop MapReduce job log	Distributed system	48.61 MB	394,308	298
OpenStack software log	Distributed system	60.01 MB	207,820	51

situation hinders collecting logs from various infrastructures to perform research and experimentation. Zhu et al. (2019) collected and released[1] an extensive assemblage of logs from 16 various infrastructures (distributed systems, supercomputers, operating systems, mobile systems, server applications, and standalone software).

Experimentation is performed on Android, BGL, Hadoop, and OpenStack logs. Hadoop and BGL are production logs released by researchers, whereas Android and OpenStack are gathered from the real-world system. Table 1 furnishes the particulars of datasets utilized during the experimentation. Statistical information such as dataset size, number of log records and number of log templates are used for further investigation.

3.3 Data Preprocessing

Engineers script logging statements at the time of software development. Furthermore, these statements get recorded in the form of logs on the occurrence of events. In the logging statements, developers specify the static part, and dynamic parts of the message get updated as per the information about components and resources involved in the event accomplishment. The principal intent is to convert unstructured logs (single-line logs included with many parameters) into structured logs (presentation of data under a specific title).

Various log parsers are available and discussed in the existing literature. Drain (He et al. 2017) parser has been adapted in this research work to parse log datasets considering execution availability, accuracy, and flexibility parameters. The drain parser employed the fixed-depth tree structure to perform and retrieve log templates. Drain stipulates acceptable accuracy for different types of system logs.

For example (Android log)

[1] GitHub—logpai/loghub: A large collection of system log datasets for AI-powered log analytics. [Online]. Available: https://github.com/logpai/loghub. Accessed: 27 June 2021.

Raw log: "03-17 16:13:38.819 1702 8671 D PowerManagerService: acquire lock = 233570404, flags = 0x1, tag = "View Lock", name = com.android.systemui, ws = null, uid = 10037, pid = 2227".

Parsed log using Drain: "LineId - 2, Date - 3/17/2022, Time - 12:13:39 AM, Pid – 1702, Tid – 8671, Level – D, Component – PowerManagerService, Content - acquire lock = 233570404, flags = 0x1, tag = "View Lock", name = com.android.systemui, ws = null, uid = 10037, pid = 2227, EventId – E10, EventTemplate - acquire lock = <*>, flags = <*>, tag = "<*>", name = <*>, ws = <*>, uid = <*>, pid = <*>".

The discussed sample log example is depicted in Fig. 1 under section two, structured log. Here, EventTemplate is retrieved from the constant part of the contents/log message of the record.

3.4 *Feature Extraction*

On execution of log parser ("Drain"), log templates are extracted as a static part of the log message. This log template is then selected as input data for further analysis. The log templates are selected from structured log data out of the multiple parameters. Thus, log template is considered a feature of log data which is determined to perform word embedding and classification.

3.5 *Word Embedding Techniques*

The data must be numerical to build any ML and DL model as they cannot be applied directly to text data. The conversion of text data into numerical vectors is called vectorization or word embedding. The extracted vectors can be provided to build the different ML and DL models. Table 2 presents the summarized view of bag of words, TF-IDF, and Word2Vec embedding techniques. Based on the critical points observed during the review of bag of words, TF-IDF, and Word2Vec techniques, they are compared using a type of embedding, suitable for which kind of data and pros-cons of the method. All studied embedding approaches work on static word embedding. Thus, in the problem, where text and document classification is the aim, bag of words and TF-IDF are the best choices. Whereas semantic relation between the words is essential, then Woc2Vec is the appropriate choice.

3.6 *Log Analysis Using Machine Learning Model*

In accordance with the "no free lunch theorem", using multiple and different classification models on the master feature vector is highly recommended to evaluate the best results (Grandgirard et al. 2002). The probability of a single classifier

Table 2 Comparative analysis of word embedding techniques/models

NLP technique	Embedding type	Suitable for	Pros	Cons
Bag of words	Static word embedding	Text classification	Easy to understand and implement, flexible to customize	Ignores context by discarding word order; sparse matrix representation is complex due to time and space complexity
TF-IDF	Static word embedding	Document	Easy to understand, fast	Unable to grab semantics, cannot reflect sequence information, useful as a lexical level feature
Word2Vec	Static word embedding	Semantic relation between word	Easy to understand and implement; little preprocessing required	Fails to learn a representation of out-of-corpus words

working better for a particular dataset is less. Hence, we selected different classifiers, logistic regression, random forest, Naïve Bayes, and support vector machine, for implementation.

4 Results and Interpretation

This section provides insight of the work performed and analyzes the results obtained. The entire experiment was done in a Python environment, taking the help of the Google Colab-pro GPU, especially for word embedding. The classification models were trained over a random selection of training and testing data from the provided datasets. Records are selected using different seeds as the 80% log entries as training data and 20% remaining as testing data. All classifiers were evaluated based on precision, recall, $F1$-score, and accuracy for each log level within our labeled dataset.

4.1 Evaluation Metrics

Using testing instances, the assembled classification model calculates which posts belong to a dataset class (i.e., "info, error, warning, fail, fatal, etc."). The performance of classification models is evaluated based on true negative, false positive, false negative, and true positive values.

The effectiveness of the proposed solution is evaluated based on the standard precision, recall, $F1$-score, and accuracy metrics. Precision is the fraction of mined ground truth edges in all ground truth edges, and recall is the fraction of the mined ground truth edges. Accuracy, which is a generally used evaluation matric to check the classification performance, is utilized. Accuracy of classifiers linear regression, SVM, Naïve Bayes, and the random forest is recorded considering the categorization of similar type log templates in the appropriate group. The definition of accuracy is as shown in the equation below.

$$F1\text{-Score} = \frac{2\,\text{Recall} * \text{Precision}}{\text{Recall} + \text{Precision}} \tag{1}$$

$$\text{Precision} = \frac{\text{TP}}{\text{TP} + \text{FP}} \tag{2}$$

$$\text{Recall} = \frac{\text{TP}}{\text{TP} + \text{FN}} \tag{3}$$

$$\text{Accuracy} = \frac{\text{TP} + \text{TN}}{\text{TP} + \text{FN} + \text{TN} + \text{FP}} \tag{4}$$

True Positive (TP): Log records having the same log template are manually categorized under the same group.

True Negative (TN): Log records having different log templates are manually categorized under the different groups.

False Positive (FP): Log records having the same log template are manually categorized under the different groups.

False Negative (FN): Log records having different log templates are manually categorized under the same groups.

4.2 Experimental Results

The classification performance based on feature extraction using NLP techniques followed by ML classifiers is evaluated by generally used metrics such as precision, recall, and $F1$-score. Evaluation metrics were calculated for Android, BGL, Hadoop, and OpenStack infrastructure logs. The highest precision, recall, and $F1$-score values are highlighted in each infrastructure to observe the accomplishment of models. The analysis shows that the random forest classifier outperforms LR and SVM in almost all types of infrastructure logs. In the case of OpenStack logs, precision, recall, and $F1$-score are recorded as 0.99 when the TF-IDF approach was applied for feature extraction. The results are overfitting due to the limited (51) number of log templates (refer to Table 1) available for processing. In 207,820 log records, multiple records carry a similar kind of nature; thus, those are categorized under the same class during classification. As the number of unique log templates increases, performance classification decreases. The precision, recall, and $F1$-score were highest when TF-IDF was used as a feature extraction technique.

Figure 2 presents the comparison of classification accuracy of different classifiers on various infrastructure logs. Figure 2a illustrates the classification accuracy acquired from the execution of the proposed model by providing an Android framework log as input. Word embedding using TF-IDF delivers the highest accuracy for LR, SVM, Naïve Bayes, and random forest classifiers. Random forest with Word2Vec embedding technique gives 94% classification accuracy. Figure 2b illustrates the classification accuracy derived from the proposed model's execution by providing Blue Gene/L Supercomputer log as input. Word embedding using TF-IDF provides equal accuracy for LR, SVM, Naïve Bayes, and random forest classifiers. Random forest with Word2Vec embedding technique gave 94% classification accuracy and 87% when polarity score was used as embedding approach. Figure 2c illustrates the classification accuracy derived from the proposed model's execution by providing Hadoop MapReduce job log as input. Word embedding using TF-IDF and Word2Vec offers adequate accuracy for LR, SVM, Naïve Bayes, and random forest classifiers. Figure 2d illustrates the classification accuracy derived from the proposed model's execution, providing the OpenStack Software log as input. Word embedding using TF-IDF and Word2VEc gives adequate accuracy to LR, SVM, Naïve Bayes, and random forest classifiers. Moreover, models trained using polarity score as embedding techniques retrieve 95% classification accuracy using LR, SVM, and random forest but much less in the case of Naïve Bayes (Table 3).

5 Conclusion

System logs hold immense information about IT infrastructure components, execution of activities, and resources involved in the communication. This rich source of information is incredibly valuable for failure detection, analyzing the cause of failure, and rectifying the problem. Due to the enormous data size and imbalance nature of the logs, it is challenging to analyze logs manually. Thus, designing a system that can classify log records based on allocated levels spontaneously and expeditiously is imperative.

The proposed system does an automatic and efficient log analysis of various types of IT infrastructure logs with the help of NLP techniques. Logs are available in the combination of text, numbers, special symbols, etc. Such data cannot be processed directly by any ML classifiers. For this reason, logs are pretended as natural language and processed to convert them into vectors or numerical form using TF-IDF, polarity score, and Word2Vec approaches. Further, SVM, random forest, logistic regression, and Naïve Bayes are provided vectors to furnish robust classification log entries. Log records categorized under "fail", "error", "fatal", etc., levels seem to be abnormal and demands obligatory attention by the system administrator. In a way, the system admin

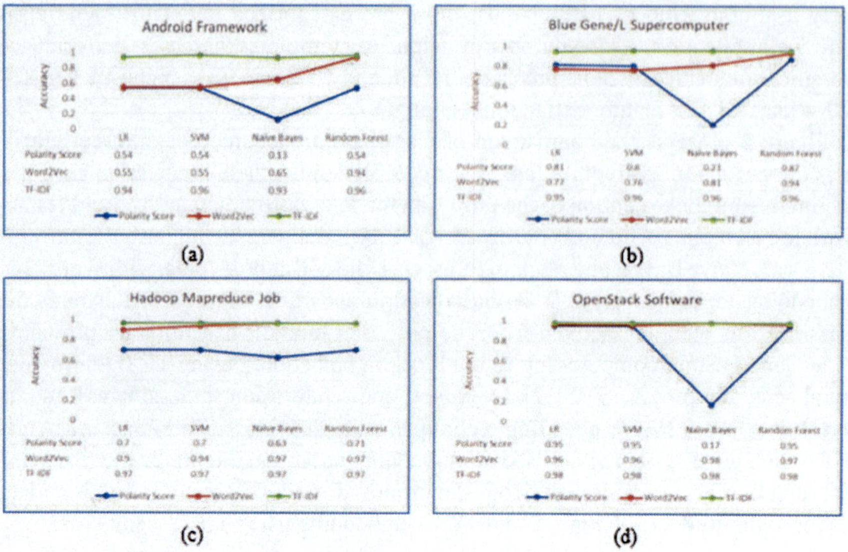

Fig. 2 Accuracy of log classification with TF-IDF, polarity score, and Word2Vec embedding technique on various IT infrastructure logs

can concentrate on limited records than multiple logs. This automated classification will save the cost and effort required for log analysis and handle the erroneous situation.

The results derived from implementing the proposed model are inconsistent on all datasets. The classification accuracy concerning TF-IDF as a word embedding technique is high, although it is not of much use in the real-time system. TF-IDF works on the frequency of words present in the text. In logs, many commonly used words appear for different purposes; thus, it is not recommended to prefer TF-IDF for sentiment analysis of logs.

These results can be improved by applying semantic analysis to understand the context of log records. Classification based on semantic analysis will be more effective in newly occurring log entries. In forthcoming research, the comprehension of logs using the BERT language model followed by classification using advanced ML models is proposed to acquire improved performance.

Table 3 Precision, recall, and $F1$-score distribution of log classification across different types of infrastructure logs

Log	ML model	Precision			Recall			$F1$-score		
		TF-IDF	Polarity score	Word2Vec	TF-IDF	Polarity score	Word2Vec	TF-IDF	Polarity score	Word2Vec
Android	LR	0.78	0.29	0.24	0.77	0.25	0.28	0.78	0.22	0.25
	SVM	0.79	0.30	0.23	0.78	0.26	0.29	0.79	0.22	0.26
	Naïve Bayes	0.80	0.21	0.57	0.97	0.23	0.53	0.81	0.10	0.54
	Random forest	0.79	0.51	0.78	0.78	0.26	0.76	0.79	0.23	0.77
BGL	LR	0.98	0.31	0.44	0.97	0.28	0.26	0.98	0.29	0.29
	SVM	0.97	0.30	0.23	0.98	0.28	0.21	0.98	0.29	0.21
	Naïve Bayes	0.95	0.32	0.56	0.98	0.35	0.49	0.97	0.18	0.52
	Random forest	0.98	0.37	0.96	0.98	0.32	0.76	0.97	0.34	0.82
Hadoop	LR	0.99	0.83	0.75	0.97	0.78	0.75	0.98	0.75	0.75
	SVM	0.98	0.65	0.73	0.98	0.59	0.73	0.99	0.57	0.75
	Naïve Bayes	0.94	0.81	0.74	0.97	0.78	0.75	0.98	0.73	0.74
	Random forest	0.99	0.87	0.75	0.98	0.79	0.75	0.99	0.78	0.75
OpenStack	LR	0.99	0.49	0.49	0.99	0.50	0.50	0.99	0.49	0.49
	SVM	0.99	0.49	0.49	0.99	0.50	0.50	0.99	0.49	0.49
	Naïve Bayes	0.99	0.52	0.97	0.99	0.59	0.98	0.99	0.18	0.98
	Random forest	0.99	0.49	0.97	0.99	0.50	0.98	0.99	0.49	0.98

References

K.A. Alharthi, A. Jhumka, S. Di, F. Cappello, E. Chuah, Sentiment analysis based error detection for large-scale systems, in *Proceedings of 51st Annual IEEE/IFIP International Conference on Dependable Systems and Networks, DSN 2021*, no. i (2021), pp. 237–249

C. Bertero, M. Roy, C. Sauvanaud, G. Tredan, Experience report: log mining using natural language processing and application to anomaly detection, in *Proceedings of International Symposium on Software Reliability Engineering ISSRE*, vol. 2017, Oct 2017, pp. 351–360

D.A. Bhanage, A.V. Pawar, K. Kotecha, IT infrastructure anomaly detection and failure handling: a systematic literature review focusing on datasets, log preprocessing, machine & deep learning approaches and automated tool. IEEE Access **9**, 156392–156421 (2021)

R. Chen et al., LogTransfer: cross-system log anomaly detection for software systems with transfer learning, in *Proceedings of International Symposium on Software Reliability Engineering ISSRE*, vol. 2020, Oct 2020, pp. 37–47

A. Das, F. Mueller, C. Siegel, A. Vishnu, Desh: deep learning for system health prediction of lead times to failure in HPC, in *HPDC 2018—Proceedings of 2018 International Symposium on High-Performance Parallel and Distributed Computing* (2018), pp. 40–51

J. Grandgirard, D. Poinsot, L. Krespi, J.P. Nénon, A.M. Cortesero, Costs of secondary parasitism in the facultative hyperparasitoid *Pachycrepoideus dubius*: does host size matter? Entomol. Exp. Appl. **103**(3), 239–248 (2002)

P. He, J. Zhu, Z. Zheng, M.R. Lyu, Drain: an online log parsing approach with fixed depth tree, in *Proceedings of 2017 IEEE 24th International Conference on Web Services ICWS 2017* (2017), pp. 33–40

S. Huang et al., HitAnomaly: hierarchical transformers for anomaly detection in system log. IEEE Trans. Netw. Serv. Manag. **17**(4), 2064–2076 (2020)

W. Meng et al., Device-agnostic log anomaly classification with partial labels, in *2018 IEEE/ACM 26th International Symposium on Quality of Service, IWQoS 2018*, no. 1 (2019a), pp. 1–6

W. Meng et al., Loganomaly: unsupervised detection of sequential and quantitative anomalies in unstructured logs, in *IJCAI International Joint Conferences on Artificial Intelligence*, vol. 2019, Aug 2019 (2019b), pp. 4739–4745

M. Platini, T. Ropars, B. Pelletier, N. De Palma, LogFlow: simplified log analysis for large scale systems, in *ACM International Conference Proceeding Series* (2021), pp. 116–125

R. Ren et al., Deep convolutional neural networks for log event classification on distributed cluster systems, in *Proceedings of 2018 IEEE International Conference on Big Data, Big Data 2018* (2019), pp. 1639–1646

Y. Tan, X. Gu, On predictability of system anomalies in real world, in *Proceedings of 18th Annual IEEE/ACM International Symposium on Modeling, Analysis, and Simulation of Computer and Telecommunication Systems, MASCOTS 2010* (2010), pp. 133–140

M. Wang, L. Xu, L. Guo, Anomaly detection of system logs based on natural language processing and deep learning, in *2018 4th International Conference on Frontiers of Signal Processing ICFSP 2018* (2018), pp. 140–144

J. Wang et al., LogEvent2vec: LogEvent-to-vector based anomaly detection for large-scale logs in internet of things. Sensors (Switzerland) **20**(9), 1–19 (2020)

J. Wang, C. Zhao, S. He, Y. Gu, O. Alfarraj, A. Abugabah, LogUAD: log unsupervised anomaly detection based on Word2Vec. Comput. Syst. Sci. Eng. **41**(3), 1207–1222 (2022)

H. Yang, X. Zhao, D. Sun, Y. Wang, W. Huang, Sprelog: log-based anomaly detection with self-matching networks and pre-trained models, vol. 2 (Springer International Publishing, 2021)

X. Zhang et al., Robust log-based anomaly detection on unstable log data, in *ESEC/FSE 2019—Proceedings of the 2019 27th ACM Joint Meeting on European Software Engineering Conference and Symposium on the Foundations of Software Engineering* (2019), pp. 807–817

J. Zhu et al., Tools and benchmarks for automated log parsing, in *Proceedings of 2019 IEEE/ACM 41st International Conference on Software Engineering: Software Engineering in Practice ICSE-SEIP 2019* (2019), pp. 121–130

Deployment and Serving ML Models Using Kubeflow and KfServing

R. J. Priyankasingh and H. Y. Vani

Abstract Network data is the data that is moving across the network. Due to the rise of network data, its utilization is rising. This made the development of predictions using machine learning algorithms. In this paper, building and deploying the trained LSTM and prophet model using Kubeflow is proposed, and the trained model is served with the help of KfServing which will quickly provide an inference of the model. By utilizing the URL generated from KfServing, the model is tested. Kubeflow and KfServing are present in the MLPaaS framework. This paper uses the key performance indicator (KPI) as the dataset that is obtained by performing the pre-processing to the counter value obtained from the network data and the paper provides affirmation on whether the framework supports both the algorithm or not. Another objective of this paper is to deploy the logs into the docker container instead of the virtual machine. The container of the pod contains the output of the training, inference of the model (URL) that has a path where the trained model is stored and the predicted dataset. A robot framework is used to automate this process, where the whole tasks are divided into subtasks. Each task is executed only after the execution of the previous task. It is monitored for continuous integration in the Jenkins tool.

Keywords Machine learning · Long short-term memory (LSTM) · Prophet · Kubeflow and KfServing · Docker container environment · Jenkins

1 Introduction

The request is sent to collect the network data from RAN and network elements. RAN is a crucial feature of a wireless communication system that uses radio links to link the devices to other areas of a network. Over a fiber or wireless backhaul

R. J. Priyankasingh (✉) · H. Y. Vani
Department of ISE, JSS Science and Technology University, Mysore, India
e-mail: priyasingh141198@gmail.com

H. Y. Vani
e-mail: vanihy@sjce.ac.in

A. B. Reddy et al. (eds.), *Proceedings of Third International Conference on Advances in Computer Engineering and Communication Systems*, Lecture Notes in Networks and Systems 612, https://doi.org/10.1007/978-981-19-9228-5_25

connection, the RAN connects user equipment. That connection leads to the core network, which controls subscriber data, location, and other things. Antennas and radios are among the crucial components that make up a RAN. Radio waves are produced from electrical signals via antennas and radios to ensure that transmissions are in the proper frequency bands. The design of the RAN has changed with the introduction of 4G LTE. To better meet the requirements of contemporary mobile devices, C-RAN was introduced, separating the remote radio head transceiver and antenna from the baseband unit. RRU interfaces with an antenna on one end and BBU on the other end it connects the BBU through a common public radio interface and converts the radio frequency signal into the data signal and vice-versa. It does filtering and amplification of RF signal and manages the whole base station including signal processing. The antenna interfaces with a cell phone wirelessly and it transmits and receives RF signal.

Network data obtained from the network elements should be processed using AI/ML technology. In this era of big data, machine learning plays a major role in predicting the data and reducing the misperception which can be seen.

ML algorithms behave well in understanding the network data. In this project, the LSTM and prophet algorithms are introduced to train the model and to get an inference. The obtained inference is utilized to test the data. The logs of the training, inferencing, and testing are deployed in a docker container. Docker offers operating system-level virtualization that can deploy applications in different environments. Docker will run on the host operating system kernel, and it is compatible with any programming language. Docker is software that allows the creation of containers that will wrap up the code, configurations, libraries, files required, and networking information to run the application quickly. The packaged application is called a docker image. Therefore, dockers are more lightweight than virtual machines.

The algorithms are trained on the MLPaaS framework that has Kubeflow which is an open-source ML platform that is built on top of K8s. It makes it easier for everyone who wants to develop and deploy portable, scalable machine learning models. It provides components such as TensorFlow model training, pipelines, notebooks, and several components which are used to build ML training models. Kubeflow pipelines provide a complete solution for delivering and controlling end-to-end ML processes. Runs can be planned, and complete reports on each run can be viewed. The trained model can be served using KfServing which is installed by default as part of Kubeflow installation. It is built with features like canary rollout, and auto-scaling, providing frameworks like TensorFlow, scikit-learn, and a clean interface making it easy to review models. It includes pre-processing and prediction and supports GPU. It will be useful for real-time forecasting.

2 Literature Review

This section provides information regarding the works done by the authors.

The authors of the paper (Xie et al. 2017) have found that traditional cloud hosts are less efficient than containers based on processing speed, disk reading, and disk writing. Container virtualization technology provides more efficient virtualization, faster deployment, faster delivery, and ease of migration. In paper (Gedia and Perigo 2018), authors have found that containerized technology has outperformed the VM in terms of throughput, performance, and memory consumption. Furthermore, in paper (Spoiala et al. 2016), it is found that VM are affecting the performance by 5–10% and they found that a container's boot time is one second whereas a virtual machine typically takes more than 10 s to boot. In paper (Lingayat et al. 2018), it has found that deployment of containers on bare metal is better than virtual machines as it affects the performance in deploying the containers that are present in docker. The docker performs 50% better than the virtual machine in bare metal.

The author of the paper (Noor Fathima and Vani 2020) has worked on building and deploying the home location register which is a database that stores the information of mobile subscriber identity and directory number of mobile station international subscribers and it serves as GSM architecture. It is concluded that the use of the virtual machine is cumbersome since they find it difficult in accessing the hardware, and it is also required to send the requests to the host to get access to hardware resources.

Rook-Ceph is the block storage that runs as a Kubernetes operator and Ceph is a shared file system. It has several components to take care of Ceph functions. The authors of the thesis (Baumann et al. 2019) have investigated a software tool that will test and monitor a Ceph cluster deployment using Rook on K8s, and in Rook, failures are recovered within a few seconds.

Every company needs to handle automation, authors of Mysari and Bejgam (2020) and Seth and Khare (2015) have used Jenkins ansible for continuous integration and development for developing and deploying the software. Jenkins for CD/CI is the best automation process since it is platform-independent, easily configurable, and saves a lot of time by helping the users for building and test the software continuously.

The authors of the paper (Jain 2020) have worked on traffic forecasting in telecom networks by using prophet (Garlapati et al. 2021) and XGBoost algorithms where the prophet algorithm helps in predicting future data by considering the seasonality and events. It can handle irregular and abnormal data, whereas the XGBoost algorithm supports the smaller dataset. In paper (Qian and Chen 2019), the authors worked on experimenting with LSTM and ARIMA model for stick price prediction and they found that LSTM can do better testing and the error rate was low when compared to the ARIMA model.

3 Machine Learning Models

3.1 LSTM Model

LSTM is mainly designed to resolve the drawback of the vanishing gradient descent problem that occurs in RNN. The structure of LSTM is shown in Fig. 1. The hidden layer in the LSTM has a memory cell, forget gate, input, and output layer. The key element to the LSTM is the cell state, it can add and discard the information from the cell state and this decision will be made by forget gate. Forget gate is composed of a sigmoid layer that helps in adding and removing the information through the cell state. The following mathematical operations are performed by the LSTM

$$f_t = \sigma\left(w_i \cdot \left[h_{t-1}, x_t\right] + b_i\right) \tag{1}$$

$$i_t = \sigma\left(w_j \cdot \left[h_{t-1}, x_t\right] + b_j\right) \tag{2}$$

$$c_t = \tanh\left(w_k \cdot \left[h_{t-1}, x_t\right] + b_k\right) \tag{3}$$

$$C_t = f_t * C_{t-1} + i_t * c_t \tag{4}$$

$$o_t = \sigma\left(w_l \cdot \left[h_{t-1}, x_t\right] + b_l\right) \tag{5}$$

$$h_t = o_t * \tanh(C_t) \tag{6}$$

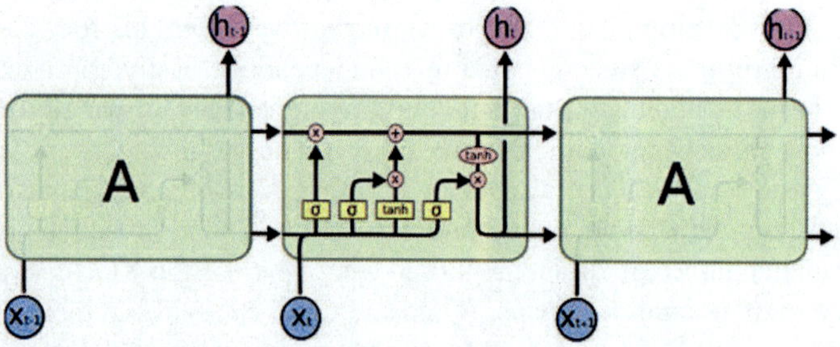

Fig. 1 LSTM architecture

3.2 Prophet Model

A prophet tool is released by Facebook's team. It is designed for forecasting time-series dataset having nonlinear trends. It will fit and predict the yearly, weekly, and daily seasonality data. Seasonality occurs over a short period, and it cannot be called a trend. It also can consider holiday effects to predict data. Prophet handles outliers well and is robust to missing data, and dramatic changes occur in the time series.

$$y(i) = g(i) + s(i) + h(i) + e(i) \tag{7}$$

The RHS of the above equation represents non-periodic changes (linear), periodic changes (yearly, monthly, weekly, and seasonality changes), and holiday effects.

4 Methodology

This project builds and deploys the machine learning models into the docker container. To deploy the models, initially, K8s is installed to maintain containerized technology and Kubeflow is installed on top of that. Cephfs and Rook-Ceph are also installed as a prerequisite to support the shared file system and block storage.

MLPaaS is an API-driven AI/ML framework on the NOKIA SEP Platform. It provides an environment to build, train, and deploy ML models. The MLPaaS framework consists of a Kubeflow pipeline for ML model development and scheduling ML pipelines for training, KfServing for model serving, Kafka for elastic data ingestion, Cassandra for Data Lake, Postgres DB for feature store.

The following steps describe the workflow of the training in the MLPaaS.

1. The data ingestion will read the data from different input sources using Kafka.
2. Storing the data to Data Lake is taken care of by Cassandra which is a NoSQL distributed database. As it uses multiple master nodes, it ensures high availability.
3. The pre-processed data is stored in the feature store which is implemented using Postgres DB which is an open-source RDBMS.
4. The ML pipeline is then executed by invoking an API call to the service control manager that schedules the ML pipeline using the APIs provided by Kubeflow.
5. The trained model is stored in the model store and deployed into the pod (Figs. 2 and 3).

The following steps describe the workflow of the inference in the MLPaaS.

1. An inferencing job is invoked by a call to the service control manager (SCM). This job will onboard the model from the model store to KF Serving, which in turn will return the prediction URL for the model.
2. The data is sent to the URL to get the predictions by invoking a call to the SCM for ML pipeline execution and predicted results are deployed.

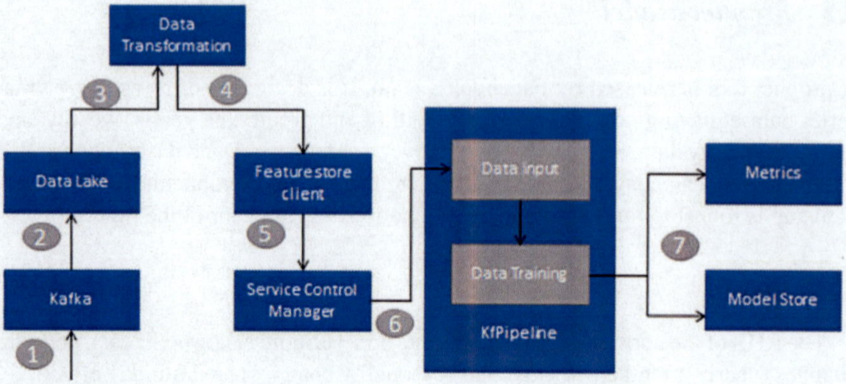

Fig. 2 Training workflow

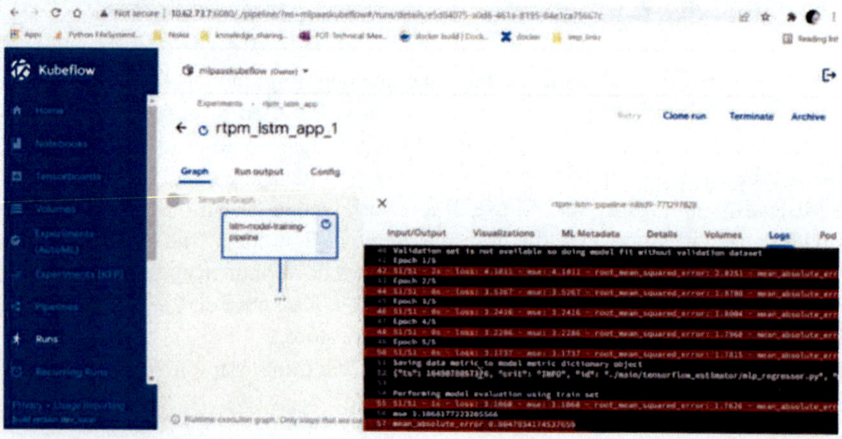

Fig. 3 Kubeflow dashboard

The logs are deployed into the docker container. As the docker concept is used the application can be deployed in a different environment by importing the docker image that packages up code, prerequisites, and libraries with the dependencies that are required to run the applications. As the docker container provides isolation capabilities the applications will be safer in the containers. Scalability is also one of the benefits of containerization that helps in scaling out by creating new containers (Figs. 4 and 5).

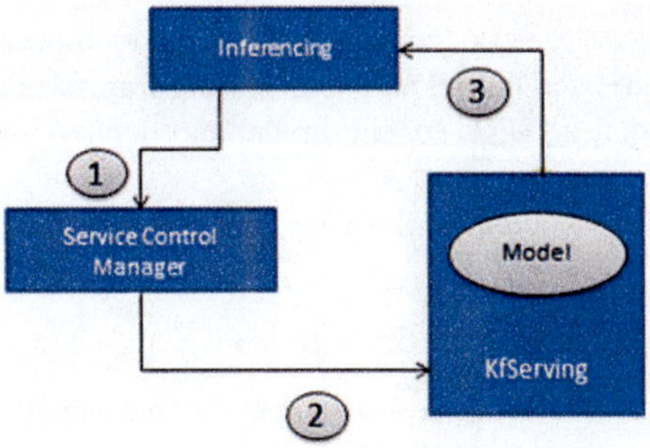

Fig. 4 Inferencing workflow

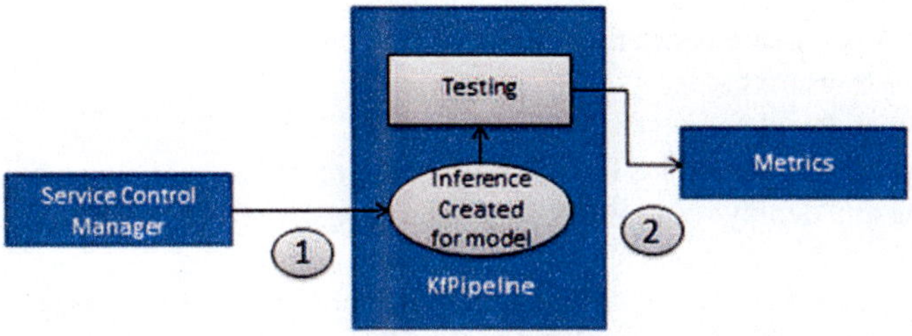

Fig. 5 Testing workflow

5 Implementation

This paper talks about building, deploying, and testing machine learning models. It can be brought up with the following tasks and all the tasks are interlinked.

Step 1: Data collection

In the previous section, an explanation regarding data collection is provided. From Kafka, counter data is collected and performed mathematical operations on that to obtain KPI data.

Step 2: Pull the mlpaas_training image and deploy prerequisites into the machine K8s, Cephfs, Rook-Ceph, and namespaces are all deployed as prerequisites. mlpaas_training image forms a base image to bring-up the container, and the artifacts required to create the model and inference are copied into the machine.

Step 3: Train and deploy the ML model

Sending the curl request to servicectrl/runpipeline component by providing the JSON file to train the LSTM and fbprophet models for KPI data in Kubeflow and deploy the logs into the container of the pod.

Step 4: Creating an inference of the model

Sending the curl request to servicectrl/inference/create component with the JSON file. JSON file has generated model path and serving framework as a TensorFlow to create the inference that is, URL for the model and deploys the inference into the pod.

But fbprophet failed to create inference since it does not support libraries of fbprophet.

Step 5: Test the data and deploy the predictions

Sending the curl request to servicectrl/runpipeline component with the JSON file. The JSON file is having Python code where the test dataset is passed to the URL to obtain predictions and deployed into the pod.

Step 6: Automate the above tasks

The automation code is written in the robot framework, and the complete tasks are divided into four test cases.

Creating and deploying a trained LSTM model.

Creating and deploying LSTM inference and predictions of unknown datasets.

Creating and deploying a trained fbprophet model.

Deploying the predictions of unknown datasets.

These test cases are monitored in the Jenkins tool. If all the test cases are passed, then the job is said to be successful. A report will be generated based on the execution that will provide detailed information on the execution of the test cases. If the test case is passed successfully, it turns the color green else it turns the color red.

6 Results and Discussion

Building and deploying the training, inference, and testing of the models into the pods are automated and integrated continuously in Jenkins. Wherein, building and deployment of the tasks are done by writing Python code and automation is done by using robot scripts.

Once the image is pulled and all the prerequisites are installed successfully, the status is checked. When it is up, the curl request is sent to the machine to train the ML models. Figure 6 shows the successful execution of fbprophet training.

Figure 7 represents the LSTM pipeline pod deployment. Pods get deployed when the training is done, and the pod should be in a completed state. If the pod is not in a completed state, scripts should be modified and merged with the code.

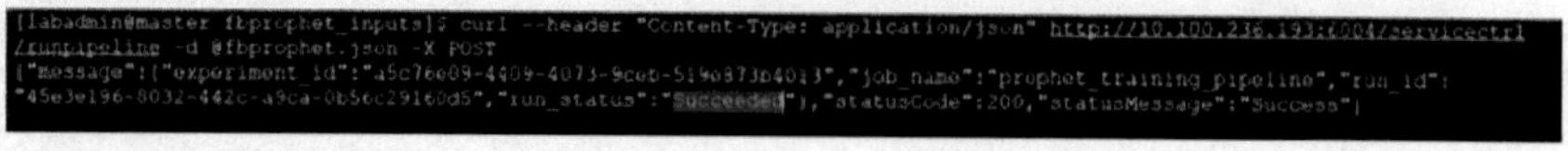

Fig. 6 Successful execution of fbprophet training

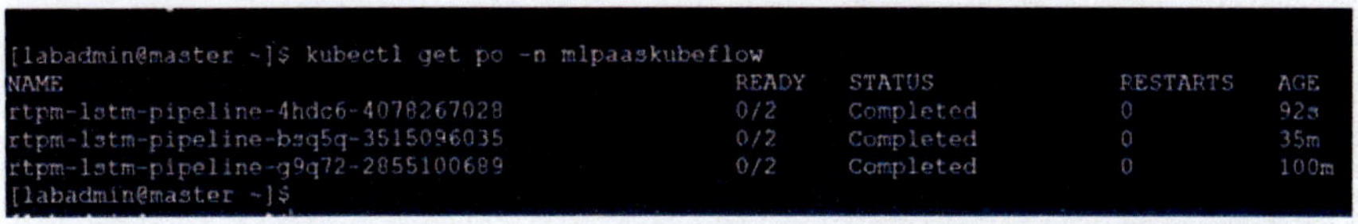

Fig. 7 Successful bring-up of LSTM-trained pod

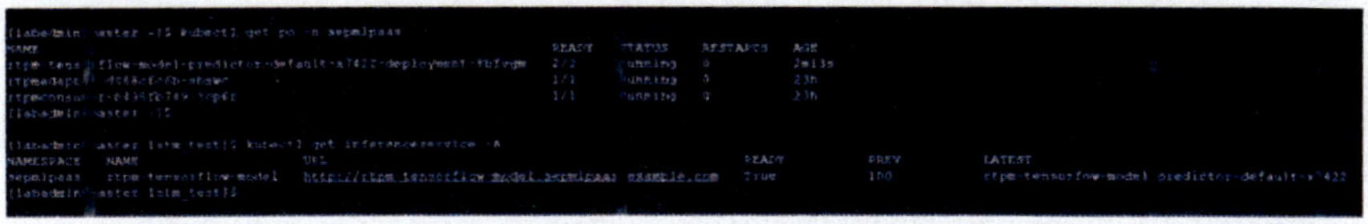

Fig. 8 Successful bring-up of LSTM inference pod and URI

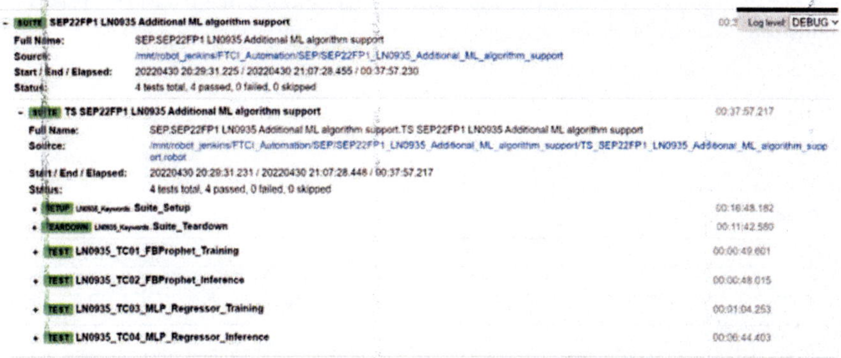

Fig. 9 Successful bring-up of automated test cases in Jenkins

Figure 8 represents creating the inference and deploying the pod for the LSTM model after doing curl request to servicectrl/inference/create component. Generated inference helps in testing the datasets.

Figure 9 shows the report generated for the above test cases. It tells that the suite is generated successfully as it appears in green color. If any of the test cases or any of the keywords in the test case fails, then it appears in red and it will send the report to the developer to identify the reason for the failures.

7 Conclusion and Future Work

The proposed paper "Deployment and Serving of Machine Learning models using Kubeflow and KfServing" is more efficient since the Kubernetes and docker concepts are used. It helps in executing the application in a different environment where K8s is running, and it supports portability, scalability, and flexibility. This paper concludes that the LSTM algorithm can be invoked via the MLPaaS framework, whereas it does not support fbprophet libraries. Using the model inference, the real-time data can be tested. Also, it provides an automated approach for deploying the trained and inference logs into the docker container. This project is also effective since it uses an automated approach to deploy the pods and it is monitored using the Jenkins tool, it helps in dividing multiple tasks into smaller tasks and it takes care of passing control of the tasks based on the status of the tasks. If any of the tasks failed, Jenkins stops the execution of the further tasks, and the report will be generated based on the execution and it will be sent to the owner of the task. As Jenkins supports continuous integration and continuous delivery that helps in the management of the project.

From the implementation and testing, it is clear that the LSTM algorithm can be invoked via the MLPaaS framework and it supports all the libraries related to LSTM. In the future, the LSTM algorithm can be utilized in MLPaaS for any of the use cases such as anomaly detection, and traffic steering to test the dataset and send the optimized value to the network elements. It helps in optimizing the RAN.

References

L. Baumann, S. Benz, L. Militano, T.M. Bohnert, Monitoring resilience in a rook-managed containerized cloud storage system (IEEE, 2019)

A. Garlapati, D.R. Krishna, K. Garlapati, Stock price prediction using Facebook prophet and Arima models, in *6th International Conference for Convergence in Technology (I2CT)*, Pune, India, 02–04 Apr 2021

D. Gedia, L. Perigo, Performance evaluation of SDN-VNF in virtual machine and container, in *2018 IEEE Conference on Network Function Virtualization and Software Defined Networks (NFV-SDN)*

G. Jain, Machine learning, prophet and XGBoost algorithm: analysis of traffic forecasting in telecom networks with time series data (2020)

A. Lingayat, R.R. Badre, A.K. Gupta, Performance evaluation for deploying docker containers on baremetal and virtual machine, in *Proceedings of the International Conference on Communication and Electronics Systems (ICCES 2018)* (2018). IEEE Xplore Part Number: CFP18AWO-ART. ISBN: 978-1-5386-4765-3

S. Mysari, V. Bejgam, Continuous integration and continuous deployment pipeline automation using Jenkins ansible, in *2020 International Conference on Emerging Trends in Information Technology and Engineering (ic-ETITE)* (2020), p. 15

F. Noor Fathima, H.Y. Vani, Building, deploying and validating a home location register (HLR) using Jenkins under the docker and container environment, in *Proceedings of the International Conference on Smart Electronics and Communication (ICOSEC 2020)* (2020). ISBN: 978-1-7281-5461-9

F. Qian, X. Chen, Stock prediction based on LSTM under different stability, in *2019 IEEE 4th International Conference on Cloud Computing and Big Data Analytics* (2019)

N. Seth, R. Khare, ACI (automated continuous integration) using Jenkins: key for successful embedded software development, in *Proceedings of 2015 RAECS*, UIET, Panjab University, Chandigarh, 21–22 Dec 2015

C.C. Spoiala, A. Calinciuc, C.O. Turcu, C. Filote, Performance comparison of a WebRTC server on docker versus virtual machine, in *13th International Conference on Development and Application Systems*, Suceava, Romania, 19–21 May 2016

X.-L. Xie, P. Wang, Q. Wang, The performance analysis of docker and rkt based on Kubernetes, in *2017 13th International Conference on Natural Computation, Fuzzy Systems and Knowledge Discovery (ICNC-FSKD 2017)* (2017)

Lung Cancer Classification Using Cross Stage Partial Network: A New Enhanced Learning Capability of CNN

Vadlapudi Likitha, Bokka Nishanth, Mandala Vamsi Krishna, Talasila Dileep Eeswara Sai, Eali. Stephen Neal Joshua, Thirupathi Rao Nakka, and Debnath Bhattacharyya

Abstract State-of-the-art approaches have been enabled by neural networks to attain accurate results on tasks such as detection of objects which are related to computer vision but the success of these approaches relies on computational resources that are costly and hinders people who prefer cheap devices to advanced technology. A network named cross stage partial network (CSPNet) is proposed in this paper to diminish the problem that requires computations based on heavy inference in the view of network architecture. This problem is caused due steeping gradient descent problem that is present within the proposed neural network optimization. The maintenance of the gradient descent problem by the proposed networks is done by combining feature maps both at starting and ending of the neural networks, the computations is reduced by 20% with equal or even greater accuracy on the image dataset of chest CT-Scan. The implementation of CSPNet is quite easy and also standard in nature to deal with architectures that are built on SparseNet and ResNet.

Keywords Cross stage partial network (CSPNet) · Convolutional neural networks (CNN) · SparseNet · ResNet

1 Introduction

The disease of cancer is a condition such that the blood cells develop more of the spread and effects the other parts of the body (Chao et al. 2019). Cancerous cell

V. Likitha · B. Nishanth · M. V. Krishna · T. D. E. Sai · Eali. S. Neal Joshua (✉) · T. R. Nakka
Department of Computer Science and Engineering, Vignan's Institute of Information Technology (A), Visakhapatnam, Andhra Pradesh, India
e-mail: stephen.eali@vignaniit.edu.in

Eali. S. Neal Joshua
Department of Computer Science and Engineering, Gitam (Deemed to be University), Visakhapatnam, India

D. Bhattacharyya
Department of Computer Science and Engineering, Koneru Lakshmaiah Education Foundation, Vaddeswaram, Guntur, Andhra Pradesh 522302, India

A. B. Reddy et al. (eds.), *Proceedings of Third International Conference on Advances in Computer Engineering and Communication Systems*, Lecture Notes in Networks and Systems 612, https://doi.org/10.1007/978-981-19-9228-5_26

starts from any part of the body among the trillions of cells present in our body. Human cells normally divide and replicate to generate new cells. This mechanism can occasionally fail (Joshua et al. 2022), resulting in the growth of aberrant cells. Tumors may arise from these cells. Cancerous tumors can spread throughout the body and invade surrounding tissues in our body. There are over 100 various forms of cancer.

However, we are concentrating on lung cancer (Huang et al. 2017) in this article. This is particularly common in smokers, but after COVID, many people's lungs were harmed. Among the most common cancer-related deaths, lung cancer stands out at the top (Law and Deng 2018). We cannot notice any symptoms at this stage. When the condition has progressed, symptoms appear. Cough, chest pain, headache, hoarseness, sometimes blood cough, shortness of breath, weight loss, bone pain, and other symptoms can occur. There are various kinds of lung cancer such as cancer carcinoma (Law et al. 2019), adenocarcinoma, and also large-cell carcinoma (Eali et al. 2018). Many lives can be saved by early prediction. Blood tests, X-rays, biopsies, and CT-Scans can all be utilized in order to diagnose it. However, identifying the nodule is difficult. The nodule is a tiny tissue tumor that appears as a lump. Earlier studies have proved that the radiologists failed to detect the nodules for positive cases. The main challenging situation in finding the nodule is because of its different size, shape, boundaries (Liu et al. 2018), and at different locations. But in the fast-growing technology, we can identify the nodules.

The dataset which we use is chest CT-Scan images (Eali et al. 2022) that is broken into three parts namely testing, training, and validation dataset. In order to train the model, training dataset is used and to optimize the model parameters, validation dataset can be utilized. The testing dataset is used to provide an unbiased estimate to the finished model. To the model we have to provide a CT-Scan image for classification (Doppala et al. 2021) as input and the model provides label or class name as output. The labels are large-cell carcinoma, squamous cell carcinoma, and adenocarcinoma (Qin et al. 2019).

2 Related Works

Design of CNN architectures: Xie et al. demonstrated that the effectivity of the cardinality can be increased than the dimensions of depth and width in ResNeXT (Bhattacharyya et al. 2020). As shown in SparseNet (Joshua et al. 2021a), it reduced the number of parameters significantly and computations because of the adoption of approach of a great number of re-claim structures. In order to maximize cardinality, feature outputs of the next layers are concatenated (Joshua et al. 2021b). The sparse connection for exponential spaced connection can be adjusted by SparseNet to improve parameter utilization resulting in good outcomes (Wang et al. 2019), further explained how to improve the network learning ability with the help of sparse connection and high cardinality and by using the concept of gradient combination. With the above technique, a partial ResNet (PRN) is developed (Joshua et al. 2021c). Ma et al.

(Doppala et al. 2021) introduced guidelines in order to improve the inference rate of CNN that are to be followed and design the ShuffleNet (Chao et al. 2019) proposed a harmonic SparseNet (HarDNet), a metric convolutional input/output (CIO), low memory traffic CNN, and an estimate of RAM traffic that is proportional to the real RAM traffic measurement.

This object is nodules in the CT-Scan of lungs. The famous object detectors in real time are YOLO (Bhattacharyya et al. 2020) and SSD (Eali et al. 2022). The novel technology on entity finding using mobile graphic processing unit (GPU) and CPU can be achieved with SSD, LRF (Wang et al. 2019), and RFBNet (Liu et al. 2018). Object detector that is anchor-free in recent times (Joshua et al. 2022; Law and Deng 2018; Law et al. 2019; Zhang et al. 2019; Neal Joshua et al. 2021) had become the main-stream system and the object detectors related to this type are CornerNet-Lite (Law et al. 2019) and CenterNet (Neal Joshua et al. 2021). These perform very efficiently and are effective. We can receive excellent performance of detection of object on mobile GPU or CPU with light-head RCNN (Eali et al. 2018) based ThunderNet (Qin et al. 2019), YOLO-based PRN (Joshua et al. 2021a), and SSD-based Pelee (Joshua et al. 2021b).

3 Method

3.1 Basic Methodology

The steps involved in this methodology is gathering the images to make the dataset, data preprocessing, building the network, training the model, validating the model, and at last testing the model. A detailed flow diagram of the model was described in Fig. 1.

Image dataset: The dataset used in this paper consists of data in three parts. They are testing data, validation data, and training data. The name of the dataset is chest CT-Scan images.

Data preprocessing: The steps to be taken are

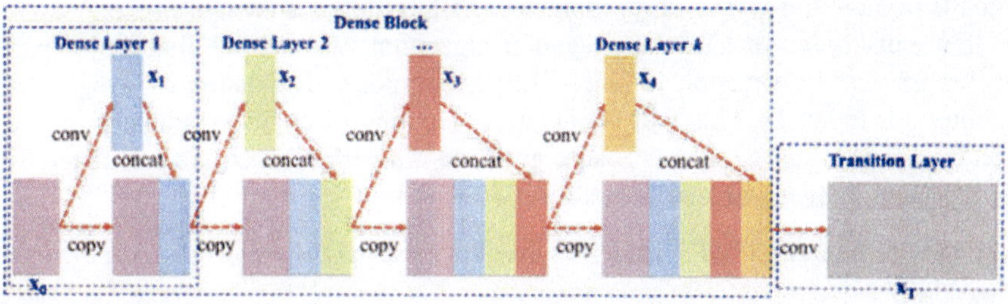

Fig. 1 Illustrations of DenseNet (Neal Joshua et al. 2020)

Image reading
Image resizing
Noise removal (denoise)
Image enhancement
Data augmentation
Morphology (smoothing edges).

Building the model: In this paper, we are using the CSPNet architecture. We will build the model by applying CSPNet on the ResNet which is the backbone to the model. The process of cross-channel pooling during the feature pyramid generating process is used to compress the feature maps and mish activation function which improves the performance of the neural network.

Training the model: After data preprocessing, the next is to train the model that we build. We train the model with training dataset using the model that we build in model building to identify patterns and make decisions.

Validating the model: After executing the classifiers, we validate the prototypical with the validation dataset which is one of the main layers in the dataset to optimize the model parameters.

Evaluating the model: We finally test the model using the testing dataset.

3.2 CASPNet-Partial Network with Cross Stage

The foremost determination of CASPNet remains toward achieve a richer gradient combination by enabling the architecture while reducing the computation amount.

SparseNet. It is utilized to solve the problem of vanishing gradient in deep neural architectures, maintain both low and high-level features together in the hierarchy of the convolutional neural layers.

A sparse layer and a sparse block are present in each stage of a neural network. Many sparse layers are present in sparse block. Let us consider, there are n sparse layers in the sparse block. Concatenation of the input of the ith sparse layer is done with the output of ith sparse layer and the output of this concatenation will be changed into the input to the $(n + 1)$th sparse layer which is clearly shown in Fig. 2.

If we make use of a backpropagation algorithm, we can see that major part of the gradient information is reused for the purpose of updation of weights of various sparse layers. This will result in the learning of copied gradient information by different sparse layers because gradients for earlier layers are calculated by multiplying gradients of later layers.

CSPNet-cross stage partial SparseNet. Cross stage partial network contains the below elements:

Partial sparse block

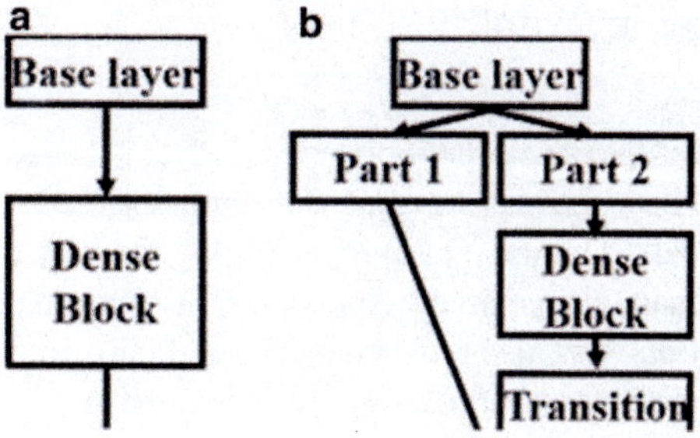

Fig. 2 **a** DenseNet consisting of solitary path, **b** proposed CASPDenseNet: transition → concatenation → transition (Neal Joshua et al. 2020)

Partial fully connected layer.

Between x0′ and x00″, the former feature map x0′ is related to the finish of the phase, and the final feature map namely x00″ will have to go through a sparse block. The sparse layers produced gradients are not duplicated. By splitting the feature map, it doubled the paths for the gradient flow during backpropagation which deals with vanishing gradient problem, reduced the computational bottleneck and reduced memory traffic.

Partial Sparse Block. This partial sparse block is designed to (a) Increase in gradient path: Because of strategy of split and merge which is shown in Fig. 2b, the number of gradient paths can be increased by twice than normal. This is solved by using cross-stage hierarchy. (b) Balance computation of every layer: In SparseNet, all the channels in the sparse layer operation involve the base layer which is shown in Fig. 2 but in partial sparse block only half of the channels of the base layer is used for operations that can successfully resolve closely partial of the computing bottlenecks, i.e., a layer which helps in reducing the number of feature maps. (c) Reduces memory traffic: Because of using cross stage hierarchy for feature maps concatenation which helps in compressing the feature maps in generating feature pyramid process.

Partial Transition Layer. This layer guides the gradient flow in order to avoid the duplication of the gradients. There are two variants of partial transition layer based on the order of transition and fusion of the feature maps which are clearly shown in Fig. 2. They are fusion first: fusion is carried out first and then the transition are performed. This significantly reduces the computational cost but causes a notable drop in the accuracy. Fusion last: this transition is performed on the output from the sparse layers and then fused with former split of the feature map. This reduces the computational cost but not as much as fusion first approach but this offers a good accuracy.

Application of CASPNet to Additional Designs. CASPNet can also be applied to architectures such as ResNet and ResNeXT. There is no need of the bottleneck layer as only partial of the extraction map goes through the ResNet blocks. This results in less consumption of energy, memory, and computations.

4 Results

In this paper, the proposed CSPNet is tested with the chest CT-Scan image dataset to know the performance. This smaller model outperformed well in conducting image classification. On light weighting the network, it enhanced the learning capability to maintain accuracy. Applying CSPNet on SparseNet is easy. The number of computations was also reduced because of partitioning of the base layer feature map into two parts. One part of the spitted feature map is allowed into the sparse block which can effectively reduce computation bottleneck and thus, reduce the unnecessary energy consumption. Applying CSPNet on ResNet and ResNeXT is easy and computations effort gets reduced because there is no need for a bottleneck layer. The result from the sparse block and the leftover part of the partitioned feature map are merged together through a cross-stage hierarchy which flattens the maps during the pyramid generation process feature. At last, the results in decreasing the memory usage and the costs because dynamic random-access memory is very expensive. We truly believe that these results were relevant to data generation applications and CSPNet is far better than CNN. The confusion matrix obtained from actual and predicted class is shown in Fig. 3. The specificity for the three classes of lung cancer along with the false positive and negative rates as mentioned in Table 1.

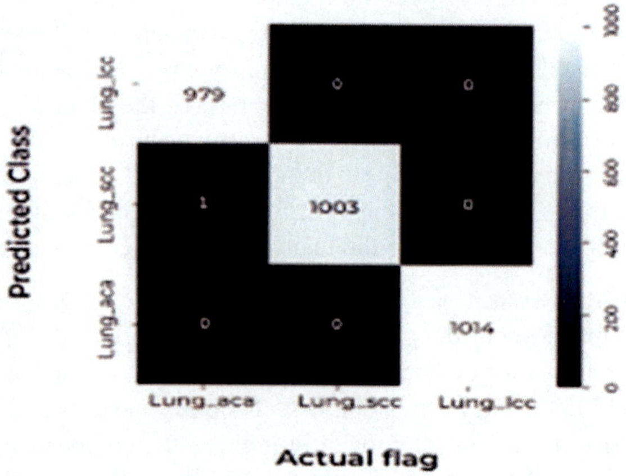

Fig. 3 Showing the confusion matrix of the proposed classifier

Table 1 Specificity and sensitivity of the three classes of lung cancer

Description	Specificity	1 − Specificity	FNR	FPR
Lung squamous cell carcinoma	1	0.99	0	0
Lung large-cell carcinoma	0.99	0.99	0	0
Lung adenocarcinoma	0.99	1	0	0

5 Conclusion

The proposed CSPNet is efficient and can be easily implemented. It is adequate to manage architectures based on the backbones such as residual neural networks called ResNet, ResNeXT, and sparsely connected convolutional networks (SparseNet). For applying the CSPNet into the backbone, we deliver dually on mobile GPU and CPUs. The major contribution made by us resulted in recognizing that the redundant problem of gradient information results in memory usage, optimization that is inefficient and also inference computations that are costly. In order to amplify the variability of cultured topographies present in various layers, the byproduct extraction synthesis approach and the slope movement have been proposed by us to outperform the entrants in the relations of correctness and memory usage. In addition to that, it also compresses the feature pyramid by incorporating the Maxout operation, thus making it compatible with devices consisting of edge computing and making them efficient.

References

D. Bhattacharyya, N.M.J. Kumari, E.S.N. Joshua, N.T. Rao, Advanced empirical studies on group governance of the novel corona virus, MERS, SARS and EBOLA: a systematic study. Int. J. Curr. Res. Rev. **12**(18), 35–41 (2020). https://doi.org/10.31782/IJCRR.2020.121828

P. Chao, C.-Y. Kao, Y.-S. Ruan, C.-H. Huang, Y.-L. Lin, HarDNet: a low memory traffic network, in *Proceedings of the IEEE International Conference on Computer Vision (ICCV)* (2019)

B.P. Doppala, S. NagaMallik Raj, E. Stephen Neal Joshua, N. Thirupathi Rao, Automatic determination of harassment in social network using machine learning (2021). https://doi.org/10.1007/978-981-16-1773-7_20. Retrieved from www.scopus.com

S.N.J. Eali, N.T. Rao, K. Swathi, K.V. Satyanarayana, D. Bhattacharyya, T. Kim, Simulated studies on the performance of intelligent transportation system using vehicular networks. Int. J. Grid Distrib. Comput. **11**(4), 27–36 (2018). https://doi.org/10.14257/ijgdc.2018.11.4.03

S.N.J. Eali, D. Bhattacharyya, T.R. Nakka, S. Hong, A novel approach in bio-medical image segmentation for analyzing brain cancer images with U-NET semantic segmentation and TPLD models using SVM. Trait. Signal **39**(2), 419–430 (2022). https://doi.org/10.18280/ts.390203

G. Huang, Z. Liu, L. Van Der Maaten, K.Q. Weinberger, Sparsely connected convolutional networks, in *Proceedings of the IEEE Conference on Computer Vision and Pattern Recognition (CVPR)* (2017), pp. 4700–4708

E.S.N. Joshua, D. Bhattacharyya, M. Chakkravarthy, Lung nodule semantic segmentation with bi-direction features using U-INET. J. Med. Pharm. Allied Sci. **10**(5), 3494–3499 (2021a). https://doi.org/10.22270/jmpas.V10I5.1454

E.S.N. Joshua, D. Bhattacharyya, M. Chakkravarthy, H. Kim, Lung cancer classification using squeeze and excitation convolutional neural networks with grad cam++ class activation function. Trait. Signal **38**(4), 1103–1112 (2021b). https://doi.org/10.18280/ts.380421

E.S.N. Joshua, M. Chakkravarthy, D. Bhattacharyya, Lung cancer detection using improvised grad-cam++ with 3D CNN class activation (2021c). https://doi.org/10.1007/978-981-16-1773-7_5. Retrieved from www.scopus.com

E.S.N. Joshua, D. Battacharyya, B.P. Doppala, M. Chakkravarthy, Extensive statistical analysis on novel coronavirus: towards worldwide health using Apache spark (2022). https://doi.org/10.1007/978-3-030-72752-9_8. Retrieved from www.scopus.com

H. Law, J. Deng, CornerNet: detecting objects as paired keypoints, in *Proceedings of the European Conference on Computer Vision (ECCV)* (2018), pp. 734–750

H. Law, Y. Teng, O. Russakovsky, J. Deng, CornerNet-Lite: efficient keypoint based object detection. arXiv preprint arXiv:1904.08900 (2019)

S. Liu, D. Huang et al., Receptive field block net for accurate and fast object detection, in *Proceedings of the European Conference on Computer Vision (ECCV)* (2018), pp. 385–400

E.S. Neal Joshua, M. Chakkravarthy, D. Bhattacharyya, An extensive review on lung cancer detection using machine learning techniques: a systematic study. Rev. d'Intell. Artif. **34**(3), 351–359 (2020). https://doi.org/10.18280/ria.340314

E.S. Neal Joshua, D. Bhattacharyya, M. Chakkravarthy, Y. Byun, 3D CNN with visual insights for early detection of lung cancer using gradient-weighted class activation. J. Healthcare Eng. **2021** (2021). https://doi.org/10.1155/2021/6695518

Z. Qin, Z. Li, Z. Zhang, Y. Bao, G. Yu, Y. Peng, J. Sun, ThunderNet: towards real-time generic object detection, in *Proceedings of the IEEE International Conference on Computer Vision (ICCV)* (2019)

T. Wang, R.M. Anwer, H. Cholakkal, F.S. Khan, Y. Pang, L. Shao, Learning rich features at high-speed for single-shot object detection, in *Proceedings of the IEEE International Conference on Computer Vision (ICCV)* (2019), pp. 1971–1980

X. Zhang, F. Wan, C. Liu, R. Ji, Q. Ye, FreeAnchor: learning to match anchors for visual object detection, in *Advances in Neural Information Processing Systems (NeurIPS)* (2019)

Emoji Creation from Facial Expression Detection Using CNN

G. Ramesh Chandra, Ravi Chandana Golla, Tejasri Gangireddy, Anusha Nagula, and Dhanya Sri Bala

Abstract This paper implements a deep learning model to classify human facial expressions and map the expression with the corresponding emojis. The model uses facial expressions that express the emotions of humans as input to the model. Then the input is classified and mapped to the respective emoji. Emojis are majorly used to represent verbal cues. Emojis have been the most important part of online interactions through messages, gathering ratings and reviews of products, and for many more purposes. The areas of computer vision and deep learning have evolved, and it is now easier to reorganize facial expressions. The interpersonal relations between humans are affected by facial expressions. The identification of human facial expressions plays an important role in computer–human interactions. It may also be used in understanding the behaviors of humans. It is easier for humans to identify emotions through facial expressions, but it is difficult for a deep learning model to detect emotions. Machine learning has advanced quickly to the point that it is now able to identify emotions in people by their facial expressions.

Keywords Convolutional neural network · Emotional expression detection · Emotional expression classification · Emoji mapping

1 Introduction

A person's facial expression identifies the feelings, which corresponds to emotions of oneself. The better method to identify a person's integrity is to understand his facial expressions. Because emojis are non-verbal ways of expressing an emotion. Emojis or avatars are ways to convey non-verbal messages. People express their emotions through their voice, hand gestures, body gestures, etc. Most of the time, emotions are expressed through facial expressions. Emojis work as visual depictions of human emotions. New-age communication applications like Facebook, WhatsApp, Twitter, chat boxes, etc., use emojis for interactions.

G. Ramesh Chandra · R. C. Golla (✉) · T. Gangireddy · A. Nagula · D. S. Bala
Department of Computer Science and Engineering, VNR VJIET, Hyderabad, India
e-mail: gollaravichandana@gmail.com

A. B. Reddy et al. (eds.), *Proceedings of Third International Conference on Advances in Computer Engineering and Communication Systems*, Lecture Notes in Networks and Systems 612, https://doi.org/10.1007/978-981-19-9228-5_27

The relations between people are affected by facial expressions. The recognition of human facial expression tasks plays a key role in computer–human interactions. Facial emotion recognition has its applications in studying behaviors. It is easier for humans to identify emotions through facial expressions, but it is difficult for a deep learning model to detect emotions. Machine learning has advanced quickly to the point that it is now able to identify emotions in people by their facial expressions. Over the past few years, the development in the techniques for the detection and classification of facial expressions has made it possible to use human facial expression identification in many modern applications. Online chat, product reviews, brand promotion, and many others have taken advantage of these cues in recent years.

It also paved the way for improving research on data science related to storytelling using emojis. Recent advancements in machine learning have made it possible to recognize human emotions in images. A breakthrough in machine learning and computer vision (Hinton 2006) made it possible to detect human facial expressions in images. In this paper, prior work and the existing system has been discussed in Sect. 2, proposed work is discussed in Sect. 3; testing and results are discussed in Sects. 4 and 5. In Sects. 6 and 7, conclusion and future scope are discussed.

2 Prior Work

Facial expressions are the most commonly used forms for expressing emotions. There is a lot of research (Kartali 2018; Li and Deng 2018) going onto improve the techniques used for facial emotion identification as it has many applications such as robotics, behavioral sciences, human–machine interactions, stress detection, sign language recognition, etc.

Ekman et al. (Hinton 2006) identified seven universal, culture-unrelated emotions in the twentieth century. The seven emotions considered are sadness, anger, surprise, fear, disgust, happiness, and neutral. Two datasets used extensively in related literature are CK+ and FER2013.[1] The CK+ dataset is small. This dataset consists of very well-centered faces. Positions of eyes, noses, and mouths are identical from picture to picture. On contrary, the FER2013 dataset is large, and faces in the FER2013 dataset vary in a great amount from one to the other. We have variability factors such as the position of the face, the tilting angle of the head, the age of the face, and different expressions in the same emotion class.

Machine learning algorithms are used to learn underlying patterns in data without external instructions. In facial expression detection, convolutional neural networks (CNN) are one of the artificial neural networks that are used very frequently, and it has shown state of art results (Lopes et al. 2017; Yu and Salzmann 2017). There are numerous issues such as different backgrounds and some unwanted objects. Those problems are resolved using convolutional networks. The advancements in

[1] Dataset is taken from https://www.kaggle.com/fer2013. Accessed: July 02, 2021.

machine learning methodologies make the detection of emotions much easier and more efficient, and accessibility to the public has been increased.

3 Proposed Work

In this paper, three steps are carried out by the proposed system. These are countenance detection, facial expression recognition, and facial expression classification. The first phase entails utilizing a video camera to capture a live human face and pinpointing its precise location using box coordinates. Face detection is carried out with an open CV library and Haar cascade detection (Pramerdorfer and Kampel 2016). Images that have been discovered include shapes, objects, landscapes, etc. In this stage, a human face is identified, and facial features are extracted and saved for expression recognition.

Human responses are classified using the CNN model (Lopes et al. 2017). The captured human responses are classified based on the facial expression in the practical sense such as Angry, fear, disgust, happiness, neutral, and surprise. Finally, the corresponding emoji is mapped with an expression.

(a) **Dataset**

The dataset used for implementing this project is the FER2013 dataset from the Kaggle. About 35,887 labeled photos total in the dataset, of which 28,709 are used to train the model and 3589 are test images. An additional 3589 private test photos make up the dataset, which is utilized to verify the trained model. The private test data is needed to check whether it overfits the public test data. FER2013 dataset consists of images having a size of 48×48 and are in black and white (Emotion Detection … 2018). Figure 1 represents some of the images from the FER2013 dataset, and Table 1 depicts the details of the dataset.

(b) **Workflow Design**

Figure 2 represents the flow of tasks carried out throughout the complete process of the implementation. If the model is trained, it takes input from facial expression, performs image processing, and outputs the detected expression emoji. Else the model is trained with FER2013 dataset which uses Deep CNN algorithms and some image processing.

4 Modules

4.1 Facial Expression Identification

The input image may have noise and are indifferent in size and color. To get more accuracy and efficient results on the model, some preprocessing steps need to be

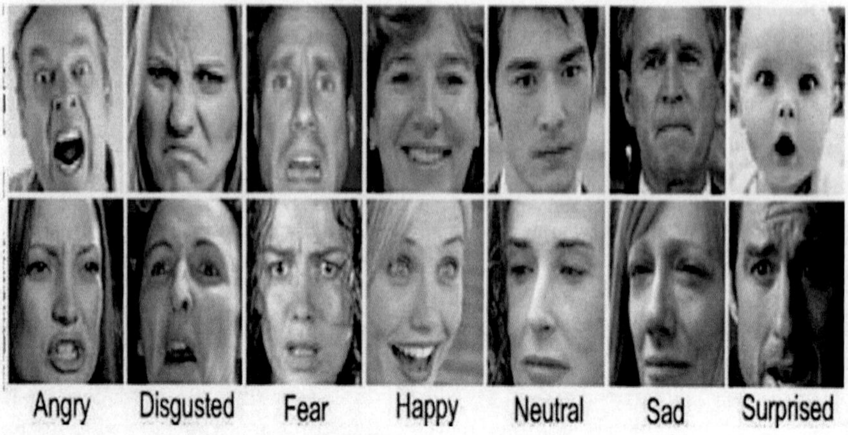

Fig. 1 Sample from FER2013 database

Table 1 Amount of data in the FER2013 dataset

Micro-expression (classification)	Validation data		Training data	Dataset total
	Public	Private		
Angry	467	491	3995	4953
Disgust	56	55	436	547
Fear	496	528	4097	5121
Happy	895	879	7215	8989
Sadness	653	594	4830	6077
Surprise	415	416	3171	4002
Contempt	607	626	4965	6198
	3589	3589	28,709	**35,887**

performed on the image. The image is converted to grayscale using preprocessing techniques, and then it is normalized and resized.

Normalization—Normalization of an image is performed to remove variations in data without changing its shape and obtain an accurate face image.

Grayscaling—Using pixel values that rely on the amount of light present in the image, gray scaling is the process that turns a colored image into a grayscale image. This is done because colored images are challenging for computers to process.

Resizing—The image is resized to a fixed size without affecting image quality. This step helps in reducing memory usage and increasing computation speed.

(i) **Face Detection**

The recognition of faces is the first step of any facial emotion identification system. The Haar cascades are an efficient method for detecting objects in images that are

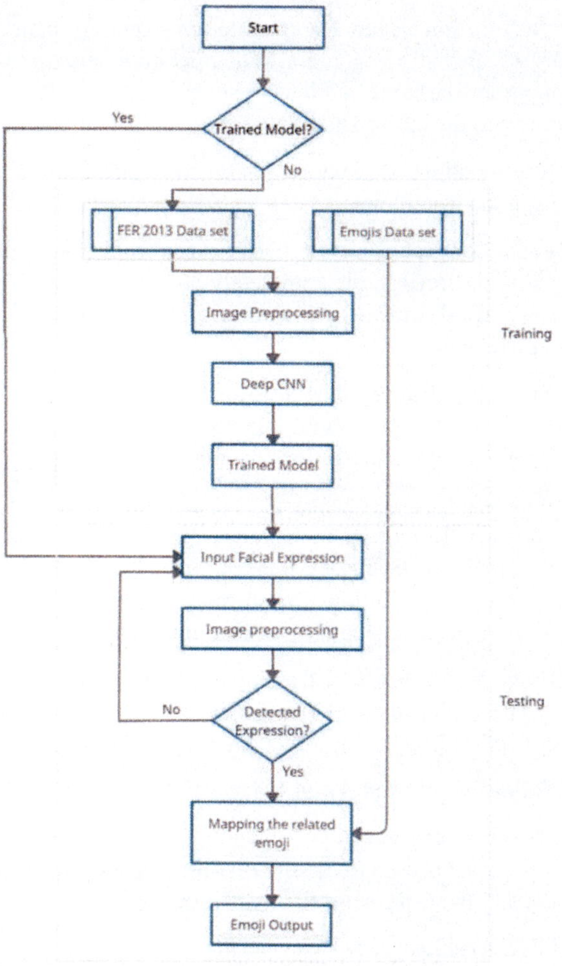

Fig. 2 Workflow design of the implementation

trained over a set of positive and negative faces. They have been widely used in detecting objects in images. Object detection using Haar cascade features detects three dark regions on the face, for example, edge features and line features.

(ii) **Facial Expression Classification**

The model assigns the image to one of the seven universal emotions that are listed in the FER2013 dataset: joyful, sad, angry, surprised, disgusted, fear, and neutral. The model is trained using the CNN classifier, a kind of neural network effective at analyzing images (Liliana et al. 2017). Training and testing datasets are the two groups into which the dataset is separated (Smith and Windeatt 2015). The emotion classification step contains the following steps Training and Generation of model.

The layers of the neural network architecture are as follows:

Convolution Layer: The convolution layer is the major building block of CNN which uses convolution operation. The operation determines the dot product between the input and the kernel or filter. The output is a feature map that is obtained after applying filters to the input. These layers are applied only to the initial stages.

Max Pooling: Max pooling helps to reduce overfitting by selecting the maximum from the feature map.

Fully Connected Layer: Each neuron from the preceding layer is connected to the output neurons in the completely connected layer. The number of classes that are categorized for the input data determines how many neurons there are in the final output layer.

Activation Function: Activation functions are mainly used to reduce overfitting. There are many activation functions available. ReLu activation function is used in CNN. The ReLu activation function has the benefit that its gradient is always equal to 1.

Softmax: The softmax function is a normalized exponential function that returns the probability of each class. It takes a vector of N real numbers and returns N numbers that sum to 1. Softmax transforms values between 0 and 1, regardless of whether they are positive, negative, or zero in the input.

Batch Normalization: Training neural networks is challenging due to initial weights and the arrangement of inputs to layers changing with each mini-batch when weights are updated. The inputs to layers are standardized via batch normalization.

Validation of Model and Using the Model to Classify Real-Time Images

The generated model is validated on a testing dataset that contains 3589 images. The model is used to capture real-time facial expressions using the web camera and classify the expression and map to the emoji.

CNN Architecture

Figure 3 represents the type of layers, output shapes, and parameters of each layer created in the deep learning model (Pramerdorfer and Kampel 2016).

5 Results

This section presents the results of the implemented model in which appropriate emoji is presented based on the live facial expression of the user. Along with the mapped emoji, the probability distribution of each class is shown, and the correct class is predicted using the probability.

- **Happy**—Figure 4 shows the mapped emoji (happy) with 94.6% probability.

Fig. 3 Convolutional neural network architecture

Layer (type)	Output Shape	Param #
conv2d_13 (Conv2D)	(None, 48, 48, 32)	320
batch_normalization_18 (Batc	(None, 48, 48, 32)	128
conv2d_14 (Conv2D)	(None, 48, 48, 32)	9248
batch_normalization_19 (Batc	(None, 48, 48, 32)	128
max_pooling2d_5 (MaxPooling2	(None, 24, 24, 32)	0
conv2d_15 (Conv2D)	(None, 24, 24, 64)	18496
batch_normalization_20 (Batc	(None, 24, 24, 64)	256
conv2d_16 (Conv2D)	(None, 24, 24, 64)	36928
batch_normalization_21 (Batc	(None, 24, 24, 64)	256
max_pooling2d_6 (MaxPooling2	(None, 12, 12, 64)	0
dropout_8 (Dropout)	(None, 12, 12, 64)	0
conv2d_17 (Conv2D)	(None, 12, 12, 128)	73856
batch_normalization_22 (Batc	(None, 12, 12, 128)	512
conv2d_18 (Conv2D)	(None, 12, 12, 128)	147584
batch_normalization_23 (Batc	(None, 12, 12, 128)	512
max_pooling2d_7 (MaxPooling2	(None, 6, 6, 128)	0
dropout_9 (Dropout)	(None, 6, 6, 128)	0
flatten_3 (Flatten)	(None, 4608)	0
dense_8 (Dense)	(None, 256)	1179904
activation_8 (Activation)	(None, 256)	0
batch_normalization_24 (Batc	(None, 256)	1024
dropout_10 (Dropout)	(None, 256)	0
dense_9 (Dense)	(None, 256)	65792
activation_9 (Activation)	(None, 256)	0
batch_normalization_25 (Batc	(None, 256)	1024
dropout_11 (Dropout)	(None, 256)	0
dense_10 (Dense)	(None, 256)	65792
activation_10 (Activation)	(None, 256)	0
batch_normalization_26 (Batc	(None, 256)	1024
dropout_12 (Dropout)	(None, 256)	0
dense_11 (Dense)	(None, 7)	1799
activation_11 (Activation)	(None, 7)	0

- **Neutral**—Figure 5 shows the mapped emoji (neutral) with 68.3% probability.
- **Fear**—Figure 6 shows the mapped emoji (neutral) with a 50.3% probability.
- **Surprised**—Figure 7 shows the mapped emoji (neutral) with a 50.3% probability.
- **Accuracy and Loss**—Figure 8 describe the model accuracy and loss graph on training and validation. With the implementation of the above model, an accuracy of 72% and a loss of 0.36 is obtained on training data, and an accuracy of 65% and a loss of 1.2 is obtained on validation data.

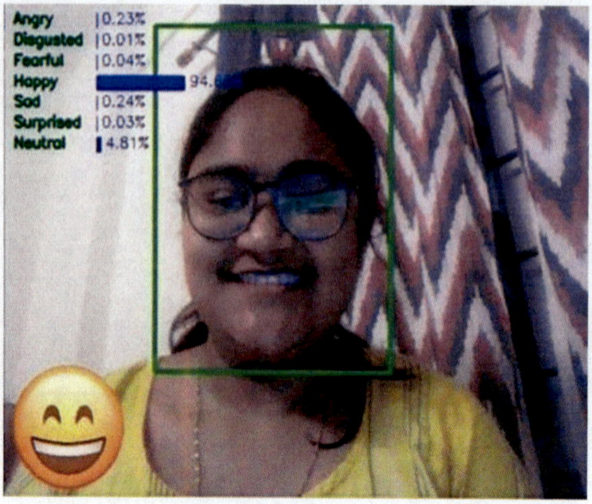

Fig. 4 Predicted emotion (happy)

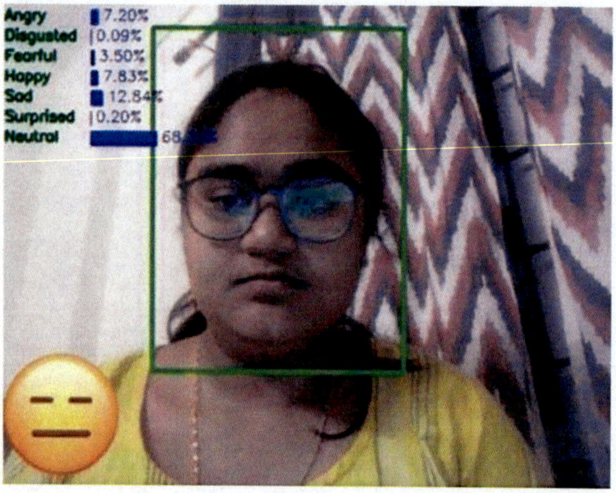

Fig. 5 Predicted emotion (neutral)

6 Conclusion

This project developed a deep learning model that maps emojis to classify facial emotions from static facial images and a live video stream. A convolution neural network is constructed, and the model is trained using the FER2013 dataset from Kaggle for emotion detection from live video. It includes capturing the human face and recognizing the facial expression of the user in the live video capture. With advances in computer vision and deep learning, it is now much more possible to identify human emotions in photos. The developed model would be able to foretell four to five different emotions, such as anger, happiness, sadness, fear, surprise, etc. With a validation accuracy of 65% and testing accuracy of 73%, the project's goal of

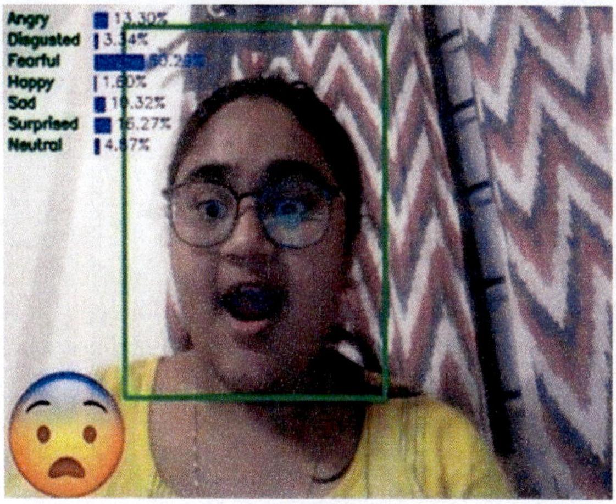

Fig. 6 Predicted emotion (fear)

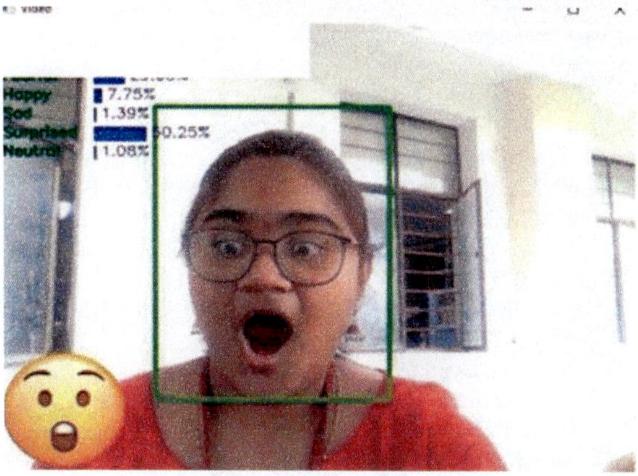

Fig. 7 Predicted emotion (surprised)

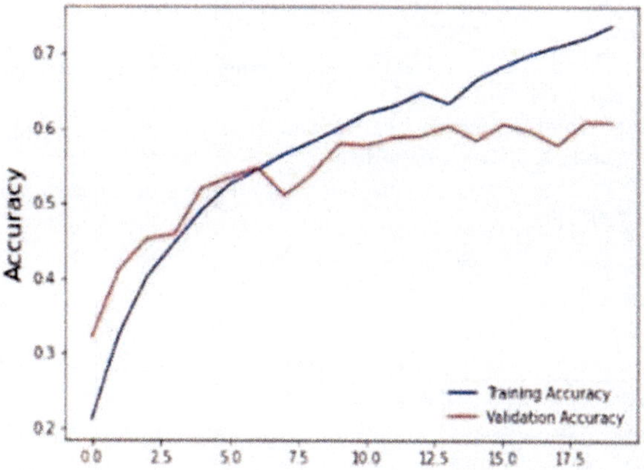

Fig. 8 Model accuracy graph

capturing and recognizing the facial emotion and displaying the appropriate emoji has therefore been effectively accomplished.

7 Future Scope

In this design, mortal facial expressions can be identified, classified, and mapped with the relating emojis. This plan can be additionally evolved by updating its accuracy. There are also many more emojis, such as those with hand movements that can be used. It is also possible to integrate this into commonly used keyboards to make facial expression recognition and emoji generating simpler and faster. It is also possible to display a personalized emoji of the person's face. The speed of bringing results can be expanded by executing it on GPU.

References

About Tensor Flow from https://www.tensorflow.org/

Emotion detection using image processing in Python, Feb 2018

G. Hinton, Reducing the dimensionality of data with neural networks (2006)

https://www.kaggle.com/code/dansbecker/rectified-linear-units-relu-in-deep-learning. Accessed 12 July 2021

A. Kartali, Real-time algorithms for facial emotion recognition: a comparison of different approaches (2018). IEEE Xplore. IEEE

S. Li, W. Deng, Deep facial expression recognition: a survey. CoRR, vol. abs/1804.08348 (2018). [Online]. Available: http://arxiv.org/abs/1804.08348

D.Y. Liliana, C. Basaruddin, M.R. Widyanto, Mix emotion recognition from facial expression using SVM-CRF sequence classifier (2017), pp. 27–31

T. Lopes, E. de Aguiar, A.F. De Souza, T. Oliveira-Santos, Facial expression recognition with convolutional neural networks: coping with few data and the training sample order. Pattern Recogn. **61**, 610–628 (2017)

C. Pramerdorfer, M. Kampel, Facial expression recognition using convolutional neural networks: state of the art. 12 (2016)

R.S. Smith, T. Windeatt, Facial action unit recognition using multi-class classification. Neurocomputing **150**, 440–448 (2015)

K. Yu, M. Salzmann, Second-order convolutional neural networks. arXiv preprint arXiv:1703.06817 (2017)

IoT-Based Telemedicine Health Monitoring System with a Fuzzy Inference-Based Medical Decision Support Module for Clinical Risk Evaluation

Analene Montesines Nagayo, Mahmood Zayid K. Al Ajmi, Naga Rama K. Guduri, and Fatma Saleh H. AlBuradai

Abstract This paper discusses the design and implementation of an Internet of Things (IoT)-based telemedicine health monitoring system (THMS) with an early warning scoring (EWS) function that reads, assesses, and logs physiological parameters of a patient such as body temperature, oxygen saturation level, systemic arterial pressure, breathing patterns, pulse (heart) rate, supplemental oxygen dependency, consciousness, and pain level using Particle Photon microcontrollers interfaced with biosensors and switches. The Mandami fuzzy inference-based medical decision support system (FI-MDSS) was also developed using MATLAB to assist medical professionals in evaluating a patient's health risk and deciding on the appropriate clinical intervention. The patient's physiological measurements, EWS, and health risk category are stored on the Particle cloud and Thing Speak cloud platforms and can be accessed remotely and in real-time via the Internet. Furthermore, a RESTful application programming interface (API) was developed using GO language and PostgreSQL database to enhance data presentation and accessibility. Based on the paired samples *t*-tests obtained from 6 sessions with 10 trials for each vital sign per session, there were no significant differences between the clinical data obtained from the designed prototype and the commercially sold medical equipment. The mean differences between the compared samples for each physiological data were not more than 0.40, the standard deviations were less than 2.3, and the *p*-values were greater than 0.05. With a 96.67% accuracy, the FI-MDSS predicted health risk levels that were comparable to conventional EWS techniques such as the Modified

A. M. Nagayo (✉) · M. Z. K. Al Ajmi · N. R. K. Guduri · F. S. H. AlBuradai
Electrical and Electronics Section, Engineering Department, University of Technology and Applied Sciences-Al Musanna, Al Muladdah, Sultanate of Oman
e-mail: analene@act.edu.om

M. Z. K. Al Ajmi
e-mail: mahmood@act.edu.om

N. R. K. Guduri
e-mail: naga@act.edu.om

F. S. H. AlBuradai
e-mail: fatmaalburadai@act.edu.om

A. B. Reddy et al. (eds.), *Proceedings of Third International Conference on Advances in Computer Engineering and Communication Systems*, Lecture Notes in Networks and Systems 612, https://doi.org/10.1007/978-981-19-9228-5_28

National Early Warning Score (m-NEWS) and NEWS2, which are used in the clinical decision-making process for managing patients with COVID-19 and other infectious illnesses.

Keywords Telemedicine · Internet of Things · Fuzzy logic · MDSS · COVID-19

1 Introduction

Following the outbreaks of the coronavirus diseases MERS-CoV in 2012 and COVID-19 in December 2019, the World Health Organization and the Oman Government issued a number of precautionary measures and recommendations to the public for infection prevention and control (World Health Organization 2015, 2022; Al Awaidy et al. 2020; Al Ghafri et al. 2021). There were over 433 million reported cases of COVID-19 as of March 6, 2022, with over 5.9 million deaths worldwide, and the number continues to rise due to virus gene mutations (World Health Organization 2022). To battle this pandemic, several strategies have been implemented, including social distancing, the utilization of face masks and personal protective equipment (PPEs), the immediate identification of suspected and probable COVID-19 cases through COVID-19 triaging, screening, and quarantine/isolation methods, and the use of technology for patient consultation via phone or videoconference (Al Ghafri et al. 2021). In situations where hospitals are overcrowded and there is a high demand for intensive care for severe cases, certain COVID-19 patients with mild symptoms may be monitored and treated at home, unless their health rapidly deteriorates (World Health Organization 2020).

To facilitate remote physiological data monitoring of patients at home or in quarantine institutions, to allow doctors in the hospital to diagnose, supervise treatment, and interact with patients remotely, and to reduce the exposure of frontline medical professionals and other people in hospitals/clinics to contagious diseases such as COVID-19, a telemedicine system was developed using IoT integrated with a knowledge-based, fuzzy inference medical decision support system. The Particle Photon-based Telemedicine Health Monitoring System (THMS) reads and processes data from biomedical sensors that measure patient physiological parameters such as core body temperature (BT), blood oxygenation level (SpO_2), systemic arterial pressure (BP), breathing patterns (RR), pulse rate (PR)/heart rate (HR), level of consciousness, and toggle switches that indicate the patient's age range, need for supplemental oxygen, and pain status. Using the Thing Speak and Particle cloud platforms, the vital sign parameters, clinical assessment, early warning score, and health risk state are then stored and retrieved remotely and in real time over the Internet. In addition, the medical data can be viewed via the built-in web dashboard and the Particle mobile apps.

The fuzzy inference-based medical decision support system (FI-MDSS) was developed in MATLAB as a rule-based clinical tool to assist medical professionals in making clinical decisions and interventions based on the patient's symptoms and health risk condition. In the medical scoring system, each physiological parameter is given a score between 0 (normal) and 3 (high risk), and the total EWS is calculated by adding all of the scores together (Choudhury and Baruah 2015). The total EWS score would determine the patient's health risk level and the clinical intervention to be administered. The output of the FI-MDSS was compared to existing medical scoring approaches like the Modified National Early Warning Score (m-NEWS) and NEWS2, which are clinically utilized to monitor the health status of patients with COVID-19 and other infectious diseases (Tagliabue et al. 2021).

In the remaining portion of this paper, Sect. 2 discusses emerging trends in Internet of Things-based physiological monitoring systems and telemedicine-related fuzzy inference systems. The developed telemedicine systems with fuzzy logic-based MDSS are described in Sect. 3, followed by experimental data and results analysis in Sect. 4. Lastly, Sect. 5 discusses the designed system's conclusions and future works.

2 Related Works

Terminologies like telemedicine, telehealth, and e-health describe the use of online platforms and other digital information and communication technologies (ICT) in the health sector (Montesines-Nagayo and Bugtai 2008). Telemedicine refers to the use of ICT for remote medical services, whereas telehealth encompasses activities not only limited to health such as education, training, and administrative functions (Fatehi and Wootton 2012). In its broadest definition, the word "e-health" refers to the application of all electronic online activities as well as the connecting of computers to the Internet in the field of medicine. These technologies encompass a wide range of products, including those that are scientific, instructional, and commercial, in addition to digital services that are given by medical professionals, non-professionals, and enterprises (Montesines-Nagayo and Bugtai 2008; Fatehi and Wootton 2012).

Several telemedicine systems have been developed and thoroughly researched in order to provide quality healthcare services when distance separates doctor and patient. Typical applications of telemedicine research include: (a) medical information systems that handle patient data and medical results (Montesines-Nagayo and Bugtai 2008; Zhang et al. 2020), (b) patient telemonitoring systems (Abo-Zahhad et al. 2014; Nagayo et al. 2021; Gutte and Vadali 2018; Athira et al. 2020; Arun et al. 2018; Viswavardhan Reddy and Kumar 2021; Mahajan and Birajdar 2019), and (c) web-based or IoT-based medical decision support systems (Choudhury and Baruah 2015; Hameed et al. 2020; Ahamad and Bharti 2021; Shatnawi et al. 2021; Al-Dmour et al. 2019; Benhamida et al. 2019).

The telemedicine information system (TMIS) developed by Montesines-Nagayo and Bugtai (2008) consists of a network of computers designed to simplify and consolidate the collecting and storage of obstetrical, gynecological, and general medical data. The TMIS interacts with clients via the Apache Web server and the MySQL relational database server in order to store user profiles, patient medical records, doctor medical diagnoses based on data and photos submitted electronically, clinical treatment, and prescriptions. The dynamic TMIS web application was created using XHTML, PHP, and JavaScript. As an application in the field of telemedicine, Zhang et al. (2020) proposed a real-time IoT framework for remotely monitoring human ECG, HR, and breathing rate using a wearable watch. Bluetooth can be used to transfer the output data to an Android application. The data was stored locally in CSV format, while the MySQL database served as a remote data storage and extraction tool. In this project, a Python program was made to send data to a remote database, while MATLAB was utilized to evaluate and process the data. Users are presented with processed data via a user interface based on the B/S architecture.

Abo-Zahhad et al. (2014) developed a remote patient monitoring system that reads and processes clinical parameters such as HR, SpO_2, BT, and BP using an E-health sensor shield V2.0 interfaced with a PIC18F458 microcontroller. In critical cases, medical data is sent to the doctor via GSM/GPRS cellular network, while long-term monitoring is done via a web-based interface. Nagayo et al. (2021) designed a vital sign monitoring system for medical emergencies that is capable of performing the following functions: (a) reads and logs six physiological parameters (BT, SpO_2, BP, HR, RR, and motion response) of a patient using Arduino microcontrollers linked to medical sensing devices; (b) transmits the physiological parameters of the patient from the medical transport vehicle to a medical practitioner in the hospital through an SMS using GSM module; (c) launches a drone system from an ambulance to inspect the accident scene, deliver the first-aid package, and enable the transmission of visual footage of the patient to the paramedic, and (d) assesses the health risks by computing the EWS using NEWS2 and the COVID-19 Severity Index methods. Gutte and Vadali (2018), on the other hand, used a Raspberry Pi microcontroller to connect with BT, HR, and ECG sensors for remote health monitoring. It processes and stores sensor read parameters in the IoT server, as well as displays the data in the LCD, using the basic MQTT protocol of IoT. The system will generate an alert message in the event of a critical condition.

Athira et al. (2020) proposed an IoT patient monitoring system for the elderly that utilizes sensing devices attached to the patient's body to gather four physical parameters (HR, RR, SpO_2, and BT). In the event of an emergency, the system uses the ESP8266 Wi-Fi module to send an email to the patient's doctors. Based on the information gathered, the physicians can subsequently provide the necessary medical care. The ECG waveform was presented on a digital signal oscilloscope, whereas the other parameters were displayed on a liquid crystal display. Another telemedicine system was developed by Arun et al. (2018) that uses NI myRIO 1900 to collect data from an ECG sensor module. LabVIEW's virtual instrumentation blocks were utilized in the processing of signals and the computation of heart rates during model implementation. The system was also configured to interface with the IBM Watson IoT platform to show the patient's heart rhythm.

Viswavardhan Reddy and Kumar (2021) developed a framework for a smart health monitoring system that enables a particle photon microcontroller to read vital signs from sensing devices that determine a patient's ECG, EMG, GSR, PR, body movements, and snoring pattern and upload the data to the Particle cloud. The cloud-based data was sent as a .csv file and imported into the e-Health web and mobile applications. Similarly, Mahajan and Birajdar (2019) devised a Particle Photon-based health monitoring system for chronic disorders such as cardiovascular diseases, osteoporosis, and asthma. The particle cloud and Adafruit web server store the medical data collected by sensors measuring BT, SpO_2, and acceleration.

As compared to the works of Montesines-Nagayo and Bugtai (2008), Zhang et al. (2020), Abo-Zahhad et al. (2014), Nagayo et al. (2021), Gutte and Vadali (2018), Athira et al. (2020), Arun et al. (2018), Viswavardhan Reddy and Kumar (2021) and Mahajan and Birajdar (2019), the designed IoT-based THMS with an EWS function reads, evaluates, and records six physiological data such as core BT, SpO_2, BP, RR, HR, and consciousness state of the patient using Particle Photon microcontrollers interfaced with biosensors. Additionally, toggle switches provide the microcontroller with input signals for the patient's age group, pain status, and the use of supplemental oxygen. Because ThingSpeak does not support applications involving large amounts of data (Payal and Singh 2020), PostgreSQL, an open-source relational database management system, was used to improve data accessibility and security. Furthermore, a representational state transfer (REST) API was developed using the Go programming language to provide real-time data presentation and remote access to patient medical data as well as health scores via the web dashboard. REST is an architectural model for developing web-based applications that communicate through HTTP request methods like GET, POST, DELETE, and PUT (Yellavula 2020).

A medical decision support system (MDSS) is an interactive computer system that helps physicians diagnose and treat patients based on the association between the patient's medical history, disease signs and symptoms, and clinical lab findings (Kumar et al. 2011). Knowledge-based and non-knowledge-based MDSS are the two fundamental types of decision support for medical diagnosis (Abbasi and Kashiyarndi 2006). Knowledge-based MDSS, such as fuzzy logic-based MDSS, contains rules and associations of collected data in the form of IF–THEN rules, whereas non-knowledge-based MDSS employs a type of artificial intelligence or deep learning that enables computing machines to learn from previous experience and/or recognize patterns in clinical data (Abbasi and Kashiyarndi 2006; Belle et al. 2013).

Various knowledge-based medical decision support systems (MDSS) have been developed by researchers for the monitoring, classification, and evaluation of patients with medical conditions. Hameed et al. (2020) presented an Arduino-based system linked to BT, PR, and BP measuring devices for assessing a patient's health. The system used a knowledge-based fuzzy logic system to identify probable diseases and treatments, with the results being kept on a web server. Ahamad and Bharti (2021) and Shatnawi et al. (2021) proposed fuzzy rule-based decision support systems to assess the possibility of COVID-19 infection based on patient symptoms and medical histories, including a history of diabetes, fever, cough, breathing difficulty, sore throat, absence of scent, or flavor. On the other hand, Al-Dmour et al. (2019) developed and implemented a wireless health monitoring and warning system based on fuzzy logic that employs RFID-based sensor modules to measure five vital signs: systolic BP, HR, SpO_2, blood sugar level, and body temperature. The data collected is sent to a server, which hosts a fuzzy logic-based system designed to be well matched with the modified early warning system (MEWS). The fuzzy system could analyze and predict patient health conditions into one of 15 health risk categories.

In comparison with previous works (Choudhury and Baruah 2015; Hameed et al. 2020; Ahamad and Bharti 2021; Shatnawi et al. 2021; Al-Dmour et al. 2019), the FI-MDSS module was designed and implemented based on a modified national early warning scoring system (m-NEWS) and medical expert recommendations. It uses physiological parameters from the telemedicine module such as RR, SpO_2, supplemental oxygen, BT, systolic BP, HR, alert/voice/pain/unconsciousness (AVPU) state, and patient age. According to Tagliabue et al. (2021), m-NEWS, a modified version of NEWS2, was used to manage COVID-19 patients clinically, with the age parameter included as one of the risk factors for survival.

3 Materials and Methods

3.1 System Block Diagram

As shown in Fig. 1, the Particle Photon microcontroller PP1 reads the patient's body temperature, blood oxygenation percentage, and motion response from sensors, and then accepts three inputs from toggle switches for the patient's age, oxygen supplement, and pain level. PP1 determines the corresponding early warning score (EWS) for each parameter using the modified National Early Warning Score (m-NEWS) algorithm. Microcontroller PP2, on the other hand, reads the BP systolic, BP diastolic, pulse (heart) rate, and breathing rate from sensors and then assigns the EWS for each vital sign. The individual EWS generated by PP2 is sent to PP1 to calculate the patient's overall health score. The conventional m-NEWS score is used as the basis for determining the health risk category and clinical intervention. The patient's physiological parameters, EWS, and health risk level are stored and accessed remotely from the Internet via the Particle cloud and Thing Speak cloud platforms. Furthermore, the medical data is migrated from the ThingSpeak cloud server to a more stable and secure data storage system capable of managing big data, such as PostgreSQL. Particle IoT mobile app and the developed RESTful API web service using the Go programming language servicing requests from a web dashboard are utilized for data visualizations.

The Mandami fuzzy inference-based medical decision support system (FI-MDSS) is installed on the doctor's computer to assist with clinical decision-making tasks and treatments based on the patient's symptoms and level of health risk. The medical data exported from the ThingSpeak cloud as a .csv file was used as inputs for the rule-based medical decision support system implemented in MATLAB. The FI-MDSS output for the patient health risk level (PHRL) is stored in the PostgreSQL database for comparison with the PHRL obtained from the conventional m-NEWS.

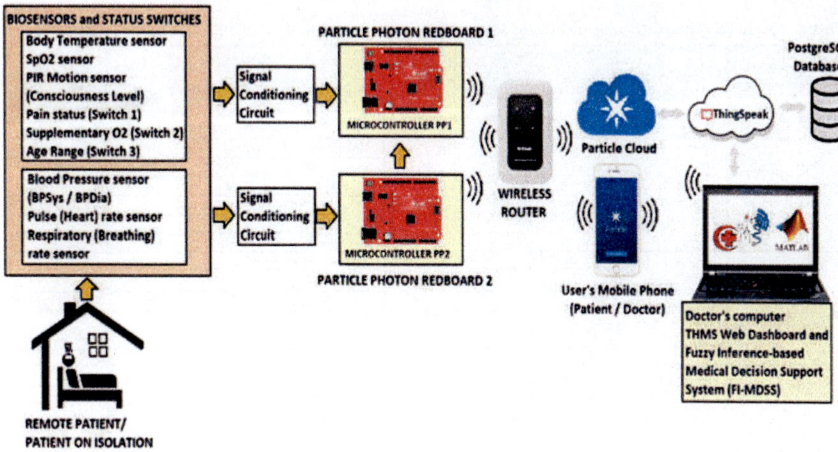

Fig. 1 Overall system block diagram

3.2 IoT-Based Telemedicine Health Monitoring System (THMS) with Physiological Data Assessment and Clinical Health Risk Detection

The THMS was designed and implemented to acquire physiological data from patients using the biosensors listed in Table 1. Particle Photon microcontrollers, which are STM32 ARM Cortex M3 microcontrollers with integrated Broadcom BCM43362 Wi-Fi chips, read and process physiological parameters from sensors as well as toggle switch inputs to provide the patient's assessed early warning score and health risk condition.

Table 1 gives the medical data assessment and early warning scoring for each physiological parameter derived from Nagayo et al. (2021), medical professional recommendations, and the m-NEWS approach presented in Tagliabue et al. (2021). The microcontrollers are programmed to assign an early warning score of 0–3 for each medical parameter read by the biosensors and from the switch inputs. The patient's overall health score will then be computed by microcontroller PP1 by summing all of the individual EWS. According to Table 2, a total EWS of 0 denotes a very low or zero health risk, and the patient should be observed every 12 h as the recommended clinical intervention. A combined EWS between 1 and 4 suggests that the patient is at low risk and should be checked every 4–6 h. An overall EWS score between 5 and 6 or a score of 3 on any of the parameters indicates that the patient is at moderate risk and needs urgent medical care every 1–2 h. A total score of 7 or higher means that the patient is at greater risk and requires continuous monitoring as well as emergency treatment.

The patient's medical data, health score, and recommended clinical intervention based on health risk level are published in the Particle cloud console and can be

Table 1 THMS physiological data assessment and early warning scoring

Physiological parameters	Component used and description	Sensor read values	Unit	Physiological data assessment [12]	EWS score based on m-NEWS [7]
Body temperature (BT)	DS18B20 temperature sensor	36.1 ≤ BT ≤ 37.2	°C	Normal BT	0
		37.3 ≤ BT ≤ 38.0		Elevated BT/low grade FEVER	0
		38.1 ≤ BT ≤ 39.0		High BT/high grade FEVER	1
		BT ≥ 39.1		Very high BT/very high grade FEVER	2
		35.1 ≤ BT ≤ 36.0		Low BT	1
		BT ≤ 35.0		Very low BT	3
Pulse/heart rate (HR)	Sunrom blood pressure sensor 1437	51 ≤ HR ≤ 90	beats per minute (bpm)	Normal HR/normal PR	0
		41 ≤ HR ≤ 50		Low HR/low PR	1
		HR ≤ 40		Very low HR/very low PR	3
		91 ≤ HR ≤ 110		Mod high HR/mod high PR	1
		111 ≤ HR ≤ 130		High HR/high PR	2
		HR ≥ 131		Very high HR/very high PR	3
Blood oxygen level (SpO2)	MAX30100 pulse oximeter and heart rate sensor	96 ≤ SpO2 ≤ 100	%	Normal Sp O2	0
		94 ≤ SpO2 ≤ 95		Moderately low SpO2	1
		92 ≤ SpO2 ≤ 93		Low SpO2	2
		SpO2 ≤ 91		Very low SpO2	3
Supplementary oxygen (SuppO2)	Toggle switch (S2)	3.3–5	V	With O2 supplement	2
		0–0.8		Without O2 supplement	0
Respiratory/breathing rate (RR)	MLX90614 non-contact IR temperature sensor for inhale and exhale temperature detection	12 ≤ RR ≤ 20	breaths per minute (brpm)	Normal RR	0
		21 ≤ RR ≤ 24		High RR	2
		RR ≥ 25		Very high RR	3
		9 ≤ RR ≤ 11		Low RR	1
		RR ≤ 8		Very low RR	3
Age range	Toggle switch (S3)	3.3–5	V	Age ≥ 65	3
		0		Age < 65	0
Pain status	Toggle switch (S1)	0–0.8	V	With pain	3
		3.3–5		No pain	0
Alertness/voice/pain/unresponsive (AVPU) status	HC-SR501 PIR motion sensor	3.3–5	V	Alert/voice/conscious	0
		0–0.8		Unconscious /unresponsive	3
Arterial blood pressure—systolic/diastolic (SBP/DBP)	Sunrom blood pressure sensor 1437	SBP < 90 or DBP < 60	mmHg	Low BP/hypotension	
		90 ≤ SBP ≤ 120 and 60 ≤ DBP ≤ 80		Normal BP	
		121 ≤ SBP ≤ 139 or 81 ≤ DBP ≤ 89		Elevated BP/pre-hypertension	
		140 ≤ SBP ≤ 159 and/or 90 ≤ DBP ≤ 99		High BP S1/hypertension stage 1	
		160 ≤ SBP ≤ 179 and/or 100 ≤ DBP ≤ 109		High BP S2/hypertension stage 2	
		180 ≤ SBP ≤ 219 and/or DBP ≥ 110		High BP S3/hypertension stage 3	
		SBP ≥ 220 and/or DBP ≥ 110		Very high BP S4 /hypertension stage 4	
Systolic BP (SBP)		SBP ≤ 90			3
		91 ≤ SBP ≤ 100			2
		101 ≤ SBP ≤ 110			1
		111 ≤ SBP ≤ 120			0
		121 ≤ SBP ≤ 219			0
		SBP ≥ 220			3

Table 2 Patient health risk evaluation and clinical intervention based on the m-NEWS and NEWS2 (Tagliabue et al. 2021; Nagayo et al. 2021)

Total m-NEWS score	Patient health risk condition	Clinical intervention
0	Zero health risk (ZHR)/very low risk	O12h/monitor every 12 h
1–4	Low health risk (LHR)	O4–6h/monitor every 4–6 h by nurse
3 in one of the physiological parameter	Medium health risk (MHR)	O1–2h/monitor every1–2 h; nurse notify doctor for urgent evaluation
5–6		
7 or more	High health risk (HHR)	Continuous monitoring/nurse inform doctor for emergency evaluation

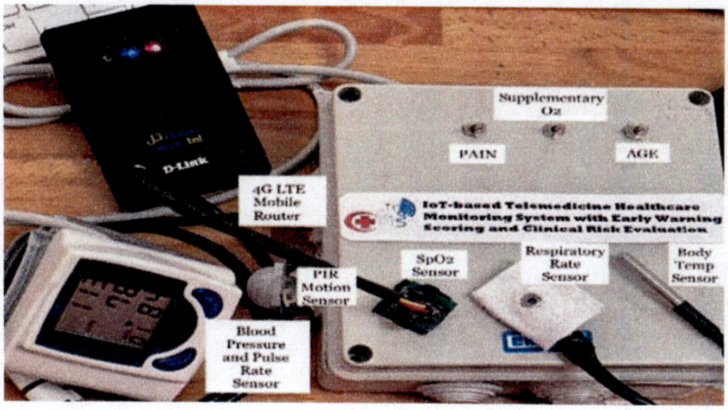

Fig. 2 THMS prototype

accessed at any time by the doctor via the Particle mobile app. The medical data is also stored in the ThingSpeak cloud and PostgreSQL. The RESTful API, created with the Go programming language, uses a GET request to retrieve medical information from the database. The THMS prototype is presented in Fig. 2.

3.3 Fuzzy Inference-Based Medical Decision Support System (FI-MDSS) for Clinical Risk Evaluation

The FI-MDSS, which is installed on the physician's computer, was developed with MATLAB R2021b to aid medical professionals in making clinical decisions and interventions based on the patient's physiological parameters. Fuzzy logic, a form of artificial intelligence, has shown potential in imitating human reasoning when dealing with complex and unstable biomedical inputs, allowing more precise decisions under uncertain conditions (Benhamida et al. 2019).

Figure 3 shows the Mandami FI-MDSS implementation, with eight input variables representing the patient's physiological parameters and a single output indicating the patient's health risk level. Table 3a gives the fuzzy sets and related ranges for each input variable based on medical doctor recommendations and m-NEWS. The membership functions of fuzzy sets were considered to be trapezoidal.

The FI-MDSS output variable represents the patient's health risk level (PHRL). Selecting one of 24 possible health risk categories determines the severity of the patient's illness. Table 3b gives the FI-MDSS output variable with its corresponding fuzzy sets and ranges. The membership functions for the health risk categories were triangularly configured.

The FI-MDSS rule base engine consists of 24,000 IF–THEN rules that provide for all possible health scenarios. The overall number of fuzzy rules, FR, is obtained by getting the product of the number of linguistic terms for the n input variables.

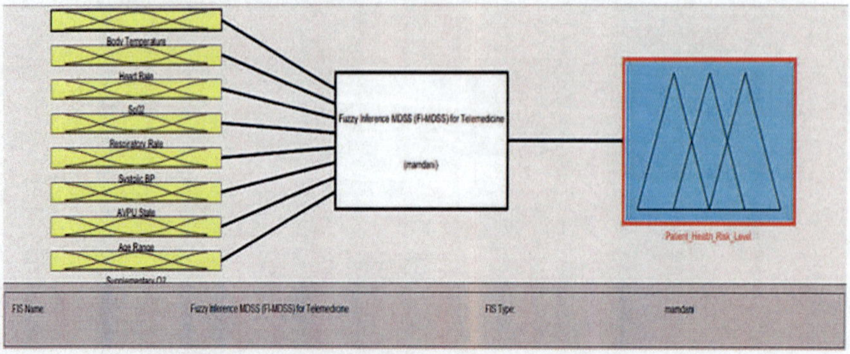

Fig. 3 Mandami fuzzy inference-based medical decision support system

Based on the number of fuzzy sets per input variable in Table 3a, FR is computed as (1):

$$FR = 5 \times 6 \times 4 \times 2 \times 5 \times 2 \times 2 \times 5 = 24{,}000\,\text{rules} \tag{1}$$

The rules were formulated using the medical expert's advice and the m-NEWS scoring method. Due to the interdependence of the physiological inputs, the AND logical operation was utilized in the rules. Using a fuzzy centroid approach, data is defuzzified to get a precise output. Figure 4 shows a sample fuzzy rule for identifying the patient's health risk level. Rule #221 states that if (Body Temperature is NRM_ElevBT_0) and (Heart Rate is NRM_HR_0) and (SpO_2 is NRMSpO2_0) and (Respiratory Rate is NRm_RR_0) and (Systolic BP is LowSBP_2) and (AVPU State is AVPU_3) and (Age Range is below 65_0) and (Supplementary O_2 is Air_0) then (Patient Health Risk Level is MHR_5).

4 Results and Discussions

4.1 Experimental Results of the IoT-Based THMS with EWS Function

Six testing sessions (S1–S6) were carried out on the THMS prototype to ensure that it provided accurate results. The physiological parameters of a single human subject who consented to act as the patient in isolation were measured for each trial per session by successively attaching the THMS prototype and commercially available medical equipment to the patient. Ten trials were performed at predefined intervals for each human test subject throughout each session, as indicated in the human test consent form approved by the UTAS-Al Musanna Research Committee and duly signed by the authors as well as the volunteer patient. Figure 5 shows the published

Table 3 **a** FI-MDSS fuzzy sets and related ranges for each input variable and **b** FI-MDSS output variable with its corresponding fuzzy sets and ranges

FI-MDSS input field (physiological parameters)	Fuzzy sets	Range
Body temperature (BT)	NRM_ElevBT_0	35.9 ≤ BT ≤ 37.4
		37.1 ≤ BT ≤ 38.2
	HighBT_1	37.9 ≤ BT ≤ 39.2
	VHighBT_2	BT ≥ 38.9
	LowBT_1	34.9 ≤ BT ≤ 36.2
	VLowBT_3	BT ≤ 35.0
Pulse/heart rate (HR)	NormHR_0	49 ≤ HR ≤ 92
	LowHR_1	39 ≤ HR ≤ 52
	VLowHR_3	HR ≤ 40
	ModHighHR_1	89 ≤ HR ≤ 112
	HighHR_2	109 ≤ HR ≤ 132
	VHighHR_3	HR ≥ 131
Blood oxygen level (SpO2)	NRMSp02_0	95 ≤ SpO2 ≤ 100
	ModLSp02_1	93 ≤ SpO2 ≤ 96
	LowSp02_2	91 ≤ SpO2 ≤ 94
	VLowSp02_3	SpO2 ≤ 91
Supplementary oxygen (SuppO2)	SuppO2_2	1.9-5
	Air_0	0-2
Respiratory/breathing rate (RR)	NRM_RR_0	10 ≤ RR ≤ 22
	HighRR_2	19 ≤ RR ≤ 26
	VHighRR_3	RR ≥ 25
	LowRR_1	7 ≤ RR ≤ 13
	VLowRR_3	RR ≤ 8
Age range	65&above_3	1.9-5
	below65_0	0-2
Alertness/voice/pain/un responsive (AVPU) status	AVPU_0	1.9-5
	AVPU_3	0-2
Systolic BP (SBP)	VLowSBP_3	SBP ≤ 91
	LowSBP_2	89 ≤ SBP ≤ 102
	ModLSBP_1	99 ≤ SBP ≤ 112
	SBP_0	109 ≤ SBP ≤ 221
	highSBP_3	SBP ≥ 220

(a)

FI-MDSS output field	Fuzzy sets	Range
Patient health risk level (PHRL)	ZHR0	0.0 < PHRL < 0.5
	LHR1	0.5 < PHRL < 1.5
	LHR2	1.5 < PHRL < 2.5
	LHR3	2.5 < PHRL < 3.5
	LHR4	3.5 < PHRL < 4.5
	MHR5	4.5 < PHRL < 5.5
	MHR6	5.5 < PHRL < 6.5
	HHR7	6.5 < PHRL < 7.5
	HHR8	7.5 < PHRL < 8.5
	HHR9	8.5 < PHRL < 9.5
	HHR10	9.5 < PHRL < 10.5
	HHR11	10.5 < PHRL < 11.5
	HHR12	11.5 < PHRL < 12.5
	HHR13	12.5 < PHRL < 13.5
	HHR14	13.5 < PHRL < 14.5
	HHR15	14.5 < PHRL < 15.5
	HHR16	15.5 < PHRL < 16.5
	HHR17	16.5 < PHRL < 17.5
	HHR18	17.5 < PHRL < 18.5
	HHR19	18.5 < PHRL < 19.5
	HHR20	19.5 < PHRL < 20.5
	HHR21	20.5 < PHRL < 21.5
	HHR22	21.5 < PHRL < 22.5
	HHR23	22.5 < PHRL < 23.5

(b)

BP, HR, and BT readings from the THMS prototype in the ThingSpeak cloud, as well as the results from Braun monitoring equipment, which were all relatively close in values. As shown in Fig. 6, the sample SpO_2 measurements obtained from the THMS prototype and the Tomorotec medical device were identical. Tables 4, 5, 6, 7, 8 and 9 compare the readings of the THMS prototype to those of commercially

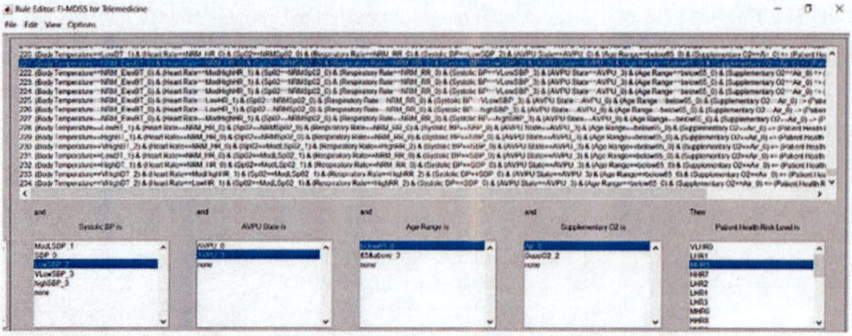

Fig. 4 Sample fuzzy rule for identifying the patient's health risk level

available medical equipment using the paired samples *t*-test method in IBM SPSS software. The bolded numbers in Tables 4, 5, 6, 7, 8 and 9 represent the maximum standard deviations and paired mean differences for the physiological parameters.

The SpO_2 readings of the designed THMS and the Tomorotec pulse oximeter did not differ significantly as given in Table 4, since the paired samples *t*-test yielded a *p*-value greater than 0.05. Furthermore, the largest mean difference (MD) across the compared devices' measurements was just 0.300. According to Table 5, the highest MD between body temperature readings of the compared medical equipment was 0.040, and the *p*-value was more than 0.05, indicating that the Braun digital thermometer readings were nearly identical to those of the designed THMS. As for the respiratory rate measurements given in Table 6, there was a maximum MD of 0.400 between the Tomorotec device and the THMS prototype. A *p*-value greater than 0.05 indicates that the readings were not statistically different.

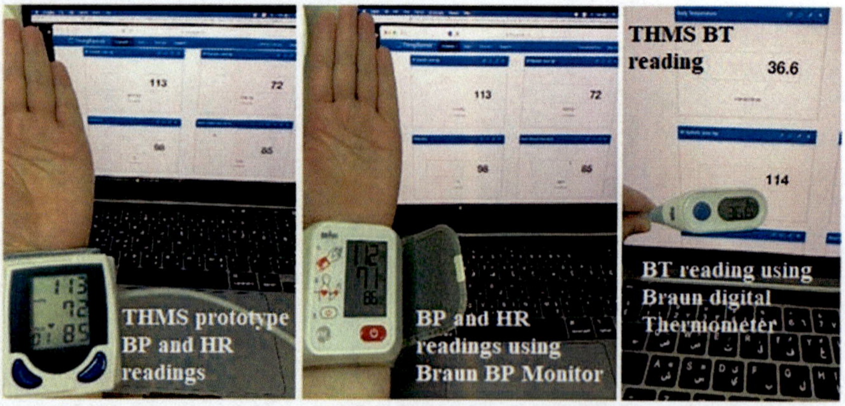

Fig. 5 Sample SBP, DBP, HR, and BT measurements using the **a** THMS prototype and **b** Braun medical equipment

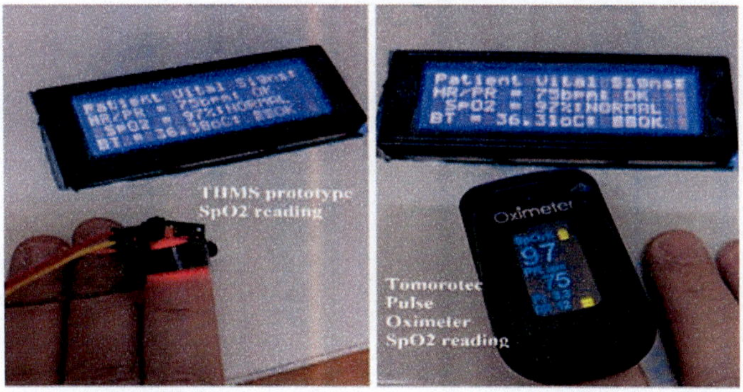

Fig. 6 Sample SpO_2 measurements using the **a** THMS prototype and **b** Tomorotec pulse oximeter device

The HR, SBP, and DBP readings obtained with the THMS prototype and the Braun BP monitor were close, with a maximum mean difference of 0.400 for each parameter, according to Tables 7, 8 and 9. Since the *p*-values obtained after using the paired samples *t*-test method for the three parameters were greater than 0.05, the readings of the two compared devices did not differ significantly. All test trials per session using both THMS and commercially available devices yielded comparable results; hence, the system may be considered accurate. The small variations in physiological parameter readings between the two systems, as given in Tables 4, 5, 6, 7, 8 and 9, were caused by movement artifacts and sensor positioning. Furthermore, the HR, SBP, and DBP values from the two devices were taken sequentially and not under the same conditions.

In addition, the standard deviation (SD) for THMS prototype readings was determined for each testing session to assess whether the results were near to the mean. As seen in Tables 4, 5, 6, 7, 8 and 9, the standard deviation (SD) of each measured parameter was minimal, indicating that the clinical data did not significantly deviate from their respective means. No measurements significantly diverge from the mean throughout all trials, demonstrating the system's reliability. The maximum standard deviations for SpO_2, BT, RR, HR, SBP, and DBP measured by THMS were 0.876, 0.383, 1.932, 1.989, 1.897, and 2.251, respectively.

Figure 7a illustrates sample THMS test results for body temperature, HR, BP, SpO_2, AVPU, and total EWS, as presented on the ThingSpeak cloud platform, along with a health risk assessment. Based on the clinical data shown in Fig. 7a, an adult patient aged 50 with an age score of 0 and no supplementary O_2 had a normal BT of 36.4 °C with an equivalent EWS of 0, a normal heart rate of 72 bpm with a corresponding EWS of 0, a low blood pressure of 97/58 with an SBP EWS score of 2, a normal SpO_2 of 97% with an EWS of 0, a normal RR of 12 brpm with an equivalent EWS of 0, and was conscious but had a severe headache at the time of testing, resulting in an AVPU EWS of 3. When the individual EWS values for each physiological parameter were added together, the total EWS was 5, indicating that

Table 4 Paired samples *t*-tests for SpO_2 level

		Mean	*N*	Std. deviation (SD)	Paired mean difference (MD)	*t*	df	Significance	
								One-sided *p*	Two-sided *p*
Patient 1 (S1)	SpO_2 Tomorotec	97.70	10	0.675	0.200	1.500	9	0.084	0.168
	SpO_2 THMS prototype	97.50	10	0.527					
Patient 2 (S2)	SpO_2 Tomorotec	99.10	10	0.738	0.200	1.500	9	0.084	0.168
	SpO_2 THMS prototype	98.90	10	**0.876**					
Patient 3 (S3)	SpO_2 Tomorotec	96.50	10	0.527	0.100	0.429	9	0.339	0.678
	SpO_2 THMS prototype	96.40	10	0.516					
Patient 4 (S4)	SpO_2 Tomorotec	98.00	10	0.816	**0.300**	1.000	9	0.172	0.343
	SpO_2 THMS prototype	97.70	10	0.823					
Patient 5 (S5)	SpO_2 Tomorotec	97.40	10	0.516	0.200	1.500	9	0.084	0.168
	SpO_2 THMS prototype	97.20	10	0.422					
Patient 6 (S6)	SpO_2 Tomorotec	98.10	10	0.738	0.200	0.802	9	0.222	0.443
	SpO_2 THMS prototype	97.90	10	**0.876**					

Table 5 Paired samples *t*-tests for body temperature (BT)

		Mean	*N*	Std. deviation (SD)	Paired mean difference (MD)	*t*	df	Significance	
								One-sided *p*	Two-sided *p*
Patient 1 (S1)	BT Braun	36.41	10	0.185	0.030	1.000	9	0.172	0.343
	BT THMS prototype	36.38	10	0.193					
Patient 2 (S2)	BT Braun	36.34	10	0.184	− 0.020	− 0.514	9	0.310	0.619
	BT THMS prototype	36.36	10	0.171					
Patient 3 (S3)	BT Braun	37.76	10	0.401	− **0.040**	− 1.500	9	0.084	0.168
	BT THMS prototype	37.80	10	**0.383**					
Patient 4 (S4)	BT Braun	36.32	10	0.063	0.020	1.500	9	0.084	0.168
	BT THMS prototype	36.30	10	0.067					
Patient 5 (S5)	BT Braun	36.55	10	0.165	0.030	1.406	9	0.097	0.193
	BT THMS prototype	36.52	10	0.162					
Patient 6 (S6)	BT Braun	36.02	10	0.132	0.010	0.318	9	0.379	0.758
	BT THMS prototype	36.01	10	0.137					

Table 6 Paired samples *t*-tests for respiratory rate (RR)

		Mean	*N*	Std. deviation (SD)	Paired mean difference (MD)	*t*	df	Significance	
								One-sided *p*	Two-sided *p*
Patient 1 (S1)	RR Tomorotec	14.70	10	2.058	− 0.100	− 0.361	9	0.363	0.726
	RR THMS prototype	14.80	10	**1.932**					
Patient 2 (S2)	RR Tomorotec	15.00	10	1.944	0.200	0.480	9	0.321	0.642
	RR THMS prototype	14.80	10	**1.932**					
Patient 3 (S3)	RR Tomorotec	19.10	10	1.101	0.100	0.557	9	0.296	0.591
	RR THMS prototype	19.00	10	1.054					
Patient 4 (S4)	RR Tomorotec	18.10	10	1.287	0.100	0.361	9	0.363	0.726
	RR THMS prototype	18.00	10	1.333					
Patient 5 (S5)	RR Tomorotec	17.30	10	1.567	0.300	1.152	9	0.139	0.279
	RR THMS prototype	17.00	10	1.414					
Patient 6 (S6)	RR Tomorotec	15.60	10	1.430	**0.400**	1.177	9	0.135	0.269
	RR THMS prototype	15.20	10	1.687					

Table 7 Paired samples t-tests for heart rate (HR)

		Mean	N	Std. deviation (SD)	Paired mean difference (MD)	t	df	Significance	
								One-sided p	Two-sided p
Patient 1 (S1)	HR Braun	78.60	10	1.955	0.200	0.375	9	0.358	0.716
	HR THMS prototype	78.40	10	1.955					
Patient 2 (S2)	HR Braun	72.20	10	1.874	0.100	0.190	9	0.427	0.853
	HR THMS prototype	72.10	10	1.912					
Patient 3 (S3)	HR Braun	71.50	10	1.958	**0.400**	0.885	9	0.200	0.399
	HR THMS prototype	71.10	10	1.969					
Patient 4 (S4)	HR Braun	85.00	10	1.333	– 0.300	– 0.669	9	0.260	0.520
	HR THMS prototype	85.30	10	1.337					
Patient 5 (S5)	HR Braun	105.00	10	2.000	– 0.200	– 0.429	9	0.339	0.678
	HR THMS prototype	105.20	10	**1.989**					
Patient 6 (S6)	HR Braun	89.50	10	1.354	0.300	0.896	9	0.197	0.394
	HR THMS prototype	89.20	10	1.398					

Table 8 Paired samples *t*-tests for systolic blood pressure (SBP)

		Mean	N	Std. deviation (SD)	Paired mean difference (MD)	t	df	Significance	
								One-sided p	Two-sided p
Patient 1 (S1)	SBP Braun	113.60	10	1.838	0.300	0.635	9	0.271	0.541
	SBP THMS prototype	113.30	10	1.767					
Patient 2 (S2)	SBP Braun	96.30	10	1.703	0.200	0.635	9	0.346	0.693
	SBP THMS prototype	96.10	10	1.663					
Patient 3 (S3)	SBP Braun	123.40	10	1.647	0.100	0.198	9	0.424	0.847
	SBP THMS prototype	123.30	10	1.636					
Patient 4 (S4)	SBP Braun	102.20	10	1.619	**– 0.400**	– 1.078	9	0.155	0.309
	SBP THMS prototype	102.60	10	1.647					
Patient 5 (S5)	SBP Braun	88.30	10	1.636	– 0.300	– 0.818	9	0.217	0.434
	SBP THMS prototype	88.60	10	1.506					
Patient 6 (S6)	SBP Braun	136.80	10	1.814	0.200	0.327	9	0.375	0.751
	SBP THMS prototype	136.60	10	**1.897**					

Table 9 Paired samples t-tests for diastolic blood pressure (DBP)

		Mean	N	Std. deviation (SD)	Paired mean difference (MD)	t	df	Significance	
								One-sided p	Two-sided p
Patient 1 (S1)	DBP Braun	66.40	10	1.897	**0.400**	0.768	9	0.231	0.462
	DBP THMS prototype	66.00	10	1.886					
Patient 2 (S2)	DBP Braun	62.00	10	1.886	0.100	0.208	9	0.420	0.840
	DBP THMS prototype	61.90	10	1.524					
Patient 3 (S3)	DBP Braun	89.50	10	2.273	0.300	1.000	9	0.172	0.343
	DBP THMS prototype	89.20	10	**2.251**					
Patient 4 (S4)	DBP Braun	67.20	10	2.098	− 0.200	− 0.514	9	0.310	0.619
	DBP THMS prototype	67.40	10	1.955					
Patient 5 (S5)	DBP Braun	63.10	10	1.969	− 0.300	− 0.669	9	0.260	0.520
	DBP THMS prototype	63.40	10	1.955					
Patient 6 (S6)	DBP Braun	97.80	10	1.932	0.200	0.318	9	0.379	0.758
	DBP THMS prototype	97.60	10	1.838					

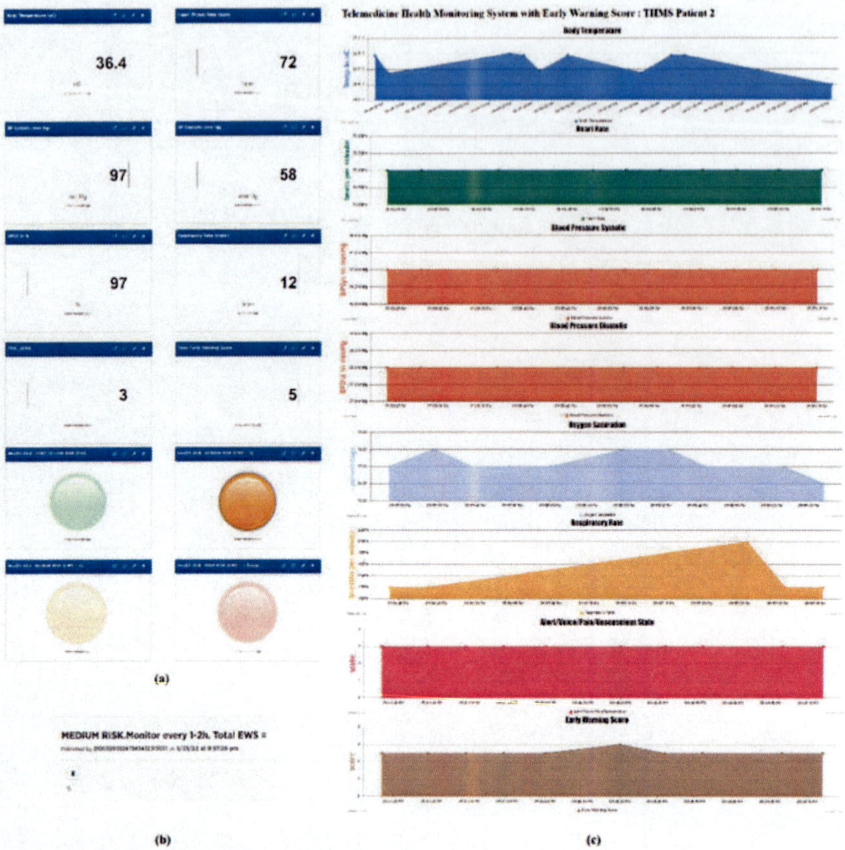

Fig. 7 Sample test results as seen in **a** ThingSpeak cloud API, **b** Particle Web console, and **c** THMS web dashboard

the patient had a medium health risk level (MHR5), and the orange light on the ThingSpeak dashboard was activated. Referring to Fig. 7b, the patient's condition should be checked by medical professionals every 1–2 h, as specified in the Particle Web console. Figure 7c also illustrates graphs of physiological parameters, which can be observed by the doctor through the developed THMS web dashboard.

4.2 *Experimental Results of the FI-MDSS for Clinical Risk Evaluation*

During the FI-MDSS testing, the patient health risk group is determined by importing the patient data from the ThingSpeak cloud as a FI-MDSS .csv file into MATLAB,

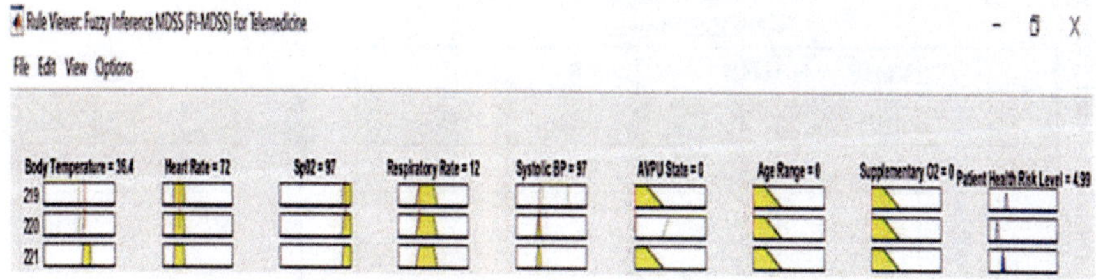

Fig. 8 FI-MDSS sample output data

Table 10 Comparison between the FI-MDSS health risk assessment and m-NEWS score based on patient's physiological data

Physiological parameters								Patient health risk evaluation			
BT (°C)	HR (bpm)	SpO2 (%)	RR (brpm)	SBP (mmHg)	EWS_ AVPU	EWS_ Age	EWS_ SuppO2	m-NEWS		FI-MDSS	
36.2	72	97	12	97	3	0	0	5	MHR	4.99	MHR
36.6	78	97	16	116	0	0	0	0	VLHR	0.26	VLHR
37.8	71	96	20	123	0	0	0	0	VLHR	1.31	LHR
38.3	72	95	20	122	0	0	0	2	LHR	1.99	LHR
39.0	80	95	19	125	3	0	0	5	MHR	4.99	MHR
39.2	80	94	21	130	3	0	0	8	HHR	8.04	HHR

evaluating the parameters with the evalfis (fis, input) function, exporting the FI-MDSS output from MATLAB as an MS Excel file with the writematrix (output, filename) function, and storing the results in the PostgreSQL database. Figure 8 is a sample output of the FI-MDSS utilizing the MATLAB fuzzy logic toolbox with the clinical parameters specified in Fig. 7a as inputs. The result shows that the patient's health risk level (PHRL) is 4.99, inferring a medium health risk category (MHR5). Using Tables 1 and 2, the conventional m-NEWS aggregated score for the medical data in Fig. 7a is 5, indicating the same health risk condition as the FI-MDSS, which presents a medium risk. Based on the data in Table 10, the FI-MDSS health risk assessment is comparable to the conventional m-NEWS approach for evaluating COVID-19 patients. According to the system's confusion matrix shown in Fig. 9, the FI-MDSS correctly detected 58 of 60 health risk categories with a 96.67% accuracy. Two patients had borderline SpO_2 and RR readings, indicating a low health risk condition rather than a very low health risk level. This shows that, in the case of borderline clinical data, the developed FI-MDSS can forecast the patient's health deterioration.

5 Conclusion

This research explored how the Internet of Things (IoT) and artificial intelligence (AI) can be used to promote telemedicine practice during infectious disease outbreaks. The Particle Photon-based health monitoring system for telemedicine application

mNEWS PHRL \ Fi-MDSS PHRL	VLHR	LHR	MHR	HHR		
VLHR	9 (15%)	2 (3.33%)	0 (0%)	0 (0%)	81.82%	18.18%
LHR	0 (0%)	30 (50%)	0 (0%)	0 (0%)	100%	0%
MHR	0 (0%)	0 (0%)	11 (18.33%)	0 (0%)	100%	0%
HHR	0 (0%)	0 (0%)	0 (0%)	8 (13.33%)	100%	0%
	100%	93.75%	100%	100%		
	0%	6.25%	0%	0%		

Fig. 9 Confusion matrix for the FI-MDSS

with an early warning scoring (EWS) feature can accurately measure, evaluate, and log a patient's physiological parameters such as body temperature, SPO_2 level, BP, RR, HR, supplementary O_2 dependency, consciousness state, and pain level. THMS enables physicians to diagnose, treat symptoms, and communicate with patients remotely, reducing the exposure of frontline medical personnel, and other individuals in hospitals/clinics to infections and diseases such as COVID-19. The Mandami fuzzy inference-based medical decision support system (FI-MDSS), which demonstrates the use of artificial intelligence, was successfully developed to aid medical professionals in assessing a patient's health risk and selecting the most appropriate clinical approach.

The medical data, EWS, and health risk status of the patient were stored on the Particle cloud and Thing Speak cloud platforms and may be viewed remotely and in real time through the Internet. In addition, a RESTful application programming interface (API) was developed utilizing the GO programming language and the PostgreSQL database to improve the presentation and accessibility of data. Based on experimental results, the physiological parameter readings obtained from the developed THMS were comparable to those obtained from commercially available medical equipment. Furthermore, the FI-MDSS predicted health risk levels matched those of the Modified National Early Warning Score (m-NEWS), an EWS method used to medically diagnose and treat COVID-19 patients.

The following recommendations can be considered as potential future advancements for a large-scale THMS implementation: (a) integrating a decision support system based on deep learning algorithms to predict the deterioration of patients under monitoring; (b) developing THMS mobile applications to improve doctor-patient interaction and patient health awareness, and (c) developing dynamic THMS web applications to store more clinical information such as medical test images as well as add more security when accessing clinical data.

Acknowledgements Special thanks to the University of Technology and Applied Sciences-Al Musanna for their unwavering support of this research endeavor, as well as The Research Council (TRC) of Oman for providing the financial resources that made it possible.

References

M.M. Abbasi, S. Kashiyarndi, *Clinical Decision Support Systems: A Discussion on Different Methodologies Used in Health Care* (Marlaedalen University, Sweden, 2006)

M. Abo-Zahhad, S.M. Ahmed, O. Elnahas, A wireless emergency telemedicine system for patients monitoring and diagnosis. Int. J. Telemed. Appl. (2014)

M.K. Ahamad, A.K. Bharti, Prevention from COVID-19 in India: fuzzy logic approach, in *2021 International Conference on Advance Computing and Innovative Technologies in Engineering (ICACITE)* (IEEE, 2021), pp. 421–426

S.T. Al Awaidy, A.A. Al Maqbali, I. Omer, S. Al Mukhaini, M.A. Al Risi, M.S. Al Maqbali, A. Al Reesi, M. Al Busaidi, F.H. Al Hashmi, T.K. Al Maqbali, V. Vaidya, The first clusters of Middle East respiratory syndrome coronavirus in Oman: time to act. J. Infect. Public Health **13**(5), 679–686 (2020)

J.A. Al-Dmour, A. Sagahyroon, A.R. Al-Ali, S. Abusnana, A fuzzy logic–based warning system for patients classification. Health Inform. J. **25**(3), 1004–1024 (2019)

T. Al Ghafri, F. Al Ajmi, L. Al Balushi, P.M. Kurup, A. Al Ghamari, Z. Al Balushi, F. Al Fahdi, H. Al Lawati, S. Al Hashmi, A. Al Manji, A. Al Sharji, Responses to the pandemic COVID-19 in primary health care in Oman: Muscat experience. Oman Med. J. **36**(1), e216 (2021)

U. Arun, S. Natarajan, R.R. Rajanna, A novel IoT cloud-based real-time cardiac monitoring approach using NI myRIO-1900 for telemedicine applications, in *2018 3rd International Conference on Circuits, Control, Communication and Computing (I4C)* (IEEE, 2018), pp. 1–4

A. Athira, T.D. Devika, K.R. Varsha, Design and development of IoT based multi-parameter patient monitoring system, in *2020 6th International Conference on Advanced Computing and Communication Systems (ICACCS)* (IEEE, 2020), pp. 862–866

A. Belle, M.A. Kon, K. Najarian, Biomedical informatics for computer-aided decision support systems: a survey. Sci. World J. (2013)

A. Benhamida, M. Takács, M. Kozlovszky, H. Redjimi, M. Ogbolu, Fuzzy model for early warning score system, in *2019 IEEE International Work Conference on Bioinspired Intelligence (IWOBI)* (IEEE, 2019), pp. 000167–000172

M.K. Choudhury, N. Baruah, A fuzzy logic-based expert system for determination of health risk level of patient. Int. J. Res. Eng. Technol. **4**(5), 261–267 (2015)

F. Fatehi, R. Wootton, Telemedicine, telehealth or e-health? A bibliometric analysis of the trends in the use of these terms. J. Telemed. Telecare **18**(8), 460–464 (2012)

A. Gutte, R. Vadali, IoT based health monitoring system using Raspberry Pi, in *2018 Fourth International Conference on Computing Communication Control and Automation (ICCUBEA)* (IEEE, 2018), pp. 1–5

K. Hameed, I.S. Bajwa, S. Ramzan, W. Anwar, A. Khan, An intelligent IoT based healthcare system using fuzzy neural networks. Sci. Program. (2020)

D.S. Kumar, G. Sathyadevi, S. Sivanesh, Decision support system for medical diagnosis using data mining. Int. J. Comput. Sci. Issues (IJCSI) **8**(3), 147 (2011)

S. Mahajan, A.M. Birajdar, IoT based smart health monitoring system for chronic diseases, in *2019 IEEE Pune Section International Conference (PuneCon)* (IEEE, 2019), pp. 1–5

A. Montesines-Nagayo, N. Bugtai, Design and implementation of a telemedicine information system (TMIS) for managing obstetrical, gynecological and general medical data, in *1st AUN/SEED-NET Regional Conference in Manufacturing Engineering* (2008)

A.M. Nagayo, M.Z.K. Al Ajmi, N.V.R. Guduri, F.S.H. AlBuradai, A.R.A. Al Kindi, A.H.A. Al Farsi, An unmanned aerial robot and physiological data monitoring system integrated into a patient transport vehicle for emergency medical services and telehealth, in *2021 5th International Conference on Information Systems and Computer Networks (ISCON)* (IEEE, 2021), pp. 1–8

R. Payal, A.P. Singh, A study of various hardware and cloud based Internet of Things platforms (No. 4335). EasyChair (2020)

M. Shatnawi, A. Shatnawi, Z. AlShara, G. Husari, Symptoms-based fuzzy-logic approach for COVID-19 diagnosis. Int. J. Adv. Comput. Sci. Appl. **12**(4) (2021)

F. Tagliabue, D. Schena, L. Galassi, M. Magni, G. Guerrazzi, A. Acerbis, C. Rinallo, D. Longhi, A. Ronzani, P. Mariani, Modified national early warning score as early predictor of outcome in COVID-19 pandemic. SN Compr. Clin. Med. **3**(9), 1863–1869 (2021)

K. Viswavardhan Reddy, N. Kumar, A framework for remote health monitoring, in *ICT Systems and Sustainability* (Springer, Singapore, 2021), pp. 101–112

World Health Organization, Middle East respiratory syndrome coronavirus (MERS-CoV): summary of current situation, literature update and risk assessment (2015). https://apps.who.int/iris/handle/10665/179184. Accessed 30 Sept 2021

World Health Organization, Home care for patients with COVID-19 presenting with mild symptoms and management of their contacts: interim guidance, 17 Mar 2020. World Health Organization (2020). https://apps.who.int/iris/handle/10665/331473. Accessed 30 Sept 2021

World Health Organization, COVID-19 weekly epidemiological update, edition 82 (2022). https://www.who.int/docs/default-source/coronaviruse/situation-reports/20220308_weekly_epi_update_82.pdf. Accessed 20 Mar 2022

N. Yellavula, *Hands-On RESTful Web Services with Go: Develop Elegant RESTful APIs with Golang for Microservices and the Cloud* (Packt Publishing Ltd., 2020)

H. Zhang, J. Li, K. Motegi, Y. Shiraishi, IoT framework of wearable device for telemedicine application, in *2020 59th Annual Conference of the Society of Instrument and Control Engineers of Japan (SICE)* (IEEE, 2020), pp. 1616–1621

HDMFRTDB: Design of a High-Efficiency Deep Learning Classification Model for Fruit Ripeness-Type Differentiation Via Bioinspired Optimization

Archana G. Said and Bharti Joshi

Abstract Fruits often alter their outward appearance in accordance with their inside makeup. As a result, immature fruits seem different from ripe fruits, and fruits that have a disease look different from fruits that are healthy. Today, artificial fruit ripening is expanding quite quickly in order to produce fruits more quickly in order to meet the needs of the rising population. However, these fruits are unhealthy and regularly consumed individuals experience a broad range of physical and mental health problems. In order to let daily customers verify their fruits before purchasing or regularly eating them, high-efficiency identification of these fruits based on their visual appearance is required. Researchers have suggested a broad range of clinical approaches to carry out this duty, but the most of them either have poor accuracy or can't be applied to many fruit varieties. The contribution of this research is the building of a unique, very effective deep learning classification model for fruit ripeness-type separation using bioinspired optimization, which is suggested in this text to get beyond these restrictions. To properly categorize fruits, the suggested model first combines a Contrast Limited Adaptive Histogram Equalization (CLAHE) enhancement technique with a CNN model based on LAB color space. A bioinspired model based on Grey Wolf Optimization (GWO) and Particle Swarm Optimization (PSO) is used to identify classes of fruit ripeness based on this categorization. Using a cloud-based subscription model, these categories—which include naturally ripened, artificially ripened, and unripe categories—will help regular consumers identify fruit classifications. The inclusion of pre-trained VGGNet-19, You Look Only Once (YoLo V2), and Mobile V Net2 models, which aid in high-efficiency classification, gives the proposed model a high processing speed. These models were chosen because they performed more accurately over a variety of dataset settings. The suggested model was evaluated using data from the ILSVRC2014 dataset, Fresh Fruit Database, and Fruits 360 databases. This examination led to the conclusion that the suggested model outperformed current techniques in terms of accuracy, delay, precision, and recall

A. G. Said (✉) · B. Joshi
Department of Computer Engineering, RAIT, D. Y. Patil University, Nerul, Navi Mumbai, India
e-mail: saidarchana07@gmail.com

B. Joshi
e-mail: bharti.joshi@rait.ac.in

A. B. Reddy et al. (eds.), *Proceedings of Third International Conference on Advances in Computer Engineering and Communication Systems*, Lecture Notes in Networks and Systems 612, https://doi.org/10.1007/978-981-19-9228-5_29

by 15.4%, 8.3%, 9.5%, and 6.2%. Additionally, it was discovered that the suggested model had a 94.5% accuracy rate when predicting the maturity type of five distinct fruit varieties. The model was also installed on the cloud for general testing and assessment, and it was shown to be capable of high-speed deployments since it could analyze fruit photos in less than 500 ms. The suggested approach is helpful for a broad range of clinical and real-time use cases because of its excellent performance.

Keywords Fruit · Ripe · Artificial · Natural · Processing · Performance · Delay · Accuracy · Precision · Recall · Histogram · Enhancement · Healthy · LAB · CNN · GWO · PSO

1 Introduction

Designing models for image pre-processing, segmentation, clustering, feature extraction, feature selection, classification, and post-processing is necessary for the multidomain problem of ripeness detection from fruit pictures. Each of these models has to be specially tuned for datasets that are disease-specific, which necessitates the creation of adaptive parameter tuning techniques needed for greater scalability. Figure 1 shows an example of a common image processing model used for strawberry ripeness detection, where ripeness levels are determined using spectral reflectance data, Strawberry Fruit Water Content (SFWC), and fruit photos (Kaur et al. 2021).

Simple thresholding, a static detection technique with few applications, is utilized in this model to identify ripeness stages. In the next portion of this book, a study of comparable models (AlFaris et al. 2021; Ringer and Blanke 2021; Ingendoh et al. 2021; Silva et al. 2022) for various fruits is covered along with their specifics, benefits, drawbacks, and potential areas for further research. This debate led to the conclusion that the accuracy of the current models is either poor or they can't be applied

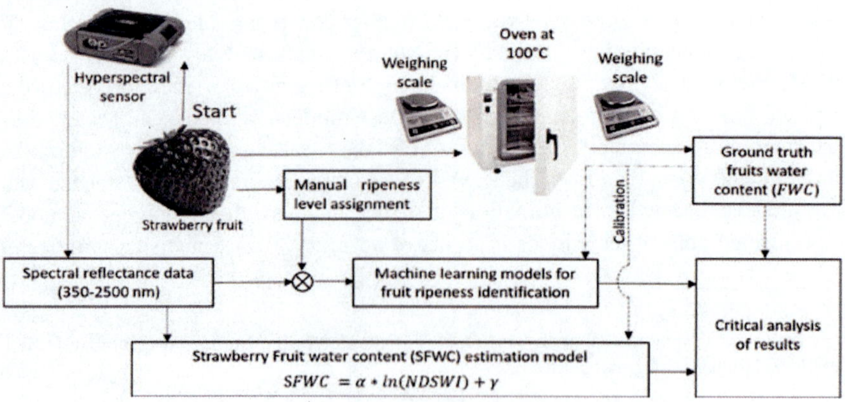

Fig. 1 A typical Strawberry Fruit ripeness detection model

to diverse varieties of fruit. Section 3 presents the building of a unique, very effective deep learning classification model for differentiating fruit ripeness types using bioinspired optimizations to address these problems. In Sect. 4, the suggested model's performance was assessed in terms of accuracy, precision, recall, and latency, and this performance was compared with several state-of-the-art techniques. This essay finishes with some thought-provoking remarks regarding the suggested model and suggestions for how to enhance it under various fruit varieties.

2 Literature Review

A wide variety of models are proposed by researchers to identify ripeness levels in fruits, and performance of these models varies widely w.r.t. their characteristics. For instance, work in Samkumar et al. (2021) proposes a model that uses abscisic acid-regulated anthocyanin biosynthesis for identification of fruit ripeness. But the model is highly complex and requires significant assembly before deployment, thus cannot be used for real-time classification applications. To overcome this limitation, work in Khodabakhshian and Baghbani (2021) proposes use of atomic force microscopy for identification of ripening using peel roughness analysis. The model is applied to banana fruit, but can be extended to multiple other fruit types. This extension can be observed from Castro et al. (2019), Abasi et al. (2021), Dao et al. (2020), wherein use of support vector machine (SVM) with cross validation (CV), multiple lightning sources with visible/near infrared (Vis/NIR) spectral analysis, and narrow-beam antenna (NBA) operated at 10.525 GHz and 35° half-power beamwidth are used for non-destructive ripeness detection of multiple fruits. Similar models that use Fuzzy Mask Recurrent Convolution Neural Network (FMRCNN) (Huang et al. 2020), combination of transfer learning models including DenseNet, AlexNet, ResNet-18, Inception-v3, and ResNet-50 (Miraei Ashtiani et al. 2021) for development of ensemble CNN (eCNN), stiffness-based ripeness detection (Xu et al. 2020), SVM with thresholding (Septiarini et al. 2019), and use of spectrophotometry with CNN (Chopra et al. 2021) are also proposed by researchers. These models assist in low complexity classification of fruit ripeness levels via use of intelligent feature reduction and classification methods.

Similar methods are discussed in Mghames et al. (2020), bin Mat Seri et al. (2021), Cardona et al. (2020), which propose use of Interactive Probabilistic Movement Primitives (I-ProMP), LiDAR Lite V3 CNN Model, and Eulerian Video Magnification (EVM) for identification of ripeness via estimation of exterior features. These features are extracted from fruit surface and assist in non-intrusive classification of different fruit ripeness types. Extensions to these models include use of Multiple Scale Context Aggregation (MSCA) (Ilyas et al. 2021), Linear Discriminant Analysis (LDA) with LAB Color Space (Septiarini et al. 2020), HSV Color Model with k Means-based clustering (Mueez 2020), and YOLOv3 Model (Selvam et al. 2021), which assist in high-speed computations along with better accuracy under different image types. These models use different datasets and introduce feature selection and reduction methods to improve classification performance during detection of ripeness levels. Similar models are discussed in Dandavate and Patodkar (2020), Rodrigues et al. (2021), Wagner et al. (2021), Pardede et al. (2019), which introduce utility of 3D Imaging, multiple augmentation and feature selection models, CNN with drone images, and augmentation of different color space models for detection of fruit ripeness levels. Extended models are discussed in Dhiman and Kumar (2017, 2018), Kaur et al. (2020), Dhiman and Kaur (2019, 2021), Chatterjee (2021), Vaishnav et al. (2021), Gupta et al. (2022), Sharma et al. (2022), Shukla et al. (2022), which further evaluate deep learning methods for various classification applications. Based on this evaluation, it can be observed that a wide variety deep learning models are proposed for fruit ripeness detection, but none of these models deploy fruit classification methods, due to which they are unable to implement fruit-dependent ripeness level classification models. This limits their scalability, which can be improved via the proposed High-Efficiency Deep Learning Classification Model for Fruit Ripeness-Type Differentiation via Bioinspired Optimization method that are described in the next section of this text. The model is evaluated on different datasets, and its efficiency is compared in terms of precision, recall, accuracy, AUC, and delay performance w.r.t. various state-of-the-art models.

3 Design of the Proposed High-Efficiency Deep Learning Classification Model for Fruit Ripeness-Type Differentiation Via Bioinspired Optimization

The examination of the available literature shows that the majority of them cannot be scaled for other fruit varieties or have poor accuracy when applied to numerous datasets. This section explains the architecture of the suggested high-efficiency deep learning classification model for fruit ripeness-type distinction using bioinspired optimization approaches in order to get beyond these restrictions. Figure 2 shows the overall flow of the proposed model, which begins by enhancing the visual quality of a picture in LAB color space using the Contrast Limited Adaptive Histogram Equalization (CLAHE) image enhancement technique. To properly categorize fruits from the processed photos, a CNN Model that incorporates VGGNet-19, You Look Only Once (YoLo V2), and Mobile V Net2 approaches is used. A bioinspired model based on Grey Wolf Optimization (GWO) and Particle Swarm Optimization (PSO) is used to identify classes of fruit ripeness based on this categorization. These classifications, which help regular customers recognize fruit types, include artificially ripened, naturally ripened, and unripe categories. In order to help researchers deploy the proposed model in part (or in whole) or as a whole for their use cases, the suggested model design is broken into several submodules and is discussed in various subsections of this text.

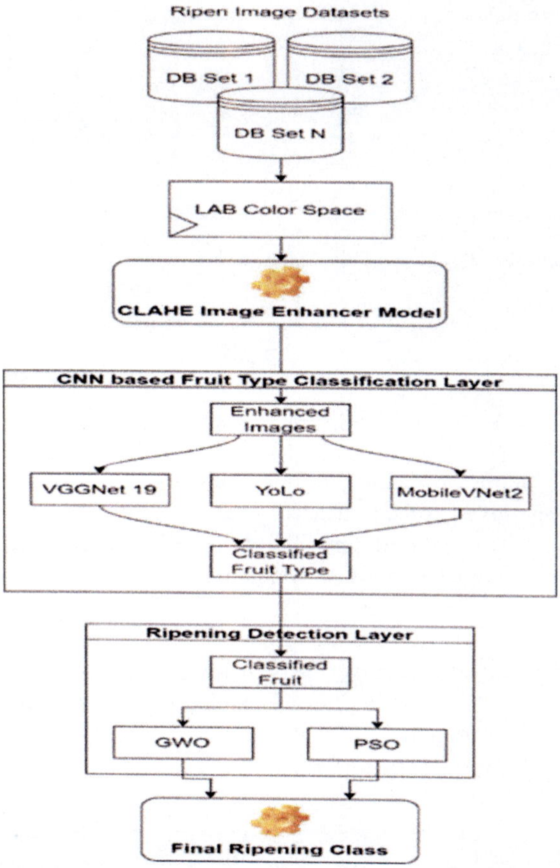

Fig. 2 Overall flow of the proposed model

3.1 Design of CLAHE Model for Image Enhancement

Fruit images from multiple datasets are initially processed via use of a CLAHE model that assists in image enhancement process. This model uses limited contrast enhancement, which assists in reducing amplification of noise during various enhancements.

Here, slope of a specialized transformation function is used to amplify contrast of different image regions. The transformation function is built via use of neighborhood cumulative distribution function (CDF), due to which histogram of that pixel value is limited via a clipping process as observed from Fig. 3 as follows.

The clipping value indicated in Fig. 3 is evaluated via Eq. 1, wherein variance between neighborhood regions during histogram normalization process is used for evaluation as follows,

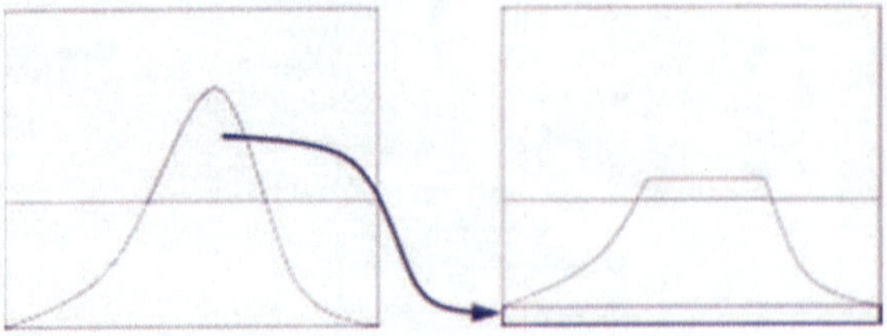

Fig. 3 Clipping process followed by CLAHE model

$$\partial = \sqrt{\sum_{r=1}^{R}\sum_{c=1}^{C}\left[\begin{array}{c}\frac{1}{R*C}* \\ I(r,c)- \\ \frac{1}{R*C}\sum_{r1=1}^{R}\sum_{c1=1}^{C} I(r1,c1)\end{array}\right]} \quad (1)$$

where *R*, *C* and *I* represents number of rows and number of columns in the image, and original image pixel levels. Enhanced pixels with intensity values above ∂ are discarded, while others are used normally by the equalization process. Results of this equalization are depicted in Fig. 4, which indicates original input image and its corresponding equalized image sets.

Based on visual quality of these images, it can be observed that CLAHE model outperforms other models in terms of visual representation. An improved visual representation assists in improving feature extraction, selection, and classification capabilities. Thus, it is used for training the proposed deep learning model, which is described in the next sub-section of this text.

Fig. 4 Results of image enhancement by various methods (**a**-original image, **b**-enhanced via adaptive thresholding, **c**-enhanced via back propagation histogram equalization, **d**-enhanced via gamma correction, **e**-enhanced via CLAHE, and **f**-enhanced via fuzzy enhancements)

In order to improve ripeness classification performance, a novel multimodal deep learning model is discussed in this section, which combines VGGNet, YoLo, and MobileVNet 2 methods. The proposed model uses a combination of these methods, because of their classification superiority w.r.t. different fruit types. In this study, it was observed that apple, kiwi, and orange were accurately classified via use of VGGNet-19 Model, while YoLo was able to classify banana and lemons, and MobileVNet2 was capable of high accuracy classification for strawberry and mango fruit types. Based on this evaluative observation, these models were selected for improving fruit classification performance under different image types. The VGGNet-19 Model is depicted in Fig. 5, wherein different layers, along with their window and stride sizes can be observed. The model uses a combination of convolutional layers with Max Pooling and Dropout layers, which assist in extraction of highly representative feature sets. The extracted sets are used to train a fully connected neural network (FCNN), that uses Soft Max layer for activation of final output fruit class. This model extracts convolutional features via Eq. 2, wherein input image (I), rows and columns (m and n) sizes for feature windows, and rows and columns (a and b) sizes for convolutional strides are used for final classification,

$$CF_{\mathrm{out}_{i,j}} = \sum_{a=-\frac{m}{2}}^{\frac{m}{2}} \sum_{b=-\frac{n}{2}}^{\frac{n}{2}} ReLU\left(\frac{m}{2} + a, \frac{n}{2} + b\right) * I(i - a, j - b) \tag{2}$$

These features are extracted for all input images, and are processed via a Max Pooling layer, which estimates feature value threshold via Eq. 3 as follows,

$$\sqrt{\sum_{r=1}^{m} \sum_{c=1}^{n} \left[\begin{array}{c} \frac{1}{m*c} * \\ CF(r, c) - \\ \frac{1}{m*n} \sum_{r1=1}^{m} \sum_{c1=1}^{n} CF(r1, c1) \end{array} \right]} \tag{3}$$

Features with values, $f \geq F_{\mathrm{th}}$ are directly used by the classification process, while others are discarded on a layer-by-layer basis.

This process is repeated for all window sizes, and the final feature vector is classified via use of Soft Max activated FCNN that works via Eq. 3 as follows, $c_{\mathrm{out}} = \mathrm{SoftMax}\left(\sum_{i=1}^{N_f} CF_i * w_i + b\right)$.

Where, N_f, w and b represents total number of extracted features, weight of the feature, and a bias value that can be tuned as per model's accuracy requirements. Similarly, the YoLo and MobileVNet2 Models are depicted in Figs. 6 and 7, respectively, wherein it can be observed that both the models use same convolutions as VGGNet and add additional feature extraction and selection layers to enhance classification performance. The YoLo model is depicted in Fig. 6, wherein different convolutional layers and Max Pooling layers can be observed.

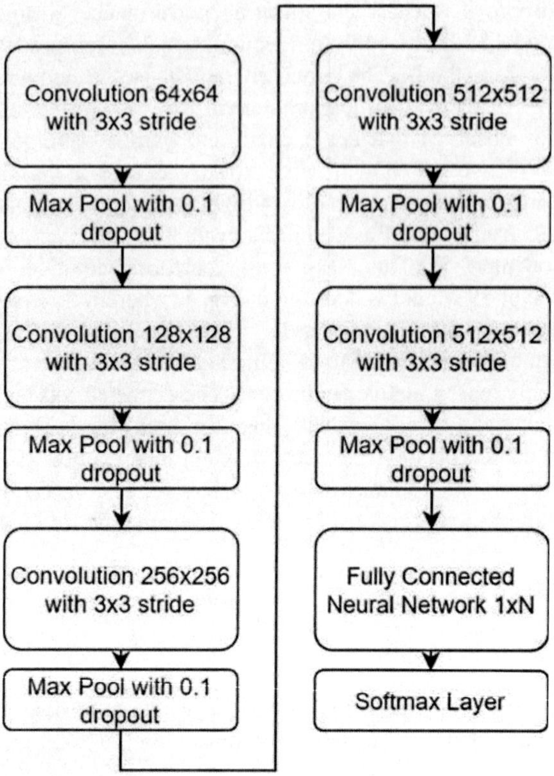

Fig. 5 Used VGGNet-19 model for classification of fruit types

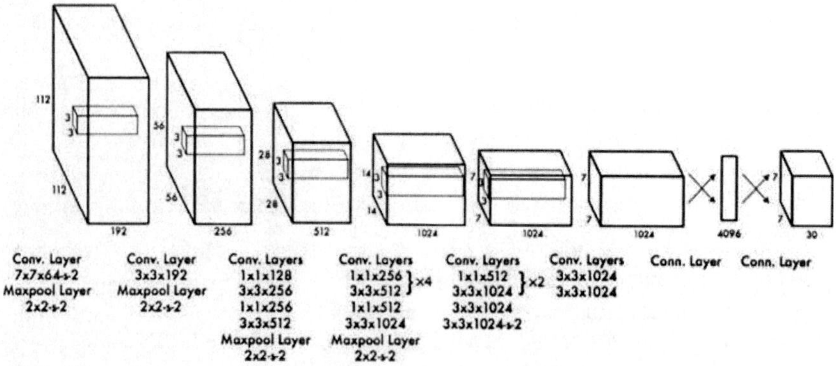

Fig. 6 Design of the YoLo model for fruit classification

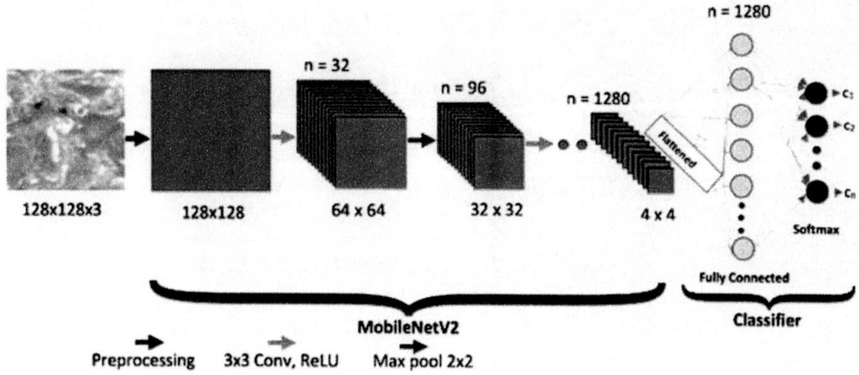

Fig. 7 Design of the MobileNetV2 model

This model uses the same mathematical formulas as VGGNet-19 but has a complex structure which assists in improving feature extraction and classification capabilities. This model is cascaded with MobileVNet2 Model, which can be observed from Fig. 7, wherein different convolutional, dropout, and pooling layers are connected.

These layers assist in improving classification speed via reduction of redundancy from feature sets. A combination of these layers is used to classify different fruit types, which assists in improving classification accuracy. Results of classification are processed via an ensemble GWO-based PSO Model for better ripeness classification efficiency. Design of this model is depicted in the next subsection of this text.

3.2 Design of Bioinspired Model for Final Detection of Fruit Ripening Levels

The classified fruit images are processed via a combination of GWO and PSO, which assists in iterative accuracy improvement for detection of ripeness levels. The PSO model is used for initial population generation and velocity update operations, while GWO is used for iteratively selecting best classifier configurations. This model works via the following process:

- Initialize GWO & PSO
- Number of iterations (N_i)
- Number of wolves/particles (N_w)
- Social Learning Rate (L_s)
- Cognitive Learning Rate (L_c)
- Wolf Learning Rate (L_w)
- Initially generate solutions via the following process: Stochastically generate classifier configurations for support vector machine (SVM), random forest (RF), and

k Nearest Neighbors (kNN) via Eqs. 4, 5, and 6 as follows:

$$\text{SVM}(e) = \text{STOCH}\left(\frac{\min(L_s, L_c, L_w)}{N_i * N_w}, \frac{1}{N_i * N_w}\right) \quad (4)$$

$$\text{NF} = \text{STOCH}(\min(L_s, L_c, L_w) * N_i * N_w, N_i * N_w) \quad (5)$$

$$k = \text{STOCH}(1, N_i * N_w * \min(L_s, L_c, L_w)) \quad (6)$$

where STOCH represents a stochastic process, which is used to generate numbers between given ranges, SVM(e) represents SVM error tolerance, NF represents number of forests for the RF classifier, and k represents number of neighbours for the kNN classiifer.

- Use these configurations for SVM, RF, and kNN to identify ripeness types via processing the classified images.
- Convolutional features extracted via the three CNN models are used as feature inputs to these ensemble classification methods.
- Identify fitness of each particle (or wolf) via Eq. 7 and mark it as particle best, as follows.

$$P\text{Best}_j = f_j = \sum_{i=1}^{N_c} \frac{A_i}{N_c} \quad (7)$$

where N_c represents number of images used for classification, A_i represents accuracy of this classifier, and f_j represents fitness of the particle.

- Identify Global Best via Eq. 8.

$$G\text{Best} = \text{Max}\left(\bigcup_{i=1}^{N_w} f_i\right) \quad (8)$$

- Mark all particles as 'delta wolves', where $f_i \leq G\text{Best} * L_w$, else mark other particles as 'Beta wolves'
- Repeat the following process for all iterations.
  - For each particle (or wolf) perform the following process,

    If the particle is not marked as 'delta wolf', then go to the next wolf
    Else, modify the particle via stochastic generation of SVM(e), NF and k values via Eqs. 4, 5, and 6, update particle fitness via Eq. 7, and identify GBest via Eq. 8, which will assist in identification of a new particle with modified fitness.
    Further modify fitness of this particle via Eq. 9,

$$f(\text{New}) = f(\text{Old}) * r + L_s * (G\text{Best} - f(\text{Old})) + L_c * (P\text{Best} - f(\text{Old})) \quad (9)$$

Based on this new fitness value, modify values of SVM(e), NF&k to obtain mutated fitness.
Modify PBest $= f(\text{New})$, if $f(\text{New}) > f(\text{Old})$, and accordingly modify the value of $GBest$
Based on these fitness values, evaluate fitness threshold via Eq. 10,

$$f_{\text{th}} = \sum_{i=1}^{N_w} f_i * \frac{L_w}{N_w} \quad (10)$$

Mark all particles as 'delta wolves', where $f_i \le f_{\text{th}}$, else mark other particles as 'Beta wolves'

- Repeat this process for all iterations, and modify solutions to obtain different wolf types.

  At the end of last iteration, select solution with maximum fitness, mark it as 'Alpha Wolf', and use its configuration for identification of fruit ripeness. Based on this process, the model is capable of identifying input images into different ripeness categories. Performance of this model is evaluated on different datasets and compared with different state-of-the-art models. This performance can be observed from the next section of this text.

4 Statistical Analysis and Comparison

- The proposed model combines multiple deep learning models with bioinspired methods to improve performance of fruit ripeness detection, which can be used under different datasets. To estimate this performance, the proposed model was tested on various fruit databases including:
- Fruits 360 (https://www.kaggle.com/datasets/moltean/fruits),
- Fresh Fruit Database (https://www.kaggle.com/datasets/sriramr/fruits-fresh-and-rotten-for-classification)
- ILSVRC2014 (https://www.image-net.org/challenges/LSVRC/2014/) datasets.

Due to use of multiple datasets, performance can be validated on multiple scenarios. These sets were combined to form a total of 15,000 images, out of which 70% were used for training the model, while remaining 30% were used to test the model in terms of accuracy (A), precision (P), recall (R), area under the curve (AUC), and delay (D) metrics. These models were also compared on multiple parameters, which makes it easier to validate its performance levels. This comparison is done with SVM CV (Castro et al. 2019), eCNN (Miraei Ashtiani et al. 2021), and YoLo V3 (N. A. Mohd Basir Selvam, Z. Ahmad and I. A. Mohtar 2021) in order to estimate its performance w.r.t currently proposed methods. Estimation of average accuracy is done based on Eq. 11,

$$\text{Acc.} = 100 * \frac{\sum_{i=1}^{N_s} \frac{\sum_{j=1}^{N_d} \frac{abs(A_p - E_p)}{\max(A_p, E_p)}}{N_d}}{N_s} \tag{11}$$

where N_s, N_d, A_p, and E_p represents number of images evaluated, number of different types for which evaluation was done, actual image class, and expected image class. Based on this experimentation, accuracy of ripeness detection w.r.t. number of images can be observed from Table 1 as follows.

This accuracy can be observed from Fig. 3, wherein accuracy is visualized against number of images for which these values were evaluated. Based on this evaluation,

Table 1 Average accuracy for ripeness type prediction

Num. images	Acc. SVM CV (Castro et al. 2019)	Acc. eCNN (Miraei Ashtiani et al. 2021)	Acc. YoLo V3 (Selvam et al. 2021)	Acc. Proposed
363	58.14	76.10	79.46	85.53
544	58.61	77.46	80.44	86.64
726	59.08	79.12	81.63	87.93
907	59.80	80.44	82.81	89.19
1089	60.27	82.10	84.00	90.44
1270	60.51	81.93	84.07	90.51
1452	60.75	81.69	84.14	90.58
1633	60.98	81.02	83.97	90.41
1815	61.22	81.73	84.51	90.98
1996	60.95	82.34	84.61	91.12
2177	60.78	82.85	84.75	91.25
2359	60.68	83.22	84.85	91.36
2540	60.92	83.39	85.08	91.63
2722	61.12	83.59	85.32	91.90
2903	61.32	83.83	85.59	92.20
3085	61.49	84.20	85.93	92.51
3266	61.63	84.54	86.20	92.78
3448	61.73	84.85	86.44	93.02
3629	61.80	85.15	86.64	93.25
3810	61.86	85.42	86.85	93.46
3992	61.93	85.76	87.05	93.66
4173	62.03	86.10	87.25	93.90
4355	62.29	86.69	87.72	94.41
4500	62.56	87.28	88.22	94.97

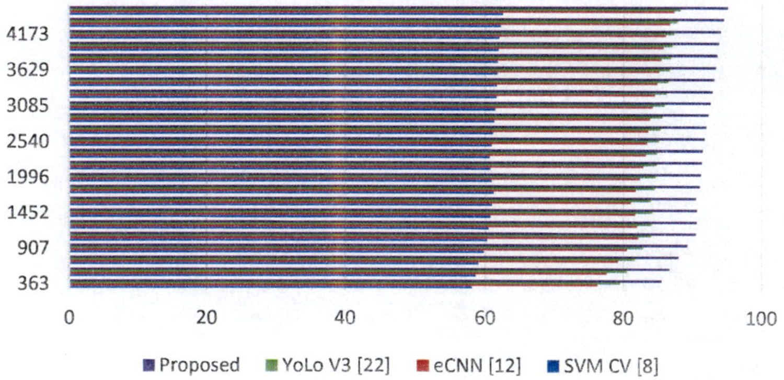

Fig. 8 Accuracy of image ripeness estimation for different models

it can be observed that the proposed model is 31% better than SVM CV (Castro et al. 2019), 8% better than eCNN (Miraei Ashtiani et al. 2021), and 5% better than YoLo V3 (Selvam et al. 2021) in terms of ripeness type prediction accuracy. This makes the proposed model applicable for real-time ripeness type prediction scenarios and can be used by agricultural firms for accurate prediction (Fig. 8).

Similarly, value of precision is estimated using Eq. 12.

$$Pre. = 100 * \frac{\sum_{i=1}^{N_s} \frac{\sum_{j=1}^{N_d} \frac{abs(Ap-Ep)}{Ap}}{N_d}}{N_s} \tag{12}$$

Value of precision is estimated using Eq. 13.

$$\text{Rec.} = 100 * \frac{\sum_{i=1}^{N_s} \frac{\sum_{j=1}^{N_d} \frac{\text{abs}(Ap-Ep)}{\min(Ap,Ep)}}{N_d}}{N_s} \tag{13}$$

And, Value of AUC is estimated using Eq. 14.

$$\text{AUC} = 100 * \frac{\sum_{i=1}^{N_s} \frac{\sum_{j=1}^{N_d} \frac{\min(Ap,Ep)}{\max(Ap,Ep)}}{N_d}}{N_s} \tag{14}$$

Based on these evaluations, observations were made for precision performance, and can be observed from Table 2, wherein precision is tabulated w.r.t. number of images for each of the image types (Fig. 9).

Based on this evaluation and Fig. 11, it can be observed that the proposed model is 19.5% better than SVM CV (Castro et al. 2019), 20.5% better than eCNN (Miraei Ashtiani et al. 2021) and 24.5% better than YoLo V3 (Selvam et al. 2021) in terms of precision of ripeness type prediction. Similar observations were made for precision

Table 2 Average precision for image ripeness prediction

Num. Images	Pre. SVM CV (Castro et al. 2019)	Pre. eCNN (Miraei Ashtiani et al. 2021)	Pre. YoLo V3 (Selvam et al. 2021)	Pre. Proposed
363	70.00	59.31	60.69	80.77
544	71.00	61.38	62.41	83.08
726	72.00	63.79	64.14	85.38
907	74.00	65.86	66.21	87.69
1089	75.50	68.62	67.93	90.00
1270	76.50	68.62	67.93	90.38
1452	77.00	68.28	67.93	90.77
1633	77.50	67.24	67.59	90.38
1815	78.00	68.28	68.62	91.54
1996	77.00	69.31	68.62	91.54
2177	76.50	70.00	68.97	91.92
2359	76.00	70.69	68.97	91.92
2540	77.00	71.03	69.66	92.69
2722	77.50	71.38	70.00	93.08
2903	78.00	71.72	70.69	93.85
3085	78.00	72.41	71.03	94.23
3266	78.50	73.10	71.38	95.00
3448	79.00	73.79	71.72	95.38
3629	79.50	74.14	72.07	96.15
3810	79.50	74.83	72.76	96.54
3992	79.50	75.17	73.10	96.92
4173	79.50	75.86	73.45	97.31
4355	80.10	76.71	74.00	98.28
4500	80.83	77.78	74.73	99.47

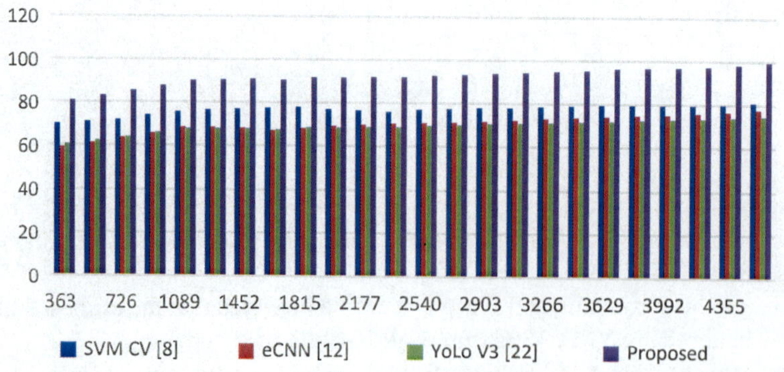

Fig. 9 Precision of image ripeness estimation for different models

performance, and can be observed from Table 3, wherein precision is tabulated w.r.t. number of observation images for each of the ripeness types (Fig. 10).

Based on this evaluation and Fig. 12, it can be observed that the proposed model is 8% better than SVM CV (Castro et al. 2019), 5% better than eCNN (Miraei Ashtiani et al. 2021), and 6% better than YoLo V3 (Selvam et al. 2021) in terms of recall of ripeness type prediction. This makes the proposed model applicable for real-time ripeness type prediction scenarios and can be used by agricultural firms for accurate prediction. Similar observations were made for area under the curve performance and can be observed from Table 4, wherein AUC is tabulated w.r.t. number of observation images for each of the ripeness types.

Based on this evaluation, it can be observed that the proposed model is 28.5% better than SVM CV (Castro et al. 2019), 23.5% better than eCNN (Miraei Ashtiani et al. 2021), and 27.4% better than YoLo V3 (Selvam et al. 2021) in terms of AUC

Table 3 Average recall for image ripeness prediction

Num. Images	Rec. SVM CV (Castro et al. 2019)	Rec. eCNN (Miraei Ashtiani et al. 2021)	Rec. YoLo V3 (Selvam et al. 2021)	Rec. Proposed
363	53.48	47.93	48.62	61.15
544	53.91	48.97	49.66	62.31
726	54.78	50.34	50.69	63.46
907	55.65	51.38	51.72	64.62
1089	56.09	52.76	52.41	65.77
1270	56.09	52.76	52.41	65.77
1452	56.52	52.41	52.41	65.77
1633	56.96	52.07	52.41	65.38
1815	57.39	52.41	52.76	65.77
1996	56.96	53.10	52.76	65.77
2177	56.96	53.45	52.76	66.15
2359	56.96	53.79	53.10	66.54
2540	57.39	53.79	53.45	66.92
2722	57.39	53.79	53.79	66.92
2903	57.39	54.14	53.79	66.92
3085	57.39	54.48	53.79	67.31
3266	57.39	54.83	54.14	67.69
3448	57.83	54.83	54.48	68.08
3629	58.26	55.17	54.83	68.08
3810	58.70	55.52	54.83	68.08
3992	58.70	55.86	54.83	68.46
4173	58.70	56.21	54.83	68.85
4355	58.93	56.67	55.21	69.45
4500	59.23	57.24	55.68	69.77

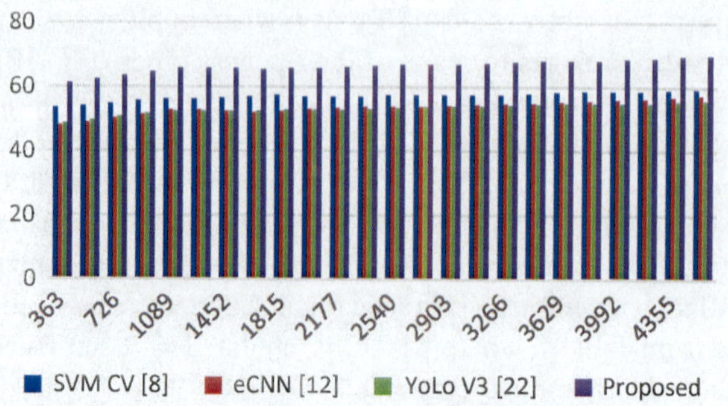

Fig. 10 Recall of image ripeness estimation for different models

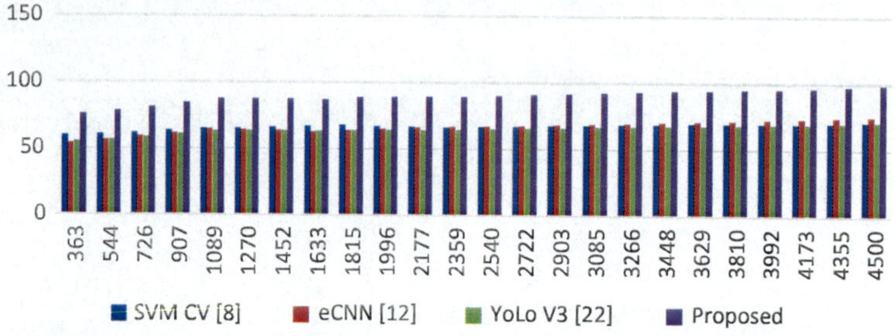

Fig. 11 Average AUC for image ripeness prediction

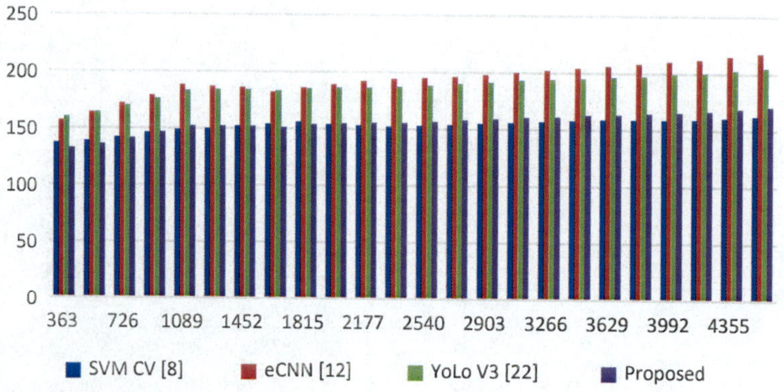

Fig. 12 Average delay for image ripeness prediction

performance for ripeness type prediction. Similar observations were made for delay performance, and can be observed from Table 5, wherein precision is tabulated w.r.t. number of observation images for each of the ripeness types.

Based on this evaluation, it can be observed that the proposed model is 0.5% faster than SVM CV (Castro et al. 2019), 24.5% faster than eCNN (Miraei Ashtiani

Table 4 Average AUC for image ripeness prediction

Num. Images	AUC. SVM CV (Castro et al. 2019)	AUC. eCNN (Miraei Ashtiani et al. 2021)	AUC. YoLo V3 (Selvam et al. 2021)	AUC. Proposed
363	59.57	54.14	55.17	76.15
544	60.43	56.55	56.55	78.46
726	61.74	59.31	58.62	81.54
907	63.48	61.72	60.69	84.62
1089	64.78	64.83	63.10	87.69
1270	65.22	64.48	63.45	87.69
1452	66.09	64.14	63.45	87.69
1633	66.96	62.76	63.10	87.31
1815	67.83	64.14	63.79	88.85
1996	66.96	65.17	64.14	89.23
2177	66.52	66.21	64.14	89.62
2359	66.09	66.90	64.48	89.62
2540	66.52	67.24	64.83	90.38
2722	66.96	67.59	65.52	91.15
2903	67.39	68.28	65.86	91.92
3085	67.83	68.97	66.55	92.69
3266	68.26	69.66	66.90	93.08
3448	68.70	70.34	67.24	93.85
3629	69.13	71.03	67.59	94.23
3810	69.13	71.72	67.93	95.00
3992	69.13	72.41	68.62	95.38
4173	69.13	73.10	68.97	96.15
4355	69.77	74.09	69.85	97.24
4500	70.53	75.00	70.58	98.60

et al. 2021), and 20.5% faster than YoLo V3 (Selvam et al. 2021) in terms of delay performance for ripeness type prediction. Due to these advantages, the proposed model is capable of deployment for a wide variety of real-time applications.

5 Analysis of Results

As observed from different result visualizations, this accuracy can be seen by looking at the number of images for which the values were evaluated. These results show the proposed model is 31% better than SVM CV (Castro et al. 2019), 8% superior to eCNN (Miraei Ashtiani et al. 2021), and 5% superior to YoLo V3 (Selvam et al.

Table 5 Average Delay for image ripeness prediction

Num. Images	D (ms) SVM CV (Castro et al. 2019)	D (ms) eCNN (Miraei Ashtiani et al. 2021)	D (ms) YoLo V3 (Selvam et al. 2021)	D (ms) Proposed
363	137.0	157.0	160.0	132.0
544	139.0	164.0	164.0	136.0
726	142.0	172.0	170.0	141.3
907	146.0	179.0	176.0	146.7
1089	149.0	188.0	183.0	152.0
1270	150.0	187.0	184.0	152.0
1452	152.0	186.0	184.0	152.0
1633	154.0	182.0	183.0	151.3
1815	156.0	186.0	185.0	154.0
1996	154.0	189.0	186.0	154.7
2177	153.0	192.0	186.0	155.3
2359	152.0	194.0	187.0	155.3
2540	153.0	195.0	188.0	156.7
2722	154.0	196.0	190.0	158.0
2903	155.0	198.0	191.0	159.3
3085	156.0	200.0	193.0	160.7
3266	157.0	202.0	194.0	161.3
3448	158.0	204.0	195.0	162.7
3629	159.0	206.0	196.0	163.3
3810	159.0	208.0	197.0	164.7
3992	159.0	210.0	199.0	165.3
4173	159.0	212.0	200.0	166.7
4355	160.5	214.9	202.6	168.6
4500	162.2	217.5	204.7	170.9

2021) in terms of the accuracy of ripeness type prediction. When applied in real time, the proposed model is capable of accurately forecasting the ripeness type of a crop. In terms of ripeness type prediction precision, the proposed model outperforms SVM CV (Castro et al. 2019), eCNN (Miraei Ashtiani et al. 2021), and YoLo V3 (Selvam et al. 2021) by 19%, 20.5%, and 24.5%, respectively. This is due to combination of different CNN Models which assist in improving classification performance for a wide variety of use cases.

According to the proposed model, ripeness type prediction is 8% better than SVM CV, 5% better than eCNN, and 6% better than YoLo V3 (Selvam et al. 2021) when compared to these other models. When applied in real time, the proposed model is capable of accurately forecasting the ripeness type of a crop. For ripeness type prediction, the proposed model outperforms SVM CV by 28.5%, eCNN by 23.5%, and YoLo V3 by 27.4% when it comes to AUC. SVM CV (Castro et al. 2019) is 0.5% faster, eCNN (Miraei Ashtiani et al. 2021) is 24.5% faster, and YoLo V3 (Selvam et al. 2021) is 20.5% faster in terms of delay performance for ripeness type prediction than the model under consideration. Many different real-time applications can be implemented with this model thanks to these advantages.

6 Conclusion and Future Work

Machine Learning Models (MLMs) and bioinspired models are combined in the proposed model in order to improve the classification of fruit ripeness. According to the results of this study, the proposed model is 31% better than SVM CV (Castro et al. 2019), 8% better than eCNN (Miraei Ashtiani et al. 2021) and 5% better than YoLo V3 (Selvam et al. 2021) in terms of ripeness type prediction accuracy, and 19.5% better than SVM CV, 20.5% better than eCNN and 24.5% better than YoLo V3 in terms of precision of ripeness type prediction. Additionally, the proposed model had a recall of ripeness type prediction that was 8% better than SVM CV (Castro et al. 2019), 5% better than eCNN (Miraei Ashtiani et al. 2021), and 6% better than YoLo V3 (Selvam et al. 2021), and performed faster due to the use of pretrained models. Consistent performance under a variety of datasets makes the model extremely useful in a wide range of applications. The model's accuracy improves inexorably when fruit classification is used before ripeness detection, making it ideal for a wide range of real-time applications.Some limitations of this work are unavailability of real-time datasets, and lack of bioinspired optimization models. Which can be taken up in future research, thus in future, researchers can deploy Q Learning Models, along with multiple deep learning methods to improve performance under different lighting conditions. It is also recommended that performance of the proposed model must be validated on multiple datasets, which will allow to evaluate its scalability under multiple scenarios.

References

S. Abasi, S. Minaei, B. Jamshidi, D. Fathi, Development of an optical smart portable instrument for fruit quality detection. IEEE Trans. Instrum. Measur. **70**, 1–9 (2021). Art no. 7000109. https://doi.org/10.1109/TIM.2020.3011334

N.A. AlFaris, J.Z. AlTamimi, L.A. AlMousa, et al., Antioxidant content determination in ripe date fruits (*Phoenix dactylifera L.*): a scoping review. Food Anal. Methods **14,** 897–921 (2021). https://doi.org/10.1007/s12161-020-01923-z

A.D.I. bin Mat Seri, M.S. bin Mohd Kassim, S.R. binti Abdul Rahman, A.A.B. Sajak, Development of Virescens fresh fruit bunch ripeness prediction using LiDAR for smart agriculture, in *2021 IEEE Region 10 Symposium (TENSYMP)* (2021), pp. 1–8. https://doi.org/10.1109/TENSYMP52854.2021.9550981

C.I. Cardona, et al., Vibration shapes identification applying Eulerian video magnification on coffee fruits to study the selective harvesting, in *2020 19th International Conference on Mechatronics—Mechatronika (ME)* (2020), pp. 1–8. https://doi.org/10.1109/ME49197.2020.9286641

W. Castro, J. Oblitas, M. De-La-Torre, C. Cotrina, K. Bazán, H. Avila-George, Classification of cape gooseberry fruit according to its level of ripeness using machine learning techniques and different color spaces. IEEE Access **7**, 27389–27400 (2019). https://doi.org/10.1109/ACCESS.2019.2898223

I. Chatterjee, Artificial intelligence and patentability: review and discussions. Int. J. Mod. Res. **1**, 15–21 (2021)

H. Chopra, et al.: Efficient fruit grading system using spectrophotometry and machine learning approaches. IEEE Sens. J. **21**(14), 16162–16169 (2021). https://doi.org/10.1109/JSEN.2021.3075465

R. Dandavate, V. Patodkar, CNN and Data Augmentation Based Fruit Classification Model," 2020 Fourth International Conference on I-SMAC (IoT in Social, Mobile, Analytics and Cloud) (I-SMAC), 2020, pp. 784–787, doi: https://doi.org/10.1109/I-SMAC49090.2020.9243440.

H.N. Dao, C. Phongcharoenpanich, M. Krairiksh, Narrow-Beam antenna for short-distance non-destructive sensor in fruit-ripeness monitoring. IEEE Access **8**, 226142–226150 (2020). https://doi.org/10.1109/ACCESS.2020.3044908

G. Dhiman, A. Kaur, STOA: a bio-inspired based optimization algorithm for industrial engineering problems. Eng. Appl. Artif. Intell. **82**, 148–174 (2019)

G. Dhiman, V. Kumar, Spotted hyena optimizer: a novel bio-inspired based metaheuristic technique for engineering applications. Adv. Eng. Softw. **114**, 48–70 (2017)

G. Dhiman, V. Kumar, Emperor penguin optimizer: a bio-inspired algorithm for engineering problems. Knowl.-Based Syst. **159**, 20–50 (2018)

V.K. Gupta, S.K. Shukla, R.S. Rawat, Crime tracking system and people's safety in India using machine learning approaches. Int. J. Mod. Res. **2**(1), 1–7 (2022)

Y.-P. Huang, T.-H. Wang, H. Basanta, Using fuzzy mask R-CNN model to automatically identify tomato ripeness. IEEE Access **8**, 207672–207682 (2020). https://doi.org/10.1109/ACCESS.2020.3038184

T. Ilyas, A. Khan, M. Umraiz, Y. Jeong, H. Kim, Multi-Scale context aggregation for strawberry fruit recognition and disease phenotyping. IEEE Access **9**, 124491–124504 (2021). https://doi.org/10.1109/ACCESS.2021.3110978

A. Ingendoh, M. Perkins, V. Langford, High-Throughput analysis of fruit ripening by SIFT-MS. Lebensmittelchemie **75**, S073–S073 (2021). https://doi.org/10.1002/lemi.202158074

S. Kaur, L.K. Awasthi, A.L. Sangal, G. Dhiman, Tunicate swarm algorithm: a new bio-inspired based metaheuristic paradigm for global optimization. Eng. Appl. Artif. Intell. **90**, 103541 (2020)

S. Kaur, S. Randhawa, A. Malhi, An efficient ANFIS based pre-harvest ripeness estimation technique for fruits. Multimed. Tools. Appl. **80**, 19459–19489 (2021). https://doi.org/10.1007/s11042-021-10741-2

R. Khodabakhshian, R. Baghbani, Classification of bananas during ripening using peel roughness analysis—an application of atomic force microscopy to food process. J. Food Process Eng. **44**(11), e13857 (2021). https://doi.org/10.1111/jfpe.13857

R. Kumar, G. Dhiman, A comparative study of fuzzy optimization through fuzzy number. Int. J. Mod. Res. **1**, 1–14 (2021)

S. Mghames, M. Hanheide, A.E. Ghalamzan, Interactive movement primitives: planning to push occluding pieces for fruit picking, in *2020 IEEE/RSJ International Conference on Intelligent Robots and Systems (IROS)* (2020), pp. 2616–2623. https://doi.org/10.1109/IROS45743.2020.9341728

S.-H. Miraei Ashtiani, S. Javanmardi, M. Jahanbanifard, A. Martynenko, F.J. Verbeek, Detection of mulberry ripeness stages using deep learning models. IEEE Access **9**, 100380–100394 (2021). https://doi.org/10.1109/ACCESS.2021.3096550.

A. Mueez, A cost-effective framework to predict the ripeness of any fruit based on color space, in 2020 IEEE Region 10 Symposium (TENSYMP) (2020), pp. 1729–1733. https://doi.org/10.1109/TENSYMP50017.2020.9231002

J. Pardede, M.G. Husada, A.N. Hermana, S.A. Rumapea, Fruit ripeness based on RGB, HSV, HSL, L a b color feature using SVM, in *2019 International Conference of Computer Science and Information Technology (ICoSNIKOM)* (2019), pp. 1–5.https://doi.org/10.1109/ICoSNIKOM48755.2019.9111486

T. Ringer, M. Blanke, Non-invasive, real time in-situ techniques to determine the ripening stage of banana. Food Measure **15**, 4426–4437 (2021). https://doi.org/10.1007/s11694-021-01009-2

B. Rodrigues, R. Kansara, S. Singh, D. Save, S. Parihar, Ripe-Unripe: machine learning based ripeness classification, in *2021 5th International Conference on Intelligent Computing and Control Systems (ICICCS)* (2021), pp. 1–5. https://doi.org/10.1109/ICICCS51141.2021.9432349

A. Samkumar, D. Jones, K. Karppinen, A.P. Dare, N. Sipari, R.V. Espley, I. Martinussen, L. Jaakola, Red and blue light treatments of ripening bilberry fruits reveal differences in signalling through abscisic acid-regulated anthocyanin biosynthesis. Plant Cell Environ. **44**(10), 3227–3245 (2021). https://doi.org/10.1111/pce.14158

N.A.M.B. Selvam, Z. Ahmad, I.A. Mohtar, Real time ripe palm oil bunch detection using YOLO V3 algorithm, in *2021 IEEE 19th Student Conference on Research and Development (SCOReD)* (2021), pp. 323–328. https://doi.org/10.1109/SCOReD53546.2021.9652752

A. Septiarini, H. Hamdani, H.R. Hatta, A.A. Kasim, Image-based processing for ripeness classification of oil palm fruit, in *2019 5th International Conference on Science in Information Technology (ICSITech)* (2019), pp. 23–26. https://doi.org/10.1109/ICSITech46713.2019.8987575

A. Septiarini, H.R. Hatta, H. Hamdani, A. Oktavia, A.A. Kasim, S. Suyanto, Maturity grading of oil palm fresh fruit bunches based on a machine learning approach, in *2020 Fifth International Conference on Informatics and Computing (ICIC)* (2020), pp. 1–4.https://doi.org/10.1109/ICIC50835.2020.9288603

T. Sharma, R. Nair, S. Gomathi, Breast cancer image classification using transfer learning and convolutional neural network. Int. J. Mod. Res. **2**(1), 8–16 (2022)

S.K. Shukla, V.K. Gupta, K. Joshi, A. Gupta, M.K. Singh, Self-aware execution environment model (SAE2) for the performance improvement of multicore systems. Int. J. Mod. Res. **2**(1), 17–27 (2022)

V.D. Silva, A.M.C. Santos, J.E. Oliveira, E.S. Medeiros. J. Appl. Polym. Sci. e52386 (2022). https://doi.org/10.1002/app.52386

P.K. Vaishnav, S. Sharma, P. Sharma, Analytical review analysis for screening COVID-19. Int. J. Mod. Res. **1**, 22–29 (2021)

N. Wagner, R. Kirk, M. Hanheide, G. Cielniak, Efficient and robust orientation estimation of strawberries for fruit picking applications, in *2021 IEEE International Conference on Robotics and Automation (ICRA)* (2021), pp. 13857–13863.https://doi.org/10.1109/ICRA48506.2021.9561848

C. Xu, H. He, S.C. Hauser, G.J. Gerling, Tactile exploration strategies with natural compliant objects elicit virtual stiffness cues. IEEE Trans. Haptics **13**(1), 4–10 (2020). https://doi.org/10.1109/TOH.2019.2959767

Emotion-Based Music Recommendation System to Generate a Dynamic Playlist

M. Sunitha, T. Adilakshmi, and Renuka

Abstract Music plays significant role in every individual life. People often get confused with the large set of music library which songs they have to listen based on current mood and this is time consuming process, very tedious, and need manual work. Different types of algorithms have been introduced for automating the music library. However, the existing algorithms used are slow and less accurate. This proposed system algorithms in view of facial expression will create a playlist consequently thereby reducing the work and time engaged with delivering the cycle physically. In terms of accuracy, emotion extraction algorithm gives around 80–90% for real-time images, 95–100% for the static pictures. In this way, it yields better exactness concerning execution and computational time and lessens the planning cost, contrasted with the algorithms utilized in the literature survey. Playlist is created, based on the detected feature.

Keywords Music recommendation · Emotion · Facial expression

1 Introduction

Looks play an important role in communicating an individuals' feelings. Music has forever been perceived to change an individual's state of mind. Catching and recognizing an individual's inclination, as well as playing music that matches the individual's state of mind, can assist with quieting the client's psyche and give an interesting piece of music. The venture will probably catch an individual's feelings through their looks. A music player is expected to record human articulation utilizing the web camera interface accessible on figuring frameworks. The product catches the client's picture and afterward utilizes picture division and picture handling procedures to separate highlights from the substance of an objective person to decide the feeling that the individual is endeavoring to communicate. By gathering the client's photo, understanding the interest of client and playing the tunes that match the client's

M. Sunitha (✉) · T. Adilakshmi · Renuka
Vasavi College of Engineering, Hyderabad, India
e-mail: m.sunithareddy@staff.vce.ac.in

A. B. Reddy et al. (eds.), *Proceedings of Third International Conference on Advances in Computer Engineering and Communication Systems*, Lecture Notes in Networks and Systems 612, https://doi.org/10.1007/978-981-19-9228-5_30

interest generally lift up the client's enthusiasm. Since antiquated times, the best sort of appearance examination known to mankind has been look acknowledgment. The most effective way to decipher or conclude the inclination, feeling, or contemplations that someone else is endeavoring to communicate is through their looks. Mind-set swings can likewise assist individuals with conquering discouragement and misery now and again. Numerous wellbeing dangers can be kept away from with the assistance of articulation investigation, and steps to further develop the client's state of mind can likewise be taken.

2 Dataset

Proposed system utilized two primary datasets to prepare the models: FER-2013 and CK+ (broadened Cohn-Kanade). The FER-2013 dataset comprises of 28,000 named pictures in the preparation set, 3500 marked pictures in the advancement set, and 3500 pictures in the test set. Each picture in FER-2013 is named as one of seven feelings: cheerful, miserable, furious, apprehensive, shock, disdain, and unbiased, with blissful being the most predominant inclination, giving a gage to irregular speculating of 24.4%. The pictures in FER-2013 comprise of both presented and unposed headshots, which are in grayscale and 48 $\times$ 48 pixels. The FER-2013 dataset was made by social occasion the consequences of a Google picture search of every inclination and equivalent words of the feelings (Tang et al. 2015; Zaanen and Kanters 2010).

The CK+ dataset has a sum of 5876 named pictures of 123 people. Out of these pictures, we involved 4113 pictures for preparing, 881 for dev, and 881 for test. Each picture is named with one of seven feelings: cheerful, miserable, irate, apprehensive, shock, disdain, and scorn. Pictures in the CK+ dataset are totally presented with comparable foundations, generally grayscale, and 640 $\times$ 490 pixels.

However, the FER-2013 and CK+ dataset both have comparably marked feelings, we found while fostering our model that it was exceptionally simple to accomplish incredibly high exactnesses on the CK+ dataset (rather than FER-2013). This is in all likelihood on the grounds that the FER-2013 dataset was presented, had more people and more variety than CK+. Thusly, we zeroed in more on working on our model's presentation on the FER-2013 dataset, as the more extensive scope of unposed pictures all the more firmly mirrored the pictures we would see while moving the abilities over to constant (He et al. 2008; Kwon and Kim 2011).

3 Literature Survey

Zaanen and Kanters (2010), the course of multi-layered decrease in which essential information is diminished to a wide range of classes for arranging or coordinating. The feeling of a client is separated by catching the client's picture with a webcam.

By following the essential information, the caught picture is upgraded through the course of layered decrease.

These qualities are switched over completely to parallel picture design, and the fisherface and Haarcascade techniques are utilized to distinguish the face. The underlying or essential information taken from the human face and decreased to numerous different classes. These classes are arranged and coordinated utilizing the techniques depicted previously. The component separated from the human face is utilized to identify feeling. The principal objective of the component removing module is to diminish the quantity of assets expected from huge informational indexes Deshmukh et al.

Fisherface Algorithm. The main part examination (PCA) technique is utilized to diminish the face space aspects, and afterward the fishers direct discriminant (FDL) or the LDA strategy (Lee and Hu 2012; Speck et al. 2011) is utilized to acquire the element of the picture qualities; the recommendation system here utilizes this strategy specifically on the grounds that it augments the division between classes during the preparation cycle. This calculation helps with the picture acknowledgment process, though the matching appearances calculation utilizes the base Euclidean space, permitting us to group the client's demeanor. In this paper, linear discriminant analysis (LDA) which is likewise called fisherface is an appearance-based strategy utilized for the dimensionality decrease and kept an extraordinary execution in face acknowledgment. This strategy chips away at a similar rule as the eigenface technique (PCA).

Haarcascade Algorithm. It is an AI calculation that is utilized to order protests in a caught picture. Its essential application is object location. The outpouring classifier has different phases of assortment that are like feeble students. These weak classifiers are the most basic type of classifier, also known as boosting. If the label range is positive, it advances to the next stage, which displays the result. These have a positive and negative viewpoint in that they recognize the pictures in light of the names. On different stages, these have a bunch of positive pictures over regrettable pictures. Images with higher resolution have a higher quantity and produce better quality results. To find the articles in the picture, we use Haarcascade front facing face default.xml. The items here are the nose, eyes, ears, and facial lips. Haarcascade, made by OpenCV, is planned to perceive the front face. It likewise can perceive attributes from the source. It works via preparing the negative picture on top of the positive picture it overlays. Positive images include only the images that you want the classifier to classify. Negative images are any images that do not contain the object you are looking for (Lameere and Pampak 2008; Hu and Downie 2010).

4 Proposed System

It can recognize and extract facial features from the user's expressions as shown in Fig. 1 and classify them in order to capture the user's specific emotions. Once the emotions are categorized, the user will see songs that match the user's emotions. Humans have an unintentional tendency to express their emotions, which are mostly reflected in their faces. We can utilize the proposed framework to give connection between the client and the music framework. This undertaking centers principally around the client's number one music, which is suggested in view of profound mindfulness. The proposed research work provided three options in the early stages of the proposed system, each with its own functionality. Research work answered with a rundown of melodies and feelings in view of spatial acknowledgment. At the point when the application fires up, it catches pictures utilizing the webcam or some other physiological gadgets. Our essential objective in fostering this framework is to make a refined music player that can work on the client's state of mind, and music is probably the best device for transforming one's temperament. Pictures procured by the framework are contrasted with informational indexes, and four feelings are chosen since people have a wide scope of feelings that are hard to foresee in light of the fact that they vary from one individual to another, so four normal and effectively recognizable states of mind are chosen, i.e., irregular choice of melodies that could end up being useful to us to light up our states of mind, and the other mode is line mode, with which we can make our own playlist, and in all modes we do not utilize past client information, however, individual client information.

Drawbacks

- Precision of the model is the main issue in the event of facial expression, because of which there are situations where anticipated feeling is off-base.
- Sometimes the music may not match with randomly played songs.
- The framework actually cannot record every one of the feelings accurately because of the less accessibility of the pictures in the picture dataset being utilized.
- The nature of the picture ought to be basically higher than 320p for the classifier to precisely anticipate the feeling of the client.

5 System Architecture

Running the primary website page in this task sets off a XML record, which then helps OpenCV in catching pictures from the webcam and playing out the execution of the OpenCV fisherface strategy for grouping. It contains two choices, one for feeling-based recognition and the other for irregular tune determination. Irregular melody choice by perceiving and joining spatial articulations, this framework depicts looks. Following component extraction, feelings are arranged into four sorts: blissful, irate,

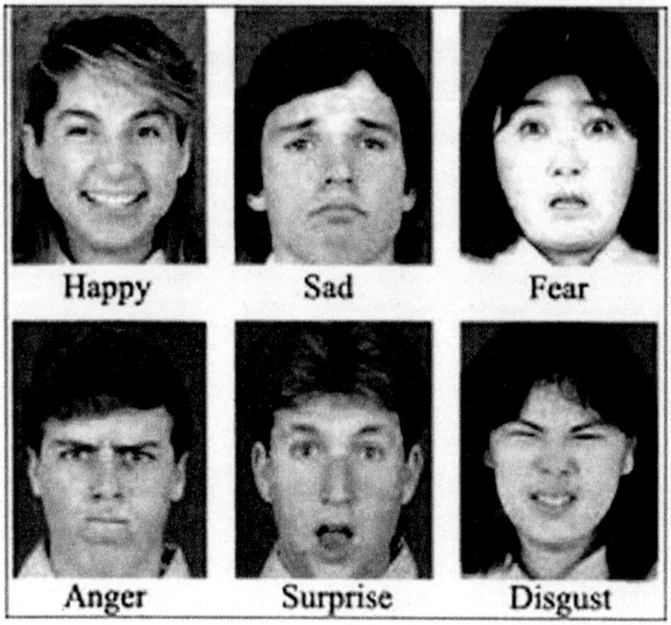

Fig. 1 Different types of emotions

miserable, and impartial. The feelings communicated in the last advance are mathematical, and music is played in light of the perceived feelings. The face acknowledgment strategy's primary objective is to recognize the casing, for example, the face. The other period of the venture is the arbitrary mode, for which we use Eel to choose sporadic tunes from the line aimlessly. The Win sound motor is utilized to get to the Windows stage's nearby strong delivering apparatus.

6 Methodology

6.1 Capturing Face

Proposed system architecture is shown in Fig. 2. The picture of the client is caught utilizing a camera/webcam. When the transformation is finished, the picture is shipped off the classifier calculation, highlight extraction strategies are applied to remove the face from the casing of the web camera feed. These pictures will be utilized to prepare the classifier so that when a totally new and obscure arrangement of pictures is introduced to it. From preparation set, facial milestones are separated based on the pictures are obtained and return the directions of the new facial tourist spots that it distinguished. Major objective of this meeting is to catch pictures, so we

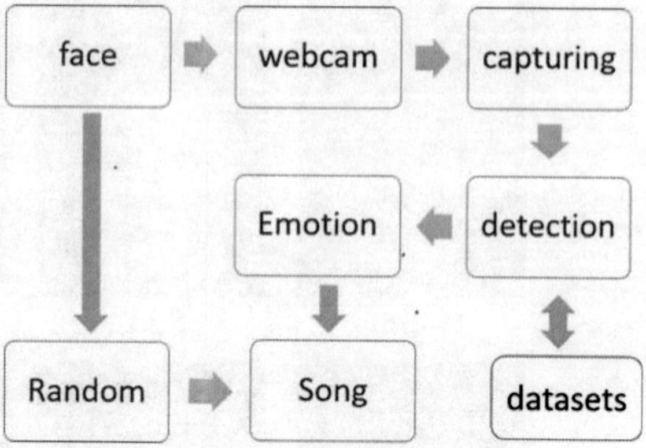

Fig. 2 System architecture

are utilizing a typical gadget, a webcam, or some other physiological gadget for that reason, and we are utilizing the PC vision library. This works with combination with different libraries that can likewise utilize NumPy, and it is fundamentally utilized for constant PC vision. At the point when the underlying system starts, it gets to the camera stream and catches around ten pictures for additional interaction and feeling discovery. Thus, in the start of this undertaking, to catch pictures and identify faces. We utilize a calculation that can take valid pictures and group them, and we really want a great deal of positive pictures that just hold back pictures with faces, as well as regrettable pictures that just hold back pictures without faces, with the goal that the classifier can be prepared. The ordered pictures are integrated into the model.

6.2 *Face Detection*

The face affirmation is seen as one of the most awe-inspiring technique for choosing a singular's perspective. This image taking care of structure as displayed in Fig. 3. is used for lessening the face space angles using the fundamental part analysis (PCA) system and a short time later it applies fishers straight discriminant (FDL) or the LDA procedure to get the component of the image credits, we especially use this since it expands the readiness cycle in the center between classes, this computation helps with taking care of for picture affirmation is done in fisherface while, matching appearances estimation we use least Euclidean it helps us with orchestrating the disposition that proposes the sensation of the client. Fisherface with OpenCV generally it essentially complement on the class unequivocal change matric subsequently, they do not acknowledge illustrative pictures as the subject and feeling is dominatingly shut by the model that the value surveyed from the connection can help us with thinking the demeanor of the user by differentiating the enlightening assortments that each tendency is differentiated and a few set aside pictures and scale gives the particular inclination so it can play the music considering the idea made by the structure by using the going with advances and strategies. Besides, it does not depend

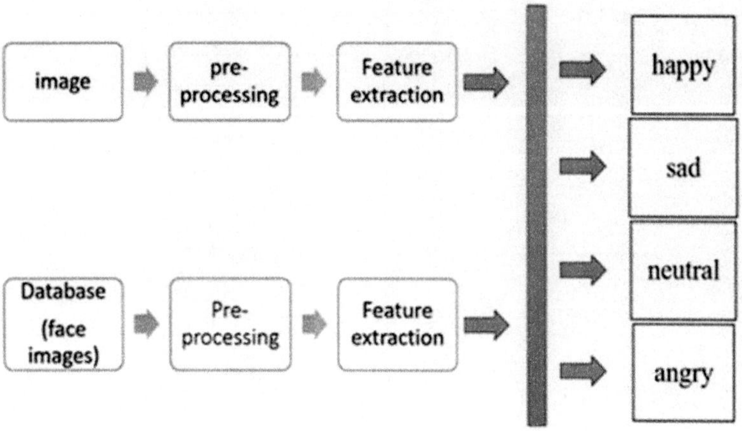

Fig. 3 Face detection

upon the other individual nuances like the other existing programming's straight gathering step in the face acknowledgment process. It helps with chipping away at the straight portrayal instead of SVM. It is for reducing the computational time so that the gathering framework will take and makes a prevalent distinguishing proof.

6.3 Classified Emotion

On the off chance that the face is effectively distinguished, the container will show up as a picture, overlaying the picture to extricate the face and further examining it in the subsequent stage as shown in Fig. 4. Recently removed pictures are handled by the capacity. The code removes the spatial face position from the face picture and uses a supporting calculation in light of the force upsides of the pixels listed at each point. It plays out an examination of the info information with the put away information, so you can foresee the class that contains feelings. If any of the four feelings of outrage, misery, impartiality, and delight are incorporated. Furthermore, feeling recognition resembles a log jam order, which is finished to dial back the wheelchair and safeguard the client from danger. It supports the disentanglement of straight grouping instead of SVM, which is for diminishing computational time with the goal that the characterization interaction requires some investment and produces a superior discover.

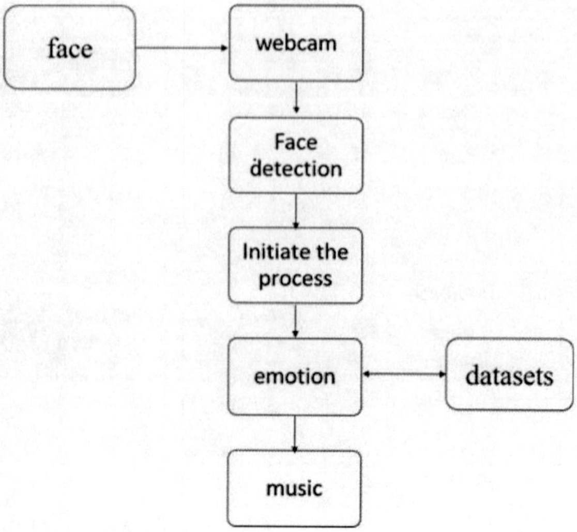

Fig. 4 Emotion classification

6.4 Music Recommendation System

The captured input image is taken from the webcam and used to capture the real-time image. There are seven main emotions here, as it is very difficult to define all emotions, and limited options can reduce compile time and result in more refined results. The value is transferred to run the web service. The song is played from the perceived emotions. Emotions are assigned to each song. When communicating emotions, each song and emotion is numbered, placed, and assigned to each song. However, many types of models are recommended for accuracy. It also uses fisherface, which includes PCA and LDA algorithms, so it is more accurate than other algorithms. The sound mechanism uses a winning sound and uses the commonly used Python library for in frequency and duration. The resulting mechanism is compared to a value that exists as a threshold and has alternative options other than emotion-based system, cue mode, and random mode. In cue mode, you can create playlists just like any other popular music software. The final mode is a random mode in which songs are selected randomly rather than in order, and it is also one of the treatments that can brighten the mood. At the same time, when you play the song based on the user's emotions, the emotions will be displayed in the form of emoji's with four different emotions. Each emotion is numbered and means music, and the emoji that each is recognized.

7 Experimental Results

SVM-The framework we attempted was sklearn's one up against one (OVO) SVM with a RBF bit. To lay out a pattern, we originally utilized crude, grayscale pixel values as the elements for the SVM. With this blend, we accomplished a precision that scarcely outperformed the strategy for picking the most widely recognized feeling like clockwork. We could prepare on the initial 5000 preparation models involving this strategy as preparing the model took excessively lengthy. Then, we had go at scaling the pixels so that each picture had a mean pixel worth of 0 and a fluctuation of 1 and involved the scaled pixel values as our new highlights. This superior our exactness to near 40%. In any case, we actually had issues preparing the model inside a sensible time span, which prevented us from utilizing more than the initial 5000 preparation models. We then, at that point, utilized principal component analysis (PCA) to endeavor to separate the main parts for our investigation. Diminishing the dimensionality of the pictures permitted us to utilize the full preparation set. We explored different avenues regarding various quantities of parts, going from 10 to 250. We observed that the best precision was with 25 parts, where we had the option to arrive at a test exactness of 43%. We then, at that point, read a paper that contended for the utilization of one versus all (OVA) SVM's in feeling acknowledgment and chose to utilize sklearn's direct bit SVM. We rehashed the method involved with attempting crude pixels and scaled pixels as elements and PCA to restrict the top parts, however, were simply ready to accomplish a most extreme dev set precision of 34.77%. We at last had a go at utilizing histogram oriented gradients (HOG) to portray the conveyance of slopes and edge bearings in the pictures prior to handling them. The motivation of applying HOG is to find different unique feelings and unmistakable slopes, especially around the mouth and eye regions. While HOG did not essentially assist with the precision of the OVO SVM, it knocks the exactness of the OVA classifier up to our most elevated SVM exactness of 45.94%. CNN consists of a collection of preprocessing strategies and model building techniques. These are applied to design a custom model set for accomplishing accuracy of 65.78% on the FER-2013 test set. For preprocessing, proposed work explored various avenues regarding focusing (i.e., reducing mean) and scaling information. It is supportive to take away the mean tracked down in the train conveyance from all sets prior to preparing/assessing. While scaling was useful with SVMs, it did not help in a CNN model. In addition to this, the research work also implemented data expansion, arbitrarily turn, shift, flip, crop, and sheer our preparation pictures. This yielded around a 10 p.p. expansion in correctness.

Table 1 Proposed system results by using FER_2013 dataset with different machine learning algorithms

Prototype	Parameters	Accuracy (training) (%)	Development (accuracy) (%)	Test set (accuracy) (%)
SVM	Pixels of scaled	42.26	37.50	41.15
SVM (OVO)	Scaled pixels PCA-25 Comps	56.70	43.74	43.18
Linear SVM (OVA)	HOG(4,4) pixels/cell	61.35	44.99	45.95
CNN	Dachapa Lly (He et al. 2008)	53.88	52.57	52.38
CNN	DeXpression	72.25	63.86	61.63
(CNN)-Linear	Custom-made	89.17	66.78	65.78

7.1 Results

Results on FER_2013 dataset

Checking out at precision as shown in Table 1. Versus test exactness, proposed generally low change shows that we stay away from overfitting as long as we stop around 100 epochs. Preparing past that reliably brings about overfitting. In spite of that, proposed emotion-based model experiences high predisposition; expanding the boundaries in the organization did not prompt seriously overfitting, insofar as it was neutralized by adding dropout.

For happy, surprised, neutral, and angry, proposed emotion-based recommendation system have a high level of precision as shown in Fig. 5. These are in accordance with our subjective discoveries from the continuous demo. Accuracy and review vary fundamentally by class. Classes like afraid have a low precision, whereas happy has a high recall. From the dataset, it is based on visual review, it is observed that several emotional combinations—surprise and fear, or disgust and surprise—are remarkably same, deluding even human workers (Fig. 6).

7.2 Results Analysis

A linear SVM with scaled pixel values to accomplish a training precision of 99.59%, and a test exactness of 98.47% on the CK + dataset (Fig. 7).

Real-Time Classification. The research proposed in the paper used OpenCV Haar fountains to distinguish and extricate an image district from a webcam video feed, then ordered it utilizing our CNN model. We identified it better to neither take away the preparation mean nor standardize the pixels in the distinguished face locale prior to characterizing it. Continuous characterization better uncovered our model's assets:

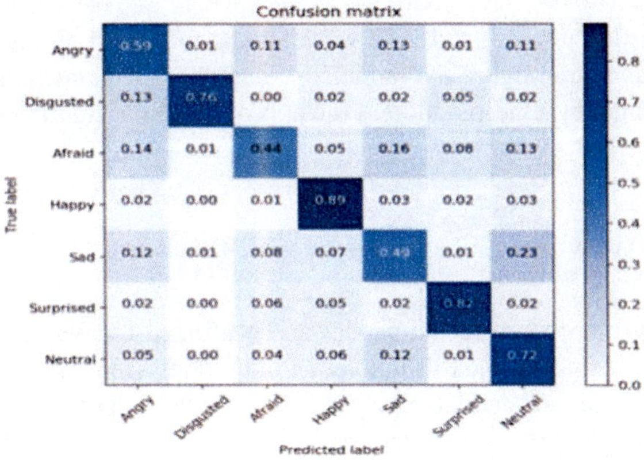

Fig. 5 Confusion matrix from FER-2013

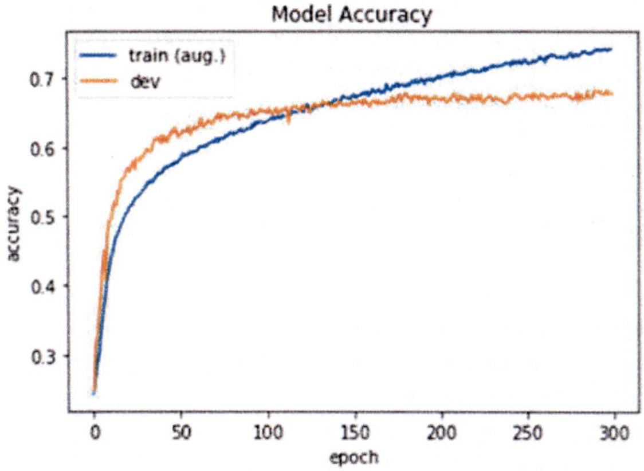

Fig. 6 Accuracy during training model

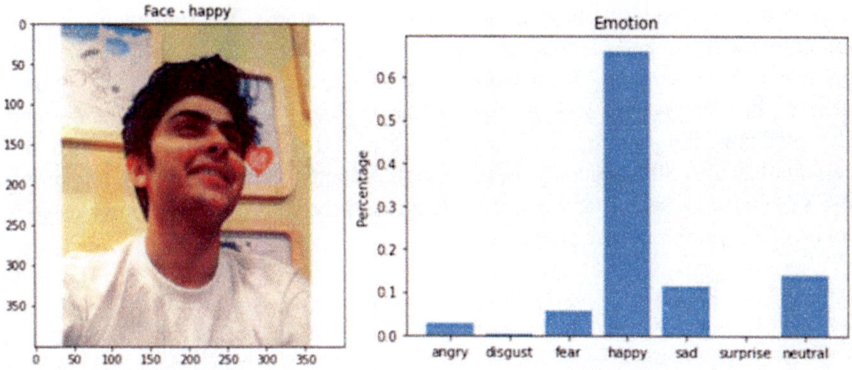

Fig. 7 Accuracy of emotion extraction

impartial, cheerful, amazed, and irate were for the most part all around identified. Brightening was a vital figure the model's presentation. This recommends that out preparing set may not honestly address the appropriation of lighting conditions.

8 Future Work

The music player proposed in the paper can be utilized locally, and all that these days is convenient and simple to ship, yet an individual's feelings can be caught by different wearable sensors, and it is easy to use rather than all the manual work that is conceivable. Physiological plethysmography sensors (PPG) and galvanic skin reaction (GSR) that would give us enough data to foresee the client's opinion precisely. Upgraded include framework will actually want to profit from the improved element framework furthermore, will require steady refreshing. Location is a technique for further developing melody auto-play. Looks are perceived utilizing the programming connection point present on the neighborhood PC. The elective strategy depends on the extra feelings that our framework does not perceive.

9 Conclusion

The music proposition model in this paper relies upon the sentiments got logically photos of the client. This adventure is intended to additionally foster collaboration between the music structure and the client, as music can assist with changing the client's temperament and can be a pressure reliever for certain individuals. Ongoing improvements demonstrate an expansive potential for fostering a feeling-based music proposal framework. Feeling acknowledgment has filled in importance in all pieces of life, and a calculation that can definitively orchestrate an individual's sentiments can be carried out, considering huge progression in the business. The framework had the option to catch a client's feeling effectively. Subsequently, the ongoing framework is a face acknowledgment framework that distinguishes feelings and plays music as needs be. The system was arranged using the facial achievement strategy and attempted in various circumstances to obtain results. It will in general be seen that the classifier has over 80% accuracy for most trials, which is very great for feeling order. It can likewise be seen that when tried progressively with the client, the classifier can precisely anticipate the client's looks.

References

G.H. He, J. Jin, Y. Xiong, B. Chen, W. Sun, L. Zhao, *Language Feature Mining for Music Emotion Classification via Supervised Learning from Lyrics*, in *International Symposium on Intelligence Computation and Applications* (Springer Berlin Heidelberg, Berlin, Heidelberg, 2008), pp. 426–435

X. Hu, J. Downie, When lyrics outperform audio for music mood classification: a feature analysis, in *ISMIR*, ed. by J. Downie, R.C. Veiltkamp (International Society for Music Information Retrieval, 2010), pp. 619–624

H.-C. Kwon, M. Kim, Lyrics-based emotion classification using feature selection by partial syntactic analysis, in *2011 IEEE 23rd International Conference on Tools with Artificial Intelligence (ICTAI 2011)* (2011), pp. 960–964

P. Lameere, E. Pampak, Social tags and music information retrieval, in *ISMIR 2008, 9th International Conference on Music Information Retrieval* (Dreexel University, Philladelphia, PA, USA, Sept. 14–18 2008), p. 24

J.H. Lee, X. Hu, Generating ground truth for music mood classification using mechanical turk, in *Proceedings of the 12th ACM/IEEE-CS Joint Conference on Digital Libraries, JCDL '12* (ACM, New York, NY, USA, 2012), pp. 129–138

J.A. Speck, E.M. Schmidtt, B.G. Morten, Y.E. Kim, A comparative study of collaborative vs. traditional musical mood annotatization, in *ISMIR*, ed. by A. Klapari, C. Leiider (University of Miaami, 2011), pp. 549–554

D. Tang, B. Qin, T. Liu, Deep learning for sentiment analysis: successful approaches and future challenges. Int. Rev. Data Min. Knowl. Disc. **5**(6), 292–303 (2015)

M. van Zaanen, P. Kanters, Automatic mood classification using tf*idf based on lyrics, in *ISMIR*, ed. by J.S. Downie, R.C. Veltkamp (International Society for Music Information Retrieval, 2010), pp. 75–80

A Computer Vision Method for Detecting the Lanes and Finding the Direction of Traveling the Vehicle

V. Sujatha, Y. Prasanthi, C. H. Pravallika, S. D. Jani Nasima, S. K. Ayesha Banu, and M. Sahithi

Abstract Road lane detection is very important in the driving assistance systems and it enhances the vehicle's active safe driving. This project is based upon the computer vision techniques. The boundaries are detected, and radius of curvature is displayed on the screen; apart from this, the direction of the vehicle is also detected from the video that is given as input. The lane is captured with the camera that is present on the top of the vehicle. By using the camera distortion coefficient, we have to rectify the distortion present in the set of images. Using this distortion coefficient converts the distorted image into undistorted image. To convert the undistorted image into binary image we use hsv (hue, saturation, color) thresholding. By using perspective transformation, the picture is converts to bird's eye from the binary undistorted picture. Later, we will find the left curvature and the right curvature and distance from the axis of the road and vehicle. So finally, we wrap the input with the detected lane and display the radius of curvature.

Keywords Camera calibration · Perspective transformation · Hough transform · Radius pf curvature

---

V. Sujatha · Y. Prasanthi · C. H. Pravallika · S. D. Jani Nasima (✉) · S. K. Ayesha Banu · M. Sahithi
CSE Department, Vignan's Nirula Institute of Technology and Science for Women, Guntur, India
e-mail: nasimasyed333@gmail.com

V. Sujatha
e-mail: sujathav.cse@gmail.com

Y. Prasanthi
e-mail: yarlagaddaprasanthi2001@gmail.com

C. H. Pravallika
e-mail: pravallikareddy581@gmail.com

S. K. Ayesha Banu
e-mail: shaikayeshabanu5a4@gmail.com

A. B. Reddy et al. (eds.), *Proceedings of Third International Conference on Advances in Computer Engineering and Communication Systems*, Lecture Notes in Networks and Systems 612, https://doi.org/10.1007/978-981-19-9228-5_31

1 Introduction

From the previous information in the (Conceicao et al. 2020) European country, they have revealed about 89% of the accidents are based on the irresponsibility, careless of the drivers. In recent years, many scientists are keeping their effects on the self-driving cars to prevent from the accidents happens on the road and also to get the safe driving, with the improved technology, large techniques were developing to warn the drivers whenever there is wrong path, many improved technologies are introduced in the 1989. This has become a sensitive issue nowaday's many people are damaging their life and their beloved people life because of these road accidents. In this, we give the image or the video of the road so that the computer vision techniques are introduced or applied on that video (Bailur et al. 2016). Before introducing this algorithm, we have simple lane detection which was introduced by Lin et al. (2020) He proposed a method that is useful for only the straight roads paths to detect the lanes using the canny edge detection and Hough transform this gives not better accuracy and it does not work in the curved lanes so to get the better accuracy Li et al. (2020) proposed CNN-based algorithm to detect the lanes but it also fails to train the data in several conditions and this takes more time so that there have used Open cv, computer vision techniques to do this project so in this paper we are going to introduce the computer-based technique to detect the lanes so that this can be used to detect the roads easily whenever there is also the curved path boundaries are detected and radius of curvature is displayed on the screen, apart from this the direction of the vehicle is also detected from the video that is given as input. The lane is captured with the camera that is present on the top of the vehicle. By using the camera distortion coefficient we have to rectify the distortion present in the set of images. Using this distortion coefficient converts the distorted image into undistorted image. To convert the undistorted image into binary image we use has (hue, saturation, color) thresholding. By using perspective transformation the picture is converted to bird's eye from the binary undistorted picture. So that the lane is detected. Finally, we have to find the radius of curvature.

2 Related Work

Conceicao et al. (2020) by using NMPC-based path control scheme building a novel visual line detecting system 2020 this system is suggested to detect and extract useful lane stroke arguments together to utilize them in a NMPC-based visual track. This method uses a visual system so that it can recognize and draw out the arguments from the lane stroke, directly from the image position, followed by various vision techniques such as robust computer vision to victual NMPC controllers. This method stops to alter a technique based on attached regions that are tough to get when displayed to worst weather conditions.

Lin et al. (2020) convolution neural network is lane detection that depends on instance segmentation that consists in edge-cloud computing. It achieves reliable identical solutions for strokes in various cases. This stroke recognition order is much better when compared to other lane-identifying models. Under critical conditions such as shadows and obstacles, it is very efficient to use. This approach fails because it projects the vanishing point to infinity whenever the ground is changed.

Li et al. (2020) lane localization and detection based on deep reinforcement learning. For accurate stroke localization and detection this method initiated into cursory lane detection prototype. In the case of curved lines, it is observed that it enhances the precision. This approach breakdown when pointing its last landmark.

Dhar et al. (2020) detecting and tracking road border lanes by using ML Approach. Beyond the road border lanes when the car leaps, we make use of ml methods to keep away from road issues and guarantees the safety driving it represents a novel approach. It gives high efficiency and precision. Identifying lanes has become more efficient but this not included the climatic conditions.

Brabandere et al. (2019) by using differentiable least-Squares fitting for end-to-end lane detection. In order to regress the stroke arguments and to guide a lane detector in a manner of end-to-end, it introduces this system. In least-squares fitting method, there is a chance to propagate back. It shows a huge effect on the shape of the curve when a small mistake in one argument value than a mistake with same magnitude in one more argument.

Besbes et al. (2017) by processing the image techniques detecting the Lane departure. For detecting lanes, it provides feature-based mechanisms. Main advantage is observation of lane boundary lines using artificial vision. There is a chance of incorrect observation regarding road strokes whenever the road strokes are altered. So that track must be enhanced.

Phung and Bouzerdoum (2020) by using filter method detecting a lane. Detecting lane as well as tracking systems are based on generalized annealed particle filter algorithm which merges annealed particle filter in the company of multiple images. For each frame, the cost of time of this filter is greatly reduces. Even though the time cost is reduced, it does not consider the road conditions.

Van Gool (2018) approach to end-to-end lane observation with instance segmentation it introduced to emit the lane detection complications as intense divisions issue-in which every lane produces its own instance. So, it can be instructed from end-to-end. It handles a variable no. of strokes and manage with whenever lane changes. The points which are close to vanishing point don't fit in a correct manner if the inclination of the ground plane is altered.

Prokhorov et al. (2016) finding marks of road boundaries and lanes with the help of intelligent vehicles. It converts to gray image from RGB road image and employed the flood-fill algorithm. Whenever there is unused area is detected then we can deduct it like other side of a way. This algorithm even has some disadvantages that include road strokes apart from white and also contour condition.

3 Proposed Work

Here in this the lane detection is detected by using thresholding perspective transformation and camera calibration. In existing system, there is only straight path detection. Here, we introduced the curved path detection, and it gives the radius of curvature. In previous systems, there is a hough transform which will combine pixel by pixel and form a straight line. But here is the camera calibration which we train the camera to take either convex or concave pictures. In existing system, there is no perspective transformation, but in the proposed work, there is a perspective transformation which will convert 3D image into 2D image (Fig. 1).

3.1 Algorithm for Proposed System

Step 1: For advanced lane detection, video is taken as input with.mp4 extension.

Step 2: Find the **camera calibration** and distorted coefficients. Camera calibration is the process of finding the parameters of a pinhole camera. These arguments are represented in a form of 3 × 4 matrixes known as camera matrix.

Radial distortion can be represented as:

$$A_{\text{distorted}} = a\left(1 + t_1 r^2 + t_2 r^4 + t_3 r^6\right) \tag{1}$$

$$B_{\text{distorted}} = b\left(1 + t_1 r^2 + t_2 r^4 + t_3 r^6\right) \tag{2}$$

Tangential distortion is represented as:

$$A_{\text{distorted}} = a + \left[2e_1 ab + e_2\left(r^2 + 2a^2\right)\right] \tag{3}$$

$$B_{\text{distorted}} = b + \left[e_1\left(r^2 + 2b^2\right) + 2e_2 ab\right] \tag{4}$$

From this, we find distortion coefficients = (t_1 $t2$ $e1$ $e2$ $t3$).
Let the focal length be (*x*, *y*) and the optical center are (c, d) then

$$\text{Camera matrix} = \begin{pmatrix} x & 0 & c \\ 0 & y & d \\ 0 & 0 & 1 \end{pmatrix}$$

Step 3: By using the function undistort() we can convert the distorted image into undistorted image.

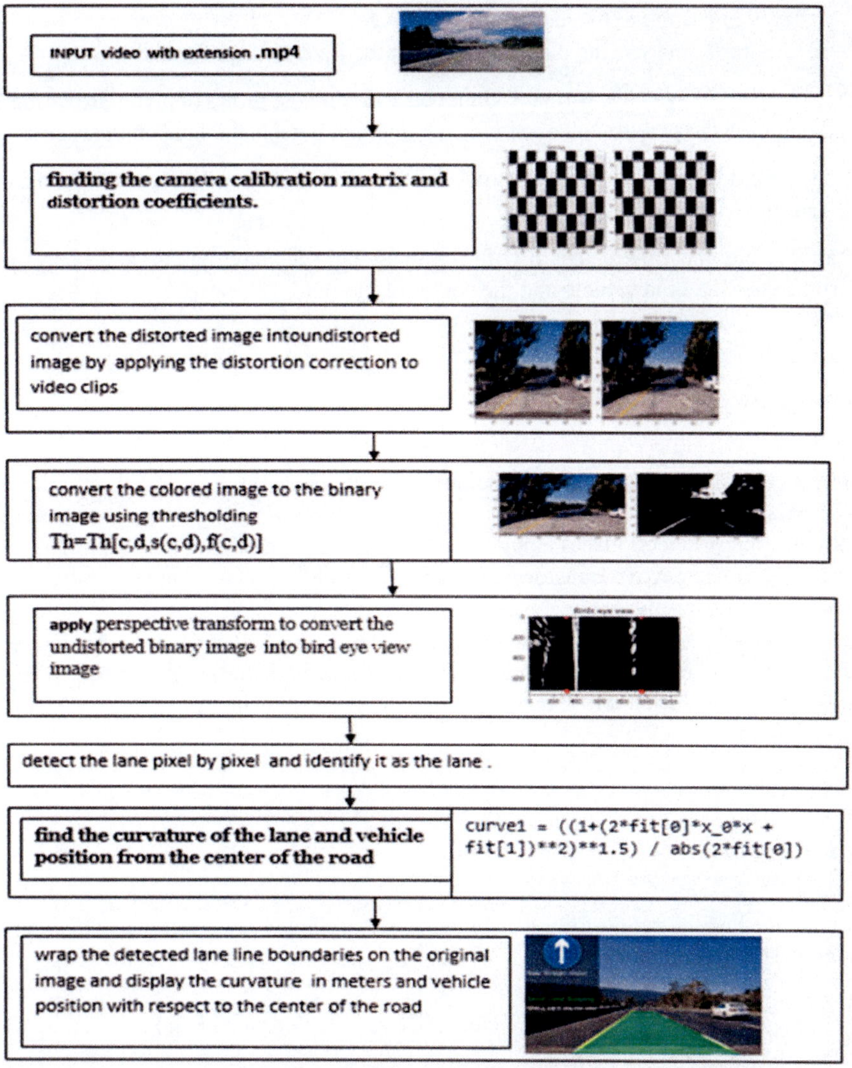

Fig. 1 System architecture of proposed computer vision technique

Step 4: Convert the colored undistorted image into binary image using the threshold() function. Th = Th[c, d, $s(c, d)$, $f(c, d)$] here (c, d) are coordinates and $s(c, d)$, $f(c, d)$ are the grayscale image pixels.

Step 5: Apply Perspective transformation to convert the 3D images into the 2D images.

The first angle is $\tan\theta = -a/f$.

The second angle is $\tan\theta = a/b$.

After comparing the two equations we can get,
$Y = -$ far/b this y is the distance at which the invert image formed.

Step 6: The bird eye view format is obtained after that the group of pixels are formed and combine the pixel by pixel and form a line that detects the lane on the road.

Step 7: **Find the curvature of the road lane and vehicle location from the axis of the road**

Step 8: Combine Line detected with the image and display the curvature and the distance between the vehicle and the center of the road.

4 Results

Here in this part, we are going to collate our proposed computer vision algorithm with existing approaches CNN by Li et al. (2020). The previous methods are having the canny edge detection and Hough transform to detect the lanes but this proposed system uses perspective transformation and thresholding.so we compare these two mechanisms using the accuracy, false positive rate, precision, range. So following are the results for the two mechanisms. We have taken 20 videos as the dataset and we apply the camera calibration to convert into undistorted image later apply perspective transformation on dataset to get bird's eye view image and find the radius of curvature.

4.1 Accuracy

It is defined as the ratio of number of precisely predicted observations to the total number of observations.

$$\text{Accuracy} = (\text{PT}_\text{R} + \text{PG}_\text{n})/(\text{PT}_\text{r} + \text{AT}_\text{r} + \text{AG}_\text{n} + \text{PG}_\text{n})$$

Our proposed model computer vision shown good results compared to existing approaches, though CNN also shown good results as the size of data increases our proposed system accuracy is close to 96% Because of low false positive rate (Fig. 2).

4.2 Precision

It is the ratio of number of precisely forecast positive observations to the total number of forecast positive observations.

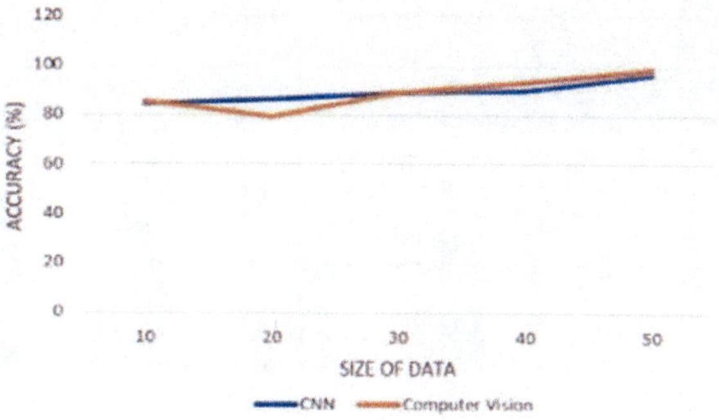

Fig. 2 Accuracy for proposed computer vision

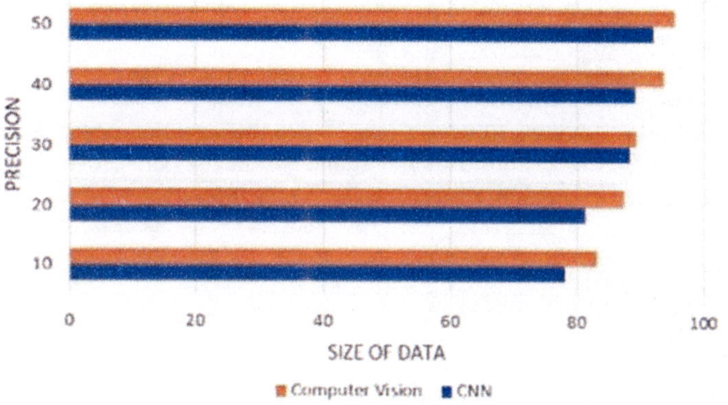

Fig. 3 Precision for proposed computer vision

$$\text{Precision} = \text{PT}_{\text{R}}/(\text{PTr} + \text{AT}_{\text{r}})$$

Here the results shows proposed approach has a better precession irrespective of size of the data compared with other approaches as the number of positive observations has no affect but for other two as the false observations increased, it effected the precession (Fig. 3).

4.3 *False Positive Rate*

It is deliberated as it is the ratio of the number of negative actions mistakenly classified as the positive and the total number of the actual negative actions occurred. It is also called as fall out or false alarm ratio. It refers to expectancy of false positive ratio.

$$\text{FPR} = F\,(U/k_0) \tag{5}$$

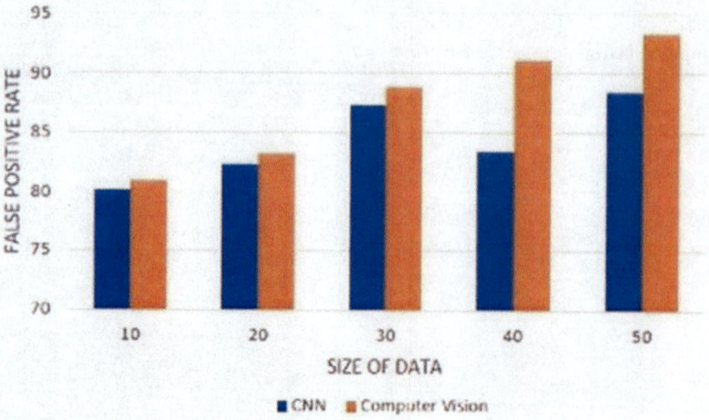

Fig. 4 False positive rates for proposed computer vision

Our proposed computer approach has shown better false positive rate when compared with CNN algorithm. It is not stable for all the data set so the computer vision technique is best with high false positive rate (Fig. 4).

4.4 Range

The range calculation is very important for any technique it is defined as the difference between the maximum no of events and the minimum number of events

$$\text{Range}(T) = \text{Maximum}(T) - \text{Minimum}(T) \tag{6}$$

Since the precision and range are good, false positive rate is also good, and consistent and it is not effected by size of the data. Overall, our proposed system has shown a very good results (Fig. 5).

5 Conclusion

This paper gives an advanced road lane detecting system mistreatment computer vision-based technologies these will accurately detect the lane boundaries of road. As we have applied different methods like perspective transform, thresholding, and preprocessing are applied on projected road lane detecting system. HLS thresholding gradient observe the lane stroke in binary footage effectively. For recognizing right and left lane on road we use window search. The cropping method performed entirely at the extreme region that consists the lane strokes. From the exploratory outcome, it is over the approach detects the lanes efficiently with any environment of the setting.

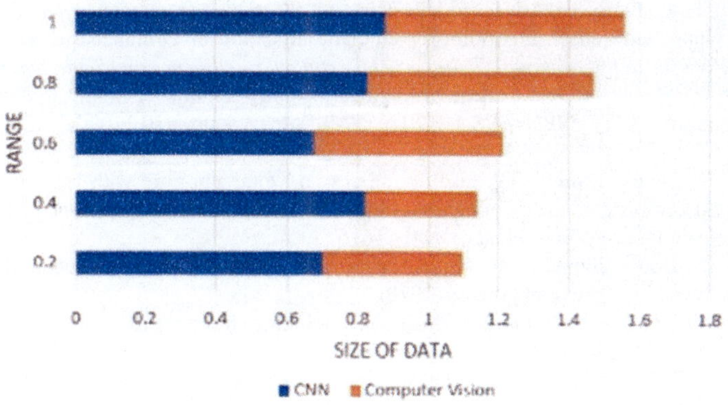

Fig. 5 Range for proposed computer vision

The approach is performed to any road consisting marked lines together enforced with the fixed system for the assistance of modern driver facilitate method and the blind people for map reading to remain them in correct pathway. In upcoming period of time the method with hardware implementation square measure progressing to be developed which can capture the photographs from the period of time state of affairs and observe the lanes supported the projected technique likewise give rise to a warning for the particular persons.

References

S.V. Bailur, A. Gurghian, V.N. M.K.J. Carey, T. Koduri, DeepLanes: estimating END-TO-END position of the lane using neural networksa (2016), pp. 38–45

K. Besbes, S. Lahouar, M. Marzougui, M. Hergli, J. Baili, J. Subash Chandra Bose, A. Sboui, Using Image Processing Techniques Lane Departures Detection (IEEE, 2017)

B. De Brabandere, M. Proesmans, D. Neven, W. VûÛÛan Gansbeke, L. Van Gool, Through differentiable least squares fitting end-to-end lane detection, in *IEEE International Conference on Computer Vision Workshop (ICCVW) 2019* (2019)

A.G. Conceicao, L.J.P.B. Franco, T.T. Ribeiro, A novel visual lane line detection system for a NMPC-based path following control scheme (2020)

P. Dhar, S.K. Satti, KS. Devi, P. Srinivasan, *Detecting and Tracking Road Boundering Lanes by Using Machine learning* (ICT express, 2020)

M.M Islam, M.R. Haque, H. Iqbal, K.S. Alam, A computer vision based lane detection approach. Int. J. Image Graph. Sig. Process. **3**, 27–34 (2019)

H. Jiang, Y. Yue, Q. Wang, L. Chen, Q. Dai, Q. Zou, By using deep neural networks detecting robust lane detection from continuous driving scenes (IEEE, 2019)

S. Li, H. Shen, X. Miao, On-Board lane detection system for intelligent vecicle based on monocular vision. **5**(4) (2012)

H. Lin, J. Wang, W. Wang, Convolution Neural Network based Lane detection with instance segmentation in edge cloud computing. J. Cloud Comput. Adv. Syst. Appl. **9**, 27 (2020)

S.-Y. Lo, P.-R. Chen, S.-W. Chan, H.-M. Hang, J.-J. Lin, Deep Learning mechanism for efficient marking detection of road lane

W.-S. Moon, H.-C. J.-H. Lee, Advanced lane detecting algorithm for unmanned vehicle (2013)
K.J. Naik, Optimized genetic algorithm for efficient allocation of virtualized resource in cloud (OGA_EAVRC), in *International Conference on Artificial Intelligence and Smart Systems (ICAIS 2021), 25–27 Mar-2021* (JCT College of Engineering and Technology, Coimbatore (T.N), India (IEEE Explore, 2021). ISBN: 978-1-7281-9537-7. https://doi.org/10.1109/ICAIS50930.2021.9395951
K.J. Naik, D.H. Naik, Minimizing deadline misses and total run-time with load balancing for a connected car systems in fog computing. Scalable Comput. Pract. Experience **21**(1), 73–84 (2020). https://doi.org/10.12694:/scpe.v21i1.1616
S.L. Phung, A. Bouzerdoum, T. A. Nguyen Pedestrian lane detection in unstructured scenes by hybrid deep learning-gaussian process (2020)
D. Prokhorov, J. Li, X. Mei, D. Tao, In traffic scene predicting structural and lane detection by deep neural network, in *2016 IEEE*
P. Singh, R. Singh, Integrated lane colorization using hough transformation and bilateral filter. **2**(10) (2013)
M.T. Smith, A. Borkar, M. Hayes, A novel lane detection system with efficient ground truth generation. IEEE Trans. Intell. Transp. Syst. **13**(1) (2012)
C.-C. Tseng, B.-S. Jeng, H.-Y. Cheng, Using geometry information and modified hough transform a algorithm for a lane detection, in *18th IPPR Conference on Computer Vision, Graphics and Image Processing (CVGIP 2005)*
L. Van Gool, S. Georgoulis, D. Neven, M. Proesmans, B. De Brabandere, K.U. Leuven, An instance segmentation approach towards end-to-end lane detection (2018)
X. Li, Z. Zhao, Q. Wang, Lane detection and localization based on deep reinforcement. Neuro Comput. **413**, 328–338 (2020)
S. Zhou, J. Xi, G. Xiong, Y. Jiang, J. Gong, H. Chen, A novel lane detection based on geometrical model and gabor filter, in *2010 IEEE Intelligent Vehicles Symposium University of California June 21–24, 2010*

Ensembled Machine Learning Techniques for DDoS Detection in SDN

P. Tarakanadha Reddy, P. V. Shalini, and V. Radha

Abstract Software Defined Networking(SDN) focuses on overcoming the drawbacks of traditional networks and offers the advantage of flexibility in managing the networks. On the other hand, this new paradigm makes networks susceptible to attacks. DDoS is one of those significant attacks. DDoS makes resources unavailable to legitimate users, and one of the mechanisms that attackers follow is the TCP-SYN flood to launch the DDoS attack. The TCP SYN flood attack takes advantage of the three-way handshake to exhaust the web server's resources. We proposed an approach to detect DDoS attacks in SDN based on an ensemble technique.Our proposed approach uses stacking model, combining bagging and boosting models as ensembled techniques. we implemented our proposed approach on dataset. We have generated our own dataset containing the required features. We show that our proposed approach gives better accuracy than existing models in the literature. We validated our proposed approach on both generated dataset and existing dataset.

Keywords Software defined networks · Ddos attacks · Ensembled techniques · Stacking model

1 Introduction

Distributed Denial of Service (DDoS) in Software Defined Networking (SDN) has become an emergent serious cyber security issue. DDoS attacks aim to drain the victim's resources by preventing legitimate users from accessing them. Even though

P. Tarakanadha Reddy
School of Computer and Information Sciences, University of Hyderabad, Hyderabad, India

P. V. Shalini
Department of Computer Science and Engineering, National Institute of Technology, Warangal, India

P. Tarakanadha Reddy (✉) · P. V. Shalini · V. Radha
Centre for Cloud Computing, Institute for Development and Research in Banking Technology, Hyderabad, India
e-mail: potturitarak@gmail.com

A. B. Reddy et al. (eds.), *Proceedings of Third International Conference on Advances in Computer Engineering and Communication Systems*, Lecture Notes in Networks and Systems 612, https://doi.org/10.1007/978-981-19-9228-5_32

SDN is a viable technology with a bright future, it is susceptible to DDoS attacks. Some of the key reasons that DDoS is hard to detect are: the attack's spreading nature, the attack's changeable time trend, the frequency of the attack, the use of spoofed IP addresses, and the complexity in recognizing traffic features (Bhushan and Gupta 2019).

In traditional networks, various strategies have been employed to mitigate the impact of DDoS attacks (Jarraya et al. 2014). Packet analysis is resource-intensive in traditional networks. As a result, sampling approaches were used to verify the packets. To detect and mitigate DDoS attacks, there are statistical, machine learning, and deep learning methodologies (Wang et al. 2015).

Many studies have used OpenFlow's flow capabilities to identify attacks in SDN, especially DDoS (Dimolianis 2019). The flow level information captured from all the switches is used to identify the existence of attacks in this study. Flow level information can be easily retrieved because SDN is built on a flow-based architecture. Flow analysis is more resource-efficient than packet analysis. TCP SYN flood consumes resources on the target web server by exploiting a portion of the regular TCP three-way handshake.

2 Related Work

In this section, we discuss schemes on security of SDN in recent years.

Banitalebi et al. (2021) introduced a approach based on both statistical and machine learning techniques for classification. They used the statistical entropy value of the collected features for classification.

In the Scheme (Ye et al. 2018) Ye, Jin presented a classification approach based on SVM for identifying DDoS attack traces in the network. The authors have generated their own dataset using both attack and benign traffic in mininet environment. The dataset which is generated for their study is too small. They got an accuracy of 95.24%.

Chen et al. (2018) proposed a boosting approach called XGBoost to detect DDoS attacks. They worked on cloud environment . Their scheme (Chen et al. 2018) got an accuracy of 98.53%

For detecting the TCP SYN flood attacks, Tuan et al. (2020) applied KNN and XGBoost. They validated their model on CAIDA 2007 data set and a testbed environment. This scheme obtained an accuracy of 98%.

Perez-Diaz et al. (2020) used six machine learning algorithms with random and grid search hyperparameter optimization techniques to detect DDoS attacks of low rate. They implemented their model on CIC DoS 2017 dataset and obtained an accuracy of 95%.

Deepa et al. (2019) used ensembled techniques to detect DDoS attacks. Their scheme SVM combined with the self-organizing maps got an accuracy of 98.12%.

Table 1 Previous studies

Previous works	Features considered
Braga et al. (2010) Self organizing maps	Average of packets, average of bytes, pair flows percentage, vverage of duration of all the flows, different ports growth, single flows growth
Jose et al. (2021) SVM classification	Count of flows, average byte count, average packet count, average duration, entropy of protocol, entropy of source IP, entropy of destination IP

Jose et al. (2021) used all the existing classification models on their generated custom data set. Out of all those classification models, SVM achieved an accuracy of 99.73%.

Braga et al. (2010) used self-organizing maps by extracting six features for detecting DDoS attacks. These six features are: Average Flow Duration, Average Packet Count, Pair Flows Percentage, Different Ports Growth, Single flows Growth, and Average Byte Count. Their scheme got an accuracy of 98.61%.

By performing a literature review, we know that most of the works are carried out on the datasets with existing features. Schemes (Braga et al. 2010, Jose et al. 2021) show that the extracted features improve the accuracy over the existing features.

Our proposed work is focused on generating the dataset having the essential six features from the Table. 1 by using a testbed environment. Apart from this, we proposed using the stacking based ensembled model that combines different models at the base and high levels to improve the accuracy of the classifying the attack. Our proposed approach is validated with "DDoS attack SDN Dataset".

3 Proposed Work

In this section, we discuss the proposed work. We created our own network topologies using mininet to run various scenarios. Pox is one of the open flow controllers used in Software Defined Networking to control the behavior of the networks. The controller connects to the OpenFlow switches in the network. These switches will have flow tables. A flow table is the collection of flow entries. OpenFlow switch acts like the forwarding medium of the packets based on the entries in the flow table. Packet in the network arrives first at the switch through one of the ports. After packet arrival, it will be matched with entries in the flow table. If a match to the incoming packet is found, then based on the actions associated with the flow entry, the packet will be forwarded. If a match to the incoming packet is not found, then that packet will be sent to the controller. These packets are called flow-initiations. Controller on receiving the flow-initiations, will associate the actions to be taken and add a flow entry to the switch's flow table as a response.

3.1 Feature Extraction from the Flows Collected

We collect the flows coming to the switches in our network. Flow is defined as the group of packets that are having the same src_ip, dst_ip, src_port, dst_port, and protocol. From these flows, we extract the required features and store them to create a dataset. Considering the previous works mentioned in Table. 1, the following six features: Average Flow Duration, Average Packet Count, Average Byte Count, Pair Flows Percentage, Different Ports Growth, and Flows Count (Braga et al. 2010, Jose et al. 2021) are taken in our proposed approach. The six features used in our proposed approach are,

Average Packet Count: During the benign traffic, the flow of packets is high as nodes are intended to communicate. But during attack traffic, flow of packets is low as the attacker's goal is to exhaust as many ports as possible by sending a very minimal count of packets to the web server on a network. Here, finding the median value will be more effective than taking up the average when the flows have a high number of packets.

Here, C represents the number of packets of each flow and N represents the total count of flows in a network.

Average Byte Count: During attack traffic, attackers send minimal bytes to the web server to use the port space. The average byte count is calculated as shown in the Fig. 1, where C represents the total number of bytes per each flow and N represents the total count of flows in a network.

Average Flow Duration: In the same way, we can calculate the median value to the duration of all flows. Flow duration will be higher if there is a benign traffic, and it will be lower if there is an attack traffic as flows from other hosts in a network increases. The average flow duration is calculated as shown in the Fig. 1, where C represents the duration of flow, and N represents the total flows in a network.

Pair Flows Percentage: Pair flows percentage helps to determine the count of pair flows during an interval in a network. A flow is said to be a pair flow only when the following three conditions satisfy,

1. Source IP of Flow A is the same as Destination IP of Flow B,
2. Destination IP of Flow A is the same as Source IP of Flow B,
3. Both Flow A and Flow B have the same protocol used for communication.

There will be packets from spoofed IP's during the attack, which decreases the percentage of pair flows. Pair flows percentage is calculated using Eq. 1.

Fig. 1 Average packet count

$$\text{Average Packet Count} = \begin{cases} \frac{C(\frac{N}{2})+C(\frac{N+1}{2})}{2} & \text{if N is even} \\ C(\frac{N+1}{2}) & \text{otherwise;} \end{cases}$$

$$\text{Pair Flows Percentage} = \frac{(2 * \text{Count of Pair Flows})}{\text{Count of Flows}} \quad (1)$$

Different Ports Growth: During attack traffic, the attackers target to exhaust the port space present on the web server, which increases the count of ports being used during the attack time. Calculating the different port's growth can be done using Eq. 2.

$$\text{Different Ports Growth} = \frac{\text{Number of Ports}}{\text{Interval}} \quad (2)$$

Flows Count: During attack traffic, count of flows increases as the flows will be generated with packets from spoofed IP's. The number of flows taken from the network during attack will give the count of flows in a interval.

3.2 Classification

Here, we proposed a stacking based ensembled approach to improve the accuracy in classifying the DDoS attacks. Stacking based ensembled approach contains base level and high level and it's architecture is given in Fig. 2. Ensembled approaches are of three types, they are Averaging, Boosting, and Bagging methods. Averaging is also known as stacking and it typically refers to the combination of different models and uses a meta-classifier to merge many basic classification models. This method uses a combination of traditional classifiers to create a generic machine learning model.

The initial stage in stacking is to aggregate every model's output in a base level which results into a new dataset. This dataset comprises the prediction of the every model in base level along with correct classification for every instance of the original dataset. The new dataset that is generated from base level models will be provided as input to the classifier in the high level. The architectures of the base and high-level classifiers are necessary for stacking-based ensemble approach. In this study, We taken the base classifiers as Adaboost and Extra Tree classifier, while Logistic Regression taken as high-level classifier.

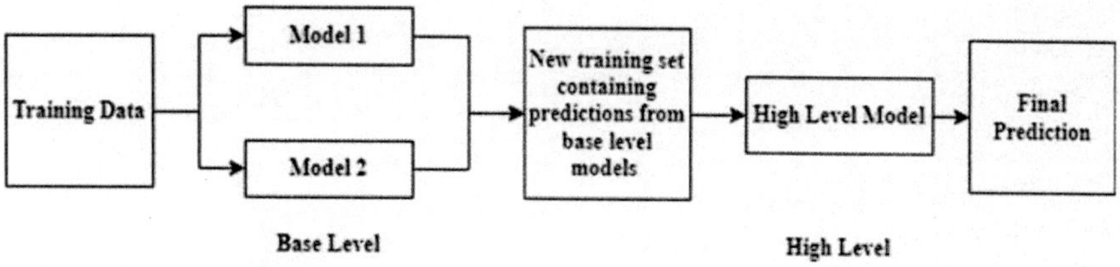

Fig. 2 Architecture of stacking model

4 Experimental Setup

The dataset generation work has been carried out on the Lenovo Thinkpad E470, having 8 GB RAM and a 64-bit processor with Ubuntu operating system.

4.1 Dataset Generation

A detailed description of the steps followed for generating dataset are given below

Topology Creation: Mininet is a emulator which builds network that is virtual in nature and similar to real world network topologies. This virtual network comprises hosts, switches, controllers, and connections. Mininet is used for R&D, learning, modeling, testing, debugging, and other activities related to the network. As part of our proposed approach, we created a few custom topologies and taken a few default topologies from Mininet. The topologies created are given in Table 2.

Benign (Normal) Traffic Generation: Curl command is a command-line tool that transfers the data from/to the server. Curl command used in a script file generates benign traffic without user involvement. The script file sends requests from hosts to the web server at regular time intervals.

Collecting Flows of Benign Traffic: The flows present in the switches of network can be collected using commands. Using the commands, flows are collected at a regular interval between 60 and 100 s (The hard time out to the flows is 60 s).

Table 2 Details of the topologies created

Topology	No of switches	No of hosts	Web server
1 (Custom)	3	5	H5
2 (Linear)	3	12	H6
3(Custom)	4	41	H29
4 (Custom)	7	61	H42
5 (Single)	1	15	H12
6 (Custom)	7	80	H72
7 (Custom)	3	5	H5
8 (Custom)	13	36	H36
9 (Linear)	3	6	H6
10 (Custom)	9	120	H98
11 (Custom)	12	160	H122
12 (Minimal)	1	2	H1

Feature Extraction from Collected Flows: After collecting the flows from the switches. The steps to extract the values of the six features are given in Sect. 3.1. As this is benign traffic, label "0" is added and stored feature values on to a csv file.

DDoS Traffic Generation: Scapy is the tool used for packet generation, and it is one of the powerful packet manipulation program written in python. As scapy can create and send packets to the web server, it is considered for our attack generation along with Hping3. Hping3 is used to flood the victim and manages the amount, size and, fragmentation of packets. By considering size and amount, the attack can be performed from any host to the web server on a network. Scapy generates TCP packets with the source IP as random from the topology and the destination as the web server. As the server resources were being used to address the connection requests, legitimate users can not communicate with the server. During the attack, benign traffic has been generated and sent to the web server. The time taken to get acknowledgment from server to legitimate users is more when compared to the normal time and even some times, legitimate users are unable to get the acknowledgement.

Collecting Flows of DDoS Traffic: The flows present in the switches of network can be collected using commands. Using the commands, flows are collected at a regular interval between 60 and 100 s (The hard time out to the flows is 60 s).

Feature Extraction from Collected Flows: After collecting the flows from the switches. The steps to extract the values of the six features are given in Sect. 3.1. As this is benign traffic, label "1" is added and stored feature values on to a csv file.

The dataset that is generated has 24,025 records. These records are extracted from an average of 5 lakh flows roughly. Out of 24,025 records, 13,753 records indicate benign traffic, and 10,272 records indicate attack traffic. With this distribution of records, It seems to be that the generated dataset is balanced. As the next step, generated dataset is used to train a model based on our proposed approach for attack detection.

4.2 Classification

During classification, partition of the dataset was 80% of the training set and 20% of the testing set. Classification algorithms like K Nearest Neighbor, Logistic Regression, Support Vector Machine, Gaussian Naive Bayes, and Linear Discriminant analysis were performed on the dataset. At last, one of model from bagging and boosting were taken as the base classifiers for the proposed stacking based ensembled approach, along with Logistic regression as the higher level classifier. We are combining base and higher-level classifiers in our proposed approach. This proposed approach evaluated with generated custom dataset. "DDoS attack SDN dataset" was evaluated with the proposed approach. This dataset contain 22 features. Chi-Square, PCA, and Mutual Information(MI) were used to select the best features and feature subset of size (12) is constructed.

5 Results and Analysis

In this section, we discussed the results and analysis of our proposed approach. Result and analysis was done in two sections. In the first Sect. 5.1, result analysis was done with generated dataset. In the second Sect. 5.2, result analysis was done with "DDoS attack SDN dataset". The training and testing of the proposed approach carried out on Google Colab, having an Intel Xeon processor with a frequency of 2.20 GHz and 13 GB of RAM. For measuring the model's performance, performance metrics like accuracy, recall, precision, and f_1 score were used.

5.1 *Result Analysis with Generated Dataset*

The dataset with features Average Packet Count, Average Byte Count, Pair Flows Percentage and Flows Count was applied with the proposed approach. The performance metrics of the proposed approach were given in Table 3. Out of all the techniques, GNB had less accuracy of 81.49%. KNN, SVM, LR, LDA were having accuracy of 99.93%, 99.61%, 99.25%, 99.09% respectively. Our Proposed approach (Base Learners—AdaBoost, Extra Tree & High level Learner—Logistic Regression) was having higher accuracy of 99.98% over the other techniques.

5.2 *Result Analysis with "DDoS Attack SDN Dataset"*

"DDoS attack SDN Dataset" was an emulated dataset. The performance metrics of the proposed approach were given in Table 4. The proposed approach with Mutual Information and Chi-Square provided an accuracy of 99.2%. By selecting all the features, the proposed approach offered an accuracy of 99.83%. Using the feature selection methods, the proposed approach given the similar results by reducing the dimensionality.

Table 3 Classifiers' performance by considering all the features

Classifier	Accuracy (%)	Recall (%)	Precision (%)	F_1 Score(%)
SVM	99.611	99.187	99.901	99.543
GNB	81.492	69.713	99.607	82.021
KNN	99.93	99.901	99.934	99.918
LDA	99.098	98.57	99.312	98.94
Logistic regression	99.25	98.733	99.509	99.119
Proposed approach	99.986	100	99.967	99.983

Table 4 Our proposed approach performance with DDoS attack SDN Dataset

Method	Accuracy(%)	Recall(%)	Precision (%)	F1 Score(%)
All features	**99.837**	**99.983**	**99.593**	**99.837**
Mutual information	99.20	99.923	97.973	98.938
Chi-Square	99.242	99.629	98.396	99.009
PCA	97.875	96.182	98.38	97.269

Bold values shows significant improvement compared to other rows in the table

6 Conclusion and Future Work

DDoS (Distributed Denial of Service) had become an emergent serious cyber security concern. Even while Software Defined Networking (SDN) makes network management more manageable, it was /prone/ subject to DDoS attacks. In traditional networks, a different range of approaches had been used to detect and mitigate attacks. In this study, flow characteristics collected from network were used to detect DDoS attacks. In our work, flow collection was done, followed by feature extraction, made an emulated dataset. This dataset was used to train the classifier. The classifier taken up for this work was based on stacking based ensembled approach. The proposed approach was an effective strategy that integrated different learning algorithms at the base level and high level with improved attack detection accuracy. The proposed approach combined several classifiers into a single composite model that was more accurate in detection. The goal of this research was to see that the proposed ensembled approach would outperform the classification models in terms of accuracy by employing extracted features from the flows. By seeing the results of the classifiers, there was an increase in the accuracy of the proposed ensembled approach when compared to the other techniques used. As part of future work, We want to include other types of mechanisms used to launch DDoS attacks in our dataset, increasing the dataset size which will help to ensure that the proposed approach performs as expected.

References

A. Banitalebi Dehkordi, M.R. Soltanaghaei, F.Z. Boroujeni, The DDoS attacks detection through machine learning and statistical methods in SDN. J. Supercomputing **77**(3), 2383–2415 (2021)

K. Bhushan, B.B. Gupta, Distributed denial of service (DDoS) attack mitigation in software defined network (SDN)-based cloud computing environment. J. Ambient Intell. Humanized Comput. **10**(5), 1985–1997 (2019)

R. Braga, E. Mota, A. Passito, Lightweight DDoS flooding attack detection using NOX/OpenFlow, in *IEEE Local Computer Network Conference* (IEEE, 2010), pp. 408-415

Z. Chen, et al., XGBoost classifier for DDoS attack detection and analysis in SDN-based cloud, in *IEEE international conference on big data and smart computing (bigcomp)* (IEEE, 2018), pp. 251–256

V. Deepa, K. Muthamil Sudar, P. Deepalakshmi, Design of ensemble learning methods for DDos detection in SDN environment, in *2019 International Conference on Vision Towards Emerging Trends in Communication and Networking (ViTECoN)* (IEEE, 2019), pp. 1–6

M. Dimolianis, Mitigation of multi-vector network attacks via orchestration of distributed rule placement, in *IFIP/IEEE Symposium on Integrated Network and Service Management (IM)* (IEEE, 2019), pp. 162–170

Y. Jarraya, T. Madi, M. Debbabi, A survey and a layered taxonomy of software-defined networking. IEEE Commun. Surv. Tutorials **16**(4), 1955–1980 (2014)

A.S. Jose, L.R. Nair, V. Paul, Towards detecting flooding DDOS attacks over software defined networks using machine learning techniques. Revista Geintec-Gestao Inovacao E Tecnologias **11**(4), 3837–3865 (2021)

J.A. Perez-Diaz et al., A flexible SDN-based architecture for identifying and mitigating low-rate DDoS attacks using machine learning. IEEE Access **8**, 155859–155872 (2020)

N.N. Tuan et al., A DDoS attack mitigation scheme in ISP networks using machine learning based on SDN. Electronics **9**(3), 413 (2020)

R. Wang, Z. Jia, J. Lei, An entropy-based distributed DDoS detection mechanism in software-defined networking, in *IEEE Trustcom/BigDataSE/ISPA*, vol. 1. (IEEE, 2015), pp. 310–317

J. Ye et al., A DDoS attack detection method based on SVM in software defined network, in *Security and Communication Networks 2018* (2018)

Classification of Cotton Leaf Diseases Using Transfer Learning-DenseNet-121

B. Arathi and Uma N. Dulhare

Abstract Farmers growing cotton will have a great help if the cotton yield is predicted accurately and helps in making decisions such as crop insurance, how much to store, investments, requirement of fertilizers, water, etc. Generally, yield is measured by means of sample surveys using destructive sampling of cotton fields and will take enormous time, cost for the labor is high. As we know that these cotton plants are affected by various bacterial and fungal diseases based on the climate conditions resulting in the decline of cotton productivity. Plants are prone to numerous diseases. In the cotton plants, the mostly affected part is the leaf that damages the plant resulting in the damage of the entire crop. In order to detect the diseases of the cotton leaf, image processing and machine learning techniques are employed. In the existing work DEEP learning technique, CNN is employed for feature extraction which is used to detect plant diseases. There is an issue with the accuracy of these traditional CNN algorithms. The Experimental results achieved showed that the proposed model i.e., DenseNet-121 pre-trained Model is capable of classifying different leaf images in the dataset with higher classification accuracy of 91%. This transfer learning technique uses ImageNet weights to detect the diseases of cotton plant accurately. The abstract should summarize the contents of the paper in short terms, i.e. 150–250 words.

Keywords Cotton leaf disease · DenseNet-121 · Convolutional neural network

B. Arathi (✉)
Computer Science and Engineering, Kamala Institute of Technology and Science, Huzurabad, Karimnagar, Telangana, India
e-mail: artibairi@kitssingapuram.ac.in

U. N. Dulhare
Computer Science and Artificial Intelligence, Muffakham Jah College of Engineering and Technology, Hyderabad, Telangana, India

A. B. Reddy et al. (eds.), *Proceedings of Third International Conference on Advances in Computer Engineering and Communication Systems*, Lecture Notes in Networks and Systems 612, https://doi.org/10.1007/978-981-19-9228-5_33

1 Introduction

Among all the fiber crops cotton accounts for 35% of the global total fiber and can be used in the production of biofuel, staple, and as a raw material for manufacturing clothes (Satraj Sohrab et al. 2014). India's economy is affected by its yield (Ashourloo et al. 2016). It is the main crop that is available economically but it is affected by various pathogens that limits its production. Our country ranks first in the cotton yield and there is a necessity to detect the diseases priorily to still further enhance the productivity (Dulhare and Gouse 2022).

According to yield, India tops rank one in area wise and rank three in the production. For India, Cotton is very vital in its economy and in its industrial activities (Bashish et al. 2011).

Temperature and humidity are vital factors which strengthens the yield of the crop and the root exudates of it improve the nutrients of the soil (Tijare et al. 2015). The plant growth and yield are affected by the plant illness and have an influence on agriculture both socio-biological and financial aspects (Zheng et al. 2019).

At various stages of plant growth, plant infections influence its development, collection, acquire, and analyze of cotton diseases manually may not be efficient and identifying the diseases becomes not that easy (Lv et al. 2012). These diseases occur because of various microorganisms, nematodes, and other agents.

Keeping all this in mind, effective methods for the management of diseases is necessary. Earlier work shows that these diseases will have an adverse affect on the country's economy.

Various cotton leaf diseases, symptoms and methods. Imbalances are caused in cotton plants because of low nutrients, environmental stress, and chemical factors which leads to low productivity of cotton. Diseases like verticillium wilt, cotton leaf curl limits the growth of cotton (Kalbande and Pati 2016) (Table 1).

2 Literature Survey

Plant diseases can be detected and classified by various algorithms and accordingly research is taking place in this area and few of them are working on the classification of cotton diseases and increase the yield of cotton. Following are the work carried out by various researchers in this area.

Rothe and Kshirsagar (2014) made use of picture preparing strategies to find the diseases and extracted the main portions of the diseased leaf images by using image enhancement and segmentation.

Li et al. (2011) developed an image processing procedure alongside edge detection that takes computerized pictures and different strategies are applied to remove RGB highlights which helps in identifying the diseased part.

Table 1 Various cotton leaf diseases, symptoms and methods

Disease Name	Image	Symptoms	Suitable methods
Angular leaf spot or black arm disease		• Small spots appear below the cotyledons, which may also dry and fall off • Elongated dark brown lesions appears on the steam and branches	Principal component analysis K-nearest neighbor (KNN)
Vascular Wilt disease		• The yellowing of cotyledons sauteing of petioles and filling of dried leaves occur at the seedling stage • In the young and grown-up plants deficiency of bloat, leaf hanging and demise of the plants takes place	Edge detection
Grey Mildew	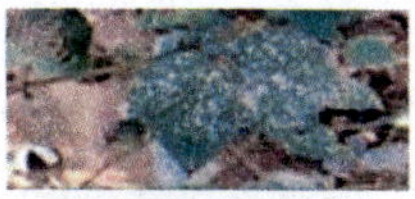	• Nonuniform rakish, pale and lackluster spots which leads the leaf color to yellowish brown	Decision tree algorithm and K-nearest neighbor
Anthracn ose disease		• Seedling stage minor, reddish round spots appear on the cotyledons and primary leaves • Bolls are circular, slightly sunken reddish-brown spots turns to black	Fuzzy feature and decision tree
Root Rot disease		• Brown sports in color on the cotyledons • At the collar area, there is dark shading, extends towards the lower parts • Rotting and shredding appear at the bark of the roots	Neural network
Boll Rot disease		• Small spots disease's which are brown or black in color which expands encloses the completed bolls	Neural network

(continued)

Table 1 (continued)

Disease Name	Image	Symptoms	Suitable methods
Leaf spot or blight disease		• Irregular round spots with tiny pale brown exits on the leaves so several spots together form blighted areas	Image processing and support vector machine

Brugger et al. (2011) employed the method of Agricultural Growers Resource Organization to find the disease of the crop in the initial stages to bridge a gap between farmers and agricultural expert.

Hannunal et al. (2011) worked on quick and exact recognition of cotton diseases using K means and KNN with a detection accuracy of 94%.

Kranthi et al. (2002) worked on cotton leaf diseases and predicted them using machine learning methods.

Wani and Ashtankar (2017) used the random forests algorithm with decision trees using MLP classifier to characterize and found from their work that multi yield regressor performed well by using a random forest system.

Rothe and Kshirsagar (2014) used preprocessing image technology to smooth the images in the detection of cotton plant diseases such as Myrothecium, Alternaria, etc.

Gulhane and Gurjar (2011) worked on the identification of cotton plants where in their leaf's are affected with diseases known as Leaf spot with the help of Support vector machine classification. It uses GLCM technique for disease detection and HSV algorithm in the identification of leaf disease.

Prashar et al. (2019) presented a paper to recognize the leaf diseases by using multilayered perceptron with overlapping pooling to find the contaminated leaves. KNN and SVM are used for morphological division, design coordinating restricting the diseased area with 96% accuracy.

Li et al. (2016) implemented CNN for segmenting the cotton bolls using cotton boll segmentation algorithm.

Images can be analyzed by using semantic segmentation. Xu et al. (2021) implemented the tool ENVI deep learning module i.e. U-Net Network model that makes use of Tensor Flow to train the model and identifies the feature in the image according to the spatial dimensions. The UAV remote sensing data images is given as input to the model.

Bargoti and Underwood (2017) and Rahnemoonfar and Sheppard (2017) implemented CNN for the extraction of the given images and finds applications in agriculture for detecting the objects.

A CNN based models needs labeled data as a training dataset, but it takes a longer time for processing if the set is bigger and sometimes the model misses the cotton bolls even they are present in the images. Therefore, its challenging for CNN to avoid misclassification.

Liu et al. (2016) used the CNN, a deep learning technique for recognizing the flower species effectively.

Dulhare et al. (2019) proposed Image fusion to produce a very high-resolution muti spectral picture by adjoining two or more images which helps in reducing the redundancy of the output.

Pérez-Zavala et al. (2018) discussed in their work regarding the detection of grape bunch using Image-based methods. Yin et al. (2018) proposed plant leaf segmentation and tracking and Stein et al. (2016) discussed mango fruit detection.

3 Methodology

The methodology for identifying the leaf diseases with its classification methods are presented in Fig. 1. The work discusses on the cotton plant leaf diseases and their classification to detect the contaminated leaves accurately.

Image Segmentation and Feature Extraction: In this classification, identification of diseases utilizing picture preparing strategies which extracts the important aspects like region, major, and minor axes, direction, etc. from the images of diseased leaves. For this purpose, two strategies image enhancement and image segmentation are used.

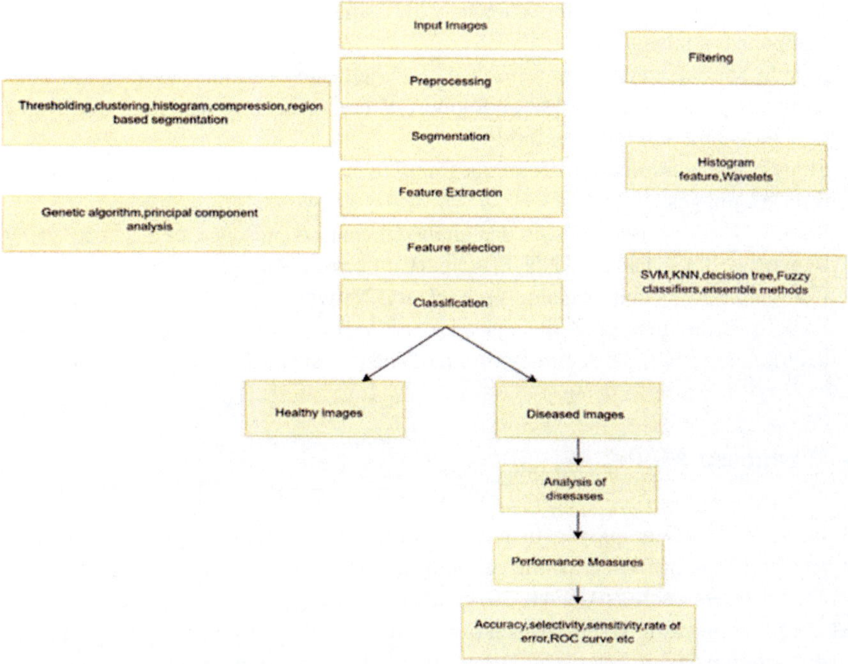

Fig. 1 Structure of cotton leaf diseases detection

The proposed method uses the CNN image classification technique to classify the leaf picture. Based on retrieved information at each convolution layer, it can detect and recognize diseases automatically. For disease identification, the system uses the image processing techniques. The image of a cotton plant leaf should be uploaded first. The algorithm will pre-process the image before the CNN algorithm is applied. The network is built using a combination of pretrained on ImageNet and the DenseNet-121 component, and it performs the other state-of-the-art techniques. Every convolution layer within dense block is tiny, so each convolution kernel is still in charge of learning the tiniest details.

From many years, humans have grown plants for various domestic needs, and the monitoring of diseases on these plants is very important to avoid shortage of food and raw material for various other needs.

The following Fig. 1 shows general structure of cotton leaf disease detection and list the various detection analysis methods.

There are various ways to detect cotton leaf disease detection:

Step 1: Leaf images are given as input for pre-processing and using filtering techniques, the noises in the images are eliminated.
Step 2: The filtered images are segmented using thresholding, clustering, histogram, compression, region-based segmentation.
Step 3: In the segmented images, the features are extracted as histogram feature, wavelets.
Step 4: The extracted features are selected using genetic algorithm and principal component analysis.
Step 5: The selected features are then classified using various classification techniques like K-nearest neighbor Support Vector Machine, decision tree, fuzzy classifiers and ensemble methods.
Step 6: The classification techniques help us in segregating healthy images and diseases images.
Step 7: The diseased images are analyzed for the presence of diseases using Support vector machine, KNN, Principal Component Analysis, Ensemble Method (Random Forest) and Decision Tree, Neural Network.
Step 8: After analyzing the diseases the various performance measures are determined to find the accuracy of the selected Transfer Learning technique.

4 Proposed Model

DenseNet-121 was employed to minimize error and maximize the performance. DenseNet-121 is proposed as one of the best performers in image classification of popular datasets such as CIFAR-10, ImageNet, etc. DenseNet-121 uses a simple pattern for connecting layers to each other directly in a feed-forward pattern, that is each layer consists of several additional inputs from previous layers and transmits its own feature maps to subsequent layers.

Fig. 2 Healthy and Unhealthy cotton plant leaves

In order to perform classification, we used the New Plant village dataset which is a hugely popular dataset available from Kaggle are downloaded which is a crowd-sourced and open-source platform in Lu et al. (2017) and Kaur et al. (2019) where they get trained, challenged to solve numerous problems. The New Plant village dataset consists of 1951 images with both healthy and diseased plant leaves in Fig. 2 Adedoja et al. (2022). The images are then split in the ratio as train and test sets. Every image is resized into 100 × 100 pixels to perform model predictions on these images. To split the data into training, testing and validation 40% data goes to testing and remaining 60% goes to training.

Table 2 lists the various parameters that are used in training DenseNet-121 from the dataset of plant village. Initially, the dataset images are loaded with the help of augmented image data store. Then, pretrained DenseNet-121 network is loaded, and the graphs from the trained network are extracted. Finally with the help of augmented training data, Dulhare and Ali (2021).The network is trained, and by validation data the classification accuracy is found out.

DenseNet-121 which is a pre- trained model which consists of 121 layers by a process known as Transfer Learning, which is re-trained by the DenseNet121 model by augmenting images, resized them to 100 × 100 size. The results were much more promising. Our main objective is to find out the higher classification accuracy.

From Tables 3 and 4 represents Transfer Learning DenseNet-121 model is trained over 20 epochs with each batch size of 32 so each epoch tells us the number of times model will be trained and achieved high Training and validation accuracy of 91% is achieved at 20 epochs. After training, the dataset consists of 781 validation images of healthy and unhealthy plant leaves.

Table 2 DenseNet-121 training validation parameters

Parameters	Value
Parameters	Value
Input layer size	100 × 100
Epochs	20
learning rate	0.001
Image augmentation	Image flipping, translations, rotations, shearing, zooming
Hardware resources	Single GPU
Batch size	32

Table 3 Performance comparison of training/validation loss for the dataset used

Training	Training Loss	Validation loss
1	1.3012	0.4892
2	0.5020	0.4098
3	0.4144	0.3534
4	0.3710	0.3762
5	0.4028	0.3717
6	0.3467	0.3108
7	0.3264	0.2777
8	0.3086	0.3112
9	0.3256	0.4516
10	0.2880	0.2641
11	0.2734	0.2673
12	0.2876	0.2823
13	0.2880	0.3221
14	0.2778	0.3294
15	0.2900	0.3515
16	0.2542	0.2750
17	0.2802	0.2369
18	0.2162	0.3127
19	0.2292	0.3209
20	0.2246	0.2629

5 Results and Discussion

The matrix shown below is the confusion matrix between predictions and test images. The matrix illustrates the four classes. The confusion matrix is used to make the summary of the prediction in a graphical way. It compares actual test images with predictions made by the model after training and running the test images through

Table 4 Performance comparison of training/validation accuracy for the dataset used

Training	Training accuracy	Validation accuracy
1	0.5940	0.8041
2	0.8051	0.8464
3	0.8453	0.8784
4	0.8701	0.8528
5	0.8479	0.8604
6	0.8632	0.8860
7	0.8769	0.9052
8	0.8795	0.8848
9	0.8744	0.8284
10	0.8872	0.9052
11	0.8974	0.9078
12	0.8863	0.9078
13	0.8957	0.8860
14	0.8966	0.8784
15	0.8855	0.8758
16	0.8991	0.9065
17	0.8991	0.9193
18	0.9265	0.8848
19	0.9103	0.8796
20	0.9060	0.9065

the DenseNet-121. Rows represent the Predicted labels, while columns represent the actual True labels.

In the below Fig. 3, Confusion Matrix can analyze the model as four class labels namely Class '0'-unhealthy cotton leaf, Class '1'-unhealthy cotton plant, Class '2'-healthy cotton leaf, Class '3'-healthy cotton plant.

From the Table 5, it is evident that model predictions are almost accurate. The experimental results achieved showed that the proposed model is capable of classifying healthy and unhealthy plant leaf images contains 781 validation images in

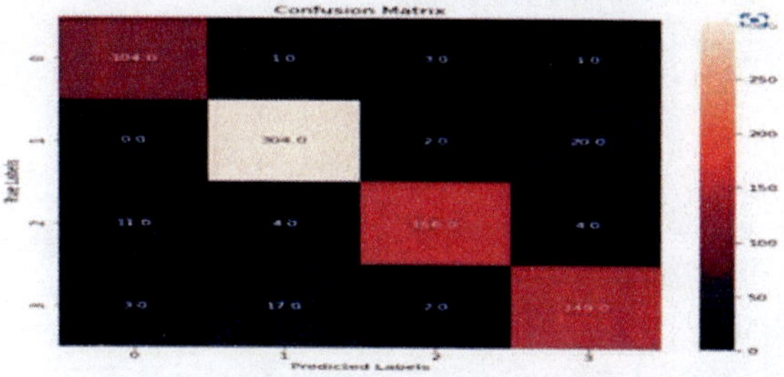

Fig. 3 DenseNet-121 Confusion Matrix

the dataset with higher classification accuracy 91% using DensNet-121 pretrained model with four classes ‘0’, ‘1’, ‘2’, ‘3’. From Fig. 3 Transfer learning DenseNet-121 Model Layer is trained and represented Figs. 4 and 5.

In Fig. 4, the graph illustrates the plot between loss and validation loss of DenseNet-121 model training. In the initial iteration 20 epochs were used, the loss is found to be very high around 1.3, validation loss is 0.48, during training, after each epoch, both loss and validation loss comes down. During the 5th epoch, there was a slight increase in Val loss, but it comes down again in the next epoch, as the training still continued.

In Fig. 5, the graph represents a validation accuracy of 91% obtained over 20 epochs of training, while a training accuracy of 89% was reported. This is an effective measure of the classification made by Transfer Learning-DenseNet-121 model illustrates the plot between Training and Validation accuracy of DenseNet-121 model training. In the initial iteration 20 epochs were used and the accuracy is found to be very low around 0.5, validation accuracy is 0.8, during training, after each epoch,

Table 5 Classification results with various metrics

Classes	Precision	Recall	F_1-score	Support
Class 0 Unhealthy cotton leaf	0.88	0.95	0.92	109
Class 1 Unhealthy cotton plant	0.93	0.93	0.93	326
Class 2 Healthy cotton leaf	0.96	0.89	0.92	175
Class 3 Healthy cotton plant	0.86	0.87	0.86	171
Macro avg	0.91	0.91	0.91	781
Weighted accuracy	0.91	0.91	0.91	781
Accuracy			0.91	781

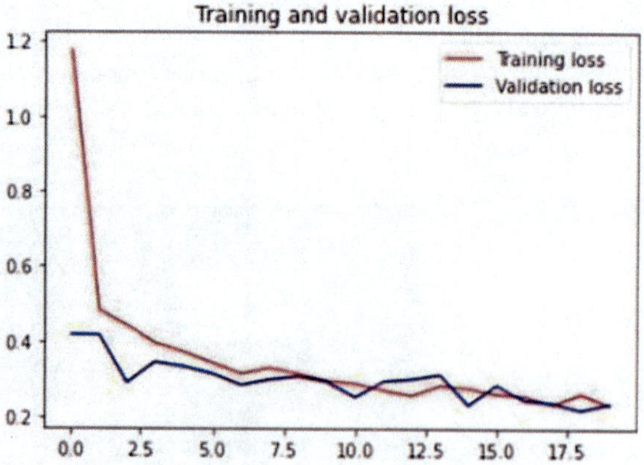

Fig. 4 Graphical representation of training and validation loss of the model

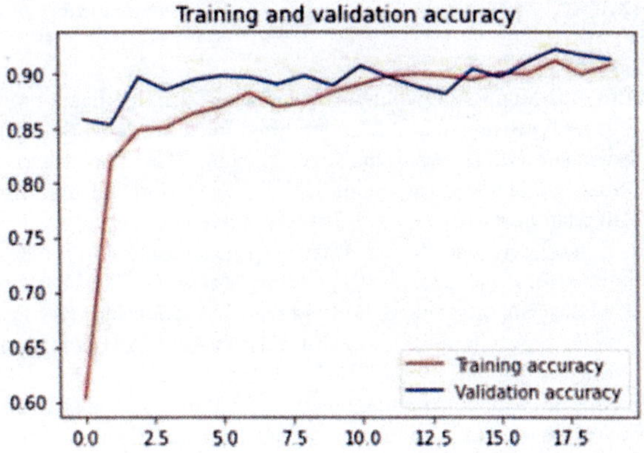

Fig. 5 Graphical representation of training and validation accuracy of the model

both accuracy and validation accuracy goes high. During the 5th epoch, there was a slight decrease in accuracy but it comes high in the next epoch, as the training still continued. After the 11th epoch the validation accuracy begins to rise.

6 Conclusion

The proposed model analyzes and provides plant village dataset of various type plant leaf images using Transfer Learning DenseNet-121 which performs the classification of healthy and unhealthy plant leaf images with an accuracy of 91%. In this system, the images are captured by the aerial vehicle are given and compared with the predefined dataset images to classify healthy and unhealthy which helps for early prediction is made so as to give prior information to farmers for the improvement of cotton yield.

7 Future Scope

In the future work, the diseases that are causing the cotton leaves unhealthy can be identified and classified according to the disease of the cotton leaf so that measures can be taken to protect the cotton crop from the disease thus enhancing the cotton yield.

References

A.O. Adedoja, P.A. Owolawi, T. Mapayi, C. Tu, Intelligent mobile plant disease diagnostic system using NASNet-Mobile deep learning. Int. J. Comput. Sci. (IJCS) (2022)

D. Ashourloo, H. Aghighi, A.A. Matkan, M.R. Mobasheri, A.M. Rad, An investigation into machine learning regression techniques for the leaf rust disease detection using hyperspectral measurement. IEEE J. Sel. Top. Appl. Earth Observations Remote Sens. 1–7 (2016)

S. Bargoti, J.P. Underwood, Image segmentation for fruit detection and yield estimation in apple orchards. J. Field Rob. (2017). https://doi.org/10.1002/rob.21699

D. Al Bashish, M. Braik, S. Bani-Ahmad, Detection and classification of leaf diseases using k-means based segmentation and neural network-based classification. Inf. Technol. J. 267–275(2011). ISSN 1812-5638

F. Brugger, Mobile Applications in Agriculture. Syngenta Foundation, Basel, Switzerland (2011)

U.N. Dulhare, M.H. Ali, Underwater human detection using faster R-CNN with data augmentation. Mater. Today Proc. (2021)

U.N. Dulhare, S. Gouse, Automation of rice cultivation from ploughing-harvesting with diseases, pests and weeds to increase the yield using AI, in *Proceedings of the 4th International Conference on Communications and Cyber Physical Engineering (ICCCE)*, vol. 828, pp. 505–513 (2022)

U.N. Dulhare, A. Mohammed, A.M. Hussam. A review on diversified mechanisms for multi focus image fusion, in *International Conference on Communication and Information Processing (ICCIP)* (2019)

V.A. Gulhane, A.A. Gurjar, Detection of diseases on cotton leaves and its possible diagnosis. Int. J. Image Process. (IJIP) **5**(5), 590–598 (2011)

S. Hannunal, N. Anantrasirichi, S. Subrmanian, S. Prashant, A. Jhunjhunwala, C.N. Canagarajah, Agriculture disease mitigation system. ICTACT J. Commun. Technol. Spec. Issue Next Gener. Wirel. Netw. Appl. **2**(2), 364–369 (2011)

H. Lv, W. Song, S. Sun, Service oriented architecture for mobile geospatial information data collection. Comput. Syst. Appl. **21**(7), 43–46 (2012)

B.B. Kalbande, A.S. Pati, Plant tissue culture independent agro bacterium tumefaciens medicated in-plant transformation strategy for upland cotton. J. Genet. Eng. Technol. **14**(1), 9–18 (2016)

S. Kaur, G. Joshi, R. Vig, Plant disease classification using deep learning google net model. Int. J. Innovative Technol. Exploring Eng. (IJITEE) **8** (2019)

K. Kranthi, D. Jadhav, S. Kranthi, R. Wanjari, S. Ali, D. Russell, Insecticide resistance in five major insect pests of cotton in India. Crop Prot. **21**(6), 449–460 (2002)

H. Li, R. Ji, J. Zhang, X. Yuan, H. Kaiqun, L. Qi, WEB- Based intelligent diagnosis system for cotton diseases control. IFIP Adv. Inf. Commun. Technol. **346**, 483–490 (2011)

Y. Li, Z. Cao, H. Lu, Y. Xiao, Y. Zhu, A.B. Cremers, In-field cotton detection via region-based semantic image segmentation. Comput. Electron. Agric. **127**, 475–486 (2016)

Y. Liu, F. Tang, D. Zhou, Y. Meng, W. Dong, Flower classification via convolutional neural network, in *International Conference on Functional- Structural Plant Growth Modeling, Simulation, Visualization and Applications (FSPMA) (Qingdao)*. https://doi.org/10.1109/FSPMA.2016.7818296 (2016)

Y. Lu, S. Yi, N. Zeng, Y. Liu, Y. Zhang, Identification of rice diseases using deep convolutional neural networks. Neurocomputing **267**, 378–384 (2017)

R. Pérez-Zavala, M. Torres-Torriti, F.A. Cheein, G. Troni, A pattern recognition strategy for visual grape bunch detection in vineyards. Comput. Electro. Agric. **151**, 136–149 (2018). https://doi.org/10.1016/j.compag.2018.05.019

K. Prashar, R. Talwar, C. Kant, CNN based on over-lapping pooling method and multi-layered learning with SVM and KNN for American cotton leaf diseases recognition, in *Proceedings of the International Conference on Automation, Computational and Technology Management (ICACTM), London, UK* (2019)

M. Rahnemoonfar, C. Sheppard, Deep count: fruit counting based on deep simulated learning. Sensors (basel). (2017). https://doi.org/10.3390/s17040905

P.R. Rothe, R.V. Kshirsagar, A study and implementation of active contour model for feature extraction: with diseased cotton leaf as example. Int. J. Curr. Eng. Technol. **4**(2) (2014a)

P.R. Rothe, R.V. Kshirsagar, A study and implementation of active contour model for feature extraction with diseased cotton leaf as example. Int. J. Current Eng. Technol. **4**(2), 812–816 (2014b)

S. Satraj Sohrab, E.I. Azhar, M.A. Kamal, P.S. Bhattacharya, D. Rana, Genetic variability of cotton leaf curl beta satellite in Northern India. Saudi J. Biol. Sci. **21**(6), 626–631(2014)

M. Stein, S. Bargoti, J. Underwood, Image based mango fruit detection, localisation and yield estimation using multiple view geometry. Sensors **16**(11), 1915 (2016)

R. Tijare, P. Khade, R. Jain, The survey of disease identification of cotton leaf. Int. J. Innovative Res. Comput. Commun. Eng. (2015)

H.K. Wani, N. Ashtankar, An appropriate model predicting pest/diseases of crops using machine learning algorithms, in *4th International Conference on Advanced Computing and Communication Systems (ICACCS)* (2017)

W. Xu, P. Chen, Y. Zhan, S. Chen, L. Zhang, Y. Lan, Cotton yield estimation model based on machine learning using time series UAV remote sensing data (2021)

X. Yin, X. Liu, J. Chen, D.M. Kramer, Joint multi-leaf segmentation, alignment, and tracking for fluorescence plant videos. IEEE Trans. (2018)

Q. Zheng, W.J. Huang, X.M. Cui, Y.Y. Dong, Y. Shi, H.Q. Ma, L.Y. Liu, Identification of wheat yellow rust sing optimal three-band spectral indices in different growth stages. Sensors **19**, 35 (2019)

Student Interaction on Moodle for a Foundation Course at a Tertiary Institute in Fiji

Komal Karishma and Krishna Raghuwaiya

Abstract In this paper, we examine which interaction(s) students engage in the most on Moodle in a Pre-degree (English Foundation) course in a tertiary institute in Fiji. A mixed-method research approach was used to collect data. Quantitative data was collected and analyzed for students' interaction on Moodle with the content, teacher, other learners, learning environment, assessment, and feedback on assessment. Qualitative data was examined to explore students' interaction with the institution. It was found that students interacted mostly with the learning environment than the other forms of interaction, as this included the total number of times students logged into the course (LLFXX) Moodle page. Their interaction with the assessment and feedback on assessment was also higher than the content, teacher, and other learners. Therefore, it can be concluded that students' interaction was high for assessment-related interactions (assessment and feedback on assessment).

Keywords e-learning · Interaction · LMS · Moodle

1 Introduction

E-learning has become a major component of the teaching–learning process. Classes are being facilitated using technology-enhanced learning tools. Due to technology-enhanced teaching, Blended mode of teaching is emerging. Learning management system (LMS) has emerged and has made great impact in ensuring that technology-enhanced learning is implemented easily. A globally adopted LMS is adopted globally is (Modular Object-Oriented Dynamic Learning Environment (MOODLE). It is used to facilitate the teaching–learning process virtually.

For teaching–learning process to be successful and productive, it is crucial for interaction to take place. The need to interact in a classroom is still crucial for the instructors and students despite the shift from a traditional classroom to a virtual classroom. The mode of teaching, whether it is Blended or Online, has not changed the

K. Karishma (✉) · K. Raghuwaiya
The University of the South Pacific, Suva, Fiji
e-mail: komal.karishma@usp.ac.fj

A. B. Reddy et al. (eds.), *Proceedings of Third International Conference on Advances in Computer Engineering and Communication Systems*, Lecture Notes in Networks and Systems 612, https://doi.org/10.1007/978-981-19-9228-5_34

expectation and/or the perception of the instructor and the students. The shift to virtual classroom strongly emphasizes on the importance of interaction in the teaching–learning process (Wagner 1994). Researchers have usually looked at students' interaction on Moodle in total or have studied students' interaction with content, teacher, and other learners (Meri 2015; Moore 1989).

It has been emphasized in the recent studies that all forms of interactions are equally important for students. Initially, student to content, student to instructor and student-to-student communication were focused on (Anderson 2003; Gunduz and Ozcan 2017; Hirumi 2002a, b, 2006; Moore 1989; Rhode 2009). Now it has been realized that in order to have a holistic approach to learning, it is essential for a learner to collaborate with his or her learning and teaching setting. Every student needs to interact with his or her assessments, response to assessments or performance and the university altogether (Hirumi 2002a, b, 2006; Meri 2015; USP Flexible Learning Policy 2019). A very interactive learning process is developed if students interact in all the seven forms.

With such holistic approach, this paper will explore learner interaction with content, teacher, other learners, learning environment, assessment, feedback on assessment, and the institution (Hirumi 2002a, b, 2006).

The study was done on one of the mandatory English Language courses at Foundation level (Pre degree) at a tertiary institute in Fiji. The institute offers courses to students from Fiji and other countries. Moodle assists in facilitating teaching and learning. This LMS has been used in the institute since 2008. When it started, it was used for certain courses only. However, over the years it has become mandatory for every course to use Moodle.

The courses are offered through either print, blended, or online mode. The presence of Moodle in the courses totally depends on the mode of teaching. For example, for an online course, Moodle facilitates teaching and thus is the virtual classroom. Since there is no set guideline for blended mode of study (Racule and Buadromo 2020), the use of Moodle differs from course to course. However, for LLFXX, Moodle is used for uploading resources, submitting assessments, providing feedback on assessments, communicating with students and teaching staff, and for sharing information.

The purpose of the research is to see which interaction students engage in the most.

2 Methodology

Qualitative or quantitative data is not able to deliver a complete comprehension of the research problem alone. As a result, a mixed method approach provides enhanced strategies to collect or analyze data (Creswell 2003; Creswell and Clark 2006).

A total of 109 students (80 Blended and 29 Print mode) participated in the research. These were Foundation students from main campus enrolled in LLFXX in the institute. The students' interaction on the Moodle page with the content, teacher, other learners, learning environment, assessment, feedback on assessment, and the institution was investigated.

From the seven types of interaction, six had quantitative data, whilst one form interaction had qualitative data to be analyzed. The quantitative data was derived from the Moodle page. The number of times students clicked onto the LLFXX Moodle page and accessed its various sections were recorded and used in the research. These were categorized under six different forms of interaction. For the seventh form of interaction, data was collected from a questionnaire that was prepared in Moodle and the students were asked to complete the questionnaire online. This collected data on students' interaction with the institution. Therefore, mixed method was the most appropriate approach for this study.

3 Dimensions of Flexibility

Student interaction has been divided into seven categories. These are known as the "Dimensions of Flexibility'. These seven categories are learner interactions with the content, teacher/instructor, other learners, learning environment, assessment activities, feedback, and with the institution. These will be discussed below.

3.1 Learner—Content Interaction

LLFXX Moodle page has a range of resources, like recordings, notes, course outline, course book, tutorial questions, assignment guidelines, and additional resources. Data was collected on the number of times these were accessed.

Table 1 shows that the most accessed content on the Moodle page was Lecture Recordings for students studying via blended mode and tutorial recordings were the least accessed. The range was also the highest for lecture recordings for blended mode students. Students studying via print mode mostly accessed additional resources and tutorial recordings and tutorial questions were the least accessed. The range for these three components was the highest for print mode students. In total content was accessed by blended mode students 661 times with a mean of 7.34 and by print mode students 141 times with a mean of 4.86.

3.2 Learner—Teacher Interaction

Four means were available for learners to interact with their facilitator and coordinator. These were through BBB, Class News and Announcement, Discussion Forum, and Messaging.

Table 1 LLFXX students' and content interaction on the Moodle page

Content	Number of clicks		Mean, μ (Standard deviation, σ)		Range	
	Blended	Print	Blended	Print	Blended	Print
Recordings of lecture	198	8	2.2 (4.9)	0.28 (0.78)	(39–0) 39	(4–0) 4
Tutorial recordings	19	3	0.21 (0.59)	0.1 (0.3)	(3–0) 3	(1–0) 1
Notes on lecture	67	7	0.74 (2.1)	0.24 (0.81)	(17–0) 17	(4–0) 4
Additional resources	69	45	0.76 (1.65)	1.55 (1.63)	(11–0) 11	(6–0) 6
Course book	133	35	1.48 (2.04)	1.2 (1.6)	(10–0) 10	(6–0) 6
Outline of course	43	16	0.48 (0.81)	0.55 (1.07)	(4–0) 4	(4–0) 4
Questions on tutorial	25	3	0.28 (1.35)	0.1 (0.3)	(10–0) 10	(1–0) 1
Assignment two guidelines	107	24	1.19 (1.67)	0.83 (1.31)	(7–0) 7	(6–0) 6
Overall	661	141	7.34 (10.66)	4.86 (4.2)	(86–0) 86	(14–0) 14

As shown in Table 2, both blended and print mode students interacted mostly by communicating on discussion forum. This is an average of 13.72 for blended mode students and 24.66 for print mode students. Messaging was the least used means of communication with blended mode students accessing it 19 times and print mode students accessing it only twice. For blended mode students the highest range was for Class News and Announcements, but for print mode students the highest range was for discussion forums.

Table 2 LLFXX students interaction with the teacher on the moodle page

Means of interacting with teacher	Number of clicks		Mean, μ (Standard deviation, σ)		Range	
	Blended	Print	Blended	Print	Blended	Print
BBB	47	5	0.52 (1.38)	0.17 (0.53)	(7–0) 7	(2–0) 2
Class news and announcement	1194	148	13.27 (24.14)	5.1 (8.05)	(153–0) 153	(28–0) 28
Discussion forum	1235	715	13.72 (24.41)	24.66 (76.43)	(150–0) 150	(420–0) 420
Messaging	19	2	0.21 (0.59)	0.07 (0.25)	(3–0) 3	(1–0) 1
Overall	2495	870	27.72 (41.73)	30 (81.57)	(232–0) 232	(448–0) 448

Table 3 LLFXX students' interaction with the other students on the moodle page

Means of interacting with other learners	Number of clicks		Mean, μ (Standard deviation, σ)		Range	
	Blended	Print	Blended	Print	Blended	Print
BBB	49	5	0.54 (1.42)	0.17 (0.53)	(7–0) 7	(2–0) 2
Class news and announcement	0	0	0 (0)	0 (0)	0	0
Forum for discussion	15	0	0.17 (0.52)	0 (0)	(3–0) 3	0
Message	0	0	0 (0)	0 (0)	0	0
Overall	64	5	0.71 (1.53)	0.17 (0.53)	(7–0) 7	(2–0) 2

3.3 *Learner—Learner Interaction*

Table 3 shows that BBB, Class News and Announcement, Discussion Forum, and Messaging were the means of interaction between learners.

Unlike learner-teacher interaction, the interaction between the learners (learner-learner) was very low. Learner-learner interaction mostly took place in BBB sessions for both blended and print mode students. There was no interaction among learners through message or Class News and Announcement.

3.4 *Learner Interaction with the Learning Environment*

On the Moodle platform, learner interaction with the learning environment was analyzed by examining the number of times students logged in into the LLFXX Moodle page. The other Moodle page component that enabled them to interact with the learning environment was by signing up for tutorials (Table 4).

Table 4 LLFXX students' interaction with the learning environment on the moodle page

Means of interacting with the learning environment	Number of clicks		Mean, μ (Standard deviation, σ)		Range	
	Blended	Print	Blended	Print	Blended	Print
Logged in into moodle	99,763	29,722	1108.48 (886.6)	1024.9 (610.64)	(7326–207) 7119	(2592–77) 2515
Tutorial signup	231	0	2.57 (2.56)	0 (0)	(14–0) 14	0
Overall	99,994	29,722	1111 (887.1)	1024.9 (610.64)	(7331–207) 7124	(2592–77) 2515

Table 5 LLFXX students' interaction with the assessment on the moodle page

Assessments	Number of clicks		Mean, μ (Standard deviation, σ)		Range	
	Blended	Print	Blended	Print	Blended	Print
Assign. 1	1567	580	17.41 (13.55)	20 (12.52)	(96–0) 96	(50–0) 50
Assign. 2	1825	619	20.28 (22.45)	21.34 (14.02)	(189–0) 189	(60–0) 60
Test 1	473	275	5.26 (3.21)	9.48 (6.7)	(17–0) 17	(37–0) 37
Test 2	337	166	3.74 (2.8)	5.72 (4.61)	(19–0) 19	(25–0) 25
Sample test 1	319	207	3.54 (3.25)	7.14 (6.1)	(14–0) 14	(30–0) 30
Sample test 2	221	176	2.46 (2.99)	6.07 (4.77)	(16–0) 16	(19–0) 19
Grammar starting quiz	347	184	3.85 (3)	6.34 (4.91)	(14–0) 14	(19–0) 19
Verb tense activity	212	138	2.36 (3.46)	4.76 (3.27)	(23–0) 23	(11–0) 11
Essay writing activity	173	0	1.92 (2.35)	0 (0)	(13–0) 13	0
Overall	5474	2345	60.82 (40.36)	80.86 (43.73)	(332–5) 327	(233–17) 216

The mean of this interaction is very high for both, Blended and Print mode students. This is due to their frequency of logins. Students had logged in into the LLFXX Moodle page quite frequently (Blended-$\mu = 1111$ and Print-$\mu = 1024.9$) for various reasons, basically for interacting with content, teacher, other students, assessment activities, feedback, and the institution.

3.5 *Learner Interaction with Assessment Activities*

LLFXX Moodle page has nine assessment activities. These are the tests, assignments, sample tests, starting test, and quizzes.

As per Table 5, the students mostly accessed the assignments (Assign.) (1 and 2). Blended mode students accessed assignment (Assign.) 2 1825 times and assignment (Assign.) 1 1567 times. Print mode students accessed assignment (Assign.) 2 619 times and assignment (Assign.) 1 580 times. The least accessed assessment activity was the essay writing quiz (173 for blended mode and 0 for print mode). The highest range for both, the blended mode (189) and print mode (60) was for assignment 2.

3.6 *Learner Interaction with Feedback*

Feedback was given to learners for Assignment 1, Assignment 2, Test, and Final exam answer.

Table 6 LLFXX students' interaction with the feedback on the moodle page

Feedback	Number of clicks		Mean, μ (Standard deviation, σ)		Range	
	Blended	Print	Blended	Print	Blended	Print
Assign. 1	234	87	2.6 (2.55)	3 (2.36)	(11–0) 11	(9–0) 9
Assign. 2	148	29	1.64 (2.83)	1 (2.02)	(20–0) 20	(9–0) 9
Quiz	104	62	1.16 (2.45)	2.14 (2.6)	(13–0) 13	(11–0) 11
Past semester final examination answer	74	48	0.82 (1.36)	1.66 (1.56)	(9–0) 9	(5–0) 5
Overall	560	226	6.22 (6.58)	7.79 (5.73)	(38–0) 38	(24–0) 24

Table 6 shows that both blended and print mode students mostly accessed the feedback for assignment (Assign.) 1 (Blended—234 and Print—87). The least accessed feedback for blended mode students was final examination answer (74), and for print mode students was assignment (Assign.) 2 (29).

3.7 *Learner Interaction with Institution*

This interaction was measured through qualitative approach, whereby students' response on how practical the use of Moodle was for studies for different learning modes, how helpful the development of ICT at the institute was for pre-degree students, and how relatable the Moodle services were for LLFXX students.

Majority of the students found Moodle to be very viable and its services very relatable. Most of the LLFXX students agree that the ICT development for pre-degree student is quite helpful.

3.8 *Summary of Findings*

The quantitative data shows the number of times students interacted with the components on the LLFXX Moodle page. These have been summarized below for six out of seven types of interaction.

Figure 1 shows that blended mode students mostly interacted with the learning environment. This includes them accessing the Moodle page. Therefore, the value is very high. Their least interaction by blended mode students was with other learners.

Figure 2 depicts that print mode students, also had a similar trend of interaction; most interaction was with the learning environment and the least interaction was with other learners.

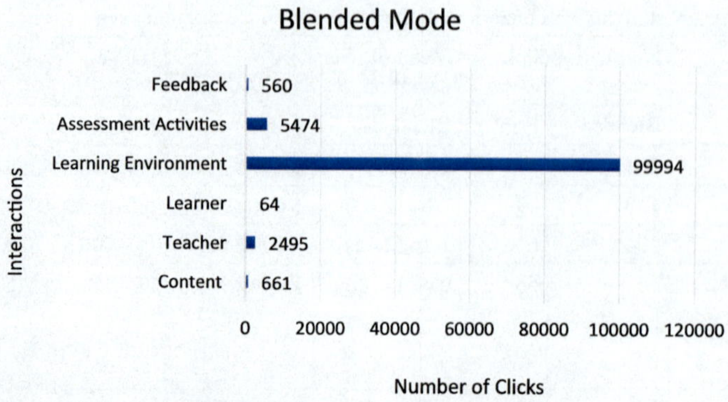

Fig. 1 Interaction on moodle by blended mode students

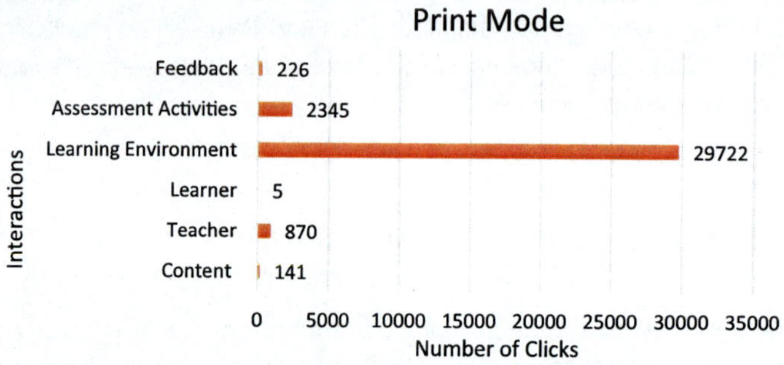

Fig. 2 Interaction on moodle by print mode students

4 Discussion

The goal of this study was to empirically examine student interaction on the Moodle page for a pre-degree (Foundation) course in a tertiary institute in Fiji. The study was done by examining student interaction on Moodle page with the content, teacher, other learners, learning environment, assessment, feedback on assessments, and the institution.

Table 1 shows that blended mode students interacted the most with the lecture recordings (198). This may be due to low attendance during lecture times. During the beginning of the semester, students' attendance in lecture is very high, but after few weeks it decreases. The possibility of students accessing lecture recordings on Moodle rather than attending the face-to-face lecture is very high.

In addition to this, Table 2 shows that students interact with the teacher the most through discussion forums. The interaction is 1235 for blended mode students and 715 for print mode students. This is the component of Moodle where students raise and clarify any confusion that they may have. They also respond to any necessary postings done by the teacher.

On the contrary, learner—learner interaction on Moodle as shown in Table 3 is very low. Discussion forum interaction amongst learners is only 15 for blended mode students and 0 for print mode students. This shows that students do not prefer to interact amongst themselves on discussion forum.

For total interaction, it was found that both blended (Fig. 1) and print (Fig. 2) mode students interacted with the learning environment. Interaction is very high for the learning environment when compared to other interactions as this interaction takes into account every time the students logged into the Moodle page for the course (LLFXX). This is regardless of how genuine or in-genuine the reasons for logging in would had been.

Figure 1 shows that blended mode students interacted with the assessment (5474) many times also and then with the feedback for the assessments (560). Print mode students (Fig. 2) similarly interacted more with the assessments (2345) than the feedback on assessments (226). This clearly shows that students are interacting only for marks or better final grades. They interaction is result centered and not merely for interaction or for communication purpose.

The least interaction for both blended mode (64) as shown in Fig. 1 and print mode (5) as shown in Fig. 2 were with other learners. These students also as shown in Figs. 1 and 2 interacted with the teacher (blended mode—2495 and print mode—870) more than with the content (blended mode—661 and print mode—14).

The qualitative data showed that majority of the students found Moodle to be very viable and its services very relatable; 'the services are relevant to my student needs'. Most of the LLFXX students agree that the ICT development at the institute for pre-degree student is quite helpful; 'the development has made me have access to things that I need for my studies and had never imagined about'.

Based on this research, it could be easily seen which components of the Moodle page students interact more on rather than analyzing just a few of them or looking at the total number of student interaction only. Such specific data can assist instructors of the Moodle page to upgrade to the complete competency of the Moodle page. This may increase student interaction with the Moodle page and assist them with their e-learning process.

5 Conclusion

It was the first time through this study that foundation students were researched on (at the institution). Their interaction on Moodle was the main component of the research. The data collected through this research can assist to make some changes to enhance interaction by the learners and the instructors. Such research can guide instructors to ensure that they are designing the Moodle page and adding resources to the Moodle page that are student friendly and learner need oriented.

The research examined foundation students interaction on the Moodle page in the main campus for blended and print mode students. This is a pre-degree course, hence, the possibility of it being the first exposure to Moodle or to any LMS. As a result, it can be said that there are variations in student interaction with the content, teacher, other learners, learning environment, assessment, feedback on assessment, and the institution.

The tertiary institute has made it mandatory for all courses to have Moodle presence despite their mode of teaching (Face to face, blended, online, or print). More courses are taught in blended mode. Therefore, it is crucial to explore learner interaction or the Moodle page. It was established in the study that of all the interactions on Moodle, student interaction was soaring for interactions associated with assessment (assessment and feedback on assessments).

References

T. Anderson, Getting the mix right again, an updated and theoretical rationale for interaction. Int. Rev. Res. Open Distance Learn. **4**(2), 1–14 (2003)

J.W. Creswell, Research design: qualitative, quantitative, and mixed methods approaches, 2nd edn. SAGE Publications, London (2003)

J.W. Creswell, V.L.P. Clark, Designing and conducting Mixed Methods research, 2nd edn. Sage Publication Inc, Los Angeles (2006)

N. Gunduz, D. Ozcan, Implementation of the moodle system into EFL classes. Sues Teacher's Prof. Dev. **19**(Suppl. 1), 51–64 (2017)

A. Hirumi, A framework for analyzing, designing and sequencing planned e-learning interactions. Q. Rev. Distance Educ. **3**(2), 141–160 (2002a)

A. Hirumi, The design and sequencing of elearning interactions: a grounded approach. Int. J. E-Learn. **1**(1), 19–27 (2002b)

A. Hirumi, Analysing and designing e-learning interactions, in *Interactions in Online Education: Implications for Theory and Practice* ed by C. Juwah (Routledge Falmer Taylor and Francis Group, London, 2006), pp. 45–69

Meri, S.: Exploring the interaction between learners and tools in e-learning environment, in *Critical CALL-Proceedings of the 2015 EUROCALL conference*, ed. by Helm, F.B. (Padova, Italy (2015), pp. 397–403

M.G. Moore, Editorial: three types of interaction. Am. J. Distance Educ **3**(2), 1–7 (1989)

E. Racule, R. Buadromo, Evaluating students' perceptions of blended learning, in *Teaching and Learning with Technology: Pushing Boundaries and Breaking downwalls* ed. by Naidu, S., Narayan, S (USP Press, Suva, 2020)

J.F. Rhode, Interaction equivalency in self-paced online learning environment: an exploration of learner preferences. Int. Rev. Res. Open Distrib. Learn. **10**(1), 1–15 (2009)

USP Flexible Learning Policy, The University of the South Pacific, http://policylib.usp.ac.fj/form.reddoc.php?id+746. Last accessed 12 June 2019

E. Wagner, In support of a functional definition of interaction. Am. J. Distance Educ **8**, 6–29 (1994)

A Privacy-Oriented Neural Collaborative Filtering-Based Framework for Recommender System

Nidhi Shukla, Ashutosh Kumar Singh, and Vijay Kumar Dwivedi

Abstract Recommender system (RS) is growing rapidly and participating in deep learning to provide high quality of recommendations. RS is used to select the best choice for user what they might like. In recent years, it is obvious that various RSs have been proposed for a little while. This field is buzzing with innovation to be sure. Deep RS is a model which is trained for recommendations based on interaction with a neural network. The properties which make neural architecture effective are end-to-end differentiable and provide learning biases to input data type. We propose a framework named as privacy preserve neural collaborative filtering (NCF) by using differential privacy (DP). In this paper, we use general matrix factorization and multi-layer perception network which is apply on real-world datasets for experimentation. We also compare of our framework with other models that are based on matrix factorization and implicit feedback. To measure the performance evaluation, we use parameters such as: MAP, NDGC, precision, and recall. For measuring the DP performance, optimal Renyi differential privacy (RPD) order is considered which shows the effectiveness of the proposed framework.

Keywords Recommender system · Collaborative filtering · Privacy · Neural network · Deep learning

1 Introduction

Deep learning-based (DL) recommender system (RS) is designed using existing techniques such as collaborative filtering. The two established approaches used in RS are model based and memory based. The model-based approaches have been especially utilized in real-world systems. The model-based technique such as matrix factorization (MF) is used for the interaction between user and item in shared latent space which reconstructed by latent vectors using user-item interaction (Bai et al. 2017). Collaborative filtering (CF) is becoming efficient and interprets for real-time

N. Shukla (✉) · A. K. Singh · V. K. Dwivedi
United College of Engineering and Research, Naini, Prayagraj, U.P. 211010, India
e-mail: nidhisona250@gmail.com

A. B. Reddy et al. (eds.), *Proceedings of Third International Conference on Advances in Computer Engineering and Communication Systems*, Lecture Notes in Networks and Systems 612, https://doi.org/10.1007/978-981-19-9228-5_35

personalized recommendations which predicts users choice from its behavior but also from others users behavior. By creating users profile based on previous interacted items, it recommends items using similarity between the items and users profile (He et al. 2018). There are two main components in CF models for learning: embedding and interaction modeling. Embedding converts users-items to vector presentation, and interaction model recreates historical interactions (Wang et al. 2019). MF straightly embeds user-item ID as vector and also its interaction with inner product (Koren et al. 2009). The MF interaction function is replaced with nonlinear neural network by neural CF models (Wang et al. 2019).

In recent years, the RS changed drastically by using deep learning and brought more scope of performance improvement. The deep learning-based RS is gaining attentiveness by overcoming from barriers of conventional model and improved recommendation quality (Zhang et al. 2019). CF technique is more popular than the other recommendation techniques as it is easy to represent a user-item data and also to model the user-item interaction based on representation. It is utilized more when integrated with latent factor model (Koren et al. 2009). In this paper, we presented a privacy-preserving neural network framework for RS using CF technique for implicit feedback. We are focusing on using complicated dataset with independent function. To increase the expressiveness of framework using both the networks linear and non-linear networks such as: standard GMF and MLP. We are applying Renyi differential privacy (RDP) for the privacy-preserving framework.

The main contribution of this paper is as follows:

- We present a privacy-preserved neural collaborative filtering framework using PyTorch library for the implicit recommendation of feedback data and applying DP through RDP.
- We construct a neural network for the interaction of user-item with embedding dimensions.
- We experiment on real-world datasets and compare the framework from the other similar models for signifying the effectiveness of proposed work.

Remainder paper is structured as follows. Section 2 includes the detailing of related work of the paper. Section 3 covers the preliminary work. Section 4 is about dataset, evaluation metrics, and baseline models to compare the work. Section 5 includes result and discussion, and last Sect. 6 includes the conclusion of the paper.

2 Related Work

Authors of He et al. (2017) presented a framework named as neural network-based CF using implicit feedback. For nonlinear modeling, a multi-layer perception is used for the interaction of user-item with GMF and neural MF. In He et al. (2018), author proposed a network architecture outer product-based CF using multi-layer neural network to acquire knowledge of high-order correlations between embedding dimensions. Authors Bai et al. (2017) of proposed neighborhood-based neural CF model (NNCF) based on implicit recommendations and also cope with data sparsity. It uses user-item interactions with neighborhood information for recommendation. The authors of Zhang et al. (2019) proposed a bipartite graph structure for user-item interaction into embedding process. For user-item graph, it provides high-order connectivity. Authors of He et al. (2018) proposed model for item-based CF using attention network, which is used for differentiating the importance of historical items for recommendation by additional parameters.

2.1 Privacy in Recommendation System

To generate any kind of recommendation, a system needs datasets which require data collection of an individual may raise worry about privacy as system need personal information. It somewhere collects other related data without knowledge of user which can generate a threat to user (Xu and Yan 2016). Issues which may lead to user in risk are sharing data with third parties, exploitation of data, generation of incorrect recommendations (Bilge et al. 2013), collection of data implicitly without user permission and knowledge may lead to serious consequences. Trusted proxy ensures users to continue open system without even ensuring authentication which may lead to malicious users to create false rating and swindle the system (Lam and Riedl 2004). There is a need of protection for smart terminals, but making of private protocols decreases the computing performance and accuracy which is considered as a challenge for developers (Shao et al. 2021). The huge amount of sharing and circulating data through social networks creates a major security risk. So, users need awareness to make a safe use of networks (Albladi and Weir 2020). Authors of Lian et al. (2018) stated that an ideal RS had to be fast, distributed, and secured. The duty of RS is to protect users, content providers, and intermediate channels.

2.2 Differential Privacy in Recommendation System

Authors of Zhu et al. (2013) proposed private neighbor CF which consist two mechanisms private neighbor selection and recommendation-aware sensitivity. The DP is used with Laplace mechanism where noise is added to protect users rating and also from K-nearest neighbor (KNN) attack. Authors of Selvaraj and Gangadharan (2021) presented a hybrid RS by combining content based (CB) and CF for deep learning. The privacy is conserved by using DP and improves accuracy by adaptive perturbation. The system uses user-provided rating and processes it by using MF; then, a global DP algorithm is applied for adding Laplace noise and processes deep neural learning for obtaining result. In Chen et al. (2022), author presented a framework for solving the 'Transfer Across the Bridge' problem where the data directly transfer without considering privacy issues. In source domain, it creates rating matrix using DP and analyzes the privacy through rating publishing methods. To publish sourced rating matrix, it use deep auto-encoder and deep neural network which preserve privacy. Authors of Wang et al. (2021) proposed a deep neural network RS formed on decentralized federated learning, and for preserving security developed a model which protects user identity from attacker, this protected model could be applied on other RS. Authors of Li et al. (2020) presented a framework using methods DP and upper confidence bound, presenting that the decision-making can be affect by privacy.

There are various model developed in this field (He et al. 2017; Wang et al. 2019). The models use user-item interaction—general matrix factorization (GMF), multi-layer perception (MLP) network and neural matrix factorization (NeuMF). There model is simple and focused on particular data with data-independent function. The real-world feedback have complicated patterns for which other neural recommender models are introduced (Bai et al. 2017) for better performance. The assumptions are made for embedding dimensions are independent from each other, and for data points prediction, it contributes equally which is impractical assumption as embedding items could be interpreted some specific properties of items (He et al. 2018).

We are presenting a NCF framework by using MLP and GMF for the prediction. By reviewing the previous work, it highlights that no privacy-preserved framework using NCF with RDP is presented. The privacy-preserving framework is using RDP with NCF to provide privacy as well as better results in terms of accuracy using neural network. We also compare our work from the other considered baseline models which are similar in terms of MF and CF.

3 Preliminaries

Neural networks are applying quite frequently to generate the recommendations but are most often applied for auxiliary data. The implicit feedback generated prevalent in RS is though clicks, purchases, and watches which are some of the usual implicit feedback that are simple to collect and utilized for the indication of users preferences which becomes the reason of privacy breaches. RS is considered as the two-way interconnection between the users and their item choices. By using nonlinearity of neural network, the accuracy of recommendation increases.

The neural matrix factorization is carried out by the combination of two different networks GMF and MLP. Final prediction is calculated through the concatenation of two networks.

- **General Matrix Factorization (GMF)**—it is a neural network version for MF where the input is one-hot encoded. The input product of user and item latent factors are obtained using point-wise matrix factorization. The user latent vector is denoted by p_u and item latent vector denoted by q_i. So, the mapping function of initial neural CF layer is in Eq. (1).

$$x = p_u \odot q_i \hat{y}_{ui} = \alpha\left(h^T x\right) \tag{1}$$

where Hadamard product of vectors is denoted by $\odot$. $P \in \mathbb{R}^{m \times k}$ and $Q \in \mathbb{R}^{n \times k}$ are user-item latent matrix. $P_u \in \mathbb{R}^k$ is $u_t h$ row of P and $P_u \in \mathbb{R}^k$ is $i_t h$ row of Q. Activation function denoted by α and h. $\hat{y}_{ui}$ is user's prediction for item.

– **Multilayer Perception Network (MLP)**—the user-item interaction may become complex. So, for higher flexibility and nonlinearity of interaction, the MLP network used, but it does not share user-item embeddings with GMF. The embeddings of user-item are concatenated for input. MLP is efficient for the estimation of user-item interaction besides the complication of interactions and nonlinear transformation. There is at least a hidden layer of neurons with nonlinear activation function. The layers of neurons are connected densely, and one layer of neuron is connected to other neuron in next layer. MLP sub-network defined in Eq. (2).

$$z^{(1)} = \phi_1 (U_u, V_i) = [U_u, V_i]$$
$$\phi^{(2)} \left(z^{(1)}\right) = \alpha^1 \left(W^{(2)} Z^{(1)} + b^{(2)}\right)$$
$$\cdots\cdots\cdots\cdots\cdots\cdots\cdots\cdots\cdots\cdots$$
$$\phi^{(L)} \left(z^{(L-1)}\right) = \alpha^L \left(W^{(L)} Z^{(L-1)} + b^{(L)}\right) \tag{2}$$

where activation function denoted by α_*, weight matrix denoted by W_*, and bias vector denoted by b_*. Function of corresponding layer denoted by ϕ_* and output of that corresponding layer denoted by z_* to obtain the results of GMF and MLP functions. Neural MF the second last layers of two sub-networks that are concatenated and form a feature vector which further passed to next layers. For the activation function that is used in MLP layer can be created using various functions such as sigmoid, hyperbolic tangent (tanh), and rectifier (ReLU). In this, the results are obtained with matrix h using ReLU activation function. Formula used for prediction layer is in Eq. (3).

$$\hat{y}_{ui} = \alpha \left(h^T \phi^L \left(z^{(L-1)}\right)\right) \tag{3}$$

– **Renyi Differential Privacy (RDP)**—RDP denoted by R has randomized mechanism $f : D \to R$ to have RDP of order (α, ϵ) or α. Hence, if for any adjacent $D, D' \in D$, it holds $D_\alpha \left(f(D) \parallel f(D')\right) \leq \epsilon$ (Mironov 2017).

The framework consists of GMF and MLP with embedding vectors of non-identical user and item. To constitute different categories of variables as continuous vectors. Neural network embeddings used, they place similar entities closer in embedding space because they are low-dimensional and learned. Neural network embedding model and supervised ML task are used to create embeddings. The result of the embeddings creates vectors, and its purpose is to find nearest neighbor in embedding space, to visualize in low dimensions, and work as an input for machine learning models. The hidden parameters in hidden layers controlled the structure of MLP, and for the activation function, ReLU is used which helps in deciding whether the neurons are activated or not.

The training of the datasets ensures that the error rate minimizes and the well generalization of framework in the new data. It is carried out by the cyclical learning rate scheduler, and then, the output is divided as training and validation sets that also, helps to check the negative sampling which is carried out by each user with the items that a user has not interacted to unobserved entries.

4 Dataset and Baselines

4.1 Dataset

We use movielens one million dataset (https://grouplens.org/datasets/movielens/1m/) which contains real-world data. The data stored include rating as an implicit feedback. The details of its statistics of dataset are number of users are 6040 and items are 3676. Total ratings for training are 750,121 in which unique users are 6040 and unique items are 3664. Total ratings for testing are 2,550,088 in which unique users are 6040 and items are 3676.

4.2 Input and Embeddings

We use Keras6 for implementation. Some authors use TensorFlow, but PyTorch improves development. All the other code is implemented in Python using PyTorch library. The $\_init\_()$ and $forward()$ functions are specifically used for framework. $\_init\_()$ used for defining the framework structure and $forward()$ used for defining the way that input tensors are forwarding.

For input using user u and an item i with its features (e.g., userID, movieID, movie category, etc.). The one-hot encoding is applied on their features. For u and i, feature vector is v_u^U and v_i^I, respectively, used to obtain embeddings p_u and q_i where $P \in \mathbb{R}^{M \times k}$ and $Q \in \mathbb{R}^{N \times k}$ denotes the embedding matrix. Embeddings size denoted by K, number denoted by M, and user feature denoted by N. The feature used in CF to identify u and i is ID. M used for number of users and N used for number of items. The embeddings of user and item are shown in Eq. (4)

$$P_u = P^T v_u^U \text{ and } Q_i = Q^T v_v^V \tag{4}$$

To create a dense network in our framework in embedding layer, the unique number of users and items is being identified for creating embedding matrix. An embedding dropout is applied after embedding layer for hidden layers. The number of integers or integer list is defined in hidden layers are identified, and after that the dropout layers rate are applied in each hidden layer to calculate the loss.

Before training loop, a cyclical learning rate (CLR) scheduler is applied to support cosine annealing with restart. PyTorch scheduler has *get_lr*() method. The *step*() method call is used to invoke the *get_lr* method each time which returns a list of learning rates, and this learning rates purely depend on current training epoch. In our framework, learning rate is same for all layers; hence, it returns a single value. For the current training epoch and learning rate base value, the single call function in CLR class returns new learning rate value.

Negative sampling is calculated for pairwise ranking loss. The item which user not interacted is calculated for each user because the model makes sure that item liked by user rank high and item not liked or interacted by user rank low at training stage.

In Fig. 1, the flowchart of NCF is shown which shows the workflow of the framework.

Fig. 1 Flowchart diagram of NCF

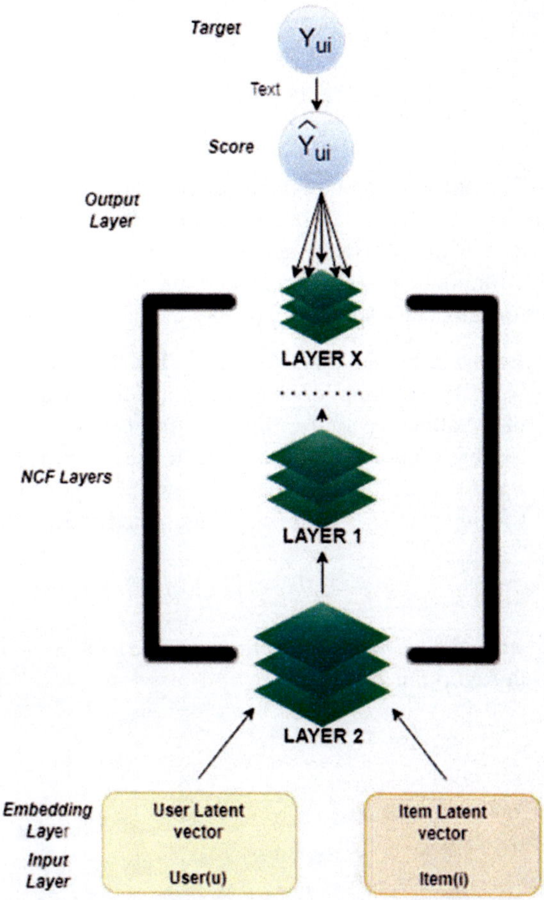

Algorithm 1 is used for the privacy-preserved NCF in which privacy is carried out by RDP. NCF includes GMF and MLP which combined further to produce results.

Algorithm 1 Pseudo-code for NCF

Input: Movielens 1m dataset
1: Load D and pre-process dataset by splitting into 1/4 as train and test sets.
2: Define hyperparameters of NCF.
3: Define hyperparameters of DP_SGD, Compute embedding using one-hot encoding.
4: Optimize using ReLU for.
5: Run CyclicLR scheduler.
6: Train NCF model.
7: Calculate train loss and epoch loss.
8: Calculate DP_SGD.
9: Evaluate the top k prediction.
Output: Accuracy, MAP, NDCG@10, Precision@10 and DP_SGD Recall@10

4.3 Baselines

We compare our proposed privacy-preserved NCF with Bayesian personalized ranking (BPR), restricted Boltzmann machines (RBMs), and simple algorithm for recommendation (SAR) by applying DP on them also. The propose model is based on MF for implicit feedback and focus on the relationship between users and items. So, we mainly considered the models for comparison which are using similar methods.

- **Bayesian Personalized Ranking (BPR)**—it optimizes the MF models with pairwise BPR ranking loss based on implicit user behavior (Rendle et al. 2012). The observations we used are only positive feedback while neglecting the real negative feedback and missing values. BPR uses item pairs i and j; for user preference, u current ranking is optimized; hence, i is positive, and j is negative items. So, training data $D_s : U \times I \times I$ determined as Eq. (5).

$$D_s = \{(u, i, j) \mid i \in I_u^+ \Lambda j \in I \setminus I_u^+\} \tag{5}$$

where i over j is prefer by the user. The BPR pseudo-code with RDP is mentioned in Algorithm 2.

Algorithm 2 Pseudo-code for BPR

Input: Movielens 1m dataset
1: Load D and pre-process dataset by splitting into 1/4 as train and test sets.
2: Define hyperparameters of DP_SGD.
3: Train (u, i, j) with user embedding u.
4: Optimize BPR by stochastic gradient descent.
5: Optimize TensorFlow privacy.
6: Update parameters.
7: Positive item i and negative item j previously interacted with embedding v_i and v_j, respectively.
8: Train data set D_s (Eq. 4).
9: Compute affinity $x_{ui} =< u, v_i >$.
10: Optimize likelihood function $p(i > v/\Theta)$ and prior probablity $p(\Theta)$.
11: Predict visible u feature vector w_u and hidden i feature vector h_i for derivating MF.
Output: Predict, Evaluate MAP, NDCG@10, Recall@10, Precision@10 and DP_SGD privacy

- **Restricted Boltzmann Machines (RBM)**—RBM is a unsupervised learning-based neural network model. It learns probability distribution between latent variables which are hidden units and input data which are visible units. By learning the distribution, the new data are obtained through sampling which are generalized. It learns user behavior instead of certain instance (Salakhutdinov et al. 2007). The model works on Boltzmann distribution defined as the least biased probability distribution on probability space $\sum$ used to obtain maximum entropy principle on space over $\sum$ which is determined as Eq. (6),

$$P = \frac{1}{Z}\mathrm{e}^{-\beta H} \tag{6}$$

where Z is partition function, β is noise parameter with inverse energy units, H is the Hamiltonian.
The training of the data is carried out by directly from joint probability distribution, conditional distributions where v is visible and h is hidden units determined as Eq. (7)

$$P(v, h) = P(v \mid h) P(h) = P(h \mid v) P(v) \tag{7}$$

The pseudo-code of RBM with DP is mentioned in Algorithm 3.

Algorithm 3 Pseudo-code for RBM

Input: Dataset movielens 1m
1: Load D and pre-process dataset by splitting into 1/4 as train and test sets.
2: Define hyperparameters of DP_SGD.
3: Define huperparamerters of RBM for joint probability distribution $P(v, h)$ where v visible and h hidden units.
4: Activate hidden layer h
5: Compute Boltzmann distribution (Eq. 5)
6: Build affinity matrix (Eq. 4) and Train (Eq. 6).
7: Compute DP_SGD
8: Compute prediction.
Output: Accuracy, MAP, NDCG@10, Recall@10, Precision@10 and DP_SGD

– **Simple Algorithm for Recommendation (SAR) for Single Node**—the algorithm applied for personalized recommendations generation though the transaction history of user. It is scaleable and works on item similarity for recommendations that user has affinity for. The similarity metrics generated are *S* for item-item relation and affinity metrics *A* for user-item relation (Sarwar et al. 2001). The similarity metrics can be obtained by co-occurrence metrics of all items $m \times m$ as C. The affinity metrics includes relationship of user and item that user has previously visited, there are two factors for affinities firstly, though differential weighting the information of type of interaction of user-item is recorded and secondly, when in past the interaction of user-item interaction take place is recorded. The two affinities for similarity are defined as Eq. (8)

$$a_{ij} = \sum_k w_k \left(\frac{1}{2}\right)^{\frac{t_0 - t_k}{T}} \tag{8}$$

where affinity a for user-item weighted k as sum of all events, w_k is weight of individual event, $\left(\frac{1}{2}\right)_n$ is scaling factor for T. The t_0 event of T takes half of the weight before as those taking place at t_0. The pseudo-code for SAR with DP is mentioned in Algorithm 4 (Table 1).

Algorithm 4 Pseudo-code for SAR

Input: Dataset movielens 1m

1: Load data set and pre-process.
2: Input data as two metrics item-item similarity metrics S and user-item affinity metrics A.
3: Define hyperparameters of DP_SGD.
4: Compute co-occurance matrix $\bar{C}$ where C_{ij} is number of time i take place within j.
   - Symmetric $C_{ij} = C_{ji}$.
   - non-negative $C_{ij} >= 0$.
   - largest element on diagonal $\forall (i, j)\, C_{ij} \left(C_{jj} \geq C_{ij}\right)$.
5: Compute Jaccard similarity matrix S_{ij}.
6: Compute affinity score a_{ij}.
7: Remove repeated seen item in training set.
8: Calculate top K item $S \times A$.

Output: Prediction, MAP, NDCG@10, Recall@10, Precision@10 and DP_SGD

5 Result and Discussion

This section includes the parameter settings of the framework. The result evaluation and discussion are based on overall performance of the framework and models, time taken by the framework for training and prediction and for privacy evaluation the comparison of RDP parameters take place.

Table 1 Evaluation metrics summary

Evaluation metrics	About	Formula
Mean Average Precision (MAP)	Recommendations that are more likely to be corrected	$\text{MAP} = \frac{1}{\min(m,N)} \sum_{k=1}^{N} P(k)$
Normalized Discounted Cumulative Gain (NDCG)	Measure the average performance of ranking of items	$\text{NDCG} = \frac{\sum_{i=1}^{P} \frac{2^{r d_i} - 1}{\log_2(i+1)}}{\sum_{i=1}^{\lvert \text{REL}_P \rvert} \frac{2^{r d_i} - 1}{\log_2(i+1)}}$
Precision@K	Ratio of numbers of recommendations that are relevant by all the possible correct item for the user	$P = \frac{\text{Correct Positive}}{\text{Predicted Positive}}$
Recall@K	Ratio of number of correct outcomes by all applicable samples	$R = \frac{\text{Correct Positive}}{\text{Condition}}$
RDP	P, Q are probability distribution over Renyi divergence R of order $\alpha > 1$	$D_\alpha (P \parallel Q) \leq \epsilon(\alpha), \forall \alpha \geq 1$

5.1 Parameter Settings

We implemented our framework on Python's library: PyTorch which considered neural network library and TensorFlow for the implementation of RDP. We split the dataset for training and test in the ratio of (0.75/0.25). The ranking metrics of the top recommended items are $K = 10$. For the activation function, ReLU is used. For RDP, the noise multiplier value used is 1.3, and delta value is exponential $(1e - 5)$ which means 1×10^{-5} or 0.00001. The baseline algorithms parameters are initialized as in the corresponding papers to compare our model.

5.2 Performance Comparison

The dataset used is splits into two parts test set and train set, and a splitting is stratified based on items. The models BPR, RBM, and framework NCF are MF-based algorithms for implicit feedback although SAR is also based on implicit feedback. The two algorithms RBM and NCF are truly neural network-based algorithms. BPR and SAR runs on CPU environment; rest of the algorithms runs on CPU as well as GPU environment. Figure 2 shows the performance of MAP, NDCG@10, recall@10, and precision@10 as for the quantity of prescient elements and predictive factors. In all baseline models, MAP consistently yields the good performance with a focus on models leveraging auxiliary data. MAP indicates the average of the maximum precision with respect to different recall values. The MAP of BPR and SAR are almost similar $(0.081 > 0.060)$; the MAP of RBM and NCF are $(0.021 < 0.063)$ in which the performance of NCF is better than the RBM. The map value is affected by the precision and recall. NCGD for '@10' specifies the threshold for the hit value. It uses optimal hyperparameter to gave the peak performances during validation by training the same amount of epochs per rounds. The NCGD@10 of baselines model shows distinct values, RBM, SAR, and NCF $(0.142 < 0.299 < 0.358)$, but BRP probability model optimizes the likelihood function and maximizes the posterior and prior probability; thus NDCG@10 is slightly higher (0.407). The other hyperparameters are learning rate used to control step size in gradient update rules which are the considered values for models (BPR 0.01, NCF 0.001, RBM 0.004 (default)) SAR learning rate is not matters because it is based on users transaction history. To train the BPR model lambda regression (value 0.01) is used, in NCF model, three layers (16, 8, 4) are applied; in RBM and SAR, affinity matrices are used. The ranking metrices precision@10 is calculated in BPR is 0.300; NCF is 0.128; SAR is 0.270, and NCF is 0.318; it clarifies that the precision of NCF is higher, and also, it is slightly similar to BPR. The ranking metrics recall@10 in BPR is 0.110; RBM is 0.050; SAR is 0.104, and NCF is 0.1294 from that the highest recall is of NCF. All the data of performance comparison is shown in Table 2.

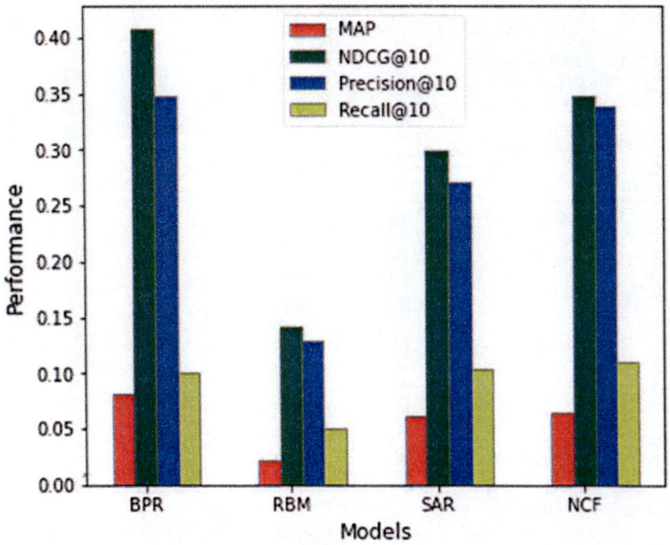

Fig. 2 Performance comparison of different models based on the parameters of MAP, NDCG@10, Precision@10 and Recall@10

Table 2 Models comparison on dataset

Model	MAP	NDCG@10	Precision@10	Recall@10
BPR	0.081621	0.407850	0.347222	0.133157
RBM	0.021648	0.142111	0.128377	0.050466
SAR	0.060519	0.299245	0.270116	0.104350
NCF	0.05922	0.210221	0.202554	0.203233

5.3 *Time Comparison*

The training time and prediction time of every model and framework is observed. The results of time of training and prediction are shown in Fig. 3. In BPR, the size of the vectors w_u and h_i and dimension of the latent space are controlled by the number factors named as k (considered, $k = 200$), and stochastic gradient descent (SGD) includes the iterations calculation which affect the training time. The prediction time is also affected by the number of iterations as well as from number of epochs. So, the optimization training time is 54.83 s, and prediction time is 27.41 s. In RBM, it use Gibbs sampling which distinguish the data into positive sampling and negative sampling; in negative, the sampling process is repeated after each epoch. The training of each epoch is increased by the sampling protocol which affect the training which is 225.15 s and prediction time is 10.52 s. In SAR, the training time is not considered because it predicts based on user preference which are need to be present on test set. The usual training time is 4.7078, and prediction time is 8.275. In NCF,

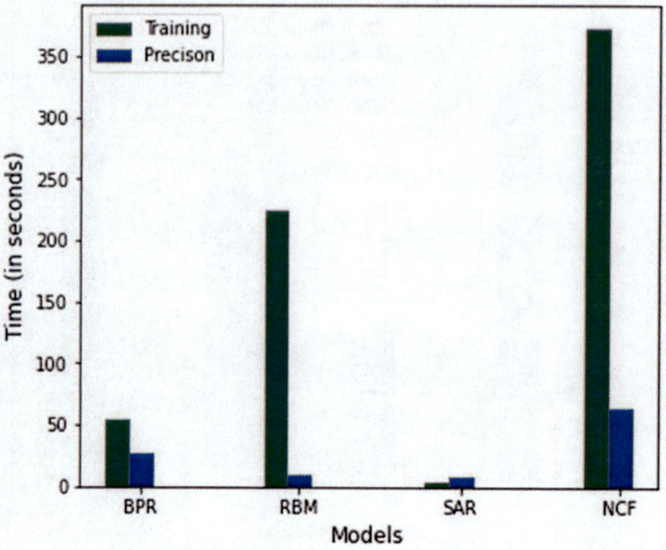

Fig. 3 Time comparison of models based on time taken for training and prediction

number of epochs control the training time and train loss; during each epoch, the train loss is calculated (epochs = train loss, 10 = 20.269, 20 = 0.263, 30 = 0.259, 40 = 0.257, 50 = 0.255), which affect the training (4686.34 s) as well as prediction time (64.49 s).

5.4 *RDP Comparison*

For the RDP, in baseline models, various parameters are considered on the basis of which optimal RPD order ϵ is calculated. The maximal the RPD order shows good practical privacy and depends on the data size and number of epochs. By using TensorFlow library, differentially private (DP-SGD) is calculated which is used to modify the gradient. The overall results of precision, recall, and DP with epochs are mentioned in Fig. 4, and the comparison of optimal RPD order is mentioned in Fig. 5. The value of noise multiplier is 1.3 which adds during the training; the increase in noise presents better result in privacy and lower result in utility and the value of δ is $(1e - 5)$ used in all models. The sampling rate is the probability of single training point used in microbatches. In BPR, the sampling rate is 0.317%; DP with epochs is 0.627 which depends on the no of epochs run in the model, and optimal RPD order calculated is 26.0, which is maximal. In RBM, the sampling rate is 0.059%, DP with epochs is 0.4958 and optimal RPD order is 24.0. In SAR, the sampling rate is 0.041%

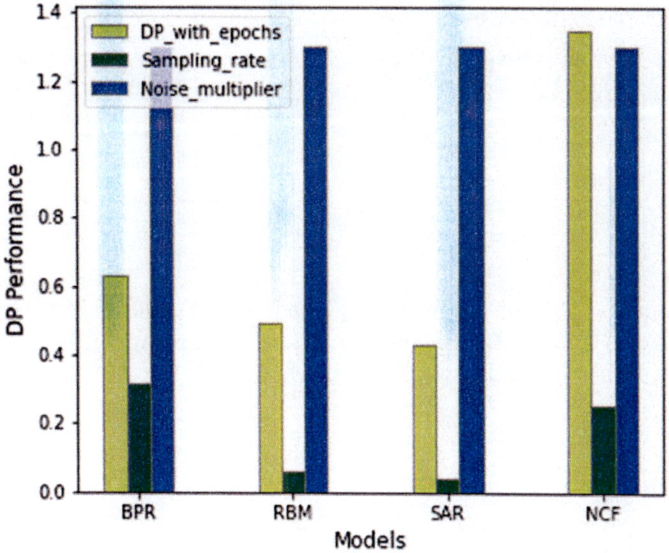

Fig. 4 Performance comparison of models for RDP based on parameters DP with epoches, sampling rate, noise multiplier

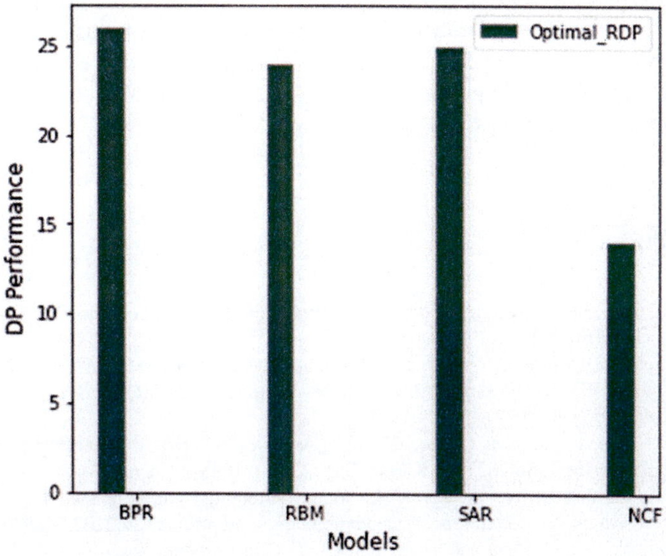

Fig. 5 Overall RPD comparison of models

Table 3 DP evaluation on models

Model	Sampling rate	DP with epochs	Delta	Noise multiplier	Optimal RPD order
BPR	0.317	0.6270	$(1e-5)$	1.3	26.0
RBM	0.0595	0.4958	$(1e-5)$	1.3	24.0
SAR	0.041	0.432	$(1e-5)$	1.3	25.0
NCF	0.254	1.3491	$(1e-5)$	1.3	14.0

which is minimal; DP with epochs is 0.432; optimal RPD order is 25.0. In NCF, the sampling rate is 0.254%; DP with epochs is 1.349, and optimal RPD order is 14.0 which is minimal. All the data of RPD is shown in Table 3.

6 Conclusion

We have presented a neural collaborative filtering with differential privacy. The framework is based on neural network and overall perform better in terms of precision@10, recall@10, and NDCG@10 as compare to other baseline models. Our framework provides attention to privacy with NCF for the secure computation and result. The interaction of privacy in framework is carried out by using RDP, which provide privacy through DP-SGD. The performance of all models is compared with framework, and the result provided is that the NCF with DP performs well and provides accurate results also maintains privacy.

References

S.M. Albladi, G.R. Weir, Predicting individuals' vulnerability to social engineering in social networks. Cybersecurity **3**(1), 1–19 (2020)

T. Bai, J.R. Wen, J. Zhang, W.X. Zhao, A neural collaborative filtering model with interaction-based neighborhood, in *Proceedings of the 2017 ACM on Conference on Information and Knowledge Management* (2017), pp. 1979–1982

A. Bilge, C. Kaleli, I. Yakut, I. Gunes, H. Polat, A survey of privacy-preserving collaborative filtering schemes. Int. J. Softw. Eng. Knowl. Eng. **23**(08), 1085–1108 (2013)

C. Chen, H. Wu, J. Su, L. Lyu, X. Zheng, L. Wang, Differential private knowledge transfer for privacy-preserving cross-domain recommendation. arXiv preprint arXiv:2202.04893 (2022)

X. He, X. Du, X. Wang, F. Tian, J. Tang, T.S. Chua, Outer product-based neural collaborative filtering. arXiv preprint arXiv:1808.03912 (2018)

X. He, Z. He, J. Song, Z. Liu, Y.G. Jiang, T.S. Chua, NAIS: neural attentive item similarity model for recommendation. IEEE Trans. Knowl. Data Eng. **30**(12), 2354–2366 (2018)

X. He, L. Liao, H. Zhang, L. Nie, X. Hu, T.S. Chua, Neural collaborative filtering, in *Proceedings of the 26th International Conference on World Wide Web* (2017), pp. 173–182

Y. Koren, R. Bell, C. Volinsky, Matrix factorization techniques for recommender systems. Computer **42**(8), 30–37 (2009)

S.K. Lam, J. Riedl, Shilling recommender systems for fun and profit, in *Proceedings of the 13th International Conference on World Wide Web* (2004), pp. 393–402

T. Li, L. Song, C. Fragouli, Federated recommendation system via differential privacy, in *2020 IEEE International Symposium on Information Theory (ISIT)* (IEEE, 2020), pp. 2592–2597

J. Lian, X. Zhou, F. Zhang, Z. Chen, X. Xie, G. Sun, xDeepFM: combining explicit and implicit feature interactions for recommender systems, in *Proceedings of the 24th ACM SIGKDD International Conference on Knowledge Discovery & Data Mining* (2018), pp. 1754–1763

I. Mironov, Rényi differential privacy, in *2017 IEEE 30th Computer Security Foundations Symposium (CSF)* (IEEE, 2017), pp. 263–275

S. Rendle, C. Freudenthaler, Z. Gantner, L. Schmidt-Thieme, BPR: Bayesian personalized ranking from implicit feedback. arXiv preprint arXiv:1205.2618 (2012)

R. Salakhutdinov, A. Mnih, G. Hinton, Restricted Boltzmann machines for collaborative filtering, in *Proceedings of the 24th International Conference on Machine Learning* (2007), pp. 791–798

B. Sarwar, G. Karypis, J. Konstan, J. Riedl, Item-based collaborative filtering recommendation algorithms, in *Proceedings of ACM World Wide Web Conference*, vol. 1, Aug 2001. https://doi.org/10.1145/371920.372071

S. Selvaraj, S.S. Gangadharan, Privacy preserving hybrid recommender system based on deep learning. Turk. J. Electr. Eng. Comput. Sci. **29**(5), 2385–2402 (2021)

B. Shao, X. Li, G. Bian, A survey of research hotspots and frontier trends of recommendation systems from the perspective of knowledge graph. Expert Syst. Appl. **165**, 113764 (2021)

Q. Wang, H. Yin, T. Chen, J. Yu, A. Zhou, X. Zhang, Fast-adapting and privacy-preserving federated recommender system. VLDB J. 1–20 (2021)

X. Wang, X. He, M. Wang, F. Feng, T.S. Chua, Neural graph collaborative filtering, in *Proceedings of the 42nd International ACM SIGIR Conference on Research and Development in Information Retrieval* (2019), pp. 165–174

K. Xu, Z. Yan, Privacy protection in mobile recommender systems: a survey, in *International Conference on Security, Privacy and Anonymity in Computation, Communication and Storage* (Springer, 2016), pp. 305–318

S. Zhang, L. Yao, A. Sun, Y. Tay, Deep learning based recommender system: a survey and new perspectives. ACM Comput. Surv. (CSUR) **52**(1), 1–38 (2019)

T. Zhu, G. Li, Y. Ren, W. Zhou, P. Xiong, Differential privacy for neighborhood-based collaborative filtering, in *Proceedings of the 2013 IEEE/ACM International Conference on Advances in Social Networks Analysis and Mining* (2013), pp. 752–759

BERT-Based Approach for Suicide and Depression Identification

S. P. Devika, M. R. Pooja, M. S. Arpitha, and Ravi Vinayakumar

Abstract With the advent of the Internet, there is an increasing requirement for sophisticated and intelligent systems that can operate efficiently handle the identification on social media about health-related issues, such as depression and suicide recognition. The data generated by social media users is unstructured and unreliable. The text representation and deep learning algorithms used, however, give only a limited amount of information's as well as expertise regarding the various user-supplied texts. We started with bidirectional long short-term memory (Bi-LSTM) with progressed toward transformer models similar as bidirectional encoder representations from transformers (BERT). In our research, we discovered that they perform considerably superior to traditional deep learning architectures such as Bi-LSTM and BERT. We have designed our own data collecting platform using Reddit, one of the majority popular social networking sites. Finally, we have used BERT plus Bi-LSTM to efficiently evaluate as well as notice indicators of depression plus suicide in social media posts.

Keywords BERT · Bi-LSTM · Depression · Suicide · Reddit

S. P. Devika (✉) · M. R. Pooja · M. S. Arpitha
Department of Computer Science and Engineering, Vidyavardhaka College of Engineering, Mysuru, Karnataka, India
e-mail: devikasp5@gmail.com

M. R. Pooja
e-mail: pooja.mr@vvce.ac.in

M. S. Arpitha
e-mail: arpitha.ms@vvce.ac.in

R. Vinayakumar
Center for Artificial Intelligence, Prince Mohammad Bin Fahd University, Khobar, Saudi Arabia
e-mail: vinayakumarr77@gmail.com

A. B. Reddy et al. (eds.), *Proceedings of Third International Conference on Advances in Computer Engineering and Communication Systems*, Lecture Notes in Networks and Systems 612, https://doi.org/10.1007/978-981-19-9228-5_36

1 Introduction

According to the latest survey, a rising number of people suffering from depression and attempting to suicide, particularly teens and young adults, are using social media to communicate thoughts. Suicide is the willful termination of one's own life. Depression is a widespread condition that affects more than one person 264 million people globally (Zeberga et al. 2022). Prolonged unhappiness and suicide might lead to self-destruction if the impacted person is not given proper treatment as well as assistance right once. Social media has lately emerged as a powerful tool for examining the psychological well-being and mental status users, in particular teenagers (Solse et al. 2022). The development of web and communication technologies, especially online informal communities, has invigorated how those engage furthermore, convey with one another automatically. Reddit, for instance, gives literary and interactive media material, yet in addition, permits clients to communicate their sentiments, feelings, and opinions about a point, subject, or issue on the web (Chiong et al. 2021). Moreover, there is a developing corpus of studies tending to the meaning of interpersonal organizations in the construction of social connections like separations, dysfunctional behavior, smoking and drinking backslide, inappropriate behavior, and self-destructive contemplations. Deep learning, which permits models to learn portrayals from information naturally, has lately resulted in a number of advances (Nisa and Muhammad 2021). Similarly, certain potential deep learning-based algorithms had been applied to some intellectual healthiness application (e.g., depression identification and suicidal ideation) and reached performance in competition. Since textual material was sequential, the sequence words and their meanings relationship in phrase, that is context, are important in drawing entire meaning from the sentence. As an outcome, there is desire to apply computationally deep techniques to textual data (Amanat et al. 2022). This paper gives a brilliant and setting mindful profound gaining system in view of bidirectional encoder portrayals to efficiently use transformers (BERT) distinguish emotional wellness related issues in online entertainment client posts with expanded order precision. As a rule, our significant commitments to this exploration study are as per the following:

(a) A novel methodology is provided for extracting massive amounts of extremely relevant depression along with suicide-related information from Reddit.
(b) BERT, a bidirectional text representation model, based on deep neural networks is utilized to incorporate textual data from mental health problems while keeping contextual and semantic implications.
(c) We present a method of knowledge distillation. It is a method of transmitting information from one source to another BERT as of a big pre-trained model on a smaller dataset one on the way to improve concert and precision.

The experiment results suggest that our model outperforms the compared approaches, even after multiple hyper-parameter tweaks.

2 Dataset

2.1 Data Collection

A dataset is a compilation in terms of a user's words, titles, and comments on other people's posts from social media, such as Reddit. The data is unidentified, and users are identifiable by user numbers. There are two types of users: those who self-harm and those who do not self-harm, represented by 1 and 0 in the data. In the mental health area, text subtask classification includes recognizing suicide notes (Asad et al. 2019).

2.2 Data Set Analysis

We investigated in our dataset terms of vocabulary, subjects, as well as other linguistic aspects to have a deeper understanding of the language signals as well as linguistic use persons who compose suicide posts. Additionally, they used term clouds to compare in a visual manner the usage of high frequency words in diverse texts. (Table 1).

There are 2 classes of user's "self-harm" and "not self-harm", denoted by 1, 0 in data.

3 Proposed Architecture

In this section, we describe the structure for retrieving, processing, evaluating, and organizing data from social networking sites concerning psychological well-being issues for example depression and suicide. To detect depression and suicidal detection, we utilized user text that was collected from the public networking site Reddit. The data collecting, preprocessing, and labeling layer are responsible for extracting useful information from data sources such as Reddit. Figure 1 depicts a generic architecture presented in this research work.

3.1 Data Collection

This paragraph goes through process of gathering information from Reddit sources. Figure 1 of step 1 depicts the whole data collecting pipeline for the proposed architecture. We concentrated our Reddit data on select subreddits that were created relevant to our intended themes after that conducted a topic-specific search query.

Table 1 Raw data and summarized data

Before	Hi. I don't know where else to go. I am devastated right now, feeling like I'm spiraling out of control, I feel so fucking numb from crying my lungs out. Even my best friend is no longer supportive, my family, parents at home are shit. I have no real friends who care, and I can't be bothered to do anything right now, I just want to sleep without waking up to see another morning. I feel so fucking pathetic and shit, I cut myself regularly now and I'm having so much fucking self pity, I'm pretending to be normal because I don't want to lose my best friend, the only person who has helped me come this far from previous suicidal attempts. But suppressing it makes it so much fucking worse I feel like nobody fucking cares, no matter how many times I shout for help. No one close to me cares. I don't know what to do. I'm leaning toward paracetamol poisoning all over again, if I die I die, If I don't I try a different method. I just feel like absolute shit, jobless. I regret everything in my life, every fucking thing, I don't know who to trust anymore, I don't even have a fucking shoulder to lean on while crying my lungs out. I'm feeling like shit. She is the best and supportive, and my best friend but I can't hurt her by harming myself so It's a fucking struggle. I just feel so enraged and angry at everything. I can't fucking find a job due to covid-19. Sometimes it feels like If I had some money to get myself kicked off I can go near her and find a job and at least stay sane. But right now it all seems so fucking impossible I feel like I'm such a fucking mess and I just wish to fucking die right now, I feel so much burden when I inhale, I just want to die. I got no where else to go. I am planning on consuming meds tonight I just can't do it anymore
After	Hi I don't really know how to phrase this situation but I'll try. My life is at a really good point right now, I'm never really depressed over stuff and 99 percent of the time my mind is clear. I'm about to graduate high school and I'm really excited. However, people in my family and friend group have tons of issues. Whether they sleep all day, hate themselves, or have no ambition to keep living on in this world, they've got problems. I don't wanna sound like I'm gloating, but usually I'm the person that a lot of these people end up going to because usually I'm able to talk people through issues and help them in the long run. Yeah sometimes their issues make me really sad and stuff because who doesn't feel sad when people are telling you they feel worthless

Fig. 1 Format of the dataset used

```
Index(['selftext', 'is_suicide'], dtype='object')
```

	selftext	is_suicide
0	Hi I don't really know how to phrase this situ...	0
1	i have been so depressed these past couple wee...	1
2	Hi..I don't know where else to go. I am devast...	0
3	The shit like "it will get better, everyone is...	1
4	I wish I was prettier. I wish I didn't feel li...	0
5	This is a weird question, but is there somepla...	1
6	I'm just venting into the void because I feel ...	0
7	Hey guys. Last week, i went through the roughe...	0
8	I don't know when I will pass. I don't know ho...	1
9	Been fighting with depression for a couple of ...	0
10	I don't talk to any of my friends I leave them...	0

3.2 Preprocessing

As shown in Fig. 1, data preparation is a process way of noise cleaning and filtering ambiguous data in order for it may subsist readily used to extract features. We must use concepts to apply machine learning and data mining extract meaning from such as text corpora before they are fed into any classification model.

(a) **Tokenization**: As indicated in Fig. 1, these are tokens used toward uncover and are patterns used as a source of information for following typical processes stemming and lemmatization in the NLP pipeline. For further analysis, all of the in the text provided a file is indicated by a collection of terminology.
(b) **Removing Stop Words**: This section discusses the elimination of terms that contain no relevant information about mental health disorders in general. We divide the text into words, convert it to phrases, and then remove any words that are on the NLTK's list of stop words in order to elimination prevent utilising terms from a phrase.
(c) **Part-of-Speech (POS) Tagging**: The primary objective for this phase is to exclude any POS that does not contribute to the diagnosis of sadness and suicide.

3.3 Representation of Text

Term embedding was the method for representing words in a corpus of text; using a vector with real values captures the words definition, so such phrase in the same is in the neighborhood expected to have the similar meaning. To retrieve the context of a particular phrase in our gathered corpus of text, we suggest the latest and effective approach, BERT.

3.4 Classifier Models

Words are represented as dense vectors using deep learning techniques, as well as sentiment categorization based on machine learning/deep learning is done by sending dense vectors to classification models. We describe a bidirectional LSTM (Bi-LSTM) that includes past and future context information from our social media data by employing two LSTM units that function in both the left and right directions.

4 Methodology

The primary purpose of this project to create a computerized system that can recognizing and analyzing psychological well-being concerns with deep learning methods like Bi-LSTM and BERT.

(a) **BERT:** To begin, BERT is an acronym that stands for bidirectional encoder representations from transformers. Every word has its own significance, which shall explore in this essay one by one. Second, BERT has been trained on a massive corpus of unlabeled text (Acheampong et al. 2021). BERT makes use of the transformer structural design, which is a device for focusing attention, detects related relationships between words. The transformer consists of two parts: a text input encoder and a text input decoder that decodes it and predicts the task. BERT uses two methods during training to seize the context in terms of words on both the right and left sides of the page orientations. A prediction is carried out in three major phases, as represented in Fig. 2:

(a) A layer of classification is applied above the encoder output.
(b) The vectors of output are multiplied as determined through the embedding matrix and transformed into lexical size.
(c) Using Softmax, compute the distribution of probabilities for all terms in the dictionary.

(b) **Bi-LSTM**: Bi-LSTM, despite its name, in fact implements and a sequence is processed using an LSTM model in both ways. To address sequence classification difficulties, a Bi-LSTM is comprised one of two such memories designed. We recommend a bidirectional LSTM (Bi-LSTM) based on our findings, which

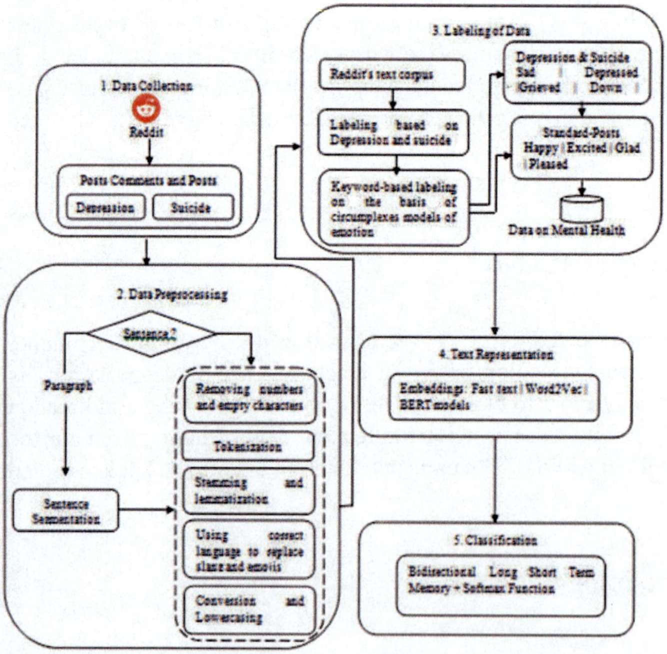

Fig. 2 Schematic designs are layered depict a well-organized perspective of the automated identification of depression and suicide based on user post

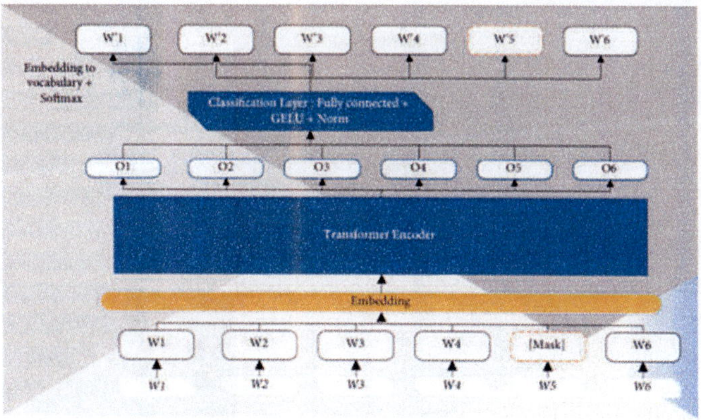

Fig. 3 Masked language modeling is a technique used to train BERT in which a model forecasts what the masked word should be using context words around a mask token

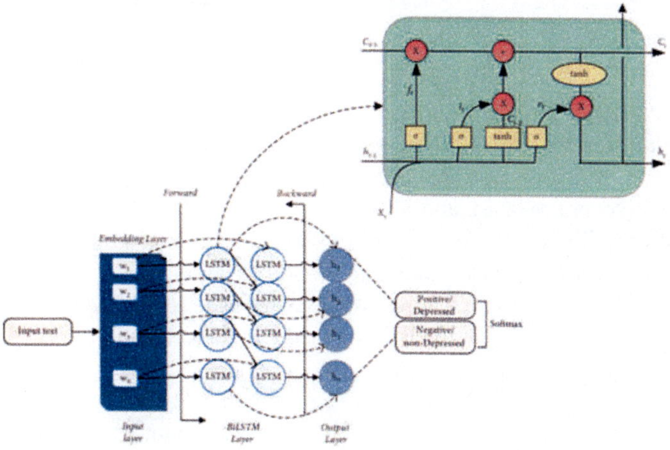

Fig. 4 Bi-LSTM technique was utilized to collect sequential characteristics

utilizes two LSTM components that function in mutually the right and left direction to include information about the our social media data provides historical and future perspective. As seen in Fig. 3, Bi-LSTM keeps a long-term link between words as well as redundant context information (Figs. 4 and 5).

5 Result and Discussion

In our experiment, we used our own data to test the suggested Bi-LSTM and BERT-based skills algorithms for recognizing depression and suicide-related sentiments,

Fig. 5 Confusion matrix

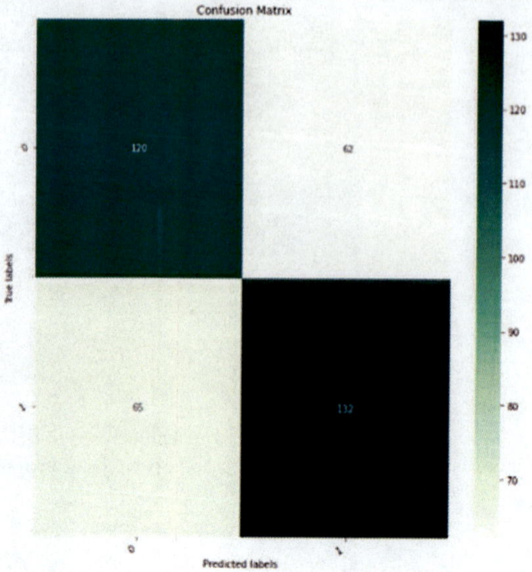

as shown in Tables 2. Based on the data we acquired, our primary goal to determine which features are extracted strategies most excellent supported the concert for depression and suicide identification. BERT and Bi-LSTM achieved considerable classification accuracy of 81 and 72%, correspondingly. In our proposed system, discovering this little model derived from a bigger pre-trained model is referred to as knowledge distillation. We can show that Bi-LSTM with BERT can reliably predict depression and suicide from a dataset. We compared the Bi-LSTM model's training and testing accuracy to that of the BERT-based text representation.

Precision: precision = (TP)/(TP + FP). The number of 1's is TP, whereas 0's is FP.

Recall: recall = (TP)/(TP + FN). When False Negative outnumbers False Positive, recall is an important metric to consider.

F_1-score

The F_1-score is calculated as the harmonic mean of recall and accuracy. The F_1-score favors classifiers with comparable accuracy and recall (Table 3).

$$F1\text{ - score}\frac{2}{\frac{1}{\text{recall}}+\frac{1}{\text{precision}}}$$

Table 2 Prediction accuracy with two different methods

Approaches	Accuracy
Bi-LSTM	81
BERT	72

Table 3 Classification report of confusion matrix

	Precision	Recall	F_1-score
0	0.65	0.66	0.65
1	0.68	0.67	0.68

A confusion matrix is far superior approach to evaluate the performance of a classifier. The fundamental concept was to count the number of examples of class A wshich are classified as class B. The confusion matrix's rows indicate actual classes, while the columns represent predicted classes.

6 Conclusion

In this study, we used BERT and Bi-LSTM are two deep learning approaches to create a robust framework for detecting depression and suicide. The suggested approach improves the smart healthcare systems' precision in detecting psychological well-being disorders, particularly depression as well as suicide. We addressed many significant elements, such as the collecting of textual data from social networks about mental health via interfaces for application programming and the preprocessing section, which focus on translation transformation of unstructured data into a structured data comprehensible shape with numerous data filtration options algorithms. Text representation model based on BERT presented converts gathered words transformed into vectors that capture the conceptual content in the gathered corpus of text to increase classification task performance utilizing an attention mechanism. Moreover, we made our own information gathering module to set up the Reddit informational index by separating the most pertinent text-based information that can be used to build a wise model for brilliant medical services frameworks. Finally, we performed a systematic experiment to demonstrate that future BERT and Bi-LSTM representation enhances the text sentiment's accuracy categorization from user comments.

References

F.A. Acheampong, H. Nunoo-Mensah, W. Chen, Transformer models for text-based emotion detection: a review of BERT-based approaches (2021)

D.A. Amanat, M. Rizwan, A.R. Javed, M. Abdelhaq, R. Alsaqour, S. Pandya, M. Uddin, Learning for depression detection from textual data (2022)

N.A. Asad, M.A. Mahmud Pranto, S. Afreen, M.M. Islam, Depression detection by analyzing social media posts of user (2019)

R.K. Bhog, S.A. Nagare, S.P Mahajan, P.S. Kor, Depression detection by analyzing social media post of user. Department of Computer Engineering. Pune (2022)

A.-M. Bucur, A. Cosma, L.P. Dinu, Early risk detection of pathological gambling, self-harm and depression using bert. Bucharest (2021)

R. Chiong, G.S. Budhi, S. Dhakal, F. Chiong, A textual-based featuring approach for depression detection using machine learning classifiers and social media texts. Australia (2021)

F. Haque, R.U. Nur, S. Al Jahan, Z. Mahmud, F.M. Shah, A transformer based approach to detect suicidal ideation using pre-trained language models. Bangladesh (2020)

P. Jain, K.R. Srinivas, A. Vichare, Depression and suicide analysis using machine learning and NLP. India (2022)

B. Jang, M. Kim, G. Harerimana, S.U. Kang, J.W. Kim, Bi-LSTM model to increase accuracy in text classification: combining word2vec CNN and attention mechanism (2020)

M.L. Joshi, N. Kanoongo, Depression detection using emotional artificial intelligence and machine learning: a closer review. India (2022)

R. Martinez-Castano, A. Htait, L. Azzopardi, Y. Moshfeghi, Early risk detection of self-harm and depression severity using BERT-based transformers (2020)

M. Munikar, S. Shakya, A. Shrestha, Fine-grained sentiment classifition using BERT (2019)

Q.U. Nisa, R. Muhammad, Towards transfer learning using BERT for early detection of self-harm of social media users. Pakistan (2021)

S. Smys, JS. Raj, Analysis of deep learning techniques for early detection of depression on social media network—a comparative study. India. (2021)

D. Solse, A. Magar, P. Harde, N. Palve, M.T. Jagatap, Depression detection by analyzing social media post in machine learning using bert algorithm. India (2022)

D. Williama, D. Suhartonob, Text-based depression detection on social media posts: a systematic literature review. Indonesia (2021)

K. Zeberga, M. Attique, B. Shah, F. Ali, Y.Z. Jembre, T.-S. Chung, A novel textmining approach for mental health prediction using Bi-LSTM and bert model. Korea (2022)

MG-Net: Multiple Person and Gadget Detection for Online Exam Proctoring System

Tejaswi Potluri and Venkatrama Phani Kumar Sistla

Abstract Recently, the education system has shifted from traditional classroom teaching to online or blended mode. Main challenge to education system is exam proctoring. One of the important limiting factors in the field of education is the inability to proctor online examinations. Manual proctoring through webcam is expensive and labor intensive. In this paper, we proposed MG-Net which automatically detects multiple persons and gadgets in the frame and alerts the proctor. The convolution operations were performed along with residual blocks to improvise the performance if the model over existing systems. Our model is trained using COCO dataset. Our model is tested on the customized dataset and performs well with 92.8 accuracy.

Keywords MG-Net · Proctoring · Residual block · Softmax · Convolution

1 Introduction

Automated Online Proctoring is the best solution that requires no human involvement. This proctoring or so-called invigilation is no longer restricted to a scheduled time and exam center anymore. The COVID-19 lockdown led to fast change of conducting exams. This is where online and remote proctoring comes to play. Nowadays, people are adopting virtual platforms to take the tests from their places. This is the modernized way of conducting exams which is scalable and can be accessed by the test takers any time.

Online exam proctoring is a simple, effective, and technology-based alternative to traditional human proctoring, where the integrity of the test is preserved with the help of AI-enabled monitoring tools to perform object detection. Object detection can be used to identify objects on screen like mobile phone, books, and unnecessary gadgets in order to avoid the malpractice during the exam.

Remote proctoring can be divided into three categories (Allen and Seaman 2015), i.e., (i) auto-proctoring, (ii) live proctoring, and (iii) record and review proctoring.

T. Potluri (✉) · V. P. K. Sistla
Department of Computer Science and Engineering, VFSTR University, Vadlamudi, Guntur, India
e-mail: tejaswi_p@vnrvjiet.in

A. B. Reddy et al. (eds.), *Proceedings of Third International Conference on Advances in Computer Engineering and Communication Systems*, Lecture Notes in Networks and Systems 612, https://doi.org/10.1007/978-981-19-9228-5_37

Auto-proctoring uses AI to identify suspicious events with the help of webcam feed and provides the vast ability to track a larger number of candidates simultaneously. Live proctoring incorporates real human with the technology in which an invigilator monitors the video feeds and can interfere in real time if necessary in a candidate's exam. Record and review proctoring method provides flexibility to candidates who do not need to fix time and coordinate with the availability of the invigilator. Instead, the audiovisual and screen-sharing feeds of candidates are recorded during the examination, a proctor then reviews the recorded exam, and marks any suspicious activity if occurs.

Further section of the paper is described as below: Sect. 2 describes the literature review, Sect. 3 describes the proposed system, Sect. 4 describes the experimental results and challenges of the model, and Sect. 5 concludes the paper.

2 Literature Survey

Most of the traditional systems process the images taken at regular intervals to identify the mischievous behaviors. Many systems have been developed for automated online proctoring like automated proctoring (Atoum et al. 2017), in which rear camera and front camera are used to capture the two views of the examinee for detecting mischievous behavior. In this system, many modules like text detection, speech recognition, gaze estimation, gadget detection are implemented. Agarwal (2020) has proposed one more system using ResNet50 Caffe model to detect the mischievous behavior of the examinee. In all the systems, the continuous video that is recorded from the webcam of the examinee is to be processed. The video should be processed further for the detection of multiple persons and gadgets (https://towardsdatascience.com/automating-online-proctoring-using-ai-e429086743c8). There are many video processing systems like (Potluri and Nitta 2016) uses dominant colors, (Potluri et al. 2017) shape features, and (Potluri and Gnaneswara Rao 2019) SURF, HARRIS, and BRISK features for extracting features from videos. The accuracy of the proctoring system depends upon the features used for the processing. The features extracted are used for the object detection.

2.1 Object Detection

Among many applications of Convolutional Neural Networks, image classification is the most popular and exciting of all the applications. Apart from simple image classifications, computer vision has many other interesting and fascinating applications. One such interesting problem is object detection. A human brain can recognize and locate the objects in a fraction of second. The main purpose of object detection is to incorporate this intelligence in a machine. Object detection is a computer vision technique for recognizing and locating the objects in images or in videos. Object detection technique is generally used in video surveillance. Authors (Tan et al. 2020) have proposed efficient for the scalable and efficient object detection from the images. The performance of the model is not remarkable. Authors (Potluri et al. 2021) have proposed Mobile Net for object detection. Object detection can also be used for controlling mouse action by processing video (Potluri et al. 2020).

Object Detection using Deep Learning. There are various techniques to perform object detection. Many object detection algorithms use machine learning and deep learning techniques to produce results. Caffe model can also use to detect the objects (Vaishali 2019) from the frame. Popular deep learning-based approaches use CNNs such as R-CNN and YOLO (Bochkovskiy and Wang 2020) which can automatically learn to detect objects.

2.2 YOLO

YOLO (Bochkovskiy and Wang 2020) is a Convolutional Neural Network (CNN) for achieving object detection. CNNs are classifier-based systems which can process input images as structured arrays of data and identify patterns between them. In comparison with other neural networks, YOLO is faster and maintains good accuracy.

The input image in YOLO is divided into S × S grids. Grids are used to detect an object when the object's center of gravity falls. For each cell in the grid, YOLO performs the classification and localization. A single composite network is used to predict multiple bounding boxes and their class probabilities at the same time. YOLO uses a 1 × 1 reduction layer and a 3 × 3 convolutional layer, consisting of 24 convolutional layers and 2 fully connected layers. A grid cell can detect only one object because the classification and search network can detect only one object.

2.3 YOLOV2

In YOLO (Du 2018), the number of localization errors is more, and it has low recall. Hence in YOLOv2, improving recall and localization are primarily focused while maintaining classification accuracy.

YOLOv2 solves complex problems with high accuracy. The author proposes here a new classification model Darknet-19 with Darknet as the backbone. Darkney-19 has 19 layers of complexity; up to five layers can achieve accuracy of 91.2% on ImageNet which is relatively better than VGG with 90% and YOLO with 88%.

2.4 YOLOV3

To perform the detection task, another 53 layers are stacked on the first 53 layers, and finally a 106-layer fully convolutional architecture is formed. There is no maximum grouping here. For each convolution operation, we have convolution, followed by batch normalization and filtered ReLu. Yolov3 (Zhao and Li 2020) achieves the best speed and accuracy in the MS COCO dataset, which is a large-scale object detection database. The YOLOv3 (Mao et al. 2019) is also used for embedded applications. Authors in Ju et al. (2019) have proposed that this model performs well in detecting multiple objects in single frame which is more useful for the proctoring application.

From the literature survey, authors have experimented in YOLOv3 in improving the performance of model for multiple person detection and gadget detection.

3 Proposed System

In the proposed system, our relevance is to focus on detecting the malpractices during online exam like usage of gadgets and appearance of more than one person. Authors have used YOLOv3 pre-trained weights trained on the COCO dataset to detect multiple examinee and gadgets in the frame.

YOLO is the most popular deep learning models widely used for object detection, semantic segmentation, and image classification. The three main variants of YOLO are YOLOv1, YOLOv2, and YOLOv3. YOLOv1 proposes a general architecture, of which YOLOv2 refines the design, uses a predefined anchor box to enhance the bounding box proposal, and YOLOv3 further refines the architecture and training process. Authors have taken the YOLOv3 as the base model.

Yolov3 uses a variant of Darknet, which is a 53-layer network trained on the ImageNet dataset. For the purpose of target detection, 53 layers have been added on top of the dark web. So YOLOV3 has 106 layers of fully convolutional underlying architecture. The architecture of yolov3 is shown in Fig. 1. Darknet-53 is used as feature extractor, and it is mainly composed of 3×3 and 1×1 filters.

Figure 2 describes the layers ad parameters of the Yolov3. Each Conv block represents a convolutional layer to perform convolution for dimensionality reduction, followed by layers of batch normalization and Leaky ReLu activation.

Figures 3 and 4 describe the operation of convolution block and residual block. The top three Conv blocks in the architecture of Yolo are modified to make the detections more accurate. In these Conv blocks, ReLu activation is replaced with softmax activation function. Softmax is generally used for classification in the output layers. Replacing ReLu with softmax is more advantageous as softmax calculates probabilities for all classes, and the most probable class is given as output. In this way, all types of gadgets and multiple persons can be properly detected.

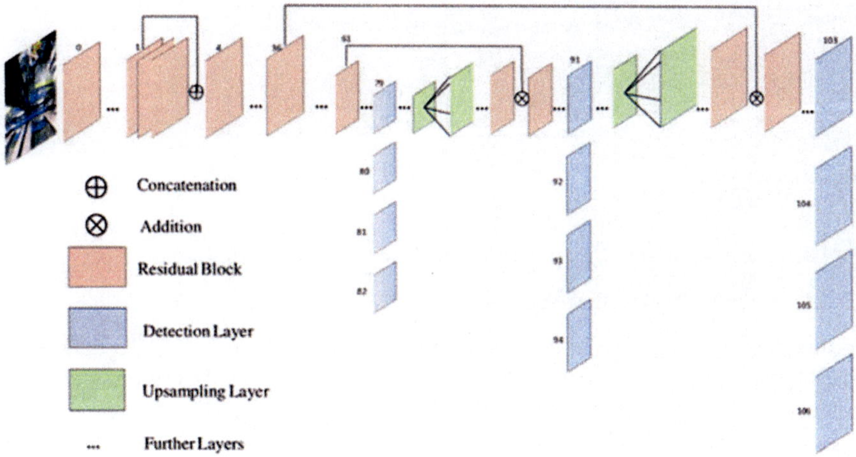

Fig. 1 YOLOv3 model

	Type	Filters	Size	Output
	Convolutional	32	3 × 3	256 × 256
	Convolutional	64	3 × 3 / 2	128 × 128
1×	Convolutional	32	1 × 1	
	Convolutional	64	3 × 3	
	Residual			128 × 128
	Convolutional	128	3 × 3 / 2	64 × 64
2×	Convolutional	64	1 × 1	
	Convolutional	128	3 × 3	
	Residual			64 × 64
	Convolutional	256	3 × 3 / 2	32 × 32
8×	Convolutional	128	1 × 1	
	Convolutional	256	3 × 3	
	Residual			32 × 32
	Convolutional	512	3 × 3 / 2	16 × 16
8×	Convolutional	256	1 × 1	
	Convolutional	512	3 × 3	
	Residual			16 × 16
	Convolutional	1024	3 × 3 / 2	8 × 8
4×	Convolutional	512	1 × 1	
	Convolutional	1024	3 × 3	
	Residual			8 × 8
	Avgpool		Global	
	Connected		1000	
	Softmax			

Fig. 2 Yolov3 architecture diagram

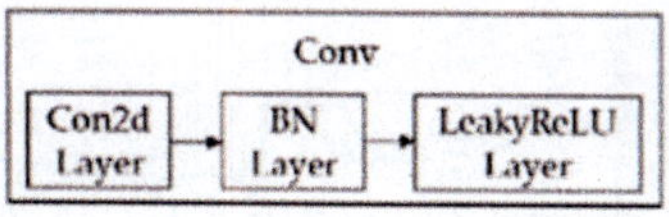

Fig. 3 Conv block

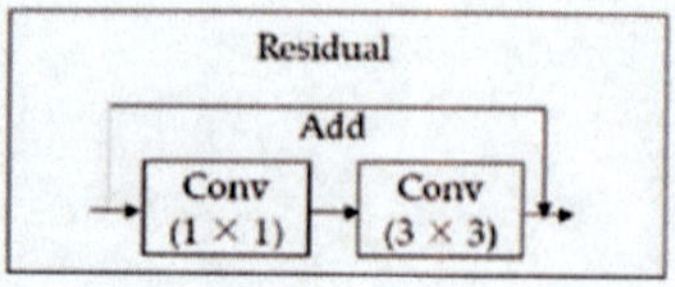

Fig. 4 Residual block

The proposed methodology to build the system includes following steps:

(1) **Download Pre-trained Weights**

In this work, the pre-trained weights of YOLOv3 trained on the COCO dataset to detect people and mobile phone in the webcam feed is used. The initial step is to download the pre-trained weights, which may be done with tools like wget or directly. Create a class names file which contains the names of classes for which the model is trained on.

(2) **Build the model**

Define the convolutional layer function and apply batch normalization to it as shown in Fig. 5. Empirical mean of batch normalization is as described below:

$$\mu_B = \frac{1}{m}\sum_{i=1}^{m} x_i \tag{1}$$

Empirical batch mean is described as below:

$$\sigma_B^2 = \frac{1}{m}\sum_{i=1}^{m} (x_i - \mu_B)^2 \tag{2}$$

Except for the top three Convolutional blocks add ReLu activation to all the Convolutional blocks. Add softmax as described below activation to top three Convolutional

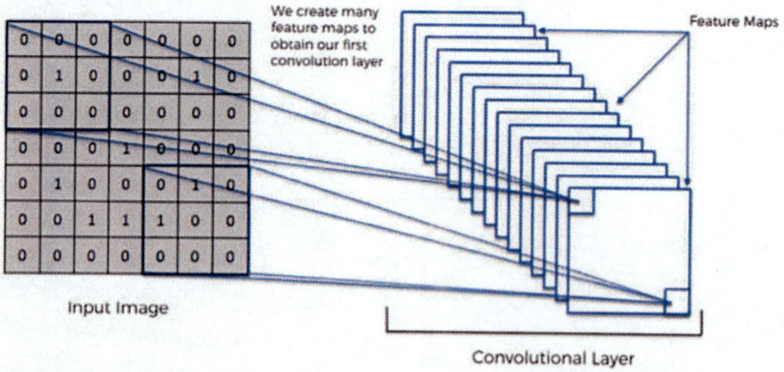

Fig. 5 Convolution operation

Fig. 6 Residual block

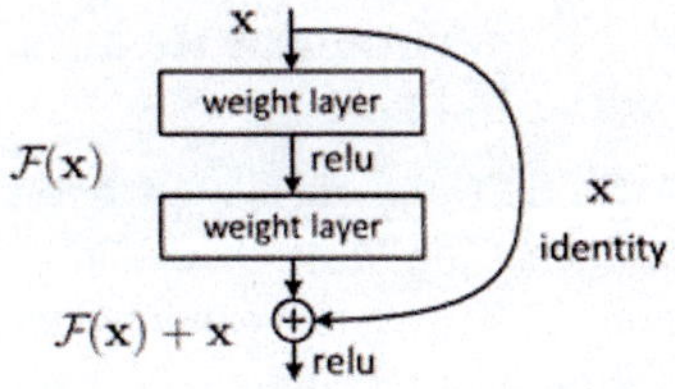

blocks.

$$\sigma(\vec{z})_i = \frac{e^{z_i}}{\sum_{j=1}^{K} e^{z_i}} \tag{3}$$

Define the Darknet residual blocks as in Figs. 4 and 6 and build the entire model. Pre-trained weights should be loaded and assigned to the model's layers.

$$(f * g)(t) \triangleq \int_{-\infty}^{\infty} f(\tau)g(t - \tau)\tau' \tag{4}$$

(3) **Data Pre-processing**

Frames are generated from the webcam feed. Instead of key frames, all frames were processed to detect the multiple people and gadgets during proctoring. Each frame of the live video should be converted to RGB and resized to 320 × 320, 416 × 416, or 608 × 608. Convert their data type to float32, expand their dimensions to four, and divide by 255.

(4) **Perform Prediction**

Pre-processed image is given as input to the trained model for making predictions. Define the anchor boxes for the detected objects and use Non-Maxima Suppression technique to find the most appropriate bounding box. The non-maximum suppression suppresses the bounding boxes with low confidence values. Use the classes file to provide the label to the detected objects.

$$s_i = \begin{cases} s_i, & \text{iou}(\mathcal{M}, b_i)) < N_t \\ s_i(1 - \text{iou}(\mathcal{M}, b_i)) & \text{iou}(\mathcal{M}, b_i)) \geq N_t \end{cases} \tag{5}$$

(5) **Detect Gadgets and Count Persons**

To detect the presence of more than one person or usage of gadgets, extract their respective indices from classes file and check whether the class is predicted or not. For example, if the person's index is 0, check to see if the anticipated class is also zero, and then increase the count.

4 Experimental Results

MS COCO dataset is used to train the YOLOv3 model and to extract the pre-trained weights. The MS COCO dataset is large-scale object detection, segmentation, key point detection, and caption dataset. This dataset consists of 328 K images, which is one of the most popular open-source object recognition databases, used to train deep learning programs. Another dataset ImageNet is used to train to improve accuracy in multiple people detection. After completion of training, our model has been tested on the customized dataset of 200 videos of various examinees trying to cheat proctor using various gadgets and seeking help from others. From Fig. 7, it is observed that authors were able to detect single examinee attempting the exam along with confidence value, mobile phone detection along with confidence value, multiple examinees in the frame and also alert the proctor if there is no examinee before the screen.

Different activation functions like Sigmoid, ReLu, and Leaky ReLu are also considered for evaluating the model. Test accuracy, precision, recall, and test loss are calculated as given in Table 1. As sigmoid is used for binary classification, it is unable to accurately classify the points when compared to softmax. From Table 1, it can be observed that test accuracy is more for softmax, and test loss is less for ReLu. The precision and recall values are almost same for ReLu and Leaky ReLu. The vanishing gradient problem can be handled by ReLu, and it is less computationally expensive. The softmax activation is considered as a better option for hidden layers in our model.

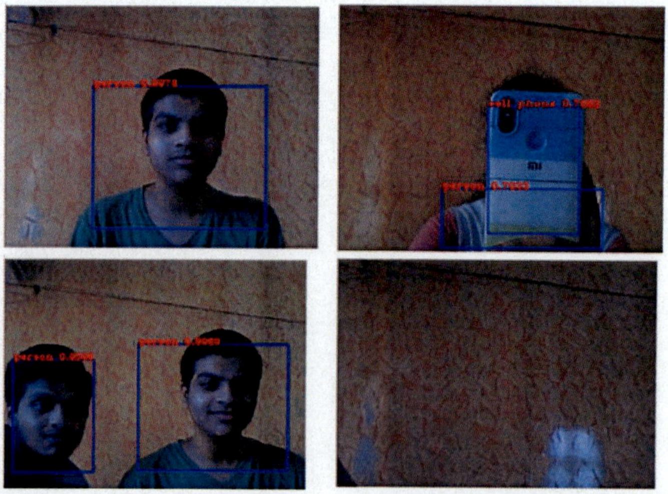

Fig. 7 Experimental results showing single person detection, person and phone detection, multiple person detection, and no person detection (from top to bottom and left to right)

Table 1 Evaluation of model with various activation functions

Activation function	Test accuracy	Precision	Recall	Test Loss
Sigmoid function	0.81	0.75	0.82	0.10
Softmax	0.92	0.86	0.95	0.05
ReLu	0.88	0.86	0.96	0.09
Leaky ReLu	0.89	0.85	0.96	0.08

Table 2 Comparing proposed system with existing systems

	Vardan AI proctoring	Automated online proctoring	Proposed system
Mean average precision	0.83	0.65	0.93
Recall	0.5	0.6	0.7
Accuracy	0.66	0.64	0.92
F_1 Score	0.62	0.64	0.8

The mean average precision (mAP) score for the proposed system is calculated using a test dataset which contains mobile phone images. The following graph images show the mAP score of the existing Yolov3 and enhanced Yolov3 with modified layers. The map score of the existing Yolov3 model is 83%, and the map score of the enhanced Yolov3 is 92% which shows that modifying the layers of the model has improved the efficiency. Comparing our model with the two existing systems like vardan and automated proctoring, as in Table 2 and Fig. 8, our model showcased better performance in the terms of mAP, recall, accuracy, and F_1 Score.

Accuracy values for various optimizers like Adam, Adagrad, Adadelta and Adamax, with various loss functions, are also used for performance evaluation of

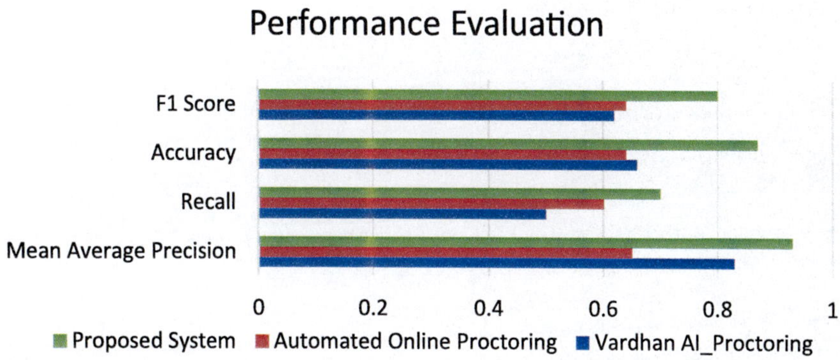

Fig. 8 Performance evaluation of proposed model

our model as shown in Fig. 9. Our model is compared with the Vardan Agarwal, and it is observed that our model performs well.

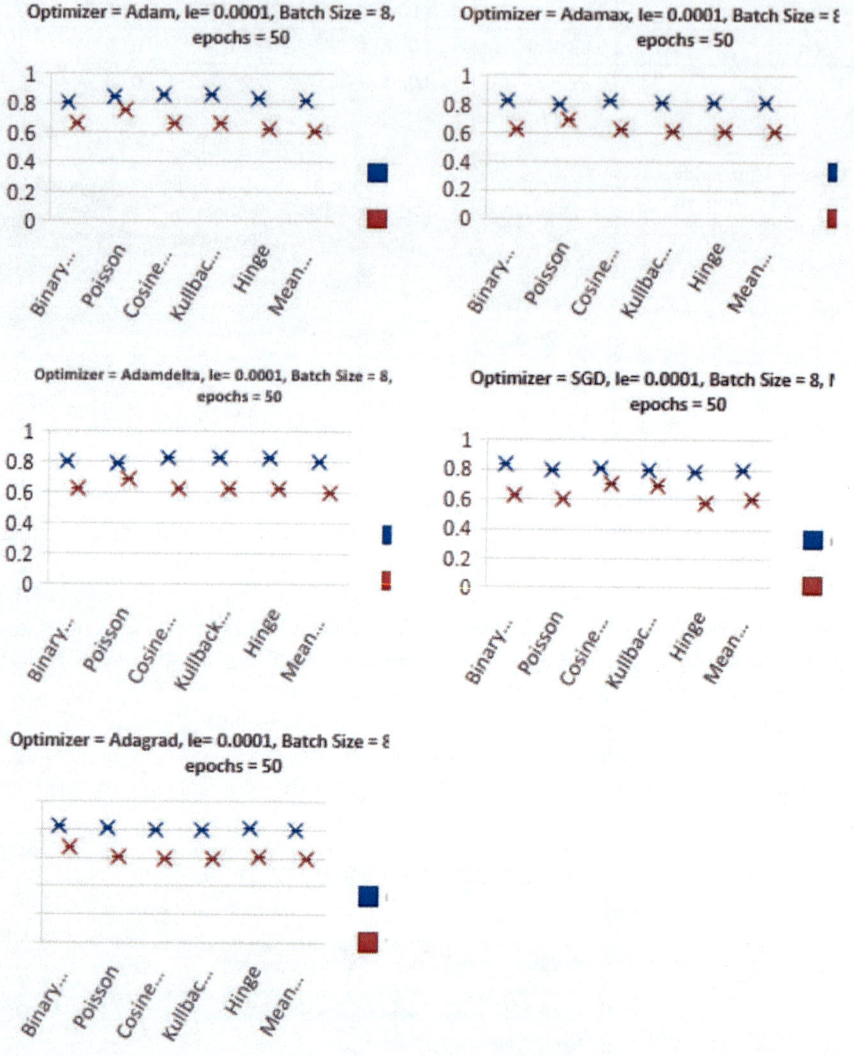

Fig. 9 Variations of accuracy with different hyper-parameters for our proposed system and Vardan AI_Proctoring

5 Conclusion

The main limitation of this system is the availability of less computational power. Provided with high computational power the model can be trained with dataset containing images only of required classes, and it can be trained from scratch to improve the efficiency of model. YOLOv3 must be replaced with other models to get good FPS on CPU. A better face spoofing model has to be made as the current accuracy is not good. Vision-based functionality must be added such that no one can replace the candidate.

We have become habituated to the online classes and video lectures. But there is a problem regarding online exams as manual invigilation is not effective. This problem can be solved using automated online exam software which involves minimal human intervention. These online exam portals are very much helpful in schools, colleges, and universities across the world.

As we know, COVID-19 pandemic has become a threat to the classroom education. However, education cannot stop. So, we need to shift toward the digital era. Proposed model can be deployed in online exam software and can be used for detecting malpractices efficiently with less effort with 92.8% accuracy.

References

V. Agarwal, Count people in webcam using pre-trainedYOLOv3 (2020)

I.E. Allen, J. Seaman, Grade level: tracking online education in the United States. Babson Survey Research Group and Quahog Research Group, LLC. Last access on 3 Feb 2016 (2015). http://www.onlinelearningsurvey.com

Y. Atoum, et al., Automated online exam proctoring. IEEE Trans. Multimedia **19**(7), 1609–1624 (2017)

A. Bochkovskiy, C.-Y. Wang, H.-Y.M. Liao, Institute of information science academia Sinica, Taiwan. YOLOV4-Optimal speed and accuracy for object detection

J. Du, Understanding of object detection based on CNN family and YOLO. J. Phys. Conf. Ser. **1004**(1) (2018)

https://towardsdatascience.com/automating-online-proctoring-using-ai-e429086743c8

M. Ju, et al., The application of improved YOLO V3 in multi-scale target detection. Appl. Sci. **9**(18), 3775 (2019)

Q.-C. Mao, et al., Mini-YOLOv3: real-time object detector for embedded applications. IEEE Access **7**, 133529–133538 (2019)

T. Potluri, N. Gnaneswara Rao, Content based video retrieval using SURF, BRISK and HARRIS features for query-by-image, in *Recent Trends in Image Processing and Pattern Recognition. RTIP2R 2018. Communications in Computer and Information Science*, vol. 1035, ed. by K. Santosh, R. Hegadi (Springer, Singapore, 2019). https://doi.org/10.1007/978-981-13-9181-1_24

T. Potluri, V. Jayaprakash, P.S. Krishna, M. Parvez, N. Rohith, Controlling mouse based game using object detection (2020)

T. Potluri, S. Jahnavi, R. Motupalli, Mobilenet V2-FCD: fake currency note detection, in *Advanced Informatics for Computing Research. ICA6ICR 2020. Communications in Computer and Information Science*, vol. 1393, ed. by A.K. Luhach, D.S. Jat, K.H. Bin Ghazali, X.Z. Gao, P. Lingrasds (Springer, Singapore, 2021). https://doi.org/10.1007/978-981-16-3660-8_26

T. Potluri, G. Nitta, Content based video retrieval using dominant color of the truncated blocks of frame. J. Theor. Appl. Inf. Technol. **85**(2), 165 (2016)

T. Potluri, et al., Content-based video retrieval using dominant color and shape feature, in *Proceedings of the First International Conference on Computational Intelligence and Informatics* (Springer, Singapore, 2017)

M. Tan, R. Pang, Q.V. Le, Efficientdet: scalable and efficient object detection, in *Proceedings of the IEEE/CVF conference on computer vision and pattern recognition* (2020)

S.S. Vaishali, Real-time object detection system using caffe model. Int. Res. J. Eng. Technol. **6**(5), 5727–5732 (2019)

L. Zhao, S. Li, Object detection algorithm based on improved YOLOv3. Electronics **9**(3), 537 (2020)

Customer Segmentation Using K-Means Clustering

V. Sujatha, Shaik Najiya, Taduvai Siva Likhitha, Malladi Sravya, and Peravali Tejaswini

Abstract This paper tells about the customer segmentation of today's fast-paced world of marketing, where the focus has shifted from product to customer; customer service management may be seen as a key to attaining revenue growth and profit. The communication system that exists between companies and their consumers is referred to as customer relationship management. In the commercial world, customer experience management is necessary. Maintaining a productive connection with business clients is crucial since a business transaction demands more decision-making and professional purchasing effort than a consumer purchase. Most scientists use data mining to examine numerous techniques to segmenting customers using a variety of clustering algorithms in a variety of businesses. In the realm of data analytics, there are a variety of approaches used in the segmentation process. It guarantees that marketers concentrate their efforts on the customers who are most likely to react. Additionally, after a successful customer segmentation procedure, businesses may be able to employ more effective marketing tactics, lowering investment risk. We utilise the *k*-means clustering technique to categorise customers based on their worth in this article. Once that, after a certain amount of time has elapsed, a change evaluation is performed. This research is unique in that it takes into account the duration and trend of customer value changes in order to increase the accuracy of forecasts based on previous customer behaviour.

Keywords K-mean algorithm · Customer relationship management · Data mining · Customer segmentation

V. Sujatha · S. Najiya · T. S. Likhitha (✉) · M. Sravya · P. Tejaswini
Department of Computer Science and Engineering, Vignan's Nirula Institute of Technology and Science for Women, Guntur, India
e-mail: likhithataduvai282@gmail.com

A. B. Reddy et al. (eds.), *Proceedings of Third International Conference on Advances in Computer Engineering and Communication Systems*, Lecture Notes in Networks and Systems 612, https://doi.org/10.1007/978-981-19-9228-5_38

1 Introduction

Consumer segmentation is the division of such a market into distinct customer groups having similar characteristics. Client segmentation is an effective strategy for uncovering unmet customer needs. Companies, on the other hand, can exceed the competition in the industry distinctive products and services. Business-customer connections have become increasingly crucial as technology has advanced. However, actions, features, and other factors can help you figure out who your most profitable customers are. Break down the big database of customers into sub-parts for this customer segmentation. Customers are segmented based on their age, gender, religion, family size, buying habits, and other factors and in accordance with their social and lifestyle choices. We might get numerous benefits as a result of this. Customer service, improved customer connections, best-performing economies, increased productivity, improved distribution channels, improved brand identification, and pricing optimization are just a few example. Some studies promote k-means clustering using customer segmentation to highlight how customer segmentation can help financial institutions. Customers' needs, on the other hand, are evolving, and they are discovering new services in the telecommunications sector that meet their needs. Because the procedure directly influences the combining consumer segmentation with a k-means clustering algorithm can result in a good segmentation process result, for accuracy is extremely important. In machine learning, there are two types of designs. Monitored and unsupervised supervision are the two types of supervision. The following are the most typical ways that firms segment their consumer base: (1) statistics that break down the population into subgroups are referred to as demographic facts. Marital status, age, race, gender, work status, income, nationality, and political orientation are all examples of demographic data. Demographic data is frequently collected on subjects for scientific research or even as part of marketing as well as public opinion surveys. The testers can better grasp how the data differs for different groups of people by using demographic data. A scientist performing an experiment, for example, would want to get whether if males and females react differently, or a corporation bringing out a new product may also want to understand how American and European opinions differ. (2) Geographic information that differs depending on the scope of the business. This data could be refer to specific cities and regions of localised enterprises. It could be a customer's residence locale, state, or even continent for larger companies. (3) Behavioural consumer categorization is based on customer behaviours that have been observed in the past and can be used to foresee future events Customers' preferred brands, as well as the times when users spend the most money. The behavioural aspect of client segmentation entails figuring out not just who your customers are, but also how they behave, why people buy things, and also how those factors change over time. (4) The analysis of cue emotional and cognitive psychographics are indeed the factors that influence a person's behaviour. This section aims to determine the factors that impact why people behave in certain ways. The actions, interests, and psychographic data are created by combining a person's thoughts and ideas (AIOs).

The rest of the paper is organised as follows: Sects. 2, 3, 4, 5, and 6.

2 Literature Review

Nandapala and Jayasena (2020) presented while analysing the quality for Big Data such as noisy, size, complexity, algorithm calculations, and cluster shape; a brief overview of the many clustering methods is categorised into partitioning, hierarchical, dense, grid-based, and model-based algorithms.

Anitha and Patil (2019) proposed "Customer segmentation utilise the K-means cluster". *K*-means, Agglomerative, and Meanshift are the three clustering techniques employed in this work. These are used to divide the customers and then compare the clustering findings. Using these techniques, five cluster segments were identified as careless, cautious, standard, target, and sensible customers. Using mean shift clustering, however, regular visits, high purchases, and occasional buyers were identified.

Kushwaha and Prajapati (2014) presented the paper "Mall Customer Segmentation Using Clustering Algorithm". The *k*-means clustering design is employed to do market-based analysis using an unsupervised machine learning technique. It also entails predicting the target clients who are easily converged among all customers. Also, develop new products that have comparable tastes to those of the clients.

Patankar (2021) investigated the need for segmentation of the brain. Clustering algorithms are used as the primary component of CRM by customers. The commonly used *k*-means and hierarchical clustering algorithms were investigated, as well as the benefits and drawbacks of various strategies were discussed. Finally, the issue of building a hybrid method is addressed by merging the above; there are two tactics that have the potential to outperform individual designs.

Ozan (2018) proposed "RFM model on customer buying behaviour using *K*-mean". They used a study using the *k*-means algorithm, we used data set classification principles based on the RFM model. Clusters are validated in this data set using the Silhouette Coefficient technique. And then factors like Sales Recency are used to compare the results of the transaction.

Jerry (2007), for recognising behavioural patterns, customer segmentation demands descriptive variables. In some cases, though, descriptive variables are insufficient. As a result, the author devised a segmentation strategy to address the issue, resulting in increased performance through the application of data mining techniques.

Chinedu (2015), intelligent technology that aids in improving business performance is referred to as corporate intelligence. Small enterprises may find it difficult to adopt this strategy due to high expenses, resource constraints, and a range of other issues. In this aspect, the author has explored not just the role of business information in major firms, and how small enterprises can benefit from the concept.

3 Methodology

3.1 K-Means Clustering

The *K*-means method is used to cluster the features. Multiple steps are involved in K-means clustering. The initial step is to consider the clustering results that will be employed. Then, at random, choose k objects to form the initial clusters. The statistic centroid is assigned after the clusters have been initialised. For this calculating the Euclidean distance between the data sets and centroid. The results of measurements of a specific centre are calculated based on their locations. Assess the price of the median from each analyses showed that remains in the cluster to update each centroid. The distances are then recalculated, and data is collected. Sets are given to the new centroid. In this step, the two previous steps are repeated. According to these conditions, the vehicles will be stopped. In this method, we use simple architectures which will not perform the better performance. It is the limitation of method. In this paper, we cluster the data so document clustering is the application used.

3.2 Algorithm

The partitioning algorithm k-means, in which the value of mean of the elements in each cluster, represents the cluster's centre.

k: total no. of data collection; *S*: a specifies collection with n no. of entities as input.

A group of k_clusters is produced as the output.

Procedure:

(1) Select *k* entities from *S* at random to serve for the early cluster_centres
(2) Redo
(3) Reallocate every entity to a group to which it belongs.
(4) After reassign the value, update the cluster value, and calculate the arithmetic value of each entity in groups.
(5) Until its changes not occurred (Fig. 1).

Data Collection. For clustering and the *k*-means technique, a data set from a retail centre is gathered. The data set contains 200 tuples and five attributes that will be used to derive information about 200 customers: age, gender, expenditure, early income, and customer id. The qualities of the data set collection are given a rating of 1–10.

Data Preprocessing. Check the null values from the data set in this data preprocessing stage. It's also crucial to detect the data set's outliers. Then, as part of the Data Preprocessing stage, eliminate all null values and outliers.

Choosing a several number of the clusters by using the *k*-means clustering. The clustering algorithm used above is shortest clustering algorithm as the algorithm

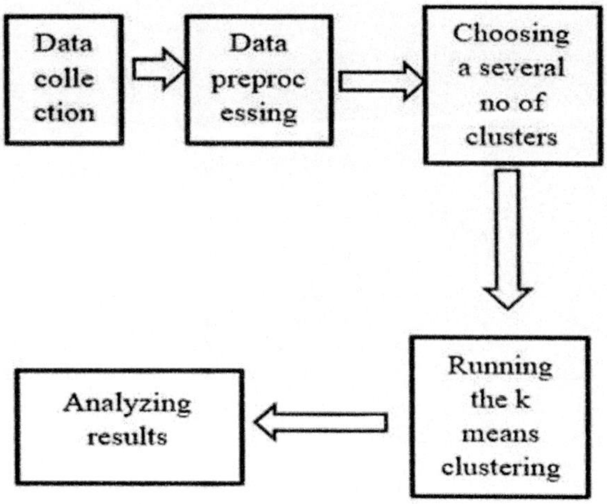

Fig. 1 Experimental design

is based on principle of segmentation; the algorithm works in a way that it will initialise the position of clustering the total number of k will be determined by the method elbow. After the determination of *k*-centroids is done with the help Euclidean distances, the data points set recognised to be closest of the centroid that will be forming the clusters. After the creation of group, the barycenters can be determined by the clustering method with the help of calculations when there will be no change in the centroid position until then the process will be repeated.

In this *k*-means clustering algorithm, here we use one method:

Elbow Method. The elbow method helps us to determine *k*-means clustering. The method's main goal is to determine the *k* value. However, determining the value for *k*, this elbow method can assist us in determining the value of *k*. The mean value is done by calculating the squared total values in between the data sets and a centroid of their associated assignment clusters.

The elbow method's primary notion is that increasing the number of subgroups can help to reduce each cluster's total of within-cluster variation. We may utilise the clustering technique for a variety of values of *k*, where *k* varies from 1 to 10 groups, and then calculate the 9 before visualising intra-cluster sum of squares based on the number of clusters to determine the ideal clusters. It also specifies the amount of clusters that are required.

Computing a squared sum of cluster errors (WCSS) requires various values of the *k* for which the WCSS starts to decline, which will be likey in the elbow method as the WCSS versus *k* representation.

The steps are used to summarise as follows:

(1) By modifying k from 1 to 10 clusters and by calculation various values of *k*-means clustering.
(2) For each value of the *k*, we will compute the cluster sum (WCSS).
(3) Plotting the WCSS curve against the *k* cluster.

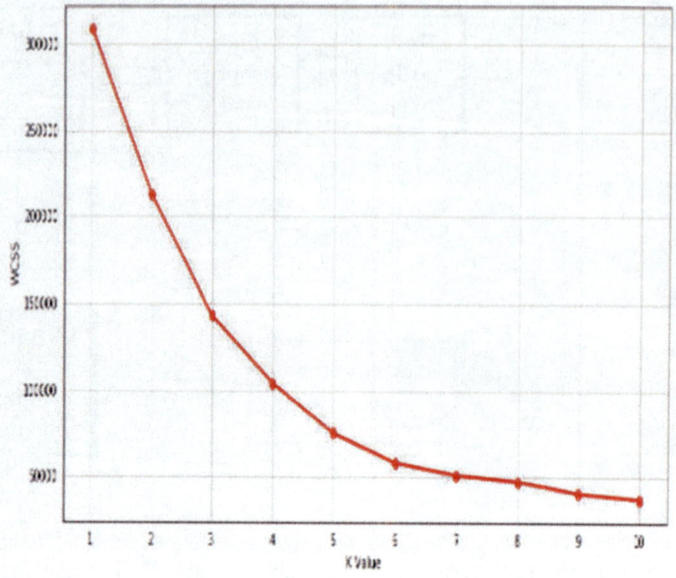

Fig. 2 Elbow method

(4) In most case, position of a bend inside of the plot is to be determined by the cluster number (Fig. 2).

3.2.1 Running the *k*-means Clustering in the Principal Components

Here running it results within the support, this allows you to see the data clusters. The *K*-means clustering leads consensus for a principle elements in this case. As a result, the data clusters can be shown.

4 Results

In Fig. 3, the id column has been removed because it does not appear to be relevant in this situation. In addition, it plotted the age distribution of customers.

In Fig. 4, then, to better comprehend the distribution range, we generated a box plot of the spending score as well as annual income. This annual income range is clearly greater than the spending score range.

In Fig. 5, then, to illustrate the number of clients based on their spending scores, we created a bar plot. The vast majority of clients have a spending score of 41–60.

In Fig. 6, then, to examine the distribution of clients by age group, a bar plot was created. The 26–35 age range clearly outnumbers all other age groups (Fig. 7).

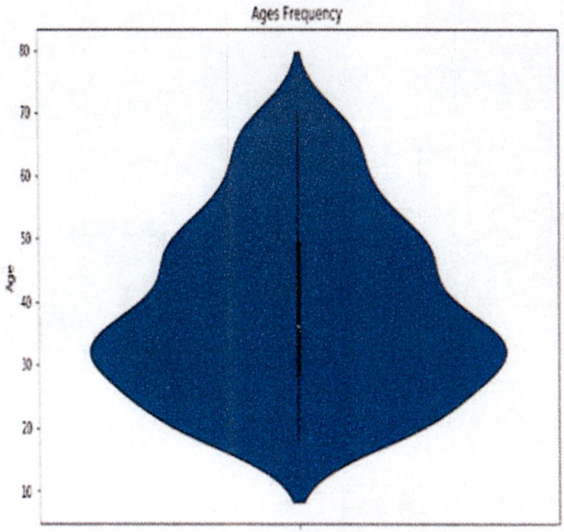

Fig. 3 Age distribution

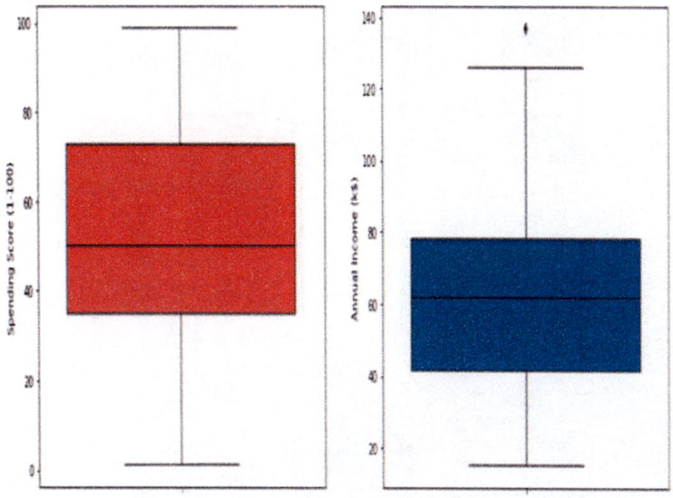

Fig. 4 Distribution range of annual income and spending score

In addition, a bar plot was created to display the number of clients based on their annual income. The bulk of clients earns between \$60,000 and \$90,000 each year.

$$\text{WCSS} = \sum_{i \in n} (X_i - Y_i)^2$$

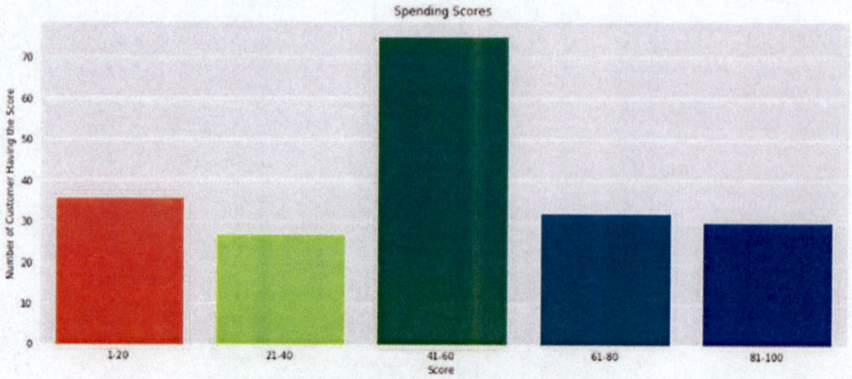

Fig. 5 Distribution range of clients 41–60

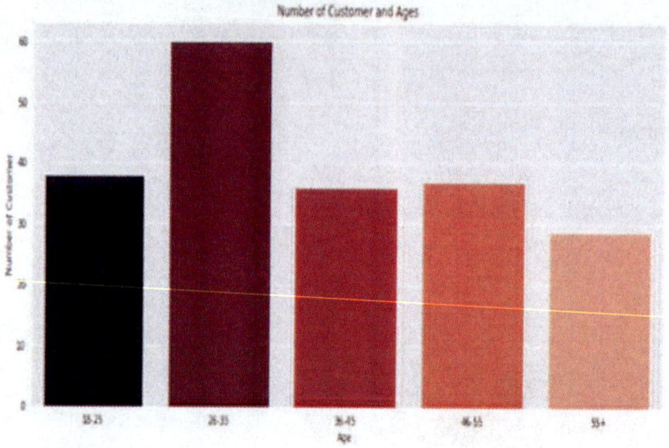

Fig. 6 Distribution range of clients 26–35

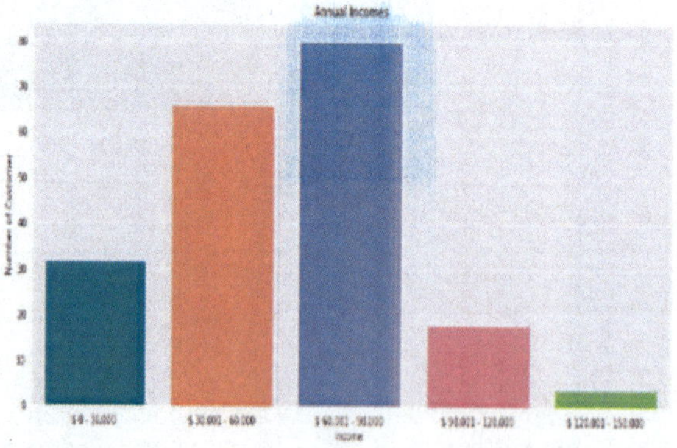

Fig.7 Annual income

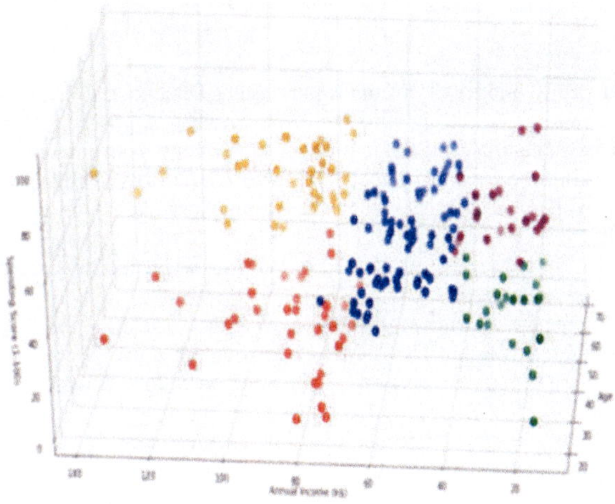

Fig. 8 Visualisation of 3D cluster

Within-Clusters Square Root (WCSS) was then plotted against the clusters (k value) and determine the appropriate number of clusters. WCSS calculates the sum of observations' distances from its cluster centroids using the formula below.

where Y_i is the observation X_i's centroid. The fundamental goal is to maximise the number of clusters, with each data point acting itself as a cluster centroid in the limiting scenario (Fig. 8).

Finally, a 3D plot depicting the consumers' expenditure score in proportion to their annual revenue was developed. The data points are categorised into five classes, which are symbolised by a various colours in the 3D visualisation.

5 Conclusion

Here, the data set taken is unlabelled. We have chosen internal clustering validation over the external clustering validation, which is depended on the external data for example labels. Segmentation of k-means is one of the most commonly used clustering methods, and it is chosen for working on the clustering problems, the people, or engineers use this clustering algorithm for structuring of data set. The evaluation of internal cluster can be used for the selection of optimal algorithm for clustering and to apply on the data set and ensure that the data is clustered accurately in opposite clusters. The most common use of k-means is the client segmentation which is used to gain a better knowledge about the clients that can be used to boost the companies income. In this paper, accuracy is high than previous papers so results are shown in better way than previous analysis.

References

P. Anitha, M. Patil, RFM model of customer purchasing behaviour using *K*-Means algorithm. Comput. Sci. Inf. J. King Saud Univ. (2019)

Chinedu, Efficient segmentation using the *K*-Means methodology, segmentation: a plan for targeted customer services. Int. J. Surv. Mach. Intell. (IJARAI) **4**(10) (2015); P. Ezenkwu, O. Simeon, K. Constance, Framework of the *K*-Means technique for efficient customer groups: a plan for directed customer care (2015)

W.T. Jerry, Segmentation of Market (2007), Recovered from www.decisionanalyst.com on 12 July 2015

Y. Kushwaha, D. Prajapati, Customer segmentation using the *k*-Means algorithms (2014)

E.Y.L. Nandapala, K.P.N Jayasena, The *K*-Mean approach are a useful methods for segmenting a customers, in *15th (IEEE) International Conference on Industrial and Information Systems (ICIIS)* (2020)

S. Ozan, A case study of customer segmentation using machine learning methods. IEEE (2018)

N. Patankar, Customer segmentation using machine learning. IOS Press and released under the conditions of Creative Commons Attribution Non-Commercial License 4.0.2021

Automated Gesture Recognition and Speech Conversion Tool for Speech Impaired

Surya Pandey

Abstract By translating hand gestures into a series of words or speech, sign language recognition (SLR) aims to improve communication between deaf and dumb people and the general public. Despite the fact that this activity has a significant societal impact, it is nevertheless challenging due to the complexity and wide variation of hand behavior. Existing SLR techniques use hand-composed pieces to characterize sign language movement and develop division models based on those aspects. It's difficult to create reliable features that respond to a wide range of hand movements. The goal of this project is to develop a sign language recognition project using CNN and predict the output in the form of text and audio.

Keywords Indian sign language · Machine learning · CNN

1 Introduction

Hearing-impaired people often use sign language to communicate because of the great variety of hand and body movements and facial expressions that can be used. Sign language recognition is still a difficult task since it is difficult to collaboratively exploit the information from hand shapes and body movement trajectory. A successful recognition model for translating sign language into text or speech is presented in this study in order to assist the hearing impaired in communicating with the general public using signs.

The most difficult part of sign language identification is generating descriptors for hand forms and motion trajectories. Tracking hand areas in the video stream, segmenting hand-shape images from the complicated background in each frame, and challenges with gesture identification are all part of the hand-shape description process. Tracking critical points and matching curves are also important aspects of motion trajectory. Many studies have been done so far, but it is still difficult to achieve

S. Pandey (✉)
Department of Computer Science and Engineering, New Horizon College of Engineering, Bangalore, India
e-mail: suryaspandey@gmail.com

A. B. Reddy et al. (eds.), *Proceedings of Third International Conference on Advances in Computer Engineering and Communication Systems*, Lecture Notes in Networks and Systems 612, https://doi.org/10.1007/978-981-19-9228-5_39

a satisfactory result for SLR because of the variance and occlusion of hands and body joints. In addition, combining the hand shape and trajectory features is not an easy task. For this reason, we have created a system that automatically incorporates hand shapes and movement trajectories, converts the image into gray scale, recognizes the sign, and converts it into audio.

Many uses of CNN (Navya and Patil 2022; Pandey et al. 2022; Pandey and Dholay 2019; Ramsamy and Eric 2022) have been seen, with the best results in most cases. We have implemented CNN in our project. This project provides output in French in addition to English, so that blind individuals may comprehend the motions more easily. In the near future, we hope to expand this project to include other languages.

The paper is designed as follows: Sect. 2 talks about algorithms implemented. In Sects. 3 and 4, we discuss about the methodology, and how we have implemented the project using CNN, respectively. The results obtained by the same are conferred in Sect. 5. Section 6 terminates the paper.

2 Related Work

Shenoy et al. (2018) proposed an Android smartphone, where touch and gestures were made by an Indian sign language (ISL) system. Client's hand pictures are caught, and their edges are moved to a server for handling. Systems like facial revelation, object adjustment, and skin shading partition are utilized for manual identification. The picture is likewise under a grid-based element extraction procedure that addresses the state of the hand as a component extraction. Hand shapes and segments are finished utilizing the k-nearest neighbors (KNN) Zhang et al. (2018) calculation. The hand signal information is given to the chains of the Hidden Markov Model relating to the 12 pre-chosen follows characterized in the ISL. Utilizing this regulation, the framework is reasonable to accomplish 97.9% exactness. With hand stances and motions as picture being input, this framework can be utilized by both hearing and discourse impeded individuals.

Abraham et al. (2019) have proposed a sign language, which acts as the means of conversation for the speech impaired community. This community cannot communicate with words and expressions. This model targets to reduce the existing gap of communication between these two communities by converting the signs into speech or text. The sensor gloves which have flex sensors assist with distinguishing the development of each finger. The gloves also have an inertial dimension unit (IMU) which analyzes the hand's orientation. This unit helps to gather information about the movement of the hand which is wirelessly transmitted and categorized into respective outputs in the speech format. LSTM Veerasamy et al. (2021) networks is used to carry out gesture classification because of their capability to study long time period dependencies. The proposed version classifies 26 gestures in total with 98% accuracy. This shows the feasibility of the use of LSTM-based NNs in sign language translation.

Sahoo (2021) provides details regarding the computerized framework acknowledgment of ISL numerical, where a conventional advanced camera was utilized to descry signals; no external predisposition is requested to catch electronic signs. The plan is to change over individual advanced images into reading material, that is, for each picture to be fitted it should contain a solitary numeric image. To view ISL token images in real-time position, a summary point is erected containing 5000 integers, 500 images of each number symbol (0–9). Image given as input can be used by both hearing and speech impaired people. The KNN grouping strategy is an example put together procedure that is based with respect to the idea of distance metric and known as one of the occasion-based method. The preparation vectors are having a place with focuses in an n-dimensional space. The result made by the KNN classifier is advanced than the Naive Bayes classifier to the extent visual delicacy. The typical accuracy of KNN is 98.36% and that of Naive Bayes Zhang et al. (2021) is 97.79%.

3 Proposed Methodology

The operation's front end is Tkinter (Tkinter is a graphical user interface (GUI) a module in Python), and it's employed with a web camera to capture hand movements. The entire signing duration is recorded that use the OpenCV video stream. Each frame from the video stream is captured and converted to gray scale images. Since the images are rather huge, there isn't much to lose by making them gray scale and downsizing them in order to accelerate the computations. Hand movements are scrutinized in the collected prints. This is a phase in the preprocessing process that takes place before the image is fed into the model for prediction.

3.1 CNN Architecture

32 filter weights are used in the first convolutional layer to process the image, which has a resolution of 64 $\times$ 64 pixels. With 62 $\times$ 62 as the input size, it is given to the following layer, where the images are down sampled using a maximum pooling of 2 $\times$ 2, and a 31 $\times$ 31 image size is obtained. These images from the first pooling layer's output are used as input for the following convolutional layer. The output size is 29 $\times$ 29, and it is processed in the following convolutional layer using 32 filter weights. These images are down sampled once more with an output size of 14 $\times$ 14 and a maximum pool of 2 $\times$ 2. Applying the flatten layer results in an output shape of (6272), or the conversion of the pooled feature map into a 1D vector. Now, these images are fed into a layer that is entirely connected using the ReLu activation function, producing an output shape of (256). The first densely connected layer's output is now fed into the next fully connected layer, which employs the softmax activation function and has seven neurons, one for each class we are classifying.

According to our proposed model as shown in Fig. 1, first the image is recorded using the user's hand gestures, then the image is preprocessed and supplied to a CNN model (Fig. 2), which predicts the output from the seven available classes as shown in Fig. 3 in text format and converted into audio.

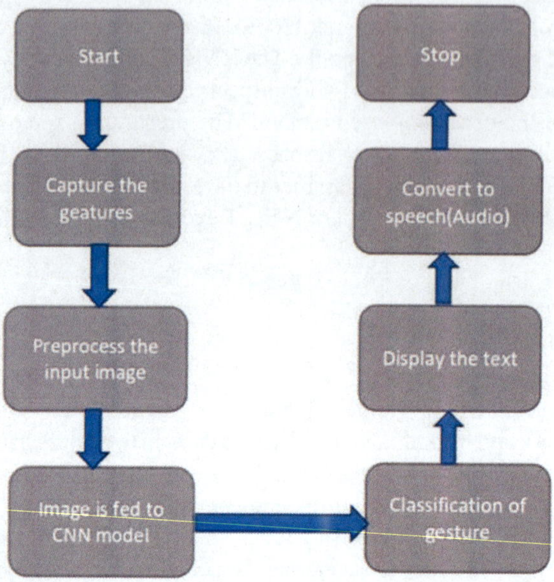

Fig. 1 Proposed model

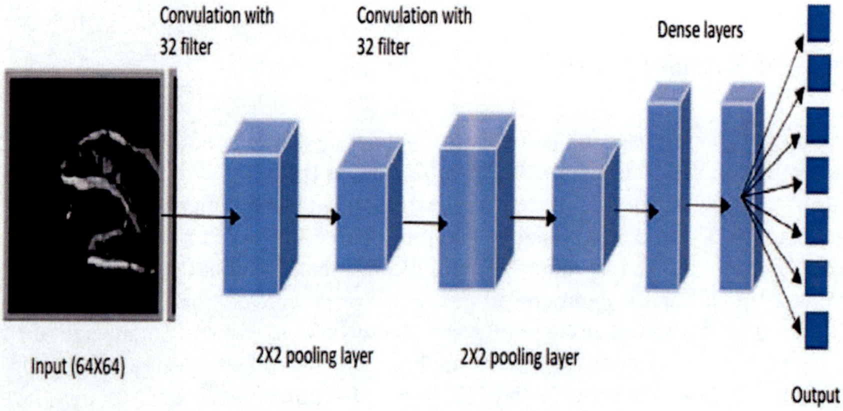

Fig. 2 CNN layers used in our model

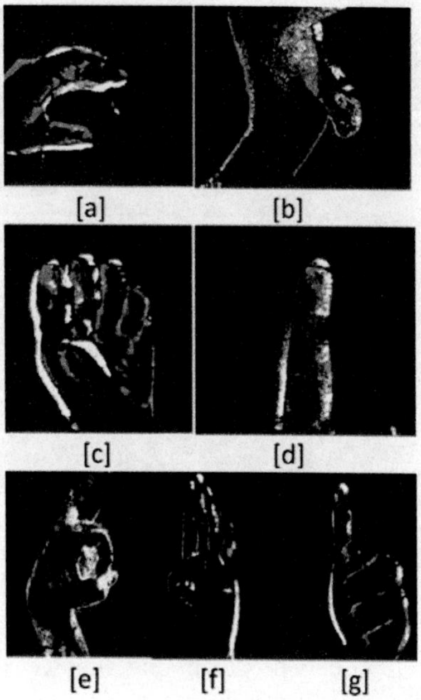

Fig. 3 Sample images present in the dataset

4 Implementation

In this project, we have classified the process of identifying the gesture from the image into three steps as described below. Our project is implemented using OpenCV environment. The actual image size is 240×215 pixels which is resized to 64×64 pixels for faster computation.

4.1 Data Acquisition and Preprocessing

For dataset acquisition, we collected a dataset from Kaggle consisting of seven different types of gestures. An average of 1600 images for each gesture are used. These gestures are namely (1) Letter I, (2) Thumps Up, (3) Palm, (4) Thumps Down, (5) OK, (6) Letter C, and (7) Fist.

The approach for detection of hands combines threshold-based color detection with background subtraction in this project, i.e., the RGB images are converted into gray scale for better prediction. We use Gaussian mixture-based background segmentation algorithm. It is a background segmentation algorithm based on a Gaussian mixture. This technique assigns a Gaussian distribution to each of the backdrop pixels. The duration of the colors' presence in the scene is what gives this distribution its weight.

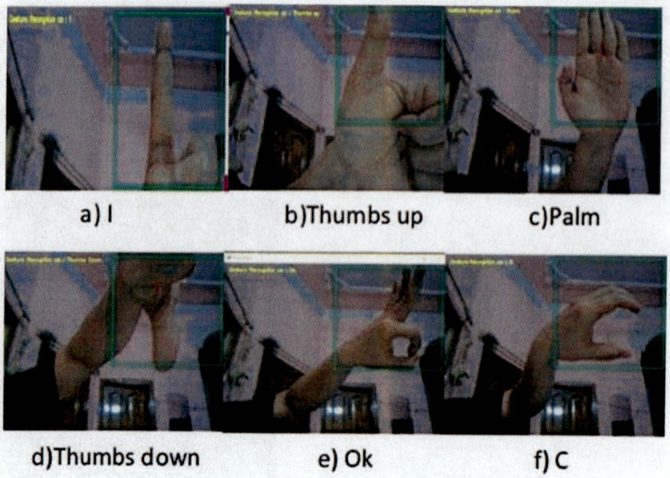

Fig. 4 Input gestures for recognition

For data preprocessing, first we capture each frame shown by the webcam of our laptop which recognizes our hand gestures. In each frame, we define a region of interest (ROI) which is signified by a bounded square as shown in Fig. 4. From this whole image, we extract our ROI and convert it into gray scale image using OpenCV. The image is then put to a Gaussian filter for image smoothing in order to lower the amount of noise in the image.

4.2 Training the Model

The preprocessed image is supplied to CNN, and the hand gestures shown by the user are classified, respectively, into the correct classes. The entire data file is separated as two parts: a training set of 80% of the images and a testing set of 20% of the images. We have used ten epochs to train our model. The batch size used is 128.

CNN uses the term "feature extraction" to describe the process of turning raw data into numerical features that can be examined, while preserving the content of the original dataset. Compared to directly applying CNN to the raw data, it yields a better result. It includes several convolutional layers followed by max-pooling and activation. We have used Adam optimizer, as it is better than every other optimization algorithm have faster computation time. It is used to modify the given model in accordance to the loss function's output.

We have used categorical cross entropy loss function Porter et al. (2021). Equation for categorical cross entropy is given as follows:

$$CE = -\sum_{i}^{C} t_i \log(s_i) \quad (1)$$

where t_i and s_i are the ground truth and the CNN score for each class i in C.

In the output screen, we need to show gesture properly, while adjusting your hands else we may get wrong predictions.

The recognized gesture is displayed as text and then converted into an audio file. The same is shown in Fig. 4. For converting text to audio, we have Google Text-to-Speech (gTTS) Hebbi et al. (2022) and CLI tool to work with Google Translates text-to-speech API. The output audio is produced in English and French language.

5 Results

From Figs. 5 and 6, we can conclude that with increase in epochs, loss is decreased, and accuracy remains same and maximum accuracy is achieved, i.e., 99.89% and a loss of 0.01.

The performance of the classification algorithm is shown and summarized using a confusion matrix as shown in Fig. 7. From the figure, we can tell that our model predicts all the classes correctly. As all the diagonal values are non-zero which represents true positive and all the non-diagonal elements are zero which represent true negative, false positive, and false negative.

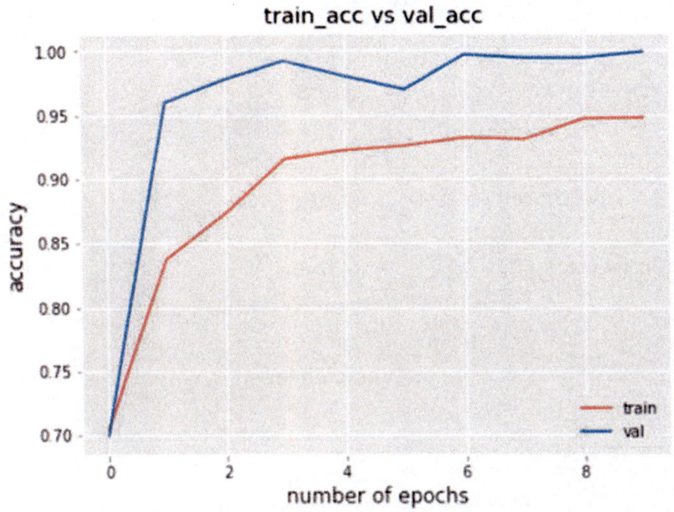

Fig. 5 Accuracy versus epoch

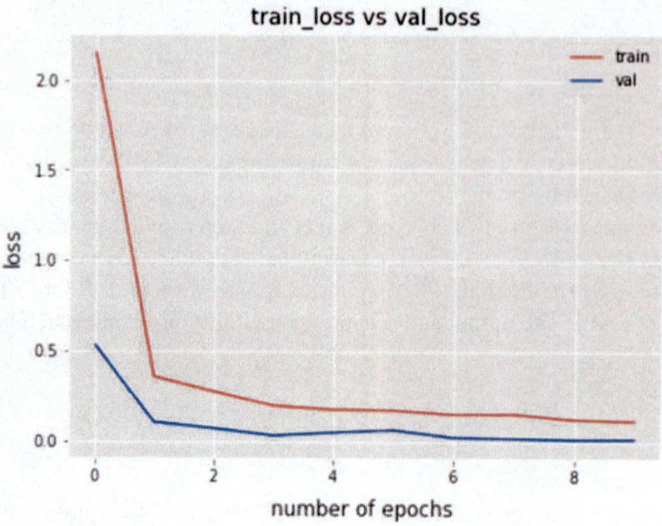

Fig. 6 Loss versus epoch

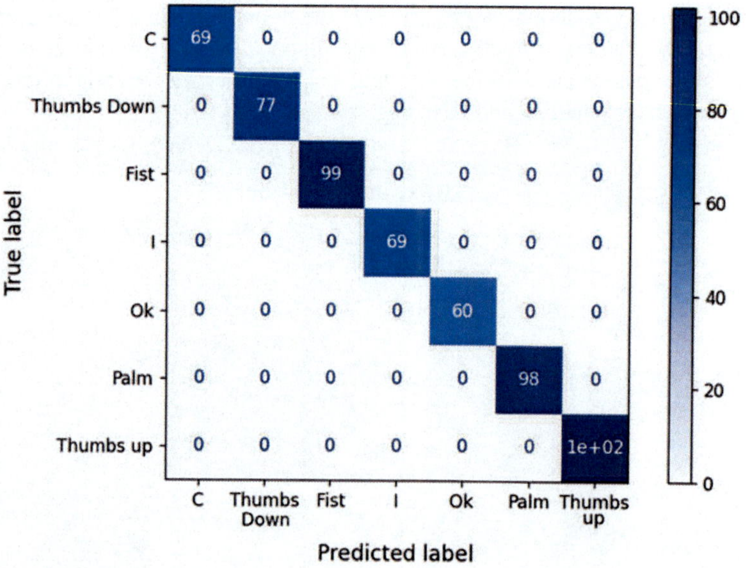

Fig. 7 Confusion matrix

From Fig. 8, we can deduce that our model's predicted positives and the actual positives are equal, i.e., precision = 1; recall score tells that our model correctly predicts the positives out of actual positives as it is equal to 1.

	precision	recall	f1-score	support
0	1.00	1.00	1.00	1
1	1.00	1.00	1.00	1
2	1.00	1.00	1.00	1
3	1.00	1.00	1.00	1
4	1.00	1.00	1.00	1
5	1.00	1.00	1.00	1
6	1.00	1.00	1.00	1

Fig. 8 Precision, f_1-score, recall, and support of the model

6 Conclusion and Future Work

We developed a CNN model for sign recognition wherein our model captures the image through webcam, converts it into gray scale image, and forecasts the output in text and speech form. The outcomes of the experiment show how effective the suggested approach is. Our model predicts all the seven classes of dataset effectively. For future work, we can collect more data and be able to recognize various other signs and also focus on recognition of sentences. In addition to the French that is being utilized here, we can also work on translating audio into other languages.

References

E. Abraham, A. Nayak, A. Iqbal, Real time translation of Indian sign language using LSTM, in *2019 Global Conference Advancement in Technology (GCAT)* (2020)

C. Hebbi, J.S. Sooraj, H.R. Mamatha, Text to speech conversion of hand- written Kannada words using various machine learning models, in *Evolution in Computational Intelligence* (Springer, Singapore, 2022), pp. 21–33

V. Navya, K.A. Patil, Identification of anomalies in images using CNN and autoencoders techniques, in *Proceedings of International Conference on Recent Trends in Computing* (Springer, Singapore, 2022)

S. Pandey, S. Dholay, An image processing approach for analyzing assessment of pavement distress. Innovations Comput. Sci. Eng. 493–502 (2019)

S. Pandey, N. Prabhu, A.S. Kumar, Survey on finger vein recognition using convolutional neural network, in *2022 Second International Conference on Artificial Intelligence and Smart Energy (ICAIS)* (IEEE, 2022)

E. Porter, et al., 10 Effect of loss functions in deep learning-based segmenta tion, in *Auto-Segmentation for Radiation Oncology: State of the Art* (2021), p. 133

M. Ramasamy, P.V. Eric, An improved deep bagging convolutional neural network classifier for efficient intrusion detection system. Bull. Electr. Eng. Inf. **11**(1), 405–413 (2022)

A.K. Sahoo, Indian sign language recognition using machine learning techniques (2021)

K. Shenoy, T. Dastane, V. Rao, D. Vyavaharkar, Real-time Indian sign language (ISL) recognition, in *International Conference on Computing and Networking Technology* (2018)

V. Veerasamy, et al., LSTM recurrent neural network classifier for high impedance fault detection in solar PV integrated power system. IEEE Access **9** (2021)

S. Zhang et al., A novel kNN algorithm with data-driven k parameter computation. Pattern Recogn. Lett. **109**, 44–54 (2018)
H. Zhang, L. Jiang, L. Yu, Attribute and instance weighted naive bayes. Pattern Recogn. **111**, 107674 (2021)

Development of Wearable and Portable Cardiac Arrest Prediction System

Saadman Kabir, Ananta Banik, Moshfiq-Us-Saleheen Chowdhury, Jannatul Ferdaus, and Md. Ashrafuzzaman

Abstract The research is founded on a medical engineering strategy for the creation of a cost-effective Cardiac Arrest Prediction Device. These gadgets have previously been invented and are in use in developed nations, but in underdeveloped nations it is difficult for families of all socioeconomic classes, from the lower to the middle class, to afford them. The paper focuses on development of a device that will primarily concentrate on medical-grade, real-time cardiac arrest prediction by continuously monitoring the heart rate of the user and alerting him to any abnormalities so that he can take the appropriate safeguards as soon as possible. The device will be based on the fundamental properties of ECG signals, will take readings from the electrodes, and will notify the user via smartphone notifications. The ST/T slope algorithm has been implemented to calculate the average values from the acquired data. The next edition of the device will emphasize a centralized monitoring system that will provide the user with access to all of his data at any time or location. Checking the status using the mobile device's application will provide several advantages. This technology will provide low-cost cardiac arrest prediction, allowing anyone to utilize it with confidence.

Keywords Cardiac arrest · Prediction · Portable

S. Kabir (✉) · A. Banik · J. Ferdaus · Md. Ashrafuzzaman
Department of Biomedical Engineering, Military Institute of Science and Technology (MIST), Mirpur Cantonment, Dhaka 1216, Bangladesh
e-mail: ashraf@bme.mist.ac.bd

M.-U.-S. Chowdhury
Department of Electrical, Electronic and Communication Engineering, Military Institute of Science and Technology (MIST), Mirpur Cantonment, Dhaka 1216, Bangladesh

A. B. Reddy et al. (eds.), *Proceedings of Third International Conference on Advances in Computer Engineering and Communication Systems*, Lecture Notes in Networks and Systems 612, https://doi.org/10.1007/978-981-19-9228-5_40

1 Introduction

Currently, heart disease is the leading cause of death worldwide. The most prevalent types of cardiac disease are stable angina, unstable angina, myocardial infarction, and sudden cardiac death (Wong, 2014). In many situations, myocardial infarction is the initial sign of heart illness. Myocardial infarction can be prevented if an accurate diagnosis is established at the earliest possible stage. Typically, the ECG test is used as an auxiliary diagnostic tool for cardiac disorders (Wang et al., 2009). The quantitative analysis of the study (Behzadi et al., 2020) conducted by Behzadi et al. (2020) revealed a good correlation between the tracings produced from 3 lead ECG and standard ECG for all of the studied segments in all three leads. Consequently, this method can be proposed as a new technique for faster, more robust, and more reliable identification of myocardial infarction that can be identified from a 3-lead ECG, particularly when standard ECG facilities are not immediately available. Consequently, the study objectives of a myocardial infarction diagnosis system based on 3 lead-based ECG images derived from 12 lead-based ECG readings. Single Shoot Detection (SSD) MobileNet v2-based Deep Neural Network architecture was utilized to detect myocardial infarction. In initial testing, the model concentrated on classifying three key classes, including Normal ECG, Myocardial Infarction, and others (ST/T Change, Conduction Disturbance, and Hypertrophy), with 98 % accuracy. This study is based on the publicly accessible PTB-XL dataset (Wagner et al., 2020), which contains a collection of 21,837 standard 12-lead-based ECG readings from 18,885 patients of 10-s length, of whom 52% are male and 48% are female, with ages spanning the entire range of 0–95 years (median 62 and interquartile range of 22).

1.1 Motivation

It has been noticed that locating a dependable and affordable gadget is difficult for all demographics. Myocardial Infarction is one of the leading causes of death, claiming hundreds of lives annually. The primary cause of death is the absence of adequate facilities for timely diagnosis and treatment. With the present rise in mobile device usage, integrating medical devices with the same impact makes it easier for both doctors and patients to identify health risks or irregularities.

With safe and convenient approaches that are also cost-effective, the quality of patient care can be multiplied significantly. The objective of the device we intend to create is to provide first aid with a low-cost, medical-grade diagnosis of myocardial infarction.

The primary objective of this development and research is to create a low-cost cardiac arrest prediction device with a central monitoring platform that can be accessed via an application on a personal mobile device for people of all socioeconomic classes, particularly in low- to middle-income countries. Even doctors will be able to use the essential information so that the device will be able to perform the basic and initial diagnostic instead of requiring a visit to the hospital, hence reducing the hospital's patient load.

1.2 Major Contribution

The invention of an accurate cardiac arrest prediction system that is reasonably priced and based on a medical engineering approach is the primary contribution of this research. In addition to this, some other contributions include the following:

- Development of an appropriate electrode equipped with a unipolar lead in order to reduce the number of electrodes required for an electrocardiogram reading of the heart.
- Extraction of the data and input it into a certain algorithm in order to compute the parameters of a cardiac arrest.
- Training the machine learning algorithms with their optimum parameters so that they can procure the data that will anticipate and inform the likely future consequence of a cardiac arrest.

2 Literature Review

The heart is the most important organ in the cardiovascular system, which can be characterized simply as a "pump." The heart's contraction generates pressure. Blood flows through the major blood arteries, such as the aorta and pulmonary trunk, as a result of this increased pressure. These vessels provide the remainder of the body with blood. Although the term "pump" connotes a mechanical device made of steel and plastic, the anatomical structure is a complex, organic muscle (Hima et al., 2010). The function of the heart in the circulatory system is comparable to that of the pump in the central heating system. Similar to how a central heating pump pushes hot water via pipes to radiators, the heart pushes blood through arteries and veins to organs, muscles, and tissues.

The cardiovascular system is governed mostly by the autonomic nervous system (sympathetic and parasympathetic). Generally, the heart rate of a healthy individual with a healthy heart is determined by the needs of the body. When a person is at rest, his or her organs, muscles, and tissues require less blood and oxygen. Due to this, blood pressure decreases along with heart rate and respiratory rate.The heart is capable of generating its own electrical impulses. Using a specific conduction pathway, it can also control the path of the impulses. There are primarily five components to this pathway (Fig. 1).

- The sinoatrial (SA) node
- The atrioventricular (AV) node

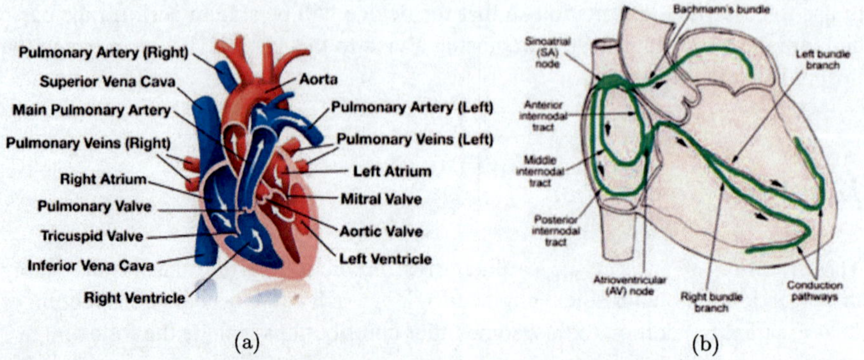

Fig. 1 **a** The passage of blood through the heart (Hima et al., 2010); **b** The cardiac conduction system (Hima et al., 2010)

- The bundle of His
- The left and right bundle branches
- The Purkinje fibers.

An ECG amplifies the heart's tiny signal and filters both internal and external noise. A differential amplifier is used for amplification, while common and differential mode filtering is used for filtering. Right Leg Drive (RLD) circuit eliminates noise and preserves common-mode voltage.

Common-mode signals from instrumentation amplifier inputs are inverted by the RLD amplifier. The output current of the right leg drive counteracts common-mode voltage changes, hence enhancing the system's common-mode rejection.The GM1 trans conductance amplifier provides the common-mode signal for the instrumentation amplifier.

3 Methodology and Implementation

The implementation of the device can be broken down to some subsections.

3.1 Designing of Data Transaction and Storing for ECG Module

The ECG module is combined with an Arduino nano, which serves as the principal control unit. The Arduino transmits data from the ECG module to an Android application via Bluetooth module (HC05).

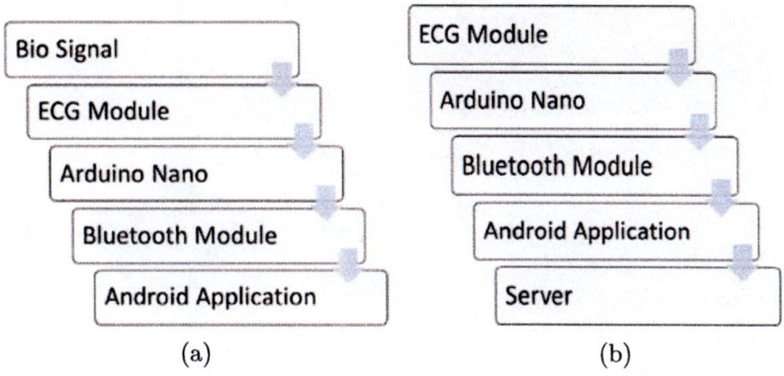

Fig. 2 **a** Block diagram for 3 lead ECG system **b** Data transaction of ECG with server

The module will be accountable for full data transfer from the patient to the doctors. The patient will send data not only to the servers, but also to the user's family members so that they can connect and contact the individual as quickly as possible. Figure 2 represents the block diagram of ECG system and Data transaction of ECG with server.

3.2 *Designing the Data Visualization Process*

Designing the Data Visualization Process: Bluetooth transmits the data acquired from the ECG module to the Android application. Our ECG module is integrated with a Bluetooth module for data display and processing in this instance. The app will connect to the website, which will be connected to the server, which will simultaneously examine the data from the database, verify the references, and then alert the individual or patient/patient's family of any deviation from a normal ECG reading.

3.3 *Collecting ECG Information and Design of a Sustainable Monitoring System*

The modules will be deployed in various medical facilities to collect ECG data. The information will be stored on a central server for myocardial infarction case diagnosis. A central monitoring system coupled to a wide area or medium area network can vastly improve the existing distribution of medical services over local areas. All of these events demonstrated the need of Central Monitoring Systems to hospitals in this region. A block diagram of data collection and sustainable monitoring system development can be comprehended from Fig. 3 and the schematic of the of the prototype can be seen from Fig. 4.

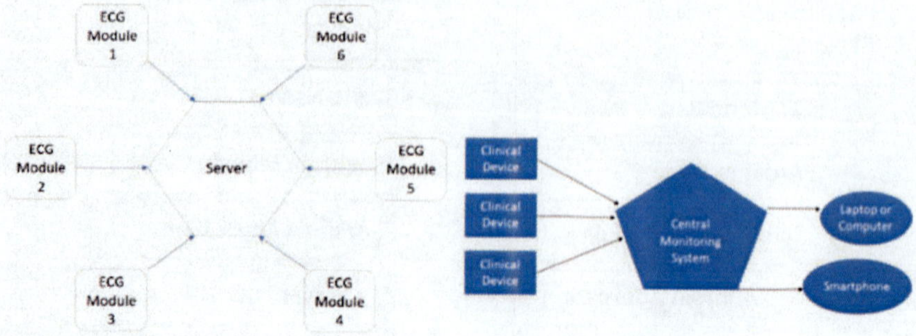

(a) (b)

Fig. 3 **a** Collection of ECG data from multiple clinic establishments **b** Central monitoring system principles

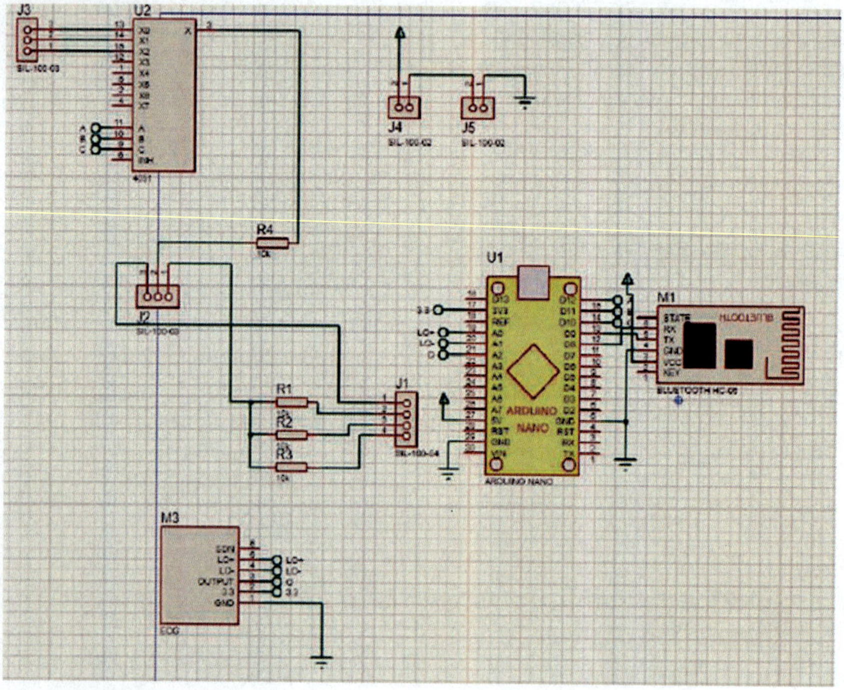

Fig. 4 Full schematic for the final prototype in proteus

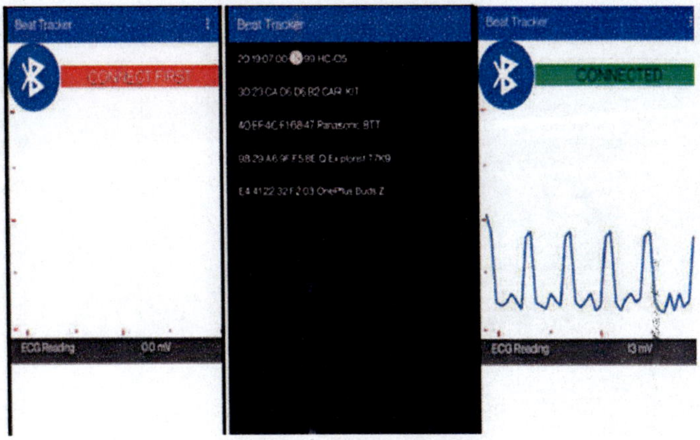

Fig. 5 View of the final readings from mobile application

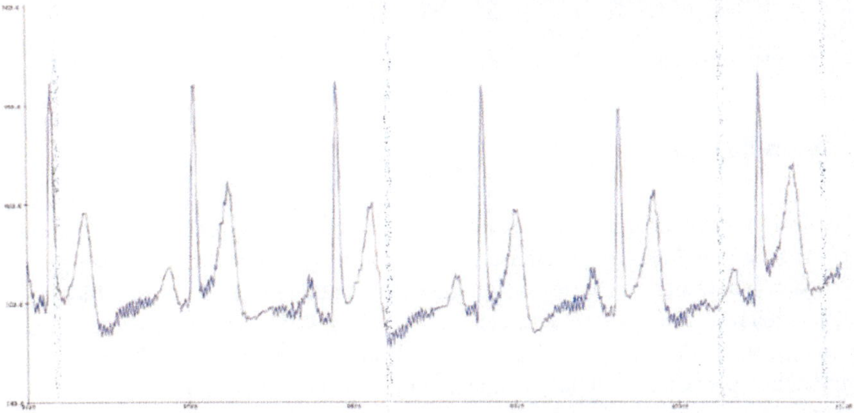

Fig. 6 Data acquisition in PC

4 Real Time Data Acquisition

For real-time data acquisition, the main PCB is Bluetooth-connected to an Android mobile device, which constantly delivers packet data to the mobile device's app. Any anomalies will be displayed on the mobile device, which can then be utilized to detect any arrhythmic disorders that can be treated.

After receiving the data from the device in the ECG system, the next objective is to enter the data into an Excel Sheet that will be stored securely for future usage and patient profiling. The Simulations can be observed from Figs. 5, 6 and 7.

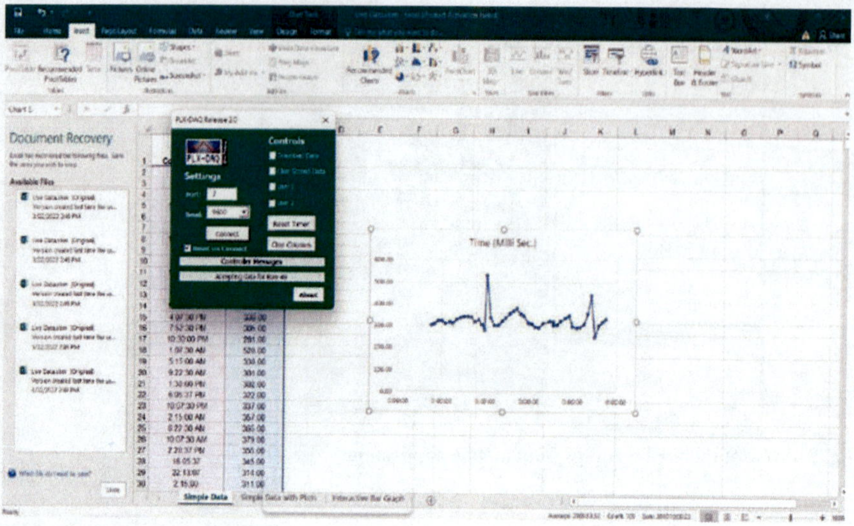

Fig. 7 Data saving in excel in PC

5 Conclusion

The development of a myocardial infarction monitoring system is expensive and involves the cooperation of specialists from a variety of professions. The development of medical equipment is a time-consuming endeavor that involves collaboration with specialized research laboratories and private vendors. Despite the drawbacks of local development, a device is developed that will alert the patient with both certified doctors and the patient's family members in the event of a sudden infarction. Even though similar technologies already exist, this platform is believed to be extremely low-cost to produce and simple to use, requiring little to no knowledge with the device and application included in the package.

The current version of the device uses Bluetooth technology, but the next generation will employ Wi-Fi for faster and more dependable data connection and data processing. In such a circumstance, the devices will be registered with distinct IP addresses, making them simpler to identify and manage centrally. Through the assistance of government and commercial groups, this low-cost platform has the ability to proceed toward original gadget development, which, if adopted countrywide in the near future, will surely contribute to saving vast populations.

References

A. Behzadi, A. Sepehri Shamloo, K. Mouratis, G. Hindricks, A. Arya, A. Bollmann, Feasibility and reliability of smartwatch to obtain 3-lead electrocardiogram recordings. Sensors **20**(18), 5074 (2020)

B.T. Hima, M. Vidyavathi, K. Kavitha, T. Sastry, K.R. Suresh, Preparation and evaluation of chitosan-gelatin composite films for wound healing activity. Trends Biomater. Artif. Organs **24**(3), 123–131 (2010)

P. Wagner, N. Strodthoff, R.D. Bousseljot, D. Kreiseler, F.I. Lunze, W. Samek, T. Schaeffter, Ptb-xl, a large publicly available electrocardiography dataset. Sci. Data **7**(1), 1–15 (2020)

L. Wang, H. Zhang, K.C. Wong, H. Liu, P. Shi, Physiological-model-constrained noninvasive reconstruction of volumetric myocardial transmembrane potentials. IEEE Trans. Biomed. Eng. **57**(2), 296–315 (2009)

N.D. Wong, Epidemiological studies of chd and the evolution of preventive cardiology. Nat. Rev. Cardiol. **11**(5), 276–289 (2014)

Political Optimizer-Based Automated Machine Learning for Skin Lesion Data

Gurram Sunitha, B. Swaroopa Rani, Shankar Nayak Bhukya, Hafeena Mohammad, and R. Hitesh Sai Vittal

Abstract Today in the age of information revolution, everything is being automated. Machine learning is needed for every industry to boost growth in business, health, agriculture, and everyday life. As machine learning is creating its importance in every industry, wherever there is scope for information generation, practicing machine learning is less than 40% in real life. This is mainly because of the challenges, while creating, implementing, and deploying machine learning models which require expertise. In this scenario, automated machine learning tools allow everyone without knowledge of coding to use insights from machine learning models. In this paper, an efficient model has been designed and developed for skin lesion imagery classification. The research work undertaken in this paper is to tune hyperparameters using political optimizer for constructing efficient classifier. The proposed model is gauged on ISIC-17 dataset, and average accuracy, sensitivity, specificity were 97.86, 0.9736, 0.9878, respectively. The proposed model has been evaluated with existing classification techniques SVM and BPN for comparative analysis, and it has been justified per the experimental outcomes that the classifier model with political optimizer for hyperparameter tuning proposed in this paper performs better than the SVM and BPN models.

Keywords Automated machine learning · Machine learning pipelines · Political optimizer · Hyperparameter tuning · Machine learning model training

---

G. Sunitha (✉)
Department of CSE, School of Computing, Mohan Babu University, Tirupati, India
e-mail: gurramsunitha@gmail.com

B. S. Rani · H. Mohammad
Department of CSE (AI&ML), CMR Technical Campus, Hyderabad, Telangana, India

S. N. Bhukya
Department of CSE (Data Science), CMR Technical Campus, Hyderabad, Telangana, India

R. H. S. Vittal
Hyundai Mobis, Hyderabad, India

A. B. Reddy et al. (eds.), *Proceedings of Third International Conference on Advances in Computer Engineering and Communication Systems*, Lecture Notes in Networks and Systems 612, https://doi.org/10.1007/978-981-19-9228-5_41

1 Introduction

"Machine Learning for Everyone as it is Future" is a new preposition in the business world from pennies to trillions of markets captive share. As machine learning is used in every industry from agriculture to astrophysics, demand for machine learning is increasing like wings of fire. This inverted theory is due to the non-capable handling of small chunks of data generated in pink share companies. But there comes the problem with machine learning because not everyone can practice machine learning due to its complex understanding and implementation nature which requires proficient engineers who should be provided high pay scales even for nitty–gritty tasks. So, providing tools required to integrate machine learning models with the user-friendly interface really helps users who don't know to code or understand complex mathematics behind the machine learning models. Anyone who are non-experts, out-of-domain practitioners who feel the need for machine learning can use the automated machine learning to construct machine learning models to the problem at hand.

Anyone with little or basic knowledge in machine learning can use the automated machine learning tools to gain insights into their data and apply the machine learning models to improve their business. A variety of issues can be addressed using the machine learning models in real world such as agriculture, food processing industry, finance industry, medical practice, entertainment, media, transportation, and communication.

The motto for the work presented in this paper is automating hyperparameter optimization problem. This is an optimization problem which deals with automating the process of hyperparameter tuning and providing insights as to the best parameter setup for the machine learning model under consideration for solving a given problem.

2 Literature Review

The development of intelligent systems depends heavily on machine learning. The machine learning pipeline comprises the steps—data processing, feature engineering, model selection, hyperparameter tuning, analyzing the models, and choosing optimal model. Experimenting with variations of these factors is labor-intensive and time-consuming process which involves high levels of human intervention at each and every step of building a machine learning model. Creating automated machine learning tools aims to reduce the interference of humans during the creation of machine learning models (Yao et al. 1810; Han et al. 2020). An automated machine learning tool will experiment with various hyperparameters, various feature engineering techniques and applies multiple models on the given data. Further, it analyzes the results thus providing insights as to the best parameter setup and most suitable machine learning model for the given data.

Many have developed AutoML tools, platforms, and technologies to encourage wide usage of machine learning to various domains and applications. Auto-WEKA tool has taken up the problems of automatic machine learning model selection and automatic hyperparameter setup (Hutter et al. 2019; Kotthoff et al. 2017; Feurer et al. 2019). Many researchers have investigated the state-of-the-art platforms and tools for AutoML and have thoroughly compared and evaluated the current tools. (Truong et al. 2019; Tuggener et al. 2019; Waring et al. 2020). Companies like Google and Amazon have been researching and developing cloud platforms to automate machine learning services (Google 2020; Amazon 2020). They have been providing cloud services to build the machine learning models, custom train, and deploy the models.

Machine learning for various healthcare applications has been an interesting area of research and investigation (Thapar et al. 2022; Khan et al. 2021; Madhavi et al. 2022; Kuraparthi et al. 2021; Ozkan and Koklu 2017).

3 Automated Machine Learning for Skin Lesion Data

The objective of automated machine learning is to unravel CASH problem. For any given problem, choosing the right machine learning model and choosing the most appropriate hyperparameter setup is a mammoth task.

The work in this paper's presentation aims to automate a part of CASH problem which is automating the process of hyperparameter tuning and providing insights as to the best parameter setup for the machine learning model under consideration for solving a given problem. A chaotic political optimizer-based method is proposed for tuning hyperparameters to determine optimal hyperparameters for the given problem and the machine learning model under consideration.

Hyperparameters are a critical component which radically affects the behavior and performance of the machine learning models. Hence, tuning of hyperparameters is integral and inescapable step to construct an optimal machine learning model. The hyperparameter space can be infinite and searching through such hyperparameter space may be a never-ending process. In such scenarios, optimization techniques play vital role in maximizing performance on the data in a timely manner by identifying the ideal mix of hyperparameter variables.

3.1 Modeling Political Optimizer for Hyperparameter Tuning

In this paper, a political optimizer is modeled to tune hyperparameters and obtain optimal values to construct an efficient machine learning model for skin lesion data. The political optimizer technique is based on the election commission process and involves five stages (a) party formation, constituency allocation (b) election promotion (b) transferring parties (4) interparty election (5) win election (Askari et al. 2020). The outline of political optimizer technique is as follows: Each individual is represented with a set of attributes; population is class labeled; initial population is generated; fitness of each individual is calculated; individual with best fitness is treated to be the party leader and constituency winner; individuals may be transferred to other parties if required; update position of each individual; dissimilar individuals are treated as members of parliament; execute parliament affairs for each constituency winner; and repeat until stopping criterion is satisfied. The political optimizer model is presented in Fig. 1.

<u>Input:</u>

P – Population

Pi – An individual in the population represented with n number of attributes

P(i) – Population of ith iteration

Sr – Switching rate threshold

t – t^{th} iteration

<u>Output:</u>

Optimal population

<u>Algorithm:</u>

```
1: Initialize initial population P(1)
2: t = 1
3: While !stopping criterion
4: for each individual Pi
5: Compute fitness of each individual Pi(t)
6: Determine the winners
7: Switch parties of individuals if required following Sr
8: Update position of each individual
9: Determine the members of parliament
10: Execute parliament affairs for each winner
11: t = t+1
12: end
13: end
14: end
```

Fig. 1 Political optimizer model for hyperparameter tuning

3.2 Proposed Methodology for Automated Machine Learning

The proposed methodology for automated machine learning model is presented in Fig. 2. The skin lesion image dataset is considered as training data. Firstly, all the skin lesion images are then converted into gray scale and then are preprocessed. Image preprocessing is to enhance quality of the images by removing body hair, smoothening images, and detecting the regions of interest. These steps will support smooth identification of lesion regions in the images. For identifying body hair in the images, the top-hat transform technique is used. Then, the pixels representing the body hair are modified by using nearest neighbor technique. Then, the skin lesions which are the regions of interest are demarked by using the steps—local entropy filtering, global thresholding, and cropping.

The next step is to perform hyperparameter tuning using political optimizer using the proposed method as shown in Fig. 1. Further, image segmentation is performed using multilevel thresholding technique, and speeded up robust features technique is used for extracting features, thus generating prioritized feature vectors. Further, these prioritized feature vectors are used by XGBoost algorithm to train the classifier. The

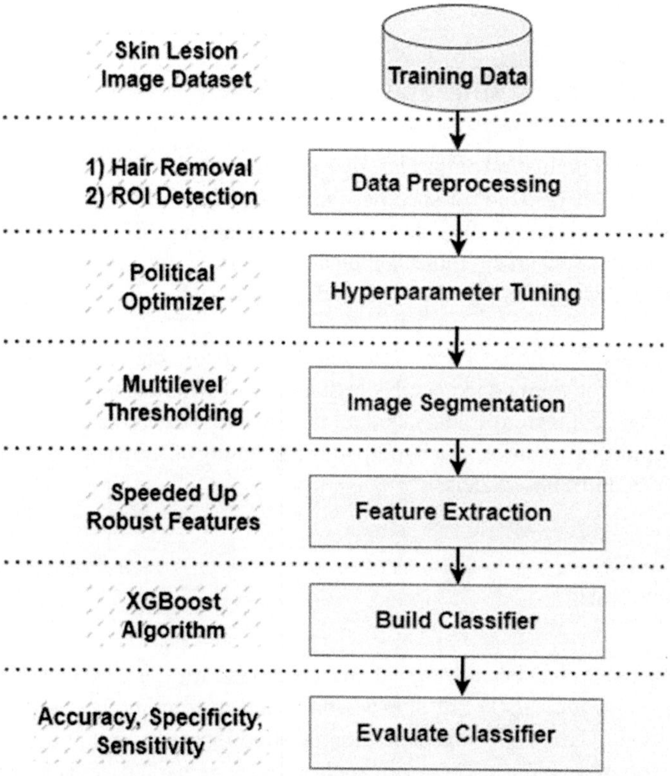

Fig. 2 Proposed learning model for classification of skin lesion dataset ISIC-17

Table 1 ISIC-17 dataset—image specifications

Item	Description	Remarks
Dermoscopic images		* All the images are tagged with a unique 7-digit identifier
Image format	JPEG	
Image size	640 × 480	
Dataset size	2000	
Ground truth segmentation images (Binary images) (Masked dermoscopic images)		
Images type	PNG	Types of Skin Lesion Diseases—Angioma, nevus, lentigo NOS, solar lentigo, melanoma, sebo keratosis, basal cell carcinoma
Image size	640 × 480	
Dataset size	2000	

advantage of XGBoost is that it manages and optimizes residuals at each iteration. Finally, the classifier model is evaluated using metrics accuracy, sensitivity, and specificity.

4 Experimental Results and Evaluation

For experimental evaluation of the classifier model, the ISIC-17 dataset is considered (ISIC-17 dataset 2017). Table 1 specifies the details of the ISIC-17 dataset. Figure 3 showcases performance of the classifier model trained on ISIC-17 dataset on 318 samples in terms of accuracy. Table 2 tabulates the performance evaluation (accuracy, sensitivity, and specificity) of the proposed classifier which is trained on ISIC-17 dataset.

The proposed model has been evaluated with existing classification techniques SVM and BPN for comparative analysis. Statistics of two runs have been presented in Table 3. The comparative analysis demonstrates that the proposed classification model with political optimizer for hyperparameter tuning of XGBoost performs better than the SVM and BPN models.

5 Conclusions

An efficient model has been designed and developed for the classification of skin lesion imagery data. Firstly, all skin lesion images are preprocessed to enhance quality of the images. The images are preprocessed by using top-hat transform and nearest neighbor techniques. Then, the skin lesions are demarked by using the steps—local entropy filtering, global thresholding, and cropping. Further, hyperparameter

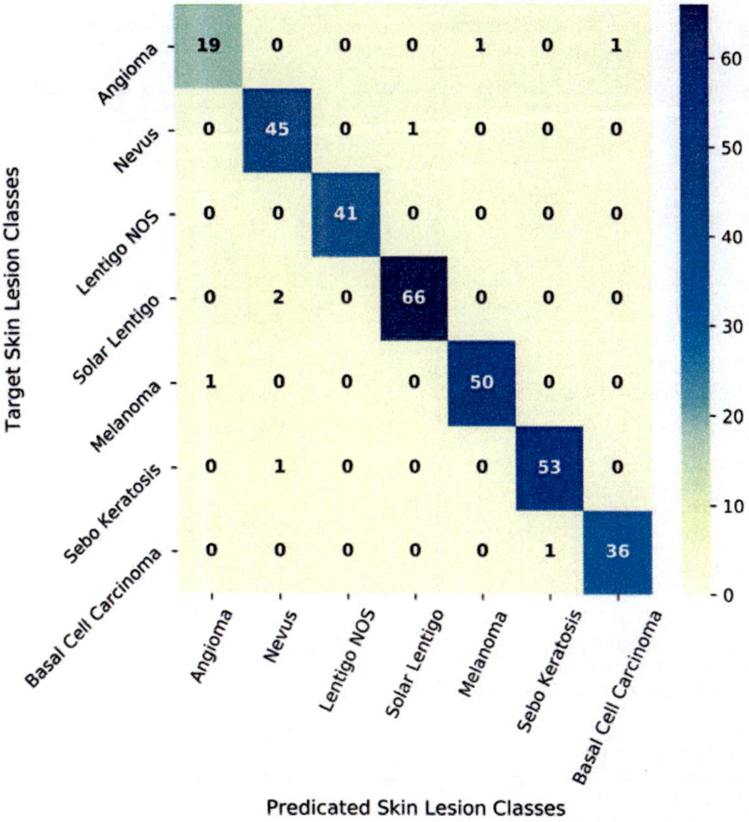

Fig. 3 Classification performance of the proposed model on skin lesion dataset ISIC-17

Table 2 Proposed model's performance evaluation of skin lesion dataset ISIC-17

Training dataset size	Accuracy (%)	Sensitivity	Specificity
321	97.484	0.954	0.986
500	97.024	0.968	0.987
1000	97. 786	0.973	0.978
1500	98.182	0.989	0.993
2000	98.751	0.984	0.995
Average	**97.86**	**0.9736**	**0.9878**

tuning is performed using political optimizer, and image segmentation is performed using multilevel thresholding technique, and speeded up robust features technique is used for extracting features, thus generating prioritized feature vectors. Finally, the extracted feature vectors are used for XGBoost algorithm for training the classifier model, and the classifier model is evaluated using the evaluation metrics. The evaluation of the proposed model is performed on ISIC-17 dataset, and average accuracy,

Table 3 Comparative performance evaluation of proposed classifier model with SVM and BPN on ISIC-17 skin lesion dataset

Training dataset size	Classifier	Accuracy (%)	Sensitivity	Specificity
100	BPN	92.180	0.912	0.936
	SVM	93.367	0.899	0.922
	Proposed	96.234	0.961	0.967
321	BPN	92.896	0.900	0.925
	SVM	94.539	0.922	0.929
	Proposed	97.484	0.954	0.986

sensitivity, specificity were 97.86, 0.9736, 0.9878, respectively. Future research will examine automated deep learning for efficacy of the classification procedures for skin lesion imagery.

References

Amazon, machine learning on AWS. https://aws.amazon.com/machine-learning/. Accessed 02 Nov 2020

Q. Askari, I. Younas, M. Saeed, Political optimizer: a novel socio-inspired meta-heuristic for global optimization. Knowl.-Based Syst. **195**, 105709 (2020)

M. Feurer, A. Klein, K. Eggensperger, J.T. Springenberg, M. Blum, F. Hutter, Auto-sklearn: efficient and robust automated machine learning, in *Automated Machine Leaning* (Springer, Cham, 2019), pp. 113–134

Google, cloud AutoML. https://cloud.google.com/automl/. Accessed 02 Nov 2020

J. Han, K.S. Park, K.M. Lee, An automated machine learning platform for non-experts, in *International Conference on Research in Adaptive and Convergent Systems* (2020), pp. 84–86

F. Hutter, L. Kotthoff, J. Vanschoren, *Automated Machine Learning: Methods, Systems, Challenges* (Springer Nature, 2019)

ISIC-17 dataset, skin lesion images. https://challenge.isic-archive.com/landing/2017/42/

M.A. Khan, M. Sharif, T. Akram, R. Damaševičius, R. Maskeliūnas, Skin lesion segmentation and multiclass classification using deep learning features and improved moth flame optimization. Diagnostics **11**(5), 811 (2021)

L. Kotthoff, C. Thornton, H.H. Hoos, F. Hutter, K. Leyton-Brown, Auto-WEKA 2.0: automatic model selection and hyperparameter optimization in WEKA. J. Mach. Learn. Res. (2017)

S. Kuraparthi, M.K. Reddy, C.N. Sujatha, H. Valiveti, C. Duggineni, M. Kollati, P. Kora, Brain tumor classification of MRI images using deep convolutional neural network. Traitement du Signal **38**(4) (2021)

K.R. Madhavi, P. Kora, L.V. Reddy, J. Avanija, K.L.S. Soujanya, P. Telagarapu, Cardiac arrhythmia detection using dual-tree wavelet transform and convolutional neural network. Soft. Comput. **26**(7), 3561–3571 (2022)

I.A. Ozkan, M. Koklu, Skin lesion classification using machine learning algorithms. Int. J. Intell. Syst. Appl. Eng. **5**(4), 285–289 (2017)

P. Thapar, M. Rakhra, G. Cazzato, M.S. Hossain, A novel hybrid deep learning approach for skin lesion segmentation and classification. J. Healthc. Eng. (2022)

A. Truong, A. Walters, J. Goodsitt, K. Hines, C.B. Bruss, R. Farivar, Towards automated machine learning: evaluation and comparison of AutoML approaches and tools, in *31st International Conference on Tools with Artificial Intelligence (ICTAI)* (IEEE, 2019), pp. 1471–1479

L. Tuggener, M. Amirian, K. Rombach, S. Lörwald, A. Varlet, C. Westermann, T. Stadelmann, Automated machine learning in practice: state of the art and recent results, in *6th Swiss Conference on Data Science (SDS)* (2019), pp. 31–36

J. Waring, C. Lindvall, R. Umeton, Automated machine learning: review of the state-of-the-art and opportunities for healthcare. Artif. Intell. Med. **104**, 101822 (2020)

Q. Yao, M. Wang, Y. Chen, W. Dai, Y.F. Li, W.W. Tu, Q. Yang, Y. Yu, Taking human out of learning applications: a survey on automated machine learning. arXiv Preprint (2018). arXiv:1810.13306

Tree Cutting Sound Detection Using Deep Learning Techniques Based on Mel Spectrogram and MFCC Features

Sallauddin Mohmmad and Suresh Kumar Sanampudi

Abstract This paper presents an approach that classifies the process of tree cutting events in a forest based on sounds detection. The forest environment consists of different types of sounds generated with different frequencies from various directions. CRNN model provides accurate results through multi-dimensional data input through which can learn various audio features for predicting the real-time scenarios. In this research, CRNN has implemented for performing sound event classification and predict the real-time scenarios in better way. Multiple features such as MFCCs and Mel Spectrogram are extracted from each sound sample for generating the prediction rate. In our research, we have taken three classes of data set related to tree cutting such as axe knocking, saw cutting, and tree falling. And the rest of the sounds is considered as noise class and is added to false ratio. The performance obtained through the proposed model has generated an accuracy of 93.4% on our newly proposed datasets which is better than existing methodologies of forest environmental sound detection.

Keywords CNN · RNN · CRNN · MFCC · Mel spectrogram

1 Introduction

The "Forest Sound Event Detection (FSED)", is considered to be the initial step in most of the applications which includes illegal deforestation monitoring, animal species identification, etc. In FSED, we have considered tree cutting with axe, tree cutting chainsaw, tree falling sound, and other forest environment sounds. The main

S. Mohmmad (✉)
JNTU, Hyderabad, India
e-mail: sallauddin.md@gmail.com

School of Computer Science and Artificial Intelligence, SR University, Warangal, TS, India

S. K. Sanampudi
Department of Information Technology, JNTUH College of Engineering, Nachupally, Kondagattu, Jagtial, TS, India
e-mail: sureshsanampudi@jntuh.ac.in

A. B. Reddy et al. (eds.), *Proceedings of Third International Conference on Advances in Computer Engineering and Communication Systems*, Lecture Notes in Networks and Systems 612, https://doi.org/10.1007/978-981-19-9228-5_42

object of this paper is to monitor illegal deforestation. As illegal deforestation is a major issue to be tackled in most of the forest areas, monitoring requires a huge human workforce. Most of the researchers in the area has suggested solutions through computer vision and ML approaches. But the computer vision solutions are not cost effective when compared to ML-based approaches. "K-Means Clustering", "Gaussian Mixture Model (GMM)", "Principal Component Analysis (PCA)", etc., have employed to classify the sounds in deferent environments based on the various feature extraction methods (Adavanne et al. 2018; Dang et al. 2018). The various feature extractions procedures such as "MFCC", "Mel Spectrogram", "RMS", and "spectral centroid" are implemented to gain better results (Deng et al. 2020). The "Multi-feature extraction process" provides more information for enhancing the prediction rate. These feature extractions become more challenging in the forest environment where multiple sounds are generated continuously and lead to noise creation. In the "acoustic event detection" process, the noise generated comprising of distinct pitch needs to be processed for detecting actual sounds. And some of the other renowned researchers has adopted sound event detection model through machine learning-based approaches implemented over proper defined datasets such as "CRNN", "SVM", "Random Forest, and K-Mean". The neural networks models such as CNN and RNN generated better results when compared with the previous studies such as integrated LSTM and BiGRU for generation of prediction rate which will be further enhanced. To detect the sound events in the forest a suitable dataset related to tree cutting is to be generated or created. Data set must include audio clips of tree knocking sound and various sounds available in a forest environment. In (Cakır et al. 2017), the authors considered deep learning techniques along with data augmentation for addressing the data scarcity for data-intensive models. The author explained the impact of data augmentation on different classes and proposed a new technique called class-based data augmentation. In (Adavanne et al. 2017), the authors considered tree cutting with a handsaw alone. And he employed Dynamic Time Wrapping (DTW) and Gauss-Bayesian classifier for prediction. Some of the researches are involved using neural network approach for providing solution through IOT-based CNN models for optimizing low-powered devices by considering "Mel Spectrogram" and "Mel Frequency Cepstral Coefficients (MFCC)" as features (Chan 2001; Adavanne et al. 1905). In our research, we have implemented the proposed model over different CRNN approaches by stacking both CNN and RNN models. The data set which is used in the model obtained through Internet-sources and self-prepared records. The tree falling sound and saw cutting sound classes are downloaded from the sources. The tree cutting axe knowing sounds are recorded by our team from different forest environmental places, where MFCCs and Mel Spectrogram are used as features in our experimental study. According to the study, different researches with respect to different data set on various environments have achieved better results. Our research has chosen forest environment and included new dataset such as axe cutting sounds to detect the acoustic events in the forest. And we got an accuracy of 93.4% along with the AUC score of 0.9915 with respect to our new dataset classes which belong to forest environmental sounds. In Sect. 2, we have discussed different studies on sound event detection with respect to various environment data sets. Section 3 presents the information about our data set collection and ratios. The proposal model execution criteria explained in Sect. 4. In Sect. 5, obtained results are analysed in different approaches.

2 Literature Survey

Adavanne et al. (2018) have explained the multichannel sound event detection (SED) using a 3D convolutional neural network (CNN) layer. The suggested C3RNN algorithm learned to distinguish overlapping sound events achieved $F1$ score increased by 7.5% and the overall ER improved by 10%. An Dang et al. (2018) used a convolutional neural network methods for acoustic scene classification in their paper. They considered the TUT Acoustic Scenes 2016 data set for SED. Before the classification in this model, they have extracted multi-feature from the sound sample such as MFCC and spectrogram to achieve the better result. Deng et al. (2020) employed a model with CRNN and implemented an enhanced Mel frequency spectrum coefficient (MFCC). According to this model, first and second order difference features of MFCC are extracted. Cakır et al. (2017) introduced an approach to identify the cardiac sounds based on CRNN enhanced with Mel band feature extractions. The data sets used were TUT SED synthetic 2016, TUT SED 2009, TUT SED 2016, CHiME-Home. Imoto et al. (2020) proposed a novel technique for sound event detection (SED) based on n multitask learning (MTL) of SED and acoustic scene categorization (ASC) using acoustic scene soft labels, which allows to model the extent to which sound events and scenes are associated. The TUT Sound Events 2016/2017 and TUT Acoustic Scenes 2016 data sets were used to prove that the suggested technique enhances SED performance by 3.80% in F-score when compared to traditional MTL-based SED. The MTL-based model was changed to include the teacher–student learning framework. The teacher network for scene classification in the proposed method is initially trained using hard scene labels, as in the traditional ASC method. The output of the trained teacher network is then used to train the soft scene label. Grondin et al. (1910), in their paper, a multichannel recording was used to propose sound event localization and detection algorithms. The researchers employed a mixture of two convolutional recurrent neural networks (CRNNs) with comparable front-end architecture. CRNN-SED is the first network, and it's been trained to detect, find, and estimate the start and offsets of sound events from a pair of microphones. CRNN-TDOA, the second network, calculates the TDOA for each pair of microphones and each sound event type. The system is tested using a four-microphone array, which combines the data of six pairs of microphones to supply a final classification and a three-dimensional direction of arrival (DOA) estimation. The suggested technique outperforms the DCASE 2019 baseline system, according to the findings. Mak et al. (2012) proposed a decent approach for sound event detection on mobile devices in this paper. He examined the power consumption of different stages, acoustic features, and kernels in the sound event detection system. He developed a low-power SVM that reduced the CPU utilization of the model by 28 times when compared to non-optimized SVM models. This approach can be extended to implement in low-powered edge computing devices. Zheng et al. (2017) made a sense of the environment by sounds which is an important research in machine learning community. In this work, a deep convolutional neural network (DCNN) model is presented to classify acoustic scenes along with a multiple spectrograms fusion method.

There are many previous works on sound event classification; few among them are considered for this experiment as a literature survey. Table 1 summarizes those previous works and describes each of the previous works briefly in the literature survey section.

3 Data Set

Our data set is a live data set made by our team. The audio files were gathered by our team. This data set consists of sounds like tree cutting sounds with axes, chainsaw tree cutting sounds, tree falling sounds, and some other sounds in forests that contribute to the false samples. The other sounds include birds sounds, animal sounds, snake sounds, wind sounds, rain sounds, water drop sounds, flowing water sounds, walking on dry leaves sounds, insect sounds, etc. We have recorded tree cutting with axes sounds by using a digital handle audio recorder in the forests nearby. Then for other sounds we have used the Internet. We have collected chainsaw tree cutting sounds, tree falling sounds, other sounds (negative samples) online from the website "freesound.org" https://freesound.org. In figure, we have a total of 862 no. of samples as a data set. All the samples are made by breaking the collected audio files into 10 s each. So, all the sample's duration will be 10 s. In this data set, we have 303 samples of chainsaw cutting sounds, 320 samples of tree cutting sounds with axes, 30 samples of tree falling sounds. Remaining 200 samples of audio clips contribute to the negative samples, i.e. forest sounds other than tree cutting sounds. Figure 1 shows the distribution of samples in the data set.

In our research, the data set classified into four categories apart from three are related to the tree cutting sounds rest of the class related to other noise in the forest. Our classes of data set are *A*, *C*, *F*, *N*. In that *A* stands for "axe tree cutting sounds", *C* stands for "chainsaw tree cutting sounds", *F* stands for "tree falling sounds", *N* stands for "negative sample sounds".

4 Proposed Model

For our research, we collected the data and followed the approach as in Fig. 2. In this approach, we first applied preprocessing techniques on audio files and split the audio clips of different lengths in to small chunks of 10 s each and then extracted features from each sample and passed to the CRNN model for training. The outputs,

Table 1 Different approaches to detect the sound events

References	Data set	Model	Feature extractions	Key technique	$F1$ score	Error rate
Adavanne et al. (2018)	TUT-SED 2017development dataset	C3RNN	Log mel-band energy + generalized cross correlation with phase transform	GRU	93.8	0.09
Dang et al. (2018)	PhysioNet	CNN	MFCC + log mel	GRU, LSTM	97.33	–
Cakır et al. (2017)	TUT SED Synthetic-2016, CHiME-Home	CRNN	MFCC	GRU	68.3	–
Adavanne et al. (2017)	TUT-SED 2016	CRNN	Log mel-band energies	LSTM	35.8	0.95
Chan (2001)	DCASE 2019	NMF-CNN	Mel-SPectrogrum	–	32.4	–
Imoto et al. (2020)	TUT sound events 2016	CRNN	Log-mel-band energy	BiGRU	49.82	0.691
Grondin et al. (1910)	Microphone-array dataset	CRNN	Log mel-band energies	GRU	87.2	0.21
Lee et al. (2017)	DCASE 2017	CNN	MFCC	–	47.62	0.71
Lin et al. (2019)	DCASE 2019-task 4 Dataset	CNN	Log mel-bank magnitudes	–	45.43	–
Cao et al. (2019)	DCASE 2019-challenge task 3	CNN	Log mel-band energies + intensity vector + GCC-PHAT	BiGRU	79.9	0.34
Lim et al. (2017)	DCASE 2017	CRNN	Log-amplitude mel-spectrogram	LSTM	96.3	0.07
Zhang et al. (2019)	DCASE 2019	CRNN	Log-amplitude mel-spectrogram + STFT	BiGRU	91.6	0.14
Xia et al. (2020)	DCASE challenge 2017	CNN	Log-mel spectrograms	LSTM	61.7	0.16
Aditya et al. (2019)	ESC-10 and ESC-50	CNN	Mel-spectrogram	–	77	–

results, and analysis are presented in the later sections of this paper. In the following sections, we will be discussing about the preprocessing and feature extractions.

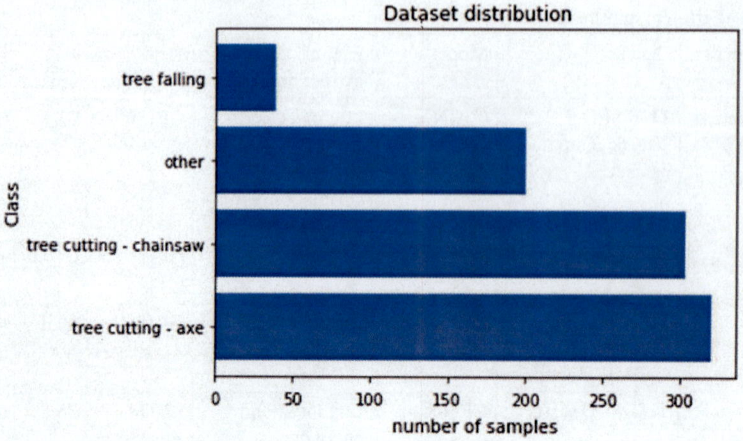

Fig. 1 Distribution of samples in the data set

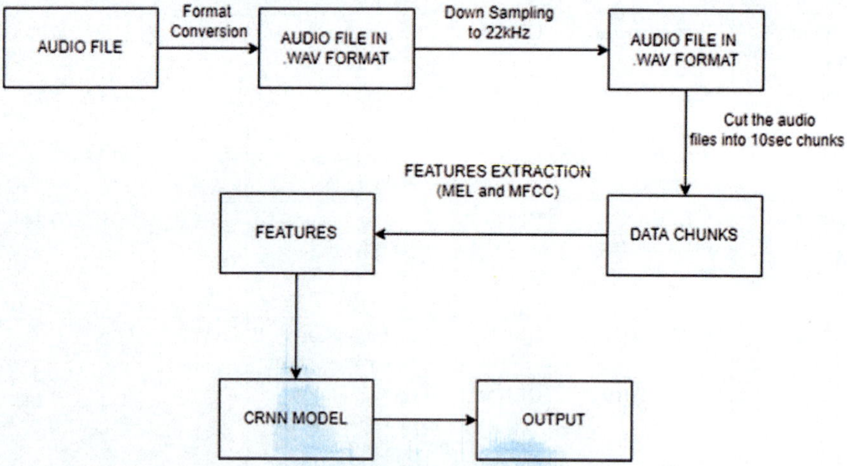

Fig. 2 Block diagram for the proposed system

4.1 Preprocessing

Data set includes audio clips of tree knocking sound and various sounds available in a forest environment. Tree knocking sounds with axes are collected with a digital handheld audio recorder at 22.05 kHz sampling frequency. We have a total of 662 audio clips of true samples with different lengths and with different file formats. As a part of preprocessing, the audio clips with different file formats are converted into a unique file format, i.e. into "wav" format. The different audio clips with various sampling frequencies are down sampled to 22.05 kHz. Samples with various lengths are broken down into small audio clips of 10 s each to keep a consistency between the features of all the samples. Thus, we preprocessed all our audio files and made a data set.

4.2 Feature Extraction

Different features are need to extract after the preprocessing on the data samples to grain the better results. In this model where we have three classes applied the two feature extractions that are MFCC and Mel Spectrogram. Mel Spectrogram is a type of spectrogram which uses a Mel scale instead of frequency on the *y*-axis. The spectrogram is a visual representation of sound intensity or energy overtime for all the frequencies present in the audio signal. It is used for analysing the different frequency components and their intensity involved in a particular audio signal. It gives information about what a human ear can perceive. MFCCs are compact representations of an audio signal. It has information about rate changes of different spectrum bands and concisely describes the shape of the spectral envelope. MFCCs are generally used in speech recognition systems and complex sound analysis. Figure 3 shows the result of chainsawing sound feature extractions and basic sound wave. Figure 4 presents the result of axe cutting sound wave and their feature extraction results. Figure 5 shows basic tree falling sound wave in the forest along with their respective feature extractions. Figure 6 conveys the result of other noise in the forest and feature extractions.

4.3 Method

We have worked on the convolutional recurrent neural network (CRNN). In this model, we used two well-known neural network approaches, i.e. recurrent neural network (RNN) and convolutional neural network (CNN). This approach sounds a bit different from the regular CRNN as we didn't stack the models one after the other; instead we built both the CNN and RNN models separately and in a later part, the output of both the models is concatenated and stacked with a few more dense layers. In this approach, image and time series-based features are extracted for an audio clip separately and passed to the model as multi-input. Most of the previous works are on either CNN or RNN, but we considered CRNN as a multi-input model to leverage the power of both CNN and RNN by taking multiple features as input in decision-making. In this approach, we have designed the model in such a way that the CNN and RNN models take a single audio clip as an input in two different kinds of representation. As discussed above, the model consists of CNN and RNN. CNN consists of convolutional layers with a ReLu activation followed by batch normalization for stabilizing the learning process of the models and a max-pooling layer for reducing the spatial size of representations and the number of parameters for computation. We have three convolutional layers with 16, 32, and 64 filters at

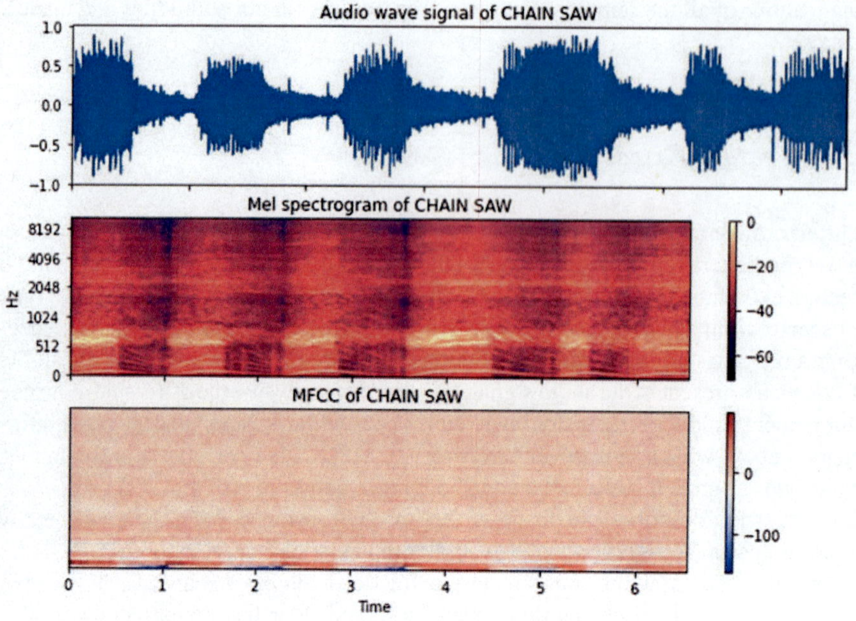

Fig. 3 Mel spectrogram and MFCC of chainsaw sound

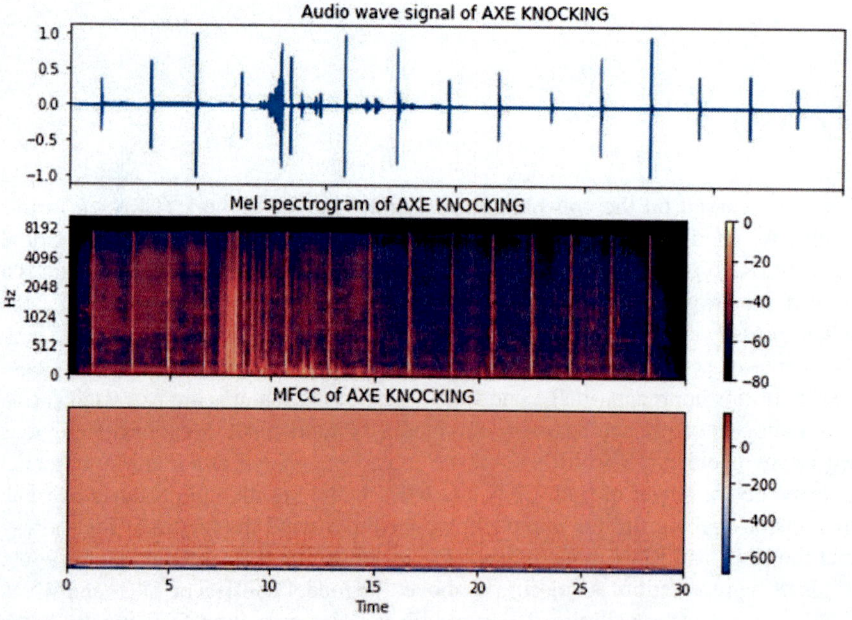

Fig. 4 Mel spectrogram and MFCC of axe sound

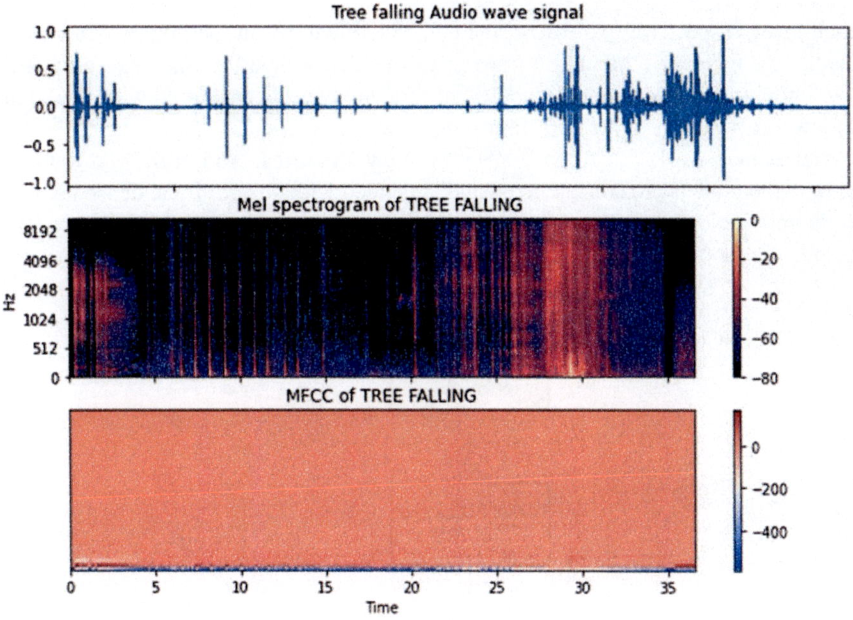

Fig. 5 Mel spectrogram and MFCC of tree falling sound

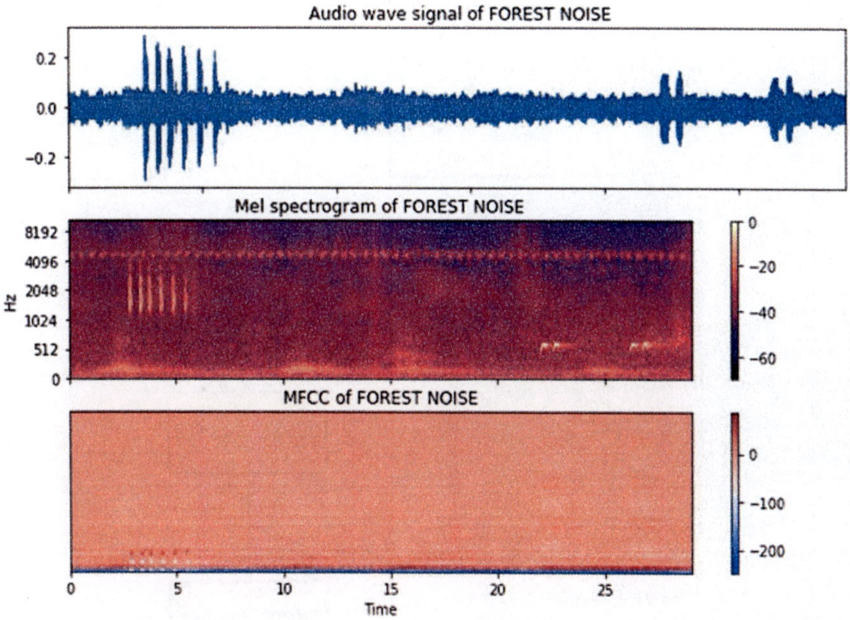

Fig. 6 Mel spectrogram and MFCC of forest noise sound

each layer, respectively, to extract the embedded features in the image. After these layers, we flatten the input and transfer to two dense layer of size 16x1 units and 4x1 units. In the RNN model, the LSTM hidden layer is considered with five units followed by a dense layer with four units. We used tanh activation function in this model as tanh worked well with RNN in many scenarios. And at the end, outputs of both the models are concatenated and appended with an additional dense layer of eight units followed by the output layer with four units and softmax activation. The graphical representation is presented in Fig. 7.

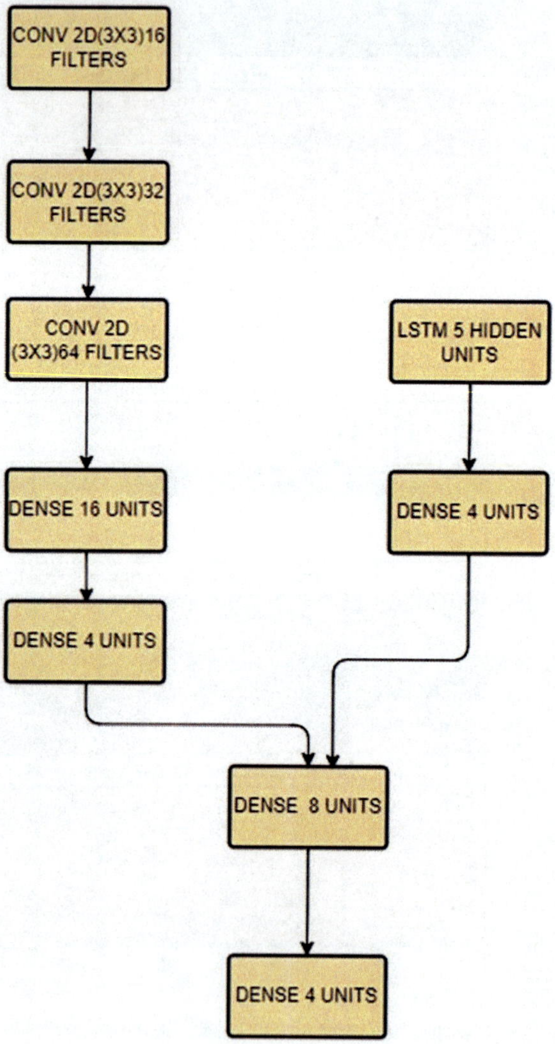

Fig. 7 Proposed model structure

5 Results and Analysis

We considered AUC score to compare the performance of models. AUC score tells the ability of the model to distinguish between the classes. The results summarized and presented with ROC curve. Higher the score better the model. In this chapter, we present the results and comparative analysis of CNN, RNN, and stacked model. If you have seen the CNN and RNN models, CNN model outperforms RNN model in most of the aspects. AUC score for RNN is 0.817553, and for CNN it is 0.975330 which proves the performance of CNN over RNN. The joint model of the CNN and RNN performed well as expected with a slight increase in the AUC score, 0.991517 when compared with CNN model. The joint model performed better than CNN on all the classes except the tree falling class. This might be because of low count of tree falling samples in the data set. Table 2 presents the metrics for individual classes with CNN model. Table 3 contains metrics for individual classes with RNN model. Table 4 presents metrics for individual classes with CRNN model.

The results are compared with different approaches such as ROC curve, confusion matrix, loss rate, and resultant wave formats individual models. According to the results of various approaches with respect to the available dataset, the axe knocking

Table 2 Metrics for individual classes with CNN model

CNN	Accuracy	Precision	Recall	$F1$ score
Negative	0.942	1.00	0.942	0.960
Tree falling	0.980	0.980	1.00	0.990
Tree cutting axe	0.914	0.902	0.969	0.934
Tree cutting chainsaw	0.953	0.981	0.946	0.963

Table 3 Metrics for individual classes with RNN model

RNN	Accuracy	Precision	Recall	$F1$ score
Negative	0.767	0.837	0.863	0.850
Tree falling	0.957	0.957	1.00	0.978
Tree cutting axe	0.903	0.930	0.913	0.922
Tree cutting chainsaw	0.712	0.804	0.737	0.769

Table 4 Metrics for individual classes with CRNN model (combined model)

Combined	Accuracy	Precision	Recall	$F1$ score
Negative	0.965	0.980	0.975	0.977
Tree falling	0.949	0.968	0.980	0.974
Tree cutting axe	0.981	0.988	0.981	0.984
Tree cutting chainsaw	0.973	0.982	0.976	0.979

and chainsawing gained the better result compared to tree falling sound. Because we have collected less amount of audio files from the sources and sound wave vary from audio file to other audio file such that all results convey the same information. The graph of individual class is represented separately. If you see the ROC curve for the RNN model, it is performing well on axe and chainsaw sounds. But its performance is low in case of falling sounds. This is because of imbalance in the data set. ROC curve for CNN presents a better performance than RNN over all the classes. We observed that when sound intensity is low, the performance of CNN and RNN models are poor. The below plot for stacked model is showing an outstanding performance for all classes. Figure 8 shows the full sound signal view of tree cutting with noise. Figure 9 presents the complete audio signal view in the forest with detection of tree cutting from the signal. Figure 10 shows the result of stacked CRNN model accuracy and loss with respect to three different dataset classes. Figure 11 extracts result of ROC curve for proposed model with respect to three data set classes. Figure 12 provides the confusion matrix with respect to the three different data set classes with other noise data which has taken as false entry for confusion matrix. Figure 13 interprets the individual results of different classes of data set in the form of confusion matrix. Figure 13a presents the chainsaw sawing true and false ratio with respect to other classes of the data. Figure 13b shows the result of axe cutting knocking sound true and false ratio. Here, the false value considered as rest value which pointed to other data classes. Figure 13c provides the result of tree falling sound, and Fig. 13d conveys the overall result true and false ratio where not wood cutting label represents the noise class.

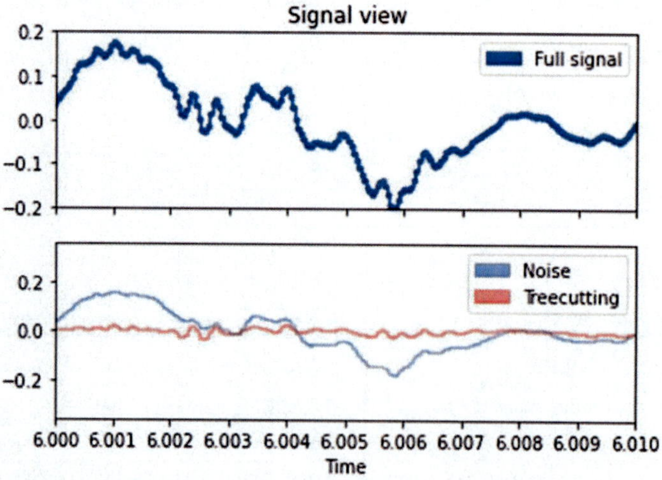

Fig. 8 Full sound signal view of tree cutting with noise

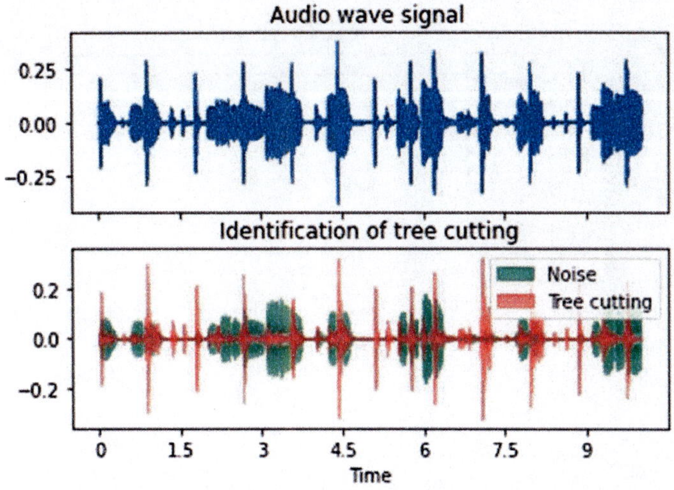

Fig. 9 Complete audio signal view and detection of tree cutting

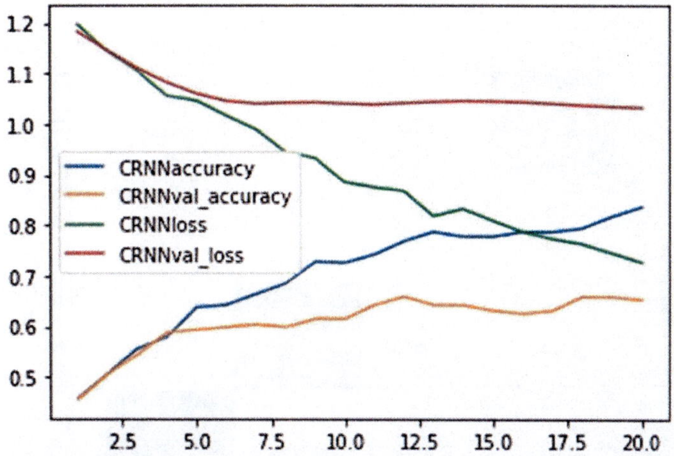

Fig. 10 Proposed CRNN model accuracy and loss

6 Conclusion

With this section we concluded that our proposed CRNN model for forest tree cutting data set performed better than CNN and RNN models on the same data set and secured the overall accuracy of 93.4% over the CNN and RNN with 87% and 66%, respectively. As a future work, we will tune the number of layers, dense layer sizes, filter sizes, and many more aspects of the network. We can also improve the data

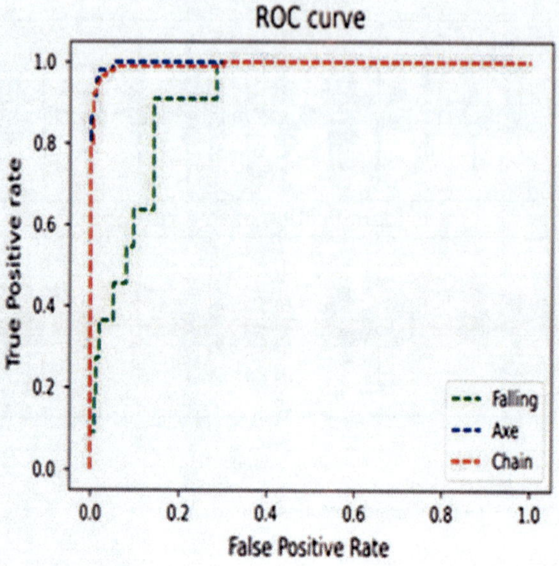

Fig. 11 ROC curve for stacked model

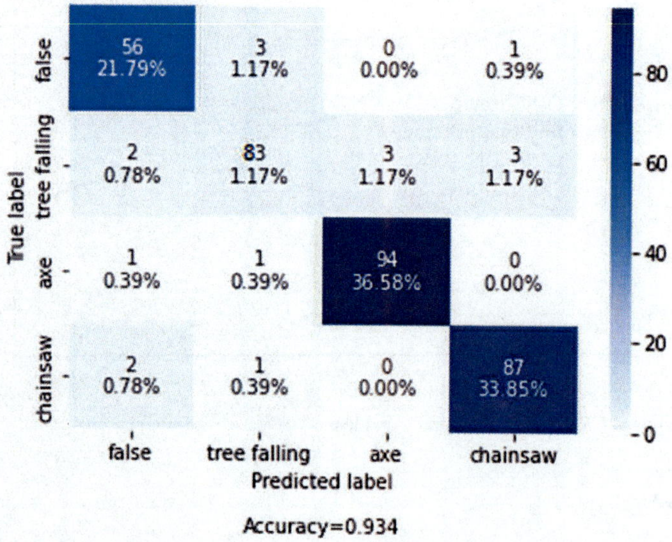

Fig. 12 Confusion matrix for all classes

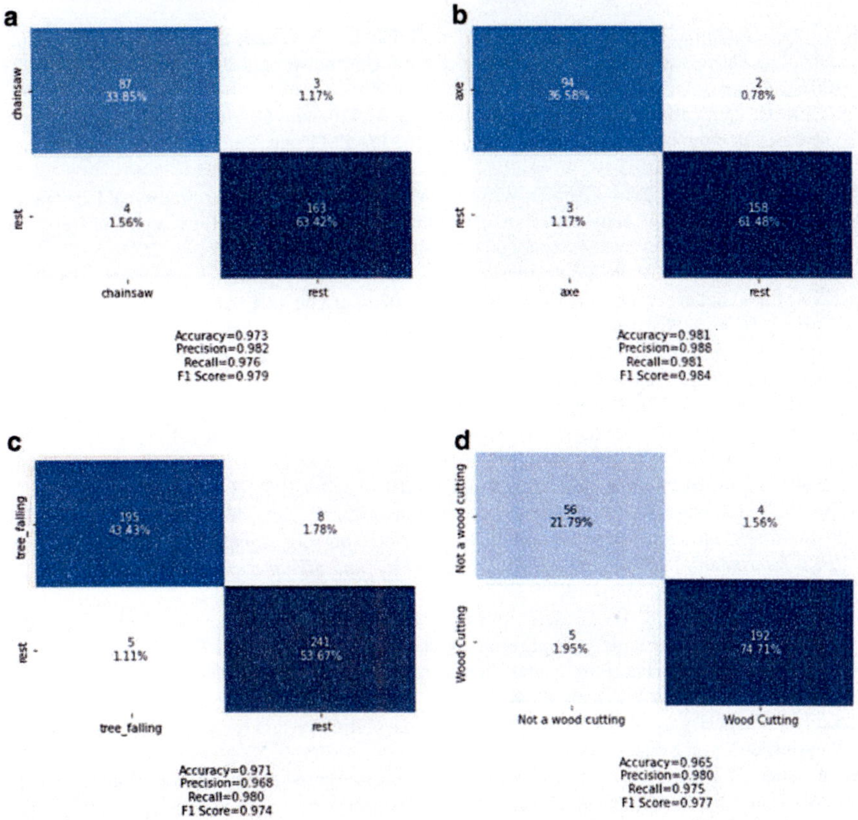

Fig. 13 Represents confusion matrix for individual classes. **a**—chainsaw, **b**—axe cutting, **c**—tree falling, and **d**—negative class

set by adding more samples specially for tree falling class for making the data set balanced and getting a better performance.

References

S. Adavanne, A. Politis, T. Virtanen, A multi-room reverberant dataset for sound event localization and detection. arXiv Preprint (2019). arXiv:1905.08546

S. Adavanne, P. Pertilä, T. Virtanen, Sound event detection using spatial features and convolutional recurrent neural network, in *2017 IEEE International Conference on Acoustics, Speech and Signal Processing (ICASSP)* (IEEE, 2017), pp. 771–775

S. Adavanne, A. Politis, T. Virtanen, Multichannel sound event detection using 3D convolutional neural networks for learning inter-channel features, in *2018 International Joint Conference on Neural Networks (IJCNN)*

K. Aditya, D. Gupta, N.G. Nguyen, A. Khanna, B. Pandey, P. Tiwari, Sound classification using convolutional neural network and tensor deep stacking network. IEEE Access **7**, 7717–7727 (2019)

E. Cakır, G. Parascandolo, T. Heittola, H. Huttunen, T. Virtanen, Convolutional recurrent neural networks for polyphonic sound event detection. IEEE/ACM Trans. Audio Speech Lang. Process. **25**(6), 1291–1303 (2017)

Y. Cao, T. Iqbal, Q. Kong, M. Galindo, W. Wang, M. Plumbley, Two-stage sound event localization and detection using intensity vector and generalized cross-correlation. Tech. report of Detection and Classification of Acoustic Scenes and Events 2019 (DCASE) Challenge (2019)

T.K. Chan, C.S. Chin, Y. Li, Non-negative matrix factorization-convolutional neural network (NMF-CNN) for sound event detection. arXiv Preprint (2020). arXiv:2001.07874

A. Dang, T.H. Vu, J.-C. Wang, Acoustic scene classification using convolutional neural networks and multi-scale multi-feature extraction, in *2018 IEEE International Conference on Consumer Electronics (ICCE)* (IEEE, 2018), pp. 1–4

M. Deng, T. Meng, J. Cao, S. Wang, J. Zhang, H. Fan, Heart sound classification based on improved MFCC features and convolutional recurrent neural networks. Neural Netw. **130**, 22–32 (2020)

F. Grondin, J. Glass, I. Sobieraj, M.D. Plumbley, Sound event localization and detection using CRNN on pairs of microphones. arXiv Preprint (2019). arXiv:1910.10049

K. Imoto, N. Tonami, Y. Koizumi, M. Yasuda, R. Yamanishi, Y. Yamashita, Sound event detection by multitask learning of sound events and scenes with soft scene labels, in *ICASSP 2020–2020 IEEE International Conference on Acoustics, Speech and Signal Processing (ICASSP)* (IEEE, 2020), pp. 621–625

D. Lee, S. Lee, Y. Han, K. Lee, Ensemble of convolutional neural networks for weakly-supervised sound event detection using multiple scale input. DCASE **1**, 14–18 (2017)

H. Lim, J.-S. Park, Y. Han, Rare sound event detection using 1D convolutional recurrent neural networks, in *DCASE* (2017), pp. 80–84

L. Lin, X. Wang, H. Liu, Y. Qian, Guided learning convolution system for dcase 2019 task 4. arXiv Preprint (2019). arXiv:1909.06178

M.-W. Mak, S.-Y. Kung, Low-power SVM classifiers for sound event classification on mobile devices, in *2012 IEEE International Conference on Acoustics, Speech and Signal Processing (ICASSP)* (IEEE, 2012), pp. 1985–1988

X. Xia, R. Togneri, F. Sohel, Y. Zhao, D.D. Huang, Sound event detection using multiple optimized kernels. IEEE/ACM Trans. Audio Speech Lang. Process. **28**, 1745–1754 (2020)

J. Zhang, W. Ding, L. He, Data augmentation and prior knowledge-based regularization for sound event localization and detection, in *DCASE 2019 Detection and Classification of Acoustic Scenes and Events 2019 Challenge* (2019)

W. Zheng, J. Yi, X. Xing, X. Liu, S. Peng, Acoustic scene classification using deep convolutional neural network and multiple spectrograms fusion, in *Detection and Classification of Acoustic Scenes and Events (DCASE)* (2017)

Vision-Based System for Detection of Petrol Pump and Charging Station

Shripad Bhatlawande, Manuja Joshi, Sakshi Nagare, Abhishek Patil, and Swati Shilaskar

Abstract This work presents a method for detection of fuel and charging stations. Automobiles may run out of fuel and leave people stranded. Detection of fuel and charging stations plays an important role in driver assistance system. This paper uses combined approach of computer vision technique and machine learning algorithms. Proposed system alerts users by sending an audio message when fuel stations are detected. Combination of Binary Robust Invariant Scalable Key points (BRISK) and Oriented FAST and Rotated BRIEF (ORB) methods is deployed for extraction of features. A blend of K-Means and principal component analysis (PCA) is used for reducing size of the feature vector. Accuracy of soft voting method with five classifiers is 77.90%. Random forest algorithm provided higher accuracy of 82.26% and 83% $F1$ score.

Keywords Driver assistance system · Detection of charging station · Detection of petrol pump · Machine learning

S. Bhatlawande (✉) · M. Joshi · S. Nagare · A. Patil · S. Shilaskar
Department of E & TC Engineering, Vishwakarma Institute of Technology, Pune 411037, India
e-mail: shripad.bhatlawande@vit.edu

M. Joshi
e-mail: manuja.joshi19@vit.edu

S. Nagare
e-mail: sakshi.nagare20@vit.edu

A. Patil
e-mail: abhishek.patil20@vit.edu

S. Shilaskar
e-mail: swati.shilaskar@vit.edu

A. B. Reddy et al. (eds.), *Proceedings of Third International Conference on Advances in Computer Engineering and Communication Systems*, Lecture Notes in Networks and Systems 612, https://doi.org/10.1007/978-981-19-9228-5_43

1 Introduction

Transportation has become an important aspect of living (Your Article Library 2022). Vehicles have changed the lives of people, by playing a significant role in transportation (CRAM 2022). People can travel as well as transfer loads easily. Majority of vehicles get stuck in the middle of the road, due to lack of fuel. Delays occur in work, and the general public is affected (Go Mechanic 2022).

Rider's unawareness about the status of fuel creates problems. Fuel stations might get skipped due to the high speed of the vehicle. Fuel station detection systems are useful for driver assistance. Drivers can easily find fuel stations if they receive alert messages.

2 Literature Survey

A good amount of work has been accomplished in the field of object detection. Roadside object detection dataset is collected by taking videos from a fixed (El Manaa et al. 2019) camera. Image is divided into several parts to extract a single object from each image. Sparse representation technique (Zhang et al. 2019) is used for building recognition. Dataset consists of samples of partial occlusion and non-uniform illumination images. Three non-parametric models (Martínez et al. 2015) are used to design a static object detection system with labeled and annotated sequences. They were used for the integral evaluation of segmentation algorithms' (LASIESTA) database. Loss of small object information is avoided using multi-attention object detection method (MA-FPN) (Ying et al. 2019). Model is designed using the dataset of object detection in aerial images (DOTA) to focus more on small objects. Vehicle-borne mobile laser scanning (MLS) based on point cloud (Fan et al. 2014) contributed to designing a manmade object detection system in urban corridors. Height of the histogram is analyzed to obtain its ground points. The color feature of images (Shinde et al. 2015) is used as a color vector for content-based classification. Machine vision-based approach (Ramík et al. 2013) is used for autonomous object detection from the surrounding environment. Microsoft Research Asia (MSRA) silent object detection model is used to check the quality of the system. Feature extraction and processing issues in computer vision (Khan and Al-Habsi 2020) have been addressed by machine learning. Canny operator is used to extract the edges of the image (Chen et al. 2020).

System provided better performance on multiple object tracking accuracy (MOTA). Block matching algorithm (Khude and Pawar 2013) is used for counting objects and vehicles in traffic, saved in static cameras. Multi-modal statistical model (Chalom and Bove 1996) of the region was developed based on user-supplied training data of fewer dataset. SIFT algorithm is parallelized on multiple accelerators and multi-core processors (Huang and Lai 2014), including FPGA and GPU. Hybrid combination of GPU provided more advantages to large images. Vision-based detection approach is used to find behavioral information of objects (Ju et al. 2014).

Full body detection (FBD) is performed using Haar-like features to scan images at different scales and locations. Variable search window technique based on feature points and color (Lim and Kang 2011) is used for object tracking. Algorithms helped in precisely detecting objects based on their color features and point features. Scale-Invariant Feature Transform (SIFT) method is used for effective image recognition systems (Zhang 2017) for mobile augmented reality games. Comparison of various features is used for visual sensor networks (VSNs). Resources are tightly controlled, as compared to SIFT and speeded-up robust features (SURF).

Deep learning-based models require a large amount of dataset to build accurate results (Bai and Qi 2010). Semi-supervised learning algorithm with deep embedded clustering (SSLDEC) approach was used earlier. Convolutional neural network (CNN) (Bai and Qi 2010; Enguehard et al. 2019) is used for building object detection systems as it provides faster and accurate results. CNN described a combination with traditional digital image processing methods (Bai and Qi 2010) to increase accuracy and speed. Combined approach of SVM method in hierarchical order helps to increase classification accuracy (Pathak et al. 2018) and to reduce power load. Gray level of color images is calculated (Demir and Ertürk 2009) including background separations as a part of an interactive segmentation method. SVM and K-nearest neighbor classifier (Pathak et al. 2018; Demir and Ertürk 2009) worked efficiently for object detection. Simple and efficient method of pixel-level classification is used for segmentation and classification.

3 Methodology

The proposed system detects petrol pumps and charging stations. It consists of a camera and a processor-based system. The camera provides inputs from the surroundings. The system interprets the detected information and translates it into an audio message. Driver is notified through this audio message. The presented model is capable to distinguish between petrol pumps and charging stations. A block diagram of the system for detection of petrol pump/charging station is shown in Fig. 1.

3.1 Dataset Description

A dataset containing 12,000 images was curated from the Internet as given in Table 1. The dataset contains a total of 6000 positive images and 6000 negative images. The positive images consist of images of petrol pump and charging station. Negative images consist of non-fuel stations (Fig. 2).

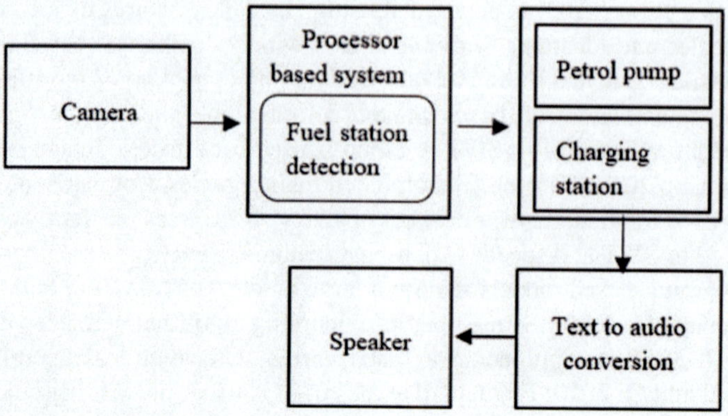

Fig. 1 Block diagram of the system for the detection of petrol pump and charging station

Table 1 Dataset description

Positive class		
S. No.	Class name	Image quantity
1	Petrol pump	3000
2	Charging station	3000
Negative class		
3	Non-fuel stations	6000
Total	Positive = 6000	Negative = 6000

Fig. 2 Sample images in the dataset

3.2 Design and Implementation of the System

All the images in the dataset are resized to 300 × 300 pixels and converted to grayscale. Prewitt operator is used for edge detection. The Prewitt better results as compared to Canny, Robert, Sobel, and Laplace. Prewitt edge detection was best in the majority cases. A fusion of BRISK and ORB is used to find the features of images. BRISK descriptor follows a predefined sampling pattern in comparison to BRIEF. Feature descriptor generated from BRISK is shown in Eq. (1).

$$a = \begin{cases} 1G\left(c_{\mathrm{y}}^{\Theta}, \sigma y\right) > G\left(c_{\mathrm{x}}^{\Theta}, \sigma x\right) \\ 0 \quad \text{otherwise} \end{cases} \tag{1}$$

where c_x^{Θ} represents the point at which cx rotates angle θ around the feature point k. $G(c_{\mathrm{x}}^{\Theta}, \sigma x)$ is gray intensity of $G(c_{\mathrm{x}}^{\Theta}, \sigma x)$ after rotating at angle θ around feature point k.

BRISK descriptor provides features of each image with dimension $Y \times 64$. Here, Y is the number of descriptors. The size of the final feature vector is 13123081 × 64. ORB is a fusion of FAST and BRIEF. Key points are found by features from accelerated segment test (FAST) method, and Harris corner detection method is used to measure topmost points. Moments of patch and moments of centroid are found using corner orientation as shown in Eqs. (2) and (3).

$$n_{\mathrm{ab}} = \sum_{x,y} x^a y^b I(x, y) \tag{2}$$

$$C = \left(\frac{n_{10}}{n_{00}}, \frac{n_{01}}{n_{00}}\right) \tag{3}$$

ORB describes features of the image in dimension $A \times 32$. Here, Y is the number of descriptors. Features of all the three classes are appended. The size of the final feature vector is 5080793 × 32. An unsupervised K-Means clustering is used to reduce large data into clusters. The value of clusters is decided by applying elbow and Silhouette method. The method provides a value of $K = 11$ for BRISK and $K = 12$ for ORB. Feature vectors are divided into K clusters depending upon Euclidian distance from centroid. The combined feature vector of ORB and BRISK has a dimension of 12,015 × 25 after applying K-Means clustering. The process of dimension reduction is presented in Algorithm 1.

Algorithm.1. Dimension Reduction

Input: Image dataset (12000 Images)
Output: Optimized feature vector (12015 x 21)

1. Data frame (data) = []
2. **For** image in image dataset (12000)
3. **Do** extract features of image -BRISK and ORB
4. DATA = append features |
   // final data frame (13123081, 64 for BRISK and 5080793, 32 for ORB)
5. **End for**
6. Applying elbow method to find value of K
7. Predict clusters – pretrained K-means [K = 11, K=12]
8. Train K-means algorithm
9. Histogram = predicted clusters
10. Combine pretrained ORB and BRISK
11. HP = normalize (histogram)
12. Scaling normalized feature vector
13. Applying PCA on scaled Feature vector
14. Selection of principal components (n=20)
15. **Return** final dimension

The number of columns in the final feature vector is 23. Principal component analysis (PCA) is performed on this feature vector to further decrease the computational complexity. PCA transformation is followed by calculation of sum of explained variance ratio. The results indicated that 96% of the variance is explained by the first 20 components of the data. A final feature vector of size 12,015 × 20 is formed.

Five classifiers, namely (i) decision tree (DT), (ii) random forest (RF), (iii) KNN, (iv) GaussianNB, and (v) support vector machine (SVM), are used to check the performance of the system. DT collects the observations from training dataset and builds a tree structure. Random forest classifier creates judgment trees from randomly selected subsets of the training set. Feature importance values from each tree are then summed and normalized as shown in Eq. (4).

$$\frac{\sum_j \text{norm} fi}{\sum_{j \in \text{all feature},\, k \in \text{all trees}} \text{norm} fijk} \tag{4}$$

This algorithm works fast and provides maximum accuracy.

Trained data are provided to seven classifiers for prediction. Final result of the classifier is in the binary form. Process of classification is shown in Fig. 3.

Soft voting classifiers work as global classifiers and improve classification accuracy. The input data are classified according to the probabilities of all the predictions made by the different classification models.

The process of voting classifier is presented in Algorithm 2.

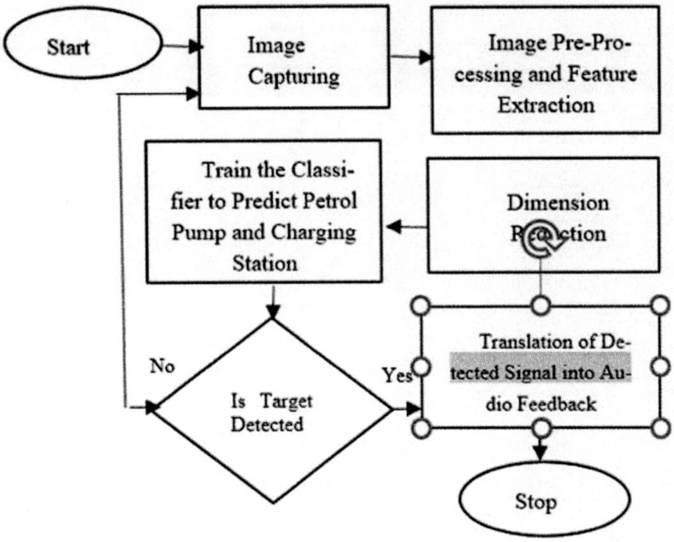

Fig. 3 Overall process of classification and detection for petrol pump and charging station detection

Algorithm 2 Voting classification.

Input: Image
Output: True class of image with highest accuracy
1. voting = Voting Classifier (Soft voting) # apply
2. soft voting on five classifiers
3. fit the model # train and fit the model
4. Call voting function to find class of given input
5. Calculate confusion matrix, precision, recall.
6. Check ROC value of classifier
7. Feed image to class which having highest probability
8. Return name of class

4 Results and Discussion

The voting classifier has an accuracy of 77.90%. The RF algorithm has maximum accuracy among seven used classifiers. The accuracy provided by RF algorithm is 86.22%. Performance evaluation of the system is accomplished based on performance parameters such as specificity, sensitivity, recall, and $F1$ score. Threshold limit is an important factor in classification, to divide the data in classes. Accuracy may get affected, because of the wrong threshold value. Receiver operating characteristic (ROC) curve is used for deciding threshold values for each classifier. x-axis shows false-positive rates, while y-axis shows true-positive rates. The ROC curve and area

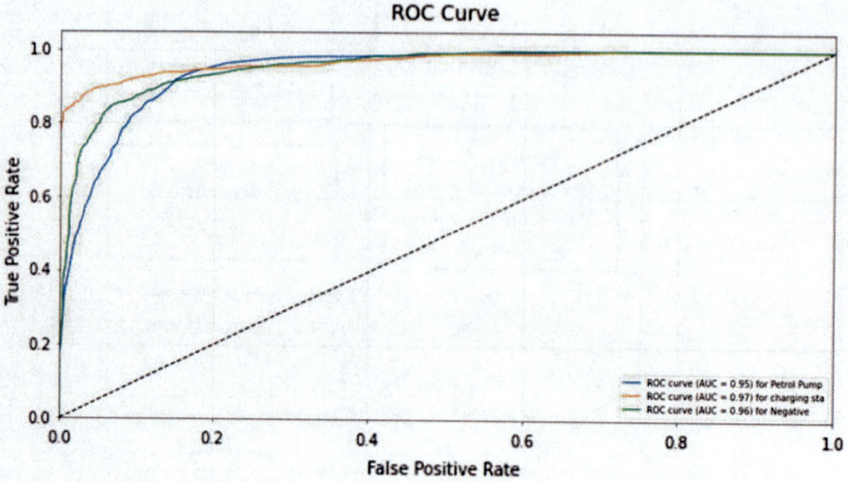

Fig. 6 Receiver operating characteristic curve of random forest

Table 2 Performance evaluation

Classifier	Training accuracy (%)	Testing accuracy (%)	Precision (%)	Recall (%)	*F*1 score (%)
Decision tree	84	70	71	70	70
Random forest	97	86	86	86	86
KNN	82	73	76	73	74
Gaussian Naive Bayes	50	49	63	49	50
SVM	57	56	56	56	56

under the ROC curve (AUC) for RF classifier are shown in Fig. 6. AUC for petrol pump and charging station class is 95% and 97%. AUC of negative class is 96%.

The accuracy, precision, recall, and *F*1 scores for five classifiers are shown in Table 2.

Metrics evaluation parameters precision, recall, and *F*1-score can be further improved in order to make better predictions in the real world. This will help drivers by notifying about fuel stations while driving.

5 Conclusion

This paper presented a solution for detection of petrol pump and charging station. Five classifiers with voting method are used for classification and detection. RF provided highest accuracy of 86.22% among all five classifiers. The accuracy of the voting-based model on test dataset is 77.90%. The RF algorithm provided 86% recall score, making it suitable for the deployment in the real world environment. This system is deployed on Jetson Nano board. The system lacks in detecting petrol pump and charging station in low light environment. In future, the system will be tested on large and improved dataset for accurate detection and classification of petrol pump and charging station.

References

Y. Bai, H. Qi, Feature-based image comparison for semantic neighbor selection in resource-constrained visual sensor networks. EURASIP J. Image and Video Process. 1–11 (2010)

Y. Bai, H. Qi, Feature-based image comparison for semantic neighbor selection in resource-constrained visual sensor networks. EURASIP J. Image and Video Process. 1–11 (2010)

Chalom, V. M. Bove, Segmentation of an image sequence using multi-dimensional image attributes, in *Proceedings of 3rd IEEE International Conference on Image Processing*, vol. 2 (1996), pp. 525–528

J. Chen, Z. Xi, C. Wei, J. Lu, Y. Niu, Z. Li, Multiple object tracking using edge multi-channel gradient model with ORB feature. IEEE Access **9**, 2294–2309 (2020)

CRAM, The importance of transportation vehicles. https://www.cram.com/essay/The-Importance-Of-Transportation-Vehicles/F3JUJ835C. Accessed 20 May 2022

B. Demir, S. Ertürk, Improving SVM classification accuracy using a hierarchical approach for hyperspectral images, in *2009 16th IEEE International Conference on Image Processing*, Nov 2009, pp. 2849–2852

El Manaa, M.A. Sabri, A. Aarab, Classification and segmentation approach for detecting moving object in road applications, in *International Conference on Intelligent Systems and Advanced Computing Sciences (ISACS)*, Dec 2019, pp. 1–5

J. Enguehard, P. O'Halloran, A. Gholipour, Semi-supervised learning with deep embedded clustering for image classification and segmentation. IEEE Access **7**, 11093–11104 (2019)

H. Fan, W. Yao, L. Tang, Identifying man-made objects along urban road corridors from mobile LiDAR data. IEEE Geosci. Remote Sens. Lett. **11**(5), 950–954 (2014)

Go Mechanic, 5 signs of a failing fuel pump that needs your attention. https://gomechanic.in/blog/signs-of-a-failing-fuel-pump/. Accessed 22 May 22

M. Huang and C. Lai, Parallelizing computer vision algorithms on acceleration technologies: a SIFT case study, in *2014 IEEE China Summit & International Conference on Signal and Information Processing*, July 2014, pp. 325–329

F. Ju, W.-M. Lu, K.-H. Chen, J.-I. Guo, Vision-based moving objects detection for intelligent automobiles and a robustness enhancing method, in *2014 IEEE International Conference on Consumer Electronics—Taiwan*, May 2014, pp. 75–76

A.I. Khana, S. Al-Habsi, Machine learning in computer vision. Procedia Comput. Sci. **16** (2020)

P.S. Khude, S. Pawar, Object detection, tracking and counting using enhanced BMA on static background videos. IEEE (2013)

H.-Y. Lim, D.-S. Kang, Object tracking system using a VSW algorithm based on color and point features. EURASIP J. Adv. Signal Process. **2** (2011). Springer

R. Martínez, C. Cuevas, D. Berjón, N. García, Detection of static moving objects using multiple nonparametric background models, in *2015 International Symposium on Consumer Electronics*, June 2015, pp. 1–2

A.R. Pathak, M. Pandey, S. Rautaray, Application of deep learning for object detection. Procedia Comput. Sci. (2018)

D.M. Ramík, C. Sabourin, R. Moreno, K. Madani, A machine learning based intelligent vision system for autonomous object detection and recognition. Appl. Intell. (2013)

S.R. Shinde, S. Sable, S. Kulkarni, D. Bhatia, Experiments on content-based image classification using color feature extraction, in *2015 International Conference on Communication, Information & Computing Technology*, Jan 2015, pp. 1–6

X. Ying et al., Multi-attention object detection model in remote sensing images based on multi-scale. IEEE Access **7**, 94508–94519 (2019)

Your Article Library, Transportations system in India: forms, significance and recent developments. https://www.yourarticlelibrary.com/transport/transportations-system-in-india-forms-significance-and-recent-developments/7552. Accessed 20 May 20

B. Zhang, Design of mobile augmented reality game based on image recognition. EURASIP J. Image Video Process. (2017)

Y. Zhang, F. Sun, B. Li, Building recognition based on sparse representation of spatial texture and color features. IEEE Access **7**, 37220–37227 (2019)

Emotion Analysis Using Convolutional Neural Network

K. Reddy Madhavi, Sandya Vooradi, P. Mounika, Satyam Yedlla, and Naresh Tangudu

Abstract In contrast to spoken words, facial expressions communicate a lot of information visually. Since there are so many different ways to portray emotions, it has become difficult to recognize them using computer vision for human–machine interaction. Nowadays, deep learning techniques own a big success in various fields including computer vision. From the background analysis, we found that the recognition of emotions is still difficult and it relies on some advancements in image preprocessing and computer vision techniques. In the proposed system, some image enhancement techniques are being used along with CNN. Our primary goal is to find an emotion that has been exhibited and identify it based on its geometry and physical characteristics. The proposed system uses convolutional neural network (CNN) variants and OpenCV to recognize seven primary human emotions—"anger, disgust, fear, happiness, sadness, surprise, and neutrality". The suggested system can be applied to programs that analyze human behavior.

Keywords Convolutional neural network (CNN) · Facial expression · Emotion analysis

1 Introduction

From the past few years, this has been an active research area in computer vision and extensively used in human–machine interaction. Facial expressions plays an important role by allowing people to communicate beyond the verbal domain. Facial expression recognition system classifies a facial expression into seven key human

K. R. Madhavi (✉)
CSE, Sree Vidyanikethan Engineering College, Tirupati, AP, India
e-mail: kreddymadhavi@gmail.com

S. Vooradi · P. Mounika · S. Yedlla
CMR Technical Campus, Hyderabad, India

N. Tangudu
IT Department, Aditya Institute of Technology and Management, Tekkali, Srikakulam, AP, India

A. B. Reddy et al. (eds.), *Proceedings of Third International Conference on Advances in Computer Engineering and Communication Systems*, Lecture Notes in Networks and Systems 612, https://doi.org/10.1007/978-981-19-9228-5_44

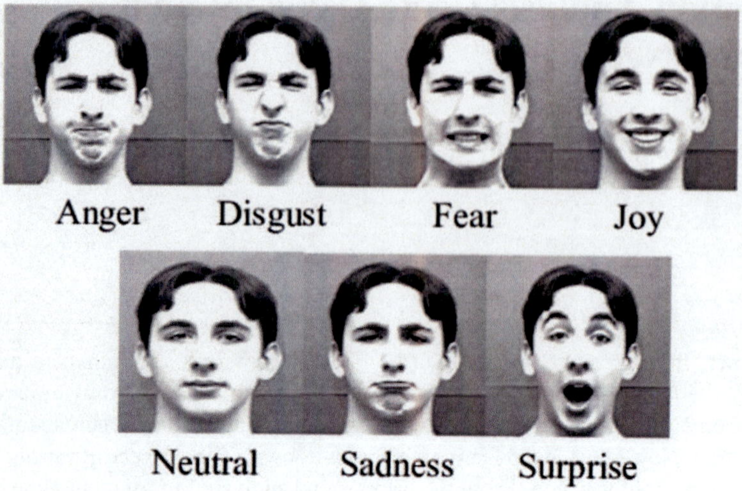

Fig. 1 Seven key human emotions

emotions such as "angry, disgust, fear, happy, neutral, sad, and surprise" which are shown in Fig. 1.

The proposed system aims to design a model that classifies the facial expression into seven human emotions using some image enhancement techniques such as histogram equalization, a preprocessing technique, which can be used to overcome the issues in previous models and to increase the recognition rate of facial images with light variations by adjusting the light contrast. The main objectives are as follows: to preprocess the images using histogram equalization, to build a CNN model that predicts human emotions, and to improve precision and accuracy. This system helps the user to detect emotions and can be effectively used in areas where human behavior plays a prominent role. It can also be used in many fields such as medical field, robotics, driver monitoring.

2 Relevant Study

When just six expressions were taken into account, the accuracy of the innovative architecture in Kabakus (2020), which used CNN, was 96% for the CK + dataset. The proposed system provided fast analysis of expressions, but the expression "happy" was misdetected as "fear" for 16% of images. The author used the deep learning techniques to facilitate the facial expression recognition which might be difficult in the context of machine learning techniques in some of the cases. The proposed system used CNN architecture with fine tuning (Fathallah et al. 2017) by Visual Geometry Group (VGG). The rate of recognition shows that this method outperformed the state-of-the-art methods. The limitation of the proposed system was that it was trained with images with faces in one position and it cannot apply on images with different face positions. Scale normalization (Zhang et al. 2019) and gray-level equalization were performed on each image before giving input into the network. For each image, the edge of each layer was extracted and was preserved by superimposing on each feature map. Softmax classifier was used to classify and recognize the expressions. The proposed model extracted the facial features and learned the pattern features effectively. But it failed for reality images with noises like blurring, occlusion, and face position. The model was built by merging deep residual depth-wise separable convolutional layers (Zhou et al. 2020).

In contrast to other methods, the one presented in (Hua et al. 2019) was straightforward and contained fewer layers. An accuracy of the system was 97.1%, CK + , 91.97%, and 92.88% for oculu-casia (Yang et al. 2017). A CNN was trained for each active region, and categorization was carried out using the frequency of each of the (Sun et al. 2018) regions. Without utilizing any preprocessing techniques, the method in (Singh and Nasoz 2020) utilized a CNN architecture. The system's preprocessing methods included grayscale picture conversion, face detection (Khemakhem and Ltifi 2019), and applying an emboss effect to photos to make them appear three dimensional. The accuracy was increased to 79.59% than the traditional model with accuracy of 71.34%. The issue with this model was that it performs more mistakes for the classes like anger in the JAFFE dataset. CK + dataset was used in the system which consists of total 3200 face images of seven expressions. The system achieved good results (Zou et al. 2019), but it failed to detect real-time expressions because of dataset size. Facial expression recognition has been performed using different methods which includes pre-trained models, different architectures of convolutional neural network, and ensemble models by using more than one model. As mentioned in above section, there are some drawbacks in the existing systems. The system was implemented on CK + dataset by considering six expressions and omitted the "contempt" expression. The existing system provided fast analysis of the facial expressions. In the existing system (Zhou et al. 2020), multi-cascaded convolutional neural networks (MTCNNs) were used for face detection. The model used global average pooling to reduce the number of parameters. The system achieved good results, but it was inaccurate for images with light variations. The existing system was developed by using customized CNN architecture with different number of layers on CK + dataset. The system was failed to detect real-time expressions because of dataset size. In the existing systems, there were problems with images having noises like light variations, blurring, occlusion; and, due to the small datasets, some of the existing systems were failed to detect real time. In some systems, "neutral" expression was excluded which resembles almost all expressions.

3 Proposed Method

Convolutional neural network model with customized architecture was developed in the proposed system to predict human emotions. In the proposed system, some techniques are used to overcome the issues of existing systems such as "neutral" expression which is also included in the proposed system. Histogram equalization, a preprocessing technique, can be used to overcome the issue with images having light variations. The system uses FER-2013 dataset with 35,887 images to overcome the issue with dataset size.

In the proposed system, histogram equalization is used in preprocessing stage to enhance the images by adjusting the light contrast. OpenCV pre-trained Haar Cascade Classifier called as Viola–Jones face detector is used to perform face detection for the facial images. CNN is applied for extracting the best of all features like eyes, nose, lips from the preprocessed images to train the model. And finally, Softmax classifier is used to classify the emotions by mapping the extracted features. Even though machine learning models obtained good accuracy, they failed to detect for images outside training. Deep learning networks extract features effectively when compared to others. So, in the proposed system, deep learning model is developed. For this problem, histogram equalization is used. It is an image enhancement technique which adjusts the light contrast in image as shown in Fig. 2. Its main idea is to transform original image histogram graph to uniform distributed histogram.

Cascading classifier: It is a "multi-stage classifier" with a classifier made out of some of the best features in each stage. If the classifier's output is positive, the input is indeed transmitted to the subsequent stage; if it is negative, it is simply rejected.

This process shortens the time needed to calculate the characteristics. As shown in Fig. 3, the cascading classifier only rejects input at the beginning if it is not a face, which helps to cut down on processing time.

Convolutional neural network (CNN). CNN architecture is built by using three layers. The general CNN architecture is represented in Fig. 4.

The proposed CNN architecture is shown in Fig. 5. It contains six convolutional layers, three max pooling layers, and two fully connected layers.

Fig. 2 Before and after histogram equalization

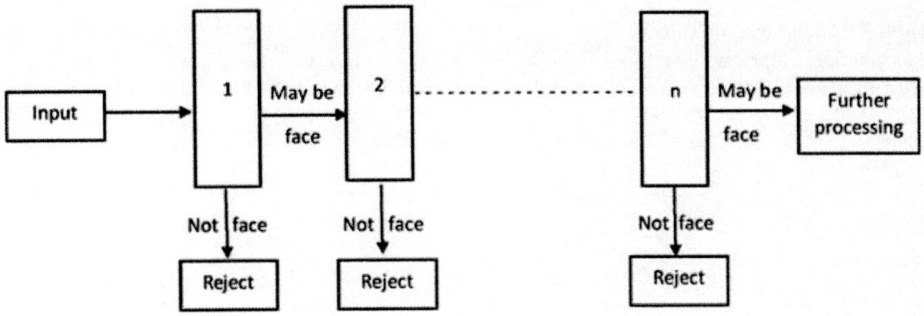

Fig. 3 Cascading classifier

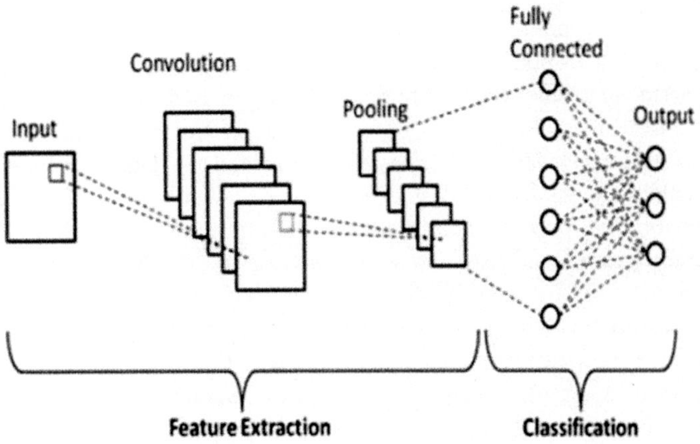

Fig. 4 General CNN architecture

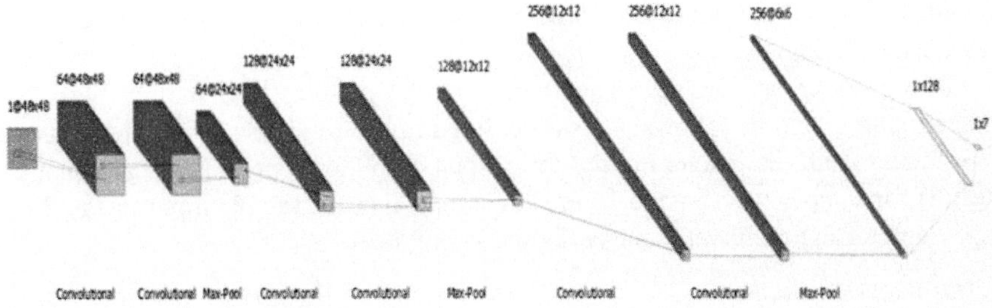

Fig. 5 Proposed CNN architecture

4 Experimental Results

Dataset description. The dataset used in the proposed system is FER2013 an open-source dataset publicly shared in Kaggle. It consists of 35,887 images, out of which

Table 1 Training and testing dataset images and emotions

Training dataset		Testing dataset	
Emotion	No. of images	Emotion	No. of images
Angry	3995	Angry	958
Disgust	436	Disgust	111
Fear	4097	Fear	1024
Happy	7215	Happy	1774
Neutral	4965	Neutral	1233
Sad	4380	Sad	1247
Surprise	3171	Surprise	831

28,907 are used for training and remaining for testing. The images of both training and testing are 48 * 48-pixel grayscale face images with seven key human emotions. Table 1 describes the emotions and number of images used for training and testing.

Preprocess the images using histogram equalization. Detection is performed using OpenCV pre-trained Haar Cascade Classifier. Train and test data are created by some data augmentation using ImageDataGenerator. A model is created using CNN layers such as convolution, pooling, and dense. Model fitting is performed by calling fit () with parameters train dataset and setting epochs to 30. This model is validated on test dataset during training. After the model is trained, evaluation of model on test data is performed. Trained model predicts the classes for the test data. The test data are also preprocessed simultaneously along with the training data. The model is tested by giving the test data as input, the model will predict for the test data using predict function, and it will return the output, which is an array of probabilities of expressions in below format and they are converted into CSV format.

["angry", "disgust", "fear", "happy", "neutral", "sad", "surprise"]

Argmax () function in numpy extracts the class with highest probability from the output array, and these values are stored in y_pred which denotes the emotion classes of respective images.

Testing on private images: The trained model predicts the emotion class of single input images as follows. Convert colored image to grayscale and detect face coordinates from the image. Predict the output from the cropped grayscale image. Results for sample input image 1 are shown in Fig. 6, for sample input image 2 in Fig. 7 and testing on private images shown in Fig. 8.

Performance Evaluation

Performance is evaluated based on the metric called accuracy shown in Fig. 9. Confusion matrix for different emotions in Table 2, classification report in Table 3 and finally comparison of results of existing and proposed method are shown in Table 4.

$$(TP + TN)/(TP + TN + FP + FN)$$

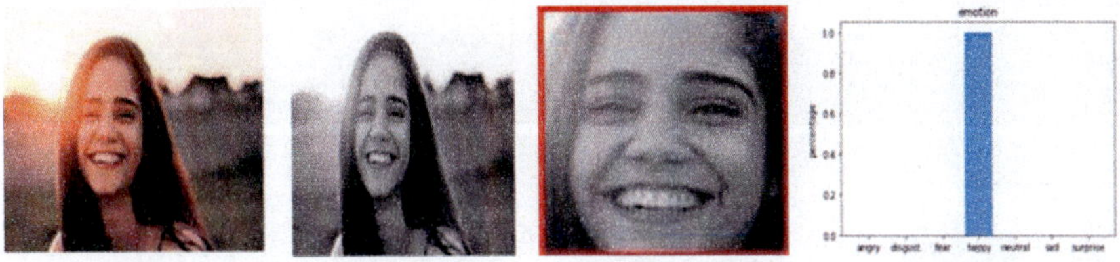

a) Input Image 1　b) Grayscale Image 1　c) Cropped Image 1　d) Classification graph 1

Fig. 6 Output for sample input image 1

Output: [2.2222705e-06, 8.1110246e-10, 5.5186774e-06, 0.99991417, 5.8390422e-05, 6.8631166e-06, 1.2846713e-05]

Classified emotion: "happy"

a) Input Image 2　b) Grayscale Image 2　c) Cropped Image 2　d) Classification graph 2

Fig. 7 Output for sample input image 2

Output: [3.1107076e-05, 1.39673e-07, 0.0026901225, 0.13357571, 5.4498538e-05, 1.0967274e-05, 0.86363745]

Classified emotion: "surprise"

Fig. 8 Testing on private images

TP—"true positives", FP—"false positives", TN—"true negatives", and FN—"false negatives". Precision = TP/(TP + FP), recall = TP/(TP + FN), and $f1$ score = (*precision * recall)/(precision + recall).

5 Conclusion and Future Work

The proposed system has been implemented to predict the emotions from the human facial images. The proposed model performed well in recognizing the seven key emotions from facial images. For the purpose of face detection, OpenCV pre-trained Haar Cascade classifier is used. An accuracy of 66.8% is obtained by using image

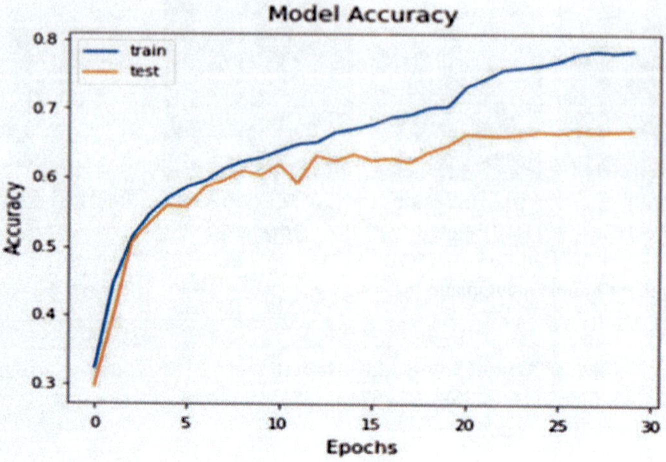

Fig. 9 Graph of train and test accuracy

Table 2 Confusion matrix

Confusion matrix							
	Angry	Disgust	Fear	Happy	Neutral	Sad	Surprise
Angry	605	10	90	34	95	109	15
Disgust	25	70	5	1	4	6	0
Fear	121	7	495	30	108	190	73
Happy	37	2	30	1537	94	29	45
Neutral	74	0	65	96	801	174	23
Sad	143	3	138	60	260	623	20
Surprise	17	1	78	35	22	12	666

Table 3 Classification report

Classification report				
	Precision	Recall	$f1$-score	Support
Angry	0.59	0.63	0.61	958
Disgust	0.75	0.63	0.69	111
Fear	0.55	0.48	0.51	1024
Happy	0.86	0.87	0.86	1774
Neutral	0.58	0.65	0.61	1233
Sad	0.55	0.50	0.52	1247
Surprise	0.79	0.80	0.80	831
Accuracy			0.67	7178
Macro avg.	0.67	0.65	0.66	7178
Weighted avg.	0.67	0.67	0.67	7178

Table 4 Comparison of existing and proposed systems

Model	Accuracy (%)	Parameters
	FER-2013	
CNN (Kabakus 2020)	58	3,934,538
CNN [20]	62	138,816
Fine-tuned xception	63	21,914,159
Proposed system	66.8	2,787,015

enhancement techniques, CNN model and a Softmax classifier is used to classify the emotions. The proposed system outperforms some of the existing systems in recognizing the facial expressions. We like to extend our model for images with noises like different face positions, blurring, and occlusion, which makes emotion recognition difficult.

References

A. Fathallah, L. Abdi, A. Douik, Face expression recognition via deep learning, in *IEEE Conference on Computer System and Applications*, Nov 2017

W. Hua, D. Fei, L. Huang, J. Xiong, G. Gui, HERO: human emotions recognition for realizing intelligent internet of things. IEEE Access **7**, 24321–24332 (2019)

A.T. Kabakus, PyFER: a facial expression recognizer based on convolutional neural networks. IEEE Access **8**, 142243–142249 (2020)

F. Khemakhem, H. Ltifi, Facial expression recognition using convolutional neural network enhancing with pre-processing stages, in *IEEE Conference on computer systems and applications*, 3–7 November, 2019

S. Singh, F. Nasoz, Facial expression recognition with convolutional neural network, in *IEEE Conference on Computing and Communication*, 6–8 Jan 2020

A. Sun, Y. Li, Y.-M. Huang, Q. Li, G. Lu, Facial expression recognition using optimized active regions. Hum. Centric Comput. Inf. Sci. (2018)

B. Yang, J. Cao, R. Ni, Y. Zhang, Facial expression recognition using weighted mixture deep neural network based on double-channel facial images. IEEE Access **6**, 4630–4640 (2017)

H. Zhang, A. Jolfaei, M. Alazab, A face emotion recognition method using convolutional neural network and image edge computing. IEEE Access **7**, 159081–159089 (2019)

N. Zhou, R. Liang, W. Shi, A lightweight convolutional neural network for real-time facial expression detection. IEEE Access **9**, 5573–5584 (2020)

J. Zou, X. Cao, S. Zhang, B. Ge, A facial expression recognition based on improved convolutional network, in *IEEE Conference on Intelligent Applied Systems of Engineering*, 26–29 April, 2019

KNOw PCOS

Sagar Yeruva, Indu Gurrala, Ramya Sri Myakala, Nimmi Agarwal, Shriya Rapolu, and Junhua Ding

Abstract Polycystic ovarian disease (PCOD), also known as polycystic ovary syndrome (PCOS), is a very common condition affecting 5–10% of women in the age group 12–45 years. Due to busy schedules, proper diet and exercise are the least things that are taken care of. Besides bad lifestyle, poor quality of food is also one of the main reasons for poor health quality. The principal features of PCOS include no ovulation, irregular periods, acne, and hirsutism. If not treated, it can cause insulin-resistant diabetes, obesity, and high cholesterol leading to heart disease. PCOS is also a common and treatable cause of infertility. Many diseases can be cured by taking a course of medication within a stipulated time, but the treatment of PCOS deals with a continuous process where diet and physical activities should be monitored. This work aims to develop a Flutter-based mobile application that predicts the presence of PCOS in women using machine learning by considering the clinical parameters or ultrasound scans. The application also gives suggestions regarding diet and exercise and also tracks the PCOS level. A user dashboard is provided where each user can sign up, create a profile, and track the progress and improvement of health status. The mobile application ensures the data integrity and confidentiality of the user's data.

Keywords PCOS · PCOD · Symptoms · Random forest algorithm · CNN · Machine learning · Suggestions

---

S. Yeruva (✉) · I. Gurrala · R. S. Myakala · N. Agarwal · S. Rapolu
Department of Computer Science and Engineering, VNR Vignana Jyothi Institute of Engineering and Technology, Pragathi Nagar, Hyderabad, Telangana 500090, India
e-mail: sagar_y@vnrvjiet.in

J. Ding
Department of Information Science, University of North Texas, Denton, TX, USA
e-mail: Junhua.Ding@unt.edu

A. B. Reddy et al. (eds.), *Proceedings of Third International Conference on Advances in Computer Engineering and Communication Systems*, Lecture Notes in Networks and Systems 612, https://doi.org/10.1007/978-981-19-9228-5_45

1 Introduction

Polycystic ovarian syndrome (PCOS) is a complicated illness marked by high testosterone levels, irregular menstruation cycles, and/or tiny cysts on one or both ovaries (https://www.ncbi.nlm.nih.gov/pmc/articles/PMC3737989/). It is the most common endocrine condition among women of reproductive age, affecting 4–6% of adolescents and 4–12% of adult women. PCOS can manifest early in life, with symptoms such as premature pubescence in children, early indicators of hyperandrogenism (acne, hirsutism), and long-term menstrual disruption in teens. Because ovarian follicles mature but do not release an egg, cysts form on and around the ovaries, which can lead to infertility and amenorrhea (Fig. 1).

Ovarian dysfunction, elevated androgen production, and abnormal gonadotropin secretion are increasingly recognized as symptoms of PCOS. Anovulatory infertility, menstrual problems, hirsutism, and acne can all be caused by these disorders.

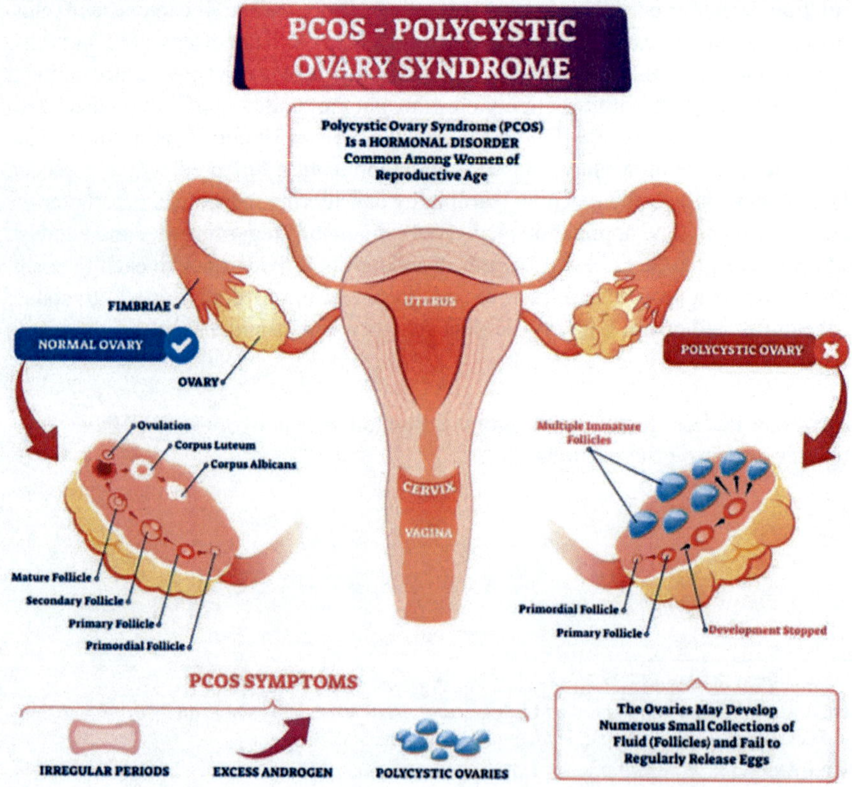

Fig. 1 Polycystic ovaries (*Source* taken from https://www.lifeextension.com/protocols/female-reproductive/polycystic-ovary-syndrome-pcos)

PCOS is also linked to metabolic problems such as insulin resistance and hyperinsulinemia. PCOS increases the risk of type 2 diabetes, dyslipidemia, and cardiovascular disease in women (https://www.sciencedirect.com/topics/medicine-and-dentistry/polycystic-ovary-syndrome).

1.1 Symptoms and Signs

PCOS clinical signs appear around the period of the first menstruation cycle in early adolescence. PCOS can occur later in life, for example, as a result of intensive weight gain (https://www.mayoclinic.org/diseases-conditions/pcos/symptoms-causes/syc-20353439). Ovulation irregularities, elevated androgen levels, and cystic ovaries are the three most prevalent variables linked with PCOS, albeit the signs and symptoms differ (Ndefo et al. 2013). As a result of PCOS, many tiny, fluid-filled sacs were formed inside the ovaries. The term "polycystic" refers to a person who has a lot of cysts. Each of these sacs is a follicle in nature, which contains an immature egg properties. Ovulation is never triggered because the eggs are never developed enough. Ovulation deficiency affects estrogen, progesterone, FSH, and LH levels. The levels of progesterone are lower than typical, while the levels of androgen are higher than usual. Extra male hormones cause the menstrual cycle to be disrupted, resulting in fewer cycles for women with PCOS (https://www.healthline.com/health/polycystic-ovary-disease#health-effects).

Irregular menstrual cycles as a result of ovulatory dysfunction are a key symptom of PCOS according to the Rotterdam criteria (Louwers and Laven 2020). Your reproductive hormones are out of balance if you have PCOS. This can cause issues with your ovaries, such as not having or not receiving your period on time (https://www.webmd.com/women/what-is-pcos). This might cause infertility issues in women. Hair growth on the face and body, as well as baldness, are also some of the symptoms of PCOS (https://www.healthline.com/health/polycystic-ovary-disease). Obese women are more likely to suffer from sleep apnea, especially, if they also have PCOS. Women with both obesity and PCOS have a 5–10 times higher risk of sleep apnea than women without PCOS (https://www.healthline.com/health/polycystic-ovary-disease#health-effects). PCOS can also affect women's mental well-being, causing low mood, stress, nervousness, low levels of self, and self-esteem issues, all of which hurt the quality of life (https://www.jeanhailes.org.au/health-a-z/pcos/symptomscauses?gclid=Cj0KCQiAkZKNBhDiARIsAPsk0WiVy7K5YgrhqBAFYmGpNRjwzuYun1vcUd9VMfToPNKhOfEc6C35gj8aAvB1EALw_wcB).

1.2 PCOS Complications

PCOS symptoms are not limited to the gynecological realm; women with PCOS characteristics have a greater prevalence of obesity, dyslipidemia, hypertension, metabolic syndrome (MS), and type 2 diabetes mellitus (DM2) than women without this illness (Rojas et al. 2014). PCOS is linked to high insulin levels in the blood. Insulin is a hormone that regulates blood glucose levels and is produced by specialized cells in the pancreas. These cells create insulin to help the body use glucose for energy when blood glucose levels rise (for example, after eating) (Louwers and Laven 2020). If blood sugar levels rise despite increasing insulin, the person may have impaired glucose tolerance, also known as "prediabetes", which could also lead to the development of diabetes (https://www.uptodate.com/contents/polycystic-ovary-syndrome-pcos-beyond-the-basics).

The uterine membrane sheds during ovulation. PCOS causes menstrual irregularities, and if women do not ovulate every month, the lining can accumulate. Endometrial cancer risk is greater of a thickened uterine lining (https://www.healthline.com/health/polycystic-ovary-disease#health-effects). Both hormonal imbalances and cognitive problems such as unwanted hair growth can have an adverse influence on your emotions. Many people with PCOS eventually develop psychological distress (https://www.healthline.com/health/polycystic-ovary-disease#health-effects). PCOS can create issues for both women and babies during pregnancy. Abortion, gestational diabetes, preeclampsia, and c-section are all more prevalent among women with PCOS. Their baby is also more likely to become obese (macrosomia) and to spend more time in a neonatal intensive care unit (NICU) (https://www.womenshealth.gov/a-z-topics/polycystic-ovary-syndrome:~:text=Having%20PCOS%20does%20not%20mean,you%20can't%20get%20pregnant).

1.3 Brief History and Statistics of PCOS

Although Stein and Leventhal are credited with being the first to investigate polycystic ovarian syndrome (PCOS), Vallisneri, an Italian scientist, described a married, infertile woman with shining, white ovaries, the size of pigeon eggs in 1721 (https://pubmed.ncbi.nlm.nih.gov/28791833/). Changes in the ovaries caused by cysts were first described in 1844 (https://en.wikipedia.org/wiki/Polycystic_ovary_syndrome#History). Later on, many scientists carried out studies to know about this disease in detail because of the increase in the count of women who got affected by the disease.

In India. PCOS affects around one in every five Indian women (20%) (https://www.thehindu.com/sci-tech/health/one-in-five-indian-women-suffers-from-pcos/article29513588.ece; https://bootcamp.uxdesign.cc/case-study-heres-why-having-pcos-should-be-normalized-dcfe5f5d66bf) as observed in Fig. 2. The prevalence of PCOS in India ranges from 3.7 to 22.5% depending on the population studied and the criteria used for diagnosis (Ganie et al. 2019).

Fig. 2 Statistical analysis of PCOS (*Source* taken from https://bootcamp.uxdesign.cc/case-study-heres-why-having-pcos-should-be-normalized-dcfe5f5d66bf)

In the world. According to the World Health Organization, it affects 116 million women globally as of 2010 (3.4% of women) (https://en.wikipedia.org/wiki/Polycystic_ovary_syndrome#History). PCOS was reported in 4 million instances in the USA in 2005, costing $4.36 billion in healthcare costs (https://en.wikipedia.org/wiki/Polycystic_ovary_syndrome#History). Globally, 1.55 million (95% UI: 1.19–2.08) incident cases of PCOS among women of reproductive age (15–49 years) were recorded in 2017, showing a 4.47% (2.86–6.37%) increase from 2007 (Liu et al. 2021).

2 Literature Survey

2.1 An Automated Detection of PCOS Using Follicle Recognition (Deshpande and Wakankar 2014)

In this source, Deshpande et al. presented that an automated detection of polycystic ovarian syndrome is given by calculating the number of follicles in the ovarian using ultrasound images and also uses bioclinical, clinical, and imaging properties to classify patients into two groups, i.e., normal and PCOS affected. The technique used for classification is the support vector machine (SVM) algorithm. The dataset contains 20 records of women with ovarian ultrasound images, BMI, cycle length, LH, and FSH values. The accuracy of this method of recognition is around 95%.

2.2 Classification of the PCOS from the Ultrasound Images Using Machine Learning Algorithms (Gopalakrishnan and Iyapparaja 2021)

In this source, Gopalakrishnan et al. have presented PCOS detection and classification using ultrasound images by analyzing the affected and unaffected areas in images. This method uses preprocessing of input images with Gaussian low-pass filter and multi-level thresholding for image segmentation process. This method has obtained an accuracy of 93.82% for SVM machine learning algorithm, 89.7 for random forest, and 88.26 for the Naive Bayes algorithm. Dataset includes 90 images in total, in which 30 are normal, 25 are cystic, and 35 are PCOS ovaries.

2.3 Classification of PCOS with Ultrasound Images and Machine Learning Algorithms (Purnama et al. 2015)

Here, Purnama et al. have used two different datasets wherein the first dataset has 40 images, of which 26 are normal images and 14 are PCOS-indicated images. The second dataset is of 40 images which consist of 34 normal images and 6 PCOS-indicated images. They have also used three classification algorithms, namely the neural network learning model with vector quantization approach, SVM classifier, and KNN algorithms. For the first dataset, greater accuracy is obtained from SVM with 82.55%, and KNN gives maximum accuracy for the second dataset with an accuracy of 78.81%.

2.4 Dr. Zio Health Fitness (https://play.google.com/store/apps/details?id=drzio.pcos.exercise.dietplan.forweightloss.pcod.cure.yoga)

This is a women PCOS treatment fitness application that supports 100 + languages and 200 + unique home-based workout exercises for PCOS cure with high-quality video support that targets all of your muscles and is specially designed for women. This application also tracks your weight loss progress and calories burned and calculates your BMI. It also gives you a workout reminder and helps you make the workout a daily habit. Final statistics on the completed training and exercises will be provided.

2.5 *Femcare Reverse PCOS Consult Doctor Online (https://play.google.com/store/apps/details?id=in.co.femcare)*

Femcare is the most customized online doctor consultation and PCOS reversal platform for women. Femcare offers holistic PCOS reversal schemes for 1, 3, and 6 months that helps the patient naturally reverse PCOS/PCOD. Gynecologists will first analyze the patient's medical information to identify the type of PCOS and the root problem of hormonal imbalance. After gaining knowledge about the body and the type of PCOS, the Femcare specialists' team builds a personalized strategy that meets an individually tailored PCOS diet plan, live PCOS exercises, changes in lifestyle, health and beauty counseling, and more.

2.6 *JGutke Designs (https://play.google.com/store/apps/details?id=com.le0fdf8af265.app)*

Every PCOS body was established as an all-encompassing reference for the latest information about the frequently misunderstood syndrome of polycystic ovary syndrome (PCOS). There was a lot of misconception nowadays, and sometimes, even experts are not always up to date on many of PCOS's complexities.

2.7 *MyAva Application (https://play.google.com/store/apps/details?id=in.myava.app)*

This application gives information about PCOS. Along with the information, it also provides suggestions to reverse the PCOS and details of expert gynecologists in the city. In this application, there is a team of people who help us to reverse PCOS by tracking their health conditions and symptoms to balance the patient's hormones. From the literature survey, it can be noted that PCOS detection is done using a combination of clinical and physical parameters and/or using ultrasound scans. Many mobile applications are also developed which bring awareness about the disease, provision to get connected to the doctors, videos related to daily workouts, etc. Such products make it easy to understand what the disease actually is and also the measures in reducing it.

3 About Dataset

3.1 Clinical Dataset

A dataset containing physical and clinical parameters used to determine PCOS is collected from Kaggle which contains 541 records and 43 attributes, of which PCOS (Y/N) is the key attribute defined as the class label of prediction. The dataset is prepared, and the key attributes are modified as PCOS Level-0, Level-1, Level-2, and Level-3 and the number of independent attributes is reduced from 43 to 12. The dataset reduction and preparation are done based on the suggestions and guidance provided by a gynecologist. The 12 finalized parameters with their detailed explanation are depicted below:

1. Age (age of the person), 2. weight (weight of the person), 3. hair growth (has excessive external hair growth), 4. skin darkening (has skin darkness in recent times), 5. pimples (has pimples in recent times), 6. fast food (consumes fast food frequently, 7. follicle number (count of the follicle in women's ovaries), 8. average follicle size (average Follicle Size), 9. FSH/LH (ratio of follicle-stimulating hormone and luteinizing hormone), 10. AMH (level of anti-mullerian hormone), 11. pregnancy (pregnant previously), and 12. cycle length (monthly menstrual cycle length).

3.2 Ultrasound Images' Dataset

A person affected with PCOS generally undergoes a lower abdomen ultrasound scanning test process. These scanned images give information about the size and number of follicles that are required to predict the presence of PCOS. For this work, real-time data from ultrasound scanned images of the lower abdomen are collected from a diagnostic center in Warangal city of Telangana State, India. The dataset contains 53 images in total, of which 23 are PCOS and 30 are non-PCOS images.

4 Design Methodology

This work consists of three major components which include a mobile application, detection of PCOS using clinical parameters with random forest classifier, and detection of PCOS using ultrasound scans with CNN as observed in Fig. 3.

A detailed explanation of each component is depicted below.

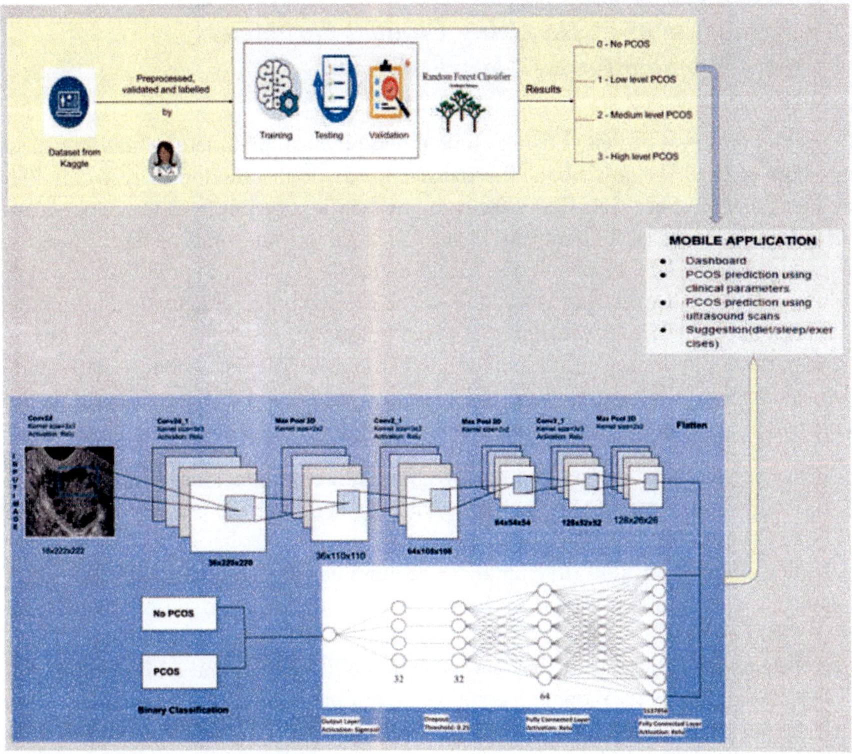

Fig. 3 Design methodology of the model

4.1 Mobile Application

The mobile application developed is Flutter based which supports both Android and IOS types of mobiles. It consists of components like the dashboard, about, profile, update profile, menstrual cycle track, suggestions to reverse the PCOS, PCOS prediction using clinical parameters and ultrasound scans, etc. Google Firebase is used to store and sync the user details such as username, email id, phone number, and password. User authentication is done by validating the stored details of the user and details provided during login time.

- One can log in or sign up for the application using the dashboard.
- It contains general information about the disease.
- The profile contains the basic information of a user, whereas the updated profile consists of some additional details.
- The menstrual cycle track consists of a calendar to mark dates that keep track of their period.

4.2 Prediction of PCOS Using Clinical Parameters with Random Forest Classifier

To predict the level of PCOS using clinical and physical parameters, random forest classifier is used. Details about the dataset used for the random forest classifier are mentioned in Sect. 3.1. Using the train_test_split function from the Scikit-learn library, the dataset is split into 80% and 20% for training and testing purposes, respectively. The data are then fed to a random forest classifier to predict the level of PCOS as Level-0, Level-1, Level-2, and Level-3 based on 12 independent attributes. The normal range of a few attributes is given below:

With the collaboration and association with the gynecologist, who has rich experience and expertise in treating the problems associated with PCOD, the classification done by the random forest algorithm is observed to follow the below pattern:

- If the person is a teenager and has symptoms like pimples, hair growth, normal FSH/LH ratio, normal AMH value, and is pregnant, then the person is not having PCOS (Level-0).
- If the person is an adult (having age > 28) and has pimples, hair growth, weight gain, and abnormal FSH/LH and AMH values, then the person is having light PCOS (Level-1).
- If the person is an adult and has hair growth, high FSH/LH and AMH values, and follicle number > 12, then the person is having moderate PCOS (Level-2).
- If the person is an adult and has pimples, hair growth, weight gain, high FSH/LH and AMH values, is not pregnant, and follicle number > 15, then she is said to be having high PCOS and must consult the doctor immediately (Level-3).

The implementation and observation of random forest classifier with the above described and identified levels of prediction with the accuracy are observed as 83% (Fig. 4).

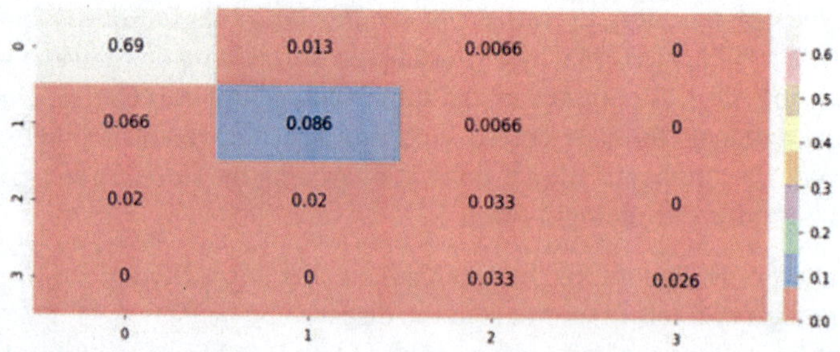

Fig. 4 Confusion matrix of random forest algorithm

4.3 Prediction of PCOS Using Ultrasound Scans with CNN

Ultrasound scanned images predict the presence of PCOS accurately when compared to physical and clinical parameters. Parameters like follicle number and follicle size can be extracted from ultrasound scans which strongly determine the presence or absence of the disease. Also, physical parameters cannot always decide and guarantee the presence of the disease. This is why ultrasound scans are used in the prediction of the presence of the disease.

Convolutional neural networks are feedforward neural networks that are fully connected. CNNs are exceptionally good at lowering the number of parameters without sacrificing model quality. The fundamental advantage of this algorithm is that it discovers essential traits without the need for human intervention. The ultrasound images are cropped using ImageDataGenerator method of keras.preprocessing.image library, and data replication is done to increase the size of the dataset. The dataset is split as 70% and 30% for training and testing, respectively.

CNN models consist of alternately stacked convolutional/hidden layers and pooling layers. The CNN model of this work consists of four convolutional layers and three max pool layers. CNN layers extract patterns from the basic level using a different number of filters whose count increases as we advance to the further layers. The pooling layers in between summarize the features in a region of the feature map created by the convolution layer. As a result, rather than precisely positioned features created by the convolution layer, the following actions are conducted on summarized information (https://www.geeksforgeeks.org/introduction-convolution-neural-network/). The activation function ReLU is simple and fast and determines whether a neuron is stimulated and transmitted to the next layer or not. In this way, training of the model is done in a faster and more precise manner as only the necessary data are passed to the subsequent layers (Fig. 5).

Training accuracy of the CNN model is obtained as 93%, whereas testing accuracy is 91% (Fig. 6).

5 Results and Discussions

The output of this model is a mobile application that has the features mentioned in Sect. 4.1. Figure 7 depicts the flow and usage of the application.

6 Conclusion

PCOS is the most common observed syndrome that affects almost 10% of women and requires immediate attention. The symptoms of this syndrome have a greater effect on health and lifestyle resulting in affecting infertility. This is an attempt

Fig. 5 Summary of the CNN model

```
Model: "sequential"
_________________________________________________________________
 Layer (type)                Output Shape              Param #
=================================================================
 conv2d (Conv2D)             (None, 222, 222, 16)      448

 conv2d_1 (Conv2D)           (None, 220, 220, 36)      5220

 max_pooling2d (MaxPooling2D  (None, 110, 110, 36)     0
 )

 conv2d_2 (Conv2D)           (None, 108, 108, 64)      20800

 max_pooling2d_1 (MaxPooling  (None, 54, 54, 64)       0
 2D)

 conv2d_3 (Conv2D)           (None, 52, 52, 128)       73856

 max_pooling2d_2 (MaxPooling  (None, 26, 26, 128)      0
 2D)

 dropout (Dropout)           (None, 26, 26, 128)       0

 flatten (Flatten)           (None, 86528)             0

 dense (Dense)               (None, 64)                5537856

 dropout_1 (Dropout)         (None, 64)                0

 dense_1 (Dense)             (None, 1)                 65

=================================================================
Total params: 5,638,245
Trainable params: 5,638,245
Non-trainable params: 0
```

```
print(model.evaluate(test_data))
print(model.evaluate(train_data))

1/1 [==============================] - 1s 1s/step - loss: 0.2114 - accuracy: 0.9333
[0.21142002940177917, 0.9333333373069763]
2/2 [==============================] - 3s 156ms/step - loss: 0.2771 - accuracy: 0.9189
[0.27707672119140625, 0.9189189076423645]
```

Fig. 6 Accuracy of the CNN model

to identify this syndrome in the early lifestyle with its severity level with guided and regular monitory through proper medication, food habits, and yoga guidance with a mobile application named "k**NO**w **PCOS**". The mobile application felicitates to predict the PCOS through clinical parameters and also through ultrasound scan images for more accuracy and integration of health policies. The prediction level of PCOS through clinical parameters is estimated as 83%, and through the ultrasound scan, images are 93%. The application provides a single platform where the detection of the disease, suggestions, and menstrual tracker features are provided. High-end mobile application technologies are used to develop a user-friendly UI, and some of the better machine learning algorithms are used in disease prediction more accurately giving better results.

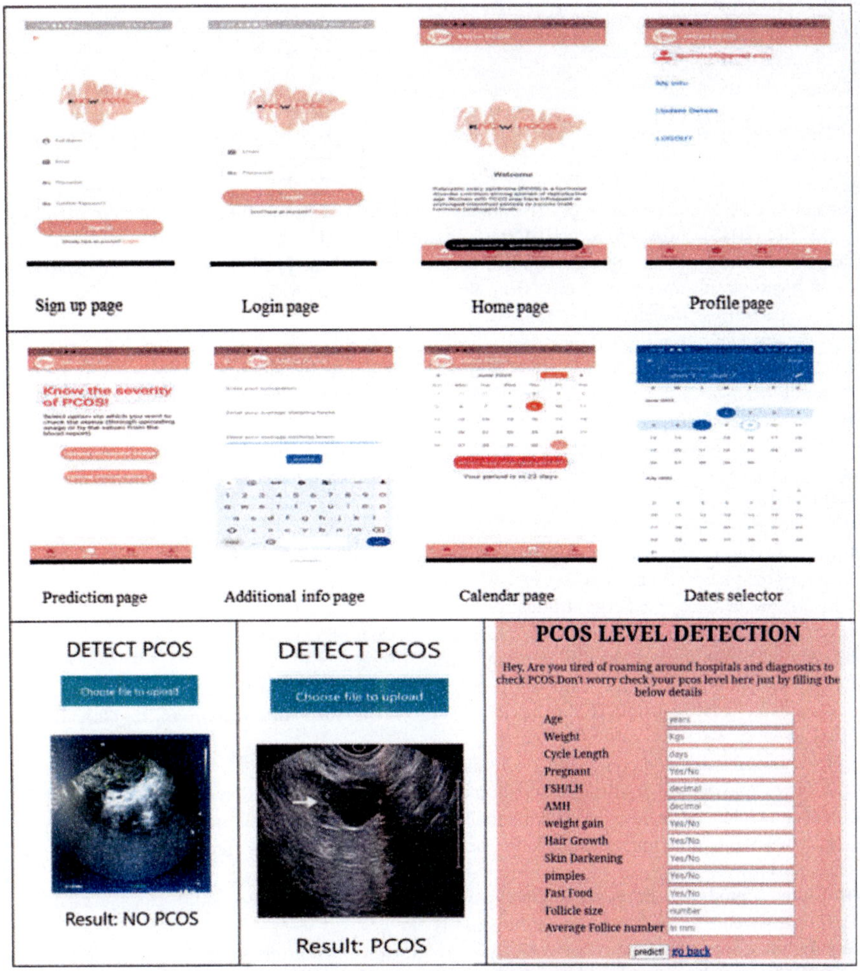

Fig. 7 Working model of the mobile application in various phases for the prediction of PCOS using two models

Acknowledgements The authors also acknowledge the Gynecologist Dr. Anuradha (MBBS, MD in Gynecology and Obstetrics) who helped in the process of understanding the PCOS symptoms and severity of the problem. Madam also helped in finding and filling the class label of the level of the PCOS based on the observations. Sincere acknowledgments to the doctor for sparing the time and support of the work.

References

S.S. Deshpande, A. Wakankar, Automated detection of polycystic ovarian syndrome using follicle recognition, in *2014 IEEE International Conference on Advanced Communications, Control and Computing Technologies* (IEEE, 2014), pp. 1341–1346

M.A. Ganie, V. Vasudevan, I.A. Wani, M.S. Baba, T. Arif, A. Rashid, Epidemiology, pathogenesis, genetics & management of polycystic ovary syndrome in India. Indian J. Med. Res. **150**(4), 333–344 (2019). https://doi.org/10.4103/ijmr.IJMR_1937_17

C. Gopalakrishnan, M. Iyapparaja, Multilevel thresholding-based follicle detection and classification of polycystic ovary syndrome from the ultrasound images using machine learning. Int. J. Syst. Assur. Eng. Manag. 1–8 (2021)

https://www.ncbi.nlm.nih.gov/pmc/articles/PMC3737989/

https://bootcamp.uxdesign.cc/case-study-heres-why-having-pcos-should-be-normalized-dcfe5f5d66bf

https://en.wikipedia.org/wiki/Polycystic_ovary_syndrome#History

https://play.google.com/store/apps/details?id=com.le0fdf8af265.app

https://play.google.com/store/apps/details?id=drzio.pcos.exercise.dietplan.forweightloss.pcod.cure.yoga

https://play.google.com/store/apps/details?id=in.co.femcare

https://play.google.com/store/apps/details?id=in.myava.app

https://pubmed.ncbi.nlm.nih.gov/28791833/

https://www.geeksforgeeks.org/introduction-convolution-neural-network/

https://www.healthline.com/health/polycystic-ovary-disease#health-effects

https://www.healthline.com/health/polycystic-ovary-disease

https://www.jeanhailes.org.au/health-a-z/pcos/symptomscauses?gclid=Cj0KCQiAkZKNBhDiARIsAPsk0WiVy7K5YgrhqBAFYmGpNRjwzuYun1vcUd9VMfToPNKhOfEc6C35gj8aAvB1EALw_wcB

https://www.lifeextension.com/protocols/female-reproductive/polycystic-ovary-syndrome-pcos

https://www.mayoclinic.org/diseases-conditions/pcos/symptoms-causes/syc-20353439

https://www.sciencedirect.com/topics/medicine-and-dentistry/polycystic-ovary-syndrome

https://www.thehindu.com/sci-tech/health/one-in-five-indian-women-suffers-from-pcos/article29513588.ece

https://www.uptodate.com/contents/polycystic-ovary-syndrome-pcos-beyond-the-basics

https://www.webmd.com/women/what-is-pcos

https://www.womenshealth.gov/a-z-topics/polycystic-ovary-syndrome:~:text=Having%20PCOS%20does%20not%20mean,you%20can't%20get%20pregnant

J. Liu, Q. Wu, Y. Hao, M. Jiao, X. Wang, S. Jiang, L. Han, Measuring the global disease burden of polycystic ovary syndrome in 194 countries: global burden of disease study 2017. Hum. Reprod. **36**(4) 1108–1119 (2021). https://doi.org/10.1093/humrep/deaa371

Y.V. Louwers, J.S.E. Laven, Characteristics of polycystic ovary syndrome throughout life. First Published March 18, 2020, Review Article. https://doi.org/10.1177/2633494120911038

U.A. Ndefo, A. Eaton, M.R. Green, Polycystic ovary syndrome: a review of treatment options with a focus on pharmacological approaches. P. T. **38**(6), 336–355 (2013)

B. Purnama, U.N. Wisesti, F. Nhita, A. Gayatri, T. Mutiah, A classification of polycystic ovary syndrome based on follicle detection of ultrasound images, in *2015 3rd International Conference on Information and Communication Technology (ICoICT)* (IEEE, 2015), pp. 396–401

J. Rojas, M. Chávez, L. Olivar, M. Rojas, J. Morillo, J. Mejías, M. Calvo, V. Bermúdez, Polycystic ovary syndrome, insulin resistance, and obesity: navigating the pathophysiologic labyrinth. Int. J. Reprod. Med. **2014**(719050), 17 (2014). https://doi.org/10.1155/2014/719050

Factors Affecting Interaction on Moodle: An Empirical Study Based on TAM

Komal Karishma and Krishna Raghuwaiya

Abstract In this paper, we looked at the factors that affect students' interaction on Moodle. These factors were assimilated into technology acceptance model (TAM) to explore if the mentioned factors actually affected the actual usage of Moodle. A qualitative method research approach was used to collect data. Students filled in a questionnaire with open-ended questions for the purpose of this research. Data were collected and the identified factors were categorized according to the segments of TAM. The factors were divided under perceived usefulness, perceived ease of use, attitude toward using, and behavioral intention to use. It was found that the factors did have an effect on students' interaction on Moodle.

Keywords E-learning · Interaction · Moodle · TAM

1 Introduction

With the emergence of e-learning in the teaching–learning process, the use of technology is undebatable. E-learning is technology-enhanced learning. It is a shift from traditional (face-to-face) classroom to a virtual classroom. Learning management system (LMS) has been introduced to accommodate the shift to a virtual learning environment (VLE).

Virtual classrooms are set up, and teaching is facilitated online. However, all these would be futile if students do not interact on Moodle. Student interaction on Moodle has been investigated on by many researchers. All dimensions of this has been explored to enhance the teaching–learning process that is facilitated by Moodle or any other LMS.

Similarly, this paper will explore the factors that affect students' interaction on Moodle. These factors have been identified by LLFXX students in the questionnaire

K. Karishma (✉) · K. Raghuwaiya
The University of the South Pacific, Suva, Fiji
e-mail: komal.karishma@usp.ac.fj

A. B. Reddy et al. (eds.), *Proceedings of Third International Conference on Advances in Computer Engineering and Communication Systems*, Lecture Notes in Networks and Systems 612, https://doi.org/10.1007/978-981-19-9228-5_46

given to them. These factors identified by the students are then matched with technology acceptance model (TAM) (Chau 1996; Davis et al. 1989). This will reflect how TAM incorporates the factors identified by the students.

The research was conducted at foundation level (pre-degree) on one of the mandatory English courses at a tertiary institute in Fiji. The institute offers courses to students from Fiji and other countries.

At the institute, courses are offered through either blended, print, or online mode. The presence of Moodle in these courses depends on the modes it is offered in. LLFXX is offered on both blended and print modes, and therefore, Moodle is used to facilitate teaching and learning. However, the interaction of students on the Moodle platform is debatable. Thus, the purpose of this research is to examine the factors influencing students' interaction on Moodle and to incorporate these factors in TAM.

2 Methodology

The participants for the research were foundation students from the main campus enrolled in LLFXX. These were a total of 109 students (80 blended and 29 print modes) from the pre-degree program. The factors affecting students' interaction on the Moodle page were investigated with the factors being included with TAM.

The factors affecting students' interaction on Moodle were identified by students in open-ended questions (Davis et al. 1989) in a questionnaire. The questionnaire was prepared in Moodle and the students completed the questionnaire online. The questionnaire was completed at the end of a semester before the covid-19 pandemic. Therefore, qualitative method of research was used for the purpose of this research.

The ability of qualitative research to enlighten 'processes and patterns of human behavior' (Tenny et al. 2022) cannot be possibly done in quantitative research as they are 'not easily handled by statistical procedures' (Bogdan and Biklen 2007). It is hard to precisely apprehend the 'experiences, attitudes, and behaviors' through the quantitative approach (Creswell 2003; Tenny et al. 2022). However, the qualitative approach enables participants to clarify 'how, why, or what they were thinking, feeling, and experiencing at a particular time or during an event' (Creswell 2003; Tenny et al. 2022).

This research needs students to answer this 'what' in open-ended questions. The factors identified by the students were analyzed by using technology acceptance model (TAM). These factors were assimilated into the different categories of TAM.

3 Technology Acceptance Model (TAM)

Technology adoption is when learners embrace an e-learning system. The most widespread theory in the area of information technology is the technology acceptance model—TAM (Alkhateeb and Abdalla 2021).

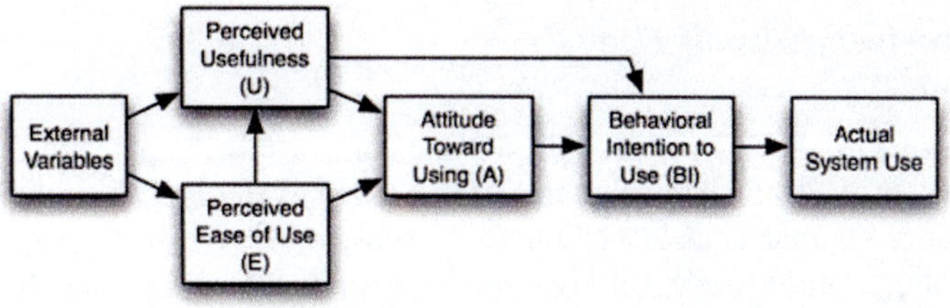

Fig. 1 Technology acceptance model adapted from Davis et al. (1989)

Since its introduction by Davis (Davis 1989), technology acceptance model (TAM) is used in research, either in its original or modified form to measure users' acceptance of information technology systems (Alkhateeb and Abdalla 2021). Davis proposed TAM to explain the possible user's behavior plans when using a technology or an improved technology, because it describes the connections between different components of TAM (the usefulness of a system and ease of use of a system) and the learners' or participants' approaches, aims, and the real use of the system (Šumak et al. 2011) (Fig. 1).

This information system theory shows user acceptance, adoption, and usability of technology. The motivation of the users of technology is portrayed by three factors: perceived usefulness, perceived ease of use, and attitude toward use (Essel and Wilson 2017; Taherdoost 2018). TAM is known to be one of the most prominent research models while researching determining factors of information systems/information technology acceptance (Chau 1996).

The TAM (Davis 1989) is widely used to describe individual behavior usage involving information technology. It implies that TAM is quite popular in the area of technology acceptance (Marangunić and Granić 2015). The principal TAM is to assist in investigating the effect of external variables that impact the perceived ease of use and perceived usefulness and impact approaches of employing technology and the behavioral aim to utilize technology as the dependent variable (Davis 1989; Davis et al. 1989; Damnjanovic et al. 2013; Essel and Wilson 2017).

The principal TAM concepts are:

3.1 *Perceived Usefulness (U)*

U is the degree to which a person believes that using a specific procedure would enhance his or her output (Šumak et al. 2011). *U* has been proven as the primary prerequisite for BI among the various variables that can possibly affect the application of the system (Šumak et al. 2011). This paper looks at *U* as the student being aware of the importance of technology.

3.2 Perceived Ease of Use (E)

E is the degree to which a person believes that using a specific procedure would be easy (Šumak et al. 2011). It refers to the effort needed by the user to get the benefit of the application (Šumak et al. 2011). For the purpose of this paper, *E* is the assurance that technology can be used easily or effortlessly and without any complications.

3.3 Attitude Toward Using (A)

An attitude is 'a summary evaluation of a psychological object captured in such attribute dimensions as good–bad, harmful–beneficial, pleasant–unpleasant, and likable–dislikable' (Šumak et al. 2011). The learners would most likely substitute the system with a new one if they do not like an e-learning system (Aizen 2001; Šumak et al. 2011). The approach of students while using technology will be focused on in this section.

3.4 Behavioral Intention to Use (BI)

BI is an indication that an individual is prepared to execute a certain behavior. It is thought to be an instant antecedent of behavior (Šumak et al. 2011). A learner's feelings about a system can determine his/her BI (Šumak et al. 2011). This is the dependent variable which measures the extent to which an individual has devised mindful strategies to execute or not execute some specific potential actions (Šumak et al. 2011). This component of TAM looks at the purpose of actions of the users known as netiquette in the modern days.

4 Factors Affecting Student Interaction on Moodle

Students (both blended and print modes) were given a questionnaire to fill in. One of the sections required them to identify factors that have affected their interaction on Moodle at pre-degree level.

4.1 Factors that Increased Student Interaction on Moodle

Figure 2 shows that at the close of the semester, the students enrolled via the blended and print modes, ensured that they completed all their assessments on Moodle. This

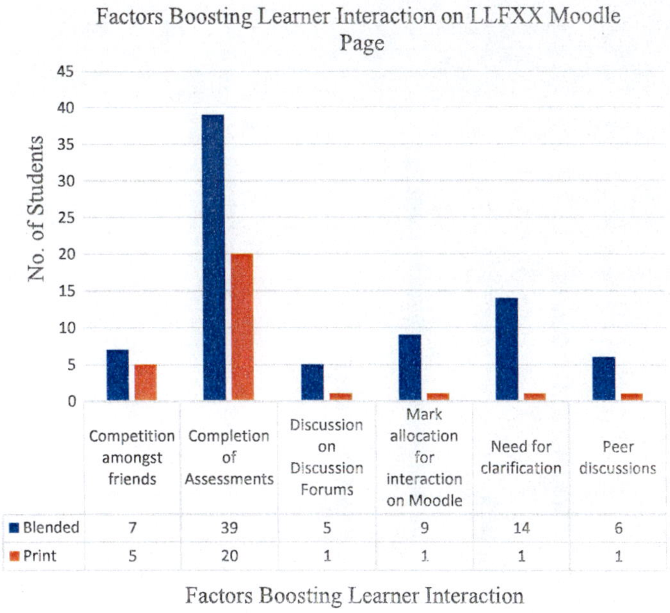

Fig. 2 Factors boosting interaction on Moodle

boosted their engagement on Moodle. This was 36% for students studying via blended mode and 18% for students studying via print mode. Blended mode students accepted that on Moodle, their interaction was the least in forums where discussions occurred (5%). The least common factors to boost interaction for the print mode students were forums for discussion, allocating marks for student engagement on Moodle, interacting with other users on Moodle to seek clarification, and for peer discussions.

4.2 *Factors that Decreased Student Interaction on Moodle*

Figure 3 depicts that blended mode students, at the close of the semester, proved that lack of time (31%) was the most common factor that decreased their interaction on Moodle. High workload (21%) and peer pressure (15%) were also notable factors decreasing interaction for blended mode students. Print mode students, on the other hand, identified high workload (13%) as the most common reason for the decrease in their interaction on Moodle.

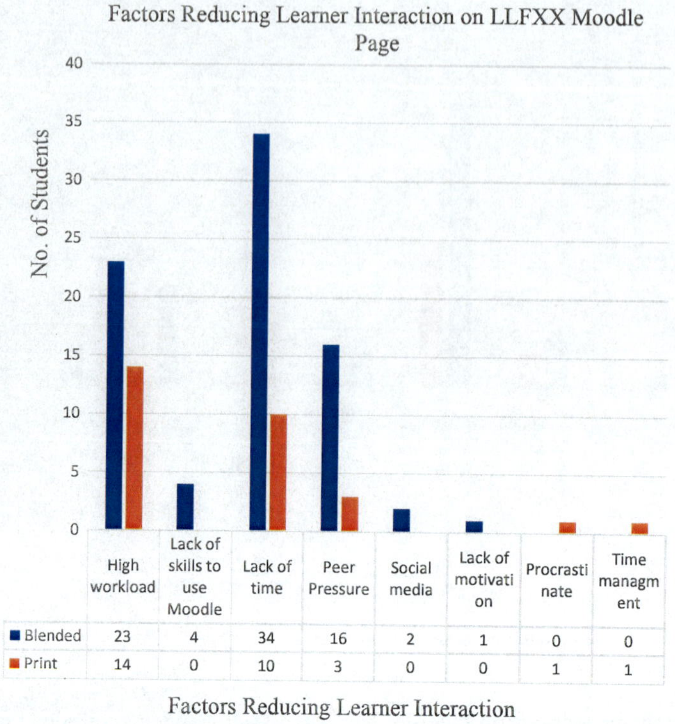

Fig. 3 Factors reducing student interaction on Moodle

5 Factors Affecting Student Interaction on Moodle and TAM

These factors that have been identified by the students in Figs. 2 and 3 can be categorized under the components of TAM. These reflect or try to explain the reasons for increasing or decreasing student interaction on Moodle.

5.1 Perceived Usefulness (U)

According to the LLFXX students, the reasons for their increased interaction on Moodle were to complete assessment, to participate on discussion or online forums, and to get clarification when need arises.

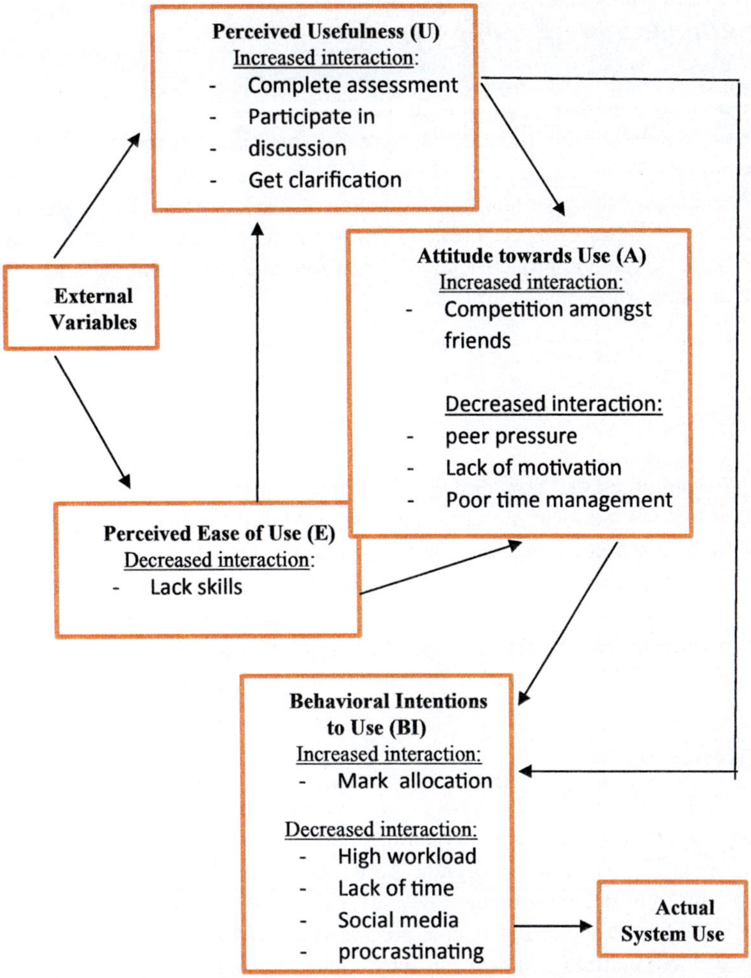

Fig. 4 Factors affecting students' interaction on Moodle tagged in TAM

5.2 *Perceived Ease of Use (E)*

Though *E* emphasizes that students feel that technology can be used effortlessly, there are students who lack skills to use Moodle. This results in a decrease in their interaction on Moodle. This is contrasting to what is perceived by technology users about technology and its use.

5.3 Attitude Toward Using (A)

The increasing interaction on Moodle is also due to competition among friends. This competitive attitude in students increases their interaction Moodle for study purposes.

There are also some factors that are negatively affecting student interaction on Moodle. These factors are peer pressure, lack of motivation, and poor time management. These adverse approach leads to little interaction on Moodle and can lead toward an undesirable final result.

5.4 Behavioral Intention to Use (BI)

Students highlighted that they prefer 'using Moodle more when there is some form of reward like mark allocation for interaction on Moodle'. This encourages them to interact regularly and not when assessments are due.

There are also factors that discourage them from interacting on Moodle. These are high workload, lack of time, social media, and procrastinating. These will lead to poor netiquettes and decreased motivation to interact on Moodle.

6 Discussion

The goal of this study was to identify factors that either increase or decrease students' interaction on Moodle and to integrate these factors with TAM to examine if TAM is practical for pre-degree students.

The findings show that if students are made aware of these factors, then their Moodle use will definitely improve as they could be able to realize the benefits that their increased interaction on Moodle will have.

According to the LLFXX students, the reasons for their increased interaction on Moodle were to complete assessment, to participate on discussion or online forums, and to get clarification when need arises. Since it is mandatory for students to submit all assessments on Moodle, LLFXX students had to interact on Moodle. Also, it is easier for students to participate in online discussion and seek clarification online if need be.

These reflect that LLFXX students are aware that technology and Moodle are crucial toward their completion of the course successfully.

This becomes more motivating as students feel that technology can be used easily. However, LLFXX students, on contrary, identified lack skills to use Moodle as being a factor for little interaction on Moodle. This re-emphasizes the fact that it is not necessary that all technology users will come with the same level of skills to use technology, and most specifically, LMS, as being pre-degree students, this is their first exposure to Moodle.

Students' attitude toward using technology also affects their usage of technology. Competitive attitude of students, for example, increases their interaction on Moodle. Competition can be in form of uploading or 'attempting assessments online to get better grade' or thriving to be one of the 'first ones to check marks' (feedback) on assessments. Such competition encourages students positively to interact on Moodle and consequently work toward a better final grade.

While the attitude of students is positively affecting their interaction on Moodle, there are some factors that are negatively affecting student interaction on Moodle. These factors are peer pressure, lack of motivation, and poor time management.

Students are easily manipulated by their peers to indulge in activities or groups that are not academic friendly. This deviates them from their focus and 'keeps them away from their studies and Moodle'.

Moreover, there is lack of motivation among students to use Moodle. One student mentioned that 'having games will be exciting'. This may motivate them to interact more on Moodle.

Furthermore, poor time management was identified to be a reason for little interaction on Moodle. Since Moodle is a platform that facilitates virtual interaction and there is an absence of face-to-face interaction, students do not find it mandatory to interact on Moodle. Interaction only happens when students find time and have to submit mandatory components like assessments. If students manage their time properly, they would be able to use Moodle more frequently.

Such attitude leads to a behavioral pattern among students. They would interact on Moodle if there are some forms of reward, like mark allocation for interaction on Moodle.

These behavioral patterns can also have negative impact on their way of looking at Moodle and its usage. For example, when students' workload is high, they spend their time meeting assessment deadlines, attending face-to-face class (tutorials, laboratories, and lectures), and participating in group work (face-to-face). This results in students having lack of time to interact on Moodle unless they need to submit assessments.

Also, some students are addicted to social media. They 'login to Facebook more than logging into Moodle'. This behavior of theirs distracts them from studies, and thus, their contact with study-related components is little.

Not surprisingly, some students said that they 'keep postponing Moodle during the day'. Such act of procrastinating hinders their chances of interacting with content that are associated with school. Since school work is postponed frequently, all materials or gazettes containing notes, assessments, and revision materials are ignored. Hence, there is a decreased interaction on Moodle.

7 Conclusion

The factors identified by the students with regard to their interaction on Moodle assimilate quite well with TAM. These factors have reflected the reasons for the actual system, which in this study is Moodle, use. The identified factors were distributed according to Davis TAM (Davis 1989). These factors asserted that it is the perceived usefulness, perceived ease of use, attitude toward using, and behavioral intention to use that decide the actual usage of Moodle, that is, whether it is used to the maximum or to the minimum.

The manner in which these factors fitted into the model was quite exciting and thought-provoking. It emphasizes that if the motivating factors are encouraged and the demotivating factors are improved upon, then students' interaction on Moodle (actual system use) will increase.

References

I. Aizen, Nature and operation of attitudes. Annu. Rev. Psychol. **52**, 27–58 (2001)

M. Alkhateeb, R. Abdalla, Factors influencing student satisfaction towards using learning management system Moodle. Int. J. Inf. Commun. Technol. Educ. **17**, 138–153 (2021)

R.C. Bogdan, S.K. Biklen, *Qualitative Research for Education: An Introduction to Theory and Methods*, 5th edn. (Pearson Education Inclusive, Boston, 2007)

P.Y.K. Chau, An empirical assessment of a modified technology acceptance model. J. Manag. Inf. Syst. **13**(2), 185–204 (1996)

J.W. Creswell, *RESEARCH DESIGN Qualitative, Quantitative, and Mixed Methods Approaches*, 2nd edn. (SAGE Publications, London, 2003)

V. Damnjanovic, S. Jednak, I. Mijatovic, Factors affecting the effectiveness and use of Moodle: students' perception. Interact. Learn. Environ. **23**, 1–19 (2013)

F.D. Davis, Perceived usefulness, perceived ease of use, and user acceptance of information technology. MIS Q. **13**(3), 318–339 (1989)

F.D. Davis, R.P. Bagozzi, P.R. Warshaw, User acceptance of computer technology: a comparison of two theoretical models. Manage. Sci. **35**, 982–1003 (1989)

D.D. Essel, O.A. Wilson, Factors affecting university students' use of Moodle: an empirical study based on TAM. Int. J. Inf. Commun. Technol. Educ. (IJICTE) **13**(1), 14–26 (2017)

N. Marangunić, A. Granić, Technology acceptance model: a literature review from 1986 to 2013. Univ. Access Inf. Soc. **14**, 81–95 (2015)

B. Šumak, M. Hericko, M. Pušnik, G. Polančič, Factors affecting acceptance and use of Moodle: an empirical study based on TAM. Informatica **35**(1), 91–100 (2011)

Taherdoost, H.: A review of technology acceptance and adoption models and theories. **22**, 960–967 (2018)

S. Tenny, G.D. Brannan, J.M. Brannan et al., *Qualitative Study* (StatPearls). https://www.ncbi.nlm.nih.gov/books/NBK470395/. Last Accessed 12 June 2022

Metacart—Decentralized Social Media Marketplace to Incentivize Creators and Ensure User Data Privacy

Sharat Chandra Manchi Sarapu, Nagaratna P. Hegde, Sireesha Vikkurty, and Krishna Priya V. S. Garimella

Abstract Metacart is an advertisement-free, tracking-free, and creator-friendly web3-based social media powered by cryptocurrencies and NFTs. Unlike current social media applications that track user activity, we ensure data privacy by leveraging decentralized IPFS storage. We use Ethereum-based tokens to facilitate content monetization through micro-tipping. This feature enables users to support their favorite creators by tipping apart from liking and sharing. We democratize digital asset ownership by minting each post into an NFT with just a click, which subsequently can be traded on our own NFT marketplace. Utilizing our proprietary dynamic pricing algorithm, which gauges the value of the assets intelligently based on virality and engagement, we determine the price of NFT. In all the transactions, we use our tokens built on Ethereum to carefully construct healthy web3 business models that leverage tokenomics to create value for every party in a positive sum manner.

Keywords Blockchain · Smart contracts · Decentralization · Web3 · Social media · NFT marketplace · Non-fungible tokens · Micro-tipping · Tokenomics · Intelligent pricing models · Intellectual property protection

---

S. C. Manchi Sarapu (✉) · N. P. Hegde · S. Vikkurty · K. P. V. S. Garimella
Vasavi College of Engineering, Hyderabad, India
e-mail: mssharatchandra@gmail.com

N. P. Hegde
e-mail: nagaratnaph@staff.vce.ac.in

S. Vikkurty
e-mail: v.sireesha@staff.vce.ac.in

K. P. V. S. Garimella
e-mail: krishnapriyagarimella9@gmail.com

A. B. Reddy et al. (eds.), *Proceedings of Third International Conference on Advances in Computer Engineering and Communication Systems*, Lecture Notes in Networks and Systems 612, https://doi.org/10.1007/978-981-19-9228-5_47

1 Introduction

The present social media landscape is not very promising as it has several issues. Problems with the web2 social media ecosystem include the vast power held by a few corporations, platform monitoring, and transparency. Moreover, the algorithms used by these organizations are for their benefit and usually guide their users into a rabbit hole of consumption and polarization.

The existing social media revenue model runs on the attention economy, which involves platforms selling their users' attention for advertising. This game of attention forces these firms to indulge in unethical practices. These practices are causing us to become addicts of the platform and making our lives worse while dividing our society by agitating us and giving different individuals various versions of the same facts.

Additionally, these platforms do not encourage creativity. We often see the content of small creators blatantly plagiarized by more popular creators without any real consequences. Even the monetization models for these creators are very centralized which forces these creators to make polarizing content to earn money.

A potential remedy is switching to a subscription-based model, which would enable businesses to control their relationships with customers, minimize addiction to participation, and stop toxic behavior like hate speech and the dissemination of false information and disinformation. But those who cannot subscribe would have less access to crucial items like job listings and business tools, interpersonal connections, and access to information.

Therefore, we want to redefine the entire landscape of traditional social media and make it a place of the future with promising web3 business models involving NFTs and cryptocurrencies. We tackle the data privacy problem by adopting a dual backend. We integrate web2 and web3 by storing user images on IPFS and keeping the hashes obtained on the blockchain. This helps small creators to stay protected as their proof of work is safe. Additionally, to prevent creators from catering to existing harmful algorithms, we have enabled monetization of their content by introducing peer-to-peer micro-tipping using Ethereum-based tokens.

To simplify the ownership of digital assets, we have democratized the NFT minting process by enabling the minting of every user post with just a click of a button. Moreover, to give the buyer and seller the best price for the NFT, we incorporated our dynamic pricing algorithm. This algorithm takes in the base price set by the user and marks up the price based on the virality and engagement of the post. This prevents the creators from underselling their NFTs. Further, these NFTs can be listed on our own NFT marketplace and will be open for commerce. The revenue model we propose is to leverage our tokens and let the network effects increase the value of the token. We are essentially removing the need for advertisements with our revenue model, thereby ensuring user data privacy.

To summarize, the main contributions of this paper include:

- Providing advertisement-free social media that ensures user data privacy.

- Leveraging smart contracts to maintain proof of work to encourage creativity among creators.
- Proposing a sustainable web3 revenue model by leveraging tokenomics.
- Democratizing the process of creating and owning NFTs and pricing them with our proposed dynamic pricing model.
- Enabling micro-tipping feature to allow users to support their favorite content creators.

2 Literature Survey

The increasing use of social networks has been enhanced by the availability of low-cost internet access to the masses. This enables the quick and easy usage of new, fascinating, and interactive applications that make users feel social empowerment and a part of the market. It has also been found that global businesses have identified the web as a promising space to run their campaigns and advertise their business products. According to the intensive survey done by Constantinides (2009) and Pinheiro et al. (2014), global companies have already started to leverage the vast distribution and daily active users on these various web2 platforms. It led to the social media marketing wave in which billions have been invested as the conversion and adoption rates have been much higher than traditional flyers and hoardings.

It has been found by Power and Phillips-Wren (2011) that the centralization of social media has a significant impact on the decision-making of the users. The firms' marketing strategies influence the users to make bad financial decisions that align with the advertising companies' behavioral change goals. The world is super connected, and the efficiency with which these networks are spreading information is unmatched. Also, Zubiaga et al. (2016) found some crucial vulnerabilities of social media and have shown the caveat of the hyperconnectivity of the world. The misinformation more often than not has widespread real-world consequences. This underlying problem was addressed by many using various approaches, but the most promising one has been from Paul et al. (2019) where it was proposed to use blockchain as a solution to detect fake news, especially on social media where the ease of constructing fake news is the apex.

Additionally, even Ochoa et al. (2019) leveraged data mining as a consensus mechanism to validate material shared on social networks using blockchain. This has streamlined our work and helped us explore web3 and its benefits for creating a sustainable social media platform. On the other hand, the exploitation of creators on such aggregator platforms is very well known and problematic. Kopf (2020) have explained this issue in detail about how these creators do not hold an empowering position in the process of monetization of their content.

We wanted to solve all these issues because we realize social media is a very integral part of our generation as it drove a lot of positive changes in society. This space for discussion of ideas and expression of one's personality should not be shrouded by the above-mentioned shortcomings.

Furthermore, an elaborate discussion by Petrescu and Krishen (2020) has indicated the growing turmoil of data manipulation, invasive personalized advertisements, and AI algorithms that censor data. There has been a call for an algorithm that does not use AI at all to censor the data and give the users the rights of their personal information and censorship. These findings have directed us to approach the whole social media problem through the promising benefits of web3 and blockchain taking advantage of immutability and trust.

From the study of Dutta and Saini (2020) and Freni et al. (2020), it was identified that many attempts have been made already by 40 + platforms like Steemit, Obsidian, Earn, Indorse, Social X, Enlte, Voice, Sapien, Sola, Ong.Social, Minds, SoMee, Smoke, Alfa, and so on. A thorough investigation has been done on the merits and shortcomings of these alternatives. It was identified that the common goals of all these platforms were to reshape the social media scenario through fair distribution of value, widespread awareness and practice of ownership, transparency, the resistance to censorship, and personal data decentralization.

Ushare by Chakravorty and Rong (2017) has proven to be efficient in dealing with the ownership problem as it lets its users have complete control of their data by harnessing user-centric blockchain, encryption techniques, and a local personal certificate authority. Likewise, Li and Palanisamy (2019) have tried to solve the value redistribution problem by introducing cryptocurrency incentives for popular content and also tried to experiment with the consensus protocol of Delegated Proof of Stake (DPoS). They elected 21 members periodically, were made in charge of synchronously operating the platform, and have found that the DPoS generates undesirable results to scale up to a highly decentralized social media.

Furthermore, it has also been identified that the reward system was being hacked by bots and is not serving its entire objective but indicating the misuse. To carefully navigate this reward system gap, we further wanted to comparatively investigate and understand various options. Guidi (2021) has done a tremendous contribution by performing a comparative study on various consensuses and blockchains and also a thorough survey on the decentralized applications (Dapps). The work was very helpful to validate the social media problem and leveraging Ethereum as our blockchain to develop smart contracts on solidity language. Many efforts are put forward to increase the efficiency of Ethereum such as Arbitrum, and proof of work (PoW) consensus is still better than the other existing alternatives that have questionable control access over the blockchain. This study further helped us understand the complexity of the blockchain systems for the environment, and hence, we have decided to modify and fine-tune the architecture of our own social media application by leveraging the benefits of web2 wherever we can. One such advancement was to use the InterPlanetary File System (IPFS) proposed by Zheng et al. (2018) to address the growing data volumes. It has to be noted that a remarkable compression ratio up to 0.0817 can be achieved by adopting IPFS in blockchain systems. Instead of the entire media, only the hash is stored on the blockchain, which provides additional security and synchronization speeds.

Simultaneously, there has been a new wave of non-fungible tokens (NFTs) being leveraged to boost the art and the frenzy of buying and selling these NFTs which has assured the artists of value creation with their contributions. From the systematic study on the current NFT markets by Wang et al. (2021), it has been established that this mushrooming NFT wave has already promised prosperity to the creators with over thousand-fold returns. But, it has also been understood that these markets are premature and people can easily get lost in the frenetic evolution. To prevent this, the need for standardization and fraud protection is necessary. The study done by Vasan et al. (2022) indicates the importance of social media following for an artist as the followers decide the valuation of the art. Hence, the decision to include digital assets into our social media application was made keeping in view, the democratization of blockchain and digital assets.

The key problem of pricing these NFTs has been identified by Dowling (2022) as the premature markets are vulnerable and many market manipulations can drive the prices irregularly if not monitored. In this process, we did not want our users and creators to be cheated; hence, the decision to create a pricing algorithm of our own was designed and developed which is purely driven by original users and their actions, leading toward promising democratic markets. And, to further incentivize our creators, various existing reward mechanisms have been studied from the report of Kukartsev et al. (2020). It has been inferred that the peer-to-peer micro-tipping will be a huge boost toward content monetization and hence its transparency if the entire transaction is facilitated by Ethereum.

Finally, the studies of Cong et al. (2021) and Potts and Rennie (2019) were very inspiring and convincing to profitably move away from agency-centric business models and establish a creator-centric business model to endure and preserve the spirit of art and also create value in the process for all the shareholders without any exploitation. This led to the empowering business model creation around tokens. Finally, the survey by Paramarta et al. (2018) added to our validation of the timing of such rapid tectonic changes in the existing social media landscape and creators' economy as the people are more aware than ever and will welcome such a driving force with open arms.

3 Architectural Design

The user can register and subscribe to other users' feeds to see their posts on the dashboard. Through comments, tips, and likes, users can also engage with the post. A crypto wallet that is connected to the user's browser is required to upload media because it enables tipping directly with Ethereum-based tokens. When the user uploads media, it is sent to IPFS which generates a hashcode for the image. This hashcode is stored on our database and smart contract to mint NFTs and enable tipping as shown in Fig. 1.

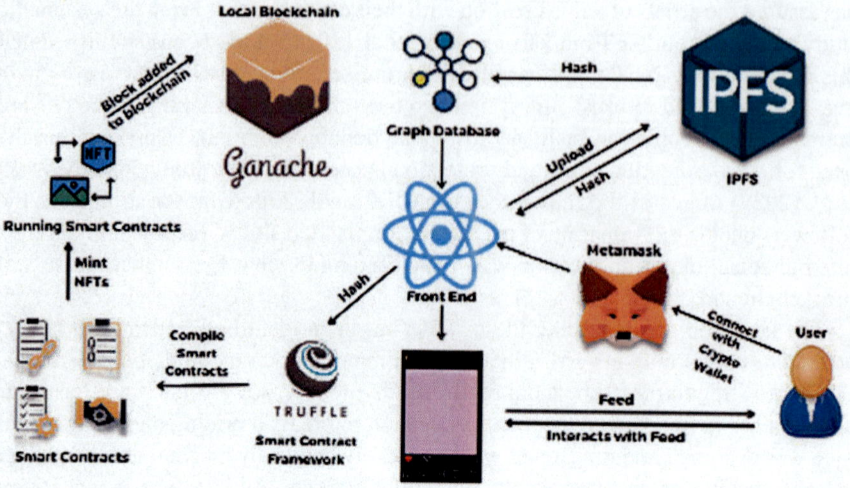

Fig. 1 Architecture of the proposed application

4 Implementation

4.1 *Smart Contracts and IPFS*

In this section, we explore the smart contracts that enable users to tip content, mint NFTs, and provide proof of work for creators. We have written two smart contracts for these purposes. They are the tipping contract and the minting contract. We additionally use InterPlanetary File System (IPFS) to store images. IPFS is a globally distributed file system that generates a hashcode for every file stored in it. This hashcode can be used to render the file on the web. The minting contract comes into play when a user wants to post content on our platform. In our efforts to democratize the minting of NFTs, we have inbuilt the minting process into the process of creating posts. For minting, we use ERC721 as it is a convention for displaying possession of non-fungible tokens. When a user creates a post, the image of the post is sent to IPFS and a hash is generated. This hash, along with the account number and base price, is used to formulate our minting smart contract. Another issue that we tackled using smart contracts was content monetization. That is where our tipping contract comes into play. If a user loves a post, they can immediately show support by clicking on the "tip" button, which will trigger a smart contract that enables them to tip the post. In this way, content creators are paid for their work, which incentivize them to make better content. Hence, we provide proof of work and security with smart contracts.

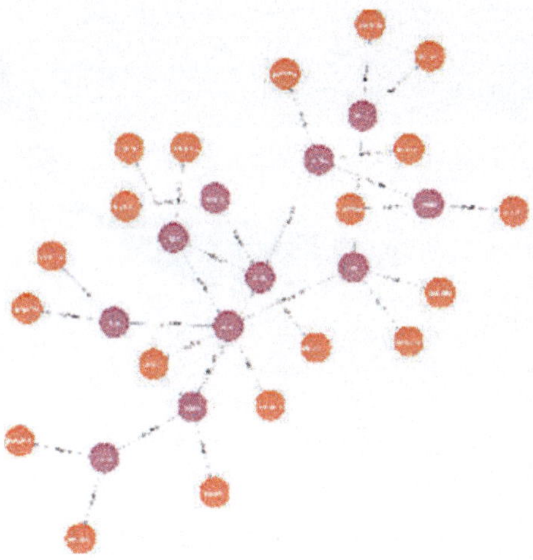

Fig. 2 Graph database model

4.2 Database

While smart contracts provide an array of benefits, it is not feasible to have smart contracts for every operation on our platform due to their high cost. Hence, we need a careful integration of web2 and web3 to provide the best of both worlds. To model relationships among various users, as in a social media setting, we decided to use a graph database to store user information. Because social media networks are already graphs, converting a graph to tables and back is pointless. There are two kinds of nodes in our database, namely, user and post. The user node stores information about each user, such as account id, username, and password. A user node can be connected to other user nodes via a follower or following relationship. These relationships among user nodes are many-to-many relationships. The post node stores information about each post, like the IPFS hash of the image, account id, and the number of likes on it. A post node is always linked to a user node via a one-to-many relationship. These relationships are visualized in Fig. 2.

4.3 Dynamic Pricing Algorithm

To tackle the problem of inconsistent pricing models and protect our creators from underselling their art and also to protect our users from getting cheated by bandwagons, we have designed an algorithm that either inflates or deflates the existing base price of the NFT according to the virality and engagement generated by the art piece.

This mathematical formula takes into account the nature of the parameters we are trying to gauge and also takes in the nature of a lot of mathematical functions that are appropriate to model real-world pricing systems without any bias or manipulation.

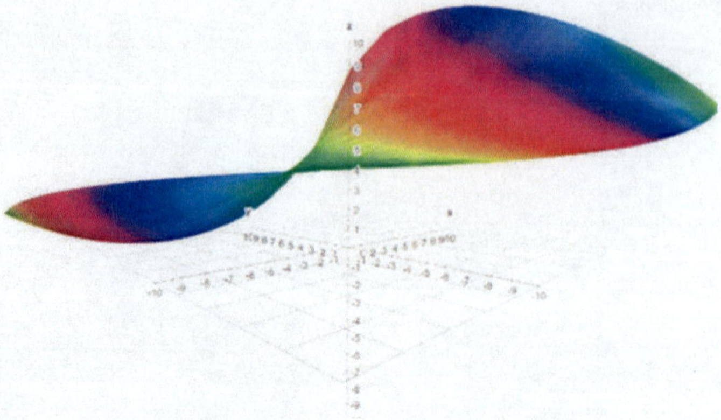

Fig. 3 Visualization of dynamic pricing algorithm

We also have ensured that the algorithm appreciates meritocracy and the perceived value of the community drives the prices unlike the perceived value of some individuals or orchestrated bandwagons.

Additionally, our algorithm is designed to turn into a boon for artists whose NFTs are mispriced as we correct their prices and ensure that they get sold to the right audience gaining maximum profit for our creators and ensuring the best prices for the users.

The formula used for dynamic pricing algorithm is

$$\text{Dynamic Pricing Algorithm Price} = \text{Base Price}\left(\log\left(\log\left(\frac{x^2}{y^{2.5}}\right)\right) + 1\right)$$

The base price is set by the creator.

x denotes the number of likes a particular post gets.

y denotes the number of hours that passed from the original post time.

Hence, we regularly update the prices and fairly encourage commerce. The visualization of the above algorithm is shown in Fig. 3.

4.4 Revenue Model

All the transactions on the platform will happen with our token. This business model adopts the standard tokenomics and is dependent on the community. We believe by creating value for the community, and they will build trust in the platform and the token and hence will continue using the tokens. Additionally, network effects and liquidity will boost the use of the token as shown in Fig. 4. Integration of various services and products can be done once there is an active community that is willing to

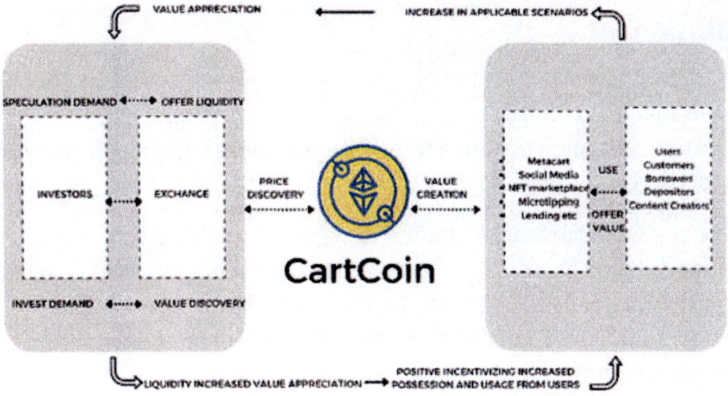

Fig. 4 Revenue model

participate in the cross-selling and upselling. This way, profitability can be achieved without having to rely on personalized advertisements and data collection.

5 Evaluation

The dynamic pricing algorithm was tested with multiple data points, and it produced desired results. One such test case is shown in Fig. 5 for an NFT quoted with a base price of 10,000 INR.

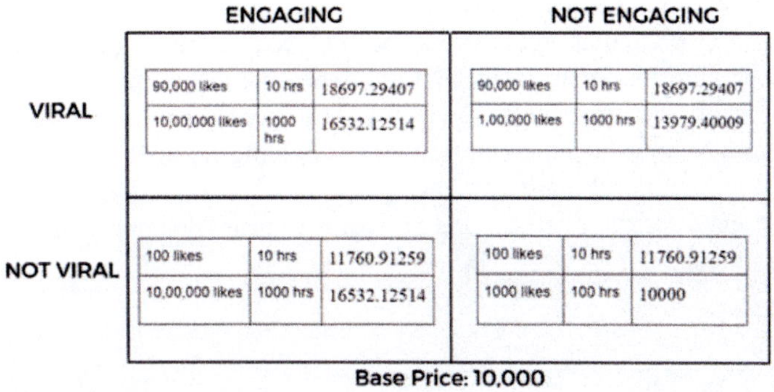

	ENGAGING			NOT ENGAGING		
VIRAL	90,000 likes	10 hrs	18697.29407	90,000 likes	10 hrs	18697.29407
	10,00,000 likes	1000 hrs	16532.12514	1,00,000 likes	1000 hrs	13979.40009
NOT VIRAL	100 likes	10 hrs	11760.91259	100 likes	10 hrs	11760.91259
	10,00,000 likes	1000 hrs	16532.12514	1000 likes	100 hrs	10000

Base Price: 10,000

Fig. 5 Measured results of dynamic pricing algorithm for various scenarios

6 Limitations

Although we managed to solve all the major issues with the existing social media platforms and paved a path toward a futuristic one, there still prevail some limitations with the proposed solution. They are

(a) Gas fee and environment: Although Ethereum turned out to be the most suitable solution for the problem, the gas fee required is relatively higher and also is not environment-friendly.
(b) Feed improvement: There is no recommended feed integrated into our platform as that goes against our data privacy policies.
(c) Plagiarism checking: Other than an intensive proof of work checking and community-based reporting, we have no robust mechanisms in place to detect and prevent plagiarism on our platform.

7 Conclusion

In this article, we covered the issues with conventional, centrally-controlled social media programs and how utilizing blockchain technology can address them. In our proposed solution, we rely heavily on smart contracts to perform operations, provide user data privacy, and guarantee transparency. We have also pointed out the predicament of artists, who are underpaid and lack appropriate avenues for monetizing their work. As the Ethereum blockchain supports the proof of work, the incorporation of NFTs into social media creates a hybrid marketplace based on trust. Additionally, the best technology stack was chosen to clear the way for social media that is truly decentralized. A particular emphasis has been placed on creating a unique, entirely mathematical algorithm that prices NFTs to overcome the anomalies in pricing the NFTs and benefit creators and users equally. Moreover, features like micro-tipping will boost and empower the creators, thereby helping them form stronger bonds with their communities. Finally, we have proposed token-based business models to generate revenue for the platform and sustain it to grow this creator-centric social media community.

8 Future Work

The aim is to enhance this platform by migrating to eco-friendly blockchain such as "Cardano". Further, the necessity of having an efficient database to store information securely stresses the need for a proprietary blockchain-based database.

This project aims to integrate blockchain-based recommendation systems that leverage artificial intelligence without compromising users' data and provide users with enhanced feed and connections.

Expansion of features is required to let the users create their own communities and interact with each other, and improving the marketplace features and allowing exclusive collectibles' deals will pave a path toward augmented reality adoption smoothly.

Finally, the same architecture can become the backbone of many pressing issues like independent journalism as the platform promises immutability, transparency, privacy, and proof of work.

References

A. Chakravorty, C. Rong, Ushare: user controlled social media based on blockchain, in *Proceedings of the 11th International Conference on Ubiquitous Information Management and Communication (IMCOM '17)*, Article 99 (Association for Computing Machinery, New York, NY, USA, 2017) pp. 1–6. https://doi.org/10.1145/3022227.3022325

L.W. Cong, Y. Li, N. Wang, Tokenomics: dynamic adoption and valuation. Rev. Fin. Stud. **34**(3), 1105–1155 (2021). https://doi.org/10.1093/rfs/hhaa089

E. Constantinides, Social media/web 2.0 As marketing parameter: an introduction, in *Proceedings of 8th International Congress Marketing trends* (2009), n. pag. Print

M. Dowling, Fertile LAND: pricing non-fungible tokens. Fin. Res. Lett. **44**, 102096 (2022). ISSN: 1544-6123. https://doi.org/10.1016/j.frl.2021.102096. https://www.sciencedirect.com/science/article/pii/S154461232100177X

S. Dutta, K. Saini, in *Blockchain and Social Media* (2020). https://www.taylorfrancis.com/books/mono/10.1201/9781003081487/blockchain-technology-applications?refId=fd86f5d0-81f5-41bf-98e6-329a46609f72&context=ubx

P. Freni, E. Ferro, G. Ceci, Fixing social media with the blockchain, in *Proceedings of the 6th EAI International Conference on Smart Objects and Technologies for Social Good (GoodTechs '20)* (Association for Computing Machinery, New York, NY, USA, 2020), pp. 175–180. https://doi.org/10.1145/3411170.3411246

B. Guidi, An overview of blockchain online social media from the technical point of view. Appl. Sci. **11**(21), 9880 (2021). https://doi.org/10.3390/app11219880

S. Kopf, Rewarding good creators': corporate social media discourse on monetization schemes for content creators. Soc. Media + Soc. (2020). https://doi.org/10.1177/2056305120969877

A.V. Kukartsev et al., Media content monetization model on the example of micropayments. J. Phys.: Conf. Ser. **1661**(1) (2020). IOP Publishing

C. Li, B. Palanisamy, Incentivized Blockchain-based social media platforms: a case study of Steemit, in *Proceedings of the 10th ACM Conference on Web Science (WebSci '19)* (Association for Computing Machinery, New York, NY, USA, 2019), pp. 145–154. https://doi.org/10.1145/3292522.3326041

I.S. Ochoa, G. de Mello, L.A. Silva, A.J.P. Gomes, A.M.R. Fernandes, V.R.Q. Leithardt, FakeChain: a blockchain architecture to ensure trust in social media networks, in *Quality of Information and Communications Technology. QUATIC 2019. Communications in Computer and Information Science*, vol. 1010, ed. by M. Piattini, P. Rupino da Cunha, I. García Rodríguez de Guzmán, R. Pérez-Castillo (Springer, Cham, 2019). https://doi.org/10.1007/978-3-030-29238-6_8

V. Paramarta, M. Jihad, A. Dharma, I.C. Hapsari, P.I. Sandhyaduhita, A.N. Hidayanto, Impact of user awareness, trust, and privacy concerns on sharing personal information on social media: Facebook, Twitter, and Instagram, in *2018 International Conference on Advanced Computer Science and Information Systems (ICACSIS)*, (2018), pp. 271–276. https://doi.org/10.1109/ICACSIS.2018.8618220

S. Paul, J.I. Joy, S. Sarker, A.-A.-H. Shakib, S. Ahmed, A.K. Das, Fake news detection in social media using blockchain, in *2019 7th International Conference on Smart Computing & Communications (ICSCC)* (2019), pp. 1–5. https://doi.org/10.1109/ICSCC.2019.8843597

M. Petrescu, A.S. Krishen, The dilemma of social media algorithms and analytics. J. Market. Anal. **8**, 187–188 (2020). https://doi.org/10.1057/s41270-020-00094-4

M.T. Pinheiro, M.B. Tiago, J.M.C. Veríssimo, Digital marketing and social media: why bother? Bus. Horiz. **57**(6), 703–708 (2014). ISSN: 0007-6813. https://doi.org/10.1016/j.bushor.2014.07.002

J. Potts, E. Rennie, Web3 and the creative industries: how blockchains are reshaping business models, in *A Research Agenda for Creative Industries* (Edward Elgar Publishing, 2019), pp. 93–111

D.J. Power, G. Phillips-Wren, Impact of social media and web 2.0 on decision-making. J. Decis. Syst. **20**(3), 249–261 (2011). https://doi.org/10.3166/jds.20.249-261

K. Vasan, M. Janosov, A.-L. Barabási, Quantifying NFT-driven networks in crypto art. Sci. Rep. **12**(1), 1–11 (2022)

Q. Wang et al., Non-fungible token (NFT): overview, evaluation, opportunities and challenges. arXiv Preprint (2021). arXiv:2105.07447

Q. Zheng, Y. Li, P. Chen, X. Dong, An innovative IPFS-based storage model for blockchain, in *2018 IEEE/WIC/ACM International Conference on Web Intelligence (WI)* (2018), pp. 704–708. https://doi.org/10.1109/WI.2018.000-8

A. Zubiaga, G. Wong Sak Hoi, M. Liakata, R. Procter, P. Tolmie, Analysing how people orient to and spread rumours in social media by looking at conversational threads. PLoS ONE **11**, n. pag. (2016)

High-Speed Low Area 2D FIR Filter Using Vedic Multiplier

Grande Nagajyothi, G. Pavan Kumar, Budati Suresh Kumar, B. P. Deepak Kumar, and A. K. Damodaram

Abstract In this paper, a novel two-dimensional (2D) finite impulse response (FIR) filter is proposed using Vedic multiplier architecture. Several multipliers, like Vedic, array, Booth, and Wallace tree, are employed in the construction of filters to reduce filter area and power consumption. The VM lowers the partial products (PP) in multiplication among various multipliers. As a result, multiplication happens more quickly. The VM uses a sutra known as "Urdhva Tiryakbhyam," which is based on ancient mathematics. Two techniques to maximize speed, area, and power are suggested in this research. The first technique makes use of a VM predictor block and the second method is based on reusable block for the VM. The proposed novelty design is coded by Verilog HDL and synthesized using Xilinx 14.7 tool. The proposed design results are compared with existing designs. The number of slices required for 2D FIR using VM is less when compared with FIR with normal multiplier.

Keywords Digital filter · Half adder · Vedic multiplier · Signal processing · Full adder first section

---

G. Nagajyothi
ECE, Golden Valley Integrated Campus, Madanapalli, India
e-mail: nagajyothisai221@gmail.com

G. P. Kumar (✉) · B. P. D. Kumar
Department of CSE, CMR Technical Campus, Hyderabad, India
e-mail: pavangurram.reddy@gmail.com

B. P. D. Kumar
e-mail: bhattudheepak@gmail.com

B. S. Kumar
Department of ECE, Chirala Engineering College, Chirala, India
e-mail: balaji2547@gmail.com

A. K. Damodaram
Department of Mechanical Engineering, Sree Vidyanikethan Engineering College (Autonomous), Tirupati, AP, India
e-mail: damodaronline991@gmail.com

A. B. Reddy et al. (eds.), *Proceedings of Third International Conference on Advances in Computer Engineering and Communication Systems*, Lecture Notes in Networks and Systems 612, https://doi.org/10.1007/978-981-19-9228-5_48

1 Introduction

A two-dimensional digital filter plays a paramount role in video and image processing applications such as image restore image development. 2D filter is also used for medical applications like ECG and EEG signals. To remove the Gaussian noises in images, these type of filters are more useful. FIR filters are generally used for image processing application rather than IIR filters due to their simple in architecture and are more stable. It is difficult for researchers to construct the 2D FIR filter architecture in VLSI. For the aforementioned applications, the low power, high speed, and compact area architecture are always preferred. Real-time applications require the implementation of specialized architectures based on Application Specific Integrated Circuit (ASIC) designs in nature.

2 Literature Survey

Many research scholars had done research on different types of FIR filters. Some of them concentrated on 2D FIR filters. In the recent years, 2D FIR filters play crucial role in signal processing. To create 2D FIR filters in hardware, three fundamental elements are needed: adder, multiplier, and delay elements. The multiplier, which is one of these three components, has the greatest effect on area, speed, and power (Saritha et al. 2020; Khan et al. 2017; Garg and Joshi 2018). The multiplier's speed affects the FIR filter processor's execution time and speed (Chavan et al. 2016). The complexity of the filter is less by using canonical signed-digit (CSD) of the filter coefficients (Murugesh et al. 2020). The area and energy are also impacted by the number representation format. Energy and area are reduced by double LSB multipliers (Chandra and Chattopadhyay 2016). Monolithic multipliers employ Booth's algorithm. It accelerates the process of multiplication (Aggarwal et al. 2016). The intricacy of the circuit required to produce a PP bit is a drawback of the Booth Multiplier. Digital multipliers' dominant critical routes can place speed constraints on the overall design. Although a high radix Booth multiplier is unsuitable for designs that priorities speed, it is still a desirable option for low power applications.

3 2D FIR Filter Formulae

The block processing idea improves the architecture's throughput in digital filters. The filter provides "L" outputs in a single iteration if the input block size is L, increasing throughput by L times. Equation 1 contains the fundamental equation for the 2D FIR filter.

$$H(z_1, z_2) = \sum_{j=0}^{N-1} \sum_{i=0}^{N-1} h_{ji} z_1^{-j} z_2^{-i} \tag{1}$$

where $[h_{ji}]$ is a 2D FIR filter's coefficient impulse response matrix for order N. Kth output block of 2D FIR filter is:

$$y_{\mathrm{m,k}} = \sum_{i=0}^{N-1} Q_{i,\mathrm{k}} \tag{2}$$

The output of the complete filter y (m, k) is expressed as follows:

$$y_{\mathrm{m,k}} = [y(m, kL) \ldots\ldots\ldots\ldots y(m, kL - L + 1)]^T \tag{3}$$

$$Q_{i,\mathrm{k}} = [Q(i, kL) Q(i, kL - 1) Qv(i, kL - L + 1). \tag{4}$$

where $Q_{i,\mathrm{k}}$ = intermediate vector.

Here, $Q_{i,\mathrm{k}}$ is product of the impulse response with the input matrix A_{k}^{m-i} and is given by Eq. 5,

$$Q_{i,\mathrm{k}} = A_{\mathrm{k}}^{m-i} \cdot h_i \tag{5}$$

where $A_{\mathrm{k}}{}^{(m-i)}$ is a component of the input matrix starting from the ($m-i$)th row of the image matrix.

$$A_{\mathrm{k}}^{m-i} = \begin{bmatrix} x(m-i, k) & \ldots & x(m-i, kL - N + 1) \\ x(m-i, kL - 1) & \ldots & x(m-i, kL - N) \\ \vdots & \vdots & \vdots \\ x(m-i, kL - L + 1) & \ldots & x(m-i, kL - N - L + 2) \end{bmatrix} \tag{6}$$

And $Q(i, kL-1) = S_{\mathrm{k}i}{}^{m-i}, hi$.

Figure 1 shows the basic diagram of direct form based FIR filter. In this adder, multiplier and delay are basic elements needed to design the circuit. Multipliers occupy more area and power consumption. To get rid of these, we go for VMs to be used in FIR filter to reduce the area and power consumption and increase the speed.

4 Vedic Multiplier Using Urdhva Tiryakbhyam Sutra

The Sanskrit word Veda, which means to know without bounds, is derived from the word Vid. This branch of Vedic mathematics was developed by Sri Bharati Krishna

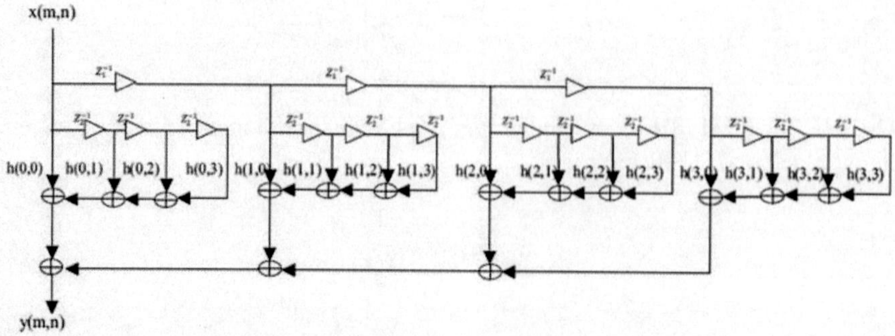

Fig. 1 2D direct form structure

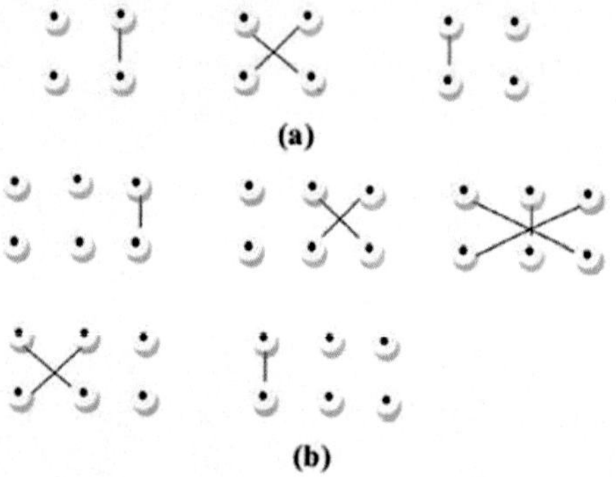

Fig. 2 VM of 2 digit and 3 digit

Tirthaji. Vedic the 16 sutras and the 13 sub-sutras form the foundation of mathematics. The 16 sutras are the source of these 13 sub-sutras. Sutras are mathematical formulas that make problem-solving simple. In Vedic mathematics, calculations are performed mentally to provide solutions more quickly. All branches of mathematics are affected by these sutras. There are 16 sutras employed, one of which is the Urdhava Tiryakbhyam (UT). Figures 2 and 3 show the VM of Urdhava Triyakbbhyam using 2 digits, 3 digits, and 4 digits. Figure 4 explains an example of the UT.

5 Proposed VM Using Predictor and Reusable

Figure 4 gives basic the VM architecture and Fig. 5 gives block diagram of VM using predictor. VM with predictor speeds up and uses less energy, but it expands the surface area. Without requiring addition, the multiplier and multiplicand are analyzed, and the outputs are anticipated using the bit values in the VM with predictor.

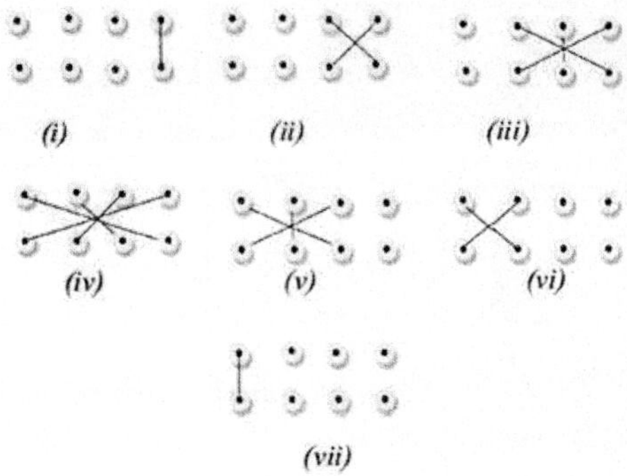

Fig. 3 4 digit UT multiplier

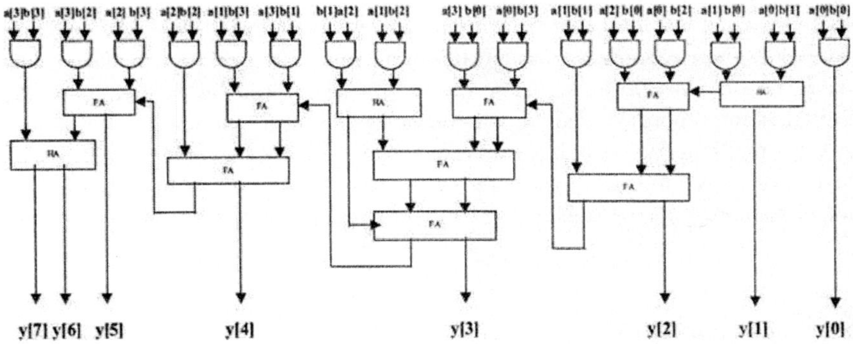

Fig. 4 VM using basic gates

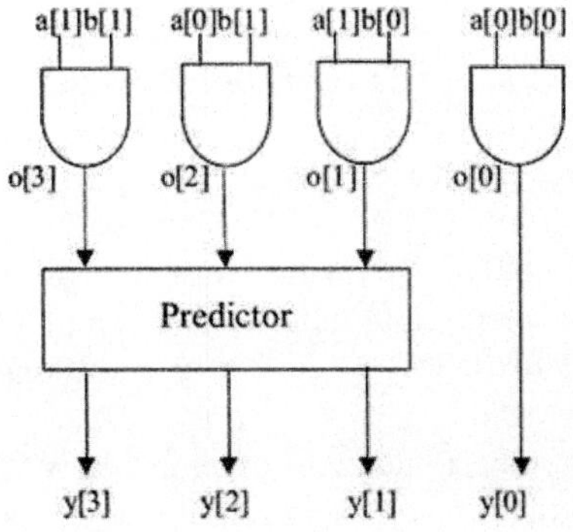

Fig. 5 VM using predictor

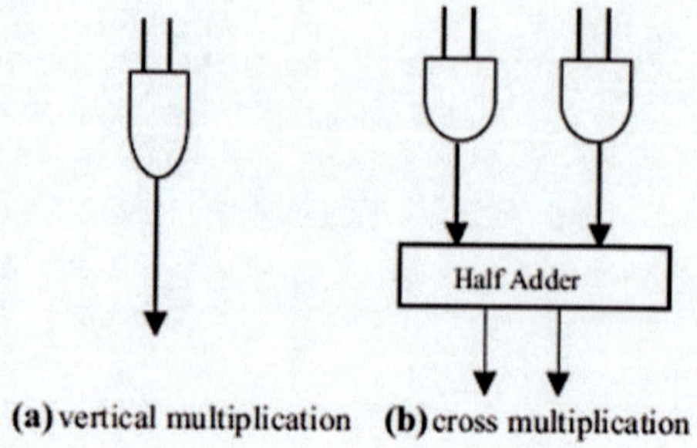

Fig. 6 VM using reusable gates

6 VM Using Reusable Logic Gate

As the bit length rises, the area in the FPGA implementation of the VM with predictor marginally increases. To get around this problem, we suggest a VM design that makes use of reusable logic. Three procedures are repeatedly carried out in the multiplication process in Vedic math. Vertical multiplication comes first, followed by cross multiplication and addition, respectively. Any bit length two numbers can be multiplied by employing these three procedures repeatedly. The fundamental VM diagram employing reusable logic gates is shown in Fig. 6.

7 Result Analysis

The proposed 2D FIR filter is implemented in Verilog HDL and synthesized in Xilinx 14.7. Table. 1 compares the area and delay of each multiplier used in the construction of 2D FIR filters. The table makes it evident that the proposed VMs using predictors and reusable logic take up less space and operate more quickly. These multipliers are used to design and implement a 2D FIR filter using Xilinx FPGA. The MATLAB FDA toolbox is used to calculate the 2D FIR filter coefficients. The comparison of the VM, VM with predictor, and VM with Reusable logic, Wallace tree multiplier, and Booth multiplier, and Array multiplier is as given in Table 2. Reusable logic-based VM-based FIR filters use less space, power, and latency than other multiplier implementations.

Figure 7 shows the synthesis diagram of proposed VM using RL and Fig. 8 shows the synthesis diagram of VM using reusable logic gate. Figure 9 shows the

Table 1 Multiplier comparison

	VM by predictor	VM by RL	Wallace	VM	Booth
No. of LUTs	14	123	115	125	222
Delay	0.945	2633	8.9	5.499	9.71

Table 2 2D FIR filters with different multiplier

	2D FIR using VM with reusable	2D FIR filter using VM with predictor	VM with Wallace in 2D FIR	2D FIR using VM	2D FIR using VM
No. of slices	31	122	321	398	456
Delay (ns)	0.933	1.0022	5.43	6.23	1.84
Power (mw)	25.44	29.78	32.5	47.99	53.28
No. of LUTs	222	322	456	565	760
No. of FF	213	332	365	456	345

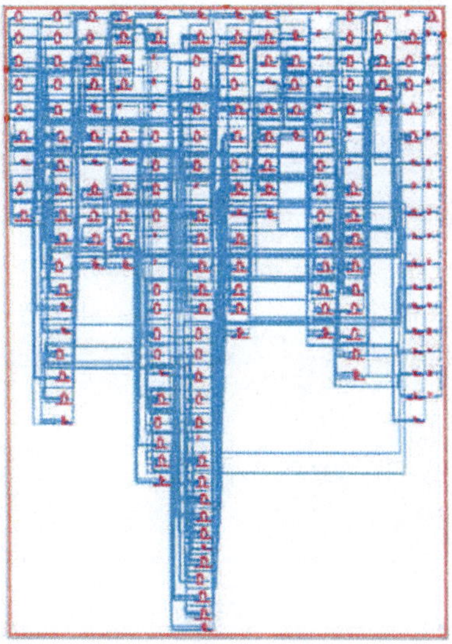

Fig. 7 VM using predictor

comparisons diagram of no. of LUTs. Figure 10 shows the throughput rate of existing and proposed work. Figure 11 shows the power consumption report.

8 Conclusion

When compared to the current design, the proposed innovative 2D FIR filter using VM uses less space and power. A 32-bit VM is built utilizing the Verilog HDL language and the Urdhva Tiryagbhyam Sutra of classical Vedic mathematics. This multiplier is then emulated using Xilinx ISE 14.7. In comparison with using the standard multiplier, the proposed multiplier uses less LUT. The proposed architecture is also particularly beneficial for applications involving the processing of medical images.

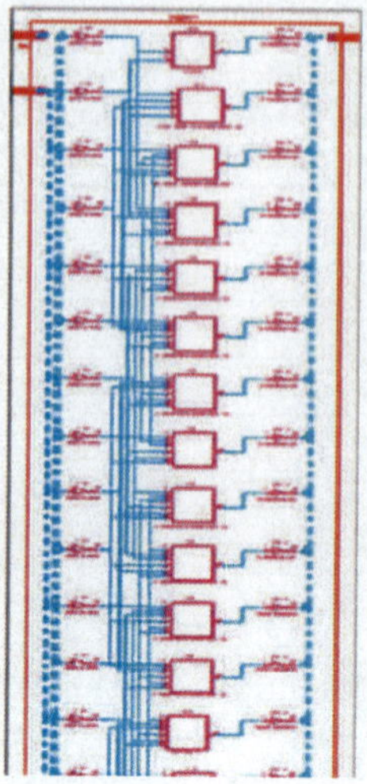

Fig. 8 VM with reusable gate

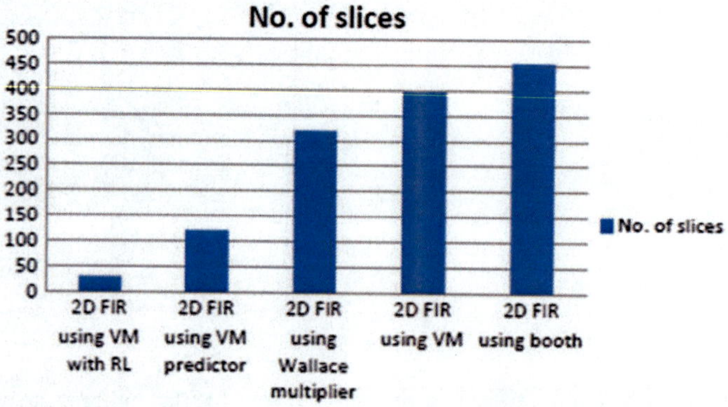

Fig. 9 No. of slices

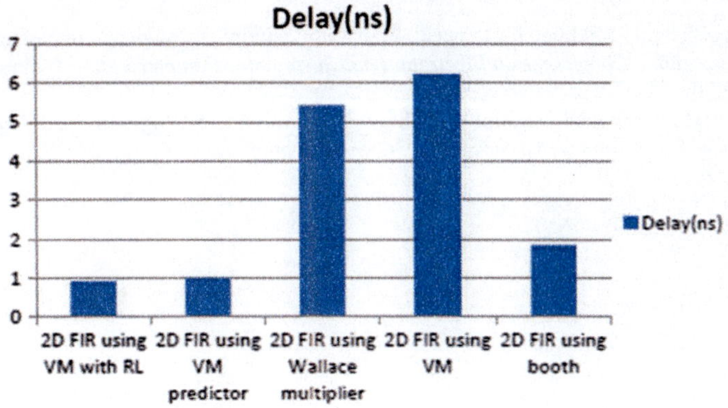

Fig. 10 Delay (ns)

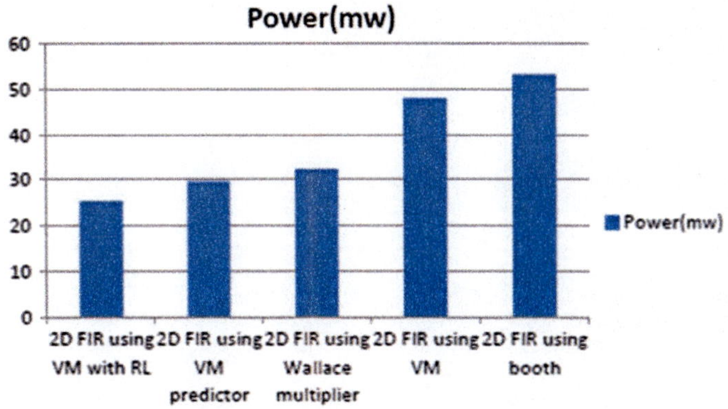

Fig. 11 Power consumption (mw)

References

A. Aggarwal et al., Optimal design of 2D FIR filters with quadrantally symmetric properties using fractional derivative constraints. Circ. Syst. Signal Process. **35**(6), 2213–2257 (2016)

A. Chandra, S. Chattopadhyay, Design of hardware efficient FIR filter: a review of the state-of-the-art approaches. Eng. Sci. Technol. Int. J. **19**(1), 212–226 (2016)

A.P. Chavan, R. Verma, N.S. Bhat, High speed 32-bit Vedic multiplier for DSP applications. Int. J. Comput. Appl. **135**(7), 35–38 (2016). Published by Foundation of Computer Science (FCS)

A. Garg, G. Joshi, Gate diffusion input based 4-bit Vedic multiplier design. IET Circ. Dev. Syst. **12**(6), 764–770 (2018)

P.A.I. Khan, S.K. Dilshad, B.K. Sree, Design of 2 × 2 Vedic multiplier using GDI technique, in *2017 International Conference on Energy, Communication, Data Analytics and Soft Computing (ICECDS)* (IEEE, 2017)

M.B. Murugesh et al., Modified high speed 32-bit Vedic multiplier design and implementation, in *2020 International Conference on Electronics and Sustainable Communication Systems (ICESC)* (IEEE, 2020)

P. Saritha et al., 4-Bit Vedic multiplier with 18 nm FinFET technology, in *2020 International Conference on Electronics and Sustainable Communication Systems (ICESC)* (IEEE, 2020)

Classification of Cotton Weeds in the Cotton Farms Using Deep Transfer Learning Techniques

Ganesan Hari Krishnan and Thiyagarajan Rajasenbagam

Abstract In current world, the automated weed control systems must important to identifying and locating the weeds from the main plant. The recognition and removal of weeds in the preliminary time is essential for increasing the productivity of the main plant. In recent years, deep learning algorithms give the better performance for the various complex tasks like image processing, audio and video processing, etc. This research paper presents the idea of classifying the various weeds from the main plant using various pre-trained models. The cotton plant was chosen as the main plant with some of its most common weeds was chosen for this work. But one of the major problems in real-world scenario is the soil, dust, etc., present in between of main plant and the weed. So, this paper proposed a methodology to segment the weeds and main plant images from others at very first stage of process. Then, the segmented images fed into the pre-trained CNN models like VGG16, ResNet50, InceptionV3, MobileNet, and InceptionResNetV2 for the classification task.

Keywords Image segmentation · Weed classification · Machine learning · Deep learning

1 Introduction

Weeds are a major source of concern in agriculture, as they reduce the production of the main crop. It uses up the main plant's nutrition, vitamins, water, sunlight, and space. It causes the crop's performance to deteriorate. Due to weed interference, yield losses in grain sorghum jumped from 37 to 61% in the United States of America, according to the findings (Ahmad et al. 2021).

G. Hari Krishnan (✉) · T. Rajasenbagam
Government College of Technology, Coimbatore 641013, India
e-mail: hari.62364@gct.ac.in

T. Rajasenbagam
e-mail: trajasenbagam@gct.ac.in

A. B. Reddy et al. (eds.), *Proceedings of Third International Conference on Advances in Computer Engineering and Communication Systems*, Lecture Notes in Networks and Systems 612, https://doi.org/10.1007/978-981-19-9228-5_49

According to field research data in Canada and the United States of America (Bakhshipour 2021), the average yield loss due to weeds increased from 31 to 94% in dry bean fields and from 61 to 83% in sugar beet farms. As a result, weeds must be removed or treated before the growth season begins. Weeds cause a 30% loss in agricultural output globally, and it is estimated that maize and soybean harvests in North America might lose $25 million each year.

Individuals pulling weeds or manually giving pesticides to weeds are the traditional technique. It is labour intensive and time demanding (Gao et al. 2018). As a result, weed identification and detection of weed sites must be quick, cost effective, and accurate in order to remove and control them (Kunz et al. 2018).

Computer vision was the first method used for object identification and detection. These systems process weed images and extract features using digital image processing techniques (Lammie et al. 2019). Various machine learning methods, including as Naive Bayes, k nearest neighbours, support vector machines (Ahmed et al. 2012), decision tree and others, are utilised to recognise the weed images.

Machine learning methods have the disadvantage of taking a long time to extract information and having low model accuracy (Osorio et al. 2020).

Deep learning then generates a superior model when compared to machine learning models. When it comes to image recognition, convolution neural networks (CNNs) beats all other machine learning methods (Smith et al. 2019). The CNN has the advantage of extracting patterns from RGB pictures automatically. However, it has several disadvantages, such as the need for a larger number of images, manual hyper-parameter tuning and huge memory (primarily GPU) (Veeranampalayam Sivakumar et al. 2020; Lammie et al. 2019).

Transfer learning is a new approach for image classification problems that involves transferring the functions of pre-trained CNN models that are all very good at one particular task (i.e. ImageNet challenge). Transfer learning approaches (Espejo-Garcia et al. 2020) provide a number of advantages. Few of them need a limited amount of input, train faster and offer fewer hyper-parameter adjusting choices, all of which improve model performance (Li et al. 2013; Olsen et al. 2019).

The segmentation algorithm for segmenting weed and cotton leaf images is proposed in this research. When the segmented weed and cotton images are given as input, the performance of the various pre-trained CNN models is discussed.

The organization of the paper divided into several parts. Section 2 discusses the existing research on weed classification and detection. Section 3 employs the data set used in this research, explaining the different pre-trained CNN models and its evaluation measures. Section 4 tells the overall results obtained from the algorithms and research findings. Finally, Sect. 5 discusses about the conclusion of this research.

2 Related Work

Previous research on weed classification and detection has been presented and analysed in this area. The type of primary plant considered, the weeds growing around that main plant, the categorisation and detection methods used, and the performance of each methodology are all included.

For weed classification in chilli (*Capsicum frutescens* L.) fields, Faisal Ahmed and colleagues (Ahmed et al. 2012) presented a machine learning classifier. They snapped the weeds with a camera that had a resolution of 1200 × 768 pixels. Colour features (red, green, and blue), size independent shape features (form factor, elongatedness, convexity, and solidity), and moment invariant features are all extracted from the captured images. Forward-selection and backward-selection methods were used to choose the best features for better classification. The extracted features and labels are then fed into an RBF kernel support vector machines (SVMs) classifier (Radial Basis Function). On a total of 224 test samples, the proposed method has a 97% accuracy.

Multiple classifiers for the classification of weeds present in cotton and tomato fields at the early growth stage were presented by Espejo-Garcia et al. (2020). Digital cameras were used to capture the weed images, which were then augmented with random rotation, blurring, and shifting. Only the feature vectors are extracted from the augmented images using pre-trained CNN models such as DenseNet, VGG16, and VGG19. For the classification task, the extracted feature vectors were fed into the SVM, logistic regression, and random forest algorithms. The F1 score was used to assess the performance of the suggested approach. Based on the results, DenseNet + SVM is the top classifier with a 99.29% $F1$ score.

Soft sets were introduced by Li et al. (2013) for classifying weeds in maize fields. A camera with a resolution of 1920 × 1040 pixels was used to take the weed photographs. The acquired images will then be segregated from the soil using image fusion and the NIR-Red technique, followed by a distance transform-based watershed segmentation algorithm. The segmented images will be used to extract shape features (rectangularity, elongation, and so on), texture features (GLCM), and fractal dimensions. Based on the evaluated results, the classification system developed Soft sets for the data set. Compared to other machine learning algorithms, such as SVM and Nave Bayes, the soft sets are outperformed well.

Olsen et al. (2019) created and contributed a multiclass weed species data set for a variety of weed species found in Australia. They gathered the images and made them available to the public as a data set. There are almost 17,509 images in total. They began by designing their own CNN architecture. The data set's performance was then evaluated using a variety of popular deep learning architectures such as Inception-V3 and ResNet-50. The data set had a classification accuracy of 95% on average based on the results.

Shaun M. Sharpe et al. made the suggestion that goosegrass could be found in tomato and strawberry fields (Sharpe et al. 2020a). For the object detection task, they used the Tiny-YOLO-v3 detection algorithm, and the $F1$ Score will be used to assess the model's performance. The suggested method produced results of 0.75 and 0.56 for strawberry and tomato fields, respectively.

The object detection method was developed by Sharpe et al. (2020b) to recognise weeds in Florida's food crops. Both indiscriminate (one class) and discriminate (three classes) vegetation, which are frequently found in Florida, were used in this study. The suggested methodology makes use of the YOLO-v3 object detection model. It proves that the model is efficient at spotting vegetation.

Several pre-trained models were employed by Ahmad et al. (2021) for the classification and detection of weeds in corn and soybean plants. They made use of the DeepWeeds (Olsen et al. 2019) open source data set for this study. For the classification task, they used pre-trained CNN models like VGG16, ResNet50, and InceptionV3. For the object detection task, the YOLO-v3 algorithm was used. The results show that the VGG16 model outperforms other models, and the YOLO-v3 model yields 54% of mAP. The models in this study were trained using both the PyTorch and Keras libraries, which is one of the most crucial things to remember.

3 Materials and Methods

The proposed work contains three modules, and overall flow of the proposed system has been shown in Fig. 1.

The following steps summarise the each modules in the proposed work.

(1) A weed leaf images collected from the real-time fields or farms.
(2) Segment the weed leaf images from the soil and other particles.
(3) Segmented images were passed into the following pre-trained CNN architectures for classification: VGG-16, ResNet50, InceptionV3, MobileNet, and InceptionResNetV2.

3.1 Weed Leaf Image Acquisition

For this research, CottonWeedID15 data set has been used. It is open source data set downloaded from https://www.kaggle.com/yuzhenlu/cottonweedid15 website (Olsen et al. 2019).

CottonWeedID15 is a data set that contains 5187 RGB photographs of 15 weeds that seem to be widespread in cotton fields in the southern United States of America.

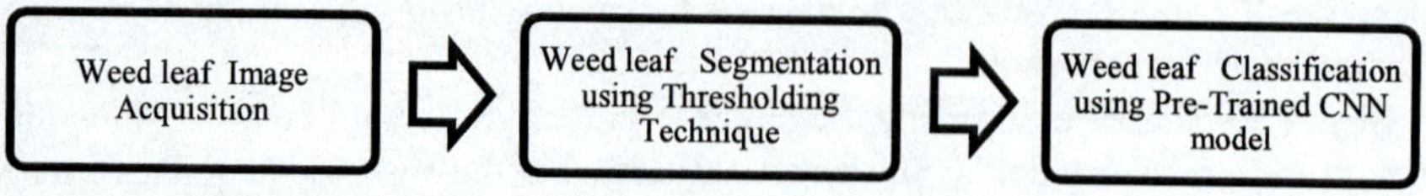

Fig. 1 Overall flow of the proposed work

Table 1 Distribution of the weed and main plant images in modified data set

S. No.	Weed name	No. of images
1	Cotton	166
2	Carpetweeds	182
3	Eclipta	123
4	Morning glory	272
5	Ragweed	92
	Total no of images	835

These pictures were taken with cell phones or handheld digital cameras. For the sake of image diversity, images were captured from different view angles, under natural field light conditions, at varying stages of weed growth in 2020 and 2021, at different locations across the U.S. cotton belt states. The images were manually labelled by weed scientists and trained individuals. Here, this research selectively focuses on four weeds of the cotton, namely Carpetweeds, Eclipta, Morning glory and Ragweed.

But one of the disadvantages of the data set is absence of the main plant (cotton) images. So for this research the cotton leaf images were captured from real-world cotton farms across the rural area at Trichy, Tamil Nadu. The modified data set formed by adding the captured cotton images with existing weed images. The distribution of the modified data set shown in Table 1.

3.2 *Weed Leaf Segmentation*

Image segmentation is the process of partitioning a digital image into multiple segments (sets of pixels, also known as image objects) or in simple terms image segmentation defined as extracting the region of interest of particular image. In this work, image segmentation applied on plant and weed images to remove the soil, dust portion of leaf image.

There are various techniques used for image segmentation process. Here, thresholding-based segmentation was applied on images (Bakhshipour 2021). The flow chart for the image segmentation process is shown in Fig. 2.

3.3 *Weed Leaf Classification*

Once segmentation of the weed images has been done, the resulting images were saved and passed into the pre-trained CNN models. In the proposed system, there are 5 pre-trained models used for the classification.

There are VGG-16, ResNet50, InceptionV3, MobileNet and InceptionResNetV2. All of the model accepts the input image as shape of 224 × 224 × 3, where first two values denote height and width of the image and third value denotes number of

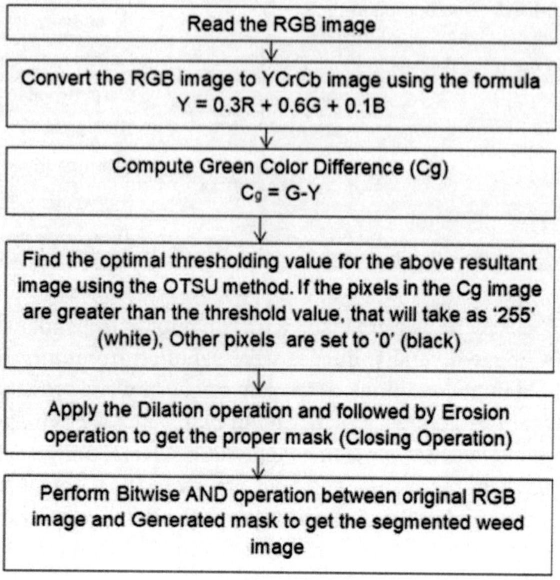

Fig. 2 Flow chart of proposed segmentation algorithm

channels (e.g. for RGB images, there are 3 channels). So all segmented images are resized into the size of 224 × 224 × 3 before fed into the models.

The segmented data set was then divided into two parts, called train and test data sets, using a 70–30 ratio. As a result, the training data set has 584 images and the testing data set has 251 images. The training data set has been used to train the models, while the testing data set was given to assess model performance.

4 Results and Discussions

Figure 3 depicts the outcomes of the segmentation algorithm. The collected results clearly demonstrate that the algorithm performs effectively in distinguishing the leaf area from others.

Then, these segmented images divided into training and testing data sets. The training images are given as input to the pre-trained models for the training process and the test images used for evaluating the trained models. One of the performance measures "accuracy" is utilised to evaluate the models in this study.

The classification accuracy is determined using the model's predicted labels and the actual labels. The models were created with the Keras neural networks library written in Python, and they were trained using Google Colaboratory's online coding platform.

The outcomes of each model are provided in Table 2. These findings clearly suggest that the VGG16 and Inception-ResNet-V2 models outperform other models in terms of accuracy.

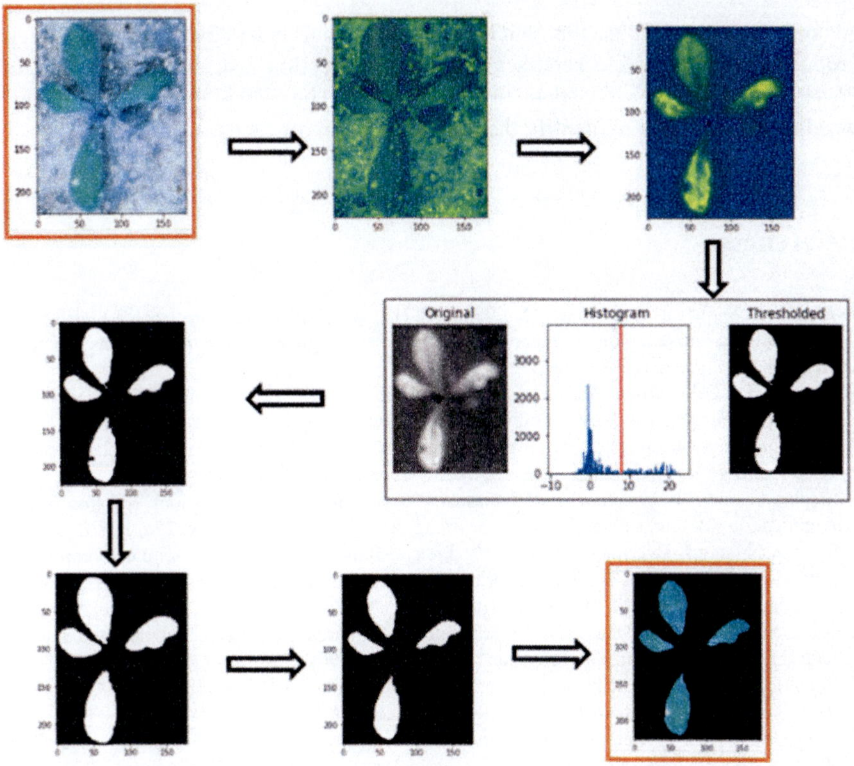

Fig. 3 Output of the proposed segmentation algorithm

Table 2 Experimental results of the proposed pre-trained models

Proposed system results		Performance measure		
		Training accuracy (%)	Testing accuracy (%)	Model size
Models	VGG16	91.44	**96.41**	356 MB
	ResNet50	70.72	81.67	680 MB
	InceptionV3	90.07	92.43	97 MB
	MobileNet	95.89	95.22	308 MB
	Inception-ResNet-V2	96.23	**96.41**	435 MB

5 Conclusion

Overall, this study proposed a new modified data set for the classification and detection of cotton and its weeds. OTSU thresholding and digital image processing processes are used to segment the collected pictures. Finally, several pre-trained models were trained using the segmented images and tested against some test images.

According to the results, the VGG16 and InceptionResnetV2 pre-trained models produced the best testing accuracy for the classification task, with an average of 96.41%. Any suitable object detection algorithm will be used in future for automated weed control systems to identify the precise location of weeds.

References

A. Ahmad, D. Saraswat, V. Aggarwal, A. Etienne, B. Hancock, Performance of deep learning models for classifying and detecting common weeds in corn and soybean production systems. Comput. Electron. Agric. **184**, 106081 (2021)

F. Ahmed, H.A. Al-Mamun, A.H. Bari, E. Hossain, P. Kwan, Classification of crops and weeds from digital images: a support vector machine approach. Crop Prot. **40**, 98–104 (2012)

A. Bakhshipour, Cascading feature filtering and boosting algorithm for plant type classification based on image features. IEEE Access **9**, 82021–82030 (2021)

B. Espejo-Garcia, N. Mylonas, L. Athanasakos, S. Fountas, I. Vasilakoglou, Towards weeds identification assistance through transfer learning. Comput. Electron. Agric. **171**, 105306 (2020)

J. Gao, D. Nuyttens, P. Lootens, Y. He, J.G. Pieters, Recognising weeds in a maize crop using a random forest machine learning algorithm and near-infrared snapshot mosaic hyperspectral imagery. Biosys. Eng. **170**, 39–50 (2018)

C. Kunz, J.F. Weber, G.G. Peteinatos, M. Sökefeld, R. Gerhards, Camera steered mechanical weed control in sugar beet, maize and soybean. Precis. Agric. **19**(4), 708–720 (2018)

C. Lammie, A. Olsen, T. Carrick, M.R. Azghadi, Low-power and high-speed deep FPGA inference engines for weed classification at the edge. IEEE Access **7**, 51171–51184 (2019)

P. Li, D. He, Y. Qiao, C. Yang, An application of soft sets in weed identification, 2013 Kansas City, Missouri, July 21–24 (American Society of Agricultural and Biological Engineers, 2013), p. 1

A. Olsen, D.A. Konovalov, B. Philippa, P. Ridd, J.C. Wood, J. Johns, W. Banks, B. Girgenti, O. Kenny, J. Whinney, B. Calvert, DeepWeeds: a multiclass weed species image dataset for deep learning. Sci. Rep. **9**(1), 1–12 (2019)

K. Osorio, A. Puerto, C. Pedraza, D. Jamaica, L. Rodríguez, A deep learning approach for weed detection in lettuce crops using multispectral images. AgriEngineering **2**(3), 471–488 (2020)

S.M. Sharpe, A.W. Schumann, N.S. Boyd, Goosegrass detection in strawberry and tomato using a convolutional neural network. Sci. Rep. **10**(1), 1–8 (2020a)

S.M. Sharpe, A.W. Schumann, J. Yu, N.S. Boyd, Vegetation detection and discrimination within vegetable plasticulture row-middles using a convolutional neural network. Precis. Agric. **21**(2), 264–277 (2020b)

L.N. Smith, A. Byrne, M.F. Hansen, W. Zhang, M.L. Smith, Weed classification in grasslands using convolutional neural networks, in *Applications of Machine Learning*, vol. 11139 (International Society for Optics and Photonics, 2019), p. 1113919

A.N. Veeranampalayam Sivakumar, J. Li, S. Scott, E. Psota, J.A. Jhala, J.D. Luck, Y. Shi, Comparison of object detection and patch-based classification deep learning models on mid to late-season weed detection in UAV imagery. Remote Sens. **12**(13), 2136 (2020)

Beam Pattern Configuration Control Unit for a 24 × 24 Reconfigurable Array of Photo Detector-Based Pixels

K. B. Sowmya, Darshan Hegde, and Pratik Desai

Abstract A versatile control unit to read 24 × 24 array of photodetector-based pixels is presented. The purpose is finding the center of the non-diffractive beam incident on the pixel. A 24 × 24 array of photodetector-based pixels is read out based on a system consisting of decoder and a multiplexer. A decoder is used to activate a particular row of pixels. All pixels of a single row are read, values are fed into multiplexer, and a single pixel value is read out. Input to decoder and select lines of multiplexer are from a sequence generator. The design of sequence generator is based on different beam patterns as required by the user. A state machine to read out user defined pattern, in a 24 × 24 array of photodetector-based pixels, is implemented using Verilog HDL. The implemented design demonstrates read out of set of pixel values for two different beam patterns.

Keywords Non-diffractive beam · Photodetector · System-on-chip

1 Introduction

Image processing applications are used in many different applications in real world today. Due to efficiency of the image processing applications developed based on FPGAs are widely used. They are faster in computation on large amount of data. But the image processing system developed is particular to a specific application. If the specific application is no longer used for any purpose, image processing system developed becomes obsolete.

K. B. Sowmya (✉) · D. Hegde · P. Desai
Department of ECE, RV College of Engineering, Bengaluru, India
e-mail: sowmyakb@rvce.edu.in

D. Hegde
e-mail: darshanhegde.lvs21@rvce.edu.in

P. Desai
e-mail: pratikbdesai.lvs21@rvce.edu.in

A. B. Reddy et al. (eds.), *Proceedings of Third International Conference on Advances in Computer Engineering and Communication Systems*, Lecture Notes in Networks and Systems 612, https://doi.org/10.1007/978-981-19-9228-5_50

The image processing system developed can be used in different applications if it can be reprogrammed as per requirements of different applications. Even though the application is no longer used, the image processing system used can be taken out and reprogrammed to use in different application. In this report, a system is proposed which detects the center of the beam incident on it, takes a user input pattern, and synthesizes it around center of the beam. Once pattern is configured, it acts as pixel-based sensor. The beam patterns can be altered at different times by user input. Re-configurability of the beam pattern makes system robust and reusable.

1.1 Literature Review

In paper (Rigoni et al. 2011), a custom CMOS active pixel sensor design for NDB optical encoders is presented. Displacement and rotation sensors based on the principle of optical function are commonly referred to as "optical encoders." These encoders capable of submicron resolution are commonly used in industrial and scientific instruments. Two different versions of the circuit having 8 and 16 annular rings photodiodes have been implemented and tested. 16 annular rings version of the chip has a chip area of 300 by 500 μm. 16 3T APS cells are connected to the analog 16 × 1 multiplexer and the output buffer. It establishes the feasibility of the design by demonstrating two different preliminary implementations.

In paper (Quintián et al. 2015), study of the progressive signal made by an optical encoder supported a non-diffractive beam (NDB) are meted out. To get NDB, a particular mechanism of diffractive optical part (DOE) is employed. The detection system consists by an application specific computer circuit (ASIC) detector. The detector consists of an array of eight concentric circular photodiodes, all given a programmable gain electronic equipment. The gain of the amplifiers is set to values, supported mathematical models for incident NDB even for 3D models. The system will synthesize non-uniform detection rates. The distinction, amplitude, and harmonic content of the curved sign is analyzed. The impact of interference between doughnut-shaped photodiodes is pronounced as a result of the distinction of the signal depends on the wavelength. Experimentally, it was verified that for different besinc^2 functions by programming different gains. It also establishes effect of different colors of light incident on sensor.

In paper (Chalimbaud and Berry 2004; Murphy et al. 2007; Birla 2006; Pearson 2009; Yan et al. 2011; Afshari et al. 2011; Fysikopoulos et al. 2012; Alqasemi et al. 2012; Werne et al. 2010; Guo and Sonkusale 2011), many different applications of image processing are presented. Image processing is vastly used in many different applications in real world. Image processing units can consist of FPGAs as their core element due to its processing capabilities. All the applications consist of a similar setup comprising of sensors and control units. The sensor inputs are read from control units and processed. If the processing computations are complex another processor can be interfaced with the sensor reading control units. The systems are designed very specific to a particular application. Color masks can be used to make systems even more specific, and power consumption is optimized based on different requirements. The systems proposed are established to be efficient in their areas of application. The control unit designs are simpler to a certain extent.

In paper (Mombello et al. 2020), a different reconfigurable array of pixel design is presented. It focusses on reusability and reconfigurability of the system. A pixel is developed with three SRAM cells and a photodetector. SRAM cells allow reconfiguring of the system. An array of 24 × 24 is implemented to realize a pixel array-based sensor. RLA lines, row shift register, and column shift register are used for control of read and write operations to SRAM. Outside the photodetector array, the RL has a gain system that allows anyone with selectable gain to connect to the output. The control unit implemented to realize a reconfigurable array. The control unit detects center of the array and synthesizes detection pattern around it. The detection pattern can be modified at any stage. Abstract idea of FSM which can be implemented using HDL is presented. It was tested by illuminating with a focused Gaussian spot. The prototype developed in this work proved to be able to find the most illuminated pixels in the entire array and perform relevant searches effectively. It has also been shown that the desired sensing pattern can be constructed around it and therefore can perform the expected function of self-alignment between the photodetector sensor and the NDB of the optical encoder.

2 Techniques and Methodology

High-level block diagram is shown in Fig. 1. The sequence of operation as per the steps of the algorithm is described below.

- System is initially in Idle state.
- Light is incident on photodetector array sensor on any pixel of the array.
- Plot graph of read line versus current in vertical direction for 12 columns as 4 pixels of 2 × 2 are provided with a single read line access to read current variations in photodetectors due to incident light.
- Similarly, plot read line versus current in horizontal direction for 12 rows.
- Find out the region where current is maximum in above two steps.
- The intersection of regions of current where it is maximum is the region where the center of the beam is located.

Fig. 1 High-level block diagram

- Since a read line indicates 4 pixels, find the pixel where current is maximum out of those 4. The pixel where current is maximum in this case is exact center of the beam.
- The user required pattern can be input to the control unit by designing a user interface. The entire setup should be able to come up with a mathematical model for variety of patterns which will be used to decide reconfiguration of pixel array.
- Around center of the beam, the pattern decided by the user has to be recognized. For this, the states of different switch control are stored in SRAM cell (West, South East and North East and Data lines).
- It is now configured in the sensor mode. The current values after light incident on sensor are selected from the cell where pattern is synthesized.
- The read values can be compared with values expected for a user defined pattern. If it matches, beam has user defined pattern around its center.
- The switch control states are stored in SRAM cells, until the user defined pattern is not changed. If user defined pattern changes, reconfiguration request will be activated and the algorithm starts from first step to reconfiguration.

Note: Row shift register is used for data line access of SRAMs for different rows, and column shift register is used to enable write/read of different SRAM for switch control and read line access.

The logic contained in the pixels is shown in Fig. 2. It can be seen that it has three MOS switches commanded by SRAM cells that allow to interconnect it with three of its neighbors: West (W), North East (NE), and South East (SE). This makes possible to configure larger independent detection zones formed by several pixels.

Figure 3 shows how the array of 2 × 2 inter-connectable pixels looks, and how the control lines of the SRAM cells are distributed. Figure 4 shows how 10 × 10 array can be implemented using 2 × 2 inter-connectable pixels. Similarly, a 24 × 24 array can be implemented.

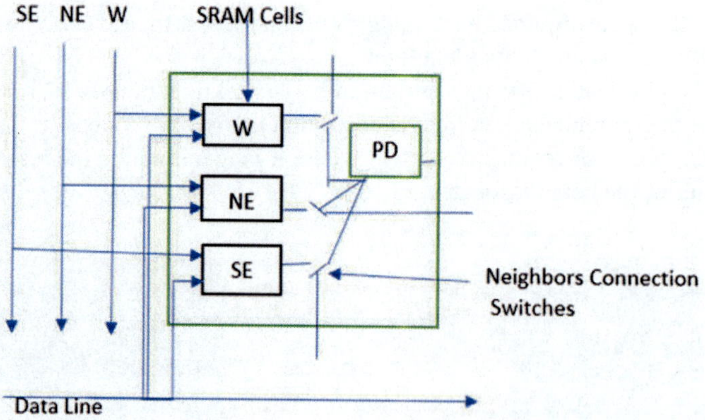

Fig. 2 SRAM-based single pixel logic

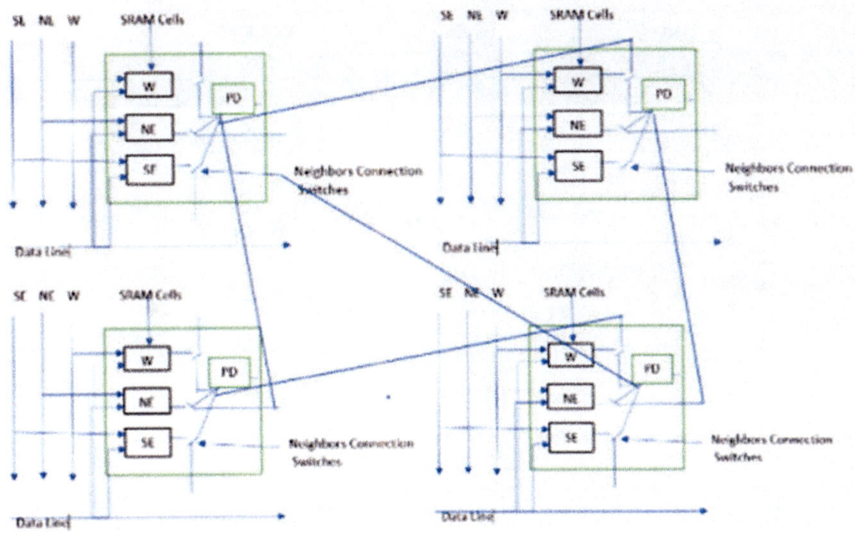

Fig. 3 2 × 2 array of inter-connectable pixel

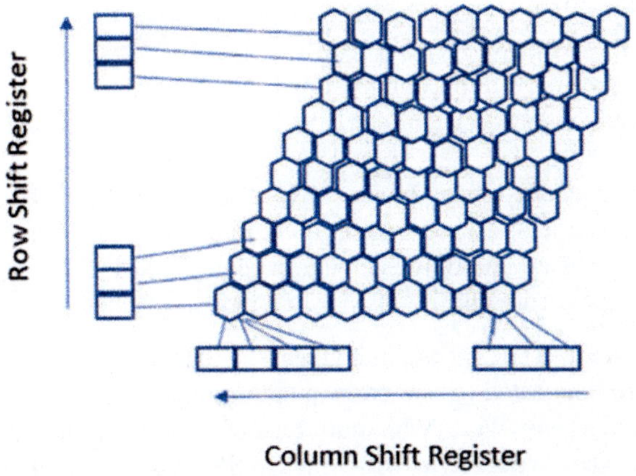

Fig. 4 10 × 10 array of pixels

The algorithm Finite State Machine of the algorithm can be implemented as per Fig. 5.

Initially, FSM is in Idle state S0 (Idle), Start = 0.

If light falls on pixel array, Start = 1, state = HS (Horizontal Search). It reads current values of each row in array of pixels and finds whichever has highest current. HS_OK = 0 till search for all rows is completed. HS_OK = 1, when Horizontal Search is completed.

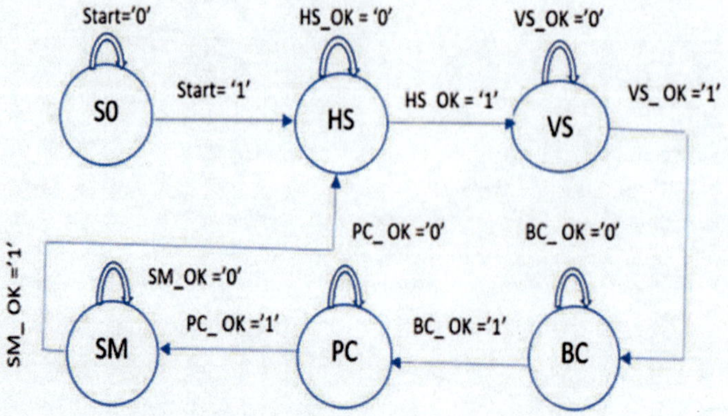

Fig. 5 FSM of pattern detection and configuration

HS_OK = 1, state = VS (Vertical Search). It follows same steps as Horizontal Search, but instead of rows, same steps are carried out for columns. VS_OK = 1 indicates completion of Vertical Search.

VS_OK = 1, state = Beam Center (BC). The intersection of HS and VS with maximum current values is located, indicating center of the beam. BC_OK = 1 after center is located.

BC_OK = 1, state = Pattern Configuration (PC). The user required pattern configuration is sent on the configuration interface. PC_OK = 1 if user required pattern is configured.

PC_OK = 1, state = Sensor Mode (SM). The required set of particular pixel values can be read as per the pattern configured. It remains in this state till there is no change in the given user pattern. If user pattern changes, SM_OK = 1, state changes to HS and the process continues.

Pattern Configuration (PC): Data line given to a SRAMs in a single row of pixels are same and are connected to one corresponding bit for one row in the row shift register. In a similar way, West (W), South East (SE), and North East (NE) of each column is same. They are given as write lines to SRAMs in column of pixels. They are connected to 3 bits in column shift register for one corresponding column, along with 1 additional bit for every alternate column for read line access. To synthesize a particular pattern, only required data lines and word lines are sent as 1.

Sensor Mode (SM): Once the required pattern is synthesized, only a set of particular pixel values has to be read. This can be done as follows. Only required rows are enabled one by one using a decoder. When a particular row is enabled, required pixels in that particular row can be read one by one, i.e., all pixels are input to a multiplexer, and select lines are varied to read out only required pixels. The control circuit for input to decoder and select lines of multiplexer is a sequence generator. The sequence to be generated is determined based on the user specific patterns. It is

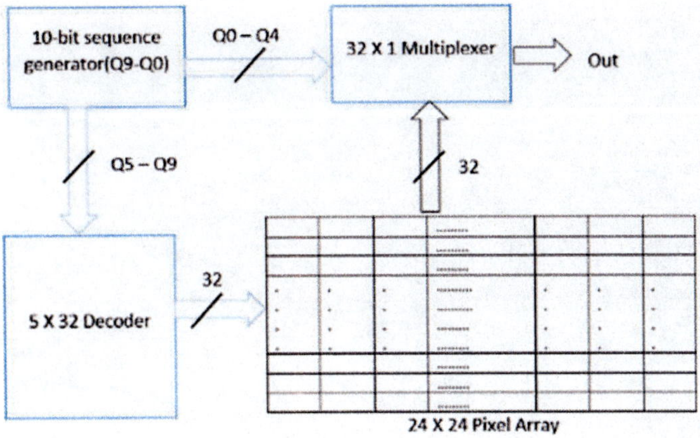

Fig. 6 Sensor mode read—circuit

demonstrated in Fig. 6. Additional multiplexer may be used to change sequences of sequence generator.

3 Result Analysis

The schematic of proposed design of 2 × 2 pixel is shown in Fig. 7.

Figure 7 shows 4 pixels connected to each other, each pixel consists of 3 6T SRAM cells, photo diode and 3 switches.

Input to the pixels is given through photo diode, and output is measured through the read line access (RLA), where the RLA is another switch.

The output waveform is plotted for various current values, and the current in each pixel is measured as shown in Figs. 8, 9 and 10.

4 Conclusion

The light incident on the pixel is input through current source. The variation of light intensities is simulated by varying current source values. Different current measured at output for variations of current sources is given in Table 1.

The proposed system, if used for a specific application, consumes high power than existing systems. But, existing systems do not provide reconfigurability of different patterns. The proposed system combines few of the existing systems, to build a single reconfigurable system, where only input patterns differ. The proposed system provides reconfigurability of patterns as required. Hence, looking at a bigger picture,

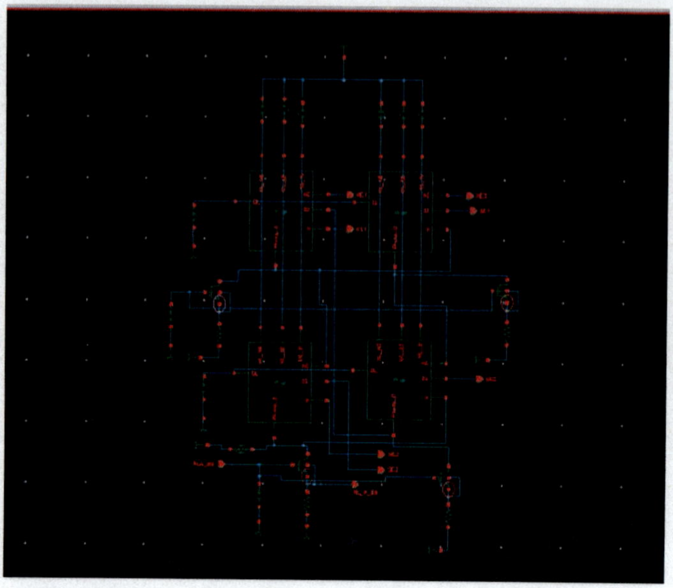

Fig. 7 Schematic of 2 × 2 array of pixels

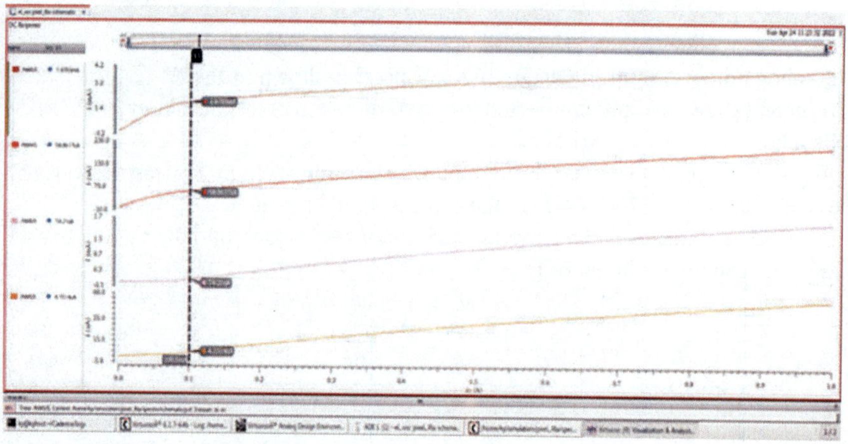

Fig. 8 Output waveforms for 100mA current

power consumption is reduced compared to having multiple systems for specific applications.

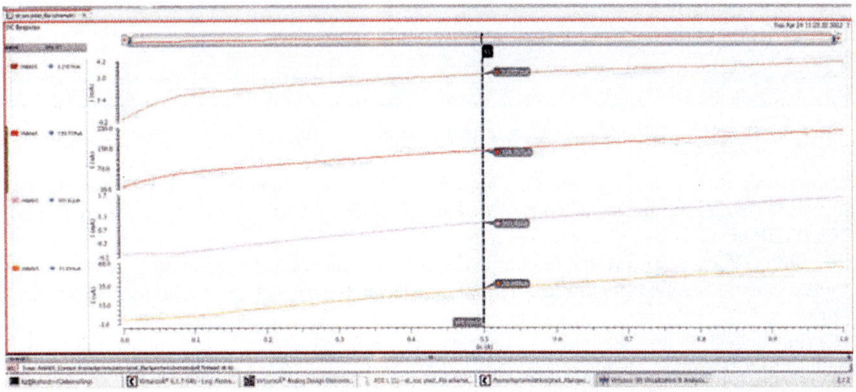

Fig. 9 Output waveforms for 500mA current

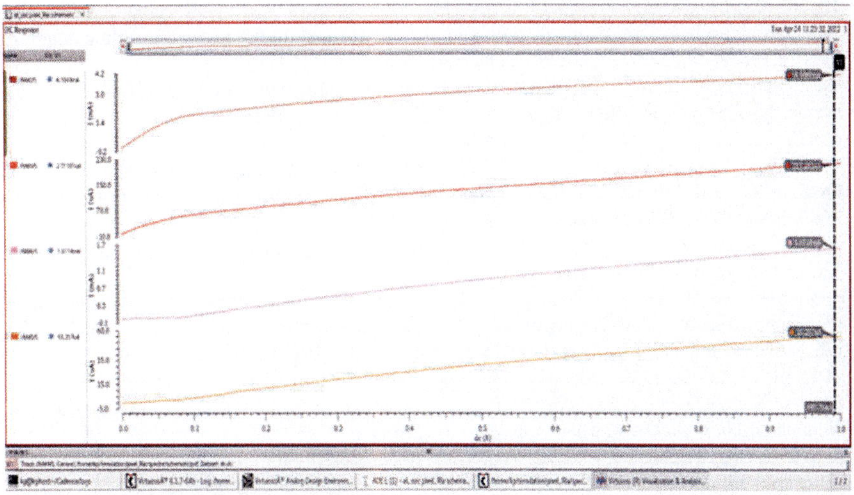

Fig. 10 Output waveforms for 990mA current

Table 1 Output current for input variations

Current value	Pixel one	Pixel two	Pixel three	Pixel four
0 A	0 A	0 A	0 A	0 A
100 mA	1.8705 mA	58.8617 μA	74.21 μA	4.1514 μA
500 mA	3.2107 mA	139.707 μA	901.82 μA	31.694 μA
990 mA	4.1049 mA	217.181 μA	1.6114 mA	55.257 μA
Existing method value	1.00 mA	400.00 μA	300.00 μA	750.00 μA

References

H. Afshari, L. Jacques, L. Bagnato, A. Schmid, P. Vandergheynst, Y. Leblebici, Hardware implementation of an omnidirectional camera with real-time 3D imaging capability, in *3DTV Conference*, Antalya (2011), pp. 1–4

U. Alqasemi, H. Li, A. Aguirreer, Q. Zhu, FPGA-based reconfigurable processor for ultrafast interlaced ultrasound and photoacoustic imaging. IEEE Trans. Ultrason. Ferroelectr. Freq. Control **59**(7) (2012)

M.K. Birla, FPGA based reconfigurable platform for complex image processing, in *2006 IEEE International Conference on Electro Information Technology*, East Lansing, MI (2006), pp. 204–209

P. Chalimbaud, F. Berry, Design of an imaging system based on FPGA technology and CMOS imager, in *IEEE International Conference on Field-Programmable Technology*, Brisbane, NSW, Australia (2004), pp. 407–411

E. Fysikopoulos, G. Loudos, M. Georgiou, S. David, G. Matsopoulos, A Spartan 6 FPGA-based data acquisition system for dedicated imagers in nuclear medicine. Meas. Sci. Technol. **23**(12), 125403–125408 (2012)

J. Guo, S. Sonkusale, A CMOS imager with digital phase readout for fluorescence lifetime imaging, in *2011 Proceedings of the ESSCIRC*, Helsinki (2011), pp. 115–118

L. Mombello, N. Calarco, F.P. Quintián, System-on-chip implementation of a self-configuration system for a programmable photodetector ASIC, in *2020 Argentine Conference on Electronics (CAE)* (2020), pp. 99–103. https://doi.org/10.1109/CAE48787.2020.9046361

C. Murphy, D. Lindquist, A.M. Rynning, T. Cecil, S. Leavitt, M.L. Chang, Low-cost stereo vision on an FPGA, in *15th Annual IEEE Symposium on Field-Programmable Custom Computing Machines (FCCM 2007)*, Napa, CA (2007), pp. 333–334

T. Pearson, Hardware-based image processing for high-speed inspection of grains. Comput. Electron. Agric. **69**, 12–18 (2009)

F.P. Quintián, N. Calarco, A. Lutenburg, J. Lipovetzky, Performance of an optical encoder based on a nondiffractive beam implemented with a specific photodetection integrated circuit and a diffractive optical element. Appl. Opt. **54**(25), 7640–7647 (2015)

N. Rigoni, R. Lugones, A. Lutenberg, J. Lipovetzky, Design of a customized CMOS active pixel sensor for a non-diffractive beam optical encoder, in *2011 Argentine School of Micro-Nanoelectronics, Technology and Applications* (2011), pp. 1–5

T.A. Werne, D.L. Bekker, P.J. Pingree, Real-time data processing for an advanced imaging system using the Xilinx Virtex-5 FPGA, in *2010 IEEE Aerospace Conference*, Big Sky, MT (2010), pp. 1–9

B. Yan, Y. Sun, F. Ding, H. Yuan, Design of CMOS image acquisition system based on FPGA, in *6th IEEE Conference on Industrial Electronics and Applications*, Beijing (2011), pp. 1726–1730

Analysis of Clustering Algorithms for Facility Location Allocation Problems

Pooja, Rakesh Kumar, Wattana Viriyasitavat, Kusum Yadav, and Gaurav Dhiman

Abstract Problems with facility location include selecting where to site a facility in order to meet the specified constraints most effectively. Choosing a production site that minimizes total weighted distances between suppliers and consumers, where weights represent the difficulty of carrying goods, is sometimes a challenging task. Numerous approaches and algorithms have been created to address location-related issues. In this research, we employ a clustering-related strategy to solve these location allocation problems. This strategy has already been applied in a variety of research projects employing clustering-based algorithms distinct from those we employed in our study. In this study, we are allocating locations for park facilities using Possibilistic C-Means clustering and Fuzzy Possibilistic C-Means clustering algorithms. To illustrate the most ideal method/algorithm, we compare the results acquired after selecting the optimal location using different clustering techniques.

Keywords Fuzzy logic · Facility location problems · Fuzzy clustering

Pooja · R. Kumar (✉)
Department of Mathematics, Lovely Professional University, Phagwara, Punjab 144411, India
e-mail: rakeshmalhan23@gmail.com

W. Viriyasitavat
Business Information Technology Division, Department of Statistics, Faculty of Commerce and Accountancy, Chulalongkorn University, Bangkok, Thailand

K. Yadav
College of Computer Science and Engineering, University of Ha'il, Ha'il, Kingdom of Saudi Arabia

G. Dhiman
Department of Computer Science, Government Bikram College of Commerce, Patiala, Punjab 147001, India

Department of Computer Science and Engineering, University Centre for Research and Development, Chandigarh University, Gharuan, Mohali, India

Department of Computer Science and Engineering, Graphic Era Deemed to be University, Dehradun, India

A. B. Reddy et al. (eds.), *Proceedings of Third International Conference on Advances in Computer Engineering and Communication Systems*, Lecture Notes in Networks and Systems 612, https://doi.org/10.1007/978-981-19-9228-5_51

1 Introduction

In 1965, Zadeh and Klaus independently created fuzzy sets as an enhancement to the standard collection instance. In an increasing variety of disciplines, including language studies, decision-making, and clustering, fuzzy relations are employed (Singh and Dhiman 2017, 2018a, b).

Allocating similar data items to the same clusters, the clustering algorithm divides a dataset into classes/clusters. The concept of fuzzy clustering is implemented when the boundary between clusters is unclear, resulting in the same data object belonging to many classes (Sharma et al. 2021; Kumar et al. 2021a). The process of allocating data points to groups in such a way that items contained within the same cluster/group are as comparable to one another as is practically conceivable, or alternatively, in such a way that groupings are as contradictory as is practically achievable (Chopra et al. 2021; Upadhyay et al. 2021).

In the subject of facility location allocation challenges, the approach of fuzzy clustering algorithms is currently applied. The Fermat–Weber problem was one of the earliest facility location issues proposed, dating back to the seventeenth century. The difficulty of facility location comprises determining where to locate a facility in order to meet the stated limits optimally (Zheng et al. 2021; Kumar et al. 2021b). Choosing a site for production that minimizes the total weighted distance between suppliers and consumers is frequently a challenge. Weights indicate the difficulty associated in transporting commodities. There are challenges with facility location in a range of industries and pursuits. Applications of the facility location problem range from public policy to telecommunications to particle physics. All of these challenges have one thing in common: distinct sites must be selected in order to meet consumer or user demand as efficiently as feasible (Singh et al. 2019).

2 Clustering Algorithms

Clustering an unlabeled dataset $X = \{x_1, x_2, x_3 \ldots x_n\} \subset \mathcal{R}^p$ is the segregating of X into $1 < c < n$ subgroups such that every single subgroup symbolizes "unusual" arrangement in X. A c-partition of X is a collection of (cn) values $\{\mu_{ik}\}$ that may be arranged as a $(c \times n)$ matrix $U = [\mu_{ik}]$. There are two partition matrices sets;

$$M_{pcn} = \left\{U \in \mathcal{R}^{cn} : 0 \leq \mu_{ik} \leq 1 \;\forall i, k; \forall k \exists i \ni \mu_{ik} > 0\right\} \tag{1}$$

$$M_{fcn} = \left\{U \in M_{pcn} : \sum_{i=1}^{c} \mu_{ik} = 1 \;\forall k; \sum_{k=1}^{n} \mu_{ik} > 0 \;\forall i\right\} \tag{2}$$

The preceding equations define the sets of possibilistic, fuzzy, and crisp c-partitions of elements set X. We observed that $M_{fcn} \subset M_{pcn}$.

2.1 Possibilistic C-Means (PCM) Clustering

PCM which stands for Possibilistic C-Means is a specific application of Krishnapuram and Keller's possibility theory (Dunn 2008; Li and Lewis 2016). It is a technique for clustering that modifies the FCM objective function (Bezdek et al. 1984; Krishnapuram and Keller 1993). The value of PCM membership might be interpreted as "degree of belongingness", whereas in FCM, it is the "sharing degree". Possibilistic C-objective Mean's function is as follows;

$$\min_{T,v}\left\{P(T,V;X,Y)=\sum_{k=1}^{n}\sum_{i=1}^{c}(t_{ix})^m\|x_k-v_i\|_A^2+\sum_{i=1}^{c}Y_i\sum_{k=1}^{n}(1-t_{ik})^m\right\} \tag{3}$$

$$P_m^{(i,j)}(T,V)=t_{ij}^m D_{ijA}^2+Y_i\left(1-t_{ij}\right)^m \tag{4}$$

where $Y_i > 0,\ 1 \le i \le c,\ u_i \in R,\ m > 1$.

And $T \in M_{pcn}$.

Where Y_i is the estimated "bandwidth, scale, or resolution" parameter. The parameter determines the distance. And is determined as,

$$Y_i = k\frac{\sum_{k=1}^{n} t_{ik}^m D_{ikA}^2}{\sum_{k=1}^{n} t_{ik}^m} \tag{5}$$

where $1 \le i \le c$; $1 \le k \le n$ and $k > 0$ (1 is the most frequent option for k).

The formula below may be used to calculate the membership value t_{ik}, which is continuously adjusted to determine the objective function,

$$t_{ik} = \frac{1}{1+\left(\frac{D_{ikA}^2}{Y_i}\right)^{\frac{1}{m-1}}} \tag{6}$$

where, $1 \le i \le c;\ 1 \le k \le n$.

Here, v_i mentioned in the above minimization formula can be calculated by following formula;

$$v_i = \frac{\sum_{k=1}^{n} t_{ik}^m x_k}{\sum_{k=1}^{n} t_{ik}^m} \tag{7}$$

where $1 \le i \le c;\ 1 \le k \le n$.

The objective function and membership value specified above satisfy the maximizing conditions as well as the fuzzy criterion (also valid for FCM) described below;

$$t_{ik} \in [0, 1];\ \forall i, j \text{ and } 0 < \sum_{j=1}^{n} t_{ik} \le 1,\ \forall i$$

Update the prototype using P^t, as indicated and then compute the value of P^{t+1} using the algorithm and hence increment in the "i" until the terminating condition satisfies, i.e., $\left\|P^t - P^{t+1}\right\| < \varepsilon$, where ε is a very small positive number.

2.2 *Fuzzy Possibilistic C-Means (FPCM) Clustering*

Keller and Bezdek introduced a revolutionary method known as FPCM which stands for Fuzzy Possibilistic C-Means in 1997 (Pal et al. 1997, 2005; Kulin and Kuenne 1962). Clustering unlabeled data with Fuzzy Possibilistic C-Means yields membership and typicality values. FPCM confines the typicality values so that the sum of a cluster's typicalities across all data elements equals one (Drezner and Hamacher 2002; Vygen 2005; Kumar et al. 2017). The row sum limitation gives inaccurate figures for typicality in large datasets. For FPCM, the following is the proposed optimization problem (Badri et al. 1998; Kumar et al. 2021b; Chandrawat et al. 2019a):

$$\min_{(U,T,V)}\left\{J_{m,\eta}(U,T,V;X)=\sum_{i=1}^{c}\sum_{k=1}^{n}\left(\mu_{ik}^{m}+t_{ik}^{\eta}\right)D_{ikA}^{2}\right\} \tag{8}$$

Depending on the constraints: $m > 1$, $\eta > 1$, $\mu_{ik} \geq 0$, $t_{ik} \leq 1$.
Where

$$D_{ik} = \|x_k - v_i\|_A \tag{9}$$

$$\sum_{i=1}^{c}\mu_{ik} = 1 \ \forall k \quad \text{i.e. } U \in M_{fcn} \tag{10}$$

$$\sum_{k=1}^{n} t_{ik} = 1 \ \forall i \quad \text{i.e. } T^t \in M_{fnc} \tag{11}$$

Members of the group M_{fnc} are the transposition of permissible T's. We can observe this by attributing typicality to object groupings. The possibilistic term $\sum_{i=1}^{c}\sum_{k=1}^{n} t_{ik}^{\eta} D_{ikA}^{2}$ will assign $\{t_{ik}\}$ for all n data points, but not for all clusters. We met the very first-class requirements for the extrema of $J_{m,\eta}$ under the normal constraints that are put on problems with C-Means approximation.

If $D_{ikA} = \|x_k - v_i\|_A > 0$ for all i, k; and $m, n > 1$, and X encompasses no less than c distinctive data points, then $(U, T, V) \in M_{fcn} \times M_{fcn} \times \mathcal{R}^p$ may minimize $J_{m,\eta}$ only if,

$$\mu_{ik} = \left[\sum_{j=1}^{c} \left(\frac{D_{ikA}}{D_{jkA}} \right)^{2/m-1} \right]^{-1}; \quad 1 \le i \le c;\ 1 \le k \le n \tag{12}$$

$$t_{ik} = \left[\sum_{j=1}^{n} \left(\frac{D_{ikA}}{D_{jkA}} \right)^{2/\eta-1} \right]^{-1}; \quad 1 \le i \le c;\ 1 \le k \le n \tag{13}$$

And

$$v_i = \frac{\sum_{k=1}^{n} (\mu_{ik}^{m} + t_{ik}^{\eta}) x_k}{\sum_{k=1}^{n} (\mu_{ik}^{m} + t_{ik}^{\eta})}; \quad 1 \le i \le c \tag{14}$$

The sum of all typicality data elements in a cluster is one due to FPCM's restriction on typicality values. When applied to huge sets of data, the row sum restriction results in typicality levels that are undesirable. The singularity form of FCM and FPCM is identical. Unlike PCM, FPCM is not always plagued by a responsiveness problem (Chandrawat et al. 2017, 2019b; Wang et al. 2020).

3 Park Allocation Illustration

We offer two distinct fuzzy clustering techniques for the facility localization problem in this research. We employed data from (Badri et al. 1998), in which twenty residential zones were extracted from an area where the government desired to build parks. Here, we intend to use this information to develop two parks (Kumar 2021; Kumar et al. 2021c). To ensure that citizens in all zones have access to parks, these two parks must be designed to maximize their accessibility from all angles. We will apply the PCM clustering method and the FPCM clustering algorithm to the data to cluster all the provided data into two clusters (because we want to develop two parks here), and then compare the outcomes of these two clustering techniques. The preliminary information is displayed in Table 1.

Positions/locations in Table 1 are represented as *X* and *Y* coordinates, respectively. The information in the preceding table is presented so that 20 residential zones are considered demand points and parks are considered feed points. As we intend to construct two parks, thus there will be two feed points. We will use the aforementioned three clustering techniques to the data in Table 1 to determine the optimal site for both parks. These algorithms will be encrypted in MATLAB. Then, we will compare the outcomes.

Table 1 Preliminary data for 20 different locations

Locations	*X*	*Y*
1	21	37
2	28	94
3	130	83
4	168	53
5	253	155
6	319	154
7	511	426
8	684	391
9	772	139
10	857	135
11	922	414
12	964	53
13	979	190
14	20	985
15	70	556
16	83	675
17	287	558
18	306	697
19	593	697
20	612	903

3.1 Comparison of Results

After applying the aforementioned two algorithms, Possibilistic and Fuzzy Possibilistic C-Means to the available data using MATLAB, we obtain the final feed points for the location of the parks. To compare the efficacy of various algorithms, we will employ the concept of the Euclidean distances between their centers and the location coordinates. Table 2 displays the final feed points (i.e., location of parks) acquired using all two techniques/algorithms.

Table 2 Feed points for all two algorithms

Clustering algorithms	1st feed point	2nd feed point
PCM	(249.8787, 374.0819)	(763.0072, 274.2252)
FPCM	(248.2809, 375.1601)	(763.9013, 275.2351)

Table 3 Euclidean distances

Clustering algorithms	Minimum Euclidean distance (1st feed point)	Minimum Euclidean distance (2nd feed point)
PCM	0.4801	0.0184
FPCM	0.4703	0.0177

With respect to different algorithms, the minimal Euclidean distances between feed points (cluster centers) and the location coordinates of 20 areas are displayed in Table 3.

In earlier parts, we discussed all three approaches, calculate the final feed points using those procedures and the available data, and calculate the Euclidean distance. Now, we will compare the two algorithms based on the outcomes of the previous sections. We utilized the concept of Euclidean distances between the centers of various algorithms and their location coordinates in order to evaluate their relative efficacy. As seen in Table 3, the FPCM algorithm yields the shortest Euclidean distance between cluster centers and location coordinates than any other approach. As a consequence of this, FPCM is an excellent contender for fuzzy rule-based system identification.

4 Conclusion

This work proposes clustering strategies for facility placement problems and evaluates the performance of clustering algorithms utilizing the concept of Euclidean distance between cluster centers (feed points) and location coordinates (demand points). We found that the FPCM clustering technique outperformed the other algorithms (i.e., PCM and FCM) we tested, while Possibilistic Fuzzy C-Mean is known for its noise sensitivity and coincident clusters problem, and it has been improved to address these issues with FPCM, resulting in the most effective gathering of cluster centers or feed points. PFCM's prototypes are less prone to outliers and can prevent coincident clusters. Even while all clustering techniques can be utilized to determine the appropriate placement for a facility, FPCM is the best option for fuzzy rule-based system identification. This research also demonstrates that in the future, clustering techniques can be applied effectively to situations involving location allocation. In addition to these fuzzy clustering algorithms, there are a variety of solutions for such facility allocation issues. Only a few fuzzy clustering algorithms have been covered in this paper, but there are many more excellent clustering techniques, such as Possibilistic Fuzzy C-Mean (PFCM), that may be used in subsequent research to address similar facility location issues.

References

M.A. Badri, A.K. Mortagy, C.A. Alsayed, A multi-objective model for locating fire stations. Eur. J. Oper. Res. **110**(2), 243–260 (1998). https://doi.org/10.1016/S0377-2217(97)00247-6

J.C. Bezdek, R. Ehrlich, W. Full, FCM: the fuzzy c-means clustering algorithm. Comput. Geosci. **10**(2–3), 191–203 (1984). https://doi.org/10.1016/0098-3004(84)90020-7

R.K. Chandrawat, R. Kumar, B.P. Garg, G. Dhiman, S. Kumar, An analysis of modeling and optimization production cost through fuzzy linear programming problem with symmetric and right angle triangular fuzzy number. Adv. Intell. Syst. Comput. **546**, 197–211 (2017). https://doi.org/10.1007/978-981-10-3322-3_18

R.K. Chandrawat, R. Kumar, V. Makkar, M. Yadav, P. Kumari, A comparative fuzzy cluster analysis of the binder's performance grades using fuzzy equivalence relation via different distance measures, vol. 955 (Springer, Singapore, 2019a). https://doi.org/10.1007/978-981-13-3140-4_11

R.K. Chandrawat, R. Kumar, V. Makkar, M. Yadav, P. Kumari, A comparative fuzzy cluster analysis of the binder's performance grades using fuzzy equivalence relation via different distance measures. Commun. Comput. Inf. Sci. **955**, 108–118 (2019b). https://doi.org/10.1007/978-981-13-3140-4_11/COVER/

S. Chopra, G. Dhiman, A. Sharma, M. Shabaz, P. Shukla, M. Arora, Taxonomy of adaptive neuro-fuzzy inference system in modern engineering sciences. Comput. Intell. Neurosci. **2021** (2021). https://doi.org/10.1155/2021/6455592

Z. Drezner, H. Hamacher, *Facility Location: Applications and Theory* (2002), p. 457

J.C. Dunn, Well-separated clusters and optimal fuzzy partitions. J. Cybern. **4**(1), 95–104 (2008). https://doi.org/10.1080/01969727408546059

R. Krishnapuram, J.M. Keller, A possibilistic approach to clustering. IEEE Trans. Fuzzy Syst. **1**(2), 98–110 (1993). https://doi.org/10.1109/91.227387

H.W. Kulin, R.E. Kuenne, An efficient algorithm for the numerical solution of the generalized Weber problem in spatial economics. J. Reg. Sci. **4**(2), 21–33 (1962). https://doi.org/10.1111/j.1467-9787.1962.tb00902.x

R. Kumar, A comparative study of fuzzy optimization through fuzzy number. Int. J. Mod. Res. **1**(1), 1–14 (2021)

R. Kumar, R.K. Chandrawat, B.P. Garg, V. Joshi, Comparison of optimized algorithms in facility location allocation problems with different distance measures. AIP Conf. Proc. **1860**(1), 20041 (2017). https://doi.org/10.1063/1.4990340

R. Kumar, G. Dhiman, N. Kumar, R.K. Chandrawat, V. Joshi, A. Kaur, A novel approach to optimize the production cost of railway coaches of India using situational-based composite triangular and trapezoidal fuzzy LPP models. Complex Intell. Syst. **7**(4), 2053–2068 (2021a). https://doi.org/10.1007/S40747-021-00313-0

R. Kumar, V. Joshi, G. Dhiman, W. Viriyasitavat, An improved exponential metric space approach for C-mean clustering analysing. Expert Syst. 1–15 (2021b). https://doi.org/10.1111/EXSY.12896

R. Kumar, R.K. Chandrawat, B. Sarkar, V. Joshi, A. Majumder, An advanced optimization technique for smart production using α-cut based quadrilateral fuzzy number. Int. J. Fuzzy Syst. **23**(1), 107–127 (2021c). https://doi.org/10.1007/s40815-020-01002-9

J. Li, H.W. Lewis, Fuzzy clustering algorithms—review of the applications, in *Proceedings of the 2016 IEEE International Conference on Smart Cloud, SmartCloud 2016* (2016), pp. 282–288. https://doi.org/10.1109/SmartCloud.2016.14

N.R. Pal, K. Pal, J.C. Bezdek, Mixed c-means clustering model, in *IEEE International Conference on Fuzzy Systems*, vol. 1 (1997), pp. 11–21. https://doi.org/10.1109/fuzzy.1997.616338

N.R. Pal, K. Pal, J.M. Keller, J.C. Bezdek, A possibilistic fuzzy c-means clustering algorithm. IEEE Trans. Fuzzy Syst. **13**(4), 517–530 (2005). https://doi.org/10.1109/TFUZZ.2004.840099

P. Sharma, N. Bhardwaj, G. Dhiman, Alexandroff soft topological spaces. Mater. Today Proc. (2021). https://doi.org/10.1016/J.MATPR.2021.01.351

P. Singh, G. Dhiman, A fuzzy-LP approach in time series forecasting, in Lecture Notes in Computer Science (including subseries Lecture Notes in Artificial Intelligence and Lecture Notes in Bioinformatics), vol. 10597. LNCS (2017), pp. 243–253. https://doi.org/10.1007/978-3-319-69900-4_31/TABLES/8

P. Singh, G. Dhiman, Uncertainty representation using fuzzy-entropy approach: special application in remotely sensed high-resolution satellite images (RSHRSIs). Appl. Soft Comput. **72**, 121–139 (2018a). https://doi.org/10.1016/J.ASOC.2018.07.038

P. Singh, G. Dhiman, A hybrid fuzzy time series forecasting model based on granular computing and bio-inspired optimization approaches. J. Comput. Sci. **27**, 370–385 (2018b). https://doi.org/10.1016/J.JOCS.2018.05.008

P. Singh et al., A hybrid fuzzy quantum time series and linear programming model: special application on TAIEX index dataset. Mod. Phys. Lett. A **34**(25) (2019). https://doi.org/10.1142/S0217732319502018

H. Upadhyay, S. Juneja, A. Juneja, G. Dhiman, S. Kautish, Evaluation of ergonomics-related disorders in online education using fuzzy AHP. Comput. Intell. Neurosci. **2021** (2021)

J. Vygen, *Approximation Algorithms for Facility Location Problems (Lecture Notes)* (2005)

Y. Wang et al., Profit optimization of products at different selling prices with fuzzy linear programming problem using situational based fuzzy triangular numbers. J. Phys. Conf. Ser. **1531**(1), 12085 (2020). https://doi.org/10.1088/1742-6596/1531/1/012085

Y. Zheng, G. Dhiman, A. Sharma, A. Sharma, M.A. Shah, An IoT-based water level detection system enabling fuzzy logic control and optical fiber sensor. Secur. Commun. Netw. **2021** (2021). https://doi.org/10.1155/2021/4229013

Biometric-Based Key Handling Using Speeded Up Robust Features

Prabhjot Kaur, Nitin Kumar, and Maheep Singh

Abstract The Biometric recognition process involves identification/verification based on individual's biometrics. During this process, the security and privacy needs to be protected against intruders and attackers. Biometric CryptoSystem (BCS) handles the issues related to security and privacy, where "Biometric" handles verification and "CryptoSystem" handles security by encryption. Almost every BCS method makes use to key for implementing their core functionality. This paper discusses a novel method for biometric-based key handling using Speeded Up Robust Features (SURF). The data is encrypted/decrypted using the derived key. The biometric key is retrieved during the enrolment stage, which then encrypts the data. After a successful biometric match, the secret is decrypted during the authentication phase. On three databases: AMI (ear), FACES (face), and UBIPr (iris), the suggested method is compared to state-of-the-art methods. The metrics used to evaluate the performance of proposed method include: Number of Pixel Change Rate, Unified Average Change Intensity, Peak Signal-to-Noise Ratio, Correlation, Mean Square Error, Mean Absolute Error, Structural Similarity, Normalised Root Mean Square Error, etc. The quantitative and qualitative results of the proposed method are better with state-of-the-art methods.

Keywords Key handling · Biometric cryptosystem · SURF

P. Kaur (✉) · N. Kumar · M. Singh
Department of Computer Science and Engineering, National Institute of Technology, Uttarakhand, Srinagar, India
e-mail: prabhjotkaur.phd19@nituk.ac.in

N. Kumar
e-mail: nitin@nituk.ac.in

M. Singh
e-mail: maheep.singh@nituk.ac.in

A. B. Reddy et al. (eds.), *Proceedings of Third International Conference on Advances in Computer Engineering and Communication Systems*, Lecture Notes in Networks and Systems 612, https://doi.org/10.1007/978-981-19-9228-5_52

1 Introduction

Along with cancelable biometrics, Biometric Cryptosystem is one of the most extensively used template protection schemes. Biometric cryptosystems combine cryptography and biometrics to maximise the capabilities of both domains (Kholmatov and Yanikoglu 2006). Biometrics adds non-repudiation and eliminates the need to memorise passwords, whilst cryptography provides high and configurable security levels in such systems. In such systems, biometrics adds non-repudiation and eliminates the need to remember passwords or carry tokens, whilst cryptography offers high and variable security levels. To protect the template in BCS, a biometric key is used. Unlike other cryptosystems that rely on passwords and other techniques to secure data, BCS employs a biometric key. In the literature, there are various ways for creating biometric keys (Rathgeb and Uhl 2011). The amount of different features extracted from the biometric sample and the key created from those features determine the performance of the biometric template matching procedure in BCS. To extract the number of distinctive features from the biometric image, this research used the Speeded Up Robust Features approach (Bay et al. 2006). SURF process predominantly follows three steps. In first step, "interest points" are chosen at prominent areas in the image, such as corners, blobs. The repeatability property of an interest point detector, is achieved by discovering the same interest points under diverse viewing situations. Second, each interest point's neighbourhood is represented by a feature vector. This description must stand out while also being resistant to noise, detection mistakes, and geometric and photometric distortions. Thirdly, the descriptor vectors are compared across images. The euclidean or mahanalobis distance between the vectors is frequently used to match images.

Motivation Traditional biometric cryptosystems based on fingerprint, fingervein, palmprint, and palmvein touch detecting devices are no longer useable due to the continuing COVID-19 epidemic. Furthermore, the use of masks makes it more difficult to identify people using a face biometric sensor, as present technologies produce more false rejections. As a result, safe biometric cryptosystems based on various biometric indicators such as ear, iris, gait, and so on are required. The goal of this project is to create a biometric cryptosystem that uses ear, face, and iris biometrics.

The remainder of the paper includes: contribution provided in Sect. 2, proposed work covered in Sect. 3, experimental results and discussion elaborated in Sect. 4, and conclusion is provided in Sect. 5.

2 Contribution

Here, on the basis of SURF, a novel biometric key management approach for biometric cryptosystems is developed. SURF is used to construct the biometric feature descriptor, which is then thresholded to generate the bio-key. At the enrolling step, this bio-key encrypts the data even more. The SURF-based feature matching is conducted at the authentication step, and after successful authentication, the bio-key decrypts the data. The suggested algorithm's resilience is tested using a Salt and Pepper noise attack on encrypted data. The encrypted data is decrypted again with the same key, and the results are compared to assess the proposed method's effectiveness using several criteria.

3 Proposed Work

One or more samples of a person are obtained, and their characteristics are retrieved and characterised using an SURF-based technique to create biometric keys for biometric cryptosystems. The SURF process is usually divided into three parts. The first stage is to choose *interest points* in significant sections of the image, such as corners, blobs. A feature vector is used to represent the neighbourhood of each interest point. The descriptor vectors are compared across images in the third step. When matching pictures, the euclidean or mahanalobis distance between the vectors is usually utilised. After that, using thresholding, a 128-bit bio-key is obtained, which is then used to encrypt the data in a biometric cryptosystem. Another biometric sample, called as the query biometric of the same individual, is utilised to extract and characterise characteristics based on SURF during the authentication phase. Between the stored descriptors and the query descriptors, a match is made. A successful match results in authentication, whereas a failed match results in the procedure being aborted. The matching saved bio-key decrypts the cryptic message to get the original message after successful authentication. The aforementioned technique is visually depicted in Fig. 1. NPCR, UACI, PSNR, Correlation, MSE, MAE, SSIM, NRMSE, and other metrics are used to assess the amount of accuracy and performance.

Based on Chen and Chandran (2007) and Loukhaoukha et al. (2012) study, the suggested algorithm's robustness is assessed using a Salt and Pepper noise attack on cryptic data. On this attacked data, the decryption procedure is carried out in order to get decrypted data that is similar to the original data. The qualitative results obtained on sample images utilising the three approaches are depicted in Figs. 2, 3, 4 and 5. The algorithm for enrollment and authentication is given in Tables 1 and 2.

Here, I is Image, S: subject, G: Gaussian, Bk: Biometric key, Dd: Decrypted data, De: Encrypted data, Cr: Correlation.

SURF-based feature extraction The SURF method generates and describes the most distinctive features from an image. This uses convolution of Gaussian kernel to locate interest points on the image. Orientations are handled using Haar-wavelet and in circular neighbourhood. Thereafter, the interest points in the image are described using sub-region handling around each interest point (Bay et al. 2006). *Biometric key handling (bio-key)* The biometric key can be generated in a variety of ways, according to the literature. By applying thresholding to the most distinctive picture characteristics, the suggested technique creates a 64-bit bio-key. *Data encryption/decryption*

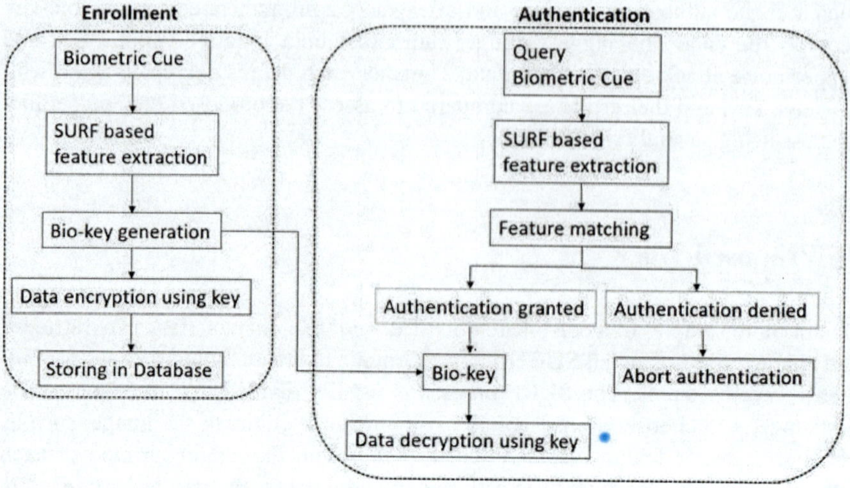

Fig. 1 SURF-based key handling

Table 1 Algorithm for enrollment

Input: Biometric images $I_1, I_2, I_3, \ldots I_n$ of subject S_1
Output: Generation of bio-key B_{k1} and encrypted data D_e
Steps:
1. Detection of interest points from biometric images $I_1, I_2, I_3, \ldots I_n$ using determinant of Hessian matrix, by applying convolution with Gaussian kernel
2. A scale-space representation of $3 \times 3 \times 3$ is applied
3. Find orientation using Haar-wavelet and in circular neighbourhood
4. Selection of most dominating descriptor and generation of 64-bit bio-key using thresholding
5. Encryption of data D_e using bio-key and XOR operation

Table 2 Algorithm for authentication

Input: Query biometric image I_q of subject S_1 and bio-key B_{k1}
Output: Decrypted data D_d
Steps:
1. Detection of interest points from biometric images using convolution with Gaussian kernel, $G(m, n) * I_q(m, n)$, where m, n represents image shape and description of interest points using 16×16 sample array
2. Feature matching based on euclidean distance; either grant authentication on successful match or abort
3. After grant authentication, decrypt the encrypted data D_e using bio-key and XOR operation to obtain the D_d

using bio-key The bio-key is used to generate cryptic data that has minimal correlation with the original data. After successful biometric authentication, the data is decrypted using the same bio-key. *Attack simulation using Salt and Pepper noise* The suggested method's strength is determined by performing a Salt and Pepper noise attack on the cryptic data (Loukhaoukha et al. 2012). The bio-key is then utilised to decode the cryptic data that has been assaulted or is noisy. The degree of difference between encrypted and original data shows the suggested BCS's resilience.

4 Experimental Results and Discussion

The proposed biometric cryptosystem key generation technique is compared to state-of-the-art approaches on three databases: AMI, FACES, and UBIPr. These methods are run on system with the configuration as: AMD Ryzen 3 2200U with Radeon Vega Mobile Gfx @ 2.50 GHz. The AMI database extends to Mathematical Analysis of Images (AMI), which is an Ear database with 700 images of 100 individuals with a resolution of 492 × 702 pixels (Gonzalez et al. 2012). The FACES database contains images of 171 people's facial expressions, totaling 2052 images at a resolution of 2835 × 3543 pixels (Ebner et al. 2010). The UBIPr database contains segmented images of both eyes with a 555 × 383 pixel image resolution (Padole and Proenca 2012).

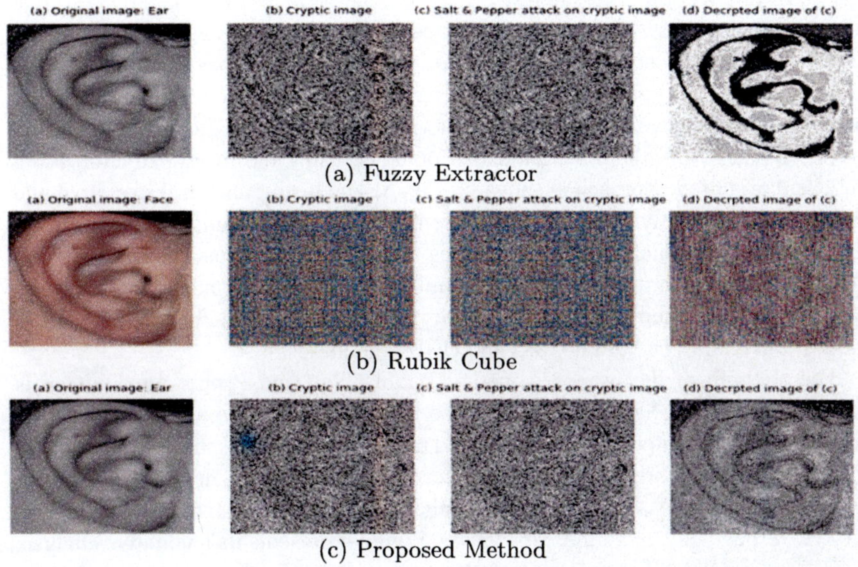

Fig. 2 Qualitative analysis of decryption of cryptic images after SALT and Pepper noise attack (AMI database)

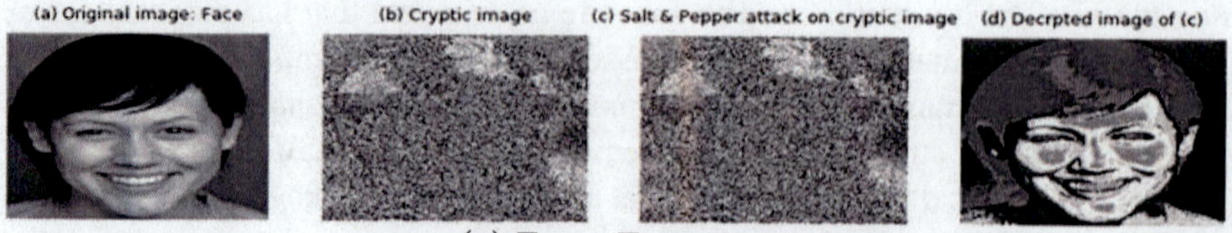

(a) Fuzzy Extractor

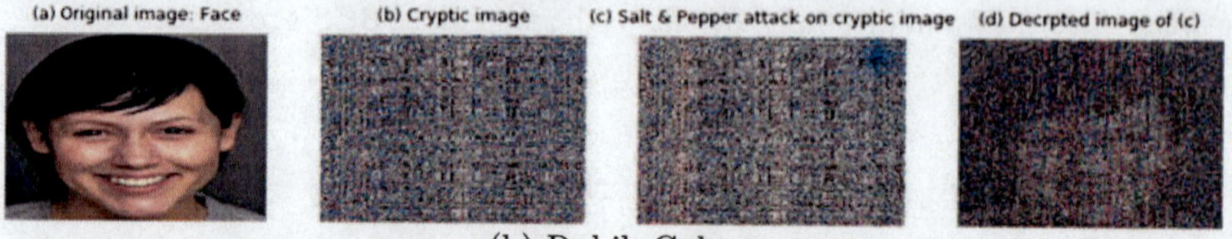

(b) Rubik Cube

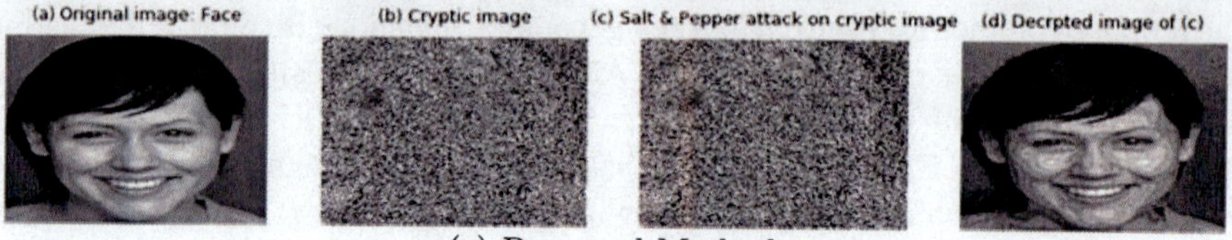

(c) Proposed Method

Fig. 3 Qualitative analysis of decryption of cryptic images after SALT and Pepper noise attack (FACES database)

The findings of the fuzzy extractor, rubik cube, and proposed approach are discussed in this section. The suggested method's performance is comparable to that of other approaches. The data is encrypted and decrypted using the key derived from the biometric cue. The ear, face, and iris are the three types of biometric cues utilised here. The cryptic image created from the original image must not reveal any of the original image's information or pattern. The histogram pattern of original and cryptic images from the three databases obtained using the fuzzy extractor, rubik cube and proposed approach is shown in Fig. 5. The histograms of the original and cryptic images plainly show no visual resemblance. Table 3 compares the quantitative aspects of original and cryptic images across three datasets. The MAE 0.4024 and PSNR 8.3050 of proposed scheme is minimum in FACES and UBIPr databases respectively. Minimum structural similarity of the proposed method is obtained as 0.0071 and 0.0089 on both FACES and UBIPr databases.

The strength of the proposed method against Salt and Pepper noise attack is also evaluated. The Salt and Pepper noise is added to the cryptic image which is subsequently decrypted using the proposed method. Table 4 presents the quantitative results between the original image and decrypted image after Salt and Pepper noise attack on three databases. The proposed scheme has lowest MAE of 0.0409, 0.4377, 0.6519, respectively, in all three databases. Figure 4 presents its qualitative analysis with other state-of-the-art methods on Faces database. The decryption result of the proposed method outperforms the other methods.

Table 3 Results comparison on AMI (ear), FACES (face) AND UBIPr (iris) DATABASES (between original and cryptic image)

Database	Ideal values/method	Cr	MSE	MAE	PSNR	SSIM	NPCR	UACI	RMSE	NRMSE
		1.00	0	0	0	0	0	0	0	0
AMI	Fuzzy extractor	−0.007	6317.71	0.0828	10.1252	0.0079	99.3585	26.1459	79.4840	0.5820
	Rubik cube	0.0409	4233.69	0.0463	11.8636	0.0195	99.4613	23.9858	65.0668	0.4728
	Proposed method	0.0583	6087.22	0.0409	10.2866	0.0116	99.5849	26.0039	78.0206	0.5713
FACES	Fuzzy extractor	−0.068	11,403.94	0.4887	7.5602	0.0104	99.4470	33.8947	106.789	0.9910
	Rubik cube	0.0055	9323.69	0.4900	8.4349	0.0102	99.5229	32.6642	96.5593	0.9358
	Proposed method	−0.0202	10,574.19	**0.4377**	7.8883	**0.0101**	99.7544	**33.1024**	102.8309	**0.9543**
UBIPr	Fuzzy extractor	−0.028	9446.59	0.6644	8.3780	0.0141	99.3777	30.9943	97.1935	1.3176
	Rubik cube	0.0200	8707.22	0.6683	8.7320	0.0099	99.5329	25.8963	93.3125	1.2302
	Proposed method	−0.0563	9408.11	0.6519	**8.3957**	**0.0054**	99.7948	31.1842	96.9954	1.3149

The bold values signifies the best results

Table 4 Results comparison on AMI (ear), FACES (face) AND UBIPr (iris) DATABASES (between original and decrypted image after Salt and Pepper noise attack)

Database	Ideal values/method	Cr	MSE	MAE	PSNR	SSIM	NPCR	UACI	RMSE	NRMSE
		1.00	0	0	0	0	0	0	0	0
AMI	Fuzzy extractor	0.6529	6544.30	0.3235	9.9721	0.0410	99.8934	29.2885	80.8968	0.5923
	Rubik cube	0.2328	5240.58	0.1677	10.9370	0.0349	99.1520	25.7550	72.3918	0.70165
	Proposed method	0.5268	**1721.73**	0.0566	15.7711	0.0844	100.0	**10.4320**	**41.4938**	**0.3038**
FACES	Fuzzy extractor	0.5224	4733.68	0.1496	11.3788	0.2198	99.9996	25.9162	68.8017	0.6385
	Rubik cube	0.2630	4948.47	0.1451	11.1860	0.0377	99.1320	25.0107	70.3454	0.6818
	Proposed method	**0.9633**	**276.80**	**0.0347**	23.7091	0.4883	**99.9999**	**6.3160**	16.6373	0.1544
UBIPr	Fuzzy extractor	−0.365	2955.52	0.0041	13.4244	0.0489	99.9807	17.8742	54.3647	0.7370
	Rubik cube	0.0117	8831.99	0.6749	8.6702	0.0096	99.5265	25.9517	93.9786	1.2389
	Proposed method	**0.5783**	**1191.93**	0.0316	17.3682	0.1320	**99.9923**	**8.3993**	34.5244	**0.4680**

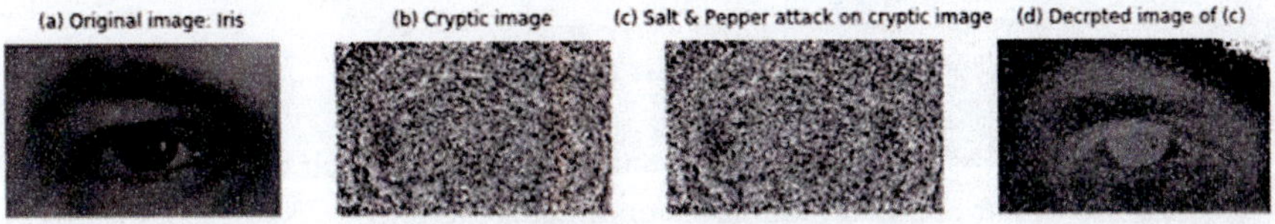

(a) Fuzzy Extractor

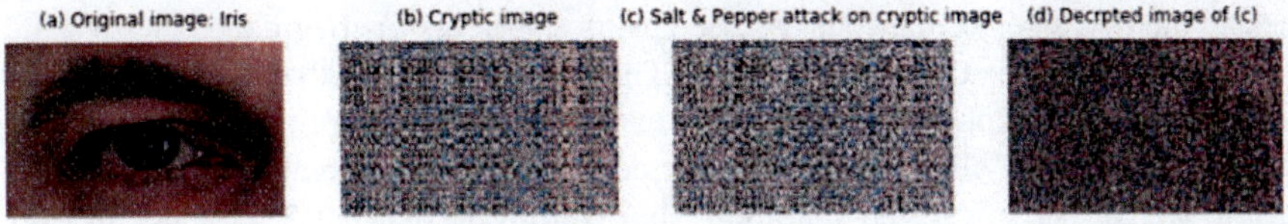

(b) Rubik Cube

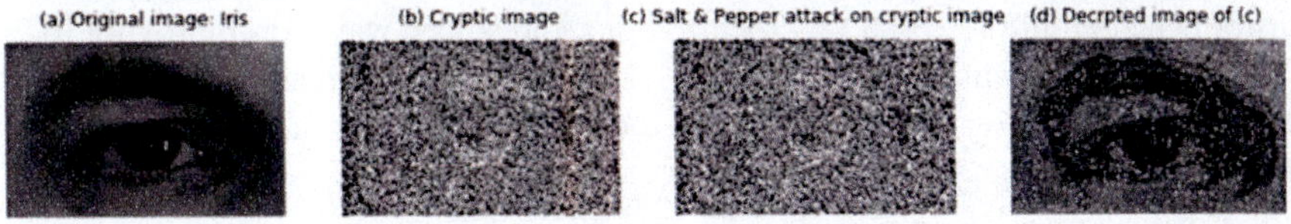

(c) Proposed Method

Fig. 4 Qualitative analysis of decryption of cryptic images after SALT and Pepper noise attack (UBIPr database)

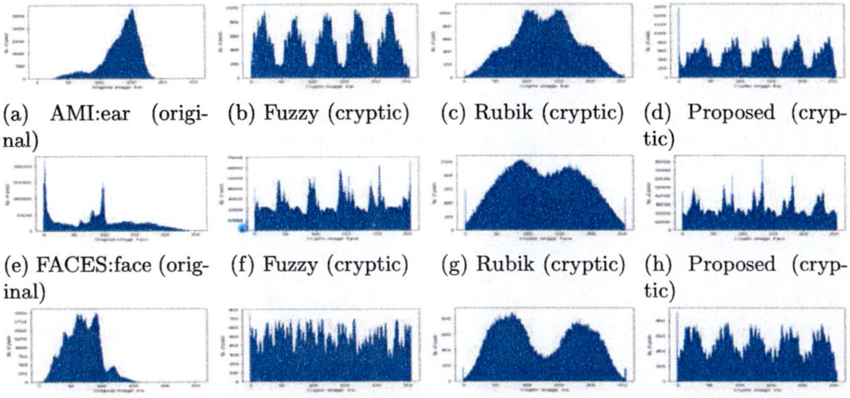

(a) AMI:ear (original) (b) Fuzzy (cryptic) (c) Rubik (cryptic) (d) Proposed (cryptic)

(e) FACES:face (original) (f) Fuzzy (cryptic) (g) Rubik (cryptic) (h) Proposed (cryptic)

(i) UBIPr:iris (original) (j) Fuzzy (cryptic) (k) Rubik (cryptic) (l) Proposed (cryptic)

Fig. 5 Histogram of original and cryptic images using proposed and state-of-the-art methods

5 Conclusion

In BCS, the SURF-based biometric key management approach excels the alternatives. The use of ear and eye biometrics eliminates the drawbacks of existing BCS that use on contact sensing biometrics like fingerprints and palmprints. The suggested method's performance is assessed using a variety of performance metrics and compared to state-of-the-art approaches. The suggested method's resilience is tested using a Salt and Pepper noise attack on encrypted data. The proposed approach is the most resilient of all the methods, as evidenced both qualitatively and numerically in the study. The suggested method's shortcoming is that it takes longer to execute high-resolution images, however this may be solved with better experimental instruments. Future research will focus on more advanced ways for improving BCS security and decryption quality. More advanced ways for increasing BCS security and image decryption quality will be the focus of future research.

References

H. Bay, T. Tuytelaars, L.V. Gool, Surf: speeded up robust features, in *European Conference on Computer Vision* (Springer, 2006), pp. 404–417

B. Chen, V. Chandran, Biometric based cryptographic key generation from faces, in *9th Biennial Conference of the Australian Pattern Recognition Society on Digital Image Computing Techniques and Applications (DICTA 2007)* (IEEE, 2007), pp. 394–401

N. Ebner, M. Riediger, U. Lindenberger, FACES—a database of facial expressions in young, middle-aged, and older women and men: development and validation. Behav. Res. Methods **42**, 351–362 (2010)

E. Gonzalez, L. Alvarez, L. Mazorra, Normalization and feature extraction on ear images, in *2012 IEEE International Carnahan Conference on Security Technology (ICCST)* (IEEE, 2012), pp. 97–104

A. Kholmatov, B. Yanikoglu, Biometric cryptosystem using online signatures, in *International Symposium on Computer and Information Sciences* (Springer, 2006), pp. 981–990

K. Loukhaoukha, J.Y. Chouinard, A. Berdai, A secure image encryption algorithm based on Rubik's cube principle. J. Electr. Comput. Eng. **2012** (2012)

C.N. Padole, H. Proenca, Periocular recognition: analysis of performance degradation factors, in *2012 5th IAPR International Conference on Biometrics (ICB)* (IEEE, 2012), pp. 439–445

C. Rathgeb, A. Uhl, A survey on biometric cryptosystems and cancelable biometrics. EURASIP J. Inf. Secur. **2011**(1), 3 (2011)

Magnitude-Based Weight-Pruned Automated Convolutional Neural Network to Detect and Classify the Plant Disease

V. Prithviraj and Sujatha Rajkumar

Abstract Agriculture systems are constantly vulnerable to pathogenic viruses and the diseases caused by them, posing a threat to a country's food security. Farmers often find it challenging to find the diseases in the plants at an early stage before it destroys the plant completely. In the proposed research work, an intelligent deep convolutional neural network for leaf image classification is developed, which can recognize 38 different types of plant diseases that are prevalent in 14 unique plant species. According to the complexity of the classification problem, various hyperparameters such as the number of epochs, batch size, hidden layers for feature extraction, dropout layers for regularization, and the number of neurons in each dense layer have been carefully designed in such a way that the model is neither overfitting nor underfitting, thus building an optimized deep CNN model. The developed CNN model for plant disease detection has an overall accuracy of 95% on the validation dataset. Further, magnitude-based weight pruning is carried out to reduce the network size by 66.7% and the overall accuracy is increased by 2%. Out of 33 test images, the model has predicted the plant diseases with an overall accuracy of 93.9% on the previously unseen test dataset. Thus, farmers would be highly benefitted from the proposed less complex weight-pruned CNN model as it predicts plant diseases using the concept of feature extraction with high accuracy, if a diseased leaf image of a plant is given as an input.

Keywords Convolutional neural network · Magnitude-based weight pruning · Plant health · Hyperparameters' tuning · Feature extraction

1 Introduction

Agriculture has always been a backbone for the Indian economy by contributing 19.9% to the GDP in the year 2020–21. But still, in this era of booming technologies, the agriculture sector is disconnected from the rapidly growing automation-based

V. Prithviraj (✉) · S. Rajkumar
Vellore Institute of Technology, Vellore, India
e-mail: prithviraj20007@gmail.com

A. B. Reddy et al. (eds.), *Proceedings of Third International Conference on Advances in Computer Engineering and Communication Systems*, Lecture Notes in Networks and Systems 612, https://doi.org/10.1007/978-981-19-9228-5_53

world. Plant diseases have always been very problematic for the farmers as the disease will destroy the plant if it is not noticed at the right time (Ristaino et al. 2021). Due to the cultivation of many crop varieties and similarities between the natures of the diseases, it is very hard and challenging for farmers and even for a pathologist to identify the diseased crops by just observing the disease-affected leaves through the naked eye. There is a high chance that even if they do so, they might end up predicting the wrong disease, and eventually, the diseased plant would not respond to the disease control treatment. On the other hand, farmers do not have enough financial support to hire experts who can help them predict the right disease. Artificial intelligence plays a major role in smart agriculture by assisting farmers in various tasks (Talaviya et al. 2020). Therefore, if we introduce the concept of artificial intelligence which can help farmers to predict plant diseases by just providing the image of the affected area of the leaf as an input, the entire process of plant disease detection can be automated, making the job simpler and easier for the farmers.

In recent times, researchers have tried to approach this problem through the concept of machine learning (Ramesh et al. 2018; Xian and Ngadiran 2021; Geetha et al. 2020) which is an extremely difficult and time-consuming technique. When it comes to applications of image classification, image segmentation, and pattern recognition, deep learning has been very efficient and successful as it can automate the process of feature extraction with high classification accuracy (Barbedo 2021; Oppenheim et al. 2019; Jenifa et al. 2019), less computational time, and low error rate. Over the past few years, the development of large neural networks has been one of the main reasons for the success of deep learning on image processing applications (Wallelign et al. 2018). Despite the improved performance of these large neural networks on various applications, they always come at the cost of a high price. Because of their large size, they demand more storage space. Hence, the process of distribution of these large models becomes a cumbersome task for the developers. These models demand costly hardware with high specifications as the execution time of large networks is more. This is especially important if a model is brought into production for a real-world application.

Therefore, in the proposed research work, as deep learning is the best suitable approach for the above-mentioned problem statement (Hasan et al. 2020), an intelligent neural network that is based on the concept of convolution and magnitude-based weight pruning is built to identify and classify various plant diseases at an early stage. The size of model is reduced by model compression, and it is made sure that there is only a minimal loss of accuracy or performance. To differentiate the plant diseases, various aspects or features of the input image are assigned learnable weights and biases (Ennadifi et al. 2020; Liu 2018) by this weight-pruned neural network. The main contributions of this paper are summarized as follows:

1. The presented paper mainly deals with the problem of overfitting and underfitting that is faced by most of the convolutional neural network models.
2. The paper focuses on tuning various hyperparameters of the CNN model in such a way that the model performs well without the problem of overfitting and underfitting.

3. A CNN model including various layers for extracting important features related to the leaf diseases is built from scratch in such a way that the model is well trained without fluctuations in its training and validation accuracies for each epoch during the training phase of the model.
4. The problem of large expensive neural networks has been addressed by building a magnitude-based weight-pruned lightweight neuronal model with reduced size and increased accuracy of detecting plant diseases than the original CNN model.
5. The proposed neuronal model can detect and classify plant diseases belonging to 38 different classes of 14 unique plants with a high accuracy despite its reduced size.

2 Related Work

The disadvantage of using ReLU as the activation function is addressed as it faces the problem of neuronal necrosis where its derivative is zero for values that are less than zero (Yadhav et al. 2020). Therefore, to improve the performance and accuracy of the CNN, a new mathematical activation function is developed and its unique characteristics are analyzed by comparing its properties with already existing activation functions. Madhulatha and Ramadevi (2020) used a pre-trained AlexNet as the disease classification model. The CNN model contains various convolution and max pooling layers, and ReLU is used as the activation function.

Ashok et al. (2020) used the concept of hierarchical feature extraction by developing a convolutional neural network algorithm. To identify the tomato plant leaf disease, the input image pixel intensities are mapped and the model classifies them into corresponding disease classes by comparing them with the images in the dataset that are already trained. Jin et al. (2021) introduced a new method that blends deep learning and image processing techniques. To detect vegetables and generate bounding boxes around them, a trained CenterNet model is employed. Weeds are the leftover green objects that fell out of the bounding boxes.

Huang et al. (2019) had come up with a model that contains two submodels in it. The first submodel is a leaf segmentation model that employs a U-Net to effectively remove interference so that the leaves can be distinguished from the background of the original image. The second submodel is used for plant disease classification. To classify diseases in the potato leaves like early blight, late blight, and healthy, the authors used a pre-trained model like VGG19 for transfer learning to extract the relevant features from the dataset (Tiwari et al. 2020). For detection and classification of various tomato leaf diseases, the authors introduced a learning vector quantization (LVQ) algorithm-based method and a CNN model (Sardogan et al. 2018).

For automatic plant disease identification, the authors developed a hybrid model based on CNN and CAE (convolutional autoencoder) (Bedi and Gole 2021). The hybrid model's training parameters are reduced using CAE. Liu et al. (2021) proposed a lightweight CNN model based on SqueezeNet. Testing results reveal that the proposed model can obtain an accuracy of 98.46% while requiring only 0.62 MB of memory. This demonstrates the technological ability to employ lightweight CNNs in an embedded system to classify plant diseases.

Rajkumar et al. (2021) have incorporated image processing techniques along with convolutional neural networks to predict tomato leaf diseases. Using the concept of data analytics, the proposed system can predict disease outbreak. An alert system has been developed to inform the tomato cultivators about the tomato leaf diseases. Gandhi et al. (2018) have used generative adversarial networks to augment the available dataset that is limited to Indian crops and diseases. They employed a class of GANs called deep convolutional generative adversarial networks (DCGANs). To stabilize the training process of a wide range of datasets, DCGAN is used which is a modified version of GAN that creates deeper generative models and higher resolution pictures.

3 Deep CNN Model for Plant Disease Prediction

A hierarchical network with convolution layers, max pooling layers, regularization layers, flatten layer, and dense layers is built to extract the important features from the diseased leaf images. The network learns from the extracted features and then helps the farmers to classify different plant diseases. As depicted in Fig. 1, the dataset containing the diseased leaf images is separated into three parts as training, validation, and test datasets for the proposed CNN model. The training dataset and validation dataset are preprocessed separately before feeding them to the dense deep neural network. The test leaf species images are converted into an array before feeding them to the network. After training the network, the proposed model can classify any unlabeled leaf images of plants. The weight-pruned model compares the test leaf images with the training leaf datasets and predicts the disease variant.

A large dataset with 87,900 images of leaves spanning 38 classes and 14 different plants is used. All the images have a resolution of 256 * 256 pixels. For the training of the CNN model, 70,295 images were separated into 38 classes, each containing 1642–2022 images. For the validation of the model, 17,572 images were split into 38 classes, each containing 410–505 images. And, a total of 33 images were used for testing purposes. The leaf images of 14 unique plants present in the dataset are corn, cherry, peach, tomato, orange, apple, squash, potato, grape, soybean, raspberry, pepper, strawberry, and blueberry.

4 Data Curation

Separate preprocessing is done for the training, validation, and test plant leaf images. Before feeding images to the model, they are scaled down by a factor of 255 (Shorten and Khoshgoftaar 2019). All the pixel values are converted from the range [0, 255] to [0, 1]. A new different perspective is given to the network by doing multiple

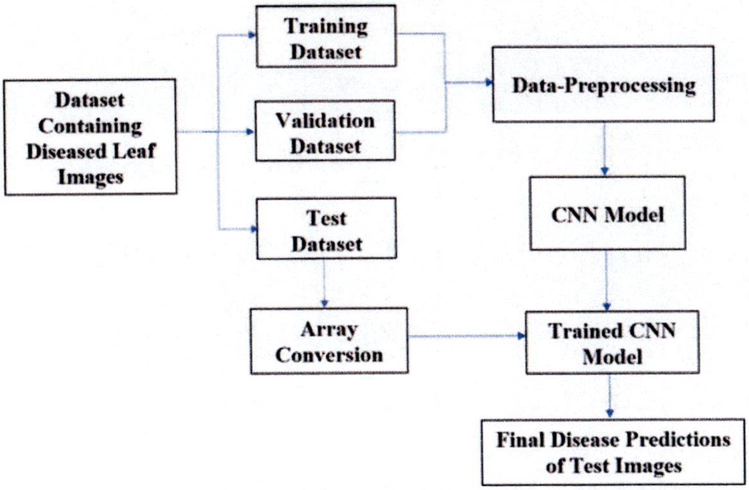

Fig. 1 Convolution-based deep AI model

transformations to the images which create various transformed copies of the same image. It is made sure that each of these transformed copies is unique in one way or the other by incorporating image augmentation techniques such as shifting, rotating, and flipping.

This technique of image augmentation improves the model's performance on the previously unseen test dataset (Frank et al. 2021). Thus, it also enhances the model's overall test accuracy. The problem of overfitting is also addressed by ensuring that the model receives unique transformed copies of the original images at each epoch (Ying 2019).

5 Deep Neuronal Model for Plant Disease Classification

5.1 Feature Extraction Using Convolution Function

The leaf image with height = 256, width = 256, and depth = 3 is fed as an input to the first convolution layer. In the first convolution layer, a total of 32 filters are learnt with a kernel size of 5 × 5. A kernel size greater than 3 is used to learn larger spatial filters and to assist in reducing spatial dimensions as the input images are greater than 128 × 128. The result of applying the filters to the input image is captured by the feature maps which highlight the relevant features of the image. Following all convolutions, a non-saturating and nonlinear activation function termed ReLU is used (Lin and Shen 2018).

$$\begin{bmatrix} I_{11} & I_{12} & I_{13} \\ I_{21} & I_{22} & I_{23} \\ I_{31} & I_{32} & I_{33} \end{bmatrix} * \begin{bmatrix} W_{11} & W_{12} \\ W_{21} & W_{22} \end{bmatrix} = \begin{bmatrix} F_{11} & F_{12} \\ F_{21} & F_{22} \end{bmatrix} \quad (1)$$

$$F[m,n] = (I * W)[m,n] = \sum_i \sum_j W[i,j] I[m-i, n-j] \quad (2)$$

$$F_{11} = I_{11} \times W_{11} + I_{12} \times W_{21} + I_{21} \times W_{12} + I_{22} \times W_{22} \quad (3)$$

$$F_{12} = I_{12} \times W_{11} + I_{13} \times W_{21} + I_{22} \times W_{12} + I_{23} \times W_{22} \quad (4)$$

$$F_{21} = I_{21} \times W_{11} + I_{22} \times W_{21} + I_{31} \times W_{12} + I_{32} \times W_{22} \quad (5)$$

$$F_{22} = I_{22} \times W_{11} + I_{23} \times W_{21} + I_{32} \times W_{12} + I_{33} \times W_{22} \quad (6)$$

The kernel is slid over the input image to capture the important features in a feature map. The above equations depict the operation of convolution on a 3×3 image with a kernel of size 2×2, which results in a feature map of size 2×2.

5.2 *Subsampling Using Max Pooling*

The feature maps that are generated by the convolutional layer are fed into a subsampling layer that reduces the spatial dimensions of the output volume for subsequent layers. Here, max pooling is used for the subsampling operation. The layer carries forward only the information that is the largest amplitude wise. Just after the max pooling layer, the feature map that is obtained as output will contain only the most significant features from the previous feature map. A pool size of 3×3 is used for the max pooling operation. Figure 2 depicts the max pooling operation, where the 4×4 matrix is subsampled into a 2×2 matrix by selecting the maximum number from each of the 2×2 matrices.

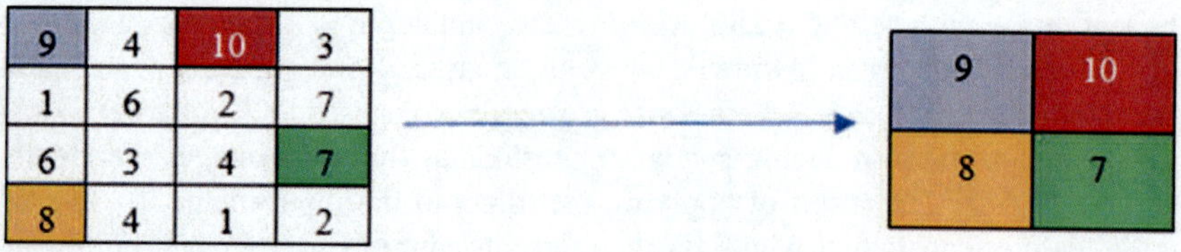

Fig. 2 Dimensionality reduction

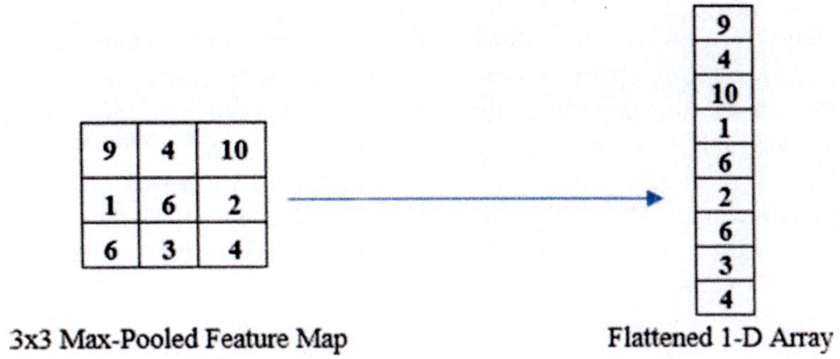

Fig. 3 Conversion of data into a single long feature vector

5.3 *Fully Connected Layers*

A flatten layer is added which acts as a connection between the Conv2D layers and the dense layer. As shown in Fig. 3, the data must be converted into a one-dimensional array that is a single long feature vector before inputting it to the dense layers. After passing the input image through the convolution layers and max pooling layers and after flattening the data into a 1D array, it is fed as an input to the dense layer containing 512 neurons. These neurons get outputs from the previous layers as input and provide a new output to the next layer.

$$G(x) = \sum_{i=1}^{n} x_i w_i + b \tag{7}$$

$$\text{ReLU function} = \text{Max}(0, G(x)) \tag{8}$$

$$\text{Output} = \begin{Bmatrix} G(x), & x > 0 \\ 0, & x \leq 0 \end{Bmatrix} \tag{9}$$

G(*x*) is the weighted sum of the input values (*x*). All the values present in the input vector are multiplied with their respective weights (*W*) and then added with a bias (*b*) and then passed on to the ReLU function.

5.4 *Optimization of the Proposed CNN Model Using Dropout*

A dropout layer has been introduced that randomly disconnects 25% of the nodes present in the current layer before passing it to the next layer to avoid overfitting. This is done by setting the value of neurons to zero at each update during the training phase (Dileep et al. 2020). This technique of dropping out nodes randomly makes sure that no single node of a particular layer is responsible for predicting a certain class. Dropout regularization prevents neurons from developing codependency among each other which avoids overfitting of the model. The output of the dropout layer is passed on to a dense layer containing 128 neurons whose outputs are passed on to the ReLU activation function. In Fig. 4, a random probability is assigned to all the 512 neurons present in the dense layer based on which dropout of neurons is carried out.

$$\text{Dropout rate} = 25\% \tag{10}$$

$$\text{Dropout probability} = 0.25 \tag{11}$$

$$\text{Retention probability} = (1 - 0.25) = 0.75 \tag{12}$$

The retention probability determines the percentage of neurons that are not dropped.

$$d^i = \{d^1, d^2, d^3, d^4 \ldots d^{512}\} \tag{13}$$

There are 512 neurons present in the dense layer.

$$P^i = \{P^1, P^2, P^3, P^4 \ldots P^{512}\}, \text{ where } 0 < P^i < 1 \tag{14}$$

P_i is the random probability assigned to the 512 neurons.

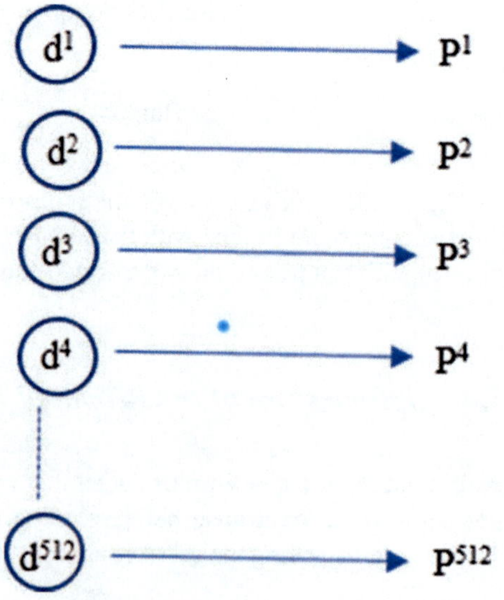

Fig. 4 Dense layer with 512 neurons

$$\text{If } P^i \leq 0.75, \text{ the neurons are retained.} \quad (15)$$

$$\text{If } P^i > 0.75, \text{ the neurons are discarded.} \quad (16)$$

The retention probability is set as the threshold probability.

5.5 *Plant Disease Classification*

A dense layer has been created that has the same number of neurons as the output class labels (38 classes). This layer acts as the output or the classification layer. The activation function for this final layer is Softmax (Deopa et al. 2019), which is used to calculate the probability of the input image belonging to one of the 38 classes. As shown in Fig. 5, the probability of the input image for each class will be returned by all 38 neurons. The image's output will be the class with the highest probability.

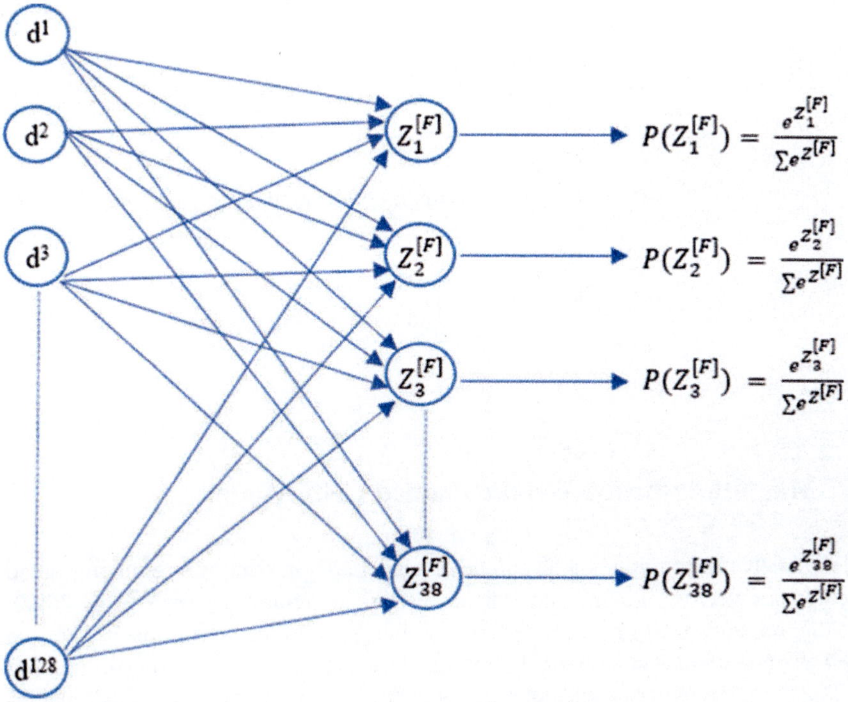

Fig. 5 One-twenty eight neurons in the dense layer are connected with the 38 neurons in the final classification layer

$$Z^{[F]} = \begin{bmatrix} Z_1^{[F]} \\ Z_2^{[F]} \\ Z_3^{[F]} \\ . \\ . \\ Z_{38}^{[F]} \end{bmatrix} = W^F * D^{F-1} + b^F \tag{17}$$

$Z[F]$ are the values that are obtained in the final (F) layer.

$$D^{F-1} = \begin{bmatrix} d^1 \\ d^2 \\ d^3 \\ . \\ . \\ d^{128} \end{bmatrix}, \text{ values at the denser layer } (F-1) \tag{18}$$

$$W^F = \begin{bmatrix} W_{1,1} & .. & W_{1,38} \\ . & .. & . \\ . & .. & . \\ W_{128,1} & .. & W_{128,38} \end{bmatrix}, \text{ weight matrix of the final layer} \tag{19}$$

$$b^F = \begin{bmatrix} b^1 \\ b^2 \\ b^3 \\ . \\ . \\ b^{38} \end{bmatrix}, \text{ bias matrix of the final layer} \tag{20}$$

$$\sum e^{Z^{[F]}} = e^{Z_1^{[F]}} + e^{Z_2^{[F]}} + e^{Z_3^{[F]}} + \cdots + e^{Z_{38}^{[F]}} \tag{21}$$

6 Magnitude-Based Weight-Pruned CNN Model

Neural network pruning is a technique of model compression where the unwanted neurons or weights of an already trained network are removed (Tessier et al. 2022). In this research work, magnitude-based weight pruning has been done where the network is made sparse by gradually zeroing out the model weights during the training process. By removing weights that are already close to zero or have a small magnitude, the effect on the network can be reduced. By discarding any weights below a given threshold, this can be accomplished. Sparse models are easily compressed. This reduces the number of parameters in the model while maintaining the same architecture. In Fig. 6, the weights present in each layer of the proposed neuronal model after magnitude-based weight pruning are shown. After pruning of the network, the total number of trainable weights is 23,860,216, out of which 11,930,502 weights are trainable and 11,929,714 weights are non-trainable whose values are not updated during the training phase of the network.

$$\text{Threshold}(W_i) = \left\{ \begin{matrix} W_i, & |W_i| > \lambda \\ 0, & |W_i| \le \lambda \end{matrix} \right\} \tag{22}$$

λ is the threshold value and W_i are the weights.

7 Hyperparameter Optimization

Hyperparameters such as the number of epochs, batch size, hidden layers for feature extraction, dropout layers for regularization, and the number of neurons in each dense layer have been tuned in such a way that the model is neither overfitting and memorizing the training dataset where the training accuracy is high but performs poorly on the validation dataset nor underfitting where it performs poorly on the training dataset itself by having low training accuracy and high training loss. In the proposed model, the validation accuracies and losses do not fluctuate for each epoch during the training of the model, thus avoiding overfitting of the model. As shown in Table 1, the images are loaded in batches with a batch size of 64 to avoid overloading of the network. The number of epochs is set as 10. Adam is used as the optimizer, and categorical cross-entropy is used as the loss function.

8 Results and Discussion

In Fig. 7, the model has an overall accuracy of 95.01% on the validation dataset with a loss of 15.88%. Figures 8 and 9 depict the accuracy and loss plots during the training and validation phases, which show that the model is neither overfitting nor underfitting. If Fig. 8 is observed, during the model's training phase, in the final epoch, the training accuracy and the validation accuracy are almost equal with a difference of only 0.18%. Similarly, if Fig. 9 is observed, in the final epoch, both the training loss and validation loss have decreased constantly and converged at a point with only a difference of 0.19%. The results show that the model is perfectly designed and trained as it performs well on the training dataset as well as on the validation dataset with high accuracy and less loss.

If Figs. 10 and 11 are observed, even after reducing the size of the proposed CNN model through magnitude-based weight pruning, the training accuracy and validation accuracy of the network are not compromised and the training loss and validation loss have also decreased further. If the network parameter values of the

```
Layer (type)                    Output Shape              Param #
=================================================================
prune_low_magnitude_conv2d      (None, 252, 252, 32)      4834
(PruneLowMagnitude)

prune_low_magnitude_max_poo     (None, 84, 84, 32)        1
ling2d (PruneLowMagnitude)

prune_low_magnitude_conv2d_     (None, 82, 82, 32)        18466
1 (PruneLowMagnitude)

prune_low_magnitude_max_poo     (None, 41, 41, 32)        1
ling2d_1 (PruneLowMagnitude
)

prune_low_magnitude_conv2d_     (None, 39, 39, 64)        36930
2 (PruneLowMagnitude)

prune_low_magnitude_max_poo     (None, 19, 19, 64)        1
ling2d_2 (PruneLowMagnitude
)

prune_low_magnitude_flatten     (None, 23104)             1
 (PruneLowMagnitude)

prune_low_magnitude_dense (     (None, 512)               23659010
PruneLowMagnitude)

prune_low_magnitude_dropout     (None, 512)               1
 (PruneLowMagnitude)

prune_low_magnitude_dense_1     (None, 128)               131202
 (PruneLowMagnitude)

prune_low_magnitude_dense_2     (None, 38)                9768
 (PruneLowMagnitude)

prune_low_magnitude_activat     (None, 38)                1
ion (PruneLowMagnitude)

=================================================================
Total params: 23,860,216
Trainable params: 11,930,502
Non-trainable params: 11,929,714
```

Fig. 6 Magnitude-based weight-pruned layers of the CNN model

Table 1 Hyperparameters used in the training phase

Parameter	Value
Number of epochs	10
Batch size	64
Optimizer	Adam
Loss function	Categorical cross-entropy
Metrics	Accuracy

```
275/275 [==============================] - 47s 172ms/step - loss: 0.1588 - accuracy: 0.9501
[0.15879225730895996, 0.9501479864120483]
```

Fig. 7 Model accuracy after training the CNN model

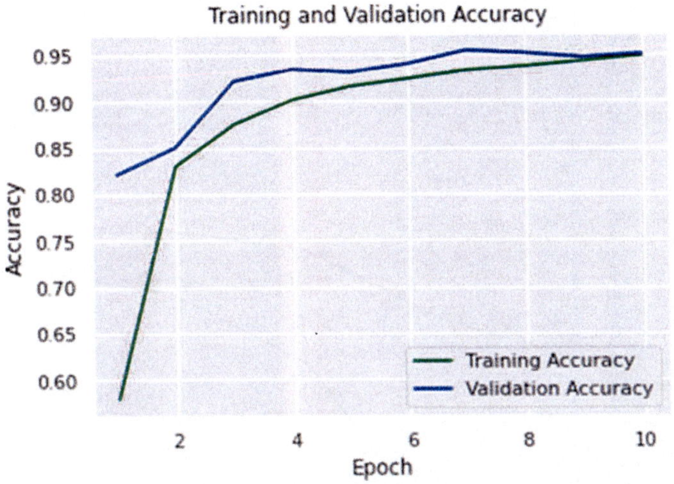

Fig. 8 Training and validation accuracy plot

final epoch in Figs. 8 and 10 are observed, after magnitude-based weight pruning, the training accuracy and validation accuracy have increased by 1.68% and 2%, respectively. Similarly, if Figs. 9 and 11 are observed, the training loss and validation loss have decreased by 31% and 39.5%, respectively. After pruning the network based on weights, the accuracy and loss plots during the training and validation phases are stable for each epoch without fluctuations. This shows that there is an increase in the overall performance of the proposed CNN model after magnitude-based weight pruning. Despite the reduction in size of the network, the proposed lightweight neuronal model predicts the plant diseases with a high accuracy and less loss.

After pruning the network based on the magnitude of weights, the model size got reduced by 66.7% from 136.6 MB in Fig. 12 to 45.5 MB in Fig. 13. The validation accuracy has increased by 2% from 95.01 to 96.87%.

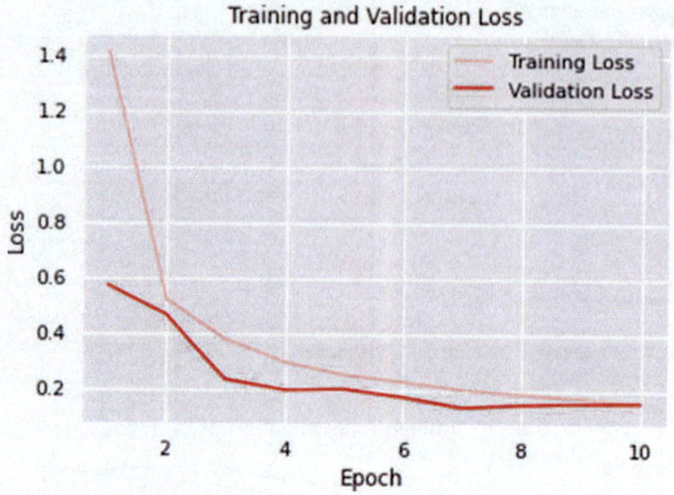

Fig. 9 Training and validation loss plot

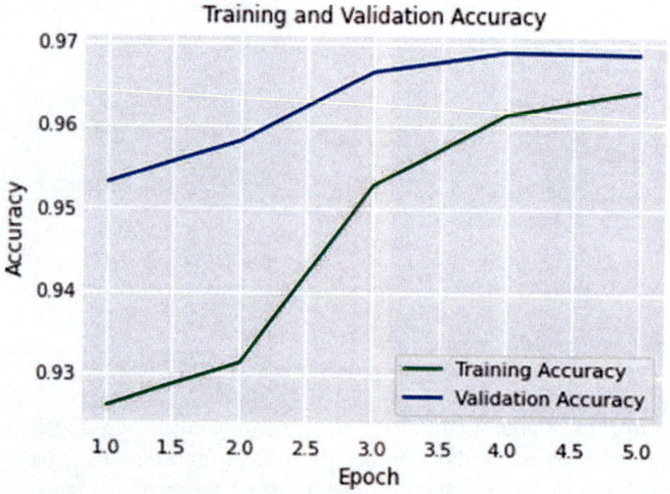

Fig. 10 Training and validation accuracy plot after weight pruning

The model has been tested with 33 different plant disease images, in which the model has not seen during the training process. In Fig. 14, out of 33 test images, the model has predicted 31 images correctly with an overall accuracy of 93.9% on the previously unseen test data. In Fig. 15, the input images of plant leaves belonging to different classes and the corresponding predictions for those classes are shown. Figure 16 shows the confusion matrix that has been plotted to visualize and understand how efficiently the network is able to classify plant diseases and to

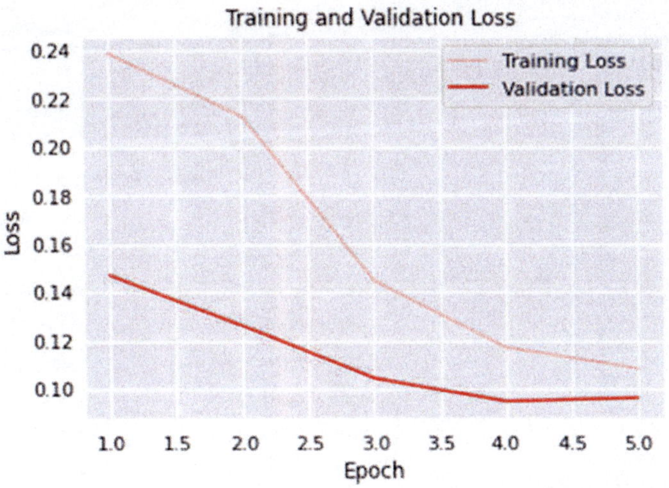

Fig. 11 Training and validation loss plot after weight pruning

Unpruned model Summary:
Model size(MB): 136.59432220458984
Time on Validation data (sec): 46.18929699199907
Accuracy on validation data: 0.9501479864120483

Fig. 12 Model summary before magnitude-based weight pruning

Pruned model Summary:
Model size(MB): 45.54493713378906
Time on Validation data (sec): 46.892362814996886
Accuracy on validation data: 0.9687002301216125

Fig. 13 Model summary after magnitude-based weight pruning

analyze the wrong predictions that are made by the network. In the confusion matrix, the diagonal values represent the number of correct predictions that are made by the model for each of the 38 different plant classes. For the proposed CNN model, the maximum number of values lies in the diagonal of the confusion matrix and only a few values are found off the diagonal. This indicates that the proposed model has been accurate most of the time and predicted the plant diseases correctly.

Fig. 14 Predicted plant disease classes

	Filename	Predicted classes
0	AppleScab1.JPG	Apple___Apple_scab
1	TomatoYellowCurlVirus1.JPG	Tomato___Tomato_Yellow_Leaf_Curl_Virus
2	TomatoHealthy1.JPG	Tomato___healthy
3	TomatoYellowCurlVirus2.JPG	Tomato___Tomato_Yellow_Leaf_Curl_Virus
4	TomatoHealthy4.JPG	Tomato___healthy
5	TomatoEarlyBlight6.JPG	Tomato___Early_blight
6	AppleCedarRust3.JPG	Apple___Cedar_apple_rust
7	TomatoEarlyBlight2.JPG	Tomato___Late_blight
8	PotatoEarlyBlight4.JPG	Potato___Early_blight
9	TomatoEarlyBlight5.JPG	Tomato___Early_blight
10	AppleCedarRust2.JPG	Apple___Cedar_apple_rust
11	AppleCedarRust4.JPG	Apple___Cedar_apple_rust
12	CornCommonRust1.JPG	Corn_(maize)___Common_rust_
13	TomatoHealthy2.JPG	Tomato___healthy
14	TomatoEarlyBlight1.JPG	Tomato___Early_blight
15	PotatoEarlyBlight5.JPG	Potato___Early_blight
16	PotatoHealthy1.JPG	Potato___healthy
17	CornCommonRust2.JPG	Corn_(maize)___Common_rust_
18	TomatoEarlyBlight3.JPG	Tomato___Early_blight
19	AppleCedarRust1.JPG	Apple___Cedar_apple_rust
20	PotatoEarlyBlight2.JPG	Potato___Early_blight
21	TomatoYellowCurlVirus5.JPG	Tomato___Tomato_Yellow_Leaf_Curl_Virus
22	AppleScab3.JPG	Apple___Apple_scab
23	TomatoYellowCurlVirus4.JPG	Tomato___Tomato_Yellow_Leaf_Curl_Virus
24	PotatoEarlyBlight1.JPG	Potato___Early_blight
25	AppleScab2.JPG	Apple___Apple_scab
26	PotatoHealthy2.JPG	Potato___healthy
27	TomatoEarlyBlight4.JPG	Tomato___Early_blight
28	TomatoHealthy3.JPG	Tomato___healthy
29	PotatoEarlyBlight3.JPG	Apple___healthy
30	TomatoYellowCurlVirus6.JPG	Tomato___Tomato_Yellow_Leaf_Curl_Virus
31	TomatoYellowCurlVirus3.JPG	Tomato___Tomato_Yellow_Leaf_Curl_Virus
32	CornCommonRust3.JPG	Corn_(maize)___Common_rust_

Fig. 15 Diseased plant leaf images and their predictions

9 Conclusion

In the proposed weight-pruned CNN model, both the training accuracy and validation accuracy are constantly increasing for each epoch without any fluctuations and they are almost equal in the final epoch of model training. Likewise, the training loss and validation loss have also decreased constantly and converged in the final epoch. Hence, the main objective of the research work has been accomplished by building an optimized lightweight CNN model for plant disease detection. By doing magnitude-based weight pruning, the model size is reduced by 66.7%, making the CNN model small and compact. The model also requires less storage space making it easier to distribute. As a result, the proposed CNN model is ideal for real-time plant disease detection applications.

Due to its smaller size and high efficiency of detecting plant diseases, it can be further developed into a web application where farmers have to just give the diseased leaf image as an input and the application will exactly predict the plant disease. The proposed CNN model helps farmers to predict plant diseases easily with high accuracy. Farmers have to be no longer dependent on field experts who are expensive to hire and take longer time than the proposed CNN model to predict plant diseases. Due to the smaller size of the proposed model, it does not require any costly hardware and it is affordable by all the farmers. If the plant diseases are accurately predicted at the right time, appropriate pest treatment can be done immediately. This improves

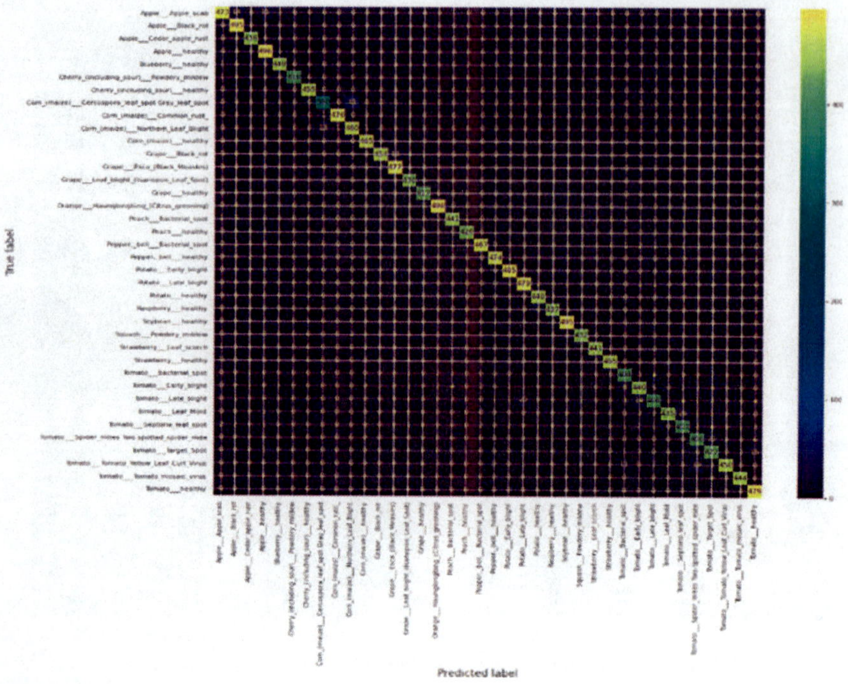

Fig. 16 Confusion matrix

the overall productivity of agricultural farms as the well-being of a plant is the most vital thing in farming. Hence, the proposed CNN model can create a huge impact on agriculture.

References

S. Ashok, G. Kishore, V. Rajesh, S. Suchitra, S.G. Sophia, B. Pavithra, Tomato leaf disease detection using deep learning techniques, in *2020 5th International Conference on Communication and Electronics Systems (ICCES)*, June 2020 (IEEE, 2020), pp. 979–983

J.G.A. Barbedo, Deep learning applied to plant pathology: the problem of data representativeness. Trop. Plant Pathol. 1–10 (2021)

P. Bedi, P. Gole, Plant disease detection using hybrid model based on convolutional autoencoder and convolutional neural network. Artif. Intell. Agric. **5**, 90–101 (2021)

A. Deopa, A. Sinha, A. Prakash, R.K. Sinha, Facial expression recognition using convolutional neural network and SoftMax function on captured images, in *2019 International Conference on Communication and Electronics Systems (ICCES)* (2019), pp. 273–279. https://doi.org/10.1109/ICCES45898.2019.9002524

P. Dileep, D. Das, P.K. Bora, Dense layer dropout based CNN architecture for automatic modulation classification, in *2020 National Conference on Communications (NCC)* (2020), pp. 1–5. https://doi.org/10.1109/NCC48643.2020.9055989

E. Ennadifi, S. Laraba, D. Vincke, B. Mercatoris, B. Gosselin, Wheat diseases classification and localization using convolutional neural networks and GradCAM visualization, in *2020 International Conference on Intelligent Systems and Computer Vision (ISCV)*, June 2020 (IEEE, 2020), pp. 1–5

L. Frank et al., Confidence-driven hierarchical classification of cultivated plant stresses, in *Proceedings of the IEEE/CVF Winter Conference on Applications of Computer Vision* (2021)

R. Gandhi, S. Nimbalkar, N. Yelamanchili, S. Ponkshe, Plant disease detection using CNNs and GANs as an augmentative approach, in *2018 IEEE International Conference on Innovative Research and Development (ICIRD)*, May 2018 (IEEE, 2018), pp. 1–5

G. Geetha, S. Samundeswari, G. Saranya, K. Meenakshi, M. Nithya, Plant leaf disease classification and detection system using machine learning. J. Phys. Conf. Ser. **1712**(1), 012012 (2020)

R.I. Hasan, S.M. Yusuf, L. Alzubaidi, Review of the state of the art of deep learning for plant diseases: a broad analysis and discussion. Plants **9**(10), 1302 (2020)

S. Huang, W. Liu, F. Qi, K. Yang, Development and validation of a deep learning algorithm for the recognition of plant disease, in *2019 IEEE 21st International Conference on High Performance Computing and Communications; IEEE 17th International Conference on Smart City; IEEE 5th International Conference on Data Science and Systems (HPCC/SmartCity/DSS)*, Aug 2019 (IEEE, 2019), pp. 1951–1957

A. Jenifa, R. Ramalakshmi, V. Ramachandran, Cotton leaf disease classification using deep convolution neural network for sustainable cotton production, in *2019 IEEE International Conference on Clean Energy and Energy Efficient Electronics Circuit for Sustainable Development (INCCES)* (IEEE, 2019)

X. Jin, J. Che, Y. Chen, Weed identification using deep learning and image processing in vegetable plantation. IEEE Access **9**, 10940–10950 (2021)

G. Lin, W. Shen, Research on convolutional neural network based on improved Relu piecewise activation function. Procedia Comput. Sci. **131**, 977–984 (2018)

Y.H. Liu, Feature extraction and image recognition with convolutional neural networks. J. Phys. Conf. Ser. **1087**(6) (2018)

Y. Liu, G. Gao, Z. Zhang, Plant disease detection based on lightweight CNN model, in *2021 4th International Conference on Information and Computer Technologies (ICICT)*, Mar 2021 (IEEE, 2021), pp. 64–68

G. Madhulatha, O. Ramadevi, Recognition of plant diseases using convolutional neural network, in *2020 Fourth International Conference on I-SMAC (IoT in Social, Mobile, Analytics and Cloud) (I-SMAC)*, Oct 2020 (IEEE, 2020), pp. 738–743

D. Oppenheim, G. Shani, O. Erlich, L. Tsror, Using deep learning for image-based potato tuber disease detection. Phytopathology **109**(6), 1083–1087 (2019). https://doi.org/10.1094/PHYTO-08-18-0288-R. Epub 2019 Apr 16. PMID: 30543489

S. Rajkumar, V. Abhyankar, P. Kurundkar, E. Ghosh, K. Kulkarni, Image processing and machine learning techniques for improvements in tomato farming, in *Soft Computing and Signal Processing* (Springer, Singapore, 2021), pp. 501–512

S. Ramesh et al., Plant disease detection using machine learning, in *2018 International Conference on Design Innovations for 3Cs Compute Communicate Control (ICDI3C)* (2018), pp. 41–45. https://doi.org/10.1109/ICDI3C.2018.00017

J.B. Ristaino, P.K. Anderson, D.P. Bebber, K.A. Brauman, N.J. Cunniffe, N.V. Fedoroff et al., The persistent threat of emerging plant disease pandemics to global food security. Proc. Natl. Acad. Sci. **118**(23), e2022239118 (2021)

M. Sardogan, A. Tuncer, Y. Ozen, Plant leaf disease detection and classification based on CNN with LVQ algorithm, in *2018 3rd International Conference on Computer Science and Engineering (UBMK)*, Sept 2018 (IEEE, 2018), pp. 382–385

C. Shorten, T.M. Khoshgoftaar, A survey on image data augmentation for deep learning. J. Big Data **6**(1), 1–48 (2019)

T. Talaviya et al., Implementation of artificial intelligence in agriculture for optimisation of irrigation and application of pesticides and herbicides. Artif. Intell. Agric. **4**, 58–73 (2020)

H. Tessier, V. Gripon, M. Léonardon, M. Arzel, T. Hannagan, D. Bertrand, Rethinking weight decay for efficient neural network pruning. J. Imaging **8**(3), 64 (2022)

D. Tiwari, M. Ashish, N. Gangwar, A. Sharma, S. Patel, S. Bhardwaj, Potato leaf diseases detection using deep learning, in *2020 4th International Conference on Intelligent Computing and Control Systems (ICICCS)*, May 2020 (IEEE, 2020), pp. 461–466

S. Wallelign, M. Polceanu, C. Buche, Soybean plant disease identification using convolutional neural network, in *The Thirty-First International Flairs Conference* (2018)

T.S. Xian, R. Ngadiran, Plant diseases classification using machine learning. J. Phys. Conf. Ser. **1962**(1) (2021)

S.Y. Yadhav, T. Senthilkumar, S. Jayanthy, J.J.A. Kovilpillai, Plant disease detection and classification using CNN model with optimized activation function, in *2020 International Conference on Electronics and Sustainable Communication Systems (ICESC)*, July 2020 (IEEE, 2020), pp. 564–569

X. Ying, An overview of overfitting and its solutions. J. Phys. Conf. Ser. **1168**(2) (2019)

Understanding the Insights of Privacy Policies Using BERT

Souvik Maitra and Dwijen Rudrapal

Abstract Privacy policies are the documents designed by an organisation to store and handle user's personal information. These documents are often long, full of legal terms and complex to understand. While reading and understanding the clauses before agreeing is mandatory in the privacy policy very few user take time to read them. User share sensitive and crucial personal information to access the content or the product of that organization without knowing policies due to the reading complexity of the privacy policy document. Rather than reading the whole policy document, an approach might be followed to know the gist of the privacy policy in least time by simply knowing the answer of specific questions users are concern about. In this work, we propose a question-answering based approach to know the answer of specific concerned questions from a privacy policy. The current work focused at the kinds of questions users are likely to pose to such a system shown by promising prior research works. We conduct experiments of our proposed approach on a corpus of privacy policies covering different types of organisations and reported our evaluation result.

Keywords Privacy policy · Summarization · Question-answering system · Machine learning · Natural language processing

1 Introduction

A privacy policy is a digital document (Ahmad et al. 2020; Ravichander et al. 2019) published by an organisation to collects, manages, uses, shares, and protects the personal information of its users. In general, user need to read and agreed to the terms and conditions include in the privacy policy (Reidenberg et al. 2015). The privacy policies are quite long in length and are frequently presented in a complicated manner (Gluck et al. 2016), with small fonts or long phrases, causing user to lose interest

S. Maitra (✉) · D. Rudrapal
National Institute of Technology Agartala, Agartala, India
e-mail: nascar.20296@gmail.com

A. B. Reddy et al. (eds.), *Proceedings of Third International Conference on Advances in Computer Engineering and Communication Systems*, Lecture Notes in Networks and Systems 612, https://doi.org/10.1007/978-981-19-9228-5_54

in the document. As a result, most of the users exchange personal information with the organisation in order to access a service or a product without reading or knowing policy terms. The lack of awareness of terms and conditions in the policy regarding the personal data being used leads to problematic situation like compromising of personal private information. Recent data misuse relevant scandal by Facebook and Cambridge Analytica (Cadwalladr and Graham-Harrison 2018), have brought serious concern to the privacy policies of various websites. This incident attracts researchers to develop such systems for quickly identifying the most salient privacy issues to the users to avoid potential security threat. In this work we propose a question-answer based approach to understand and review specific sections of policies that the users are most concerned in by presenting answers of specific questions from the privacy policies. Our work focused on set of questions drawn from prior promising research works to fetch best possible answer for quick understanding of the crucial terms in privacy policies.

2 Related Works

Prior research works make privacy policies understandable by summarizing the whole policy document quickly or answering the specific questions on policy terms. Recent research works (Liu et al. 2015; Oltramari et al. 2018; Sathyendra et al. 2017) primarily focused on understanding privacy policies by applying various Natural Language Processing (NLP) techniques. The approach proposed in Tomuro et al. (2016) generates summary of a policy document by extracting important sentences from the document. It combines important keywords defined by the human expert and machine learning algorithm in an ensemble architecture to generate the required summary. The work proposed in Nejad et al. (2019) and Harkous et al. (2018) analysed the policy documents and used NLP methods to classify the documents into different risk level section like green, yellow and red according the risk weightage. Their work is extended in the work (Keymanesh et al. 2020) and used a pre-trained encoder and convolutional neural network to classify sentences of the policy into different risk levels. A recent work in Alabduljabbar et al. (2021) proposed a machine learning-based automated classifier for classifying the content of the policy into nine classes and detects missing information if any in the privacy policies. Question answer based approaches explore the user concerned salient content from the document through specific queries. These approaches are further divided into two types: Open question-answer based (Caballero 2021; Zhu et al. 2021) systems and closed question-answer (QA) based (Caballero 2021; Lende and Raghuwanshi 2016) systems. Question answering system proposed in Kim et al. (2015) replies in yes or no for a specific user query. The system follows 3 phase architecture to serve the reply. The first phase uses a combination of ranking SVM and TF-IDF for information retrieval. Phase 2 utilizes features based word embeddings, syntactic similarities and identification of antonym relation to process the retrieved information. And the third phase answers in yes or no for unseen queries by tallying with meanings of previously extracted queries. Passage level question answering has been obtained in this work (Tan et al. 2016) by using both convolutional and recurrent neural networks. The approach in Sathyendra et al. (2017) incorporates both open and closed QA system. It segments the policy documents into sections and then ranks them based on BM25 scores by matching the user's query. The work in Zaeem et al. (2018) extracts answers of the questions and predicts the class of risk based on the keywords like email, third-party involvement, sharing of data etc. The system "PrivacyQA" built in Ahmad et al. (2020) predict the answer of question based on GloVe (Pennington et al. 2014) and BERT (Devlin et al. 2018) encoding technique. Each question was described in the experimental setting using three sets of features based on SVM: Bag-of-words (SVM-BOW), length of question in words (SVM-BOW + LEN), and parts-of-speech (SVM-BOW + LEN + POS). A recent work explained in Lamba et al. (2021) presents a transformer and sequence-to-sequence mapping approach to generate questions for understanding the privacy policies. Rather than focusing on a fixed set of questionnaire the approach generates questions to enhanced the understanding the privacy policies.

3 Proposed Methodology

We have proposed an extractive question-answering model to precise the terms in privacy policy for better understanding the risk of personal data sharing. The proposed model is trained on a dataset containing the content of privacy policy and possible question-answer pairs. At initial stage the corpus is pre-processed by tokenizing and removing unnecessary tokens. If the policy contains a sentence that is most likely to be the answer for the question it extracts and returns by the model it to the user, else it produces a suitable message to the query posted by the user.

3.1 Dataset

We have created an annotated dataset comprising of 100 randomly chosen privacy policies which amount to 6500+ sentences. The corpus has policies containing an average of 61.6 sentences where the shortest policy comprises of only 39 lines which goes up till the longest comprising of 132 sentences. The policies are collected from a wide rage of business websites like that of e-commerce, online payment platform, infotainment etc. Each context comprising of the policy is accompanied by a set of 6 question-answer pairs and *answer_start* attribute that indicates the starting character position of the answer in the context. Which means that the corpus contains 100 privacy policies and 600 question/answer pairs. As this corpus is comparatively small to train a BERT language model, hence we have also considered training the *bookcorpus* (Zhu et al. 2015) along with our primary dataset to increase the vocabulary size of the model.

3.2 Methodology Used

BERT, a form of neural network that receives a sequence as input and creates some output, has been used for the question-answering job. Unlike classic recurrent networks like LSTMs, which process each sequence element in turn, the Transformer processes all components at the same time by using an attention mechanism to build direct connections between individual elements. This not only allows for more parallelization, but it also improves accuracy across a wide range of tasks.

Here Transfer learning comes into play which is the process of training a model on one task and then using that information to improve the model's performance on another activity. BERT is first trained on two unsupervised tasks: masked language modelling (Taylor 1953) (predicting a sentence's missing word) and next sentence prediction (predicting if one sentence naturally follows another). Then BERT can work on any new assignment with a firm foundation in the workings of the English language thanks to extensive pre-training over a broad corpus (Fig. 1).

The multi-head attention of BERT allows the model to capture a wider range of word associations than a single attention mechanism would allow. As it progresses further into the model, BERT is able to generate highly rich representations through a repetitive synthesis of word embeddings. Because the attention heads have different settings, each one develops its own attention pattern. BERT Large, the version we're looking at here, contains 24 layers and 16 heads, for a total of $24 \times 16 = 384$ unique attention processes.

A compatibility function used by BERT gives each pair of words a score indicating how closely linked they should be. The model gives each word a query vector and a key vector in order to determine compatibility. These vectors are a form of word embedding similar to the value vectors, but they're built expressly for detecting word compatibility. In this scenario, the compatibility score is just the dot product of one word's query vector and the other's key vector(the query and key vectors are dynamically calculated based on the preceding layer's output). These compatibility scores are normalized to sum to one in order to convert them into valid attention weights.

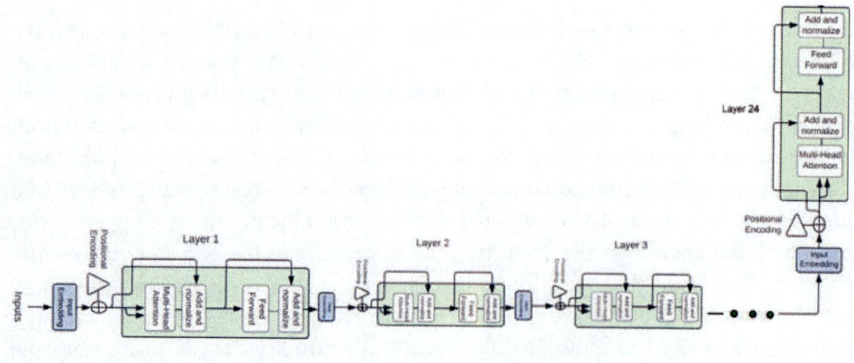

Fig. 1 BERT large language model

The softmax function is applied to the scores for a given word to achieve this. To split text into words, a basic tokenization strategy is employed where terms that are not in the lexicon are considered as "unknown". WordPiece tokenization is used to tokenize text into subword units in modern NLP models, which typically maintain linguistic meaning (i.e., morphemes). Even if the model does not recognise a word, individual subword tokens may contain enough information for the model to infer the meaning to some extent. Then we simply mask some fraction of the input tokens at random and then forecast those masked tokens to train a deep bidirectional representation. The final hidden vectors for the masked tokens are sent into an output softmax over the vocabulary, much like in a regular LM. We mask 15% of all WordPiece tokens in each sequence at random. We simply forecast the masked words rather than recreating the complete input, unlike de-noising auto-encoders (Vincent et al. 2008). The [MASK] token does not arise during fine-tuning, resulting in a mismatch between pre-training and fine-tuning, even though this enables us to construct a bidirectional pre-trained model. We don't always substitute "masked" words with the [MASK] token to prevent this. The training data generator randomly selects 15% of the token placements for prediction. If the ith token is chosen, we either (1) replace the ith token with the [MASK] token 80% of the time, (2) a random token 10% of the time, or (3) the unchanged ith token 10% of the time.

The model then learns to predict whether the second sentence in a pair is the next sentence in the original text by receiving pairs of sentences as input. Input pairs with the second sentence being the subsequent one in the original text make up 50% of the inputs during training, while the remaining 50% are randomly selected phrases from the corpus. It is considered that the random statement has nothing to do with the first.

3.3 Training the Model

Training the model with the annotated dataset needs constructing parallel lists of contexts, questions, and responses (the contexts are repeated since each privacy policy has 6 questions each). The answers are dictionaries that contain the right answer's sub-sequence as well as an integer indicating where the answer begins. To train a model on this data, we'll firstly need tokenized context/question pairings, and secondly numbers showing where the response begins and stops at each token location.

4 Result Analysis

This is where we support our experimental setup with the evaluation metrics for the 6 risk factors (1—first-party collection, 2—third-party sharing, 3—data security, 4—user control, 5—data retention, 6—policy change intimation). The evaluation process

Table 1 Confusion matrix for risk factors

Risk factor	Actual count	Predicted positive	Predicted negative
1	Positive	75	3
	Negative	17	5
2	Positive	74	6
	Negative	18	2
3	Positive	69	2
	Negative	21	8
4	Positive	42	16
	Negative	33	9
5	Positive	51	13
	Negative	24	12
6	Positive	73	16
	Negative	2	9

Table 2 BERTLarge versus BERTBase

Model	Average F1
BERTLarge	0.82
BERTBase	0.79

gets a better estimation of the accuracy of the model as a K-fold cross validation is employed. With $K = 5$ we created 5 batches of data (20 policies in each batch) out of the 100 policies in the dataset. Each batch is utilised as the test set in each fold of the process with the other 4 sets as the train set. By observing the results as shown in Table 1 we can consider the experimental setup to be fairly accurate with an average F1 0.82, but there are still scopes for improvement.

The model based on BERTLarge performs better than a similar model designed with BERTBase as per experiment results shown in Table 2. The extra layers of encoders and the attention heads of the BERTLarge come into play to exhibit better accuracy.

5 Error Analysis

For most of the queries, the proposed model generates most relevant answer from the policy document. However, in certain scenarios, the model could not generated expected answer. For example, according to the gold standard dataset this question is unanswerable in correspondence to the context, but as we can see it has produced an answer. One of the main problem with this setup is the small size of the custom dataset of 100 privacy policies which results in a lesser number of word embeddings related to this domain. A bigger dataset would have definitely produced far more accurate

```
You can review and edit your personal information at any time by logging in to
your account and reviewing your account settings and profile. You can also close
your account through the PayPal website. If you close your PayPal account, we
will mark your account in our database as Closed, but may retain personal
information from your account to collect any fees owed, resolve disputes,
troubleshoot problems, assist with any investigations, prevent fraud, enforce
our User Agreement, or take other actions as required or permitted by law.

Question: How long user information will be stored?
Answer : Sorry!, Not able to answer your question. Try a different question related to the context.
```

Fig. 2 Wrong prediction result

results. The limited number of vectored tokens for this domain fails to identify the result correctly from the context and produces an output that states the model is unable to find an answer for the question. But the gold standard dataset contains an answer for this question (Fig. 2).

6 Conclusion

Studying the works can always bring refinement in the future approaches to come. This model has a fairly high accuracy score F1 0.82 for the privacy policy question-answering but could perform better with a larger dataset specifically for this domain. The current work also developed a gold standard corpus for privacy policies with human generated answers for standard queries. However, as the open QA approaches need to address more user queries than the closed QA approaches, the scope of the present work may further explore more user's queries relevant to personal critical information and can significantly contribute to the improvement of the proposed model.

References

W.U. Ahmad, J. Chi, Y. Tian, K.W. Chang, Policyqa: a reading comprehension dataset for privacy policies. arXiv preprint arXiv:2010.02557 (2020)

A. Alabduljabbar, A. Abusnaina, U. Meteriz-Yildiran, D. Mohaisen, *TLDR: Deep Learning-Based Automated Privacy Policy Annotation with Key Policy Highlights* (Association for Computing Machinery, New York, NY, USA, 2021), pp. 103–118. https://doi.org/10.1145/3463676.3485608

M. Caballero, A brief survey of question answering systems. Int. J. Artif. Intell. Appl. **12**, 01–07 (2021)

C. Cadwalladr, E. Graham-Harrison, Revealed: 50 million Facebook profiles harvested for Cambridge Analytica in major data breach. Guardian **17**, 22 (2018)

J. Devlin, M.W. Chang, K. Lee, K. Toutanova, Bert: pre-training of deep bidirectional transformers for language understanding. arXiv preprint arXiv:1810.04805 (2018)

J. Gluck, F. Schaub, A. Friedman, H. Habib, N. Sadeh, L.F. Cranor, Y. Agarwal, How short is too short? Implications of length and framing on the effectiveness of privacy notices, in *Twelfth Symposium on Usable Privacy and Security (SOUPS 2016)* (2016), pp. 321–340

H. Harkous, K. Fawaz, R. Lebret, F. Schaub, K.G. Shin, K. Aberer, Polisis: automated analysis and presentation of privacy policies using deep learning, in *27th USENIX Security Symposium (USENIX Security 18)* (2018), pp. 531–548

M. Keymanesh, M. Elsner, S. Sarthasarathy, Toward domain-guided controllable summarization of privacy policies, in *NLLP@ KDD* (2020), pp. 18–24

M.Y. Kim, Y. Xu, R. Goebel, Applying a convolutional neural network to legal question answering, in *JSAI International Symposium on Artificial Intelligence* (Springer, 2015), pp. 282–294

D. Lamba, W.H. Hsu, Answer-agnostic question generation in privacy policy domain using sequence-to-sequence and transformer models, in *2021 2nd International Conference on Electronics, Communications and Information Technology (CECIT)* (2021), pp. 256–261

S. Lende, M. Raghuwanshi, Closed domain question answering system using NLP techniques (2016)

Y.H. Liu, Y.L. Chen, W.L. Ho, Predicting associated statutes for legal problems. Inf. Process. Manag. **51**(1), 194–211 (2015)

N.M. Nejad, D. Graux, D. Collarana, Towards measuring risk factors in privacy policies, in *AIAS@ ICAIL* (2019), pp. 18–20

A. Oltramari, D. Piraviperumal, F. Schaub, S. Wilson, S. Cherivirala, T.B. Norton, N.C. Russell, P. Story, J. Reidenberg, N. Sadeh, Privonto: a semantic framework for the analysis of privacy policies. Semant. Web **9**(2), 185–203 (2018)

J. Pennington, R. Socher, C.D. Manning, Glove: global vectors for word representation, in *Proceedings of the 2014 Conference on Empirical Methods in Natural Language Processing (EMNLP)* (2014), pp. 1532–1543

A. Ravichander, A.W. Black, S. Wilson, T. Norton, N. Sadeh, Question answering for privacy policies: combining computational and legal perspectives. arXiv preprint arXiv:1911.00841 (2019)

J.R. Reidenberg, T. Breaux, L.F. Cranor, B. French, A. Grannis, J.T. Graves, F. Liu, A. McDonald, T.B. Norton, R. Ramanath, Disagreeable privacy policies: mismatches between meaning and users' understanding. Berkeley Technol. Law J. **30**, 39 (2015)

K.M. Sathyendra, A. Ravichander, P.G. Story, A.W. Black, N. Sadeh, *Helping Users Understand Privacy Notices with Automated Query Answering Functionality: An Exploratory Study*, Technical report, Carnegie Mellon University, 2017

K.M. Sathyendra, S. Wilson, F. Schaub, S. Zimmeck, N. Sadeh, Identifying the provision of choices in privacy policy text, in *Proceedings of the 2017 Conference on Empirical Methods in Natural Language Processing* (2017), pp. 2774–2779

M. Tan, C. Dos Santos, B. Xiang, B. Zhou, Improved representation learning for question answer matching, in *Proceedings of the 54th Annual Meeting of the Association for Computational Linguistics (Volume 1: Long Papers)* (2016), pp. 464–473

W.L. Taylor, "Cloze procedure": a new tool for measuring readability. J. Q. **30**(4), 415–433 (1953)

N. Tomuro, S. Lytinen, K. Hornsburg, Automatic summarization of privacy policies using ensemble learning, in *Proceedings of the Sixth ACM Conference on Data and Application Security and Privacy. CODASPY '16* (Association for Computing Machinery, New York, NY, USA, 2016), pp. 133–135. https://doi.org/10.1145/2857705.2857741

P. Vincent, H. Larochelle, Y. Bengio, P.A. Manzagol, Extracting and composing robust features with denoising autoencoders, in *Proceedings of the 25th International Conference on Machine Learning* (2008), pp. 1096–1103

R.N. Zaeem, R.L. German, K.S. Barber, Privacycheck: Automatic summarization of privacy policies using data mining. ACM Trans. Internet Technol. (TOIT) **18**(4), 1–18 (2018)

F. Zhu, W. Lei, C. Wang, J. Zheng, S. Poria, T. Chua, Retrieving and reading: a comprehensive survey on open-domain question answering. CoRR abs/2101.00774 (2021), https://arxiv.org/abs/2101.00774

Y. Zhu, R. Kiros, R. Zemel, R. Salakhutdinov, R. Urtasun, A. Torralba, S. Fidler, Aligning books and movies: towards story-like visual explanations by watching movies and reading books. arXiv preprint arXiv:1506.06724 (2015)

Classification of Family Domain of Amino Acid Sequences Using CNN-LSTM

G. S. Mahalakshmi, Gokul Sunilkumar, Steven Fredrick Gilbert, and S. Sendhilkumar

Abstract As diseases become more and more resistant and adaptive, protein classification is crucial to identify and create cures or vaccines for the diseases. Proteins interact with other macromolecules, playing a central role in many biological processes. Investigating protein functionality is cumbersome as it mandates structural/biochemical studies. Virus protein classification thus helps researchers to determine the basis of the disease, which in turn elucidates methods to prevent and treat the disease. The proposed approach uses a CNN-LSTM architecture for the main purpose of protein identification. LSTMs have an advantage over RNNs in that they do not suffer from the vanishing gradient problem. LSTMs are very good at processing long sequences without losing any gradient value. Coupled with this, a CNN is employed to deep extract the features present in the protein sequence. The data set used here is based on real-time data of researchers, who conduct various tests on proteins and the results are archived together as a whole data set. The same classification task is performed using a FastText architecture and the performances are compared and analysed to infer the most efficient and accurate architecture.

Keywords Protein sequences · LSTM · CNN

1 Introduction

Protein classification is an important part of modern biotechnology as proteins are the macromolecules which play a major role in the biological processes of a cell. The amino acid sequences of proteins shall be derived from DNA sequences. Since

G. S. Mahalakshmi (✉) · G. Sunilkumar · S. F. Gilbert
Department of Computer Science and Engineering, College of Engineering Guindy, Anna University, Chennai, Tamil Nadu, India
e-mail: gsmaha@annauniv.edu

S. Sendhilkumar
Department of Information Science and Technology, College of Engineering Guindy, Anna University, Chennai, Tamil Nadu, India
e-mail: thamaraikumar@annauniv.edu

A. B. Reddy et al. (eds.), *Proceedings of Third International Conference on Advances in Computer Engineering and Communication Systems*, Lecture Notes in Networks and Systems 612, https://doi.org/10.1007/978-981-19-9228-5_55

amino acid sequences serve as the foremost for all biological functions, the study of proteins are crucial as different and new diseases can be studied from the proteins they are made up of and new drugs can be synthesized based on the proteins to come up with a cure. Protein functionality is embedded in the amino acid sequences. Hence, interpreting the sequence-function pair is the key issue to be resolved. Proteins with similar activities are grouped as protein families. In addition, classifying a type of protein can bring insight to its functional properties. Investigating protein functionality is a very lengthy process and involves huge research studies. Virus protein classification thus helps researchers to determine the basis of the disease, which in turn elucidates methods to prevent and treat the disease. The proposed work uses a CNN-LSTM architecture for the main purpose of identification of Class of Family domain of amino acid sequences.

2 Related Work

Saidi et al. (2010) adopted the classic methodology of filtered classification by pre-processing the input amino acid sequence through motif extraction before subsequent classification using a downstream architecture. The parallel 2D wavelet denoising algorithm processes the extracted feature vectors prior to classification (Wang and Wang 2019). Position specific scoring matrix is introduced for extracting the respective feature vectors of underlying protein sequence (Bhasin and Raghava 2006). Partitioning the protein sequence as equal-sized chunks is assumed as the first feature set. The second feature set looked at domains with the same conservation ratio. For such probing residues, physiochemical properties were extracted and considered as the third feature set. The last set is defined by accommodating all the 3 features together. The idea is evaluated using 4 baseline approaches for classification. The highest level of categorization accuracy was 72.5%. Due to a significant percentage of misclassification, the accuracy was low. However, extraction of protein sequences with better significance would increase the accuracy to a greater extent.

Rizzo et al. (2015) provide a model that is a modified version of LeCun et al.'s (1998) LeNet-5. Two lower levels of CNN are modified, followed by max-pooling with two fully connected layers, totalling six layers for processing. The max-pooling algorithm performs 1D convolutions to produce a feature map, which is subsequently down sampled. The completely connected layers have a hidden layer with 500 units and an output layer with one unit for each class. The model they have here is a CNN-based model that has used 3000 16s rRNA (Cole et al. 2009) sequences in total.

La Trobe University's Wang and Huang (2005) took a different strategy and used extreme learning machine (Huang et al. 2004) to implement the categorization. This is a single hidden layer feed-forward network (SLFNs). The inverse operation of hidden layer is used to determine SFLN's operation as a linear system.

Bosco et al. (Caruana 1997) suggested multi-task learning implementation (Bosco and Di Gangi 2016). They present a model for both a CNN and an RNN network that is a multi-task learning variation. The variation is created by reproducing the network's dense layers. Each work is replicated, with each replica having 500 units. Due to the increased network size, the model is only trained once rather than five times, with the expectation that it will perform better than the single task version.

The study provided by Bileschi et al. (Cao and Xiong 2014) introduces the ProtoCNN model, which is a CNN model based on the ResNet model. The advantage of utilizing a ResNet is that it may obtain improved accuracy as depth is increased without causing bad effects. The model comprises an initial convolution block connected to residual blocks. Each residual block is made up of batch normalized ReLU layers.

Feature encoding is performed by 1-gram method (Bileschi et al. 2019). For this, 20 amino acids are selected to define the feature set. The feature set is constructed in such a manner that every amino acid appears in some or other protein sequence. To identify super family prototypes, variable-length fuzzy genetic clustering is used followed by nearest neighbour classification. The research used three super families of proteins: globin, RAS and trypsin. On the previously indicated data set, the classification accuracy was 81.3%. The classification accuracy of a variable-length protein sequence can be improved by using more informative and more relevant characteristics. Bileschi et al. (2019) brought into light an approach known as ProtENN enabling the addition of 6.8 million entries to the PFAM data set in itself. This is possible as the deep learning technique is reliably able to predict the function of certain patterns and their mapping to the target protein family.

Caragea et al. used hash keys to convert high-dimensional characteristics to low-dimensional features using the feature hashing technique. The standard-gram encoding was used to obtain the high-dimensional characteristics. The feature vector size was lowered from 222 to 210 without a significant reduction in classification accuracy. Using 1–4 gramme sequence features, the suggested technique resulted in best feature reduction and the classification accuracy of 82.83% was achieved. Though considerable feature reduction is already performed, further reduction is possible and desirable as the size of resulting features is still larger for processing. Feature encoding and selection would improve the classification accuracy.

Some algorithms like BLAST (Altschul et al. 1990)[1] work on finding the local sequence similarities to compare biological sequence information. It explores all possible amino acid sequence databases to retrieve dependency information. The above works were applied on the benchmark PFAM protein sequence data set[2,3] in order to document their performance against a global scale.

To represent a protein sequence, a string dictionary approach is proposed. Repeated strings were only taken into account once and the frequency was observed. The frequency/probability matrix was factorized using singular value decomposition (SVD), which accurately represented each protein sequence. This resulted in better

[1] https://blast.ncbi.nlm.nih.gov/Blast.cgi?PAGE=Proteins.

[2] http://pfam.xfam.org/.

[3] https://www.kaggle.com/googleai/pfam-seed-random-split.

feature set size reduction. Hence, to overcome the above shortcomings, the following architecture is proposed.

3 Model for Classification of Family Domain of Amino Acid Sequences

The input amino acid sequence is a string of upper case characters that cannot be fundamentally embedded to any vector space using pre-trained embedding like BERT, FastText, Glove or by training a custom embedding layer as all the above are only trained on a corpus of words that are part of a natural language set (English here). Hence, the architecture is formulated in such a way that it directly one-hot encodes the input sequence and zero pads it to maintain consistent dimensions. Now, we define a batch convolution layer whose contents are a dilated convolution layer, a max-pooling layer, a batch normalization layer followed by a ReLU activation layer. The intent here is to convolve the input sequence using a dilated layer so that long-term dependencies can be accurately captured. This, when max pooled, reduces the dimensions by a factor of the stride of the filter (Fig. 1).

Batch normalization speeds up training by scaling down the penultimate layer neurons values to the same range post which, ReLU activation boosts back-propagation by reducing the vanishing gradients problem. A sequence of these layers is experimented in different permutations and the best performing order is fixed. Some batch convolution layers contain dropout layers to bring regularization into effect. The intermediate output is then subjected to an LSTM layer that helps in the contextual representation of the input sequence as every character would have a relation with every other in the sequence. Once the LSTM decoded, output is affine transformed using a multi-layer perceptron, and final exponential normalization is

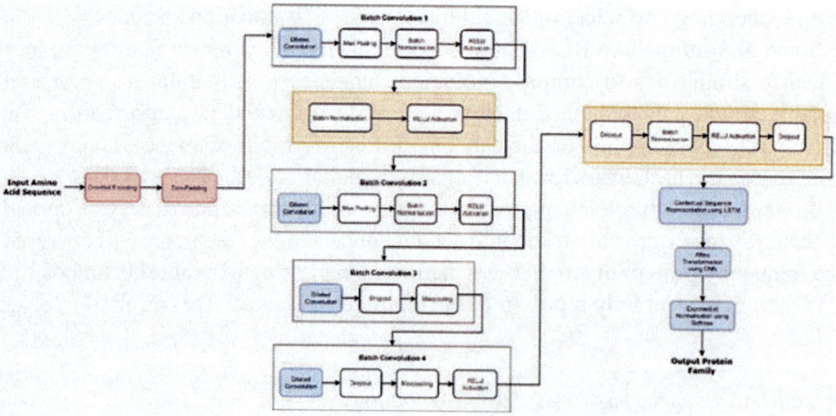

Fig. 1 Architecture for classification of amino acid sequences for family domain

performed using a Softmax activation function. This gives out the required output protein family. Although the task is classification, a classic framework has not been employed as the input to the pipeline is not a typical textual corpus or an image corpus. It is a character sequence that does not have any natural language semantic meaning associated with it.

4 Results and Discussion

The PFAM database is a collection of families of protein domains (Saidi et al. 2010). Proteins comprise multiple functional regions known as domains. Different domains in varying combinations can produce a diverse category of proteins. This data set posted by Bileschi et al. on Kaggle called the seed random split (Wang and Wang 2019; Bhasin and Raghava 2006). This data set was provided by Bileschi et al. by taking the highly curated protein families and splitting each family with at least 10 seed sequences. This means that only all the protein sequences available in this data set correspond to a single domain. The data set (Table 1) consists of sequences of amino acids along with other metadata such as sequence name, aligned_sequence, family_id and family_accession. Family accession is used here to identify the different families within the domain and is of the form PFxxxxx.y, where xxxxx is the family accession and y is the version number. This data set consists of over 1 million sequences with 18,000 unique families. Table 2 lists the hyper-parameters for the proposed model. Table 3 lists the overall loss and accuracy of the proposed system.

Taking a closer look at the data set, there are 17,929 unique family accessions. After the pre-processing of data, the dedicated train, test and validation data sets were considered for further training. This is given by train—500,000 and test—25,000 and validation—25,000 data points. The data is also shuffled so as to reduce variance and to make the model more general and to avoid any over-fitting. The overall accuracy and loss curve across epochs are seen in Fig. 2a, b.

The model was compared with other baseline approaches like ProtCNN and FastText CNN. The performance comparison is seen in Table 4.

The model took about 9 h for pre-processing (cleaning, one-hot encoding and count vectorising) and 4 h to train with a training accuracy of 94.73% and loss of 0.3454 with validation accuracy of 94.64% and loss of 0.3524. Other metrics for performance are calculated as weighted average of all classes. The time stated above is on an Nvidia Quadro P5000 16 GB RAM GPU Specifications: Cuda—11.2CuDNN—7.0.1 (Tables 5 and 6).

The proposed model performs better than the BLASTP (Rizzo et al. 2015) model which is based on a Basic Local Alignment Search Tool Query (LeCun et al. 1998) which has a much higher error rate. In FastText model, under-sampling the majority classes to 400 samples and up-sampling the minority classes to the same size has helped the model to learn a good representation of the PFAM protein database by achieving a P@1 and R@1 score of 0.908.

Table 1 PFAM data set details

Label	Data
sequence	HWLQMRDSMNTYNNMVNRCFATCIRSFQEKKVNAEEMDCTKRCVTKFVGYSQRVALRFAE
family_accession	PF02953.15
sequence_name	C5K6N5_PERM5/28-87
aligned_sequence	…HWLQMRDSMNTYNNMVNRCFATCI………RS.F…QEKKVNAEE…MDCT…KRCVTKFVGYSQRVALRFAE
family_id	zf-Tim10_DDP

Table 2 Parameters used in model training

Epochs	Optimizer	Learning rate	Loss function	Batch size
50	Adam	0.001	Categorical cross entropy	256

Table 3 Loss and accuracy of the proposed model—deep CNN-LSTM

Validation		Training	
Loss	Accuracy	Loss	Accuracy
0.3454	0.9473	0.0355	0.9892

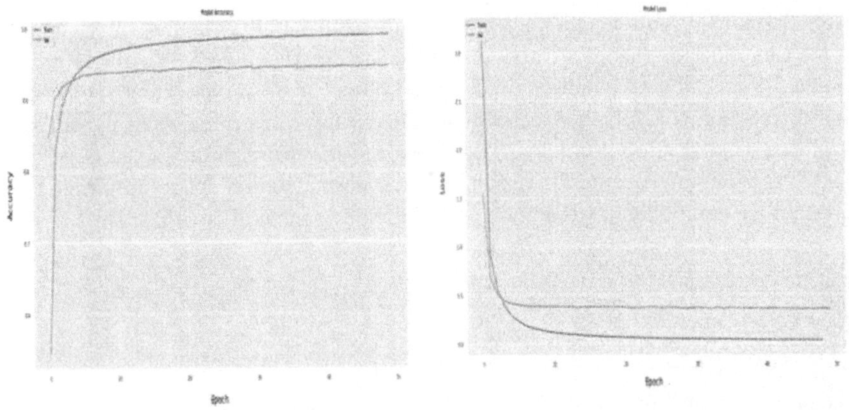

Fig. 2 **a** Accuracy curve of deep CNN-LSTM. **b** Loss curve of deep CNN-LSTM

Table 4 Comparisons between models

Model	Accuracy	*F*1 score
ProtCNN model	0.73	0.68
FastText model		0.9071
Deep CNN-LSTM model	0.9473	0.94

In CNN-LSTM, using a dilated CNN instead of a normal one proves to be more efficient as the field of reception is spread. This results in better operational perspective for same kernel parameters and computational complexity. In order to combat the problem of a highly imbalanced data set, we first over-sample the minority classes

Table 5 Performance metrics of proposed model

Classes	Accuracy %	Precision	Recall	*F*1 score
17,929	95.00	0.95	0.95	0.94

Table 6 Comparing performance metric of different models

Model	Classes	Error rate %	Number of errors
ProtCNN by Bileschi et al. (Bhasin and Raghava 2006)	17,929	0.159	201
Blastp (Bhasin and Raghava 2006; Rizzo et al. 2015)	17,929	1.645	2087
RNN (LSTM)	100	0.63	219
Proposed model deep CNN-LSTM	*17,929*	*0.52*	*180*

Significance of italics is to differentiate between our models and previous models

using Random Upsampler and then under-sample the majority classes with an appropriate sampling rate using a Random Undersampler. Using an LSTM layer post 4 sequences of batch convolutions helped the model learn a contextual representation of the amino acid sequence, which is extremely important when it comes to sequence classification problems. This improved the test accuracy from 0.8936 to 0.9473 (5.37% improvement). Usage of batch normalization speed up training by normalizing the hidden layer activation and handling internal covariate shift.

5 Conclusion

The proposed work focused in formulating two separate architectures and compared their performances on the benchmark PFAM data set. First one focuses on applying a CNN-LSTM assorted architecture while the second type uses a FastText methodology. From the experiments, the CNN-LSTM architecture outperformed in classifying the amino acid sequences to their dedicated protein families. In future, an automated classifier like this could be used to map out the vaccine generation schema of a new virus strain in a much quicker and efficient manner than the current manual mapping approach.

References

S.F. Altschul, W. Gish, W. Miller, E.W. Myers, D.J. Lipman, Basic local alignment search tool. J. Mol. Biol. **215**(3), 403–410 (1990)

M. Bhasin, G.P.S. Raghava, Computational methods in genome research. Appl. Mycol. Biotechnol. **6**, 179–207 (2006). ISSN 1874-5334. ISBN 9780444518071. https://doi.org/10.1016/S1874-5334(06)80011-0

M.L. Bileschi, D. Belanger, D. Bryant, T. Sanderson, B. Carter, D. Sculley, M.A. DePristo, L.J. Colwell, Using deep learning to annotate the protein universe. bioRxiv 626507 (2019)

G.L. Bosco, M.A. Di Gangi, Deep learning architectures for DNA sequence classification, in *International Workshop on Fuzzy Logic and Applications* (Springer, Cham, 2016), pp. 162–171

J. Cao, L. Xiong, Protein sequence classification with improved extreme learning machine algorithms. BioMed Res. Int. **2014** (2014)

R. Caruana, Multi-task learning: a knowledge-based source of inductive bias. Mach. Learn. **28**, 41–75 (1997)

J.R. Cole, Q. Wang, E. Cardenas, J. Fish, B. Chai, R.J. Farris, A.S. Kulam-Syed Mohideen, D.M. McGarrell, T. Marsh, G.M. Garrity, J.M. Tiedje, The Ribosomal Database Project: improved alignments and new tools for rRNA analysis. Nucleic Acids Res. **37**(Database), D141–D145 (2009)

G.-B. Huang, Q.-Y. Zhu, C.-K. Siew, Extreme learning machine: a new learning scheme of feed-forward neural networks, in *Proceedings of International Joint Conference on Neural Networks (IJCNN2004)*, Budapest, Hungary, 25–29 July 2004

Y. LeCun, L. Bottou, Y. Bengio, P. Haffner, Gradient-based learning applied to document recognition. Proc. IEEE **86**(11), 2278–2324 (1998)

R. Rizzo, A. Fiannaca, M. La Rosa, A. Urso, A deep learning approach to DNA sequence classification, in *International Meeting on Computational Intelligence Methods for Bioinformatics and Biostatistics* (Springer, Cham, 2015), pp. 129–140

R. Saidi, M. Maddouri, E. Mephu Nguifo, Protein sequences classification by means of feature extraction with substitution matrices. BMC Bioinform. **11**, 175 (2010). https://doi.org/10.1186/1471-2105-11-175

D. Wang, G.-B. Huang, Protein sequence classification using extreme learning machine, in *Proceedings. 2005 IEEE International Joint Conference on Neural Networks*, vol. 3 (IEEE, 2005), pp. 1406–1411

S. Wang, X. Wang, Prediction of protein structural classes by different feature expressions based on 2-D wavelet denoising and fusion. BMC Bioinform. **20**, 701 (2019). https://doi.org/10.1186/s12859-019-3276-5

A Performance Evaluation of Situational-Based Fuzzy Linear Programming Problem for Job Assessment

Shivali Slathia, Rakesh Kumar, Mudassir Lone, Wattana Viriyasitavat, Amandeep Kaur, and Gaurav Dhiman

Abstract The motive of job evaluation is to elucidate the relative work that the role of different jobs makes towards different organisational objectives. Various methods are used to solve problems related to job evaluation as ranking method, factor comparison method, and many more. In the present paper, various jobs are evaluated in fuzzy environment. Fuzzy mathematical model is developed to solve various problems in job evaluation by using suitable methods. The objective of the paper is to locate various problems in evaluation of jobs and then solve the problems using fuzzy mathematical method. An endeavour has been done to solve the job-related problems by numerical and equations.

Keywords Job evaluation · Fuzzy number · Fuzzy linear programming problems · Optimization

---

S. Slathia · R. Kumar (✉)
Department of Mathematics, Lovely Professional University, Phagwara, Punjab 144411, India
e-mail: rakeshmalhan23@gmail.com

M. Lone
Department of Mathematical Sciences, Baba Ghulam Shah Badshah University, Rajouri ,
J&K 185234, India
e-mail: mudassirlone@bgsbu.ac.in

W. Viriyasitavat
Business Information Technology Division, Department of Statistics, Faculty of Commerce and Accountancy, Chulalongkorn University, Bangkok, Thailand

G. Dhiman
Department of Computer Science, Government Bikram College of Commerce, Patiala,
Punjab 147001, India

A. Kaur · G. Dhiman
Department of Computer Science and Engineering, University Centre for Research and Development, Chandigarh University, Gharuan, Mohali, India

G. Dhiman
Department of Computer Science and Engineering, Graphic Era Deemed to be University,
Dehradun, India

A. B. Reddy et al. (eds.), *Proceedings of Third International Conference on Advances in Computer Engineering and Communication Systems*, Lecture Notes in Networks and Systems 612, https://doi.org/10.1007/978-981-19-9228-5_56

1 Introduction

Job evaluation plays a vital role in creating and keeping a compensation structure by looking at the similarities as well as the differences in the substance and worth of occupation. If we focus on an association or organization and the salary structure is confusing, compensation might come into play. The pivotal role of job evaluation is to solve existing problems or inequalities and eliminate salary-related issues (Gupta and Ahmed, 1998). The various values of the job profile can be determined through ranking, comparing, or setting points. Job evaluation is a logical method for guaranteying consistency in choosing responsibility and salary for the workers. The value of a job is somewhere determined on the basis of pay. Job evaluations are validated on the basis of a selected example such as benchmarks or on the other hand key positions (Charnes et al. 1955). What follows is to contemplate the factor which will decide the value of the job. A variety of reasons are responsible that differ from job to job. Many efforts were made to minimize the issues related to job elevation and various methods were introduced to solve the problem proposed a linear programming method to solve problems in the execution of salary structure. Llewellen considers a comparable but modified sort of method, giving many more factors and pointers in the point system of evaluation (Llewellyn, 1964). All these modifications, however, could not find a resolution because various standard objectives are involved in practical job evaluation. In every organization, the leader wants to fix the desired objectives for every target. Ahmed and Gupta broadened the task and created a model of linear programming for deciding the actual value of the different levels of factors of a job (Gupta and Ahmed, 1998; Rubin and Narasimhan, 1981). Later it was recommended that the main problem which is multiple-criteria decision-making (MCDM) can be explained precisely as a Fuzzy model where we might consider goals and imperatives as Fuzzy goals. Bellman and Zadeh introduced the idea of decision-making in the Fuzzy model. In the coming times, many workers have tried to modify this concept of decision-making problems such as Rubin and Narasimha, Zimmermann, Yager, and more. (Rubin and Narasimhan, 1981) Zimmermann created a model called the Fuzzy mathematical model to overcome the problems with functions. He was the pioneer to develop an operational FLP for maximum problems of linear vectors (Zimmermann 1981; Zimmermann, 1978). Yager developed programming with fuzzy constraints and Goal programming in a fuzzy environment was developed by Narasimha (Yager, 1979; Narasimhan, 1980). Ying–Yun investigated vector problems and tackle them by using fuzzy programming (Ying-Yung, 1983).

Since it is tough for the opinion former, decision makers to tackle the objectives completely for the benchmark's jobs, hence by this paper, we consider goals as fuzzy in nature and an attempt is made to settle down the job evaluation problems and issues as fuzzy models in the environment for those jobs having various factors and in turn, those factors having various levels.

2 Literature Review

There is an improvement in (Wang et al. 2022) to the susceptible multimodal petrol store network in terms of economy, energy, and climate change by Chia Nan Wang, Nhat Luong Nhieu, Kim Phong Tran, and Yen Hui Wang (3E evaluation). This work provides a novel method for combining a heuristic computation with a precise arrangement by means of fuzzy optimization. Practical energy production has been one of contemporary society's primary concerns. To reduce carbon dioxide emissions while generating the necessary demand for energy, it is important to evaluate various energy innovations and production systems. In (Emmanuel 2020), fuzzy optimization generated considerable results for elements that were novel, ecological, and cost effective. These discoveries assist researchers in determining where to concentrate their energy innovation simplifying research in order to make it more manageable. Using waste from various sectors and fuzzy optimization, (Aviso et al. 2022) presented a study on the development of weather networking with the aid of fuzzy optimization. Comparative research on fuzzy optimization with fuzzy numbers is presented in (Kumar and Dhiman 2021a). Instead of complex arithmetic and logistics, a simple fuzzy mathematical method is employed. In the (Kumar and Joshi 2021) study, fuzzy linear programming problems are also solved by satisfying various fuzzy constraints. The conclusion of the research is that optimization may be achieved by determining the constraint levels. In (Kumar et al. 2021a), trapezoidal and triangular fuzzy LLP models were employed to solve the optimization cost problem of Indian railway coach production. The purpose of the article was to minimise production costs and maximise output. In this article, the influence of trapezoidal on composite fuzzy LLP models is investigated. A method for improving a fuzzy linear programming (FLPP) problem (Kumar et al. 2021b) in which fuzzy integers represent right-side bounds. To address fuzzy linear programming and the basic procedure on the proposed number, the relative evaluation of exhibiting and enhancing creation cost via another cut-based quadrilateral fuzzy number is proposed. Using TFLPP-(s, l, r) and triangular fuzzy linear programming, the paper (Chandrawat et al. xxxx) offers a solution to the RCF Kapurthala manufacturing challenge. The data gathered on the cost of manufacture of the coaches from 2009 to 2010 was collected in an effort to reduce the cost of production of the coaches and optimise the total cost of production. The primary objective of the assembly firm is to maximise profit by minimising production costs and increasing the selling prices of the numerous products they produce. Eventually, reaching the optimal incentive for the company's production costs will involve increasing the selling price of goods. Nevertheless, given the circumstances, the market's volatility and intensity, the provider sells their products at unique prices. Consequently, their benefit varies. The company's most pressing issue is to locate (Kumar, et al. 2020) the enhancement of the advantage, and this elimination can be mitigated with the aid of fuzzy linear programming problem. This study presents (Garg and Kumar 2018) a fuzzy c-implies clustering algorithm that intends to simultaneously improve the areas for rectilinear min-aggregate office area problems. The punitive measures specify that the safe outcomes are of speculative interest, but that the established methods may allow for much faster calculations. This work (Kumar et al. 2020) describes a method for achieving optimization using a fuzzy linear programming problem in which fuzzy integers represent the right-side bounds. To address fuzzy linear programming and the basic method on the proposed number, the relative examination of exhibiting and enhancing creation cost via a new cut-based quadrilateral fuzzy number is proposed. Time currently plays an essential role in our lives; accordingly, time zones have emerged. There are numerous time zones around the entire globe. Essentially, everything, regardless of category, is subject to time. Few countries contain at least two double cross zones. India observes Indian Standard Time (IST). It is the primary time standard, with + 5.30UTC preceding UTC, and it persists on the Indian subcontinent. After attaining independence, India, a vast nation, is grappling with a problem inside its own time zone; specifically, north-east Indians. Consequently, in this study (Kumar et al. 2019), we created a methodology based on fuzzy grouping and fuzzy *c*-means calculation and determined the issue based on the perception of dawn and dusk in several metropolitan communities in India. An ineffectively characterised work evaluation framework creates the conundrum of workers and their capabilities for their obligations and, if necessary, their compensation. This result in representative dissatisfaction, which ultimately worsens reduction, is exorbitant due to the dearth of talented representatives. This article (Ahmed et al. 2013; Kumar et al. 2022b; Rao et al. 2022; Singh et al. 2022) proposes a VIKOR computation as an innovative approach to labour evaluations. The designing-related roles in a global avionics' organisation were analysed to determine the adequacy of the proposed strategy for addressing the work assessment issue. A model (Alidrisi 2021) is developed for any organisation where performance evaluation is important for staff motivation, improvement of demeanour and behaviour, communicating and aligning individual and hierarchical points, and fostering healthy relationships between management and staff (Tiwari et al. 2022; Bhoi et al. 2022; Mekala et al. 2022). Fuzzy control is utilised to determine the overall exhibition record by combining the results of performance in specified standards and presenting them in mathematical characteristics, which will unquestionably secure the ease of the relevant human asset faculty while calculating execution ratings. The continuing paper (Manoj Kumar Mandal 2021; Yadav et al. 2022; Sumathy et al. 2022; Alferaidi et al. 2022) was the product of the job promotion framework's numerical model. The proposed technique is a combination of quadratic programming and fuzzy logic (Vaishnav et al. 2021; Kumar and Dhiman 2021b), in which quadratic writing of computer programmes (Singh et al. 2022; Gupta et al. 2022; Shukla et al. 2022) is used to enhance objective capacity with associated limitations, such as non-direct definition. Using fuzzy logic to evaluate unknown limits, vulnerability-related data is controlled. This work (Nojavan 2021; Kumar et al. 2022a) proposes another fuzzy methodology based on administration quality for the evaluation of instructional unit performance. To determine the productivity of instructional units in the light of their administrative quality ratings, a fuzzy DEA technique is utilised.

3 Assumptions

Let us assume a job evaluation problem with factor $y_i (i = 1, 2, 3, 4 \ldots a)$ and we will consider the infinite levels as $y_{ij} (i = 1, 2, 3 \ldots a; \quad j = 1, 2 \ldots b)$ for every factor of y_i.

Let us suppose that as j increases job complexity which is written as:

$$y_{ij} Q y_{ij} - 1 = 0, \quad i = 1, 2, 3, 4 \ldots a; \quad j = 1, 2, 3 \ldots b, \tag{1}$$

where we will consider Q as the relation, more complex.

Now let us consider the benchmarks job with levels and the constraints of every individual for smallest (y_{i1}) and for biggest (y_{in}) each factor for each level and that can be written as

$$y_{i1} \geq a_{i,} \tag{2}$$

$$y_{in} \leq b_{i,} \tag{3}$$

$i = 1, 2, 3 \ldots a$; where a_i and b_i represents the score.

It is assumed that value for factor at a given point must be higher by the amount as e_i as

$$y_{ij} - y_{ij} - 1 > e_i, \quad i = 1, 2 \ldots a; \quad j = 2, 3 \ldots b \tag{4}$$

We motive to find the exact values and by those values we could find actual score for each job.

Model:

We modified a model of fuzzy environment for job evaluation.

$$\underline{Z} = (yij : i = 1, 2, \ldots a, \quad j = 1, 2, 3, \ldots b) \tag{5}$$

$$s.t X_l(\underline{Z}) \cong Q_i, \tag{6}$$

$$y_{i1} \geq a_{i,} \tag{7}$$

$$y_{in} \leq b_{i,} \tag{8}$$

$$y_{ij} - y_{ij} - 1 > e_i, \quad (i = 1, 2 \ldots a; \quad j = 2, 3 \ldots b; \quad l = 1, 2 \ldots k; \quad a_i > c_i) \tag{9}$$

$$y_{ij} > 0(i = 1, 2 \ldots a, \quad j = 1, 2 \ldots b) \tag{10}$$

where $\cong$ is fuzzy equality.

The equality function in fuzzy in Eq. 5 can be represented equivalently as follows,

$$X_l(Z) \leq Q_l \tag{11}$$

$$X_l(Z) \geq Q_l \tag{12}$$

We consider benchmark jobs as $X_l(Z)(l = 1, 2 \ldots k)$.

Let us consider the basic equation as

$$u(z) = \begin{cases} \frac{z - z_l}{z_s - z_l} & z_l \leq z \leq z_s \\ 1 & z = z_s \\ \frac{z - z_s}{z_u - z_s} & z_s \leq z \leq z_u \end{cases} \tag{13}$$

So, the maximum and minimum function can be considered as (Fig. 1).

$$u_1(z) = \begin{cases} z - (z_s - z_l)\lambda \geq z_l \\ z - (z_u - z_l)\lambda \geq z_l \\ z - (z_u - z_s)\lambda \geq z_s \end{cases} \tag{14}$$

$$u_2(z) = \begin{cases} z + (z_s - z_l) \leq z_s \\ z + (z_u - z_l) \leq z_u \\ z + (z_u - z_u) \leq z_u \end{cases} \tag{15}$$

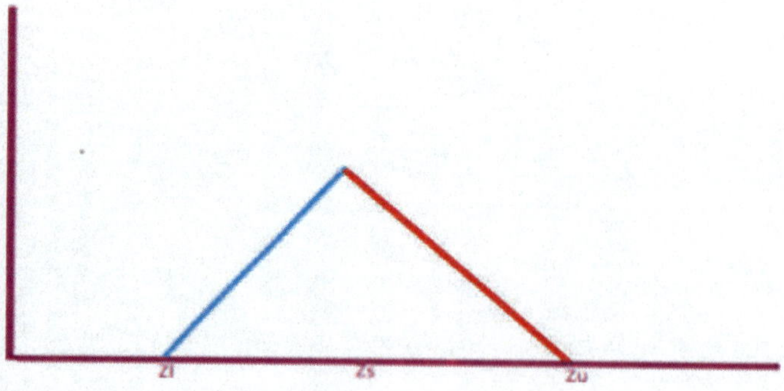

Fig. 1 Fuzzy triangular number used in Eq. 13

In the above Fig. 1, we used the triangle law to develop the equations.

We reduce the functions to simple LLP model as.

s.t

$$Z_u - (p_u - \underline{Z})\lambda \geq \underline{Z_l} \tag{16}$$

$$Z_l + (\overline{Z_l} - p_l)\lambda \leq \overline{Z_l} \tag{17}$$

$$y_{i1} \geq a_{i,} \tag{18}$$

$$y_{in} \leq b_{i,} \tag{19}$$

$$y_{ij} - y_{ij} - 1 > e_i, \quad (i = 1, 2 \ldots a; \quad j = 2, 3 \ldots b; \quad l = 1, 2 \ldots k; \quad a_i > c_i) \tag{20}$$

$$y_{ij} > 0(i = 1, 2 \ldots a, \quad j = 1, 2 \ldots b) \tag{21}$$

We will solve using suitable methods.

4 Mathematical Example

Consider an illustration of a mathematical model that explains the problem of job appraisal more clearly. Consider the topic of employment evaluation with five variables such as [$F = 5$], which may be used to pick candidates based on their performance in various factors such as nature, education, and flexibility, among others. In addition, the elements are categorised into six unique levels, such as [$L = 6$], and the different levels have been identified by particular management in the process of selecting candidates based on the various criteria also considered that the management has five benchmark jobs as [$K = 5$] and taking into account all the criteria and different levels, job evaluation benchmarks have been described. In the following equation, the maximum benchmark is deemed to be 105 for a particular candidate, while the minimum benchmark is deemed to be 50.7. The following equations reflect the numerous levels, aspects, and benchmarks associated with the job evaluation issue.

Let us consider all of them as

Factors—Nature

z_1—Education

z_2—Planning skills

z_3—Supervision quality

z_4—Duties complexity

z_5—Flexibility

Levels

$$z_{11}, z_{12}, z_{13}, z_{14}, z_{15}, z_{16}$$

$$z_{21}, z_{22}, z_{23}, z_{24}, z_{25}, z_{26}$$

$$z_{31}, z_{32}, z_{33}, z_{34}, z_{35}, z_{36}$$

$$z_{41}, z_{42}, z_{43}, z_{44}, z_{45}, z_{46}$$

$$z_{51}, z_{52}, z_{53}, z_{54}, z_{55}, z_{56}$$

Let us consider five identified benchmark jobs with worth's using Eqs. (11) and (12)

$$X_1(z) \equiv z_{16} + z_{26} + z_{36} + z_{46} \cong (95, 100, 105) \tag{22}$$

$$X_2(z) \equiv z_{15} + z_{25} + z_{35} + z_{45} \cong (80, 89, 98) \tag{23}$$

$$X_3(z) \equiv z_{14} + z_{24} + z_{34} + z_{43} \cong (70, 74.1, 78.2) \tag{24}$$

$$X_4(z) \equiv z_{13} + z_{23} + z_{33} + z_{43} \cong (60.8, 64.8, 68.8) \tag{25}$$

$$X_5(z) \equiv z_{11} + z_{22} + z_{33} + z_{42} \cong (50.7, 58.7, 66.8) \tag{26}$$

where $X_1(z)$ is the most profound job in the particular field.

The left part in the equation indicates the score for each job. So as score for the lowest and highest level for each individual is considered as

$$z_{i1} \geq 5, \tag{27}$$

$$z_{i6} \leq 40, \tag{28}$$

$$z_{ij} - z_{ij-1} \geq 2, \tag{29}$$

In the model *S*, we will investigate multiple benchmarks for job evaluation by plugging alternative values into Eqs. (16) and (17), and then evaluate the equation to get a conclusion. Putting the values into the equations and then contemplating how to address the problem of job appraisal. By using the simplex method for the values in Table 1 and 2. We conclude the values in Table 3 for the job selection.

Table 1 Compensation factors

Level	1	2	3	4	5	6
z_1	17.6	18.3	20.1	28.4	35	29.6
z_2	6	5	8	11	26.4	28.2
z_3	4	9	28.5	32	11.8	35
z_4	5	8	9	12	23.2	15

Table 2 Scores of benchmark jobs

Benchmark jobs	Scores
1	100.93
2	88.45
3	78.1
4	65.4
5	57.98

Max λ
s.t.

$$z_{14} + z_{24} + z_{34} + z_{43} - 5\lambda \geq 100, \tag{30}$$

$$z_{16} + z_{25} + z_{36} + z_{46} - 5\lambda \geq 100, \tag{31}$$

$$z_{16} + z_{25} + z_{36} + z_{46} - 10\lambda \geq 95, \tag{32}$$

$$z_{15} + z_{25} + z_{34} + z_{45} - 20\lambda \geq 89, \tag{33}$$

$$z_{15} + z_{25} + z_{34} + z_{43} - 5\lambda \geq 89, \tag{34}$$

$$z_{13} + z_{23} + z_{33} + z_{43} - 18\lambda \geq 80, \tag{35}$$

$$z_{11} + z_{22} + z_{32} + z_{43} - 4.1\lambda \geq 74.1, \tag{36}$$

$$z_{14} + z_{24} + z_{34} + z_{43} - 4.1\lambda \geq 74.1 \tag{37}$$

$$z_{11} + z_{22} + z_{33} + z_{42} - 8.2\lambda \geq 76, \tag{38}$$

$$z_{13} + z_{23} + z_{33} + z_{43} - 4\lambda \geq 64.8, \tag{39}$$

$$z_{16} + z_{25} + z_{36} + z_{46} - 4\lambda \geq 64.8, \tag{40}$$

$$z_{15} + z_{25} + z_{34} + z_{43} - 8\lambda \geq 60.8, \tag{41}$$

$$z_{15} + z_{25} + z_{34} + z_{43} - 8\lambda \geq 58.7, \tag{42}$$

$$z_{16} + z_{25} + z_{36} + z_{46} - 8\lambda \geq 58.7, \tag{43}$$

$$z_{11} + z_{22} + z_{32} + z_{43} - 16.1\lambda \geq 50.7, \tag{44}$$

$$z_{14} + z_{24} + z_{34} + z_{43} + 5\lambda \leq 95, \tag{45}$$

$$z_{16} + z_{25} + z_{36} + z_{46} + 5\lambda \leq 95, \tag{46}$$

$$z_{16} + z_{25} + z_{36} + z_{46} + 10\lambda \leq 105, \tag{47}$$

$$z_{15} + z_{25} + z_{34} + z_{45} + 20\lambda \leq 80, \tag{48}$$

$$z_{15} + z_{25} + z_{34} + z_{43} + 5\lambda \leq 98, \tag{49}$$

$$z_{13} + z_{23} + z_{33} + z_{43} + 18\lambda \leq 98, \tag{50}$$

$$z_{11} + z_{22} + z_{32} + z_{43} + 4.1\lambda \leq 70, \tag{51}$$

$$z_{14} + z_{24} + z_{34} + z_{43} + 4.1\lambda \leq 78.2, \tag{52}$$

$$z_{11} + z_{22} + z_{33} + z_{42} + 8.2\lambda \leq 78.2, \tag{53}$$

$$z_{13} + z_{23} + z_{33} + z_{43} + 4\lambda \leq 60.8, \tag{54}$$

$$z_{16} + z_{25} + z_{36} + z_{46} + 4\lambda \leq 68.8, \tag{55}$$

$$z_{15} + z_{25} + z_{34} + z_{43} + 8\lambda \leq 68.8, \tag{56}$$

$$z_{15} + z_{25} + z_{34} + z_{43} + 8\lambda \leq 50.7, \tag{57}$$

Table 3 Show the optimal values of the problem

Level	1	2	3	4	5	6
z_1	16.11104	16.75182	18.39954	25.99736	32.039	27.09584
z_2	6.126	5.87	8.008	11.011	26.4264	28.2282
z_3	4.92	11.07	35.055	39.36	14.514	43.05
z_4	4.88	7.808	8.784	11.712	22.6432	14.64

$$z_{16} + z_{25} + z_{36} + z_{46} + 8\lambda \leq 66.8, \quad (58)$$

$$z_{11} + z_{22} + z_{32} + z_{43} + 16.1\lambda \leq 66.8 \quad (59)$$

$$z_{i1} \geq 5 \text{ for } i = 1, 2, 3, 4, \quad (60)$$

$$z_{i6} \leq 40 \text{ for } i = 1, 2, 3, 4, \quad (61)$$

By keeping the above example in mind, we solve the problem as by taking all the values under consideration and Table 3 describes the different values of z_{11}, z_{12}, z_{13} and rest of the other values for the selection of the candidate for the job.

5 Conclusion

Thereby I conclude the paper by solving the problem of by using more efficient methods from earlier. The answers are more precise and correct as it helps to design various pay structure according to the need of the company. In future, we will try to solve more problems using various fuzzy numbers and try to tackle problems in job efficiency. In the paper, we have used fuzzy triangular number to solve the problem. Many papers are done to try and solve the issue, but we tried to make the method simpler to understand and to deal with. Gupta's paper helped us to understand the concept and with the help of fuzzy numbers and methods we tried and solve the problem efficiently. A lot of work is done in pasts and will be continued in future; we will also try to solve various problems of job evaluation in future by solving and constructing various methodology.

References

I. Ahmed, I. Sultana, S. Kumar, A. Azeem, Int. J. Prod. Perform. Manag. **62**(7), 718–735 (2013)

A. Alferaidi, K. Yadav, Y. Alharbi, N. Razmjooy, W. Viriyasitavat, K. Gulati, S. Kautish, G. Dhiman, Distributed deep CNN-LSTM model for intrusion detection method in IoT-based vehicles. Math. Prob. Eng. (2022)

H. Alidrisi, J. Risk Financial Manage. (2021)

E.R. Arriola, A.T. Ubando, Int. J. Energy Res. (2020)

K.B. Aviso, J.-Y. Lee, A.T. Ubando, R.R. Tan, Clean Technol. Environ. Policy **24**, 21–37 (2022)

A. Bhoi, R.C. Balabantaray, D. Sahoo, G. Dhiman, M. Khare, F. Narducci, A. Kaur, Mining social media text for disaster resource management using a feature selection based on forest optimization. Comput. Indus. Eng. 108280 (2022)

R.K. Chandrawat, R. Kumar, B.P. Garg, G. Dhiman, S. Kumar

A. Charnes, W.W. Cooper, R.O. Ferguson, Optimal estimation of executive compensation by linear programming. Management Sci. **1**(1), 138–151 (1955)

B.P. Garg, R. Kumar, R. K. Chandrawat, J. Adv. Res. Comput. Technol. Softw. Appl. **2**(1), 10–12 (2018)

J.N.D. Gupta, N.U. Ahmed, A goal programming approach to job evaluation. Comput. Eng. **14**, 147–152 (1998)

S. Gupta, M. Chakraborty, Job evaluation in fuzzy environment. Fuzzy Sets Syst. **100**, 71–76 (1998)

V.K. Gupta, S.K. Shukla, R.S. Rawat, Crime tracking system and people's safety in India using machine learning approaches. Int. J. Mod. Res. **2**(1), 1–7 (2022)

R. Kumar, G. Dhiman, Int. J. Mod. Res. **1**(1), 1–14 (2021a)

R. Kumar, G. Dhiman, A comparative study of fuzzy optimization through fuzzy number. Int. J. Mod. Res. **1**(1), 1–14 (2021b)

R. Kumar, G. Dhiman, N. Kumar, *Rajesh Kumar Chandrawat* (Varun Joshi and Amandeep kaur, Complex and Intelligent Systems, 2021a), pp.2053–2068

R. Kumar, J. Khepar, K. Yadav et al., A systematic review on generalized fuzzy numbers and its applications: past, present and future. Arch. Comput. Methods Eng. (2022a). https://doi.org/10.1007/s11831-022-09779-8

B.M. Kumar, R.K.R. Guduru, A. Srinivas, F. Ana, K. Ramudu, G. Dhiman, Wavelength assignment in optical fiber with intelligent optimization and assignment scheme for static and dynamic traffic intensity based photonic networks. Opt. Quant. Electron. **54**(8), 1–20 (2022b)

R. Kumar, V. Joshi, Eur. J. Mol. Clin. Med. 07 (2021)

R. Kumar, R. K. Chandrawat, R. Sen, V. Joshi, J. Gujarat Res. Soc. **21**(6) (2019) ISSN:0374-8588

R. Kumar, R.K. Chandrawat, B. Sarkar, V. Joshi, A. Majumder, Int. J. Fuzzy Syst. (2021b) ISSN: 1562-2479

R. Kumar, R.K. Chandrawat, B. Sarkar, V. Joshi, A. Majumder, Int. J. Fuzzy Syst. (2020)

R. Kumar et al., J. Phys. Conf. Ser. **1531**, 012085 (2020)

R.W. Llewellen, *Linear Programming, Holt* (Rinehart & Winston, New York, 1964)

M.K. Mandal, A.P. Burnwal, N. Dubey, O.P. Dubey, Int. J. Students' Res. Technol. Manage. **9**(2), 25–29. eISSN: 2321-2543

M.S. Mekala, G. Dhiman, G. Srivastava, Z. Nain, H. Zhang, W. Viriyasitavat, G.P.S. Varma, A DRL-based service offloading approach using DAG for edge computational orchestration. IEEE Trans. Comput. Soc. Syst. (2022)

R. Narasimhan, Goal programming in fuzzy environment. Decision Sci. **11**, 326 (1980)

M. Nojavan, A. Heidari, D. Mohammaditabar, Socio-economic planning sciences **73**, 100816 (February 2021)

G.S. Rao, J. Armstrong Joseph, G. Dhiman, H.S. Mohammed, S. Degadwala, R. Bhavani, Novel big data networking framework using multihoming optimization for distributed stream computing. Wirel. Commun. Mob. Comput. (2022)

P.A. Rubin, R. Narasimhan, Fuzzy goal programming with nested priorities, (Michigan State University, 1981)

T. Sharma, R. Nair, S. Gomathi, Breast cancer image classification using transfer learning and convolutional neural network. Int. J. Mod. Res. **2**(1), 8–16 (2022)

S.K. Shukla, V.K. Gupta, K. Joshi, A. Gupta, M.K. Singh, Self-aware execution environment model (SAE2) for the performance improvement of multicore systems. Int. J. Mod. Res. **2**(1), 17–27 (2022)

P. Singh, R. Kaur, J. Rashid, S. Juneja, G. Dhiman, J. Kim, M. Ouaissa, A Fog-Cluster Based Load-Balancing Technique. Sustainability **14**(13), 7961 (2022)

B. Sumathy, A. Chakrabarty, S. Gupta, S.S. Hishan, B. Raj, K. Gulati, G. Dhiman, Prediction of diabetic retinopathy using health records with machine learning classifiers and data science. Int. J. Reliable Qual. E-Healthc. (IJRQEH) **11**(2), 1–16 (2022)

P. Tiwari, B. Pant, M.M. Elarabawy, M. Abd-Elnaby, N. Mohd, G. Dhiman, S. Sharma CNN based multiclass brain tumor detection using medical imaging. Comput. Intell. Neurosci. (2022)

P.K. Vaishnav, S. Sharma, P. Sharma, Analytical review analysis for screening COVID-19 disease. Int. J. Mod. Res. **1**(1), 22–29 (2021)

C.N. Wang, N.L. Nhieu, K.P. Tran, Y.H. Wang, Mathematical fuzzy logic in the emerging fields of engineering. Finance Comput. Sci. (2022)

K. Yadav, A. Jain, N.M. Osman Sid Ahmed, S.A. Saad Hamad, G. Dhiman, S.D. Alotaibi, Internet of thing based koch fractal curve fractal antennas for wireless applications. IETE J. Res. 1–10 (2022)

R.R. Yager, Mathematical programming with fuzzy constraints and preferences on the objectives. Kybernetes **8**, 285–291 (1979)

F. Ying-Yung, A method using fuzzy mathematics to solve vector maximum problems. Fuzzy Sets Syst. **9**, 129 (1983)

H.J. Zimmermann, Fuzzy programming and linear programming with several objective functions. Fuzzy Sets Syst. **1**, 45 (1978)

H.J. Zimmermann, Fuzzy mathematical programming. Comput. Oper. Res. **4**, 291–298 (1981)

Breast Cancer Detection Using Deep Learning Model

Aliya Thaseen, Raheem Unnisa, Naheed Sultana, K. Reddy Madhavi, Grande. NagaJyothi, and S. Kirubakaran

Abstract Early detection of breast cancer is a challenging task. It is the major source of death among the leading and underdeveloped countries. The detection of breast cancer at earlier stages helps patients suffering from the disease to get proper treatment. Considering this aspect, the approach using deep learning architectures had been proposed to detect the presence of breast cancer given the cytology images. Transfer learning method is used in terms of building pre-learned model to improve the performance of the proposed approach. Using average pooling classification, the images were divided into malignant and benign cells. Three architectures of CNN combined in the proposed work for better prediction and accuracy.

Keywords Convolutional neural network · Breast cancer · Deep learning · Transfer learning

---

A. Thaseen
Department of IT, Shadan Womens College of Engineering and Technology, Hyderabad, India
e-mail: aliya.tehseen86@gmail.com

R. Unnisa (✉)
Department of CSE, CMR Technical Campus, Hyderabad, India
e-mail: raheemunnisa.hb@gmail.com

N. Sultana
Department of IT, Stanley College of Engineering & Technology, Hyderabad, India
e-mail: naheed.ruquiya@gmail.com

K. R. Madhavi
CSE, Sree Vidyanikethan Engineering College, Tirupati, AP, India
e-mail: kreddymadhavi@gmail.com

Grande. NagaJyothi
ECE, Golden Valley Integrated Campus, Madanapalli, AP, India
e-mail: nagajyothisai221@gmail.com

S. Kirubakaran
Department of CSE, CMR College of Engineering and Technology, Hyderabad, India
e-mail: dr.s.kirubakaran@gmail.com

A. B. Reddy et al. (eds.), *Proceedings of Third International Conference on Advances in Computer Engineering and Communication Systems*, Lecture Notes in Networks and Systems 612, https://doi.org/10.1007/978-981-19-9228-5_57

1 Introduction

Breast cancer is the most common problem among women aged 20–59 years. Identifying this cancer in prior can reduce the death rates. There are two kinds of breast cancer detection methods in the preliminary stages such as Biopsy and Mammography. Analyzing the cytology images (Sunitha et al. 2022) and classifying whether the cancer is malignant or benign can help in identifying breast cancer at earlier stages. The objective of the proposed work is to develop a mechanism to identify breast cancer combining deep learning architectures. The breast cancer detection and classification system is useful in detecting the cancer tissues for the given cytology images. It classifies the lesions as benign and malignant. It can also be used for classification of various categories of cancer (Kavitha et al. 2021). The proposed method can be used to detect the cancerous tissues from cytology images and classify the lesions in the early stages as benign and malignant. Accuracy will be better when applied for image classification than existing methods.

2 Related Work

This section focuses on the background analysis that were done in this domain. Medical imaging data is now frequently acquired, handled, and stored digitally. Additionally, high-content photographs of the processed samples of breast tissue are kept in databases.

There are numerous picture processing and data handling processes in the total computer-aided diagnostic pipeline. The images are first gathered and digitally stored. Then, reduction of noise, contrast enhancement, and artifacts reduction are applied for improvement. Images are divided into set of homogeneous segments after being enhanced, and these segments are then examined in subsequent processes (Bharath Simha Reddy and Rana 2021; Daly et al. 2021; Yadlapalli et al. 2021). Through the use of virtual microscopy, telepathology, automatic image processing, and slide data administration could all be made easier. By yielding more reliable results, automatic analysis could minimize pathologist workload or boost the accuracy of diagnoses. These goals demand the use of pattern recognition methods that can accurately recognize and classify objects in huge imagery. Feature segmentation is used in several pattern recognition techniques for cell-sized objects in histopathology images (Güzel and Bİlgin 2020). The right segmentation takes time, and the feature findings could be quite segmentation sensitive.

The fastest-growing sectors of the health sector today are "artificial intelligence, machine learning, and convolutional neural networks" (Bharath Simha Reddy and Rana 2021; Güzel and Bİlgin 2020; Spanhol et al. 2016). Research on AI and ML focuses on enhancing technological systems that can tackle complicated problems without the need for human intelligence (Taifi et al. 2014; Hassan and DeRosa 2020). Artificial neural networks, for deep learning, a member of the machine learning family. DL algorithms in particular can be used to increase the effectiveness and diagnostic precision of cancer detection (Pandian (2019),Sung, et al. (2020)).

Gravina et al. identified that the simple use of CNNs may not be better, since "medical images are incomparable than normal images" (Reshma et al. (2022)). Mammographic lesion separation has indeed been demonstrated to be a valuable source of information since it may aid in both the extraction of shape-related features and the precise localization of lesions. Malathi et al. used a computer-aided diagnostic (CAD) system for mammography to enable the initial detection, evaluation, and treatment of breast cancer. And, also talked about investigating a breast CAD design that fuses traits using deep learning (Kuraparthi et al. (2021), Yang et al. (2012)). Random forest algorithm (RFA) was shown to have higher precision and lower error than the CNN classifier (94.95%).

3 Proposed Method

The proposed system uses different convolutional neural network architectures such as GoogleNet, ResNet, and VGGNet with transfer learning for the accurate prediction of breast cancer. In this approach, different attributes of the images were drawn out by the architectures likely Google Net, VGG16 Net, and ResNet (Taifi et al. 2014), and then these are merged using a fully connected layer. The proposed approach helps doctors in predicting the disease in early stage and can give proper treatment. In terms of accuracy, this approach produces better results than the existing framework.

The architecture of the proposed framework is shown in Fig. 1. Three of the most recent and cutting-edge deep CNN architectures are employed in the proposed framework as feature extractors for the identification of breast cancer in cytology images: "Google Network (GoogleNet)," "Visual Geometry Group Network (VGGNet)," Dense Network (DenseNet), and "Residual Networks (ResNet)." We developed a model based on the design of "GoogleNet" that integrates multiple convolution filters of varied widths into a single new filter, reducing both the number of perimeters and the computational complexity (Yu et al. (2017)).

Steps of the Proposed Approach

- **Step 1**: Collect the data set consisting of the breast cytology images required to train and test the model.
- **Step 2**: Perform preprocessing to rescale and normalize the images.
- **Step 3**: Use data augmentation to expand the dataset's size in order to combat the overfitting issue.
- **Step 4**: Extract the features using pre-trained CNN architectures and apply transfer learning to each of the network. Train the CNN architectures with 25 epochs and calculate the evaluation metrics.
- **Step 5**: Plot the accuracy and loss graphs for each CNN against the train and validation data.

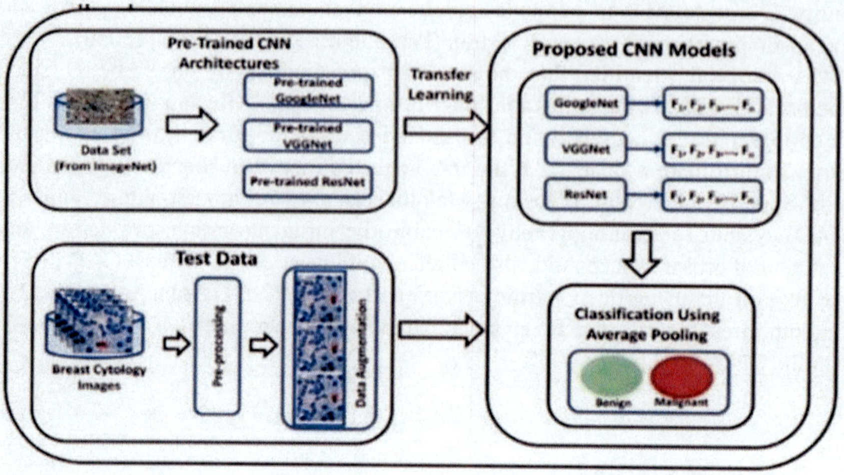

Fig. 1 Architecture of proposed deep learning framework

- **Step 6**: Ensemble Google Network, Visual Geometric Group Network, Dense network, Mobile Network, and Residual Network.
- **Step 7**: Evaluate an accuracy of the ensembled model and compare with the prevailing models.

4 Experiments

The dataset used in the proposed system is breast cancer dataset consisting of breast cytology images from Kaggle. The CNN architectures displayed a range of detection and classification accuracy rates for breast cancer. Performance measures for each classifier, including recall, precision, $f1$ score, and support, were assessed and compared for "VGG16," "GoogleNet," "ResNet," "DenseNet," and "MobileNet." To rescale and transport the images, image augmentation is used. Out of 1905 images belonging to 2 classes, 1350 images taken for training and remaining for testing. After loading VGG model, we obtained an of accuracy = 70.27%, and loss = 1.2768, shown in Fig. 2 (Tables 1, 2, 3 and 4).

We obtained an accuracy = 68.51%, loss = 2.6144 with "GoogleNet" model, and shown in Fig. 3. Similarly, we obtained accuracy = 70.95%, loss 0.6221, with ResNet model, shown in Fig. 4 (Fig. 5).

We obtained an accuracy = 70.41%, loss 3.3819 with MobileNet model as shown in Fig. 6. The Table 5 reveals the comparison of existing architectures with the proposed approach based on accuracy scores. The accuracy of the proposed approach combining different deep learning architectures is better when compared to the existing methods as specified in Table 6.

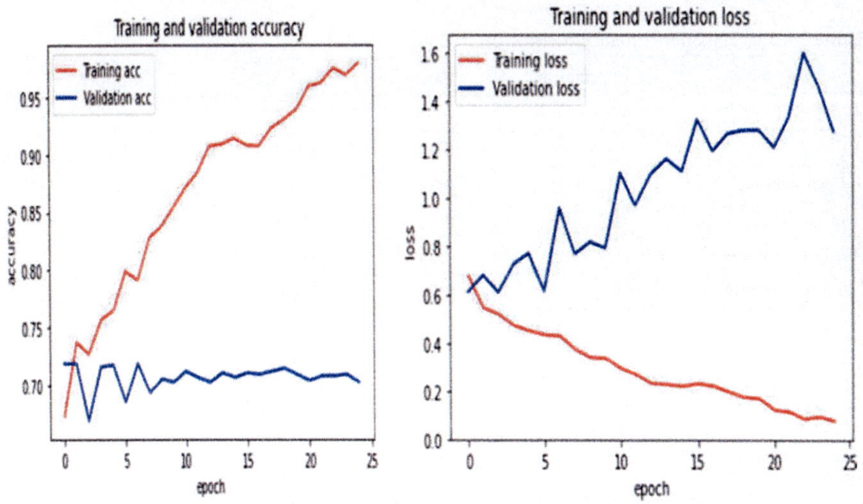

Fig. 2 VGG16 accuracy and loss graph

Table 1 VGG P

	Precision	Recall	$F1$-score	Support
0	0.73	0.92	0.82	532
1	0.4 1	0.14	0.2 1	208
Micro average	0.70	0.70	0.70	740
Macro average	0.57	0.53	0.5 1	740
Weighted average	0.64	0.70	0.65	740
Samples average	0.70	0.70	0.70	740

Table 2 GoogleNet performance metrics

	Precision	Recall	$F1$-score	Support
0	0.76	0.82	0.79	532
1	0.43	0.35	0.39	208
Micro average	0.69	0.69	0.69	740
Macro average	0.59	0.58	0.59	740
Weighted average	0.67	0.69	0.68	740
Samples average	0.69	0.69	0.69	740

5 Conclusion

The main aim of the proposed work is to detect breast cancer on the given dataset consisting of patient's cytology images. The proposed model uses CNN with different

Table 3 DenseNet performance metrics

	Precision	Recall	$F1$-score	Support
0	0.73	0.95	0.82	532
1	0.43	0.11	0.17	208
Micro average	0.71	0.71	0.71	740
Macro average	0.58	0.53	0.50	740
Weighted average	0.65	0.71	0.64	740
Samples average	0.71	0.71	0.71	740

Table 4 DenseNet performance metrics

	Precision	Recall	$F1$-score	Support
0	0.73	0.91	0.81	532
1	0.34	0. 12	0.1 8	208
Micro average	0.69	0.69	0.69	740
Macro average	0.53	0.51	0.49	740
Weighted average	0.62	0.69	0.63	740
Samples average	0.69	0.69	0.69	740

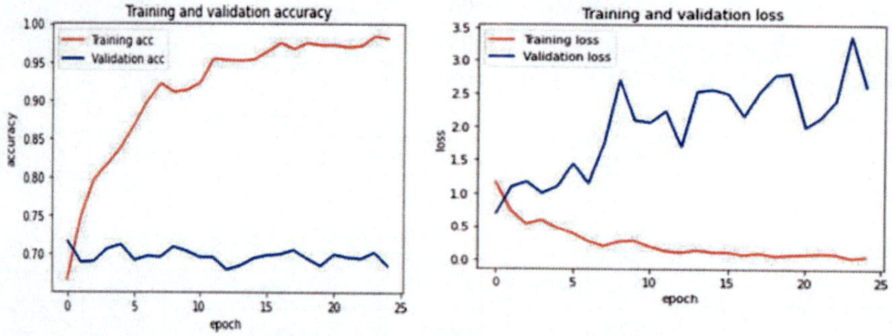

Fig. 3 GoogleNet accuracy and loss graph

deep learning models to classify the images. After preprocessing, data augmentation is applied to enhance the training of CNN and also to overcome the overfitting problem in case of smaller datasets. In the proposed method, the features were retrieved from the breast cytology images using different CNN architectures, including "VGG16," "ResNet," and "GoogleNet." Finally, all three architectures were merged through ensembling to provide improved accuracy. The findings of an individual accuracy correlation between each architecture and the proposed framework are presented in validation graphs. The transfer learning method will be used

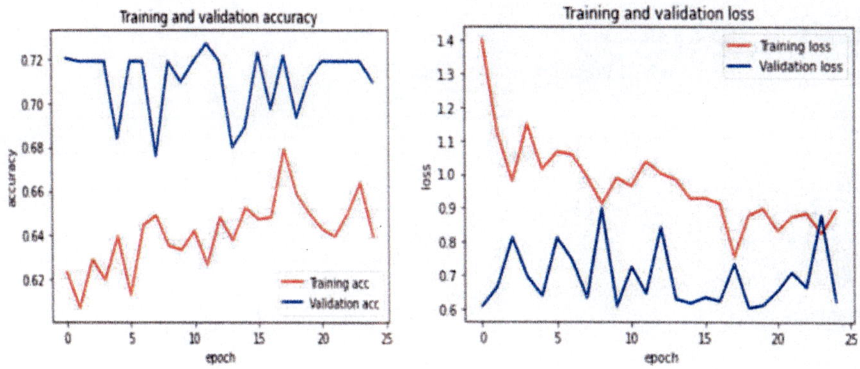

Fig. 4 ResNet accuracy and loss graph

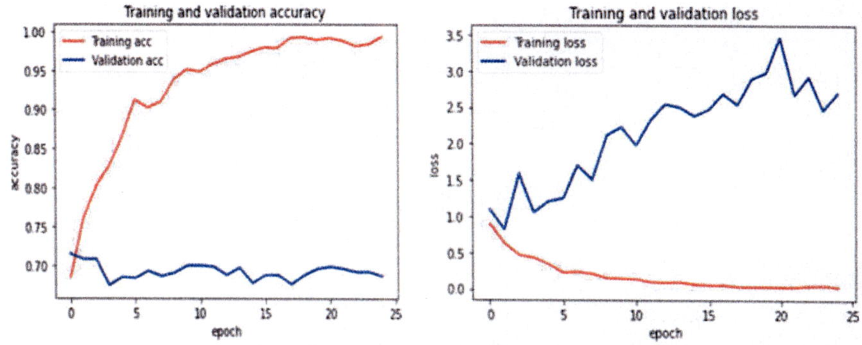

Fig. 5 DenseNet accuracy and loss graph

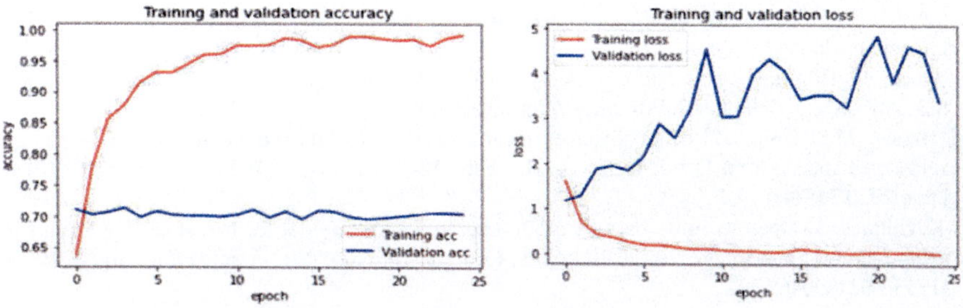

Fig. 6 MobileNet accuracy and loss

in the future to categorize different types of cancers in order to increase the model's applicability.

Table 5 MobileNet performance metrics graph

	Precision	Recall	*F*1-score	Support
0	0.74	0.92	0.82	532
1	0.43	0. 16	0.23	208
Micro average	0.70	0.70	0.70	740
Macro average	0.58	0.54	0.52	740
Weighted average	0.65	0.70	0.65	740
Samples average	0.70	0.70	0.70	740

Table 6 Comparison of architectures based on accuracy

Method	Accuracy (%)
DenseNet	68.6
VGG16 net	70.2
GoogleNet	68.5
ResNet	70.9
MobileNet	70.4
Ensembling (VGG16, Google, ResNets)	71.8

References

M. Bharath Simha Reddy, P. Rana, Biomedical image classification using deep convolutional neural networks—overview. in IOP Conference Series: Materials Science and Engineering, vol. 1022, (Jan. 2021), p. 012020. https://doi.org/10.1088/1757-899x/1022/1/012020

A.C. Daly, K.J. Geras, R. Bonneau, A convolutional neural network for common coordinate registration of high-resolution histology images. Bioinformatics (2021). https://doi.org/10.1093/bioinformatics/btab447

K. Guzel, G. Bilgin, Classification of breast cancer images using ensembles of transfer learning. Sakarya Univ. J. Sci. (Jun 2020). https://doi.org/10.16984/saufenbilder.720693

M. Hassan, M.C. DeRosa, Recent advances in cancer early detection and diagnosis: role of nucleic acid based aptasensors. TrAC, Trends Anal. Chem. **124**, 115806 (2020). https://doi.org/10.1016/j.trac.2020.115806

T. Kavitha et al., Deep learning based capsule neural network model for breast cancer diagnosis using mammogram images. Interdiscip. Sci. Comput. Life Sci. (2021). https://doi.org/10.1007/s12539-021-00467-y

S. Kuraparthi et al., Brain tumor classification of MRI images using deep convolutional neural network. Traitement Du Signal **38**(4), 1171–1179 (2021). https://doi.org/10.18280/ts.380428.J

P. Pandian, Identification and classification of cancer cells using capsule network with pathological images. J. Artif. Intell. Capsule Netw. **01**(01), 37–44 (2019). https://doi.org/10.36548/jaicn.2019.1.005

G. Reshma et al., Deep learning-based skin lesion diagnosis model using dermoscopic images. Intell. Auto. Soft Comput. **31**(1), 621–634 (2022). https://doi.org/10.32604/iasc.2022.019117

F.A. Spanhol, L.S. Oliveira, C. Petitjean, L. Heutte, A dataset for breast cancer histopathological image classification. IEEE Trans. Biomed. Eng. **63**(7), 1455–1462 (2016). https://doi.org/10.1109/tbme.2015.2496264

H. Sung et al., Global cancer statistics 2020: GLOBOCAN estimates of incidence and mortality worldwide for 36 cancers in 185 countries. CA: Cancer J. Clin. **71**(3), 209–249 (2021). https://doi.org/10.3322/caac.21660

G. Sunitha, K. Geetha, S. Neelakandan, A.K.S. Pundir, S. Hemalatha, V. Kumar, Intelligent deep learning-based ethnicity recognition and classification using facial images. Image vis. Comput. **121**, 104404 (2022). https://doi.org/10.1016/j.imavis.2022.104404

K. Taifi, S. Safi, M. Fakir, A. Elbalaoui, Breast cancer diagnosis system based on wavelet analysis and neural networks. Int. J. Comput. Vis. Image Process. **4**(1), 1–16 (2014). https://doi.org/10.4018/ijcvip.2014010101

P. Yadlapalli, M.K. Reddy, S. Gurram, J. Avanija, K. Meenakshi, P. Kora, Breast thermograms asymmetry analysis using gabor filters. in E3S Web of Conferences, vol. 309, (2021), p. 01109. https://doi.org/10.1051/e3sconf/202130901109

L. Yang, S. Hanneke, J. Carbonell, A theory of transfer learning with applications to active learning. Mach. Learn. **90**(2), 161–189 (2012). https://doi.org/10.1007/s10994-012-5310-y

Y. Yu, H. Lin, J. Meng, X. Wei, H. Guo, Z. Zhao, Deep transfer learning for modality classification of medical images. Information **8**(3), 91 (2017). https://doi.org/10.3390/info8030091

A Survey on Automatic Text Summarisation

G. L. Anand Babu and Srinivasu Badugu

Abstract Due to the massive volume of textual data which is produced by the Internet and in other archives of like news stories, academic papers, legal documents, etc., automatic text summarisation (ATS) is becoming increasingly crucial. The process of text summarisation is nothing but retrieving the significant data from the input document and then summarising it. While increasing the data's magnitude, it is critical to summarise the data manually and time consuming, so numerous applications require ATS. Hybrid, extractive, and abstractive are the various method in ATS. In extractive method, the summary is formed by choosing the significant sentences from the source document. The source document is represented using the abstractive technique in an intermediate representation, which then provides summary using sentences different from the source phrases. The hybrid technique combines abstractive and extrapolative methods. Despite all of the suggested strategies, the produced summaries still differ significantly from the summaries made manually. Most research concentrates on the extractive strategy. Techniques that are abstract and hybrid must be given more attention. This paper provides a full evaluation for researchers by covering the several ATS components—approaches, techniques, building blocks, processes, and future research goals.

Keywords Automatic text summarisation · Natural language processing · Abstractive technique · Intermediate representation · Automated summarisation

1 Introduction

In today's world, the Internet is growing enormously every day due to that the volume of data produced in Internet are also rising tremendously. The data produced is

G. L. A. Babu (✉)
University College of Engineering, Osmania University, Hyderabad, India
e-mail: anandbabu77@gmail.com

S. Badugu
Department of CSE, Stanley College of Engineering and Technology for Women, Abids, Hyderabad, India

A. B. Reddy et al. (eds.), *Proceedings of Third International Conference on Advances in Computer Engineering and Communication Systems*, Lecture Notes in Networks and Systems 612, https://doi.org/10.1007/978-981-19-9228-5_58

mostly unstructured; hence, extracting significant data manually is a critical process (Chen and You 2020). For humans, it is not easy for extracting significant data from huge volume of data. They take a while to understand the point of the information. Therefore, automated summarisation is a well-known method of overcoming such difficulties (Dave and Jaswal 2015; Widyassari et al. 2022). The field of Natural Language Processing (NLP) is now conducting intensive research in Automatic Text Summarisation (ATS). An ATS assignment aims to automatically glean the most important data from a source text and duplicate it in a more condensed manner (Guo et al. 2022; Almutiri and Nadeem 2022). Extractive and abstractive summarisations are the two kinds of summarisation methodologies. Using an extractive technique, a summary is created by concatenating the most important sentences from the original material. The abstract ones quote the original text (Gupta and Lehal 2010).

Summarisation may involve a single text or a collection of documents. While multi-document-based techniques condense the information of several documents into a single concise paragraph, mono-document-based approaches summarise one text at a time. Summarisation might be query-based or general (Kallimani 2018). Generic summaries compile data from the whole document, whereas query-based ones respond to a particular inquiry about the material. Till today, various methodologies are introduced for summarising the text (Boorugu and Ramesh 2020). Depending on the quantity of documents, there are choices for single and multi-document summaries. In the meanwhile, the summary findings serve as the foundation for the extractive and abstractive outputs. Generic and query-based, on the other hand, rely on the purpose (Widyassari et al. 2020). On the other hand, depending on the content, it is separated into indicative and informative categories (Rajasekaran and Varalakshmi 2018).

2 Literature Review

2.1 Text Summarisation Approaches in Telugu Language

Shashikanth and Sanghavi (2019) have published a survey on text summarising techniques for Telugu and other languages. In their article, the extractive summarisation approach is covered in the study and is based on the Telugu language ranking system. Telugu text is tokenised, tagged, and the stop words are eliminated during pre-processing. Following that, low-frequency sentences are eliminated before calculating the word frequency. The sentences are sorted according to frequency, and a summary is produced using the rank. For the Telugu language, Khanam and Sravani (2016) have presented a diversity-based extractive summarisation model using k-means clustering. Telugu text is tokenised, tagged, and the stop words are eliminated during the pre-processing operation. The frequency of the terms is then calculated, and clusters within the phrases are created. Low-frequency sentences are eliminated once the sentences are rated and based on frequency. According to rank, a summary is generated. Redundancy elimination in summary sentences has not been taken into account by these extractive text summarising approaches. In addition, performance is not evaluated in order to demonstrate the model's efficacy.

A method of extractive text summarisation utilising the Text-Rank algorithm has been put out by Manjari (2020). Utilising extractive text summarisation on several Telugu text documents is the method used here. To conduct word embedding, a continuous bag of words (CBOW) model has been used. The cosine similarity technique is then used to construct a similarity matrix. This helps to create a graph, and the Text-Rank method is used to rank the phrases. To evaluate the effectiveness of the model, the ROUGE score is computed. This technique has an extremely low ROUGE score. In Telugu, Kallimani et al. (2011) studied several text summarising techniques. Telugu text is tokenised, parts of speech are identified, and stop words are eliminated during the pre-processing stage. The TFIDF technique is then used to extract the keywords. Based on the score, the crucial phrases are chosen. The resulting summary is post-processed using rephrasing and summary refining techniques, together with the extracted sentences from the text. This method's effectiveness was not evaluated based on how well it performed. A multi-document abstractive text summarising method for Telugu language using a semantic similarity matrix has been suggested by Latha and Sudha (2020). Semantic concepts, Jiang similarity, and RNN are used to compute similarity. ROUGE score is used to assess the effectiveness of this particular strategy. In this method, a lower ROUGE score is created.

An extractive summarising technique based on a ranking system for Telugu language was proposed by See et al. (2017). The Telugu text is tokenised, tagged, and the stop words are eliminated during pre-processing. Following that, low-frequency sentences are eliminated before calculating the word frequency. The sentences are sorted according to how frequently they occur, and the summary is produced using the rank. This approach has the drawback of producing extracting summaries that are already included in the source material. Additionally, it does not evaluate the effectiveness of the suggested paradigm.

With *k*-means clustering, Pan et al. (2019) presented an efficient diversity-based extractive summarisation model for Telugu. The Telugu text is tokenised, tagged, and the stop words are eliminated during the pre-processing operation. After that, word clusters are created inside the phrases based on their frequency. Low-frequency sentences are eliminated once the sentences are ordered according to frequency. Sentences from clusters are picked to create a summary based on the rank. This approach has two main drawbacks: (i) additional characteristics are not taken into account when extracting Telugu sentences, and (ii) it only extracts the most crucial phrases from the original huge text corpus. Additionally, the performance is not assessed to demonstrate the model's efficacy.

In Telugu, Shi et al. (2021) studied several methods of abstractive text summarising. The Telugu text is tokenised, part of speech is identified, and stop words are eliminated during the pre-processing stage. The TFIDF technique is then used to extract the keywords. Based on the score, the crucial phrases are chosen. The content is subsequently processed using summary refining and rephrasing procedures, and the extracted phrases are used to construct the final summary. This method's performance has not been tested to determine its effectiveness. A multi-document abstractive text summarising method for Telugu language with a semantic similarity matrix was suggested by Boorugu and Ramesh (2020). With the use of RNN, Jiang similarity, and semantic ideas, the similarity is estimated. The ROUGE score is used to assess the model's performance. The produced Rouge score for this method is rather low.

2.2 *Text Summarisation Approaches in Other Language*

For ATS, a variety of strategies has been used. The majority of techniques to text summary are written in English. The method of automated text summarising that most frequently employed is graph-based extractive text summarisation which was developed by Uçkan and Karc (2020). SUMMONS, created at Columbia University, is the first multi-document summarising system (Radev and McKeown 1998).

An approach for extractive text summarisation (ETS) that is independent of language and domain has been presented by Hernadez-Castaneda et al. (2020). With the use of evolutionary and clustering techniques, this strategy enhanced the extractive summarisation methodology. The crucial sentences are recognised in the absence of any prior knowledge. Different feature-generation methods, including LDA, One Hot Encoding, TFIDF, and Doc2vec, have been suggested to map the text to a numeric value.

For the summary of Arabic literature, Elbarougy et al. (2019) suggested an ETS method using a modified Page-Rank algorithm. The initial score assigned to each node in the graph by the modified Page-Rank algorithm was determined by the nouns in the sentences. Using a cosine similarity approach, this technique produced a similarity matrix. Through the use of the Lex-Rank, Text-Rank, and modified Page-Rank algorithms, they have evaluated the effectiveness of their strategy.

Deep learning-based algorithms have outperformed rule-based ones for abstractive text summarisation (Pan et al. 2019). Deep learning is utilised for the first time in English abstractive summarisation by Rush et al. (2015). As the deep learning model for ATS, the seq2seq technique has also been extensively used. Since the seq2seq framework has been successful and can produce better results than a human-written summary, abstractive approaches are preferable to extractive ones. To improve the performance of seq2seq models, several advanced studies have been conducted in various languages, such as English. Sutskever et al. (2014) provide an encoder-decoder approach to address machine translation issues in seq2seq tasks.

Coverage mechanisms, which might lessen the issue of recurrence in text summary, are still another crucial mechanism (See et al. 2017). By incorporating the pointer generation network into the seq2seq architecture, OOV words may be produced. Pointer-generator networks were suggested by Vinyals et al. (2015). The pointer generating network was used by Nallapatti et al. (2016) to create OOV in a machine translation job with a temporal attention mechanism. Later, Gu et al. (2016) utilised this pointer generating network in their text summarising research. The performance of the abstract model is enhanced by the implementation of the fundamental seq2seq architectures with attention, coverage, and pointer generating mechanisms. However, the abstractive model's produced summaries lack fluency and are difficult for humans to read. Table 1 gives various automatic text summarisation techniques with its limitations.

3 Conclusion

There is a glut of knowledge as a result of the rapid advancement of technology and Internet use. If there were powerful text summarisers that produced a user-friendly summary of documents, this issue might be resolved. The last several decades have seen a rise in interest in the research area of automatic text summarisation from a variety of scientific disciplines. In this article, majority of the approaches largely concentrated on extractive text summary for Telugu. The abstractive text summary using Telugu text was only briefly explored in a few studies. Bi-LSTM encoder and LSTM decoder models are utilised in the suggested model to provide summaries for Telugu text using the news data set. A temporal attention mechanism, coverage mechanism, pointer-generator network, and diversified beam-search inference decoder are applied to the suggested model to strengthen the produced summary quality. Yet, there are not any published studies using the Telugu ATS method with the seq2seq model.

Table 1 Automatic text summarisation techniques and its limitations

References	Language	Authors and year	Proposed method	Advantages	Limitations	Identification of research gap
Shi et al. 2021)	English	Shi,T, Keneshloo,Y, Ramakrishnan,N, and Reddy,CK 2021 (Extractive multi-document text summarisation based on graph independent sets)	Neural abstractive text summarisation with seq2seq methodology	Extract salient phrases and fragments from the source articles to create a summary	Depend on beam-search technique ROUGE and BERT score metrics are not supported	Improve the method to be flexible to all techniques and have to support the standard ROUGE and BERT score metrics
Uçkan and Karcı 2020)	Arabic	Uçkan, T., & Karcı, A. 2020 (Egyptian Informatics Journal)	Text extraction from many documents using graph independent sets	Very effective and consistent in summarising the text	Requires pre-processing Need correct graph transformation	Enhance by introducing a graph transformation method
Radev and McKeown 1998)	English	Radev,DR., McKeown,K.R. 1998 (Comput. Linguist.,)	Summarising multiple online sources in Natural Language	When using summaries of various lengths and when combining many independent summary operators to create more sophisticated summaries with more detailed details	Entity cross-referencing was not solved	Solve entity cross-referencing
Hernandez-Castaneda et al. 2020)	English	Hernandez-Castaneda,A, Garcia-Hernandez,R.A, Ledeneva,Y, and Millan-Hernandez, C.E 2020 (IEEE)	Extorting text summarisation by partitional clustering using genetic approach	This model is independent of both language and domain	Language dependent works better for English language	Make flexible to all languages

(continued)

Table 1 (continued)

References	Language	Authors and year	Proposed method	Advantages	Limitations	Identification of research gap
Elbarougy et al. 2019)	Arabic	Elbarougy, R., Behery, G., & El Khatib, A. 2019 (Egyptian Informatics Journal)	Extractive Arabic text summarisation using modified Page-Rank algorithm	The amount of redundant information is removed from the derived summary based on the compression ratio and sentence overlap	If summary sentence is large, then it was neglected hence there was a chance of missing significant data	Avoid neglecting the sentences so that the data are not missed
Pan et al. 2019)	English	Pan,HX., Liu,H, Tang,Y. 2019 (Springer)	A topic-based attention mechanism for Seq2Seq Text summarisation	Provides a better text summarisation outcomes	Only text features were considered	Consider all the features in the document
Rush et al. 2015)	Arabic	Rush,AM., Chopra,S, and Weston,J 2015 (Conference on Empirical Methods in Natural Language Processing)	Abstractive phrase summarisation using the attention concept	This method produces accurate abstractive summaries	Grammaticality of the summaries were not considered Does not generate paragraph-level summaries	Consider grammaticality of the summaries and also generate paragraph-level summaries
Nallapati et al. 2016)	English	Nallapati,R, Zhou,B, Santos, CD, Gulçehre,CG, and Xiang,B. 2016 (CoNLL)	Using Seq2Seq RNNs for abstractive text summarisation	Proposes a novel data set for multi sentence summaries	Multiple sentences were not processed	Process all the sentences in the document

(continued)

Table 1 (continued)

References	Language	Authors and year	Proposed method	Advantages	Limitations	Identification of research gap
Sutskever et al. 2014)	English French	Sutskever,I, Vinyals,O, and Le,QV. 2014 (Adv. Neural Inf. Process. Syst.)	Learning seq2seq using neural networks	Simple, straightforward, and a relatively unoptimised approach	Long texts are difficult to summarise, performance degrades as input text length increases	Improve the performance when the text size becomes big
Bahdanau et al. 2014)	English French	Bahdanau,D, Cho,K., Bengio,Y, 2014 (ArXiv)	Translations using neural machine for alignment and translation	This approach gives higher translation performance	The length of the input or output is limited to 15–40 words Cannot summarise the unknown or rare words	Summarise all the words in the document and improve the length of the output
See et al. 2017)	English	See,A., Liu,PJ., and Manning,CD., 2017 (Association for Computational Linguistics)	Summarisation by pointer-generator approach	Reduces inaccuracies and repetition	The coverage objective interfered with the main objective reduced the overall performance	Improve the performance
Vinyals et al. 2015)	English	Vinyals,O, Fortunato,M, and Jaitly,N. 2015 (Adv. Neural Inf. Process. Syst.,)	Pointer networks	This method generalises to variable size output dictionaries in addition to being an improvement over sequence-to-sequence with input attention	Have greater complexity	Reduce the complexity

(continued)

Table 1 (continued)

References	Language	Authors and year	Proposed method	Advantages	Limitations	Identification of research gap
Gu et al. 2016)	English	Gu,J, Lu,Z, Li,H, and Li,VOK. 2016 (arxiv)	Incorporating copying mechanism in seq2seq learning	Best method to manage the mixture of two modes	Some unseen words are not caught	Caught all the words in the document
Krause et al. 2017)	English with Image	Krause,J, Johnson,J, Krishna,R, and Fei-Fei,L. 2017 (IEEE)	Methodology for producing descriptive image paragraphs that is hierarchical	Have wider vocabulary, simple, and accurate	Resulting summary lacks diversity	Improve the diversity
Cibils et al. 2018)	English	Cibils,A, Musat,C, Hossman,A, and Baeriswyl,M. 2018 (arXiv)	Abstractive summarisation novelty search with diverse beams	Despite being less extractive, the approach raises the overall total score of traditional techniques by at least two points	Selecting the candidates is a crucial step	Choose correct candidates with degrading the performance

References

T. Almutiri, F. Nadeem, Markov models applications in natural language processing: a survey. I.J. Inf. Technol. Comput. Sci. **2**, 1–16 (2022). https://doi.org/10.5815/ijitcs.2022.02.01

D. Bahdanau, K. Cho, Y. Bengio, Neural machine translation by jointly learning to align and translate. ArXiv:1409–0473 (2014)

R. Boorugu, G. Ramesh, A survey on NLP based text summarization for summarizing product reviews. in *2020 Second International Conference on Inventive Research in Computing Applications (ICIRCA)*, (IEEE, 2020, July), pp. 352–356

J. Chen, F. You, Text summarization generation based on semantic similarity. in *2020 International Conference on Intelligent Transportation, Big Data & Smart City (ICITBS)*, (IEEE, 2020, January), pp. 946–949

A. Cibils, C. Musat, A. Hossman, M. Baeriswyl, Diverse beam search for increased novelty in abstractive summarization. arXiv preprint arXiv:1802.01457 (2018)

H. Dave, S. Jaswal, Multiple text document summarization system using hybrid summarization technique. in *2015 1st International Conference on Next Generation Computing Technologies (NGCT)*, (IEEE, 2015, September), pp. 804–808

R. Elbarougy, G. Behery, A. El Khatib, Extractive arabic text summarization using modified pagerank algorithm. Egypt. Inf. J. (2019). https://doi.org/10.1016/j.eij.2019.11.001

J. Gu, Z. Lu, H. Li, V.O.K. Li, Incorporating copying mechanism in sequence-to-sequence learning (2016). https://arxiv.org/abs/1603.06393

X. Guo, K. Tan, Y. Liu, M. Jin, H. Lu, LPG–PCFG: an improved probabilistic context-free grammar to hit low-probability passwords. Sensors **22**, 4604 (2022). https://doi.org/10.3390/s22124604

V. Gupta, G.S. Lehal, A survey of text summarization extractive techniques. J. Emerg. Technol. Web Intell. **2**(3), 258–268 (2010)

A. Hernandez-Castaneda, R.A. Garcia-Hernandez, Y. Ledeneva, C.E. Millan-Hernandez, Extractive automatic text summarization based on lexical-semantic keywords. IEEE Access 1–1 (2020). https://doi.org/10.1109/access.2020.2980226

J.S. Kallimani, Survey on extractive text summarization methods with multidocument datasets. in *2018 International Conference on Advances in Computing, Communications and Informatics (ICACCI)*, (IEEE, 2018, September), pp. 2113–2119

J.S. Kallimani, B. Eswara, Information extraction by an abstractive text summarization for an Indian regional language, (2011), pp. 319–322

M.H. Khanam, S. Sravani, Text summarization for Telugu document, IOSR-JSE, 18(6) (2016)

J. Krause, J. Johnson, R. Krishna, L. Fei-Fei, A hierarchical approach for generating descriptive image paragraphs. in *Proceedings of the IEEE conference on computer vision and pattern recognition*, (2017) pp. 317–325

Y.M. Latha, D.N. Sudha, Multi-document abstractive text summarization through semantic similarity matrix for Telugu language. Int. J. Adv. Sci. Technol. **29**(1), 513–521 (2020)

K.U. Manjari, Extractive summarization of Telugu documents using textrank algorithm. in *2020 Fourth International Conference on I-SMAC (IoT in Social, Mobile, Analytics and Cloud) (I-SMAC)*, (2020), pp. 678–683. https://doi.org/10.1109/I-SMAC49090.2020.9243568

R. Nallapati, B. Zhou, C.D. Santos, C.G. Gulçehre, B. Xiang, Abstractive text summarization using sequence-to-sequence RNNs and beyond. CoNLL **2016**, 280 (2016)

H.X. Pan, H. Liu, Y. Tang A sequence-to-sequence text summarization model with topic based attention mechanism. In: W. Ni, X. Wang, W. Song, Y. Li eds. by *Web Information Systems and Applications*, WISA 2019. Lecture Notes in Computer Science, vol. 11817 (Springer, Cham, 2019). https://doi.org/10.1007/978-3-030-30952-7_29

D.R. Radev, K.R. McKeown, Generating natural language summaries from multiple on-line sources. Comput. Linguist. **24**, 469–475 (1998)

A. Rajasekaran, R. Varalakshmi, Review on automatic text summarization. Inter. J. Eng. Technol **7**, 456–460 (2018)

A.M. Rush, S. Chopra, J. Weston, A neural attention model for abstractive sentence summarization. in *Proceedings of the 2015 Conference on Empirical Methods in Natural Language Processing*, (2015), pp. 379–389

A. See, P.J. Liu, C.D. Manning, Get to the point: summarization with pointer-generator networks. in *Proceedings of the 55th Annual Meeting of the Association for Computational Linguistics*, vol. 1: Long Papers. (Association for Computational Linguistics, 2017) pp. 1073–1083

S. Shashikanth, S. Sanghavi, Text summarization techniques survey on Telugu and foreign languages. Int. J. Res. Eng. Sci. Manage. **2**(1) (2019). https://doi.org/10.1109/NLPKE.2011.6138217

T. Shi, Y. Keneshloo, N. Ramakrishnan, C.K. Reddy, Neural abstractive text summarization with sequence-to-sequence models. ACM Trans. Data Sci. (2021)

I. Sutskever, O. Vinyals, Q.V. Le, Sequence to sequence learning with neural networks. Proc. Adv. Neural Inf. Process. Syst. 3104–3112 (2014)

T. Uçkan, A. Karcı, Extractive multi-document text summarization based on graph independent sets. Egypt. Inf. J. (2020). https://doi.org/10.1016/j.eij.2019.12.002

O. Vinyals, M. Fortunato, N. Jaitly, Pointer networks. in Proc. Adv. Neural Inf. Process. Syst. 2692–2700 (2015)

A.P. Widyassari, S. Rustad, G.F. Shidik, E. Noersasongko, A. Syukur, A. Affandy, Review of automatic text summarization techniques & methods. J. King Saud Univ. Comput. Inf. Sci. (2020)

A.P. Widyassari, S. Rustad, G.F. Shidik, E. Noersasongko, A. Syukur, A. Affandy, D.R.I. Moses Setiadi, Review of automatic text summarization techniques and methods. J. King Saud Univ. Comput. Inf. Sci. **34**(4), 1029–1046 (2022). https://doi.org/10.1016/j.jksuci.2020.05.006

The Adaptive Strategies Improving Design in Internet of Things

Suresh Kallam, A. Veerender, K. Shilpa, K. Ranjith Reddy, K. Reddy Madhavi, and Jonnadula Narasimharao

Abstract The Internet of Things can bring a lot of economic, social and technical benefits, but it also raises essential challenges that could stand in the way. Issues like lack of privacy and security need to be controlled to ensure societal acceptance of IOT services, otherwise, it can undermine the user's confidence to fully enjoy the technology and result in a smaller than expected adoption. This paper has only presented an overview of few issues that come with IOT and a lot more is still needs to be researched in more details. There are issues that people and companies are still not comfortable with or do not know how to properly address. For instance, even after the product is developed, besides meeting the needs of the user, products will also have to provide seamless integration with other enterprise data, applications and environment. And how to monitor something you no longer own? Yet however, the company is still responsible for their performance. Besides all that there is also the need to meet the needs of the user and designed products will have to provide seamless integration with other enterprise data, applications and environment. Also overcome battery challenges that limit computation, display resolution and connectivity. The

S. Kallam
Department of CSE, Sree Vidyanikethan Engineering College, Tirupati, AP, India
e-mail: sureshkallam@gmail.com

A. Veerender
Department of CSE—Data Science, CMR Technical Campus, Hyderabad, India
e-mail: veerender57@gmail.com

K. Shilpa · K. R. Reddy · J. Narasimharao (✉)
Department of CSE, CMR Technical Campus, Hyderabad, India
e-mail: jonnadula.narasimharao@gmail.com

K. Shilpa
e-mail: shilpamtech555@gmail.com

K. R. Reddy
e-mail: ranjithreddy.cse@cmrtc.ac.in

K. R. Madhavi
School of Computing, Mohan Babu University, Tirupati, India
e-mail: kreddymadhavi@gmail.com

A. B. Reddy et al. (eds.), *Proceedings of Third International Conference on Advances in Computer Engineering and Communication Systems*, Lecture Notes in Networks and Systems 612, https://doi.org/10.1007/978-981-19-9228-5_59

term Internet of Things might have been present for a long time now, but still there are plenty of work ahead to be done to transform this "term" into this success that everybody has been expecting.

Keywords Internet of Things · Digital twin · Recommendation system · Virtual and health care

1 Introduction

The idea of the Internet of Things is that, rather than having few very powerful devices such as laptops and smartphones, the user will possess a large number of devices, sometimes less powerful, but they will enable them to always stay connected and informed. (Saddik 2018; Steinbach et al. 2018) discussed that products are now composed of physical, smart and connectivity components. They have become complex systems composed by hardware, software, sensors, data storage, microprocessors and connectivity. Subsequent paragraphs, however, are indented.

Design is changing the focus from only physical products to data-centric physical products. The role of designers is also changing. Now, they have to advance human experience, design for user's experiences. For example, connected products will communicate to companies when something is not working properly, so they can start acting instantly when they receive the information and address before it becomes a bigger problem. And in the process, by knowing how the product is being used, companies can also start working proactively and start preventing failures before they happen, creating a more loyal customer base. "Listening" to the product will change perception of what constitutes a product and how to improve it. Information can be used to modify the company's own products to meet customer needs or combine it with other shared data enabling a new level for the next generation of products and services (Fig. 1).

A key component of the digital twin vision is the optimization of multi-modal communication/interaction from the user perspective, which did not get any attention from industrial and academic stakeholders until so far. As a result, the demands to

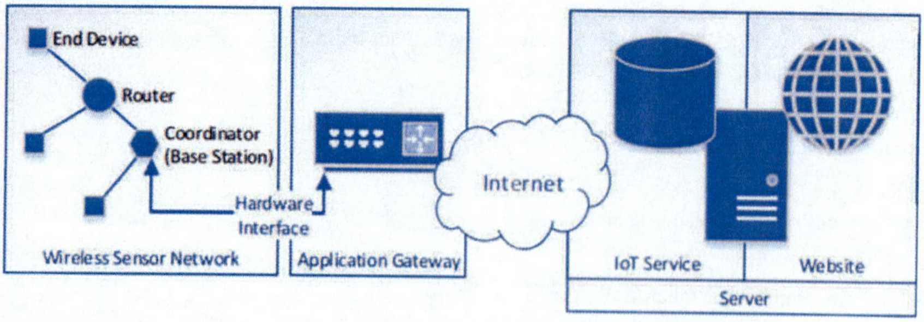

Fig. 1 Internet of things structure

conduct extensive researches and developments in this domain are still growing and infancy stages.

2 Problem Statement

Design for easy replacement of outdated hardware: Some components may have a replacement cycle longer than others may, for example, Bluetooth components have longer cycle life than the battery. Therefore, companies should consider ways to facilitate the upgrade or replacement of individual components without having to upgrade the entire product (Saddik 2018). Design for interoperability: Sharing hardware components between different devices and operational systems. Interfaces should not be limited to one single screen—or even a single device. Modular systems must be able to work in slightly different environments, infrastructures and regions. It is extremely important to utilize a software that can interact with several devices and platforms in different places and devices (Steinbach et al. 2018). Add-on connectivity to attach to "old" products: retrofitting add-on. It may cost more, but it is a great way to transform a "regular" product into a smart one, get to market faster and stay competitive (Arima et al. 2017). Build data-integration layers into existing modules facilitating flow of common data: Older systems may not be able to support gigantic data transfer from networked products, but there are several integration models that can be explored (Hassen and Steinbach 2018). Design products so that the user can customize features according with their values and willingness to pay: Customers have different needs and those needs might change many times during a short period of time or occasion. Being able to customize products either in level of engagement or technical feasibility in different scenarios is probably the future with IOT products. Design parts to be multi-functional, invest in smart algorithms that can identify required functionality from a single hardware: Look for alternatives and sophisticated algorithms that can make one unique sensor to provide different. Design parts for multiuse, create custom algorithms to explore older generation of products: Although an older product might not be able to take advantage of IOT applications, it still can provide useful systems data for the creation of new functions and features. Software applications develop agile software: Devices are always connected; huge amounts of data will be generated, and response should be instantaneous. Pursue a continuous delivery model: Companies should design for easy and continuous delivery of software updates focusing on supporting the customer. Applications must be quickly and frequently updated ensuring stability and security. Open-sourced software: Explore and develop open-sourced software to expand number of partners and accelerate the process. Consider developing open-source software for components that can stand alone. Create an open-source-based platform that can be further developed by different systems integrators.

Offloading data: Develop smart algorithms that can manage to identify the optimal point in time where data will be compressed, saved, or transmitted; and at the same time, it will reduce bandwidth requirements. Sending only the most relevant data to cloud-based servers can be an alternative to reduce cost of transmission and increase battery life. Adaptable software system: Systems must be able to deploy in different countries, with specific settings, in multiple regions and infrastructures, so flexibility is essential.

Power efficiency, always-on technology: An always-on technology can be delivered to IOT products to support context awareness, gesture recognition and indoor navigation. Users will expect to have an all-seeing awareness of the environment, ready for every change of context that needs to be informed. Developers must limit the time that components such as the display and application processor are in an inactive state. At the same time, the system must allow specific sensors to remain in an always-on state. Moreover, the challenge is that the product must do all of it without draining too much battery.

Minimize power consumption: Combining sensors with a sensor fusion algorithm can reduce power consumption. Sensors and processors are always on, but with a sensor fusion device, the main processor can go to sleep and the sensor fusion device will be on but with a very low power processor. When there is a change of context, the sensor fusion takes action and activate the main processor. Privacy and Security: Be transparent: Allow users to customize their permitted uses and allow them to opt-in and opt-out at any time. Interoperability creates space for breaches and risks increases exponentially; therefore, customers need to see a strong value proposition in the product to feel safe in sharing personal information. It will be critical that companies are transparent about the collection and usage of data, providing the highest possible levels of data security.

Create methods of protecting critical information: Information must be shared freely among companies, customers, departments and functions. New security schemes may be required and tight integration between software developers and IT operations will be needed.

3 Proposed Work

Metrics need to be aligned to companies' goals and objectives. With traditional goods, manufacturing metrics should improve quality, efficiency and reduce costs. Traditionally, metrics are determined and evaluated prior to product launch, but with a connected device, it becomes possible to monitor performance remotely even after the product is in the market. By doing so, it is possible to better understand and address a defect or a bad experience through collection and analysis of data and improve not only the current product but also the next generation. Next, are a few metrics that address some of IOT issues and can help designers to improve the smart components in cost and operational efficiency.

Cost efficiency in embedded communication is essential for the product to collect data and transmit information, interacting with other objects, people and environment. Applications1 show the data gathered from sensors in a friendlier way for the user. It would be impossible to get information and make decisions if it was not for the applications.

We can use sensors to sense any type of phenomenon in the environment. For example, we can use microphones to sense types of sound signals, a gyroscope to tell us about position and angles or a thermometer to detect temperature and humidity. As we can see from Fig. 2, sensor's cost is dropping every year and consequently, the number of sensors installed in products keeps getting bigger every day. To maintain businesses competitive, it will be necessary to have a differentiator, deliver the expected experience the user desires and also have low operating costs.

For example, a Smartphone has at least twenty sensors: gesture sensor, accelerometer, proximity sensor, gyro sensor, temperature sensor and geomagnetic sensor, to name a few. Individually, each one of them has a function such as detect the rotation or movement based on axes and change the image accordingly or display temperature and humidity levels for the user, but the sensors can also be combined to provide more valuable information. By combining a few sensors and an array of recognition technologies, Samsung can better understand users' behavior and deliver effortless user experience. Experiences like having an all-in-one companion that offers positive support during your work out by calculating how many calories are burned accurately or, understanding behavior by stopping and resuming a video when the user looks away. Therefore, when evaluating a Smartphone by its sensors and applications, this

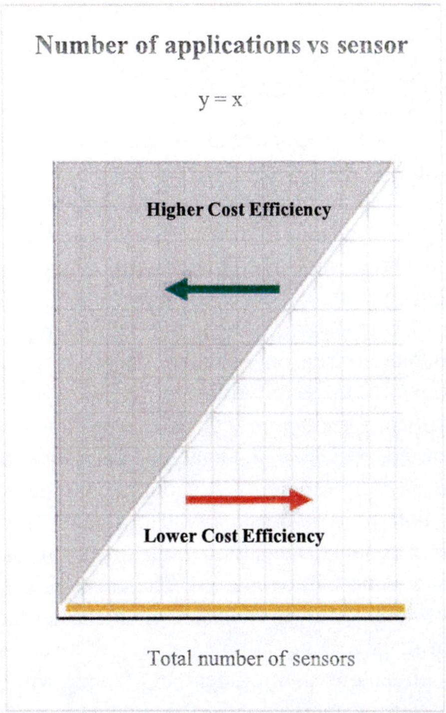

Fig. 2 Number of applications versus number of sensors

product would be positioned above the linear line on Fig. 4 and it would probably be mosre distant from the x-axis and closer to the y-axis. So, even though a Smartphone is considered a high-cost product, it would be considered highly efficient in terms of cost of sensor and applications.

Critical sensors are demonstrated that individual sensor readings are valuable, but its value increases exponentially when integrated with other data such as inventory locations, commodities prices, service histories and traffic patterns. This ability of achieving full value of data becomes a key source of competitive advantage. However, with that comes an issue of scale, and that is where one of the IOT problems begins. The volume of data produced on a device-by-device basis does not warrant concern, but once considered in aggregate, this data poses a significant threat for systems and networks. For example, according to Stephen (Hassen and Steinbach 2018, 3D systems 2021), a Boeing jet generates 10 terabytes of data per engine every 30 min of flight. Therefore, for a single six-hour flight, the total amount of data generated would be 240 terabytes. Multiply this number by the number of flights in the sky, by weeks, months and years, and the scale of sensor data gets massive, and it becomes a huge concern.

To collect all of the necessary data, manufacturers will have to equip products with faster processors and sophisticated and complex algorithms. Additionally, advanced analytics models and tools will be needed to enable data-driven optimization or predictions. Especially because part of the critical information will reside outside companies and it will probably be in an unstructured format such as social-network conversations, videos and audios, companies will have to integrate internal and external data. Making data useful will require a large investment in data capabilities (Narumi et al. 2011).

Even though there are solutions already in the market like SAP HANA that allows filtering, aggregating and enriching raw data before storing into your database, the less amount of data is there to clean, less time will be needed for the information to be delivered, less space for storage will be needed and consequently less money will be spent. For all that, minimizing the number of critical2 sensors that provides information that needs to be periodically collected will be essential.

Critical Ratio (CR) is the ratio of the number of critical sensors over the total3 number of sensors installed in a product. The Critical Ratio indicates how complex the infrastructure will need to be to support the required amount of data. The more critical sensors a product has, more information will need to be collected, cleaned, stored, protected and analyzed and more software and architecture complexity will need to be handled. Reducing the number of critical sensors can propagate into the whole process drastically. The smaller is the quantity of critical sensors deployed in a product, the less complexity in infrastructure for storage, security, analytics and so on will be needed. The closer CR is to 100% the more data will have to be gathered, analyzed and stored and the harder will be to get important information out of it. To manage all this data, more investment of analytical models and tools, data architecture, commitment and money will be required. In addition, if CR is closer to zero, it indicates that the product is more efficient regarding its sensors and the related information.

Max=X/Y

Subiect Y > X; X > 0

Where: X= Number of applications

Y= Number of sensors

Fig. 3 Number of applications versus number of sensors

Critical Ratio (CR) = Number of Critical sensors/Number of Total sensors.

Although most of the data produced is being thrown away, or when kept, companies do not know what to do with it, manufacturers should not be limited when deciding what data to collect and store. The costs and risks for not collecting certain types of data just because there is no clear use or application are too high. Companies should think about their stakeholders and what type of data will be valuable for them. With that in mind, companies can decide not to collect data at all, collect only the fields of data necessary to the product or service being offered, collect data that is less sensitive or identify the data they collect.

For example, (3D systems 2021) from Rapid7 tested 9 baby monitors and exposed that all of them had critical vulnerabilities. They all fail in all and exhibited several common vulnerabilities and exposures like easy guessed passwords. For this product, "*X*"-number of vulnerabilities would be 10. And the baby monitor deals with casual and personal data, so, the constant "*A*" is equal 0.3. With this numbers, the probability of the product to be hacked "*P*" is equal 90% (Fig. 4).

From Fig. 3, it is possible to see that the probability of a product to be hacked increases as the number of vulnerabilities varies. The baby monitor is definitely in the red line zone. Probability also increases considerable when dealing with more attractive data, if a product offers the opportunity to access long shelf life information, hackers will prefer to have the trouble to attack this one and gain more money out of it. Therefore, reducing the number of vulnerabilities in a product is the best guarantee that the product is safe from being hacked. However, manufactures cannot only depend on reducing vulnerabilities.

4 Conclusion

With the design for manufacturability and assembly principles, designers can improve the manufacturing process reducing time and cost for the physical product. With IOT product, manufacturers need to think not just about the technical details, but also of

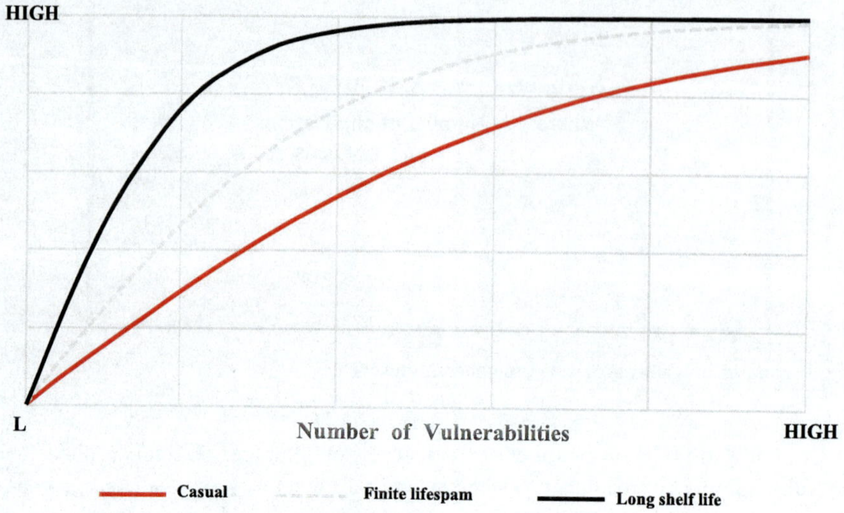

Fig. 4 Probability of an IOT product be hacked

how it will fit into the broader context of the user's life. The main components in a IOT product are its smart components such as connectivity and sensors, and these are the ones that need attention since they are usually the most expensive and complex.

The guidelines presented in this paper are going to be useful in helping designers to better understand how technologies work and how they are being used. It will be a valuable tool to identify the best way technology can work in our benefit, what are the alternatives for the smart components to deliver the same function and help the designer to decide what is the best fit for the product. These guidelines help manufactures to optimize and adapt product design to reduce resources, time and cost during mass production.

References

R. Arima, M. Sithu, Y. Ishibashi et al., Qoe assessment of fairness between players in networked virtual 3d objects identification game using haptic, olfactory, and auditory senses. Int. J. Commun. Netw. Syst. Sci. **10**(07), 129 (2017)

A. El Saddik, Digital twins: the convergence of multimedia technologies. IEEE Multimedia **25**(2), 87–92 (2018)

R. Hassen, E. Steinbach, Hssim: an objective haptic quality assessment measure for force-feedback signals. in *2018 Tenth International Conference on Quality of Multimedia Experience (QoMEX)*, (IEEE, 2018), pp. 1–6

T. Narumi, T. Kajinami, S. Nishizaka, T. Tanikawa, M. Hirose, Pseudo-gustatory display system based on cross-modal integration of vision, olfaction, and gustation. in *Virtual Reality Conference (VR)*, (IEEE, 2011), pp. 127–130

E. Steinbach, M. Strese, M. Eid, X. Liu, A. Bhardwaj, Q. Liu, M. Al-Ja'afreh, T. Mahmoodi, R. Hassen, A. El Saddik et al., Haptic codecs for the tactile internet. Proc. IEEE **107**(2), 447–470 (2018)

3D systems. Touch haptic device 2021. Online accessed: 6 May 2021